SMART SYSTEMS ENGINEERING:

COMPUTATIONAL INTELLIGENCE IN ARCHITECTING
COMPLEX ENGINEERING SYSTEMS

VOLUME 17

ASME PRESS SERIES ON INTELLIGENT ENGINEERING SYSTEMS THROUGH ARTIFICIAL NEURAL NETWORKS

EDITOR

C.H. Dagli, Editor, University of Missouri-Rolla, Rolla, Missouri, USA

Intelligent Engineering Systems Through Artificial Neural Networks, Volume 1, edited by Cihan H. Dagli, Soundar R.T. Kumara, and Yung C. Shin, 1991

Intelligent Engineering Systems Through Artificial Neural Networks, Volume 2, edited by Cihan H. Dagli, Laura I. Burke, and Yung C. Shin, 1992

Intelligent Engineering Systems Through Artificial Neural Networks, Volume 3, edited by Cihan H. Dagli, Laura I. Burke, Benito Fernandez, and Joydeep Ghosh, 1993

Intelligent Engineering Systems Through Artificial Neural Networks, Volume 4, edited by Cihan H. Dagli, Benito Fernandez, Joydeep Ghosh, and R.T. Soundar Kumara, 1994

Intelligent Engineering Systems Through Artificial Neural Networks, Volume 5, edited by Cihan H. Dagli, Metin Akay, C.L. Phillip Chen, Benito Fernandez, and Joydeep Ghosh, 1995

Intelligent Engineering Systems Through Artificial Neural Networks, Volume 6, edited by Cihan H. Dagli, Metin Akay, C.L. Phillip Chen, Benito Fernandez, and Joydeep Ghosh, 1996

Intelligent Engineering Systems Through Artificial Neural Networks: Smart Engineering Systems: Neural Networks, Fuzzy Logic, Data Mining and Evolutionary Programming, Volume 7, edited by Cihan H. Dagli, Metin Akay, Okan Ersoy, Benito Fernandez, and Alice Smith, 1997

Intelligent Engineering Systems Through Artificial Neural Networks: Smart Engineering Systems: Neural Networks, Fuzzy Logic, Evolutionary Programming, Data Mining and Rough Sets, Volume 8, edited by Cihan H. Dagli, Metin Akay, Anna L. Buczak, Okan Ersoy, and Benito Fernandez, 1998

Intelligent Engineering Systems Through Artificial Neural Networks: Smart Engineering System Design: Neural Networks, Fuzzy Logic, Evolutionary Programming, Data Mining and Complex Systems, Volume 9, edited by Cihan H. Dagli, Anna L. Buczak, Joydeep Ghosh, Mark Embrechts and Okan Ersoy, 1999

Intelligent Engineering Systems Through Artificial Neural Networks: Smart Engineering System Design: Neural Networks, Fuzzy Logic, Evolutionary Programming, Data Mining and Complex Systems, Volume 10, edited by Cihan H. Dagli, Anna L. Buczak, Joydeep Ghosh, Mark Embrechts, Okan Ersoy and Stephen Kercel, 2000

Intelligent Engineering Systems Through Artificial Neural Networks: Smart Engineering System Design: Neural Networks, Fuzzy Logic, Evolutionary Programming, Data Mining and Complex Systems, Volume 11, edited by Cihan H. Dagli, Anna L. Buczak, Joydeep Ghosh, Mark Embrechts, Okan Ersoy and Stephen Kercel, 2001

Intelligent Engineering Systems Through Artificial Neural Networks: Smart Engineering System Design: Neural Networks, Fuzzy Logic, Evolutionary Programming, Complex Systems and Artificial Life, Volume 12, edited by Cihan H. Dagli, Anna L. Buczak, Joydeep Ghosh, Mark Embrechts, Okan Ersoy and Stephen Kercel, 2002

Intelligent Engineering Systems Through Artificial Neural Networks: Smart Engineering System Design: Neural Networks, Fuzzy Logic, Evolutionary Programming, Complex Systems, and Artificial Life, Volume 13, edited by Cihan H. Dagli, Anna L. Buczak, Joydeep Ghosh, Mark Embrechts, and Okan Ersoy, 2003

Intelligent Engineering Systems Through Artificial Neural Networks: Smart Engineering System Design: Neural Networks, Fuzzy Logic, Evolutionary Programming, Complex Systems and Artificial Life, Volume 14, edited by Cihan H. Dagli, Anna L. Buczak, David L. Enke, Mark Embrechts and Okan Ersoy, 2004

Intelligent Engineering Systems Through Artificial Neural Networks: Smart Engineering System Design: Neural Networks, Evolutionary Programming, Data Mining, and Artificial Life, Volume 15, edited by Cihan H. Dagli, Anna L. Buczak, David L. Enke, Mark Embrechts, and Okan Ersoy, 2005

Intelligent Engineering Systems Through Artificial Neural Networks: Smart Systems Engineering: Infra-Structure Systems Engineering, Bio-Informatics and Computational Biology, Evolutionary Computation, Volume 16, edited by Cihan H. Dagli, Anna L. Buczak, David L. Enke, Mark Embrechts, and Okan Ersoy, 2006

Intelligent Engineering Systems Through Artificial Neural Networks: Smart Systems Engineering: Computational Intelligence in Architecting Complex Engineering Systems, Volume 17, edited by Cihan H. Dagli, Anna L. Buczak, David L. Enke, Mark Embrechts, and Okan Ersoy, 2007

SMART SYSTEMS ENGINEERING:

Computational Intelligence in Architecting Complex Engineering Systems

VOLUME 17

Proceedings of the Artificial Neural Networks in Engineering Conference ANNIE 2007) held November 11-14, 2007, in St. Louis, Missouri, U.S.A.

EDITORS
Cihan H. Dagli
University of Missouri-Rolla, Missouri
Rolla, Missouri

Anna L. Buczak
Sarnoff Corp.
Princeton, New Jersey

David L. Enke
University of Tulsa, Oklahoma
Tulsa, Oklahoma

Mark Embrechts
Rensselaer Polytechnic Institute, RPI
Troy, New York

Okan Ersoy
Purdue University
West Lafayette, Indiana

SME PRESS NEW YORK 2007

Library of Congress 92-30949
ISBN: 0-7918-0265-5
ASME Order No. 802655

Contents

PREFACE

As a follow up to the previous sixteen volumes of Intelligent Engineering Systems Through Artificial Neural Networks, edited by Dagli, C.H. et al., this volume contains the edited versions of the technical presentations of ANNIE 2007. The seventeenth international gathering of researchers interested in architecting complex engineering systems through computational intelligence was held from November 11-14, 2007 in St. Louis, Missouri, USA. The papers included in this volume provide a forum for researchers in computational intelligence field to exchange ideas on smart engineering systems architecting and design.

An extended version of each paper selected for inclusion was reviewed by two referees, then revised, edited, and condensed to the format herein. The papers in this edited book are grouped into seven categories:

- Bio-Informatics
- Infrastructure Systems
- Evolutionary Computation
- Financial Engineering and Data Mining
- Architecting and Engineering Complex Systems
- Fuzzy Systems and Machine Learning
- Smart Engineering Systems

There were six plenary sessions scheduled for ANNIE 2007. Dr. Matthia Scheutz began the conference with the Monday Morning Plenary talk titled: "Swarm Intelligence and Agent-Based Modeling". The Monday Luncheon Plenary talk was presented by Dr. Mitsuo Gen on "Evolutionary Scheduling in Manufacturing Systems". Dr. Robert Kozma opened the Tuesday presentations with his Tuesday Morning Plenary talk on "Cognitive Phase Transition in Brains and Brain Models and Applications". Dr. Gürsel Süer presented the Tuesday Luncheon Plenary on "Human-Profile Based Intelligent Systems". Dr. Lonnie Welch started Wednesday with his Wednesday Morning Plenary Talk "New Frontiers in Computational Genomics:. Dr. Mark Embrechts presented the Wednesday Luncheon Plenary with his talk titled: "An Introduction to Text Mining".

The ANNIE 2007 Banquet Plenary Speakers were K. Mark Bryden and Bill Fulkerson. Their presentation was titled "Looking Back . . .Stepping Forward: Open Systems and Narrative in Engineering".

Dr. Kenneth "Mark" Bryden is an Associate Professor of Mechanical Engineering, Iowa State University, Ames, Iowa, USA. Dr. Bryden has written more than 100 peer reviewed articles in the areas of modeling, complex systems and decision science. Currently he is an associate professor of the Mechanical Engineering Department and chair of the Complex Adaptive Systems Group at Iowa State University. He is the Program Director for the Simulation, Modeling and Decision Science Program in the DOE's Ames Laboratory and heads the Virtual Engineering Research Laboratory with the Virtual Reality Applications Center. The Virtual Engineering Research Laboratory focuses on integration of information technologies and cognition into the engineering process to support decision making for and the realization of complex systems. Prior to his arrival at ISU Dr. Bryden worked 14 years in a wide range of engineering positions at Westinghouse

Electric Corporation within the Naval Reactors Program. This included 8 years in power plant operations and testing and 6 years in engineering support. Dr. Bryden's primary research interests are in the integration of high performance computing, new computational algorithms, and new user interfaces to solve complex, tightly coupled engineering and decision analysis problems. Dr. Bryden is the recipient of numerous awards including a 2006 R&D 100 award for the development of the software package VE-Suite.

Bill Fulkerson serves as Technology Architect in the Information Systems group of Deere & Company, Moline IL. In this role, Bill facilitates the alignment of business and technology strategy and advises on organizational change initiatives. In his thirty year career with Deere, Bill has served in many areas including Information Technology, Business Development, Engineering& Technology, and Finance. Prior to joining Deere, Bill was employed as a military operations research analyst for the US Army and as instructor of mathematics at the University of Central Missouri. Bill received his BA and MA in Mathematics from the University of Central Missouri.

The editors would like to once again thank the plenary speakers, and the authors for their contributions. We would also like to recognize the organizing committee members of ANNIE 2007 for their excellent support in promoting the conference internationally. Further, we wish to express our gratitude to all the referees for their excellent and timely review efforts, which made this edited book possible within a short period of time.

We would like to mention our appreciation to Tina Dunn. Her ownership of the conference, excellent communication skills, her superb organizing capabilities, her excellent knowledge of computers, skills on databases and spreadsheets, quick adaptability to new internet portal for paper submission and review, and her attention to detail made her an integral part of the ANNIE 2007 conference and this book.

Lastly, but most importantly, we would like to thank all of our families for their patience and support during the many long hours that it took to create yet another ANNIE, ANNIE 2007, and subsequently, this book.

Cihan H. Dagli
University of Missouri-Rolla

David L. Enke
University of Tulsa

Okan Ersoy
Purdue University

Anna Buczak
Sarnoff Corp.

Mark Embrechts
Rensselaer Polytechnic Institute, RPI

PART I:

BIO-INFORMATICS

MODELING BRAIN ELECTRICAL ACTIVITY INVOLVING VAGUS NERVE STIMULATION

MARK H. MYERS
Computational Neurodynamics Laboratory
Department of CS, University of Memphis , Memphis, TN 38152
mhmyers @memphis.edu

ROBERT KOZMA
Computational Neurodynamics Laboratory
Department of CS, University of Memphis, Memphis TN 38152
rkozma@memphis.edu

ABSTRACT
Experiments involving electroencephalographic (EEG) analysis of spatiotemporal structures identified various characteristic features in the measured brain activity of human subjects, while awake, at rest, in sleep, and preceding and during epileptic seizure. Gamma power of EEG signals (20Hz to 80 Hz) varies with changes in the cognitive activity and medical conditions of the subject. Qualitative characterization of the measured effects involves the Lyapunov analysis in time domain and power spectral analysis in frequency domain. Positive Lyapunov exponents and scale free behavior in spectra are indications of nonlinear chaotic dynamics underlying brains behavior. The present work introduces results obtained by the KIV model, which represents a hierarchical approach to the spatio-temporal dynamic of the limbic system. Our approach models some aspects of the neurosurgical treatment of medically intractable epilepsy using Vagus Nerve Stimulation (VNS). Several properties of brain dynamics during VNS treatment has been reproduced by KIV simulations, which indicates a potential for future interpretation of VNS experimental data.

INTRODUCTION

In this paper, we study the electrical activity of the limbic system, which is a crucial component of the human. In normal healthy state, brains operate in a balanced manner and exhibit high-dimensional complex oscillations, which can be characterized as chaotic (Freeman, 2000). According to Freeman, the formation of global amplitude modulation patterns that are integrated into the entire hemisphere of the brain causes the brain to remain in homogeneous neuronal electrical firing states (Freeman, 2006, Freeman, Holmes, West, Vanhatalo, 2005). When the brain is subjected to seizure or stroke, its electrical activity becomes imbalanced which results in constraints of the original highdimensional EEG oscillations.

We model the balanced, homogenous chaotic background activity in the brain through the KIV model, which is a biologically inspired neural network. The KIV network is comprised of several KIII models linked with distributed and delayed feedback (Kozma, 2003, Kozma, Freeman, Erdi, 2003, Kozma, Myers,

2005). We will illustrate the imbalance of electrical brain activity through the KIV such as what is found in an EEG during the emergence of a seizure, and then reinstate the balanced state of electrical activity through external stimulation which models VNS treatment. Vagus Nerve Stimulation (VNS) is an adjunctive treatment for certain types of epilepsy and clinical depression (Wheless JW, Baumgartner J. 2004). VNS involves sending electric impulses to the vagus nerve in the neck via a lead implanted under the skin. VNS treatment returns the imbalanced electrical activity of the brain back to a homogeneous state (Adamolekun, Jouny, Franaszczuk, 2004). In this work, initial results of modeling VNS treatment are introduced.

THE KIV MODEL

The KIV model is a biologically inspired neural network (Kozma, Freeman, Erdi, 2003). K-set family includes hierarchy of K models, including K0, KI, KII, KIII, and KIV, see Fig. 1. K-sets represent family of models with increasing complexity. They represent different aspects of the vertebrate brain. K0 consists of a single node with input and output and a nonlinear transfer function. KI is a coupling of either exhibitory or inhibitory K0 units. KII is a double layer of both exhibitory and inhibitory KI units. KIII is three-layer architecture with one or more KIIs connected by feed forward and delayed feed back neural network. KIII model is a working example of the chaotic principles .The KIII model is used for robust pattern recognition. Finally, KIV is the mathematical model of the brain.

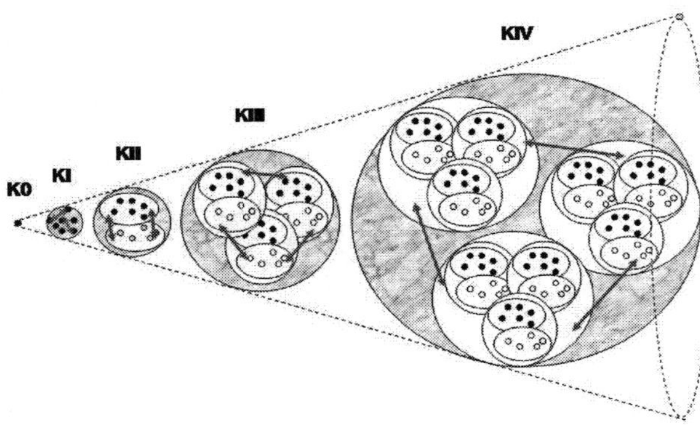

Figure 1 – Illustration of K0, KI, KII, KIII, and KIV sets. Solid and open cycles signify excitatory and inhibitory populations, respectively.

KIV model consists of Hippocampus, Cortical and Amygdala. .It has the functionality of sensory perception and action selection. KIV is a chaotic dynamic memory which encodes sensory information in the form of aperiodic spatial-temporal oscillations of non-linear processing elements. Figure 2 provides a KIV model of the hemisphere, Fig. 2a and b. It has three major parts, cortex, hippocampus and enthorhinal cortex with the amygdala, following the limbic system, Fig. 2a. Hippocampus is a KIII neural network which models proprioception including navigation functions. The cortex is another KIII neural network which models sensory processing and pattern recognition in various sensory modalities. The amygdala is a KII neural network which is incorporated into the KIV neural network. It is the unit where the activations from the both KIII are taken and decision is made concerning the next action, based on the fusion of the signals from the KIII sets. Amygdala is linked with both KIIIs as shown in the following figure with some weighted matrix. With proper weight selection the KIV neural network can maintain some non convergent chaotic oscillations among all the components of the system. The activations from Hippocampus and Cortical KIII's are transferred between them through connection weight matrix WA. The activations between Hippocampus and Amygdala are passed through weighted matrix WB. Finally, the activations between Cortex and Amygdala are passed through weight matrix WC.

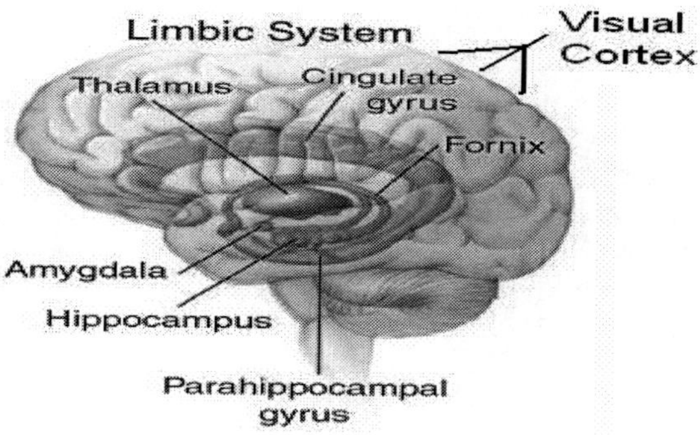

Figure 2a. Schematic illustration of the components of the limbic system.

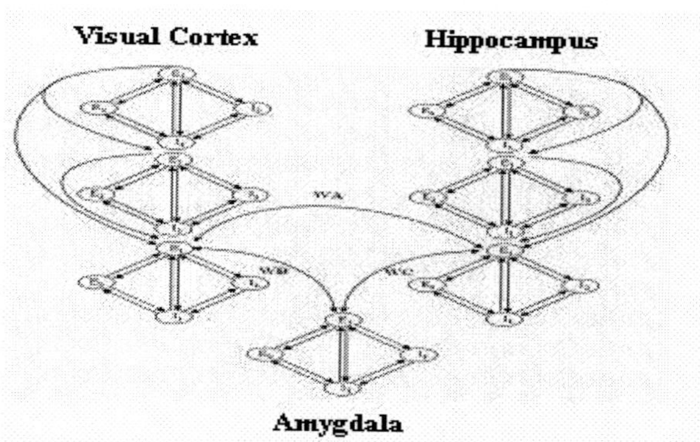

Figure 2b - The KIV model of the limbic system, consisting of two KIII's and one KII interconnected sets.

The weights between the KIII and KII subcomponents provide the manner in which we can model various states of electrical neuronal activity found in a human EEG. In this paper, will illustrate how additive inhibitory bias in the

Amygdala can compensate for simulated imbalance in electrical activity and reinstate it to a normal and balanced state.

MODELING VAGUS NERVE STIMULATION

We modify the activity in the KIV model by adjusting the weight connections between the Cortex, Hippocampus, and Amygdala. A small perturbation in initial conditions can greatly affect the evolution of chaotic systems through time, but once those conditions are estiblished, the future of a chaotic system is just as deterministic as a non-chaotic system (Kozma, 2003). By introducing a greater bias to the KII signal we can cause the signal to change its chaotic nature.

Figures 3 display the basal state A with high dimensional oscillations for a healthy patient. The Lyapunov exponent is 0.10, showing well-developed chaos. Figure 4 exhibits state B with increase inhibitory connection weights. This state simulates epileptic conditions. The trajectory becomes less chaotic, and the Lyapunov exponent is 0.02. Figure 5 shows the case of simulated VNS treatment, when highly chaotic behavior is restored in the model by applying external bias and a dynamic component. This is called state C. The calculated Lyapunov exponent is 0.12 indicating well-developed chaos. State C is similar to conditions that the VNS treatment induces on the EEG activity of the seizure patient. Namely, it forces the system back to an average system state.

Linear Regression is performed on the PSD of each data channel and the slope of the PSD has been calculated in log-log coordinates. A relatively linear log-log behavior indicates scale-free properties in the chaotic time series. The slope of the PSD is estimated within the 20-80 Hz range. Our model shows that the slope is -3.59 in experiment A, -4.98 in Experiment B, and -2.90 in Experiment C. This behavior is qualitatively agrees with experimental findings (Freeman, Holmes, West, Vanhatalo, 2005).

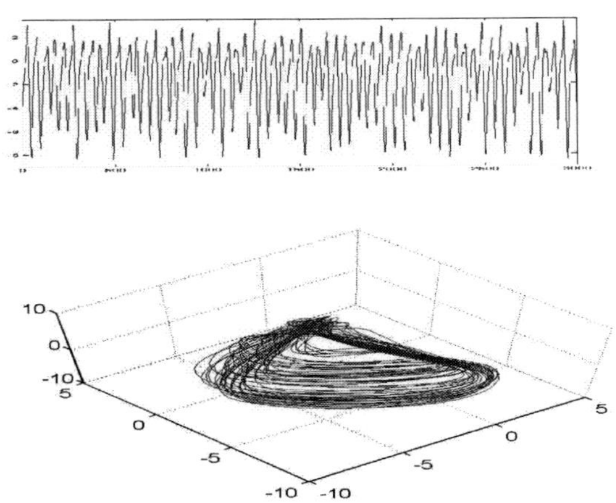

Figure 3 - Simulations with KIV model in the basal normal state A; upper panel: activation of a cortical node for a 3 s period; lower panel: return plot in the time-delayed phase space.

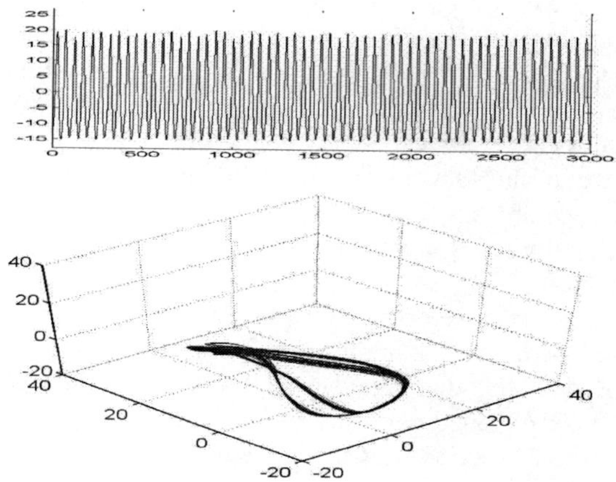

Figure 4 - Simulations with KIV model in the simulated seizure state B; notations are the same as in Figure 3.

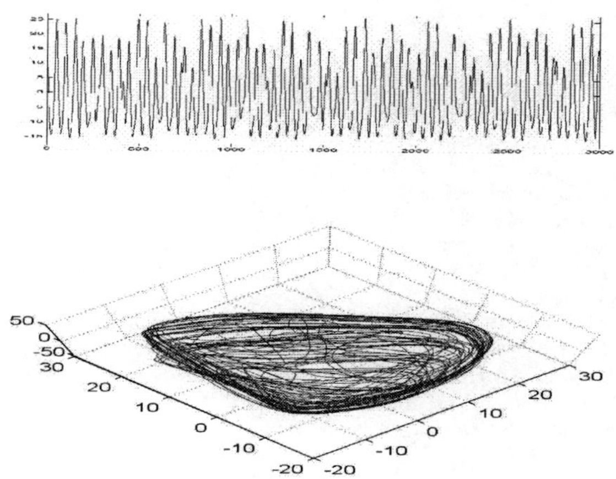

Figure 5 - Simulations with KIV model in the simulated VNS treatment state C. Note that higher complexity chaotic state is restored, as in Fig. 3.

CONCLUSIONS AND FUTURE PERSPECTIVES

The KIV model can be used to simulate the brain electrical activity as observed in EEG experiments and characterize the various cognitive states, i.e., awake, normal, or seizure. The KIV model is a biologically-inspired, non-linear dynamic neural network that can model EEG activity through incorporating the normal and pathological electrical states found in human EEG recordings. Dynamical properties of the simulated data exhibit qualitatively similar behavior as displayed in EEG recordings on seizure patients with a Vagus Nerve Implant. This characterization includes Lyapunov exponent and PSD slope over the gamma band 20Hz – 80 Hz.

Our model can aid in developing therapies through the characterization of various normal and pathological brain states, and potentially predict seizure before its onset. Eventually, this study can provide a customized therapy per patient which can be developed as a titration method for dampening a seizure occurrence while preserving/less influencing the overall electrical state of the brain.

ACKNOWLEDGEMENTS

Experimental assistance by Hima Puppala, University of Memphis is greatly appreciated.

REFERENCES

Jouny CC, Adamolekun B, Franaszczuk PJ, et al. ,2004,, Propagation of partial seizures is associated with increased signal complexity near seizure focus Epilepsia 45: 61-61 Suppl. 7

Freeman, Walter J., 2000, *How Brains Make Up Their Minds*. Columbia University Press, 37-60

Freeman WJ, 2006, Origin, structure, and role of background EEG activity. Part 4. Neural Frame Simulation. Neurophysiol. 117: 572-589.

Freeman WJ, Holmes MD, West GA, Vanhatalo S.2005, Dynamics of human neocortex that optimizes its stability and flexibility. Int. J. Intelligent Systems, Vol 21:9 pp 881- 901.

Kozma R, 2003 Sept, On the constructive role of noise in stabilizing itinerant trajectories in chaotic dynamical systems. Chaos, Vol. 13:3, pp 1078 – 1089.

Kozma R, Freeman WJ, Erdi P, 2003, The KIV Model – Nonlinear Spatio-temporal Dynamics of the Primordial Vertebrate Forebrain. Neurocomputing, 52-54, pp. 819-826.

Kozma R, Myers M, 2005, Analysis of Phase Transitions in KIV with Amygdale during Simulated Navigational Control. IEEE Inf. Joint Conf. Neural Netw., Montreal, Canada.

Wheless JW, Baumgartner J., 2004, Vagus nerve stimulation therapy. Drugs Today (Barc). 40(6):501-15.

PHASE SYNCHRONIZATION IN MESOSCOPIC ELECTROENCEPHALOGRAM ARRAYS

JOSE M. RODRIGUEZ
Department of Computer Science,
The University of Memphis,
Memphis, TN.

ROBERT KOZMA
Department of Computer Science,
The University of Memphis,
Memphis, TN.

ABSTRACT
Coordination of neural oscillations generated at different frequencies has been conjectured to be an important feature of neural activity. In this paper we investigate the phenomena of phase synchronization in electroencephalogram mesoscopic arrays. The electrical activity of the neurons in the visual cortex of rabbits is recorded and analyzed. An 8*8 array of electrodes is placed on the visual cortex of the subject. We consider each electrode as a variable in the phase space of a dynamical system. We decompose each variable in the phase space into narrow frequency signals using finite impulse filters according to standard clinical bands. We measure the instantaneous amplitude and phase of each electrode signal using the continuous time analytical signal concept. We conjecture that the dynamics of the visual cortex can be considered a metastable system with periods of absolute and relative synchronization.

INTRODUCTION

In this section we introduce the fundamental concepts of dynamical systems DS in order to provide the theoretical background for the analysis of phase synchronization in DS. A DS provides a mathematical description for evolving the state of a system forward in time. The DS can be considered as a flow $dx(t) / dt = F(x(t))$ where $\boldsymbol{x}$ is a n-dimensional vector. Given the initial conditions $\boldsymbol{x}(0)$, we can solve the equations and obtain the future state of the system $\boldsymbol{x}(t)$. The n-dimensional state space described by the DS is called the phase space, and the path followed by the system from a given initial state in the phase space is called trajectory or orbit. The phase space of the DS can be characterized using the concept of attractors and Fourier analyis. Types of attractors are fixed points, periodic attractors and chaotic or strange attractors (Strogatz 1994). In Fourier analysis Chutchfield et al. (Crutchfield, Farmer et al. 1980) shows that a power spectrum with large number of peaks corresponds to attractors with complicated geometrical structure. A power spectrum with few peaks corresponds to attractors with less complicated geometrical properties.

Biological neural networks can be considered a particular type of DS. We use electrodes to measure the electrical activity of the brain in the phase space. Each electrode samples the electrical activity of approximately 200 hundreds of biological neurons. The phase space obtained in this way has been used in many ways. Pikovsky (Pikovsky, Rosenblum et al. 2001) use the phase space to measure the strength of phase synchronization among the objects that compose the DS. Freeman (Freeman 2004) uses the phase space in order to analyze the mesoscopic activity of biological neural networks. Kelso (Kelso 1995) uses the concept of metastability to characterize the coordination dynamics of the brain. Metastability is the realization of two opposing tendencies. The tendencies in the DS components to couple together (phase trapping) and the tendency of the components to perform their own innate behavior (phase scattering).

In this paper, we study phase synchronization in mesoscopic arrays. In the first section we introduce the fundamental concepts related to phase synchronization. Hilbert transform and Fourier filtering allow us to characterize the dynamics of the neurons. Next we apply these techniques to the study the electrical activity of biological neurons in the visual cortex. We use metastability concepts to characterize the phase transitions. We use the term signal or channel meaning a variable in the DS phase space.

PHASE SYNCHRONIZATION IN DYNACAL SYSTEMS

To measure the phase coupling between two variables in a DS, we have (Pikovsky, Rosenblum et al. 2001)

$$\psi_{n,m}(t) = (n\phi_1(t) - m\phi_2(t)) \bmod (2\pi) \quad (1)$$

Here n and m are integers. $\phi_1(t)$ and $\phi_2(t)$ are the phases of each DS variable. For uncoupled variables the distribution of $\psi_{n,m}$ is uniform, whereas the interaction between the variables makes the distribution unimodal. To measure the phase synchronization strength between the two variables in the DS, we use the phase coupling index based on the Shannon entropy of the phase distribution differences (Pikovsky, Rosenblum et al. 1999). Having computed p_k in the histogram of $\psi_{n,m}$ the Shannon entropy index $\rho_{n,m}$ is computed as

$$\rho_{n,m} = (\ln N + \sum_{k=1}^{N} p_k \ln p_k) / \ln N \quad (2)$$

where N is the number of bins in the histogram and p_k is the relative frequency of finding $\psi_{n,m}$ within the k-th bin.

The first step in order to analyze the relationships between phases in the DS is to estimate the phases. We consider the method based in the construction of the complex analytical signal AS *z(t)* (Boashash 1992; Marple and Lawrance 1999) using the Hilbert transform HT. To implement this method, one has to construct from a scalar signal s(t), a complex signal z(t)

$$z(t) = s(t) + iH(s(t)) = A(t)e^{i\phi(t)} \quad (3)$$

where H(s(t)) is the HT of s(t), A(t) is the analytic amplitude, and $\phi(t)$ is the analytic phase. The HT of s(t) is defined as

$$H(s(t)) = (1/\pi) P \int_{-\infty}^{\infty} (s(\tau)/(t-\tau)) d\tau \quad (4)$$

where P means that integral is taken in the sense of the Cauchy Principal Value. Practical hints in order to compute and the usage of the HT can be found in (Boashash 1992; Rosenblum, Pikovsky et al. 2001). Using the straightened phases of $\phi(t)$, we can compute the analytic frequency for each variable in the DS phase space as follows

$$\frac{d\phi(t)}{dt} = 2\pi \frac{df(t)}{dt} \quad (5)$$

RESULTS – THE THETA BAND (3-7 HZ)

The electrical activity of neurons on the surface of the primary visual cortex in rabbits was measured using an 8*8 array of electrodes. Details of the experiments are given by (Freeman 2004); here we summarize the main conditions. The space between the electrodes is 0.79 mm covering an entire area of 5.6mm*5.6mm. Data were collected during 6 seconds using a 500 Hz sampling frequency. The data were lowpass filtered at 100 Hz. We use finite impulse filters FIR (Porat 1997) to decompose the DS phase space. We use the standard frequencies defined by the brain clinical community (Freeman 2006). Particularly we use the least-square linear phase FIR filters. Choosing correctly the order of the filter N, these filters have the property of reduce the energy around a peak. We concentrate our studies in the theta band.

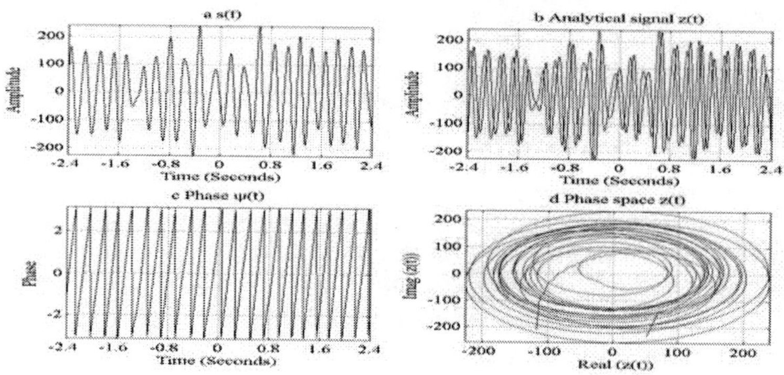

Fig 1. Computing the AS via the HT (a) Signal s(t). (b) Analytical signal z(t). (c) Analytical phase φ(t) and (d) phase space for z(t).

We decompose each electrical signal in the array of electrodes using FIR filters. The peak frequency is determined as the frequency having the largest spectral value in the range 3-7 Hz. Theta oscillations are extracted by using a finite impulse filter FIR bandpass filter with cutoff frequencies of $\Phi_{peak} \pm 2$ Hz where Φ_{peak} represent a peak. The order of the filter N is set to 181 and the attenuation in the passband and stopband are 1 dB and 40 dB respectively. We use least-square FIR filters in the decomposition process (Porat 1997). The firls Matlab function implement the least-square FIR filters (Losada 2004). Next the AS is calculated for each filtered signal. Figure 1 shows the result. Plots a,b,c and d show one channel in the time domain, the real and the imaginary part of the analytical signal, the wrapped phases and the real part of *z(t)* versus the imaginary part of *z(t)*. To reduce boundary effects, one half of a second of data is eliminated at the beginning and ending of each signal. The signal in plot (a) resembles a sine-like oscillating signal with well defined frequency. Observe that the phase has a well defined behavior as shown in plot (c). Plot d shows that all the phase space of *z(t)* is well defined.

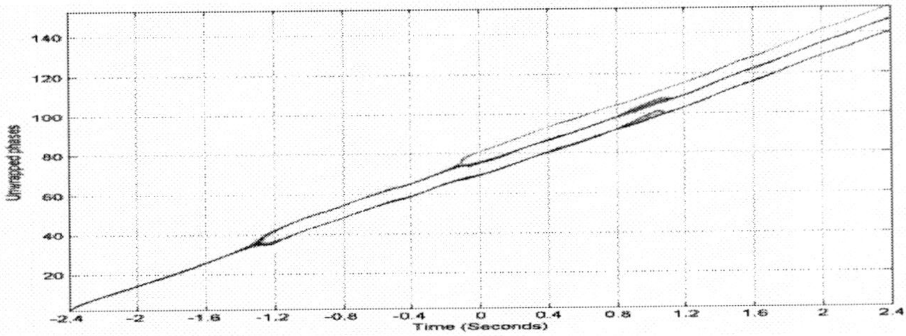

Fig 2. Unwrapped phases for all the 64 unwrapped phases. Observe periods of synchronization and desynchronization.

UNWRAPING THE PHASES

The analytic phase is computed for all the 64 channels using the AS concept. In order to track the behavior of the phase over time, the discontinuous phases are straightened by adding 2π to the arctangent2 function to get the unwrapped analytical phase. The slope of the unwrapped phases gives the average frequency over time. The upward and downward deviations show variations in the frequency content in theta oscillations. Figure 2 shows the 64 unwrapped phases. It is interesting to point out in this plot that phase deviations happen in the array at particular times (See for example times -1.3 seconds and -0.1 seconds). Here we classify the phase segments in two categories: segments without phase scatering (absolute phase synchroniation) and segments with phase scatering (relative phase synchronization). See Kelso 1995.

SEGMENTS OF ABSOLUTE PHASE SYNCHRONIZATION

Figure 3 shows a segment with absolute phase synchronization. The plot in the top-left shows the 64 unwrapped phases. The unwrapped phases look like straight lines, resembling sine-like waves with well defined peak in the frequency domain. In the top-rigth the unwrapped phase differences are shown. Here, the phase of channel j ($j = 1,..,5,6$) has been subtracted from channels $k = j + lag$, with lag = 8. Given this difference, we calculate and plot the residues after 2π division. The distribution of the phase differences has a peak in zero, showing statistical phase synchronization. The Shannon synchronization index is 0.90. In the bottom-left the analytic frequency is computed for all the channels. The bottom-right shows the distribution of the analytic frequencies. The frequencies are smooth out using a moving average filter size 11 samples (44 milliseconds). We lose resolution in time but the analytic frequencies exhibit more regular behavior.

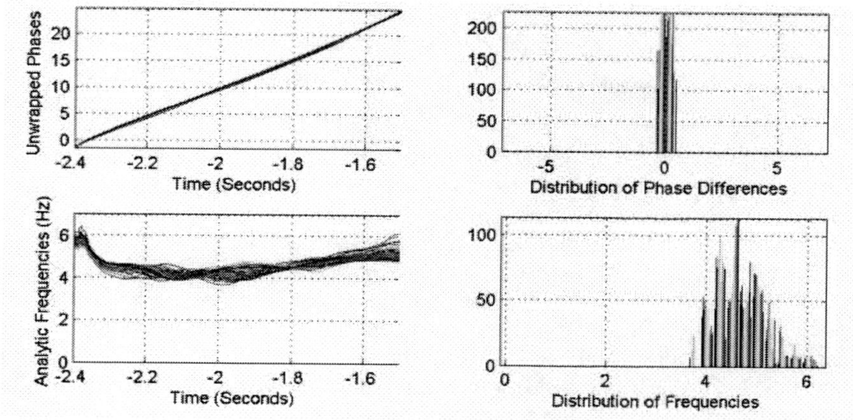

Fig 3. Absolute synchronization segment. Top-left the unwrapped phases. Top-right the distribution of phase differences. Bottom-left the frequencies and bottom-right the distribution of frequencies.

SEGMENTS OF RELATIVE PHASE SYNCHRONIZATION

Figure 4 shows a segment with relative phase synchronization. The desynchronization periods appear from -1.3 to -1.1 seconds approximately. The plots in the top show the unwrapped phases (left) and the distribution of the phase's differences (right). The desynchronization periods last for about 250 milliseconds, about one period of oscillation. The Shannon synchronization index is 0.75. In the bottom of Figure 4 the analytic frequency (left) and the distribution of frequencies (right) is shown. Notice that the analytic frequency plot shows negative values. Thus, frequencies $f[t] \geq 0$ is allowed, while $f[t] < 0$ is due to the high sensitivity of the HT (Freeman 2006).

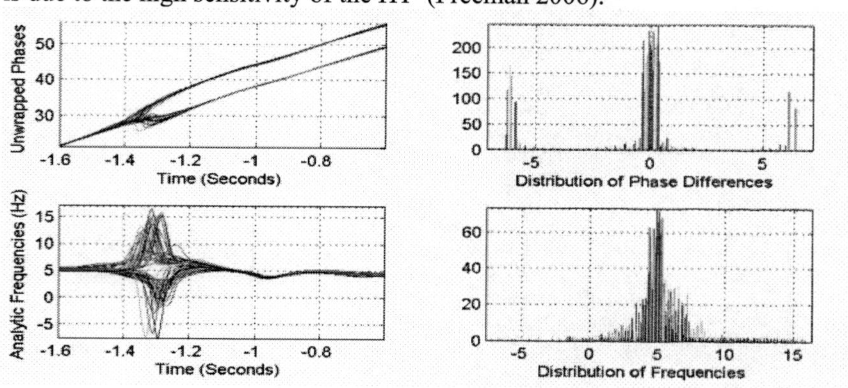

Fig 4. A relative desynchronization segment. Top-left the unwrapped phases. Top-right the distribution of the phase's differences. Bottom-left the analytic frequencies. Bottom-right the distribution of the analytic frequencies.

DISCUSSION

Two types of phase synchronization are present in the theta band according with our findings: Absolute phase synchronization and relative phase synchronization. In absolute phase synchronization periods, the distribution of the phase's differences has a well defined peak and high Shannon synchronization index. The distribution of phases differences and the distribution of frequencies are narrow. The distribution of phases differences does not cover the entire phase interval -2π to 2π. These metastable dynamics expresses the tendency of the neurons in the theta band to couple together. In relative phase synchronization the distribution of the phase's differences has a well defined peak. The distribution of differences in phase cover the entire interval of phase differences -2π to 2π. These dynamics correspond to a lower Shannon synchronization index and broad distribution of phases and frequencies. The results provide support for the hypothesis that neurons at the mesoscopic level organize their dynamics by alternating periods of absolute and relative phase synchronization. Thus, neuron oscillating in the theta band can be considered as metastable systems. The computation of the AS via the HT turns out to be a very useful approach. It allows us to compute the instantaneous phase and the instantaneous amplitude without assumptions about the linearity and stationary in the EEG records. The method is fairly robust since it is able to capture the analytic frequencies in signals with well defined peak in the power spectrum. The AS approach allows us to identify two regions in the up unwrapped phase plot: Straight phases and scattered phases. Straight phases correspond to absolute phase synchronization while scattered phases correspond to relative phase synchronization.

REFERENCES

Boashash, B. (1992). "Estimating and Interpreting the instantaneous frequency of a signal. I. Fundamentals." Proceedings of the IEEE Vol 80(14): 520 - 538

Crutchfield, J., D. Farmer, et al. (1980). "Power spectral analysis of dynamical system." Physics Letters Vol 76 A (1).

Freeman, W. (2004). "Origin, structure and role of the background EEG activity. Part 3. Neural frame classification." Clinical Neurophysilogy(116): 1118-1129.

Freeman, W. (2006). Hilbert transform for brain waves. Berkeley University of California at Berkeley. Available on line at www.scholarpedia.org.

Kelso, S. (1995). Dynamic patterns: The self-organization of brain and behavior. Cambridge, MA, MIT Press.

Losada, R. (2004). Practical FIR filter design in Matlab, The Math Works, Inc.

Pikovsky, A., M. Rosenblum, et al. (1999). "Phase syncronization In regular and chaotic systems " International Journal of Bifurcation and Chaos Vol 10(10): 2291-2305.

Pikovsky, A., M. Rosenblum, et al. (2001). Synchronization. A universal concept in nonlinear sciences, Cambridge University Press.

Porat, B. (1997). A course in digital signal processing Jhon Wiley & Sons, Inc.

Rosenblum, M., A. Pikovsky, et al. (2001). Phase synchronization: From theory to data analysis. In F. Moss and S. Gielen, Editors, Handbook of biological physics. E. science. Vol 4: 297.

Strogatz, S. (1994). Nonlinear dynamics and chaos, Addison Wesley Reading.

SIMULATION OF CELL COLLISION AND AGGREGATION USING A THREE-DIMENSIONAL COMPUTATIONAL MODEL FOR MULTICELLULAR TISSUE GROWTH

BELGACEM BEN YOUSSEF
School of Interactive Arts & Technology
Simon Fraser University
Surrey, British Columbia, Canada

LENNY TANG
School of Interactive Arts & Technology
Simon Fraser University
Surrey, British Columbia, Canada

ABSTRACT

We developed a computational model to simulate the growth of multicellular tissues using a discrete approach based on cellular automata. The model allows us to study the tissue growth rates and population dynamics of different populations of migrating and proliferating mammalian cells. Cell migration is modeled using a discrete-time Markov chain approach. We report simulation results describing cell collision and aggregation for two cell populations each having its own division and motion characteristics. Heterotypic and homotypic cell-cell interactions play important roles in cell and tissue functions. The temporal evolution of the frequency of cell collision and aggregation and their relations to other variables that quantify the dynamics of cell populations can be predicted by this model for different cell seeding topologies and cell heterogeneity ratios.

INTRODUCTION

Each year, millions of surgical procedures are performed to relieve patients who are affected by tissue loss, due to burns, injuries, or organ failure. Operations treating patients using tissue reconstruction and organ transplantation have been highly successful. However, the number of patients treated by these therapies is small due to the limited number of donors available. The primary focus of tissue engineering is the growth of three-dimensional tissues with proper structure and function. Tissue engineers combine knowledge from the areas of biochemistry, medical sciences, and engineering to develop bioartificial tissue substitutes or to induce tissue remodelling in order to repair, replace, or enhance tissue functions (Langer and Vacanti, 1993; Mooney and Mikos, 1999).

Tissue growth is a complex process affected by many contributing factors such as the type of cells, initial seeding densities, spatial distribution of the seed cells and the culture conditions (Ratcliffe and Niklason, 2002). Scaffold properties, cell activities like adhesion or migration, and external stimuli that modulate cellular functions are among the many factors that affect the growth rate of tissues. Hence, the development of bio-artificial tissue substitutes involves extensive and time-consuming experimentation. The availability of computational models with predictive abilities will greatly speed up progress in this area.

This research focuses on the development of computational models to simulate the growth of three-dimensional tissues consisting of more than one cell type. Our specific objective here is to use such models to study the dynamics of cell-cell interactions. The success of models that describe the dynamic behaviour of cell populations, however, will depend on our ability to accurately describe the collision and aggregation of different cell types in various environments.

RELATED WORK

Various modeling approaches have been used to simulate the population dynamics of proliferating cells. Early attempts to model cell population growth were limited to nonmotile cells and the study of contact inhibition phenomena on the proliferation of anchorage-dependant endothelial cells. These early models considered nonmotile cells proliferating in two dimensions (Zygourakis et al., 1991), or on microcarriers (Hawboldt et al., 1994). In disregarding cell locomotion, the cell growth rates in these early models are restricted by the effects of contact inhibition. Lee et al. (1995) showed the importance of cell motility and cell-cell interaction in describing the cell proliferation rates. Any comprehensive model for tissue growth must consider these processes and account for the growth factors that regulate their rates. Later, Ben Youssef (1999) developed the first three-dimensional cellular automata model for tissue growth. This work also utilizes a Markov chain approach to model the trajectories of migrating cells and is focused primarily on the study of a single population of proliferating and migrating cells. Cheng et al. (2006) enhanced this model by accounting for the slowing of cells resulting from cell aggregation while using one cell type. The latter two works form the basis for this model.

MODELING OF CELL COLLISION AND AGGREGATION

To model our biological system, we focused our attention on four cellular processes: cell division, cell migration, cell collision, and cell aggregation. The first two processes are described elsewhere (Cheng et al., 2006). The latter two are described in the next two sections, respectively. These four processes are illustrated on Figure 1. Mother cell 1, at the top left of the cellular space, migrates downwards, and divides into two daughter cells. These two daughter cells, in this scenario, move around until they encounter each other, at which time aggregation occurs. If one of these daughter cells moves upwards and encounters a cell of a different type (type 2), then collision occurs.

Cell Collision

Mammalian cells move in a certain direction for some period of time and, then, they turn and migrate in another direction. Cell collision occurs when a cell moving in a certain direction encounters another cell of a different type in its path. When a cell of one type collides with a cell of a different type, the cell slows down for a period of time and then changes its direction. The duration of time a cell stays in the stationary state after a cell collision is based on its type (Lee et al., 1994; Kouvroukoglou et al., 1998). Once the cell is ready to resume its motion, it migrates in a new direction. The modeling steps used in the cell collision process are described below:

- The cell stops for a number of steps.
- The cell changes its direction and resumes motion.

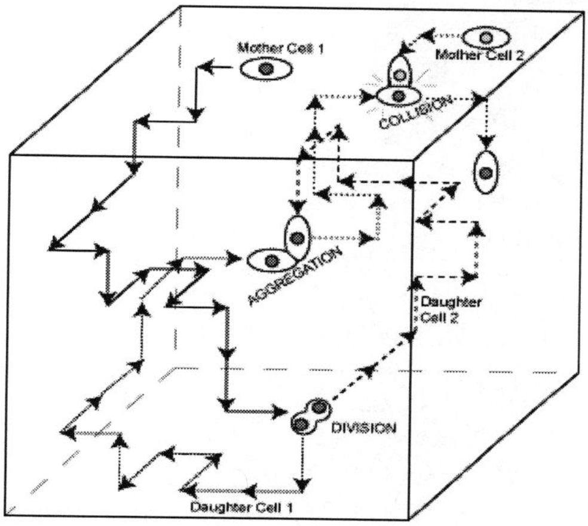

Figure 1 – An illustration showing the persistent random walk of two mother cells of different types (1 and 2) during their division cycle. The two daughter cells move away from each other and may aggregate or collide with other cells.

Cell Aggregation

Cell aggregation is a feature of tissue formation that allows the binding of cells of the same type. It is this specific grouping of cells that enables the tissue to perform its intended purpose. Cell aggregation is the combination of two cellular functions: cell-to-cell recognition and cell adhesion. The self-recognition quality lets cells identify cells of the same type. When cells of the same type encounter each other, they adhere to one another and form a cellular aggregate (Wolf, 1983). As more cells of the same cell type encounter the cellular aggregate, the cellular aggregate becomes larger forming a cluster of cells. The adhesion can be strengthened by aggregation factors that are sometimes secreted by the cells (Dyson, 1978). As a result of aggregation, the cell slows down, "sticks" to another cell of the same type, and changes its direction of motion. The basic steps used to model the process of cell aggregation are listed below:

- The two cells stop for a number of steps, thereby entering an aggregation state.
- The two cells change their direction and resume locomotion.

Cell State Information

To characterize a heterotypic cell-cell interaction, our model uses a collision stationary state (denoted by $l = 0$). If a cell c collides with another cell of a different type, cell *c* stops moving and enters what is known as the collision stationary state. It remains in that state for a defined period equal to its expected waiting time, $E(T_0)$. When $E(T_0)$ expires, cell *c* can move away in a randomly assigned direction.

In the case of a homotypic cell-cell interaction, we added a second stationary state (denoted by $l = 7$). Two colliding cells of the same type will enter the aggregation state and stick together for a defined period equal to $E(T_7)$. The value of $E(T_7)$ indicates the likelihood for the cells to form multicellular aggregates. When a cell collides with another cell of the same type, that is already in the aggregation state, then the first cell will enter the aggregation state and both cells will reset their persistence counters to $E(T_7)/\Delta t$. Once the waiting time $E(T_7)$ has expired, the two cells can move away in randomly assigned directions.

The state of a cell takes values from the following set of eight digit integer numbers: $\Omega = \{klmnpqrs \mid k, l, m, n, p, q, r, \text{and } s \in \aleph\}$, where k is the cell type. The direction of motion is identified by the direction index l. When l is equal to 0, the cell is in the collision stationary state. When the value of l ranges from 1 - 6, it represents one of six directions the cell is currently moving in (east, north, west, south, above, and below). When the value of l is 7, it enters an aggregation stationary state where it "sticks" to another cell of the same type potentially forming cellular aggregates. The digits mn denote the persistence counter. It is the time remaining until the next change in the direction of cell movement. The cell phase counter is represented by the digits $pqrs$. The cell phase counter is the time remaining before the cell divides.

Initial Conditions

Two cell populations are used to study collision and aggregation. Each of these two cell populations, referred to as cell population 1 and cell population 2, has its own division and migration chacteristics. The division time distributions for these two cell populations are given in Table 1. For instance, we observe that for cell population 1, 64% of the living cells have division times between 12 and 18 hours. Moreover, cells from population 1 are fast-moving cells that migrate at 10 μm/hr while cells from population 2 are slow-moving cells migrating at a speed of only 1 μm/hr.

Table 1 - Cell Division Time Distributions for the Two Cell Populations

	Cell Populations	
Division Times (hrs)	Cell Population 1	Cell Population 2
[12,18)	64%	4%
[18,24)	32%	32%
[24,30)	4%	64%

Cell Seeding Distributions

The uniform cell seeding topolgy is considered here. This topology was chosen for its applicability to experimental research conducted in laboratories. In this case, cells are randomly seeded in the cellular space. The uniform topology simulates the migration and proliferation of cells in a sparsely and uniformly seeded environment with the objective of simulating tissue regeneration. We have associated two types of cell distributions with this topology, a segmented distribution and a mixed one.

In the segmented distribution, each cell type is seeded in a separate area of the cellular space. During the simulation, cells can migrate freely in the cellular space, and can enter areas that were originally designated for a particular type of cell during the initial cell seeding. In the mixed distribution, the different cell types are seeded together in a random order according to the topology of cell seeding. Figure 2 shows an illustration of these two cell distributions. In this topology, the seeding density indicates the percentage of sites occupied by cells at the start of the simulation.

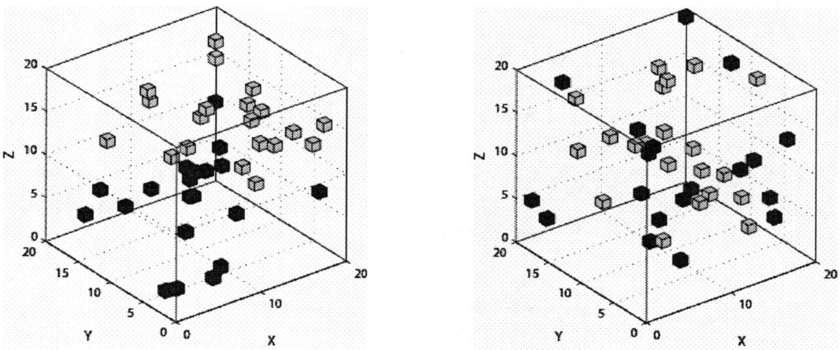

Figure 2 – A uniform seeding topology using two cell populations in (a) a segmented and (b) a mixed distribution. In each case, a total seeding density of 0.5% is used. This example is based on a 20×20×20 cellular array.

Dynamics of Cell-Cell Interactions

At the start of each simulation, the cellular space is seeded with a total number of seed cells equal to N_0. These seed cells migrate and proliferate in the cellular array based on the cellular automata rules to simulate the dynamic process of tissue growth. After some time t, $N_c(t)$ sites of the cellular automaton are occupied by cells. In addition, the frequency of cell-cell interactions, $F_c(t)$, which reflects the total number of interactions for a given cell per simulation step Δt, is provided by Equation (4). Thus, $F_c(t)$ yields the sum of the number of cell collisions, or heterotypic cell interactions, and the number of cell aggregations, also known as homotypic cell interactions.

$$F_c(t) = \frac{\text{Total number of cell-cell interactions in interval } [t, t+\Delta t]}{N_c(t) \times \Delta t} \qquad (4)$$

We also define the heterogeneity measure, H, as the ratio of the initially seeded number of cells from population 1 to that from population 2. The ratio H is given by:

$$H = \frac{\text{initial number of cells from population 1}}{\text{initial number of cells from population 2}} \quad . \tag{5}$$

That is, when $H = 9$, there are 9 cells from population 1 for every cell from population 2 in the cell seeding density.

SIMULATION RESULTS AND DISCUSSION

The presented simulation results were obtained using a 200×200×200 cellular array, a 99.99% confluence parameter, and average waiting times of 2 hours for the six directional states and 1 hour each for the aggregation state and the collision state, respectively.

Segmented Uniform Distribution

The frequency of cell-cell interactions versus time for both cell population 1 and cell population 2 for various values of H is shown in Figure 3. For cell population 1, as H is increased the frequency of cell-cell interactions increases and reaches a higher peak value due to an increase in the number of fast cells in its segment of the cellular space, which leads to more cell-to-cell interactions. Because cells in population 2 have a lower speed of 1 μm/hr, they tend to form more localized clusters. When combined with the fact that an increase in the ratio H also reduces the size of the segment for this population (and the number of slow cells), this results in fewer and fewer cell-cell interactions as H is increased.

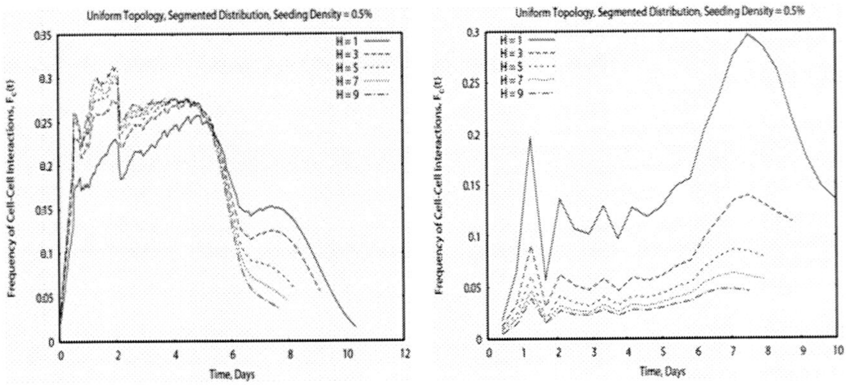

Figure 3 - The effect of varying the ratio *H* on the frequency of cell-cell interactions for cell population 1 (left) and cell population 2 (right).

Figures 4 displays the temporal evolution of the frequency of cell-cell interactions for $H = 1$ and $H = 9$. The homotypic and hetrotypic cell interactions are also shown in both cases. We observe that cell collision dominates early in the proliferation process. Cell aggregation becomes important only later in the process, even as H is increased. These results are also confirmed for the mixed uniform distribution and are not be shown here.

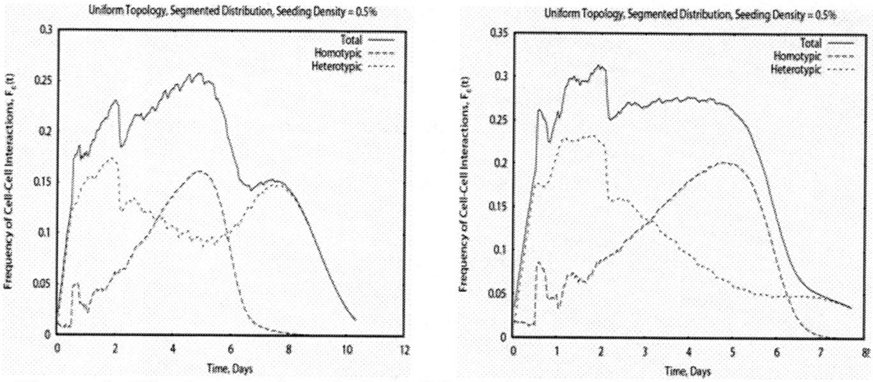

Figure 4 - The temporal evolution of the frequency of cell-cell interactions and its two components for cell population 1 when *H*=1 (left) and *H*=9 (right).

Mixed Uniform Distribution

Figure 5 shows the frequency of cell-cell interactions for populations 1 and 2 for different values of H given a fixed cell seeding density of 0.5%. For cells in population 1, during the first two days and as H is increased, the frequency of cell-cell interactions increases due to an increase in interactions between fast and slow moving cells early in the proliferation process. This is facilitated by the nature of the mixed seeding distribution.

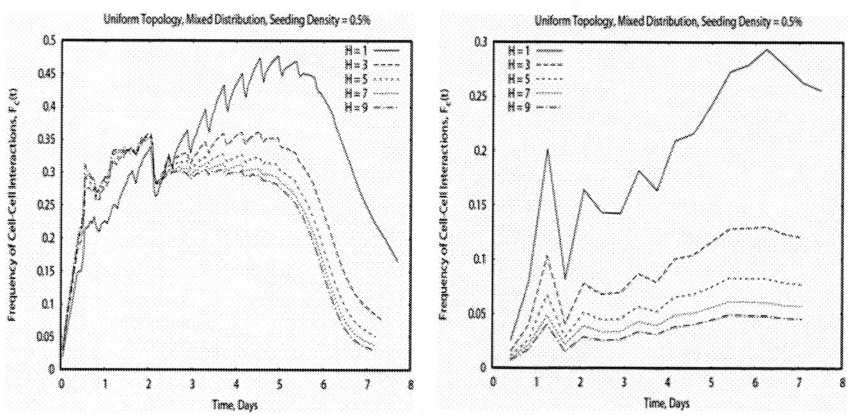

Figure 5 - The effect of varying the ratio *H* on the frequency of cell-cell interactions for cell population 1 (left) and cell population 2 (right).

This phenomenon is clearly illustrated in Figure 6. The figure depicts both the frequency of cell aggregations as well as the frequency of cell collisions for

H =1 and H =9. After the first two days, the frequency of cell collisions decreases rapidly as H is increased. Given that the frequency of homotypic cell interactions between fast cells of the same type increases with time until it reaches a maximum for all values of H, the combined effect of these two observations leads to a reversal of behaviour. That is, the frequency of cell-cell interactions decreases as H is increased after the first two days for cells in population 1. For cells in population 2, the frequency of cell-cell interactions decreases as H is increased due to the decrease in the seeded number of slow cells. This results in a smaller number of cell clusters, which in turn impacts negatively the number of homotypic cell interactions between slow cells.

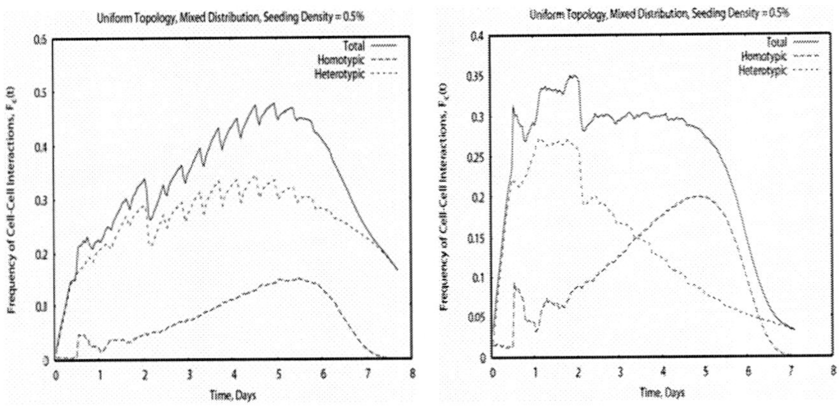

Figure 6 - The temporal evolution of the frequency of cell-cell interactions and its two components for cell population 1 when *H*=1 (left) and *H*=9 (right).

CONCLUSIONS

In this paper, we presented simulation results for cell collision and cell aggregation obtained using a three-dimensional computational model for multicellular tissue growth. When employing two types of cell populations with different migration speeds and division time distributions, our simulations showed that increasing the cell heterogeneity ratio H of fast to slow cells enhanced the frequency of cell-cell interactions in a uniform seeding topology. It appears that the frequency of cell-cell interactions may depend at least primarily on the dynamics of the cell populations present in the cellular space. Throughout these results, we noted that a rapid increase in the fraction of surrounded cells is accompanied by a fast decrease in the frequency of cell-cell interactions. Moreover, the segmented uniform distribution yielded a smaller overall frequency of cell-cell interactions for the population of fast cells. In the case of slow-moving cells, however, nearly similar simulation results were obtained by both cell seeding distributions. These conclusions may have important implications for the design of experiments that can test the efficacy of cell heterogeneity and cell seeding distributions designed to enhance or minimize the frequency of cell-cell interactions in a tissue growth environment.

To achieve the desired goals, it is recommended that assays based on their corresponding design factors be adopted.

REFERENCES

Ben Youssef, B., 1999, *Cell proliferation and migration: 3-D modelling using cellular automata, development of a parallel algorithm and its parallel implementation on an IBM SP2*, Ph.D. Dissertation, University of Houston.

Ben Youssef, B., Cheng, G., Zygourakis, K. and Markenscoff, P., 2007, Parallel implementation of a cellular automaton modeling the growth of three-dimensional tissues. *International Journal of High Performance Computing Applications,* Vol. 21, No. 2, pp. 196-209.

Cheng, G., Ben Youssef, B., Markenscoff, P. and Zygourakis, K., 2006, Cell population dynamics modulate the rates of tissue growth processes. *Biophysical journal,* Vol. 90, No. 3, pp. 713-724.

Dyson, R., 1978, Essentials of cell biology, Allyn and Bacon, Inc., Boston, MA.

Hawboldt, K.A., Kalogerakis, N. and Behie, L.A., 1994, A cellular automaton model for microcarrier cultures. *Biotechnology and bioengineering,* Vol. 43, No. 1, pp. 90-100.

Kouvroukoglou, S., Lakkis, C., Wallace, D., Zygourakis, K. and Epner, D., 1998, Bioenergetics of rat prostate cancer cell migration. *The Prostate,* Vol. 34, No. 2, pp. 137-144.

Langer, R. and Vacanti, J.P., 1993, Tissue engineering. *Science,* Vol. 260, pp. 920-926.

Lee, Y., Kouvroukoglou, S., McIntire, L.V. and Zygourakis, K., 1995, A cellular automaton model for the proliferation of migrating contact-inhibited cells. *Biophysical journal,* Vol. 69, No. 10, pp. 1284-1298.

Lee, Y., McIntire, L.V. and Zygourakis, K., 1994, Analysis of endothelial cell locomotion: Differential effects of motility and contact inhibition. *Biotechnology and bioengineering,* Vol. 43, No. 7, pp. 622-634.

Marée, A.F. and Hogeweg, P., 2001, How amoeboids self-organize into a fruiting body: Multicellular coordination in dictyostelium discoideum. *Proceedings of the National Academy of Sciences,* Vol. 98, No. 7, pp. 3879-3883.

Mooney, D.J. and Mikos, A.G., 1999, Growing new organs. *Scientific American,* Vol. 280, pp. 60-65.

Ratcliffe, A. and Niklason, L.E., 2002, Bioreactors and bioprocessing for tissue engineering. *Annals of the New York Academy of Sciences,* Vol. 961, No. 1, pp. 210-215.

Wolfe, S., 1983, Introduction to cell biology, Wadsworth Publishing Co., Belmont, CA.

Zygourakis, K., Bizios, R. and Markenscoff, P., 1991, Proliferation of anchorage-dependent contact-inhibited cells: I. development of theoretical models based on cellular automata. *Biotechnology and bioengineering,* Vol. 38, No. 5, pp. 459-470.

CYSTEINES ON AMINO TERMINUS INFER OXIDATION STATES OF OTHER CYSTEINES ON PROTEIN CHAINS

AIGUO DU
Georgia State University
34 Peachtree Street, Suite 1450
Atlanta, Georgia

YI PAN
Georgia State University
34 Peachtree Street, Suite 1450
Atlanta, Georgia

ABSTRACT
The binding states of cysteine residues on protein carry important information about the function and structure of proteins. In this research, support vector machine (SVM) was used for the prediction of bonding states of cysteines. We found that the oxidation states of amino terminus cysteines infer the oxidation states of other cysteines in the same protein chain. By including the oxidations states of up to two cysteines on the amino terminus and the flanking sequences around the cysteine as well as the global protein information, accuracy of 95% is achieved for the data set consisting of 4136 cysteine-containing segments extracted from 969 nonhomologous proteins.

INTRODUCTION

Cysteine is one of the few amino acids that contain sulfur. This allows cysteine to bond in a special way though disulfide bond and help to maintain/stabilize the 3-D structure of proteins. In addition, cysteines play very important roles in the function, properties as well as aging process of proteins. As is illustrated in Figure 1, cysteine residues have two possible chemical states in proteins: oxidized state and reduced state. These two states are interchangeable when proper condition is met. In its reduced form, cysteine can undergo chemical reactions such as alkylation, oxidation or forming complex compounds with metal ions. These chemical reactions carry critical biological meanings such as activation, deactivation of the active sites of enzymes and changes in the local environment of the protein. In their oxidized form, two cysteine residues from the same protein chain form intra-chain disulfide bond, which links distant portion of a protein chain and provides strong structural constraints in the form of long-range interactions. Two cysteines from different proteins form inter-chain disulfide bond and enable more complicated protein structures and functions. Interchange between its reduced form and oxidized form also serve as regulation switches for enzymes and other proteins [1,2]. Therefore, accurate prediction of the oxidation states of cysteine is essential to the studies of protein stability, protein function and three-dimensional structure[3~6] of proteins.

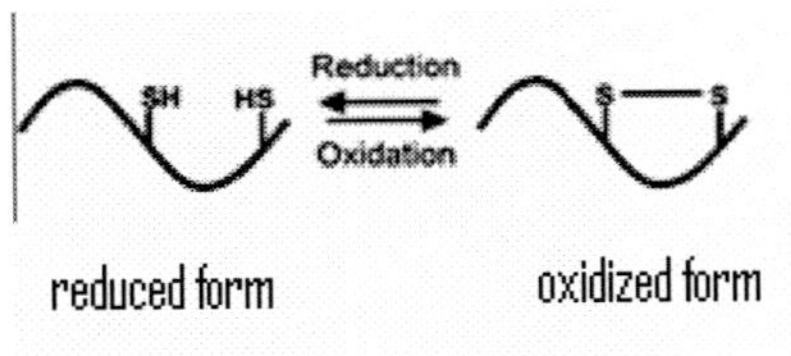

Fig 1: oxidation states of cysteine

There have been a number of studies in the prediction of the bonding states of cysteines. Muskal et al [7] obtained 81% accuracy by using neural network and the sequence information around the cysteines. Fiser et al.[8] performed statistical analysis on the amino acid frequencies in the sequence environment of cysteines and achieved 71% accuracy. Mucchielli-Giorgi et al.[9] assessed the relative efficiency of different descriptors in predicting the cysteine disulfide bonding states and concluded that the amino acid content of the whole protein is more informative than the flanking sequences. This approach has an accuracy of 84%. Martelli et al.[10,11] implemented a new hybrid system that combines a neural network and a hidden Markov model (hidden neural network) and obtained an overall accuracy of 88%. Frasconi et al.[12] introduced a support vector machines (SVM) based predictor which operates in two stages and incorporated both information at the protein level and local sequences. The prediction accuracy of this approach is 83.6%. By using the SVM based on multiple feature vectors and assuming that the transition probability of cysteines' oxidation states in the testing dataset was known beforehand, Chen et al.[13] achieved the record high accuracy of 90%. They pointed out that cysteine state sequence in a protein chain has significant influence on the prediction accuracy. Most recently, Ceroni et al. [14] employed SVM binary classifier to predict the bonding states of each cysteine, followed by a refinement step that take the overall bonding state assignment of the entire chain into consideration and reported an accuracy of 88%.

In this research, we found that the oxidation states of amino terminal cysteines infer the oxidation states of other cysteines in the same protein chain. The encodings includes the ordinal number (the order of the cysteine in the whole protein chain in comparison to other cysteines) of the cysteine in consideration, the oxidation states pattern of up to two amino terminal cysteines on the protein as well as the flanking sequence information and global information of protein chain. It was found that SVM based on this set of encodings gave improved overall accuracy (94.8%) for the data set consisting of 4136 cysteine-containing segments extracted from 969 nonhomologous proteins. This is 4.8% higher than the current record of 90% by Chen et al.[13]. The influence of the local flanking sequence of cysteines, the global information such as the number of cysteines present and amino acids composition of the whole protein chain were also studied.

METHOD

In this work, SVM were used to predict the potential bonding state of cysteines. Support Vector Machine (SVM) is a learning system that uses a hypothesis space of linear functions in a high dimensional feature space, trained with a learning algorithm from optimization theory [15]. This learning scheme was designed by Vapnik and his co-workers [16] and proved to be very powerful machine learning technique since its introduction. Compared with other machine learning techniques, SVM has a number of superior properties, such as effective avoidance of over-fitting, the ability to handle large feature spaces and information condensing of the given data set etc. It has been successfully applied to a wide range of pattern recognition problems, including isolated handwritten digit recognition, object recognition, speaker identification, and text categorization, etc.

Given data $x_1, \ldots, x_l$, the label for these data was set as +1 if x_i is in class 1 (in our case, +1 if the cysteine residue is intra-chain bounded with another cysteine) and as -1 if x_i belongs to class 2 (in our case, -1 if the cysteine residue is non-bounded or is inter-chain bounded with other cysteines). Then with these training data, SVM solves an optimization problem for binary classification:

$$\min_{\omega,b,\xi} \frac{1}{2}\omega^T\omega + C\sum_{i=1}^{l}\xi_i \qquad (1)$$

and

$$y_i(\omega^T\varphi(x_i)+b) \geq 1-\xi_i,\ \xi_i \geq 0 \qquad (2)$$

Where x_i is mapped to a higher dimensional space by the function φ; ξ_i is the allowable training error; C is the cost of error. The software SVM^{light} [17] was used in our experiments.

We first train an SVM to recognize the bounded and unbounded cysteines. Based on the training dataset, *SVM*^{light} returns a value regarding the bonding state of the cysteines. The decision was made based on the value returned by SVM. The cysteine is considered intra-chain bounded if the returned value was larger than 0. Otherwise, the cysteine was treated as non-bounded or inter-chain bounded.

Multiple feature vectors composed of information on flanking sequences, global information of the protein chain, oxidation states pattern of cysteines and ordinal number of the cysteine being considered were used in this study. We followed the previous literatures [7~14] in that only those cysteines that form intra-chain disulfide bonds were treated as in oxidized states and all free cysteines and those that form inter-chain disulfide bonds are treated as in reduced form. The global information vectors consist of two set of vectors. One corresponds to the total number of cysteines and the total number of amino acids on the chain. The other corresponds to the amino acids composition (calculated by number of particular amino acid/total number of amino acid). The ordinal number vector denotes where the cysteine is located relative to the other cysteines. The flanking sequence vectors are obtained from the adjacent amino acids of the interested cysteines. Representations developed by Meiler et al. [18] were used in this study, which accounted for graph shape index, polarizability, volume, hydrophobicity and isoelectric point of amino acids. Finally the vectors for oxidation state pattern of up to two amino terminus cysteines were used. When it is the first or second cysteine's oxidations state is to be predicted, then the oxidation state of that cysteine is not encoded.

The dataset in Martelli et al. [9] was used for testing purposes in this study. The dataset is comprised of 4136 cysteine-containing segments. There are 1446 segments in the disulfide bonded states and 2690 segments in the non-bonded states. The segments were extracted from 969 non homologous proteins from Protein Data Bank. The sequence identity of the set is less than 25%.

The prediction accuracy was calculated by following the standard conventions[19]: Accuracy for prediction is calculated by:

$$Q_2 = N_c/N_0 \qquad (3)$$

where N_c is the total number of correctly predicted cysteines and N_0 is the total number of cysteines.

Specificity is calculated by:

$$\text{Spec} = \frac{TN_x}{\text{TN}_\text{x} + \text{FP}_x} \tag{4}$$

And sensitivity is calculated by:

$$\text{Sens.} = \frac{TP_x}{\text{TP}_\text{x} + \text{FN}_\text{x}} \tag{5}$$

where x denotes the bonded cysteines or non-bonded cysteines, FP_x is the number of false negatives in the prediction and TN_x is the number of true negative predictions for bonding state x, FN_x is the number of false negatives for bonding state x. and TP_x is the number of true negative predictions for bonding state x

The Matthews correlation coefficient (MCC) [20] is calculated as:

$$\text{MCC}=\frac{TP_xTN_x - FP_xFN_x}{\sqrt{(TP_x+FN_x)(TP_x+FP_x)(TN_x+FP_x)(TN_x+FN_x)}} \tag{6}$$

The value of MCC is an indication of how good is the prediction. The closer the MCC is to 1, the closer the prediction is to a perfect prediction.

RESULTS AND DISCUSSION

Prediction performance of different combinations of input vectors was compared. The testing results are shown in Table 2. When the first cysteine's oxidation state is encoded, the overall accuracy is 87.4%. Encoding the first two cysteines on the N-terminus increases the accuracy to 94.7%. The encodings with the oxidation states of two cysteines on the amino terminus and with the global information on the protein chain yielded the best accuracy among all the tests. Accuracy of 94.8% and a MCC value of 0.89 are obtained. This might be explained by the critical biological findings that the terminals of proteins usually carry critical signals for the whole protein chain.

Table 1. Prediction performance of SVM based on different combination of input vectors

SVM vectors in encoding	SS-sens.	SS-spec.	SH-sens.	SH-spec.	Accu.	MCC
1 st cys	0.72	0.90	0.96	0.87	0.87	0.72
1,2 cys known	0.90	0.94	0.97	0.95	0.95	0.88
Global info +1st ,2nd cys	0.91	0.94	0.97	0.95	**0.95**	**0.89**
Flanking seq. + 1st, 2nd cys	0.90	0.94	0.97	0.94	0.95	0.88
1st ,2nd cys and flanking seq. and global info	0.91	0.94	0.97	0.95	**0.95**	**0.89**

Adding vectors for flanking sequences does not improve the accuracy further. However, by doing so, the sensitivity of free cysteines and specificity of the bonded cysteine were compromised slightly.

Between the 1st and 2nd cysteine on the amino end of the chain, the oxidation state of second cysteine is more influential to the oxidation states of other cysteines. Knowledge of the oxidation state of the second cysteine alone combined with the global information and flanking sequence information give a 91% accuracy, which has already outperform the best prediction result in literature [13], as is shown in Table 3.

Table 3: The relative importance of oxidation states of first cysteine and second cysteines from the amino side of protein chain. (In each testing, general information about the protein hain such as number of cysteines and number of amino acids on the chain were also included in the encoding)

SVM vectors used	SS-Sens.	SS-spec	SH-sens.	SH-spec	Accu.	MCC
1st cys only	0.72	0.9	0.96	0.87	0.87	0.72
2nd cys only	0.78	0.91	0.97	0.9	0.9	0.77
2nd cys & flanking	0.8	0.91	0.97	0.9	0.9	0.78
2nd cys & flanking & global	0.81	0.92	0.96	0.9	0.91	0.8
1st and 2nd cys & flanking & global	0.91	0.94	0.97	0.95	0.95	0.89

The way how flanking sequences was represented took a role too. As is shown in Table 4, when represented by frequency, the accuracy is about 2 % higher when represented by the order they appear.

Table 4: Influence of how flankcing sequences were represented in the encoding. (Also included in the encoding are 1st and 2nd cysteines' oxidation state, global information and general information about the chain, such as number of cysteines and number of amino acids on the chain)

SVM vectors used	SS sens.	SS spec.	SH sens	SH spec	accu.	MCC
flanking by frequency	0.91	0.94	0.97	0.95	0.95	0.89
flanking by order	0.86	0.94	0.97	0.93	0.93	0.85

CONCLUSIONS

In this research, support vector machine was used for the prediction of bonding states of cysteines. We found that the oxidation states of amino terminal cysteines infer the oxidation states of other cysteines in the same protein chain. By including the oxidations states of up to two cysteines on the amino terminus, accuracy of 95% is achieved. This might be explained by biological findings that the terminals of proteins usually carry critical signals. Between the 1st and 2nd cysteine on the amino end of the chain, the oxidation state of second cysteine is more influential to the oxidation states of other cysteines. Higher accuracy is obtained when the flanking sequences are represented by the frequency of each amino acid.

ACKNOWLEDGEMENT

This research was supported in part by the National Science Foundation (NSF) under Grants CCF-0514750, and CCF-0646102, and a GSU Research Program Enhancement Grant.

REFERENCES

[1] Nagahara, N., Yoshii, T., Abe, Y., Matsumura, T., Thioredoxin-dependent enzymatic activation of mercaptopyruvate sulfurtransferase: An intersubunit disulfide bond serves as a redox switch for activation.J Biol Chem. 2006 Nov 27; [Epub ahead of print]

[2] Mannick, JB., Regulation of apoptosis by protein S-nitrosylation. Amino Acids. 2006 Nov 30; [Epub ahead of print].

[3] Vielle, C., Zeikus, G., Hyperthermophilic enzymes: sources, uses and molecular mechanisms for thermostability, Microbiol Mol Biol Rew 2001; 65:1-43

[4] Hogg, P. J., Disulfide bonds as switches for protein function, Trends Biochem Sci. (2003) 210-214

[5] Mitsunori, F., Eiko, K., and Katsuhiko, M., Conserved N-terminal Cysteine Motif Is Essential for Homo- and Heterodimer Formation of Synaptotagmins III, V, VI, and X, J Biol Chem, (1999) 31421-31427

[6] Chuang, CC., Chen, CY., Yang, J-M., Lyu, PC., Hwang, JK., Relationship between protein structures and disulfide-bonding patterns, Proteins (2003) 1-5

[7] Muskal, S., Holbrook, S. and Kim, S., Prediction of the disulfide-bonding state of cysteine in proteins, Prot. Eng., (1990) 667-672

[8] Fiser, A., Cserzo, M., Tudos, E., Simon, I., Different sequence environments of cysteines and half cystines in proteins. Application to predict disulfide forming residues., FEBS Lett. (1992) 117-20

[9] Muccielli-Giorgi, MH., Hazout, S., Tuffery, P.. Predicting the disulfide bonding state of cysteines using protein descriptors. Proteins, (2002) 243-249

[10] Martelli, P.L., Fariselli, P., Malaguti, L. and Casadio, R., Prediction of the disulfide-bonding state of cysteines in proteins at 88% accuracy, Protein Science (2002), 2735-2739

[11] Martell, P. L., Fariselli, P., Malaguti, L. and Casadio, R., Prediction of the disulfide bonding state of cysteines in proteins with hidden neural networks, Protein Engineering, (2002) 951-953

[12] Frasconi, P., Passerini, A., Vullo, A. , A two-stage SVM architecture for predicting the disulfide bonding state of cysteines, Neural Networks for Signal Processing, 2002. Proceedings of the 2002 12th IEEE Workshop on, 25-34

[13] Chen, YC., Lin, YS., Lin, CJ. and Hwang, JK., Prediction of the bonding states of cysteines using the support vector machines based on multiple feature vectors and cysteine state sequences, Proteins: structure, function and bioinformatics (2004) 1036-1042

[14] Ceroni, A., Passerini, A., Vullo, A., and Frasconi, P., DISULFIND: a disulfide bonding state and cysteine connectivity prediction server, nucleic acids research, (2006) W177-W181

[15] Cristianini, N.., and Shawe-Taylor, J., An Introduction to Support Vector Machines. Cambridge University Press (2000)

[16] Vapnik, V. and Cortes, C., Support vector networks. Machine Learning (1995) 273-293

[17] Joachims T., Making large-Scale SVM Learning Practical. Advances in Kernel Methods - Support Vector Learning, B. Schölkopf and C. Burges and A. Smola (ed.), MIT-Press, (1999)

[18] Meiler, J., Muller, M., Zeidler, A., Schmaschke, F. Generation and evaluation of dimension-reduced amino acid parameter representations by artificial neural networks, J Mol Model (2001) 360:369

[19] Vapnik, V., The nature of statistical learning theory. New York: Springer; (1995)

[20] Matthews, B. W., Comparison of the predicted and observed secondary structure of T4 phage lysozyme. Biochim Biophys Acta (1975) 442-451

FUZZY NEURAL NETWORKS FOR DIAGNOSIS OF MALIGNANT MESOTHELIOMA

ARUN KULKARNI
Computer Science Department
The University of Texas at Tyler
Tyler, TX 75799
Arun_Kulkarni@UTTyler.edu

MADHUKAR BANDI
Computer Science Department
The University of Texas at Tyler,
Tyler, TX 75799
madhukarbandi@gmail.com

ABSTRACT

Many computer based diagnostic systems are being used in practice in areas such as the mammography, analysis of magnetic resonance imaging (MRI), and computer tomography. However, the diagnosis of pathological images is still more of an art, and it is based on the pathologist's knowledge, experience, and intuition. In pathology, suspect tissue samples are removed from the host and are stained before they are observed under the microscope. In this paper, we propose a computer based diagnostic system to analyze images of stained samples. The proposed system tries to mimic the human vision and decision making process. During preprocessing, pixels are mapped from the red-blue-green (RGB) color space to the hue-saturation-intensity (HIS) feature space. The feature extraction stage deals with obtaining color features from the segmented images. During the decision making, the images are separated in two classes positive and negatives.

BACKGROUND

Computer based diagnostic systems are commonly used by radiologists in areas such as the mammography, analysis of magnetic resonance images (MRI), and computer tomography. There are many advantages of using a computer diagnostic system. First, it yields consistent, objective, and repeatable results that are free from intra and inter observer variations. Secondly, unlike the human eye, which can differentiate only limited number of color intensity strengths, a computer diagnostic system can analyze the image pixel-by-pixel with a large number of color intensity values. Thirdly, a computer diagnostic system can handle a large amount of data with out fatigue and the disdain for repetitiveness. Until recently, most computer diagnostic systems have been designed for radiology to assist radiologists in decision-making. Computer diagnostic systems have not made their way to diagnosis of pathological images. The diagnosis of pathological images is based on the pathologist's knowledge, experience, and intuition. One of the significant differences between images in radiology and images in pathology is that most images in radiology are black-and-white and are obtained by some scanning mechanism. However, most pathological images are color images and are obtained by observing stained tissue samples under the microscope. Tissue samples are removed from the host and are stained before they are observed under the microscope. The process of staining samples modifies the color of the tissue sample. Images of stained tissue samples are used for diagnosis. In this paper, we considere a problem of designing a computer - based system to analyze pathological images for diagnosis of malignant mesothelioma (MM). Malignant mesothelioma is an aggressive neoplasm for which current therapy is unsatisfactory. This tumor causes up to 4000 deaths per year in the United States and it is a world wide health problem associated with asbestos exposure (Idell et al., 1995). In our approach, we try to mimics pathologists' vision and decision process. The key factor that pathologists use in decision making is the difference between the color of cell images and the background. In our implementation, we use color clustering for segmentation. Jiang, et al. (2006) have used color clustering for segmentation. We first segment images and then extract color features that distinguish cell areas and the background. Finally, we use a fuzzy neural network for decision making. We train the network using training samples. We have used twenty microscopic images of stained tissue samples obtained from the pathological laboratory at The University of Texas Health Center at Tyler.

METHODOLOGY

The proposed system consists of three stages the preprocessing, feature extraction, and decision making. During the preprocessing stage we map pixels from the red, blue, green (RBG) color space to hue, saturation, intensity (HIS) color space. In a color image, each pixel is characterized by a great number of combinations of RGB chromatic components. However, more complicated segmentation techniques are required to deal with rich chromatic information in the segmentation of color images. A number of color spaces have been proposed for segmentation of color images. The HIS color space is more suitable for segmentation because the RGB color space is far from exhibiting the perceptual uniformity and it does not model the way human perceive the color (Dong and Xie, 2005). In feature extraction stage, we segment the color image using unsupervised clustering. We use a fuzzy neural network with competitive learning algorithm for segmentation, and a perceptron model for decision making..

RGB to HIS Mapping

The RGB to HIS color space transformation is described by following equations (Gonzalez and Woods, 2002).

The intensity and saturation components are given by

$$I = \frac{1}{3}(R + G + B) \tag{1}$$

The saturation component S is given by

$$S = 1 - \frac{3}{(R + G + B)}[\min(R, G, B)] \tag{2}$$

The hue component H is given by

$$H = \begin{Bmatrix} \theta & \text{if } \mathrm{B} \leq G \\ 2\pi\text{-}\theta & \text{if } \mathrm{B} > \mathrm{G} \end{Bmatrix} \tag{3}$$

$$\text{where } \theta = \cos^{-1}\left[\frac{\frac{1}{2}[(R-G)+(R-B)]}{\left[(R-G)^2+(R-B)(G-B)\right]^{1/2}}\right]$$

Segmentation with Fuzzy Competitive Learning

In order to recognize objects in a given image, it is often necessary to subdivide the image into meaningful regions. The process of segmentation deals with partitioning the image into a set of non-overlapping regions whose union is the entire image. It is very difficult to find a criterion that yields meaningful segmentation. The segmentation problem can be informally described as the task of partitioning an image into homogeneous regions. Clustering is a good solution for unsupervised segmentation. We have used a fuzzy competitive network for segmentation.

A model for fuzzy competitive learning is shown in Figure 1. However, with the fuzzy competitive learning algorithm a sample can be assigned to ore than one class with the partial memberships. The partial memberships are used for updating weights. The model for fuzzy competitive learning consists of three layers: the input layer, fuzzifier layer, and output layer. Layer L_1 is the input layer, L_2 is the fuzzifier layer that maps input crisp values to fuzzy membership functions. We used triangular fuzzy membership functions in our implementation. Layers L_2-L_3 implements the competitive learning algorithm. To start with the number of clusters is set to zero. The input vector is first fuzzified. The first fuzzified vector is set as a prototype for the first cluster. With each input vector, the unit in the recognition layer with the highest weighted sum is declared as the winner. If the input vector satisfies the distance criterion the input vector is assigned to that cluster and the weights are updated using partial membership values. If no unit satisfies the distance criterion, a new cluster with the input vector as the centroid is formed. The process is described in the steps shown below (Kulkani and Muniganti, 1996; Kulkarni, 2001).

Let $\mathbf{X}_j$ represents the *i*th input vector, such that

$$\sum_{j=1}^{N} x_{ij} = 1 \qquad (4)$$

where *n* represents the number of features.

Step 1: To start with the number of clusters *M* is assumed to be zero. Let $\mathbf{x}_1$ be the first input vector. Assume, the first input vector as the first prototype for the first cluster and assign the weights accordingly, i.e. $\mathbf{x}_1 = \mathbf{v}_1$, where $\mathbf{v}_1$ represents weights connecting units in layer L_1 to unit i in layer L_2.

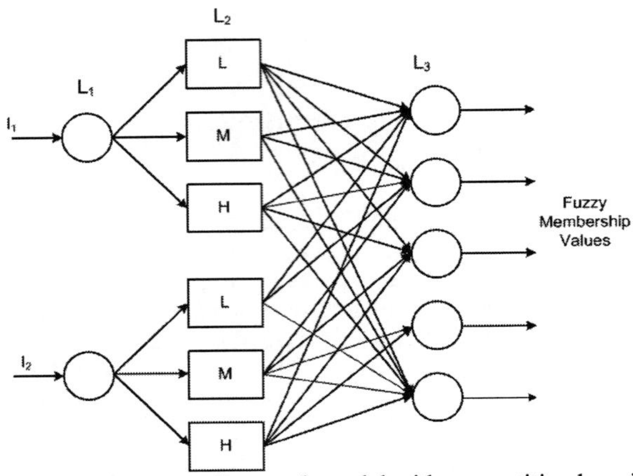

Figure 1. Fuzzy neural network model with competitive learning

Step 2: Let $\mathbf{x}_j$ be the next input vector. Find the winner unit in layer L_2 using the minimum distance (or the maximum dot product) as the criterion.

$$d_{\min} = \min \left\| \mathbf{x}_{jj} - \mathbf{v}_i \right\|^2 \qquad \text{for } i \in M$$

$$\text{where } \left\| \mathbf{x}_j - \mathbf{v}_i \right\|^2 = \sum_{k=1}^{M} \left(x_{jk} - v_{ik} \right)^2 \tag{5}$$

In Equation (5), M represents the number of clusters.

Step 3: If the winner unit does not satisfy the distance criterion given in Equation (6), then create a new cluster, and make its prototype vector be equal to $\mathbf{x}_j$.

$$R_i = \frac{\left\| \mathbf{x}_j - \mathbf{v}_i \right\|}{\frac{1}{N_i} \sum_{k=1}^{Ni} \left\| \mathbf{x}_j - \mathbf{v}_k \right\|} < \tau \tag{6}$$

where N_i is the number of samples in cluster i.

Step 4: Otherwise, update the winner cluster prototype associated with y_i by calculating the new centroid and membership values using Equations (7) and (8).

$$\mu_{ij} = \frac{1}{\sum_{k=1}^{M} \left(\frac{\left\| \mathbf{x}_j - \mathbf{v}_i \right\|^2}{\left\| \mathbf{x}_j - \mathbf{v}_k \right\|^2} \right)} \tag{7}$$

where $1 \leq i \leq M$ and $1 \leq j \leq N$.

$$\mathbf{v}_i = \frac{1}{\sum_{j=1}^{Ni} \mu_{ij}} \left(\sum_{j=1}^{Ni} (\mu_{ij})^m x_j \right) \tag{8}$$

where $1 \leq i \leq M$.

Step 5. If no more input samples then stop, else go to Step 2.

In decision making stage we used the perceptron model for classification. The model learns from training samples.

RESULTS

We developed software for the system that maps pixels from the RGB color space to HIS color space and used fuzzy competitive learning algorithm for segmentation. The segmented images are classified using the characteristics that separate cell images from the background. Figures 2 and 3 show the R,G,B and H, S, I components of an image of a positive tissue sample. The segmented image is shown in Figure 4. We used a set of 20 images that contained 10 positive and 10 negative. Five images from each set were used for training and remaining five samples from each set were used for testing.

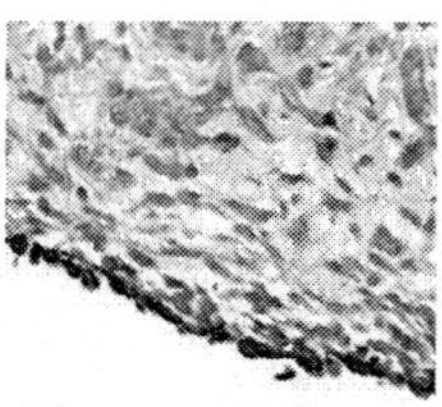

Figure 2a. R- Image

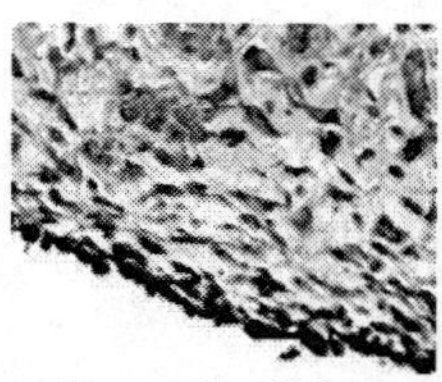

Figure 2b. G-Image

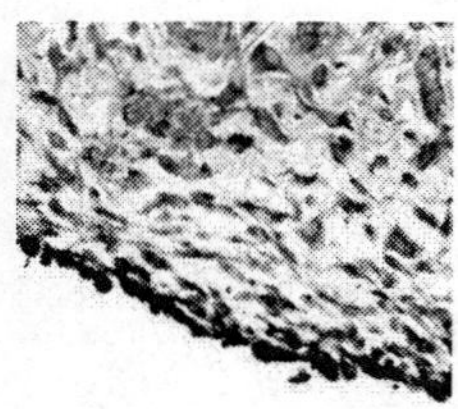

Figure 2c. B-Image

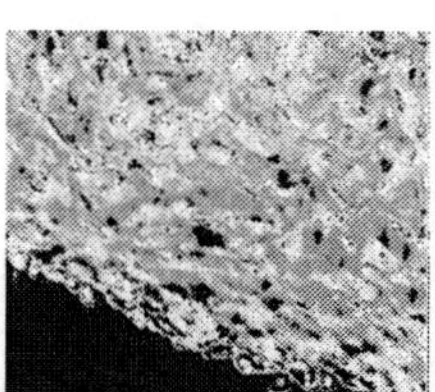

Figure 3a. H- Image

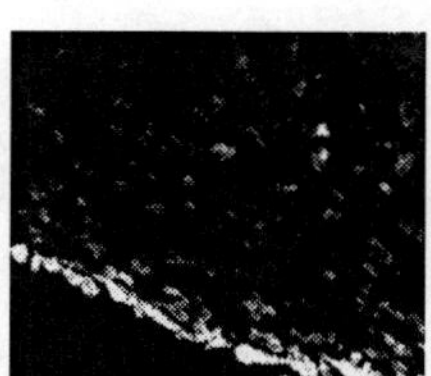

Figure 3b. S-Image

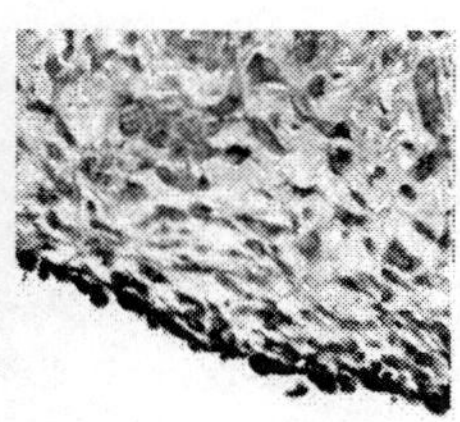

Figure 3c. S-Image

CONCLUSIONS

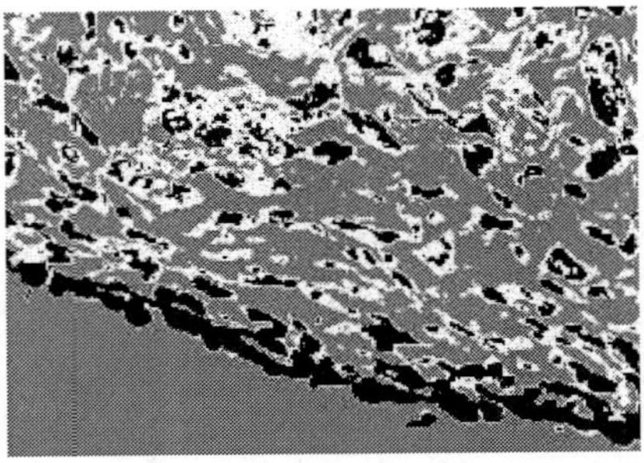

Figure 4. Segmented image

A novel approach for diagnosis of malignant mesothelioma using microscopic images of stained tissue samples is suggested. An algorithm is implemented for segmentation using clustering technique that identifies stained cells in the Malignant Mesothelioma tissue samples. In the decision making stage we used a perceptron model that learns from training samples. As an illustration the algorithm was tested with a set of twenty images obtained from a pathology laboratory. The threshold values for clustering were heuristically chosen.

Acknowledgements: The authors are thankful to Dr. Tim Allen for providing us with images of stained tissue samples.

References

Gonzalez, R. C. and Woods, R. C., 2002. *Digital image processing.* Prentice Hall, Upper Saddle River, NJ.

Jiang K., Liao Q. M., and Xiong, Y. 2006. A novel white blood cell segmentation scheme based on feature space clustering, *Soft Computing,* vol. 10, pp 12-19.

Idell, S., Pendurthi, U., Pueblitz, S., Koenig, K., Williams, L., and Rao. V. M. 1998. Tissue factor pathway inhibitor in tetracycline-induced pleuritis in rabbits. *Thromb Haemost*, vol. 79, pp 649-655.

Kulkarni, A. D., 2001. *Computer vision and fuzzy neural systems.* Prentice Hall PTR, Upper Saddle River, NJ.

Kulkarni, A. D. and Muniganti, V. K. 1996. Fuzzy neural network models for clustering. *Proceedings of the Symposium on Applied Computing*. Philadelphia, PA, pp 523-528.

PERFORMANCE VALIDATION USING SEVERAL STATISTICAL LEARNING THEORY PARADIGMS FOR MAMMOGRAM SCREEN FILM AND CLINICAL DATA FEATURES

WALKER H. LAND, JR
Dept. Of Bioengineering,
Binghamton University,
Binghamton, NY, 13902.

JOHN HEINE
H. Lee Moffitt Cancer Center &
Research Institute, Tampa, FL
33612

GEORGE TOMKO
Dept. Of Bioengineering,
Binghamton University,
Binghamton, NY, 13902.

ABSTRACT
In an ANNIE 2006 paper "Comparison of Logistics Regression (LR) and Evolutionary Programming derived Support Vector Machines (EP-SVMs) and CHI Squared derived results for breast cancer diagnosis", pp267-272 of ANNIE volume 16, it was demonstrated for the first time that LR and Evolutionary Programming derived SVMs provided essentially the same diagnostic accuracy using and ROC analysis. However, in a surprising preliminary finding, it was also shown that breast composition, age and family history are not significant diagnostic discriminators. This paper extends that research by investigating the performance results, employing the same data set and statistical cross validation methods, using, in addition to the above two paradigms, the probabilistic neural network as well as a Kernelized- partial least squares approach.

INTRODUCTION

This paper evaluates two machine intelligence emerging technologies: Kernel–Partial Least Squares (K-PLS) and Suport Vector Machines (SVMs) for diagnostic mammography applications and the classification of breast cancer. Both the Probablistic Neural Network (PNN) and Logistic Regression (LR) classifiers are used to validate the K-PLS and SVM results. The eventual intent is to develop a second opinion diagnostic aid for use in the clinical enviroment. Several applications of computational intelligence to breast cancer screening have been previously published. Earlier work by Floyd et al. (1994) and Lo et al. (1999) demostrated the applicability of backpropagation-trained Multiple Layer Feedforward Neural Networks (MLFNs) to the task of classifying lesions. EP-based feedforward networks have been applied by Fogel et al. (1995, 1995a, 1998) and Land et al. (2000) .They were able to achieve similar or better results than those trained by backpropagation, but with much simpler architectures (i.e., fewer nodes in the hidden layer). Evolutionary-Programming (EP) based linear discriminate analysis (Fogel et. al. 1995,1998a), and an adaptive boosting/EP hybrid (Land et. al. 2000a) also have been investigated as potential classification mechanisms. The original paper , and to the author's knowledge, the first to explore the use of two novel approaches called Partial Least Squares (PLS) and Kernel-PLS (K-PLS) to the diagnosis of breast cancer is described in (Land, et. al. 2005). This paper extends the research initially described in (Land, et. al. 2006) to validate the previous surprising findings using four classification paradigms.

THEORETICAL BACKGROUND

This section summarizes the approaches used to validate the results initially found in (Land, et. al. 2006) , as well as briefly comparing neural networks to SVM operation. Covered are summaries of the Probabilistic Neural Network, Statistical Learning Theory, Partial Least Squares (PLS) and Kernealized-PLS (K-PLS) , and EP-derived support vector machines. The reader may find a discussion of theory of logistics regression, as applied to computer aided diagnosis (CAD), described in (Land, et. al. 2007), and is not repeated here.

Probabilistic Neural Network (PNN)

The PNN is a classifier that makes effective use of the shapes of population distributions (Masters,1995). Given a point that previously has not been classified, the PNN determines the probability that the point belongs to a particular class based on populations of points whose class identity is known *a priori*. The PNN makes this determination in two steps. First, it uses the multivariate Parzen density to estimate (a constant multiple of) the density function of each population at the known point. It then computes the Bayesian probability of membership in each population and assigns the unknown to the class having the highest probability. The PNN has three useful properties: (1.) under reasonable assumptions, and as the training set size increases, the PNN is an asymptotically optimal classifier, (2.) the PNN is fully nonlinear. It neither imposes constraints of linearity on the model (as does ordinary discriminate analysis), nor requires linear separability like many other classifiers, and ,finally, (3.) the PNN has the ability to compute Bayesian probabilities, including use of priors if desired, which is a great advantage in many practical applications.

Statistical Learning Theory Kernels

One of the main concepts of statistical learning theory is kernel methods (Land, et. al. 2006) , which embody the following precepts: (1.) Data items are embedded into a vector space called a feature space, (2.) One attempts to ascertain linear relations developed from the data items in a feature space, (3.) Through the use of kernels, one attempts to design and implement algorithms so that the coordinates of the imbedded points are not directly used. Only the pair wise inner products of these are used, and (4.) And, from the above item, the pair wise inner products can be computed efficiently using a kernel function directly from the original data items.

Partial Least Squares (PLS) and Kernalized- PLS

PLS Regression was developed by the Swedish statistician Herman Wold (Lindgren, Geladi and Wold 1993) for econometrics modeling of multivariate time series. PLS can be viewed as a better principal components, where the data are first transformed into a different and non-orthogonal basis, and only a few (the most important) **PLS components** (or latent variables) are considered for building a regression model. PLS regression is one of the most powerful data mining tools for large data sets with an overabundance of descriptive features. What makes PLS especially interesting for computer aided diagnosis (CAD) is a recent extension to non-linear PLS or kernel PLS, which incorporates a kernel transform, similar to SVM's. The K-PLS method can be reformulated to resemble support vector machines, but it can also be interpreted as a kernel and centering transformation of the descriptor data followed by a regular PLS method. K-PLS was first introduced by Lindgren, Geladi and Wold (1993) in the context of working with linear kernels on data sets with more descriptor fields than data, in order to make the PLS modeling more efficient. The elegance and advantage of such a direct kernel method is that the non-linear aspects of the problem are captured entirely in the kernel and are transparent to the applied algorithm. If a linear algorithm was used before introducing the kernel transformation, the required mathematical operations remain linear. Consequently, linear methods such as principal component regression, ridge regression, and PLS can be turned into non-linear direct kernel methods, by using exactly the same algorithm and code: only the data are different, and we operate on the kernel transformation of the data rather than the data themselves.

Neural Networks (NN) and Evolutionary Programming Derived Support Vector Machines (EP-SVMs)

In using artificial neural networks, it is well known that the risk function is typically minimized by minimizing the empirical risk through adjustment of weights and biases, which generally results in a local minimum. We address this problem by using an iterative mutation-tourmentant selection process to ascertain the kernel mapping as well as the regularization parameters required to implement a SVM (Land, et.al. 2000). SVM implementations simultaneously minimize the empirical risk as well as the risk associated with the VC confidence interval, thereby addressing the local minimum problem mentioned above. That is, as the sample size $N \rightarrow \infty$, the empirical risk approaches the true risk because the VC confidence interval risk approaches zero. The reader may recall that obtaining larger and larger sets of valid training data would sometimes produce (with a great deal of training experience) a better performing neural network (NN), which resulted from classical training methods. This restriction is not incumbent on the EP-SRM principle and is the fundamental difference between training NNs and training SVMs.

SCALED MOFFITT DATA SET

This data set consists of 175 distinct individuals with 55 malignant cases, 40 benign controls determined by biopsy, and 80 benign follow-up controls without biopsy. The data were collected from the diagnostic center at the Moffitt Cancer Center & Research Institute. The cases are women with biopsy verified malignant masses, and the controls are women with benign breast lesions. For this work breast masses and architectural distortions define the lesion types. The benign controls consist of two distinct classes: those masses determined benign by biopsy and those determined benign after two-years follow-up without biopsy. The standard BIRADS descriptors used as input features were mass shape, mass margin, and density (relative attenuation of the mass), and the overall breast composition descriptors (defined below). In addition the presence of a halo (yes or no) around the tumor, family history (yes or no), and age were also used as input features. Mass shape is normally four-category rating: round, oval, lobulated, and irregular. The architectural distortion descriptor was added to the mass shape category making it into a five category rating for situations when there was no definite mass. Mass margin is a five-category rating: circumscribed-well defined microlobulated, obscured, indistinct, and spiculated. Density is a four-category rating of the relative attenuation of the mass volume relative to the same volume of fibroglandular tissue: high, low, equal, and fat-containing-radiolucent. Overall breast composition is a four-category relative possibility rating that the lesion could be obscured: entirely fat, scattered fibroglandular tissue, heterogeneously dense, and extremely dense.

RESULTS

In Land et. al. (2006), a logistics regression classifier showed that, for this data set, the features Halo, Family history, and Breast Composition did not contribute to classification accuracy. Consequently, for this validation study, Halo, Family History and Breast composition were removed from the original dataset and this subset, containing the **features mass shape and margin, density and age**, was evaluated by all of the separate classifiers previously mentioned. Results are delineated in tables 1 through 4. The Az results increased slightly for Logistic Regression, the PNN, and EP- SVM classifiers, and remained essentially the same for K-PLS, thereby verifying, **as least for this data set**, that these features do not contribute to classification performance. Note from table 1 that both the Wald and p statistical tests show the four remaining variables are important contributors to diagnostic accuracy.

LR Results	Fold1	Fold2	Fold3	Fold4	Fold5	Avg Az
Az	0.890152	0.890152	0.92029	0.950758	0.942308	0.918732
Model Fit	80.33496	84.46325	83.06561	88.09459	88.09459	84.8106

Wald Statistic Results

Variable	Fold1	Fold2	Fold3	Fold4	Fold5	Average
Mass shape	11.32577	11.32577	3.371042	9.481926	3.443567	7.789615
Mass margin	10.62018	10.62018	19.88363	8.83416	15.83919	13.15947
Density	11.2781	11.2781	9.959336	10.83089	9.669664	10.60322
Age	3.932637	3.932637	4.637525	7.271257	3.412639	4.637339

p values

Variable	Fold1	Fold2	Fold3	Fold4	Fold5	Avg p
Mass shape	0.063498	0.002075	0.066352	0.000764	0.000764	0.026691
Mass margin	6.90E-05	0.002956	8.23E-06	0.001119	0.001119	0.001054
Density	0.001873	0.000998	0.0016	0.000784	0.000784	0.001208

Age	0.064699	0.007007	0.03128	0.047358	0.047358	0.03954

Table 1: Logistic Regression Results for data subset for features

PNN

Fold		Basic	GCNN	SepClass	SepVar
	1	0.776515	0.719697	0.782197	0.702652
	2	0.887681	0.938406	0.880435	0.873188
	3	0.942029	0.971014	0.934783	0.934783
	4	0.935606	0.94697	0.931818	0.939394
	5	0.948718	0.991453	0.844017	0.961538
Average		0.89811	0.913508	0.87465	0.882311
Delta		-0.01051	0.053498	0.044474	0.021071

Table 2: PNN Results for different PNN configurations

EP-SVM Process

Selection Method	Kernel	avgAz
Rank	Rbf	0.92748
Rank	s2000	0.92688
Rank	Spline	0.92598
Rank	Tanh	0.92979
Tourn w/o Repl	Rbf	0.92748
Tourn w/o Repl	s2000	0.92688
Tourn w/o Repl	Spline	0.92598
Tourn w/o Repl	Tanh	0.9273
Tourn w/ Repl	Rbf	0.92748
Tourn w/ Repl	s2000	0.92688
Tourn w/ Repl	Spline	0.92598
Tourn w/ Repl	Tanh	0.9273

Table 3: SVM Results for different kernels

KPLS

Options:	Outliers	Latent Variables	Sigma	Az
	No	5	6.8	0.904924

Table 4 KPLS Results for data subset

CONCLUSIONS

This study evaluated the performance of several CAD classifiers using a reduced set of mammogrm BIRADS and clinical variable features to validate a previous conclusion that halo, family history and overall breast composition descriptors are not significant contributors to benign / malignant classification performance. Logistics Regression (LR), Probalistic Neural Networks (PNN) , Kernealized-Partial Least Squares (K-PLS), and Evolutionary Programming derived Support Vector Machines (EP-SVMs) all showed that ,at least for this data set, only mass shape and margin, density (attenuation) and age are the important indicators. Specificlly, the AZ performance results were, using this reduced data set: LR = 0.919, average PNN value of 0.892 , average EP-SVM performance using several kernels and selection methods = 0.927, and K-PLS = 0.905, which are comperable to the results obtained in Land et. al. (2006).

ACKNOWLEDGEMENT This work was in part funded by NCI grant # R21CA082639.

REFERENCES

Floyd, C.E., J.Y. Lo, A.J. Yun, D.C. Sullivan, and P.J. Kornguth. (1994) Prediction of Breast Cancer Malignancy Using an Artificial Neural Network. *Cancer*, 74:2944-98.

Fogel, D.B. (1995) Evolutionary Computation: Toward a New Philosophy of Machine Intelligence. IEEE Press, Piscataway, NJ.

Fogel, D.B., E.C. Wasson, and E.M. Boughton. (1995a) Evolving Neural Networks for Detecting Breast Cancer. *Cancer Letters*. 96:49-53.

Fogel, D.B., E.C. Wasson, E.M. Boughton, and V.W. Porto. (1998) Evolving Artificial Neural Networks for Screening Features from Mammograms. *Artificial Intelligence in Medicine*. 14:317-326.

Fogel, D.B., E.C. Wasson, E.M. Boughton, V.W. Porto, and P.J. Angeline. (1998a) Linear and Neural Models for Classifying Breast Masses. *IEEE Trans. Medical Imaging*, 17(3):485-488.

Freund, Y. (1995) Boosting a Weak Learning Algorithm by Majority. *Information and Computation*. 121(2):256-85.

Land, W. H., T. Masters, J. Y. Lo, and D. McKee. (2000) Using Evolutionary Computation to Develop Neural Network Breast Cancer Benign/Malignant Classification Models. *4th World Conference on Systemics, Cybernetics and Informatics*, 10:343-7.

Land, W. H., Jr., T. Masters, and J. Y. Lo. (2000a) Application of a New Evolutionary Programming/ Adaptive Boosting Hybrid to Breast Cancer Diagnosis. *IEEE Congress on Evolutionary Computation Proceedings*

Land, W.H., Heine, J., Embrechts, M., Smith, T., Choma, R., Wong, L., " New Approach to breast cancer diagnosis using partial least squares and kernel partial least squares", Medical Imaging, 2005, Image Processing, volume 5747,ISBN 0-8194-5721-3, pp 48-58

Land, W.H., Heine, J and Tomko, G., (2006) "Comparison of Logistics Regression (LR) and Evolutionary Programming derived Support Vector Machines (EP-SVMs) and CHI Squared derived results for breast cancer diagnosis", pp267-272 ANNIE volume 16

Land, W.H.,Jr., Heine, C., Bernam, C., Mizaku ,A, Raway, T., Kovalchuk, N., (2007) "Advancer in Automated Diagnostic mammography", IEEE BIBE07, (In press)

F. Lindgren, P. Geladi and S. Wold, (1993) ,"The kernel algorithm for PLS," *Journal of Chemometrics*, **7**, pp. 45-59.

Lo, J. Y., J. A. Baker, P. J. Kornguth, and C. E. Floyd, Jr. (1999) Effect of Patient History Data on the Prediction of Breast Cancer from Mammographic Findings with Artificial Neural Networks. *Acad. Radiology*, 6:10-15.

Masters, T., "Advanced Algorithms for Neural Networks", John Wiley and Sons, pp 157-220,1995

PREDICTION OF DISORDERED REGIONS OF PROTEINS WITH NEW N-PIECES NAÏVE BAYES ALGORITHM

UMUT ORHAN
Computer Center, Gaziosmanpasa University, Tokat, TURKIYE
umutorhan@gop.edu.tr

TURGAY IBRIKCI
Computer Engineering Dept, Cukurova University, Adana, TURKIYE
ibrikci@cukurova.edu.tr

Irem ERSÖZ
Technical Education Dept, Mersin University, Tarsus, TURKIYE
iremer@mersin.edu.tr

ABSTRACT

In this paper, Bayesian Learning with bioinformatics is utilized. Data is partitioned into a different numbers of pieces for Naïve Learning, which is based on binary Naïve Bayes algorithm. The algorithm is applied on the dataset R80 which consists of order/disorder protein structures. In our algorithm, the data is heuristically partitioned into 2, 5, 11, 101, 133, 201, 301, and undivided. The experimental results have indicated that the partition of the data into 133 pieces has given better results according to correct classification and specificity.

INTRODUCTION

The protein macromolecule represents much of the bulk of an organism and accomplishes almost all of its biochemical activities. It has been widely believed that the 3D structure of a protein is a prerequisite for its function. However, the last 50 years of research has revealed that some proteins or protein regions that lack a 3-D structure in the intrinsic state are also involved in a variety of important biological functions [1-4]. These proteins are now called natively unfolded or intrinsically disordered proteins [5]. The research has shown the importance and advantages of disordered proteins for many functional activities such as cell cycle regulation, molecular recognition, enzyme catalysis. Numerous tools and approaches for recognizing and predicting disordered regions of proteins have been developed [6-9]. Identification of these disordered proteins or regions can be done by several experimental techniques such as X-ray crystallography, nuclear magnetic resonance spectroscopy (NMR) but the methods require skilled experts to operate the equipment. This can be quite expensive and time consuming. Thus research has concentrated on study of computational methods including ANNs [10, 11], SVMs [12], logistic regression [13]. Surveys have demonstrated that amino acid composition and other attributes such as hydropathy, flexibility, sequence complexity, net charge, amino acid volumes, and secondary structure can be used as a discriminators for disorder/order propensity [14].

The application of machine learning techniques in bioinformatics problems has become increasingly known in recent years. There is a plethora of intelligent algorithms which are used in various fields. However, many of these algorithms can be used for only a restricted field. Neural networks and probabilistic prediction algorithms are common functions which can be implemented into a various fields. Bayesian networks are the subset of probabilistic prediction.

Bayesian networks are a useful tool for statistical modeling. They have been widely used in the biological sciences for the tasks of inferring cellular networks [15], modeling protein-signaling pathways [16], data integration, classification, and genetic data analysis [17]. Bayesian networks provide a neat compact representation for expressing joint probability distributions and for inference. The representation and use of probability theory makes Bayesian networks suitable for learning from incomplete datasets, expressing causal relationships, combining domain knowledge and data, and avoid over-fitting a model to training data.

A Bayesian network includes a classifier to show relationship among features. The classifier is a function f that maps the input feature vectors $x \in X$ to the output class labels $y \in \{1, . . . ,C\}$, where X is the feature space. It has been typically assumed that $X = R^D$ or $X = \{0, 1\}^D$, i.e. that the feature vector is a vector of D real numbers or D binary bits, but in general, discrete and continuous features have also been included. The class labels were taken as unordered and mutually exclusive in this study. If an input belongs to multiple classes, this is called a multi-label problem. The goal is to learn f from a labeled training set of N input-output pairs , (x_n, y_n) $n = 1 : N$ which has been known as supervised learning [18].

The probabilistic classifiers that recall $p(x \mid y)$ are investigated. The advantages of obtaining probabilities are discussed below. There has been two main ways to attain probability. The first is to learn directly the function that computes the class posterior $p(x \mid y)$, named a discriminative model, since it discriminates between different classes given the input. The alternative is to learn the class-conditional density $p(x \mid y)$ for each value of y and to estimate the class priors $p(y)$; therefore, the Bayesian Rule has been applied to compute the posterior,

$$p(x \mid y) = \frac{p(x, y)}{p(x)} = \frac{p(x \mid y)p(y)}{\sum_{y'=1}^{C} p(x \mid y')p(y')} \tag{1}$$

Naive Bayes is the simplest form of Bayesian network, in which all are independent given value of class variable. This is called conditional independence. The paper has proposed the N-pieces Naïve for constructing Bayesian networks from prior knowledge. This algorithm has been repeated for different N values (2, 5, 11, 101, 133, 201, 301 pieces and undivided data) in order for validation.

DATASET

In this study, only one dataset named R80 for training and validating processes is used. The dataset R80 including 80 protein taken from the study of Yang et al.[6] has been recreated via the PSSMP method by Su et al. [7].The details of the dataset are given at the Table 1. Each protein chain of the data contains a region of at least 21 consecutive disordered residues and all were collected from the PDB database. The method has been implemented on the R80 by conserving it without changes to compare the evaluations and measure the performance of Bayesian Learning on protein disorder prediction against the others.

Table 1: Summary of the dataset employed in the study

	Training - Testing Data
	R80
Number of chains	80
Number of ordered regions	151
Number of disordered regions	183
Number of residues in the ordered regions	29909
Number of residues in the disordered regions	3649
Total residues in the dataset	33558

VALIDATION

Predicting a residue in a given protein sequence as order or disorder is a binary classification problem, and many measures have been introduced for validation issues [19, 20]. Table 2 lists five widely used indices defined by previous related works [2-14]. The most important value is the correct classification in this paper. Also mistaken results have also been considered. Sensitivity represents the fraction of disordered residues correctly identified in a prediction method, while specificity indicates the fraction of ordered residues correctly identified. The Matthews' correlation coefficient is a popular measure in many bioinformatics problems. However, sensitivity, specificity, and the Matthews' correlation coefficient are seriously affected by the relative frequency of the target class. Therefore, the above three measures are not suitable for evaluating the performance in isolation. The probability excess is independent of the relative class frequency, and this measure can be reduced to sensitivity + specificity - 1 concisely [6, 7]. In this paper, since these five measures have the same tendency with each other, we are adopted specificity and correct classification. The other methods, however, are shown for comparison purposes.

Table 2: The definition of measures employed in the study

Measure	Abbreviation	Equation
Sensitivity	Sens.	$\frac{TP}{TP+FN}$
Specificity	Spec.	$\frac{TN}{TN+FP}$
Matthew's correlation coefficient	MCC	$\frac{TP\times TN-FP\times FN}{\sqrt{(TP+FN)\times(TN+FN)\times(TP+FN)\times(TN+FP)}}$
Probability excess	Prob. Excess	$\frac{TP\times TN-FP\times FN}{(TP+FN)\times(TN+FP)}$
Correct classification	Corr. Class.	$\frac{TP+TN}{TP+TN+FP+FN}$

The definition of the abbreviations used: *TP* is the number of correctly classified disordered residues; *FP* is the number of ordered residues incorrectly classified as disordered; *TN* is the number of correctly classified ordered residues; and *FN* is the number of disordered residues incorrectly classified as ordered.

N-PIECES NAÏVE BAYES ALGORITHM AND ITS APPLICATION

In fact, this is a simple idea that for digitalization, data is put into N number parts. The algorithm includes two phrases, which are partition of data and to apply classic Naive Bayes Learning classifier to partitioned data. The partition of data is based on heuristical values for N number pieces. Each piece of data is normalized and balanced itself. The data into two pieces is called binary data. These are then put into N balanced part. Thus, we obtain equal weighted N part are obtained. In second phase, the classic Naïve Bayesian Classifier method is applied to each piece of data.

This process was tried on the some different N values (2, 5, 11, 101, 133, 201 and 301). For each partition, the first step is to find the average error for digitalization. The results are shown in Figure 1 that shows relation between the average error and number of N pieces. When N is equal to be $6.3x10^6$, the average error is zero as the raw data. Different pieces data applied with the algorithm to follow their performances of correct classification to compare. This figure is used for finding the optimum N value with Figure 2.

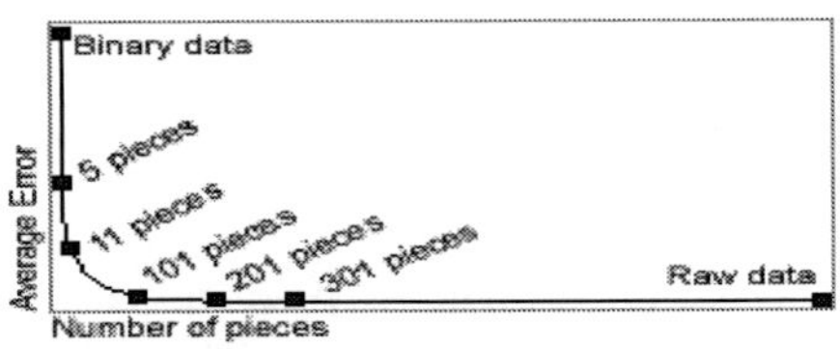

Figure 1: Relation between Average Error and N pieces values

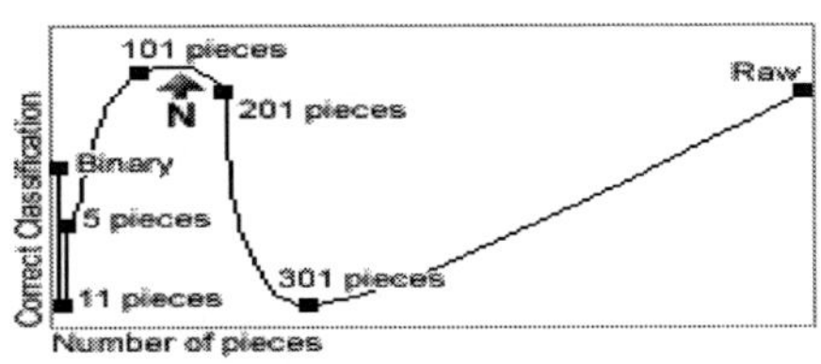

Figure 2: Relation between Correct Classification and N pieces values

When N is equal to 101, the results are promising (Figure 2). But these are not optimum values. The optimum N-pieces was looked sought for the dataset (Figure 3). From Figure 2, it appears that the optimum N value is between 101 and 201, and it is reasonable to use these values. Hence a better point can be obtained as shown by Figure 3.

The optimum N value cannot be 201, since its some other validation measurements are 0. Even 201 value for the correct classification (corr. class.) is better value then others.

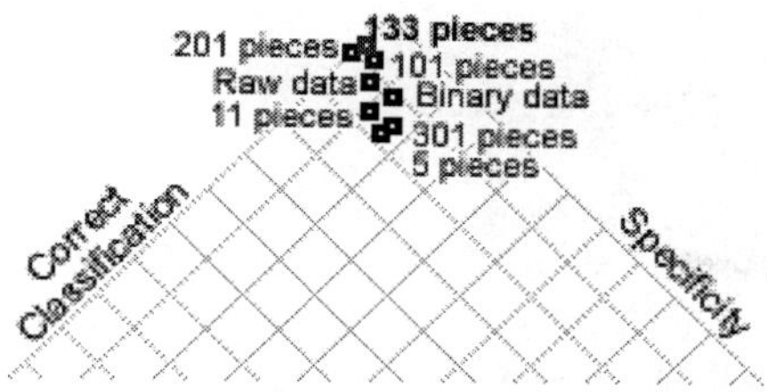

Figure 3: Relation between Correct Classification and Specificity

A study of gradient of line in Figure 2 is useful for the determination of the optimum N value. While the gradient is greater than a threshold, the optimum N value is sought. In Figure 2, the gradient between 201 and 301 is nearly equal to the gradient between 301 and $6.3x10^6$. Hence, the optimum N value cannot be in these regions. It may be between 101 and 201, but what can be an acceptable threshold value? The answer to this question is hidden in Figure 2. The N shows the optimal value which is 133 for the dataset.

Table 3: The performance of each N values on dataset

	Sens.	Spec.	MCC	Prob. Excess	Corr. Class.
Binary	0,61	0,91	0,5	0,52	0,87
5 Pieces	0,61	0,87	0,4	0,48	0,84
11 Pieces	0,57	0,90	0,4	0,47	0,86
101 Pieces	0,31	0,97	0,4	0,28	0,90
133 Pieces	0,09	1	0,08	0,09	0,89
201 Pieces	0	1	0	0	0,89
301 Pieces	0,61	0,87	0,4	0,48	0,84
Raw	0,45	0,95	0,4	0,40	0,89

CONCLUSION

This paper presented a new Naive Bayesian Learning algorithm which is called NP-NBL. The algorithm is tested on bioinformatics data with good results. The algorithm is applied on the prediction of disorder region of proteins. Many proteins have disorder structures that affect their functional activities. That is why the prediction of disorder structures of proteins are important.

The protein data is partitioned into 2, 5, 11, 101, 201, 301, and raw data. Actually, when we calculate the possibility of partitions which becomes $6.3x10^6$ pieces. According to Figure 2 that shows the critical values are between 101 thru 201. Then we focused on the gap to find the optimal partition of the data. The number of 133 for partition is optimal value with the specificity and the correct classification. However, we obtain the understanding that the optimal N value depending on individual applications.

ACKNOWLEDGMENT

This research was supported by TUBITAK-TBAG-104T505 and Cukurova University Research Funds Project Number MMF2006D11.

REFERENCES

[1] C. Klee, G. Draetta, M. Hubbard. Calcineurin. In Meister, A. (ed.) Advances in Enzymology, Vol. 61, 149–200, 1988.

[2] A.K. Dunker, Z. Obradovic, P. Romero, C. Kissinger, E. Villafranca. On the Importance of Being Disordered. PDB Newsletter, 81, 3-5, 1997.

[3] P.E. WRIGHT and H.J. DYSON. Intrinsically Unstructured Proteins: Re-assessing The Protein Structure-Function Paradigm. J. Mol. Biol., 293, 321-331, 1999.

[4] A.K. Dunker, P. Romero, Z. Obradovic, E.C. Garner, C.J. Brown. Intrinsic Protein Disorder in Complete Genomes. Genome Informatics, 11, 161-171, 2000

[5] A.K. Dunker, J.D. Lawson, C.J. Brown, R.M. Williams, P. Romero, J.S. Oh, C.S. Oldfield, A.M. Campen, C.M. Ratliff, K.W. Hipps, J. Ausio, M.S. Nissen, R. Reeves, C. Kang, C.R. Kissinger, R.W. Bailey, M.D. Griswold, W. Chiu, E.C. Garner, Z. Obradovic. Intrinsically Disordered Protein. Journal of Molecular Graphics and Modeling, 19, 26-59, 2001.

[6] Z. R. Yang, R. Thomson, P. McNeil and R. M. Esnouf. RONN: The Bio-Basis Function Neural Network Technique Applied To The Detection of Natively Disordered. Bioinformatics, 21, 3369-3376, 2005

[7] C.T. Su, C.Y. Chen and Y.Y. Ou. Protein Disorder Prediction by Condensed PSSM Considering Propensity for Order or Disorder. BMC Bioinformatics, 7:319, 2006.

[8] J.J.Ward, L.J. McGuffin, K. Bryson, B.F. Buxton, D.T. Jones. The Disopred Server for the Prediction of Protein Disorder. Bioinformatics, 20, 2138-2139, 2004.

[9] P. Romero, Z. Obradovic, A.K. Dunker. Sequence Data Analysis for Long Disordered Regions Prediction in the Calcineurin Family. Genome Informatics, 8, 110-124, 1997.

[10] P. Romero, Z. Obradovic, C.R. Kissinger, E. Villafranca, A.K. Dunker. Identifying Disordered Regions in Proteins from Amino Asid Sequence. In Proc. IEEE Int. Conf. On Neural Networks, 1, 90-95, 1997

[11] D.T. Jones, J.J. Ward. Prediction of Disordered Regions in Proteins from Position Specific Scoring Matrices. Proteins, 53, 573-578, 2003.

[12] K. Shimizu, S. Hirose, T. Noguchi, Y. Muraoka. Predicting The Protein Disordered Region Using Modified Position Specific Scoring Matrix. Genome Informatics, P150, 2004.

[13] S. Vucetic S, C.J. Brown, A.K. Dunker, Z. Obradovic. Flavors of Protein Disorder. Proteins, 52:573-584, 2003.

[14] P. Romero, Z. Obradovic, X. Li, E.C. Garner, C.J. Brown, A.K. Dunker. Sequence Complexity of Disordered Protein. Proteins: Structure. Function and Genetics, Vol. 42, 38-48, 2001.

[15] N. Friedman. Inferring Cellular Networks Using Probabilistic Graphical Models. 303, 709-805, 2004.

[16] K. Sach, O. Perez, D. Pe'er, D.A. Lauffenburger, and G.P. Nolan. Causal Protein Signaling Networks Derived from Multiparameter Singlecell Data. Science, 308, 523-529, 2005.

[17] M.A. Beaumont and B. Rannala. The Bayesian Revolution in Genetics. Nature Reviews Genetics, 5, 251-261.

[18] K. P. Murphy. Naive Bayes Classifiers. Technical Report, October 2006.

[19] E. Melamud and J. Moult. Evaluation of Disorder Predictions in CASP5. Proteins, 53:561-565, 2003.

[20] Y. Jin and R.L. Dunbrack. Assessment of Disorder Predictions in CASP6. Proteins, Early View, 2005.

PROTEIN SECONDARY STRUCTURE PREDICTION WITH HYDROPHOBICITY AND HYDROPHOBIC MOMENT

TZU-CHENG CHUANG
School of Electrical and Computer Engineering, Purdue University, West Lafayette, Indiana 47907
SAUL B. GELFAND
School of Electrical and Computer Engineering, Purdue University, West Lafayette, Indiana 47907
OKAN K. ERSOY
School of Electrical and Computer Engineering, Purdue University, West Lafayette, Indiana 47907

ABSTRACT
Protein secondary structure prediction has been satisfactorily performed by machine learning techniques such as support vector machines (SVM's). We discuss a special technique to include hyrophobicity information to further improve the classification results. Hydrophobicity or hydrophobic moment measure of each amino acid is included within a given window length in the protein secondary structure prediction using support vector machines. The input data is divided into two groups, which is subsequently classified by an SVM. By including hydrophobicity or hydrophobic moment, the classification accuracy is increased. Comparing the accuracy between using 1 SVM and 2 SVMs. 2 SVMs method has 3-9% higher accuracy than 1 SVM method.

INTRODUCTION

Protein structure prediction is a research topic of growing interest. The number of protein sequences deposited in the Protein Data Bank (PDB) grows faster than the numberof known protein structures. It is very time-consuming to crystallize each protein and use X-ray or nuclear magnetic resonance to analyze its structure.

From the tertiary structure, the shape of a protein can be estimated and which part of the protein binds to other proteins or enzymes can be predicted. Such information is important to design drugs. Before tertiary structure prediction, secondary structure prediction is usually carried out. Higher accuracy in secondary structure prediction may help predict tertiary structure more precisely.

Such research is usually initiated by using basic local alignment search tool (BLAST) or position-specific iterated BLAST (PSIBLAST) to find similar homologous proteins. Then, the sequences are aligned, and the position-specific scoring matrix (PSSM) is calculated. This is typically followed by a machine learning algorithm such as a neural network (NN) or a support vector machine (SVM) to do secondary structure prediction.

It is known that discriminating features may help improve classification accuracy. In this paper, we investigate how best to include new features to improve secondary structure prediction. We focus on several membrane proteins which have special distributions with respect to input features. In a previous paper on secondary structure prediction with support vector machines (Ibrikci et al., 2005), the protein structure was classified into 8 classes. In another paper (Rost et al., 2003), the protein structure was classified into 3 classes as alpha-helix, beta sheet and coil. In this paper, we first classify the data into 2 classes, such as alpha-helix and non-alpha-helix. Later by combining the results of three binary classifiers, the three-class classifier is obtained. The window size was investigated in a previous paper (Ibrikci et al., 2005). We pick window length equal to 7 in this paper.

Amino acids have different characteristics in terms of hydrophobicity and hydrophobic moment information. We first segment input data into two groups based on hydrophobicity or hydrophobic moment. Then, the data in each group is classified by an SVM. Below we refer to hydrophobicity unless otherwise specified since it tends to be more significant . A threshold is picked by maximizing the difference of the composition. The input pattern with hydrophobicity greater than the threshold is sent to SVM1. The input pattern with hydrophobicity smaller than the threshold is sent to SVM2. In this way, by segmenting the original dataset into 2 groups, each SVM is trained with similar composition of structural types, and the overall classification accuracy is expected to be higher. The overall structure of the learning algorithm is shown in Fig. 1. It is possible to include the hydrophobicity and/or hydrophobic moment information as part of the input vector. However, the method proposed here was more successful in increasing classification accuracy.

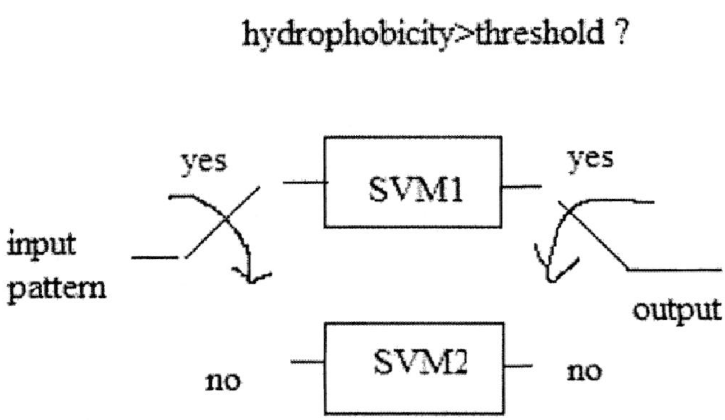

Figure 1. Hydrophobicity decision on which input pattern goes to which SVM.

DATASET

We focused upon membrane proteins since they are less understood than other proteins. The protein sequence and protein structure data for membrane proteins were obtained from the database of secondary structure assignments (DSSP) and the Protein Data Bank (PDB). The decision on which proteins are membrane proteins is obtained by using the information given in the Stephen White laboratory at UC Irvine (Stephen White laboratory). The protein id's for these membrane proteins are 2PPS, 1JB0, 1FE1, 1S5L, 2AXT, 1IZL, 1VF5 and 1Q90. Some proteins have more than one chain, resulting in 119 chains.

In the DSSP data, there are 8 classes: alpha-helix (H), beta-sheet (E), residue in isolated beta-bridge (B), 3-helix (G), 5 helix (I), hydrogen bonded turn (T), bend (S) and Coil (C). We can treat H as alpha-helix, E as beta-sheet and all the others as coil.We can also treat {H, G} as alpha-helix, {E, B} as beta-sheet and all the others as coil. The second approach was used in this paper since it is more commonly used in other literature as well.

HYDROPHOBICITY AND HYDROPHOBIC MOMENT

Hydrophobicity and hydrophobic moment were previously used to classify membrane and surface protein sequences (Eisenberg et al., 1984), (Yang, 2005). In Eisenberg hydrophobicity, the scale is normalized with mean value equal to zero and standard deviation equal to unity. The hydrophobic moment is calculated as follows :

$$\mu_H = \left\{ \left[\sum_{n=1}^{N} H_n \sin(\delta n) \right]^2 + \left[\sum_{n=1}^{N} H_n \cos(\delta n) \right]^2 \right\}^{1/2} \quad (1)$$

where μ_H is the hydrophobic moment, H_n is the hydrophobicity of the amino acid number *n*, *N* is the window length used (7 in this paper), and δ is the angle of the turn between two amino acids. For classical helices as discussed here, the angle was set to 100.

The following 2 figures show that alpha-helix, beta-sheet and coil have different distributions of hydrophobicity and hydrophobic moment. We included this information into our training and testing method.

Each amino acid has its own hydrophobicity. We calculated the sum of hydrophobicity scale in a given window length. We also calculated the hydrophobic moment in a given window length.

Based on this information, the threshold values were estimated. By including one or the other or both of these attributes, we decide which input pattern is classified by which SVM.

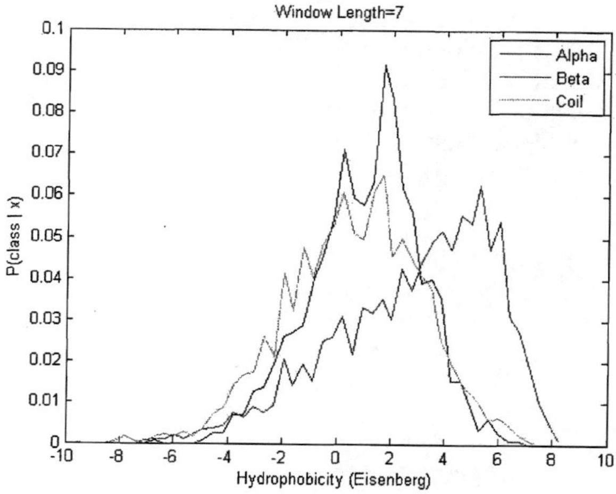

Figure 2: Hydrophobicity distribution for membrane photosystem protein with window length equal to 7.

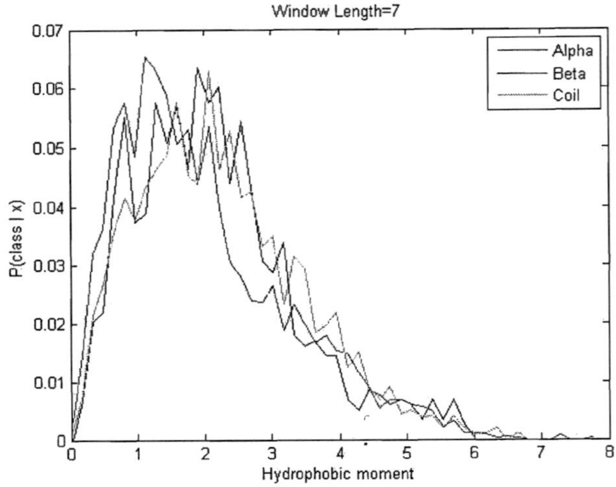

Figure 3: Hydrophobic moment distribution for membrane photosystem protein with window length equal to 7.

INPUT CODING METHOD

The class of an amino acid has some relation to its nearby neighbor. To take this into account, the window length was chosen as an odd number. For a reference amino acid, the amino acids which are half window length in front of the amino acid and behind it make up the input pattern in the window length. For example, for a window length equal to 7, the window includes the previous 3 and subsequent 3 positions of the amino acid. Each amino acid is represented as a 1x 20 vector in one-of-m representation like [0 1 0 ... 0]. If the window length is 7, the pattern within a window is represented as a 1x140 vector.

In DSSP files, some representations of amino acid are B, Z or X. B means the amino acid at that position could be Aspartic acid or Asparagine. Z means the amino acid at that position could be Glutamic acid or glutamine. X means the amino acid at that position is unknown or any amino acid. In such cases, we use 1 x 23 vector to represent an amino acid. In the experiments to be described, the 23-bit representation was used.

SUPPORT VECTOR MACHINES

Vapnik invented Support vector machines (SVM's) with a kernel function in the 1990s (Boser et al., 1992). This algorithm is initially designed for the two-class classification problem. One class output is marked as 1, and another class output is marked as -1. SVM learning involves optimization resulting in support vectors, which are vectors near the boundaries between two classes.

The algorithm trys to find the best separating hyperplane with the largest margin width. The hyperplane of the non-separable case is determined by solving the following equation:

$$\min \frac{1}{2}\|w\|^2 + C\sum_i \xi_i$$
$$subject.to. \qquad (2)$$
$$y_i(x_i^T w + b) \geq 1 - \xi_i$$
$$\xi_i \geq 0$$

where x_i is the ith data vector, y_i is the binary (-1 or 1) class label of the ith data vector, ξ_i is the slack variable, w is the weight vector normal to the hyperplane, C is the regularization parameter and b is the bias. It can be shown that the margin width is equal to $2/|w|$.

Usually the original data is mapped by using a kernel function to a higher dimensional representation before classification. Some common kernel functions are linear, polynomial, radial basis and sigmoid functions. From previous research (Ibrikci et al., 2005), radial basis function kernel is known to give good results. Hence, we used this kernel function in the experiments. Its kernel function is given by

$$K(x, x_i) = C * \exp(-\gamma * | x - x_i |^2) \qquad (3)$$

We chose γ equal to 1 and C equal to 1. SVM-light software package was used in the experiments (Joachims, 2004).

EXPERIMENTS AND RESULTS

In each experiment, we randomly picked some proteins for training and the others for testing so that the training set and the testing set do not overlap with each other. Since SVM is initially designed for 2 class classification problem, we classified the proteins into H/not H, E/ not E and C/not C. Later the results were combined to obtain 3-class classification. We set the hydrophobicity threshold equal to 0.42. We also tried several different thresholds near 0.42, and the accuracies were quite similar. Table 1 shows the results for H / not H classification.

We also investigated the distribution of hydrophobic moment to improve classification accuracy. In these experiments, we set the threshold equal to 1.7. The classification accuracy results are shown in Table 2.

In subsequent experiments, we combined three two-class SVM classifiers of H/ not H, E/ not E and C/ not C to obtain 3-class classification. We assumed the class belongs to the class which gives the largest output value. In these experiments, we used hydrophobicity to decide which input pattern goes to which SVM. The threshold is again chosen as 0.42.

The accuracy rate for the 3 classes is denoted as *Qtotal*, and is given by

$$Q_{total} = \frac{CorrectClasses}{N}\% \qquad (4)$$

where N is the total number of predicted residues and CorrectClasses is the number of correct classifications in which the prediction class is the same as the actual class. Table 3 shows the results.

Table 1. Comparison of the testing classification accuracy for H/ not H between one SVM only, and with two SVMs by considering hydrophobicity.

Experiment	1	2	3	4	5
Only 1 SVM	73.49%	81.19%	58.08%	71.69%	62.55%
SVM1	86.78%	88.63%	74.89%	82.61%	73.59%
SVM2	76.57%	83.52%	67.41%	81.22%	65.39%
2 SVMs overall	80.07%	85.29%	70.06%	81.72%	68.4%

Table 2. Comparison of the testing classification accuracy for H/ not H between one SVM only, and with two SVMs by considering hydromoment.

Experiment	1	2	3	4	5
Only 1 SVM	73.49%	81.19%	71.69%	62.55%	72.81%
SVM1	72.99%	82.19%	75.42%	60.57%	72.74%
SVM2	79.46%	83.79%	76.87%	60.64%	83.04%
2 SVMs overall	75.75%	82.87%	76.05%	60.6%	77.2%

Table 3. Comparison of Qtotal with one SVM only, and with two SVMs.

Experiment	1	2	3	4	5
Only 1 SVM	76.02%	54.09%	60.35%	54.09%	69.05%
2 SVMs overall	80.66%	63.04%	63.74%	63.04%	76.95%

DISCUSSION

This method is most useful when the distributions of hydrophobicity of the three classes are significantly separable. From the hydromoment experiment, Table 2, we see that one of the experiments actually gives worse result with 2 SVMs than with one SVM. The differentiability of the three classes with respect to the characteristic information used is really important to improve classification accuracy. Our results suggest that, since a binary classifier is used, other characteristics which differentiate beta-sheet versus non-beta-sheet or coil versus non-coil can be used with three different thresholds for different characteristics within different binary classifiers. Each classifier is specialized with the characteristic assigned to it. By combining their outputs with a consensus method, classification results can be improved further.

In the experiments conducted, the ratio of support vectors was almost half of the training vectors. This means the patterns are really difficult to differentiate.

The distribution of membrane photosystem protein is quite different from the other proteins.With some other protein data, such as the CB513 dataset, the three classes of structure have slightly different distribution in hydrophobicity (Fig. 4) and in hydrophobic moment (Fig. 5). In such cases, using hydrophobicity to classify E/not E and using hydrophobic moment to classify H/not H would give better results.

CONCLUSIONS

In this study, we showed a special approach to include hydrophobicity and/ or hydrophobic moment to improve classification accuracy in classifying the membrane proteins. In the future, we plan to search for some other natural discriminating features in alpha-helix, beta-sheet and coil to further increase classification accuracy. We also plan to use similar approaches with other non-homologous data sets, such as CB513 and RS126.

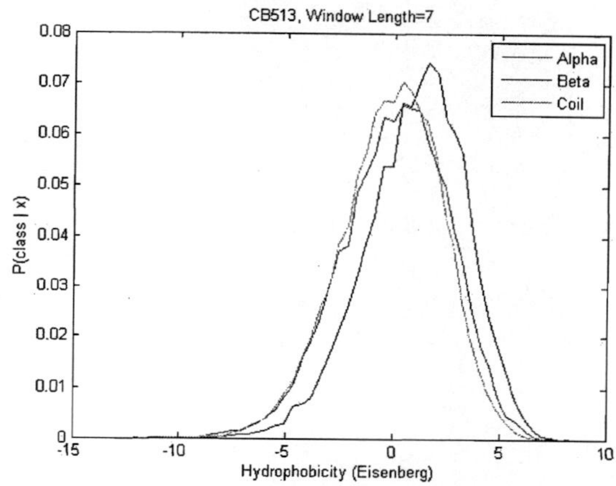

Figure 4: Hydrophobicity distribution for CB513 protein dataset with window length equal to 7.

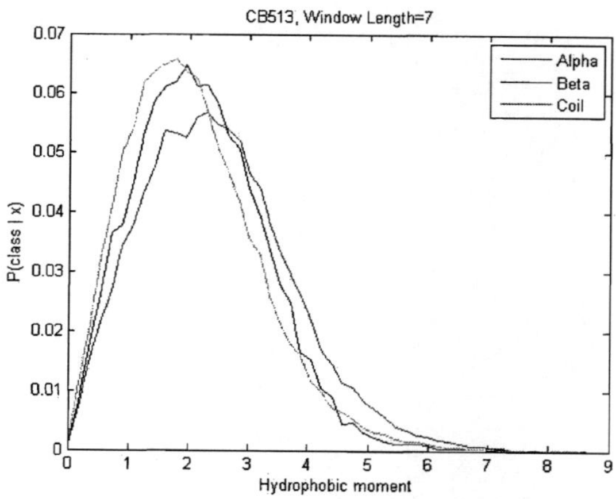

Figure 5: Hydrophobic moment distribution for CB513 protein dataset with window length equal to 7.

ACKNOWLEDGEMENT

This research was supported by NSF Grant MCB-9873139 and partly by NSF Grant #0325544.

REFERENCES

T.Ibrikci, Ayca Cakmak, Irem Ersoz and Okan K. Ersoy, "Hemoglobin secondarystructure prediction with four kernels on support vector machines," *Computational Intelligence Methods and Applications,* 2005 ICSC Congress Digital ObjectIdentifier 10.1109/CIMA.2005.1662310.

B Rost, G Yachdav and J Liu, "The PredictProtein Server," *Nucleic Acids Research* 32(Web Server issue), 2003 W321-W326.

Stephen White laboratory http://blanco.biomol.uci.edu/Membrane_Proteins_xtal.html#Latest

D. Eisenberg, E. Schwarz, M. Kmaromy and R. Wall, "Analysis of membrane and surface protein sequences with the hydrophobic moment plot," *Journal of Molecular Biology*, Volume 179, Issue 1, 15 October 1984, Pages 125-142.

J.Y. Yang, M.Q. Yang, O. Ersoy, "Datamining and knowledge discovery from membrane proteins," *Proceeding of IEEE Intelligence: Methods and Applications Conference*, Istanbul, June 2005.

B. E. Boser, I. M. Guyon, and V. N. Vapnik, "A training algorithm for optimal margin classifiers," D. Haussler, editor, *5th Annual ACM Workshop on COLT*, pages 144-152, Pittsburgh, PA, 1992. ACM Press.

Thorsten Joachims, 2004, http://www.cs.cornell.edu/People/tj/svm_light/

CLASSIFICATION OF ELECTROCARDIOGRAM ARRHYTHMIAS USING NEURAL NETWORKS

DAN ALSUP, BRIAN T. HEMMELMAN
South Dakota School of Mines and Technology
Electrical and Computer Engineering Department
Rapid City, South Dakota

ABSTRACT

The heart will beat over a billion times throughout a lifetime; however, a series of abnormal beats could result in death. Premature ventricular contraction (PVC), left bundle branch block (LBBB), and right bundle branch block (RBBB), are three types of abnormal heart beats which can lead to or indicate the risk of heart failure. A neural network has been trained and tested using signals from the MIT-BIH Arrhythmia Database to detect PVC, LBBB, RBBB, normal, and other arrhythmias. Wavelet, frequency, and temporal analysis of the QRS complex have been used to create neural network feature vectors used for classification. The network showed a specificity of 98.12% and a sensitivity of 93.35% for normal beats, 83.72% and 90.06% for PVC beats, 88.18% and 89.09% for LBBB beats, 92.85% and 96.14% for RBBB beats, and 77.21% and 89.78% for other arrhythmias. Overall, the network classified 97,545 true positives of the 105,287 testing beats, for an overall positive predictability of 92.65%.

1. INTRODUCTION

The heart is a muscular pump that controls the flow of blood throughout the human body. The pumping of the heart is controlled by electrical charges distributed through the heart's conduction system. The impulses conducted during the cardiac cycle generate electrical charges that spread through the surrounding tissues, and to the surface of the skin. From the skin surface, these charges display the electrical conduction of the heart, and can be measured with an electrocardiograph. An electrocardiogram (ECG) is not a record of the contractions of the heart, but of the electrical impulses which precede these contractions. [8]

A normal ECG, as shown in Figure 1, is composed of a P wave, QRS complex, and a T wave. The QRS complex usually ranges between 80 and 120ms in duration and is easily located by the R-peak. One interval of note is the RR interval, which is the distance between R-peaks and corresponds to the heart rate. [4]

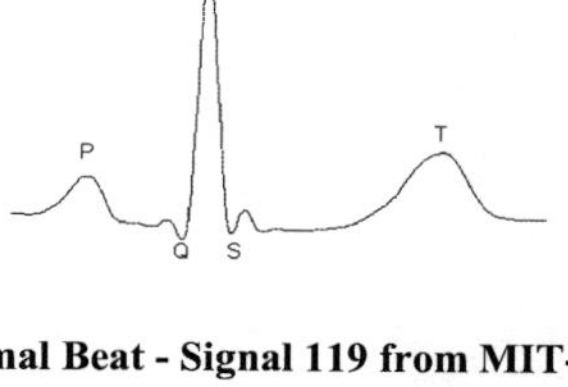

Figure 1. ECG Output for Normal Beat - Signal 119 from MIT-BIH Arrhythmia Database[2]

Sinus rhythm is defined as the regular rhythm of the heart as controlled by the SA node's rate of firing. An arrhythmia is an abnormal cardiac rhythm, or deviation from sinus rhythm. [4] Premature ventricular contraction (PVC), left bundle branch block (LBBB), and right bundle branch block (RBBB), are three arrhythmias which can lead to or indicate the risk of heart failure.

2. NEURAL NETWORK FEATURES

2.1 RR Interval

A PVC beat is a contraction of the ventricles prior to receiving the contract signal from the SA node. This premature contraction is displayed in an ECG signal as a heart beat that is earlier than the normal sinus rhythm followed by a pause, with a wider than normal QRS complex.

Because a PVC beat has an irregular beat rhythm, the RR interval is useful in classifying a PVC beat. [1] The RR interval from a normal beat to a PVC beat is shorter than the RR interval between two normal beats. Also, the RR interval following a PVC beat is longer than normal RR intervals. Therefore, the following and preceding RR intervals were used as two features for the neural network classification of PVC beats.

2.2 Form Factor

In addition to the variation in the RR interval, the shape of the QRS complex of a PVC beat is different than a normal heartbeat. The form factor is a calculation that can assist in differentiating between the shapes of a waveform. This has been proven to be useful in differentiating between the shape of a normal beat and a PVC beat. [1]

The form factor determines the measure of signal activity using the square of the variance of a signal. [1] The variance, σ_x^2, is the measure of the deviation of a signal from the mean. As a signal varies from the mean, the value of the form factor increases. This is often used in statistics, as the square root of the variance is the standard deviation. [5]

$$\sigma_x^2 = E\left[(x-\mu_x)^2\right] \tag{1}$$

where $\mu_x = E(x)$ is the mean of the signal x. The mobility, M_x, of a signal is defined as the square root of the ratio of the activity of the derivative of the signal and the activity of the signal. The form factor, FF_x, is the ratio of the mobility of the derivative of a signal and the mobility of the signal. [1]

$$FF_x = \frac{M_{x'}}{M_x} = \frac{\sqrt{\sigma_{x''}^2 / \sigma_{x'}^2}}{\sqrt{\sigma_{x'}^2 / \sigma_x^2}} = \frac{\sigma_{x''} / \sigma_{x'}}{\sigma_{x'} / \sigma_x} \tag{2}$$

where x is the input ECG signal, x' is the first derivative, and x'' is the second derivative of the ECG signal. The form factor was calculated using the region from 80ms before the R-peak to 80ms after the R-peak as the input.

2.3 QRS Frequency Spectrum

The frequency spectrum of the QRS complex has been proven to be useful in classifying PVC beats. A normal beat is narrower and contains more sharp peaks than a PVC beat. These sharp peaks result in more of the power higher in the frequency spectrum for a normal beat in comparison to a PVC beat. [6]

The Fast Fourier Transform (FFT) was used to calculate the frequency spectrum of the QRS complex. For this frequency analysis, the portion of the QRS complex used for analysis was defined as the region from 100ms before the R-peak to 200ms after the R-peak. The spectrum was then broken into eight regions for analysis, each 4 Hz wide. The first region was centered on 4 Hz, which consisted of frequencies from 2 – 6 Hz, and the remaining seven regions were centered on 8, 12, 16, 20, 24, 28, and 32 Hz. The power of each region was computed, and the eight regions were normalized giving five feature inputs for the neural network. Figures 2a and 2b show the normalized frequency regions up to 32 Hz of normal and PVC beats. It can be seen in these figures that the higher frequency regions contain more power for a normal beat in comparison to a PVC beat.

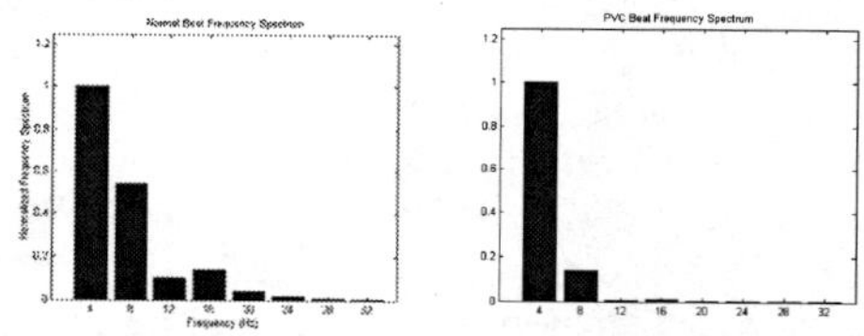

Figure 2 - (a) Frequency Spectrum of Normal Beat. (b) Frequency Spectrum of PVC Beat.

Similar to the frequency spectrum of a PVC beat, the frequency spectrum of an LBBB beat contains more power in the lower frequencies when compared to a normal beat. This is due to the fact that a normal beat contains more sharp peaks than an LBBB beat. Figures 3a and 3b, which are plots of the frequency spectra of a normal beat and an LBBB beat, show the attenuation of the higher frequencies in an LBBB beat. The frequency spectrum was broken into two regions, from 2 – 12 Hz and from 12 – 23 Hz. For these calculations, the portion of the QRS complex used for analysis was defined as the region from 25ms before the R-peak to 85ms after the R-peak. The total power contained in these regions was calculated and normalized, providing an additional two features as inputs to the neural network.

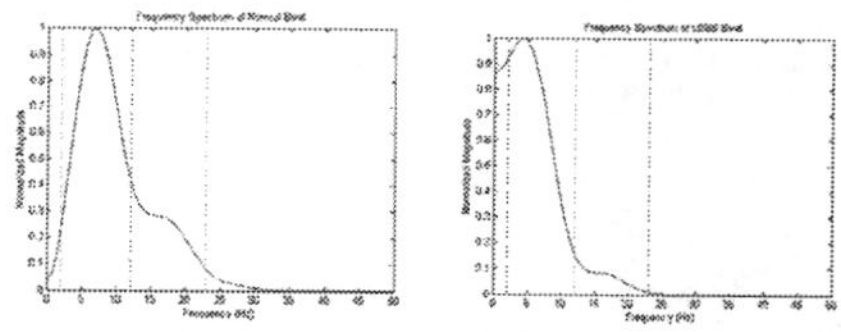

Figure 3 - (a) Frequency Spectrum of Normal Beat. (b) Frequency Spectrum of LBBB Beat

2.4 A.R. Coefficients

Autoregressive (AR) modeling is a process by which the current value of a signal is determined from previous values of itself. AR modeling has been used for modeling ECG signals, heart rate variability, as well as classification of arrhythmias. [8] The QRS complex was characterized using AR modeling with an order of 7, and the seven AR coefficients were used as neural network features. The region used as the QRS complex for AR modeling was from 50ms before the R-peak to 100ms after. The AR coefficients from the modeling of a signal vary based on the shape and properties of the signal. The AR coefficients were calculated using Burg's algorithm. [5]

2.5 Continuous Wavelet Transform

The The Continuous Wavelet Transform (CWT) is a method of decomposing a signal into a combination of the time and frequency domains. This method is based on the Short Time Fourier Transform, and was developed to work around the lack of time-frequency resolution with other transforms.[3] The CWT is defined as

$$C(a,b) = \int f(t) \cdot \Psi(a \cdot t + b)dt \tag{3}$$

where f(t) is the input signal and $\Psi(a \cdot t + b)$ is the mother wavelet. The mother wavelet defines the transform space, and the output of the CWT is dependant on the morphology of the mother wavelet. Note that the output of the CWT is a function of both a and b, where a is the scale and b is the translation of the transform space. The scale, a, corresponds to the frequency information, and the translation, b, corresponds to the time information in the CWT domain. [7]

When comparing Figures 4a and 4b, it can be seen that more power, indicated by a lighter color, is located in the region from 25 -35 in a normal beat in comparison to an LBBB beat. Testing the various scales of the CWT showed that this difference in power was maximized at scale 12. The output from the CWT at this scale was divided into two regions, from 5 - 15 and 25 – 35 (Figure 4c). The two regions were normalized, providing two feature inputs to the network. These computations were calculated using the range of 25ms before the R-peak to 85ms following the R-peak with the mexican hat mother wavelet.

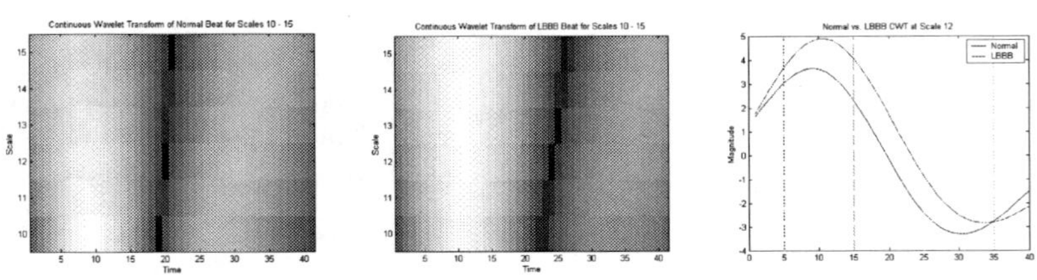

Figure 4 – (a) CWT of Normal Beat. (b) CWT of LBBB Beat. (c) CWT of Normal and LBBB Beats at Scale 12

2.6 Mean and RMS

An LBBB beat is wider than normal, and is often displayed by an inverted QRS complex. Because of the inverted QRS complex, the mean and root mean square (RMS) values of the QRS complex are useful in classifying LBBB beats.

The RMS is a measure of the magnitude of a varying signal, and is useful in determining the average power of a signal as a function of time. [1] These QRS computations were calculated using the range of 25ms before the R-peak to 85ms following the R-peak.

2.7 Network Topology

The features described above were combined to make a vector of 26 features per beat as an input to the neural network. A feed-forward backprop neural network was used to classify these features as a normal, PVC, LBBB, RBBB, or other beat. The network architecture contained three total layers of nodes; 14 nodes in the first layer, 8 in the second, and the third layer contained 5 output nodes. Each of the five output nodes was dedicated for each of the five beat classifications, where a value greater than 0 indicated a positive, and a value less than 0 indicated a negative for the given classification.

3. NEURAL NETWORK TESTING

The performance of a neural network is defined by how the network classifies inputs that have not been seen by the network. In this study, the performance of neural networks is evaluated by the sensitivity, specificity, and correct classification of unseen data. These values were computed given the number of true positives (TP), true negatives (TN), false positives (FP), and false negatives (FN). A TP is defined as a real event classified as a real event by the network, a TN is a non-event classified as a non-event, a FP is a non-event classified as an event, and a FN is an event classified as a non-event. [6] For example, a PVC input to a network would be defined as an event, while a non-PVC input is a non-event. A TP is a PVC beat classified as a PVC beat, a TN is a non-PVC classified as a non-PVC, a FP is a non-PVC classified as a PVC, and FN is a PVC classified as non-PVC.

The sensitivity is the percentage of real events that are correctly detected out of all real events.

$$Sensitivity = \frac{TP}{TP + FN} \tag{4}$$

The specificity is the percentage of non-events that are correctly rejected out of all non-events.

$$Specificity = \frac{TN}{TN + FP} \tag{5}$$

The correct classification (CC) is the overall percentage of beats classified correctly.

$$CC = \frac{TP + TN}{TP + TN + FN + FP} \tag{6}$$

4. NEURAL NETWORK RESULTS

A total of 109,287 beats from 48 files in the MIT-BIH Arrhythmia Database were used for training and testing of this network. The neural network was trained using 1,500 normal, 500 PVC, 500 LBBB, 500 RBBB, and 1,000 other beats; a total of 4,000 beats. The feature inputs, along with the annotations from the MIT-BIH Arrhythmia Database were used to train the neural network to classify beats using supervised learning.

After the network was trained, 97,545 beats of the 105,287 remaining beats were classified by the network as true positives, for an overall positive predictability of 92.65%. Table 1 shows the results for all 105,287 testing beats for each of the five output nodes.

Table 1 - Neural Network Testing Results

Classification	TP	TN	FP	FN	Sensitivity	Specificity	CC
Normal	68540	30580	1286	4881	93.35%	98.16%	94.14
PVC	5955	97517	1158	657	90.06%	83.72%	98.28
LBBB	6745	96812	904	826	89.09%	88.18%	98.36
RBBB	6496	98030	500	261	96.14%	92.85%	99.28
Other	9809	91466	2895	1117	89.78%	77.21%	96.19

5. CONCLUSIONS

An ECG signal can be measured on the skin in various locations on the body and describes the electrical charges which precede the pumping of the heart. ECG signal analysis is useful in detecting cardiac arrhythmias which can lead to or indicate the risk of heart failure. This paper has presented that arrhythmias can be detected using neural networks and information about the frequency spectrum of the QRS complex, the location of power in the signal at a given CWT scale, and various temporal features such as the RR interval, form factor, mean, and RMS values. The designed neural network classified PVC, LBBB, RBBB, normal beats and other arrhythmias at an overall positive predictability of 92.65%.

6. REFERENCES

[1] R. M. Rangayyan, Biomedical Signal Analysis – A Case-Study Approach. Piscataway, NJ: IEEE Press, 2002.

[2] R. Mark and G. Moody, MIT-BIH Arrhythmia Database Directory. Massachusetts Institute of Technology, 1988.

[3] R. Polikar, "The Engineer's Ultimate Guide to Wavelet Analysis – The Wavelet Tutorial," January 2007, http://users.rowan.edu/~polikar/WAVELETS/WTtutorial.html.

[4] G. D. Clifford, F. Azuaje, and P. E. McSharry, Advanced Methods and Tools for ECG Data Analysis. Norwood, MA: Artech House, Inc, 2006.

[5] E. A. Robinson and M. T. Silvia, Digital Foundations of Time Series Analysis – The Box-Jenkins Approach. San Francisco: Holden-Day, Inc, 1979.

[6] M. A. Chikh, "Application of artificial neural networks to identify the premature ventricular contraction (PVC) beat," Electronic Journal -Technical Acoustics, March 2004.

[7] B. B. Hubbard, The World According to Wavelets. AK Peters, Ltd, 1996.

[8] C. P. Anthony and G. A. Thibodeas, Textbook of Anatomy and Physiology, 10th Edition. St Louis: The C.V. Mosby Company, 1979.

MODELING AND REALIZATION ON A CHIP OF GROWTH HORMONE SECRETION MECHANISM

JOHN R. SHELL
Southern Illinois University

NAZEIH M. BOTROS
Southern Illinois University

YONGLIAN WANG
Southern Illinois University

ABSTRACT
In this paper, we present a mathematical model for the Growth Hormone (GH) secretion mechanism. The model is simulated using Hardware Description Language; the simulation is then synthesized on a compact electronic chip. This paper is an extension of an extension of previous work [1]. After synthesis, the chip is a stand-alone electronic device that mimics the mechanism. The chip is a Xilinx Field Programmable Gate Array (FPGA). The chip, because it is compact, can be attached to the skin of the patient and monitor, through miniature sensors, the concentration of the hormones involved in the GH secretion. The mathematical model is developed based on a modified bio-mathematical nonlinear delay differential equations. The model explores the effects of Growth Hormone Releasing Hormone (GHRH), Somatotropin Release-Inhibiting factor (SRIF), and Insulin-like growth hormone (IGF-1) on Growth Hormone (GH). Testing of the chip indicates that the chip correctly mimics the GH secretion mechanism.

INTRODUCTION

To understand the function of biological mechanisms in the human's body and to develop drugs to treat maladies in a particular mechanism, scientists and researchers need to carry out extensive investigation of the mechanisms. Medical experimentation on humans and animals is extremely limited and in some instances prohibited due to ethical issues or other constraints. An alternative approach to investigate the mechanism is to simulate it and then realize (synthesize) it on a compact electronic chip. The chip after synthesis mimics the function of the mechanism. The synthesized chip offers researchers and physicians a portal to explore and/or correct the functions of the mechanism. Also, because the chip is compact, it can be attached to the skin or be implanted in the body of the subjects. An example of such innovation is reported by Wolpert [Wolpert, 2005] to realize a simple function of the pancreas, the secretion of insulin. In this paper we focus on the Growth Hormone (GH) mechanism since it is one of the most important mechanisms in the human's body. The technique presented here is applicable to other biological mechanisms. The goal of this work is to model a simplified GH secretion mechanism and to realize the model on an integrated compact electronic chip such as the Field Programmable Gate Array (FPGA). After realization of the mechanism, the chip mimics the function of the realized mechanism. The chip offers a unique opportunity for researchers to explore the function of the mechanism.

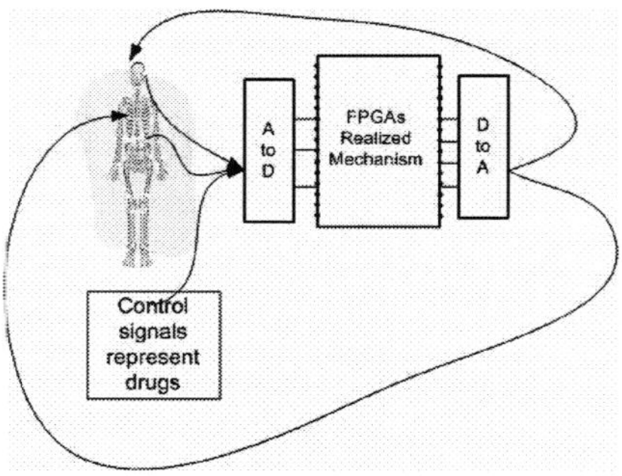

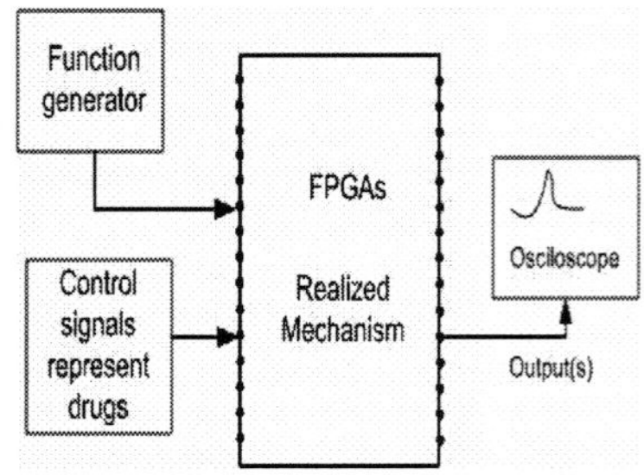

Figure 1, In-vivo Environment. Figure 2, Laboratory environment

The chip can be implemented in two environments, in-vivo and laboratory. Fig. 1 shows a general diagram of a possible in-vivo interface between the chip and the human body. The chip is attached to the body of the subject. The chip inputs are electrical signals fed from sensors attached to a selected region of the body such as chest, head, or muscle. In our case the signals represent the concentration of the hormones that control the function of the GH mechanism. The chip, through the sensors, can be programmed to monitor and analyze, in real time, the behavior (such as concentration level, or rate of change) of one or more of the hormones that control the GH mechanism; after analyzing the behavior of the hormones the chip can make a decision on certain programmed action such as turning on a pump, calling a doctor, or transmitting the data to a physician. The laboratory environment implementation is shown in Figure 2. Function generator or alike may replace the sensors to provide input signals simulating the concentration of the hormones. The effect of drugs on the mechanism can be investigated by providing control signals that represent concentration of drugs and observing the output signals. In this paper we have modeled, simulated, and synthesized the mechanism on a compact electronic chip. We have selected an FPGA to be the compact chip; other compact devices such as programmable Integrated Circuits (PIC) or a VLSI chip can be used if desired.

THE GROWTH HORMONE MECHANISM

Growth hormone (GH) or somatotropin is a peptide hormone synthesized or stored in the anterior pituitary gland. Growth hormone plays a vital role in stimulating linear growth during childhood, and may affect: lipid and carbohydrate metabolism, net muscle protein synthesis, reproductive functioning, renal functioning, and aging.

To model the mechanism taking into consideration every hormone that may affect the function of the mechanism is extremely difficult task if not impossible. In our paper here four hormones are considered: GH, Growth Hormone Releasing Hormone (GHRH), Somatotrophin Releasing Inhibiting Factor (SRIF) and Insulin-like growth hormone (IGF-1). Growth Hormone Releasing Hormone (GHRH) stimulates the secretion of Growth Hormone (GH) through binding to stimulatory protein $G_{s;}$ the binding initiates a chain of activation of enzymes Adenylate cyclase (AC), cyclic adenosine 3', 5' mono-phosphate (cAMP), and protein kinase A (PKA). These activations result in synthesizing growth hormone. On the other hand, Somatostatin or Somatotrophin Releasing Inhibiting Factor (SRIF) inhibits the secretion of GH by binding to the inhibitatory protein G_i; this binding results in deactivation of PKA. Releasing hormones are secreted in response to neural inputs from the central nervous system (CNS). Figure 3 depicts the mechanism. Insulin-like Growth Factor (IGF-1) is produced, primarily in response to GH stimulation, by the liver and some is also produced by skeletal muscles. It mediates many of the actions of GH and so stimulates the growth of bones and other tissues and promotes the production of lean muscle mass. IGF-1 provides an antagonistic feedback path to GH synthesis [1], [2]. Thus GH, SRIF, GHRH, GHS and IGF-1 form a feedback and feed-forward network.

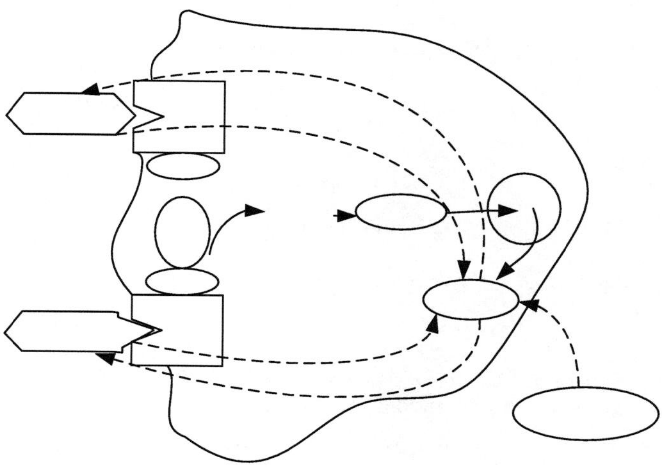

Figure 3, Simplified Growth Hormone Mechanism

THE MATHEMATICAL MODEL

In this section, we develop mathematical model that describes the relationship between the hormones that control the GH mechanism. The growth hormone mechanism, as many other biological mechanisms, is extremely complex. Developing a mathematical model that describes the complete mechanism is an extremely difficult task. Our focus here in this paper is to simulate and realize the mechanism rather than developing a comprehensive mathematical model. Several literatures have investigated the mathematical relationship between GH, SRIF, and GHRH [1],[2]. In our paper here we modify these mathematical formulae to include the contribution of IGF-1 to the GH mechanism. The equations are developed under several assumptions to simplify the model. Some of the major assumptions are: a) only four hormones, GH, GHRH, SRIF, and IGF-1 regulate the secretion of GH, b) GH after reaching a threshold, exerts delayed feed forward on the concentration of SRIF and also exerts negative delayed feedback on GHRH [2], c) IGF-1 exerts negative feedback on GH, d) each of the hormones has an infusion rate and an elimination rate. The rate of change of a hormone concentration ($\frac{dH}{dt}$) can be expressed as [2]:

$$\frac{dH}{dt} = -kH + H_i \tag{1}$$

where H is the instantaneous concentration of the hormone, k is the elimination rate constant, and H_i is the hormone infusion rate. Using equation (1), the rate of change of GH concentration can be written as:

$$\frac{dGH}{dt} = -k_1 GH + k_{r,1}[F_1^+(GHRH) F_2^-(SRIF) F_3^-(IGF\text{-}I)]$$

where $F_3^-(IGF\text{-}I) = \dfrac{1}{(IGF\ /\ t_8)^{n8} + 1}$

$F_1^+(GHRH) = (GHRH/t_1)^{n1} / (GHRH/t_1)^{n1} + 1$ and

$F_2^-(SRIF) = (SRIF / t_2)^{n2} / (SRIF / t_2)^{n2} + 1$.

The rate of change over time of the SRIF concentration into the portal blood after a certain time delay specified as *D* is:

$$\frac{dSRIF}{dt} = -k_2\, SRIF + k_{r,2}\, \{F_1^+[GH(t-D)]\}$$

The elimination constant is expressed as k_2 and the release constant is expressed as $k_{r,2}$. An approximation represented by a Michaelis-Menten equation emulates an upward regulatory function [1], [2]:

$$F_1^+[GH(t-D)] = \frac{S_{\min} - 1}{1 + GH(t-D)/t_3} + 1 .$$

$S_{\min}$ is the minimal concentration level of the regulatory function which is considered to be the basal SRIF concentration and t_3 is the Michaelis-Menten constant. It is assumed that the rate of change of the GHRH concentration in portal blood is affected by the suppression of GHRH by GH and SRIF and that

GH suppression of GHRH is time delayed. This dual suppression is combined to form the down regulatory function such that release of GHRH occurs only when GH, before a time delay (*T*) and SRIF are low values and is indicated in the following expression:

$$\frac{dGHRH}{dt} = - k_3 GHRH + k_{r,3} \{F_4^- [GH (t - T)] F_5^- (SRIF)\} \quad (2)$$

The roles of the rate of change constants are functionally similar to the equations above and the corresponding regulatory functions are;

$$F_4^- (GH) = \frac{1}{[GH(t-T)/t_4]^{n_4} + 1} \quad (3)$$

$$\text{and } F_5^- (SRIF) = \frac{1}{(SRIF/t_5)^{n_5} + 1} \quad (4)$$

Equations (2), (3) and (4) form the basis for coupled nonlinear ordinary differential equations model. A fourth equation based on the assumption that the releasable pituitary GH pool (GHS) is depleted partially and saturated fully (parameter *M*) and that GHRH-driven synthesis of GH expands the releasable pool in inverse proportion to nearness of saturation, is given by Equation 5.

$$\frac{GHS}{dt} = kr_5 \frac{(GHRH t_7)^{n_7}}{(GHRH t_7)^{n_7} + 1} [M - GHS] - kr_1 [\frac{(GHS t_0)^{n_0}}{(GHS t_0)^{n_0} + 1} \frac{(GHRH t_1)^{n_1}}{(GHRH t_1)^{n_1} + 1} \frac{1}{(SRIF t_2)^{n_2} + 1}] \quad (5)$$

From the above equations 3-5, the relationship expressing the rate of change for the GH axis with respect to time can be expressed as shown in equations 6-8.

$$\frac{dGH}{dt} = -k_1 GH + \alpha kr_1 [\frac{(GHS/t_0)^{n_0}}{(GHS/t_0)^{n_0} + 1} \frac{(GHRH/t_1)^{n_1}}{(GHRH/t_1)^{n_1} + 1} \frac{1}{(SRIF/t_2)^{n_2} + 1} \frac{1}{(IGF/t_8)^{n_8} + 1}] \quad (6)$$

$$\frac{dSRIF}{dt} = -k_2 SRIF + S_{\min} + kr_2 [\frac{GH(t-D)/t_3}{1 + GH(t-D)/t_3} + kr_3 \frac{(GHRH(t-T)/t_6)^{n_6}}{(GHRH(t-T)/t_6)^{n_6} + 1}] \quad (7)$$

$$\frac{dGHRH}{dt} = -k_3 GHRH + kr_4 \left[\frac{1}{(SRIF/t_4)^{n_4} + 1} \frac{1}{(GH/t_5)^{n_5} + 1} \right] \quad (8)$$

GHS is the amount of releasable GH concentration levels stored in the pituitary gland. *M* defines the maximum amount of GH in the stored pool. *Smin* is the minimum baseline required for SRIF release, α is a scaling constant that is used to relate the pituitary and circulatory distribution volume. The constants k_1, k_2

and k_3 describe the rates of hormonal elimination. The other constants kr_1, kr_2, kr_3, kr_4 and kr_5 signify the rate of release; t_1, t_2, t_4, t_5, t_6 , t_7 and t_8 are the threshold levels eliciting a response. The constants n0, n1, n2, n4, n5, n6, n7 and n8 are the Hill Coefficients. D is the time delay for the GH feedback effect on SRIF and T is the time delay for the GHRH feed-forward effect on SRIF.

Mathlab® package is implemented to solve the above delay differential equations (6-8); all constants necessary to solve the equations have been obtained from published in-vitro experimental data [2].

SIMULATION AND REALIZATION OF THE MECHANISM

After developing the mathematical model for the mechanism (equations 5-8), Computer Aided Design (CAD) tools have been implemented to simulate and realize the mechanism onto an FPGA [3]. The FPGA chip implemented in this paper is Xilinx-XC2S50 mounted on XSA-50 board. .

RESULTS

A Verilog behavioral code, based on case statements, has been written to describe the rate of change of GH, GHRH, SRIF and IGF-1. After simulation, a Xilinx synthesizer is implemented to synthesize the Verilog code onto the FPGA. The output bit file of the synthesizer is downloaded onto the Xilinx FPGA board XSA-50.

CONCLUSIONS

We have presented a technique to simulate and realize a simplified growth hormone secretion mechanism on an electronic chip (FPGA). The technique is based on synthesizing the HDL behavioral description of the mechanism. To write the behavioral description, a mathematical model for the mechanism has been developed. After successful realization, the chip mimics the mechanism. The chip can be attached to the skin and can be interfaced with the body of the subject through appropriate sensors. The synthesized mechanism occupies a small portion of the chip-that allows for upgrading the complexity of the mechanism without the need to use a new FPGA chip. The technique presented here can be applied for other biological mechanisms

REFERENCES

[1] Botros, N., Shell, J.R., "Modeling and Hardware Realization of Biological Mechanisms," Proceedings of the Artificial Neural Networks in Engineering Conference, St. Louis, MO, USA, November 7-9, 2005.

[2] Farhy, L. S., and Veldhuis, J. D., "Joint Pituitary and Intrahypothalamic Auto-feedback Construct of Pulsatile Growth Hormone Secretion",Am J Physiol Regulatory Integrative Comp Physiol, R00086, 2003.

[3] Botros, N. , "HDL Programming Fundamentals: VHDL and Verilog," Thomson Learning publisher, November 16, 2005.

[4] Wolpert, H. A., M.D., "Insulin Pumps and Glucose Sensor Devices", TIME Magazine, March 28, 2005.

MODELING, SIMULATING, AND SYNTHESIZING OF SIMPLIFIED RENAL MECHANISM

M. ALBANNA
Southern Illinois University

N. BOTROS
Southern Illinois University

ABSTRACT

Due to ethical constrains, medical experiments on humans and animals are not allowed, or extremely limited. To investigate the renal mechanism we propose implementing a mathematical model for the mechanism, simulate the model using Hardware Description Language (HDL), and synthesizing the simulation on a compact electronic chip. The chip mimics the function of the renal system. The function can be explored by testing the chip with input signals representing the input parameters of the system such as concentration of hormones; the signal on the output pins of the chip represents the output of the system. The mathematical model takes into consideration the interaction between the cardiovascular, the nervous, and the hormonal systems. The model is simulated on the computer using hardware description language (HDL). After verifying the simulation results, Computer Aided Design (CAD) tools are implemented to download the simulation into Xilinx FPGA chip.

INTRODUCTION

To understand the function of biological mechanisms in the human body and to develop drugs to treat maladies in a particular mechanism, scientists and researchers need to carry out extensive investigation of the mechanisms. Medical experimentation on humans and animals is extremely limited and in some instances prohibited due to ethical issues or other constraints.

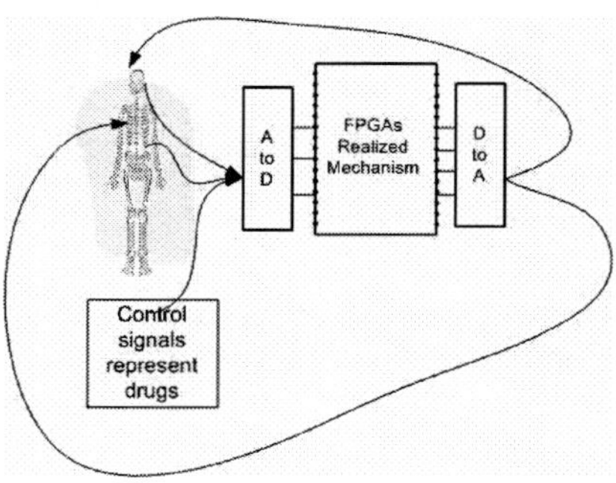

Figure 1, In-vivo Environment.

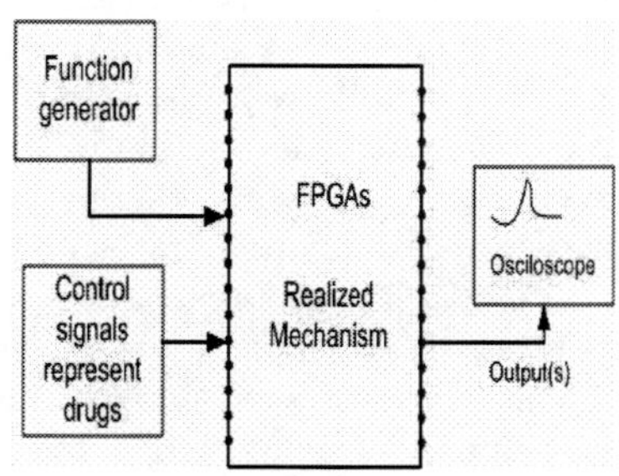

Figure 2, Laboratory environment

Researchers in the medical and biological fields are in constant search of innovative techniques to test various drugs but are restricted by the harmful side effects these drugs may have on the test subject. An alternative approach to investigate the mechanism in-vivo is to simulate it and then realize (synthesize) it on a compact electronic chip. The chip after synthesis mimics the function of the mechanism. The synthesized chip offers researchers and physicians a portal to explore and/or correct the functions of the mechanism. Also, because the chip is compact, it can be attached to the skin or can be implanted in the body of the subject. Recent advances in compact microchip technology have allowed researchers to realize complex systems such as biological mechanisms on compact chips. An example of such innovation is reported by Wolpert et al. [3] where a simple function of the pancreas, the secretion of insulin is realized. A compact sensor with a compact control unit is implanted under the skin. The unit monitors glucose level. If the level reaches a threshold, a pump worn as a watch automatically delivers insulin.

The goal of this paper is to model a simplified renal mechanism and to realize the model on an integrated compact electronic chip such as the Field Programmable Gate Array (FPGA). After realization of the mechanism, the chip mimics the function of the realized mechanism. The chip offers a unique opportunity for researchers to explore the function of the mechanism. The chip can be implemented in two environments, in-vivo and laboratory. Figure 1 shows a general diagram of a possible in-vivo interface between the chip and the human body. The chip is attached to the body of the subject. The chip inputs are electrical signals fed from sensors attached to a selected region of the body such as chest, head, or muscle. These input signals represent the input parameters that control the function of the mechanism such as blood volume, sodium concentration; the output signals represent the action of the mechanism to the input signals; an example of such action is the rate of urine flow. The laboratory environment implementation is shown in Figure 2. Function generator or alike may replace the sensors to provide input signals simulating the input parameters. The effect of drugs on the mechanism can be investigated by providing control signals that represent concentration of drugs and observing the output signals. The steps taken to simulate and realize the mechanism are summarized as follows a) develop a mathematical model for the mechanism, b) write a behavioral description of the mechanism using HDL, c) simulate the behavioral model to verify that the model correctly describes the mechanism, d) synthesize the simulation into components and gates that can be downloaded onto a FPGA chip, e) simulate the outcome of synthesis to verify the accuracy of synthesis, f) download the synthesized model onto a FPGA chip, g) test the chip by inputting selected signals on the input pins of the chip; the output signals on the output pins should reflect the discrete numerical values obtained from the mathematical mode

THE MATHEMATICAL MODEL

Biological mechanisms are in general very complex to be entirely modeled; accordingly few functions of the mechanism are usually selected. The renal mechanism is a major biological mechanism in the human body. Several comprehensive studies and analysis have been carried out to present a complete mathematical model for the mechanism [1]. Mathematical formulas have been proposed to describe the function of the renal mechanism [1],[2]. Uttamsingh et al. have introduced extensive mathematical formulas for the function of the components of the renal system. We follow the same approach proposed by Uttamsingh et al [1] by dividing the renal system into subsystems and finding the mathematical formulas for each subsystem. In this section we introduce the major mathematical formulas (many of them have been implemented in Uttamsingh et al. [1]) of two subsystems of the renal system: the Cardiovascular and the kidney. All the constants presented in this section are based on experimental data obtained from Uttamsingh et al. [1].

I: The Cardiovascular System Equations:

The relationship between the blood volume (BV) and the extracellular fluid volume (E) can be written as:

$BV=0.33E$ ………. (If $E<21.0l$)………. (1)

$BV=0.015E + 6.6$………. (If $E\geq 21.0l$)………. (2)

The mean systematic pressure (MSP) can be written as::

$MSP=3.5(BV-3)$………. (3)

The third parameter which is related to the cardiovascular system is the total peripheral resistance (TPR), and this parameter depends mainly on plasma angiotensin II concentrations (A).

$TPR=19.0 +.037A$………. (If $A\leq 27ng/l$) ………. (4)

$TPR=12.2 +5.44 \log A$………. (If $A>27ng/l$)………. (5)

The return venous resistance (RVR) is related strongly to the total peripheral resistance (TPR)

$RVR=0.07\ TRP$………. (6)

One of the most important parameters, which are related to the cardiovascular system and the kidney function, is the cardiac effectiveness (CE). The cardiac effectiveness depends mainly on extracellular concentration of sodium (PNA), and extracellular concentration of potassium (PK). To find the total cardiac effectiveness, the cardiac effective due to abnormal in sodium concentration level (CENA) and potassium concentration level (CEK) should be found as follows:

$CENA=1.0$………. (If $PNA<148.0$)………. (7)

$CENA= -0.0125\ PNA+2.85$………. (If $PNA\geq 148.0$)………. (8)

$CEK=1.0$………. (If $PK<6.5$)………. (9)

$CEK= -0.065\ PK +1.43$………. (If $PK \geq 6.5$)………. (10)

$CE=0.5(CENA+CEK)$………… (11)

The arterial pressure (AP) can be found from the total peripheral resistance (TPR) and the cardiac Output (CO) as follow
AP=CO*TPR.......... (12)

II: The Kidney:

The rate of formation of glomerular filtrate (GFR) is dependent on the net pressure forcing fluid across the glomerular capillary membranes in the Bowman's capsules of the nephron.
GFR=0.0.......... (If AP≤20.0 torr).......... (13)
GFR=1.92AP-38.4.......... (If 20.0<AP≤75.0 torr).......... (14)
GFR=-0.00808AP^2 +2.195AP-13.6........... (If 75.0<AP≤120.0 torr)........... (15)
GFR=0.035AP+129.2........... (If >120.0 torr).......... (16)

The rate of filtration of sodium into the proximal tubule (FNA) is given by
FNA=GFR*PNA/1000.0.......... (17)

The rate of reabsorption of sodium in the proximal tubule (SPTR) and of flow of sodium into the loop of Henle SFLH are given by
SPTR=GTB*FNA.......... (18)
SFLH=FNA-SPTR........... (19)
The glomerular tubular balance relationship is assumed as first order approximation, since it is known to be function of sodium concentration, which is equal to the concentration of sodium in plasma (PNA)
GTB=5.815-.0357PNA.......... (20)

The rate of water absorption in the proximal tubule (EPTR) and its rate in the next segment (loop of Henle) EFLH are given by

EPTR=GTB*GFR.......... (21)
EFLH=GFR-EPTR.......... (22)

Extrapolation to the human kidney yields the following relationships for the rates of reabsorption of water ELHR and Sodium SLHR in this nephron segment
EBLH=-0.01EFLH+0.65.......... (23)
ELHR=EBLH*EFLH.......... (24)
SLHR=0.8SFLH.......... (25)

The rate of Flow of water EFDT and sodium SFDT into the distal tubules are then given by
EFDT=EFLH-ELHR.......... (26)
SFDT-SFLH-SLHR.......... (27)
Moving to the next segment which is the distal and collecting segments to find the fraction of water load reabsorbed in the distal nephron segments (EBDT)
EBDT=0.0.......... (If ADH≤0.765 munits/l).......... (28)
EBDT=0.383 ADH - 0.293 .... (If 0.765<ADH≤3.0 munits/l).......... (29)

$EBDT=-0.0383ADH^2 + 0.364ADH+0.109$.... (If $3.0<ADH\leq5.0$ munits/l).......... (30)
$EBDT=0.0012ADH+0.9653$.......... (If ADH>5.0 munits /l).......... (31)

$EDTR=EBDT*EFDT$.......... (32)
The urine flow (UFL) can be written as:
$UFL=EFDT-EDTR$.......... (33)

SIMULATION OF THE MATHEMATICAL MODEL

To write a HDL-behavioral code for the mathematical model, we draw a flowchart of the mechanism; the flowchart facilitates the writing of the code, Figure 3 shows a simplified flow chart for the first three equations (1-3). The simulation has been done by using Xilinx CAD with Hardware Description Language (HDL). Figure 4 shows the result of simulation. From the values shown in Figure 4 we conclude that the simulation correctly represents equations 1-3.

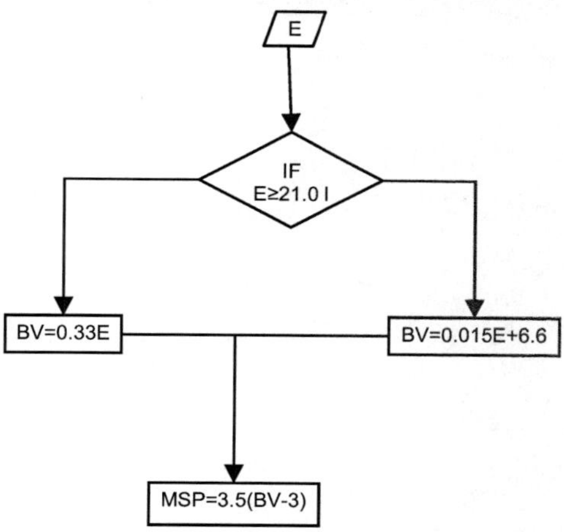

Figure 3 Flow chart for some cardiovascular parameters shows the implementation of the equations.

The next step is to synthesize the HDL code on FPGAs; the configuration of the FPGA is shown in Figure 5. The synthesis stage of Figure 3 has been completed and verified using Xilinx synthesizer

CONCLUSION

In this paper, we have presented an approach for modeling and realizing of a simplified human renal mechanism using VHDL and FPGA's. The flow chart for part of the model is developed and VHDL behavioral modeling is written to simulate this part of the mechanism. After verification of the simulation, the HDL code is realized on FPGAs chip using Xilinx CAD tools. The inputs and outputs of the chip are checked to verify that the chip mimics the simulated part of the mechanism. Our results show that the chip mimics the function of Figure 3.

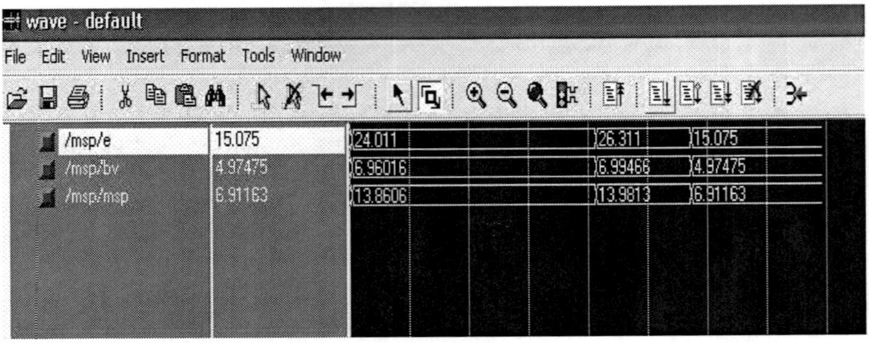

Figure 4 Simulation of equations 1-3.

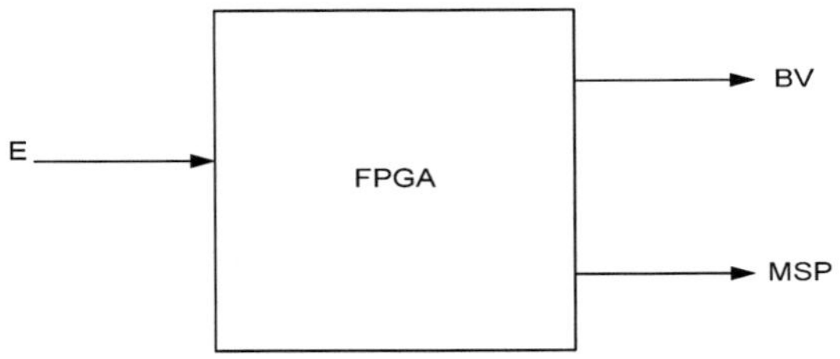

Figure 5 Configuration of FPGA for equations (1-3).

REFERENCES

1. R. J. Utamsighn, M.S.Leaning, J.A.Bushman, E.R.Carson, and L.Fimkelstein. Mathematical model of the human renal system. Med.Biol.Eng.Comput.23:525-535.1985.

2. Fatih Karaaslan,Yagmur Deizhan, Abidin Kayserilioglu, and H.Ozcan Gulcur. Long term Mathematical model involving renal sympathetic nerve activity, arterial pressure, and sodium excretion. Annals of Biomedical Engineering, vol.33. No.11, November 2005, pp.1607-16030.

3. Wolpert, H. A., M.D., "Insulin Pumps and Glucose Sensor Devices", TIME Magazine, March 28, 2005.

ADVANCEMENTS IN AUTOMATED DIAGNOSTIC MAMMOGRAPHY USING K-PLS NON-LINEAR MAPPINGS

WALKER H. LAND, JR
Dept. Of Bioengineering,
Binghamton University,
Binghamton, NY 13902.

JOHN J. HEINE
H. Lee Moffitt Cancer Center & Research Institute, Tampa, FL 33612

CLAUDIA BERMAN
Dept. of Radiology, College of Medicine, University of South Florida, and H. Lee Moffitt Cancer Center & Research Institute, Tampa, FL 33612.

ALDA MIZAKU
Dept. Of Bioengineering,
Binghamton University,
Binghamton, NY 13902.

THOMAS RAWAY
Dept. Of Bioengineering,
Binghamton University,
Binghamton, NY 13902.

NATALIYA KOVALCHUK
Dept. of Physics, University of South Florida, and H. Lee Moffitt Cancer Center & Research Institute, Tampa, FL 33612.

ABSTRACT

This paper evaluated the contribution of both magnetic resonance mammography (MRM) image and non-image (NI) features (breast cancer risk factors) to benign / malignant diagnostic accuracy using a new data set. This paper focus was using a new Kernealized-Partial Least Squares (K-PLS) and EP-SVM paradigms for this evaluation. An important initial finding is that (only) the non-imag , non-linear mapping region of the feature to diagnostic solution space contains significant diagnostic information, with a resultant overall A_Z of 0.77. The MRI image features of mass size, mass shape and mass margin yielded an A_Z of 0.96, which is excellent diagnostic performance. The EP-SVM A_Z diagnostic results for the same basic data set and statistical cross validation methods were slightly lower because "outliers" were included in these analyses.

INTRODUCTION

Breast cancer is the most prevalent form of cancer and the second leading tumor-related cause of death in women. The American Cancer Society estimates that this year, 217,440 new cases of breast cancer will be diagnosed, and 40,580 breast cancer-related deaths will occur (2004). Between 1973 and 1998, the incidence rates of breast cancer rose over 40% (Howe, et. al., 1973). The percentage of women over 40 years of age who reported having had a mammogram within the past two years, moreover, also has increased, from 29% in 1987 to 67% in 1998 (National Center for Health Statistics, 2001). This increase in the use of mammography has resulted in earlier detection of cancers and improved survival.

While mammography is very sensitive to the detection of breast cancer, it also is sensitive to the detection of non-cancerous breast lesions. As a result, only 15-34% of women referred for biopsy on the basis of mammographic findings do, in fact, have cancer (Lo, et. al, 1997). The results reported in this study are part of an ongoing effort to reduce the number of benign cases referred for biopsy by developing tools to aid physicians in classifying suspicious lesions. Previous applications of computational intelligence to this problem by the authors and others, which serve as a foundation for this study, include Evolutionary Programming (EP)-derived neural networks (Land, et. al., 2000), and EP-derived SVMs (Land, et. al., 2003).

There is a well known false positive biopsy artifact in diagnostic mammography, which is a consequence of maintaining a high true positive breast cancer classification rate. It is estimated that about 80% of the biopsies performed in current diagnostic mammography prove to be false positive. To put the current state of diagnostic mammography in perspective we quote verbatim the findings of a recent report sponsored by the Agency for Healthcare Research and Quality (AHRQ) (see citations for report web-address (2006). This report reviewed diagnostic ultrasound, magnetic resonance, position emission scanning, and scinti-mammography (it included 81 published reports) that resulted in these findings based on assuming a 2% false negative rate: (1) an average women with abnormal biopsy due to a mammogram has 20% risk of breast cancer, (2) for 1000 women with negative PET scans 924 would avoid a needless biopsy, but 76 would have missed cancers, (3) for scintimammogram the corresponding numbers are 907 avoided biopsies and 93-missed cancers, (4) for MR there are 962 avoided biopsies and 38-missed cancers, and (5) for ultrasound 950 avoided biopsies and 50 missed cancers, respectively. The conclusion indicates most women sent for biopsy do not have cancer, and none of these adjunctive tests can replace biopsies if a less than 2% false negative rate is deemed acceptable.

In reference (Land, et.al., in press ,2007) we performed a linear analysis of this problem using a new feature selection – classification process that integrates logistics regression, Wald statistical testing, p statistics and converting logistic regression probability density functions (PDFs) into ROC curves. For this paper the research is extended using a Kernealized - Partial Least Squares approach (K-PLS) to establish if any diagnostic information is contained in the nonlinear feature space. This body of work, in part, forms the foundation for developing non-invasive methods (diagnostic tool) for reducing the FP biopsy rate in conventional x-ray mammography, which defines our motivation.

The remainder of this paper summarizes the new data set used for this research, briefly discusses K-PLS mappings to ascertain the diagnostic feature information contained in the non-linear regions of the solution space, summarizes Evolutionary Programming derived Support Vector Machines (EP-SVMs) , and the results found using this new data set.

DATA DESCRIPTION

This new data set consisted of 200 samples patients undergoing magnetic resonance mammography (MRM), approximately equality divided between cases and controls. All cancer-positive cases were pathology verified and the controls were MRM patients with either negative or benign findings. The patient population represents a subset of those patients encountered in diagnostic mammography that are undergoing diagnostic MRM. The patient data used as input features for the K-PLS modeling may be divided into two categories (I) non-image type (NI), and (II) and image type (IM) features observable in the MRM image. The NI features were (1) patient age, (2) race/ethnicity, (3) weight, (4) age of menarche, (5) duration of menstruation, (6) age of menopause, (7) number of full term deliveries, (8) duration of hormone replacement therapy, (9) family history of breast cancer (yes or no), and (10) cigarette smoking (yes or no). The NI factors are related to breast cancer risk. The IM based features were lesion (1) size, (2) shape, and (3) margin. Since MRM requires contrast injection, the lesion description is dependent upon its enhancement. The shape and margins for masses are normally assessed from the first post-contrast image to avoid washout or enhancement (Morris, et. al., 2005). It should be clear that the IM features do not directly support our intended x-ray mammography application-goals.

KERNEL DEFINED FEATURE SPACE

Consider kernel defined feature mappings (Land, et. al. 2006). Non-linear relationships between data variables may be found by embedding this data into a kernel induced feature space. This kernel "trick" is used in PLS and is called Kernel –PLS,(K-PLS). That is, consider now the following embedding map:

$$\phi : \mathbf{X} \in \Re^n \Rightarrow \phi(\mathbf{X}) \in \mathbf{F} \subseteq \Re^N.$$

The objective of the map is to convert the non-linear relations into linear ones. This map will therefore reflect the expectations regarding the relation y= g(**X**), which must be learned. That is, the mapping will "recode" the data set S as:

$$S = \{\phi((\mathbf{X}_1), y_1), \phi((\mathbf{X}_2), y_2), \ldots\ldots\ldots\ldots \phi((\mathbf{X}_i), y_i)\}$$

This mapping of the data set is from nonlinear input space to a linear feature space. That is, the environment data representation in the input **X** space is **non-linear**. However, after the $\mathbf{X_i}$ data are processed by the $\phi((\mathbf{X}_i), y_i)$ function, the data characterized by this mapping is linear, with the happy result that linear techniques may be used **on the mapped** data while preserving the non-linear properties represented in the input space. This mapping is accomplished, as previously stated, by using a valid kernel function. Therefore, the purpose of this kernel function is to map the data into a feature space where the original non linear pattern (or environment) now **appears** as linear. The practical consequence is that linear techniques may be applied to the feature space data without in any way disturbing the nonlinear environment properties in the input space. Adding this kernel induced capability to the PLS approach means that a real time, non-linear optimal training intelligent method now exists which can be used to perform CAD diagnosis. A second advantage of this approach is that the Kernel computes inner products in the feature space directly from inputs.Computationally, kernel mappings have the following important properties: (1) they enable access to exceptionally high dimensional and, consequently, very flexible feature space, with a correspondly low time and space computational cost, (2) they solve the convex optimization problem without becoming "trapped" in local minima and, more importantly, (3) the approach decouples the design of the algorithms from the specifications of the feature space. Therefore, both learning algorithms and specific kernel designs are not as difficult to analyze.

EVOLUTIONARY PROGRAMMING

Based on the work of Fogel (1995), the evolutionary process as implemented in this study evolves the parameters for a collection of support vector machines. A generic description of this process is as follows: (1) a population of candidate solutions (SVMs) is randomly configured. (2) Mutation: Each of these candidate solutions then is copied and mutated, yielding a solution pool of twice the original size Mutation process involves adding both random and bias components to each SVM parameter and scaling these values based on both a Cauchy distribution as well as the inverse square root of the generation number. (3) Selection: All elements of this pool are scored using an objective function. These objective function scores then are used to order the candidate solutions from the "most fit" to the "least fit." Better results usually are obtained from using tournament selection methodologies. Each candidate solution competes against a random subset of the remaining solutions. Finally, the upper 50% of the solution pool is selected to continue as the basis for the next generation and the remaining 50% is "killed off" (discarded) to reduce the pool to the original population size. Stopping criteria was the number of generations. The objective function used was the number of correct classification for each fold of the cross validation process.

RESULTS

Both NI (non–image) and image type (IM) features are evaluated relative to their contribution to diagnostic accuracy (see data section for description). Table 1 depicts results for two types of K-PLS paradigms: auto-K-PLS and K-PLS. Auto K-PLS is an automated process which finds the optimal sigma value for the Gaussian Radial Basis Function (GRBF) mapping kernel as well as eliminating the outliers by an automated process. K-PLS has neither of these features. The EP-SVM process employs the same GRBF mapping kernel and used the EP process to optimize both the sigma (σ) value and regularization parameter(C). Both auto-K-PLS and K-PLS result in essentially the same ROC A_Z value (average of 0.76), while the EP-EVM result of 0.64 is 18.7% lower. This difference is attributed to the inclusion of the data set "outliers." The MRM Image features size, shape and margin yielded a much better K-PLS A_Z result, as expected, of 0.96. The EP-SVM process produced and expected lower A_Z of

0.91, again because of the outliers. These ROC A_Z curves are depicted in Figure 1.

Data Set	Auto K-PLS Az	KPLS Az	EP-SVM Az
NI	0.77	0.75	0.643052
IM	0.962742	0.9549	0.906607

Table 1. A_Z results for NI and IM data sets.

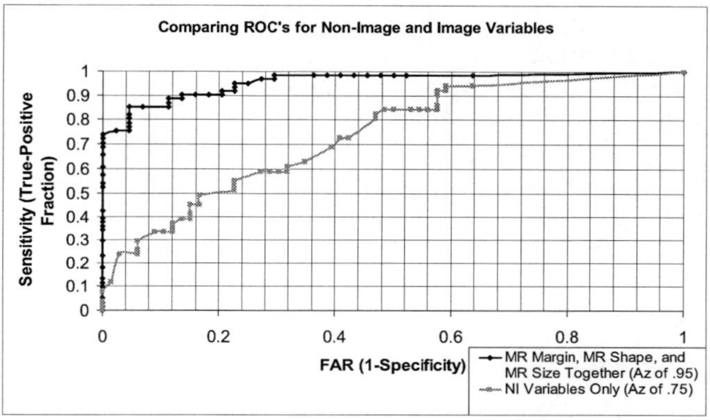

Figure 1. Comparing Non Image and Image ROC curves.

Finally, Figure 2 demonstrates the K-PLS diagnostic performance improvement for some of the more significant non-image features, when compared to the logistics linear diagnostic process (Land et. al. , 2007) for the development of the logistics regression results). Note that in all cases a significant amount of diagnostic information in contained in the non-linear mapping region of the features to the diagnostic result. This is an important new finding, and demonstrates the importance of collecting and using this clinical information in the diagnostic process.

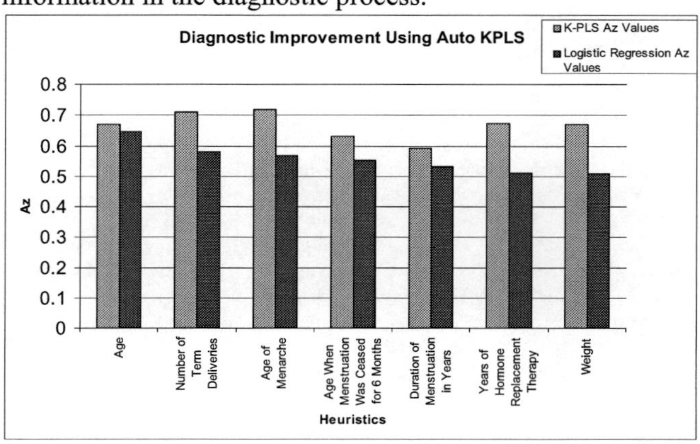

Figure 2. Diagnostic Improvement of Using Auto K-PLS in Place of LR

CONCLUSIONS

This paper evaluated the contribution of both MRI mammography image (IM) and non-image (NI) features to beingn / malignant diagnostic accuracy using a new data set. This paper focus was using a new kernealized-partial least squares (K-PLS) paradigm for this evaluation. A secondary objective was asertaining the individual non-linear contribution of several of the important NI features as well as measuring the performance improvement as compared to a logistics regression linear mapping as described in Land et.al. (2007). In all NI cases studied, additional significant information is contained in these non-linear mapping regions (see Figure 2). Finally, the collectave K-PLS ROC Az value for the NI features was 0.77, which was a significant finding, while the K-PLS MRI image features Az was 0.96. The work provides preliminary evidence suggesting that the same breast cancer risk factors that elevate a woman's risk for disease incidence may also be useful for diagnostic purposes. Similarly, the combination of IM and NI features is complementary; the IM factors suggest that it may be useful to investigate x-ray mammograms for spatial features with automated methods in meeting our FP reduction goals.

ACKNOWLEDGMENTS

This work was supported in part by NIH grant # K25 CA106799

REFERENCES

American Cancer Society, 2004, "Cancer Facts & Figures 2004" [On-line], http://www.cancer.org/downloads/STT/CAFF2004PWSecured.pdf.

Fogel, D.B., 1995, Evolutionary Computation: Toward a New Philosophy of Machine Intelligence. Piscataway, NJ: IEEE Press.

Howe, H.L., Wingo, P.A., Thun, M.J. et al., 2000 "Annual report to the Nation on the Status of Cancer (1973 through 1998), Featuring Cancers with Recent Increasing Trends." *J Natl Cancer Inst.* **93**, pp. 824-842.

HRQ Report, 2006, "Effectiveness of Noninvasive Diagnostic Tests for Breast Abnormalities," http://effectivehealthcare.ahrq.gov/reports/topic.cfm?topic=2&sid=32&rType=3.

Lo, J.Y., Baker, J.A., Kornguth, P.J., Iglehart, J.D., and Floyd, C.E., 1997, "Predicting Breast Cancer Invasion with Artificial Neural Networks on the Basis of Mammographic Features," *Radiology*, **203**, pp. 159-163.

Land, W. H., Masters, T., Lo, J. Y., and McKee, D., 2000, "Using Evolutionary Computation to Develop Neural Network Breast Cancer Benign/Malignant Classification Models." *4th World Conference on Systemics, Cybernetics and Informatics*, **10**, pp. 343-7.

Land, W. H., McKee, D. W., Velazquez, R., Wong, L., Lo, J. Y., and Anderson, F. R., 2003, "Application of Support Vector Machines to breast cancer screening using mammogram and clinical history data." *SPIE Medical Imaging 2003*, San Diego, CA, 15-20 February.

Land, W.H., Heine, J., Embrechts, M., Smith, T., Choma, R., Wong, L., " New Approach to breast cancer diagnosis using partial least squares and kernel partial least squares", Medical Imaging, 2005, Image Processing, volume 5747,ISBN 0-8194-5721-3, pp 48-58

Land, W.H.,Jr., Heine, C., Bernam, C., Mizaku, A., Raway, T., Kovalchuk, N., (2007) In Press, "Advancer in Automated Diagnostic mammography", IEEE BIBE07.

Morris E. A., L. Liberman (Editors), 2005, Breast MRI Diagnosis and Intervention, Pages 51-55, Publisher: Springer,Year",NY, NY.

National Center for Health Statistics, 2001, *Health, United States, 2001 with Urban and Rural Health Chartbool*, Hyattsville (MD), National Center for Health Statistics.

THE IMPLEMENTATION OF THE ALOPEX-B TRAINING ALGORITHM IN PARALLEL

MICHAEL DE RIDDER and STEFAN C. KREMER
Dept. of Computing and Information Science, University of Guelph
Guelph, ON N1G 2W1, CANADA {mderidde|skremer}@uoguelph.ca

ABSTRACT
The proliferation of automated testing equipment and methods has led to an increase in the number of large data sets. As a result, difficulties of analyzing the data have increased and the time to science has increased. Artificial Neural Networks (ANNs) have the ability to identify and model complex relationships, as well as identify patterns in data. ALOPEX-B is a general-purpose optimization algorithm that has been successfully used to train ANNs. We intend to demonstrate the effectiveness of the ALOPEX-B algorithm to efficiently train an ANN in parallel. The ALOPEX-B naturally lends itself to a data parallel model because of the method calculates updates. The *N*-bit parity problem is a well documented problem that has posed a challenge for ANNs in the past(Minsky and Papert, 1969) and was used to test ALOPEX-B in parallel. Our goal is to demonstrate that a parallel implementation of the ALOPEX-B training algorithm results in a notable improvement in training time.

1 INTRODUCTION

The number of data points, test values, and information generated has increased as experimental methods have become more efficient. Sampling equipment has become faster and increased in sensitivity, allowing researchers to perform larger experiments that test for a greater amount of features. Interpretation of the results has become more difficult because of the increased size and complexity. ANNs are useful for their ability to infer function from data that is presented to them(Haykin, 1999), as well as being able to perform as classifiers for a variety of problems. Training the network is the most important part of an ANN. There are many different algorithms that are used for training neural networks, the most common being vanilla backpropagation(Rumelhart et al., 1986)(Werbos, 1974). ALOPEX-B(Bia, 2001) is a general purpose optimization algorithm derived from the original ALOPEX algorithm published by Harth and Tzanakou in 1973(Harth and Tzanakou, 1974). ALOPEX is an acronym for "**AL**gorithms **O**f **P**attern **EX**traction"(Harth and Tzanakou, 1974). It lends itself to a data parallel design because of the way it calculates updates independent of neighbouring connections. Advantages of the data parallel implementation of ALOPEX-B are time savings from the exploration of the training space. Problems that were too large are an option because of time savings from the parallel training of the neural network. This enables the use of larger training sets and a fuller representation can be presented during training. A key benefit of faster training times is a decrease in the time for analysis. The network space needs to be explored to find a network configuration and training parameters that best suit the problem. Experimentation of the network space needs to be repeated several times to ensure a proper configuration. The accumulation of time savings from shorter training times will shorten the time to science of experiments. The purpose of this investigation is to examine the abilities and limitations of a data parallel implementation of the ALOPEX-B algorithm.

2 BACKGROUND

ALOPEX is a general-purpose optimization algorithm used to solve complex problems where the connections between inputs are not explicitly known. ALOPEX was first proposed in 1973 by Harth and Tzanakou(Harth and Tzanakou, 1974) and was used for determining visual receptive fields. ALOPEX uses a stochastic function to avoid local minima during training, which other algorithms such as vanilla backpropagation are susceptible to. ALOPEX has demonstrated its usefulness for solving complex problems such as modeling, prediction, diagnosis, and pattern recognition(Skurikhin and Surkan, 1994). ALOPEX-B is a derivative of the original ALOPEX developed by Alejandro Bía (Bia, 2001). ALOPEX-B functions by making random changes to weight and bias values between neurons based on the error of the network. If the network has a high degree of error then there will be many of the updates to the weights and bias values will be undone. If the network contains a low degree of error, then the weight updates will continue in the same direction as the previous step. ALOPEX-B is a simplification of the ALOPEX algorithm published by Unnikrishnan and Venugopal (Unnikrishnan and Venugopal, 1994) in 1994. The Unikrishnan and Venugopal(Unnikrishnan and Venugopal, 1994) paper is an elaborate version of Harth and Tzanakou(Harth and Tzanakou, 1974) publication, and is especially suited for the training of ANNs(Bia, 2001). ALOPEX-B differs from the original ALOPEX in that it does not contain an annealing scheme. ALOPEX uses the annealing scheme to gradually decrease the entropy of the random walk. This is to prevent the network from starting in a position of local minima. The elimination of the annealing scheme allows the correlations to be self-adjusted from the start. This also simplifies the weight update functions and saves on computation expenses. The following is a description of the ALOPEX-B algorithm and its various parts. The correlation for a connection in ALOPEX-B between two neurons i and j at iteration n is defined by equation 1.

$$C_{ij}(n) = \Delta w_{ij}(n) \cdot \Delta E(n) \tag{1}$$

$w_{ij}(n)$ is the weight of the connection and ΔE is the total error of the neural network at the nth iteration. Weight updates are determined by the previous values at iterations $n-1$ and $n-2$ as shown in equations 2 and 3.

$$\Delta w_{ij}(n) = w_{ij}(n-1) - w_{ij}(n-2) \tag{2}$$

$$\Delta E(n) = E(n-1) - E(n-2) \tag{3}$$

If $\Delta w > 0$ then the weight at n advances by a constant value of delta(δ). If $\Delta w < 0$ then the weight at n retreats by the same value of δ. δ is the step size for the weight and bias updates. It is a constant and is integral in the random walk of the space. The weights are adjusted using equation 4.

$$w_{ij}(n) = w_{ij}(n-1) + \delta \operatorname{sgn}(r - p_{ij}) \tag{4}$$

The term $r_i(n)$ is a random number between 0 and 1, and p_{ij} is the probability of change which is determined by the correlation between $\Delta w_{ij}(n)$ and $\Delta E(n)$.

$$p_{ij} = \delta(C_{ij}) \tag{5}$$

The probability of change p_{ij} is determined by the correlation between the change in error and the direction of weight change. The value $r_i(n)$ is responsible for the stochasticity of ALOPEX-B. A random decision is made based on the value of r which is generated separately for each neuron. p_{ij} acts as a threshold, separating the decision into two regions. For example if p_{ij} was 0, all the weights and biases will be incremented but if p_{ij} is equal to 1, then the weights will be decremented.

$$C_{ij} = \frac{\operatorname{sgn}\left(\Delta w_{ij}\left(n\right)\right)\Delta E\left(n\right)}{\sum_{n'=2}^{n'=n} \lambda\left(\lambda - 1\right)^{n-n'} \left|\Delta E\left(n' - 1\right)\right|} \tag{6}$$

ALOPEX-B is able to recall the effects of decisions up to two steps back. The correlation between the error of the network $\Delta E(n)$ and the direction of change $\operatorname{sgn}(\Delta w_{ij}(n))$ of the previous two steps creates a "memory" effect. The influence of the two previous steps is determined by lambda (λ), which is the "memory" variable. λ influences how fast ALOPEX-B forgets the results of the previous two steps and creates a "memory" window of past weight changes. The range of λ is between 0 and 1. The larger the value of λ, the faster the network forgets the previous two steps.

Parallel programming is the simultaneous execution of processes on multiple processors in order to achieve a speed up in computation time. It is the division of the program into many smaller executions that allows for a faster execution. The parallel design pattern that was selected for this paper was a data parallel model. Data parallelism is the performance of many identical repetitive operations on a set of data(Quinn, 2004). This allows for a decomposition of the execution of a program over multiple processors by dividing the data amongst the processors. This is in contrast to task parallelism, which refers to performing different operations on the same or different piece of data(Quinn, 2004) on each processor. The independent design of weight updates in ALOPEX-B allows for a parallel implementation with very little modification of the original algorithm(Bia, 2001). Unlike vanilla backpropagation(Rumelhart et al., 1986)(Werbos, 1974), weight updates are calculated independent of values from neighbouring neurons. Weight updates are based on the global error of the network as outlined in equation 4. The result is a divide and conquer approach to training the network so the work is distributed over many processors, reducing the amount of computation on each processor.

The *N*-bit parity problem is a classic one that posed difficulty for ANNs to "learn"(Minsky and Papert, 1969). Minsky and Papert also hypothesized that it would be impossible for a multi-layer network to "learn" the XOR data set, also known as 2-bit parity(Minsky and Papert, 1969). It was not until the introduction of adaptive training algorithms such as vanilla backpropagation, that ANNs were able to successfully "learn" the pattern. *N*-bit parity is defined by the number of 1's present in a string of *N* characters in length. If the number of 1's is odd, then the output is true and if the number of 1's is even, then the output is false, where true and false are defined by 1 and 0. The growth of the *N*-bit parity data set is defined by the function 2^N and grows in a exponential fashion. This makes the *N*-bit parity problem ideal to test our data parallel implementation, as a data set can easily be generated and with defined outputs. Another advantage of using the *N*-bit parity data set is that it can be easily inspected visually.

3 METHOD

ALOPEX-B was implemented in parallel using a data parallel model with a master slave design pattern. The training set was split up over x processors, with each of the processors having a copy of the neural network. To ensure the synchronization and allow all the networks to operate as one, the master node broadcasted a global seed to all slave nodes. The master node was also responsible for the splitting and distribution of the N-bit data and the calculation of the global error at the end of every epoch. Unsupervised batch training was selected, as it allowed for a large amount of computation before each synchronization event. Training continued until the global error was less than 0.0001 or the maximum epoch of 1,000,000 was reached. At the end of each epoch, the error from each of the x processors was collected by the head node which calculated the global error and transmitted it to all the x processors.

The neural network was trained using the parallel implementation of ALOPEX-B on the University of Guelph's super-computer Narwhal(SHA, 2007). Narwhal is a HP Opteron super-computer cluster, with 1068 processors, 2 gigabytes of ram per CPU, and runs HP's Linux XC 3.0 operating system. The boards are inter-connected by a Myrinet switch which has an approximate 1 gigabyte transfer rate. A value of 10 was selected for N, as it produced a large size data set of 1024 unique patterns, but small enough to allow for the training to be completed in a reasonable amount of time.The number of epochs, training time and percent of correctness was recorded. Each test was repeated 10 times to ensure accuracy and to limit the effect of the stochastic nature of the ALOPEX-B algorithm. The network was considered to be stuck in local minima or maxima when it reached the 1,000,000 epoch ceiling and training was halted and restarted. The time for the retraining of the network was added onto the time it took to reach the maximum epoch and the total training time was reported. The value of the forget rate, λ was 0.6, and the update step size, δ was 0.005. The values of the training parameters δ and λ were not optimized, but were selected because they worked consistently. The structure of the network was 10 inputs, 20 neurons in the hidden layer, and 2 outputs. It is important to note that the network was found to have performed consistently when the hidden layer size was 2N.

4 RESULTS

The parallel implementation of ALOPEX-B behaved as hypothesized and a quantifiable speed up in training times was observed. As more processors were added, the time for training decreased until a lower plateau was reached. This plateau appeared to be the fastest speed before the cost of parallelization effected execution time and the amount of computation. Table 4 contains the average number of epochs, seconds, and the time that was reported.

# of processors	epoch	Time (seconds)	Time(hour:minute:sec)
2	520684	1490	0:24:50
4	477157	692	0:11:32
8	605291	454	0:07:34
16	666702	273	0:04:33
32	455997	112	0:01:52
64	568158	101	0:01:41

Table 1: Table of observed results for the 10-bit parity data set

The downward trend in training time is shown in table 4. Training times steadily decrease until a lower boundary of approximately 0:01:40 is reached. It is important to note that the stochastic nature of the ALOPEX-B algorithm creates the range in the average number of epochs. In table 4, the average number of epochs ranged from 455,007 to 605,291.

There did not appear to be any relationship between the number of epoch and the number of processors. In figure 1, the drop in training time is can be seen once again. The time for

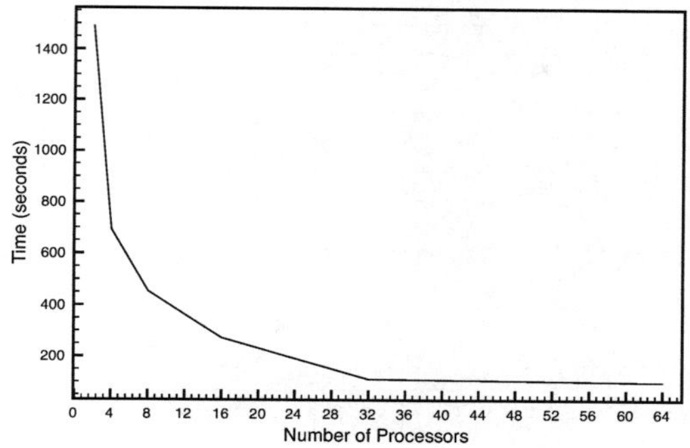

Figure 1: A plot of the number of the average training time in seconds versus the number processors

training drops dramatically to almost half when the number of processors increases from 2 to 4 processors. The training time continues to drop sharply until it begins to plateau at approximately 32 processors. From the shape of the curve in figure 1, the fastest possible training time for this data set was approximately 0:01:30.

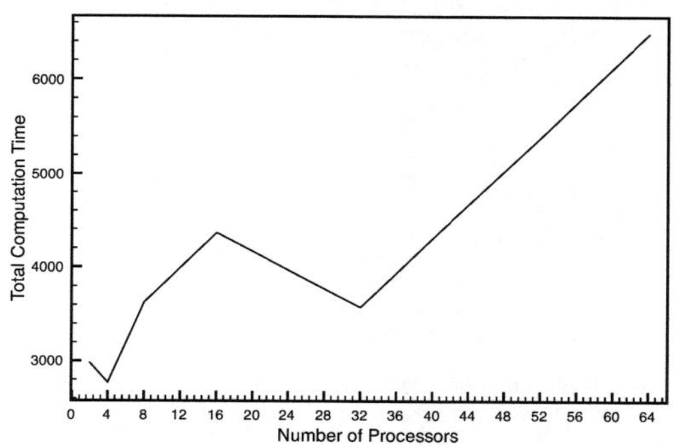

Figure 2: A plot of the total computation time relative to the number of processors used

Figure 2 plots the total computation time versus the number of processors. The total computation was calculated by multiplying the average time of computation by the number of processors. If the problem is perfectly parallelized the graph should contain a flat line straight across from beginning to end. In figure 2 it is clear that the line is not straight across, but rises and falls. Two possible reasons for this is the stochasticity of ALOPEX-B and the overhead of parallelization. It is interesting to note the sharp increase of total computation time from 32 processors to 64 processors. The total computation time is significantly more expensive than any of the other tests. This is most likely due to the data having been decomposed too far where a significant amount of computational expense is from the communication between processors. Table 4 contains the number of observed "failures" or the number of times that ALOPEX-B reached the maximum epoch before converging.

# of processors	# of restarts
2	1
4	1
8	2
16	2
32	1
64	2

Table 2: Table listing the number of restarts observed for each of the test conditions

The average number of restarts observed was approximately 1.5 ande ach test case had approximately the same number of restarts. Therefore the impact of ALOPEX-B's stochastic nature was minimal.

5 CONCLUSION

In conclusion, a quantifiable speed up in training was observed from a parallel implementation of ALOPEX-B. The time gains were significant enough to allow for the use of large data sets. The overhead required for the implementation was minimal and very little change to the original algorithm was required. The time for training with 2 processors was 0:24:50 and only 0:01:41 with 64 processors. The addition of 62 processors to the original 2 resulted in a time savings of approximately 23 minutes, a speed up of almost 14 times. Like all neural networks, the training of the network was sensitive to the completeness of the training set. The data parallel implementation of ALOPEX-B demonstrated itself as being an effective way to speed up training times. This is significant in increasing the window of problems that the ALOPEX-B can be used to analyze. The parallel implementation of ALOPEX-B helps to overcome some of the barriers related to training times and the usage of neural networks as a feasible option for large data sets.

REFERENCES

(2007). Sharcnet: Shared hierarchical academic research computing network.

Bia, A. (2001). Alopex-b: A new, simpler, but yet faster version of the alopex training algorithm. *International Journal of Neural Systems*, 11(6):497–507.

Harth, E. and Tzanakou, E. (1974). Alopex: A stochastic method for determining visual receptive fields. *Vision Research*, 14:1475–1482.

Haykin, S. (1999). *Neural Networks: A Comprehensive Foundation*. Prentice Hall.

Minsky, M. L. and Papert, S. A. (1969). *Perceptrons*. The MIT Press.

Quinn, M. J. (2004). *Parallel Programming in C with MPI and OpenMP*. McGraw-Hill Inc.

Rumelhart, D. E., Hinton, G. E., and Williams, R. J. (1986). Learning internal representations by error propagation. *Nature*, 323:533–536.

Skurikhin, A. N. and Surkan, A. J. (1994). A parallel correlation-based algorithm in j learns neural network connections. *APL Quote Quad*, 25(1):194–199.

Unnikrishnan, K. P. and Venugopal, K. P. (1994). Alopex: A correlation-based learning algorithm for feedforward and recurrent neural networks. *Neural Computation*, 6(3):469–490.

Werbos, P. (1974). *Beyond Regression: New Tools for Prediction and Analysis in the Behavioral Sciences*. PhD thesis, Harvard University, Cambridge, MA.

PROTEIN STRUCTURE ALIGNMENT USING FUZZY MEASURES

SUCHITHA SUBRAMANIAN
University of Cincinnati
Department of ECECS, ML 30
Cincinnati, Ohio 45221-0030
Email: subramsc@email.uc.edu

CARLA PURDY
University of Cincinnati
Department of ECECS, ML 30
Cincinnati, Ohio 45221-0030
Email: carla.purdy@uc.edu

ABSTRACT

The structural alignment of proteins is very important because structure of a protein is more closely related to function than sequence and also makes clear the common ancestry of proteins. There have been a number of tools developed to study three-dimensional protein structure. Most of these tools are vector based or distance based and use the Euclidean distance measure. In this project we use fuzzy alignment measures, along with powerful pattern-matching heuristics, to understand the functional and structural similarities of proteins whose underlying amino acid sequences may be different. Our datasets are the protein structures available in the Protein Data Bank (PDB), represented by their PDB ids.

1 INTRODUCTION

Proteins play an important role in every biological process. They are responsible for physiological processes such as enzymatic catalysis, structural support, and immune protection. Proteins are made up of building blocks called amino acids. There are 20 amino acids which are used to make up proteins in biological systems. These sequences of amino acids form the primary structure of proteins.

Today there are a number of ongoing projects seeking to understand the sequence information of proteins. But structural knowledge of the proteins is necessary for a detailed functional characterization of these sequences. It is also essential to study the three-dimensional structure of proteins because proteins that are structurally similar reveal a common evolutionary origin. Experimental determination of protein structures by X-ray crystallography or nuclear magnetic resonance spectroscopy (NMR) is time-consuming. Hence there is a sizable gap between the available protein sequences and experimentally derived protein structures. So researchers are working to automate the process of determining the 3D structure of proteins, using computational approaches called prediction methods, which attempt to predict a protein structure from its sequence. There are many algorithms available to do this, but they suffer from several shortcomings and can generate contradictory results which are difficult to interpret. Hence it is important to develop an algorithm that can better interpret structural alignment and discover protein structure similarity.

The structural alignment of proteins refers to comparing the two 3D protein structures and finding similarities between them. This problem may look like sequence alignment of proteins, but it is not the same, because the sequence alignment deals with the amino acid sequences, whereas structural alignment refers to 3D alignment of proteins based on their shape. Also, from the algorithmic point of view, sequence alignment can be solved in polynomial time using dynamic programming approaches (Lathrop, 1984). This is not the case in structural alignments because structural alignments refer to the rigid bodies of the protein structure that are to be matched. An optimal solution for structural alignment is difficult to obtain due to the experimental error inherent in protein structure determination. So far only an approximate polynomial-time algorithm for structural alignment of proteins has been developed (Kolodny and Linial, 2004). However, at O (n^{10} / ε^6) for a protein of sequence length n, the algorithm is still too computationally expensive for practical use. As a consequence, practical algorithms that converge to global solutions for protein alignment do not exist. Most algorithms are, therefore, heuristic. We use one such heuristic to determine an optimal structural alignment between two proteins and minimize the distance between the aligned structures.

In the past two decades many algorithms have been developed to automate protein structural alignment. These algorithms are categorized into two main types. They are

1. Intermolecular: The transformation of atomic coordinates is done by rotating and translating one of the two structures, keeping the other one stationary so that relevant parts are superimposed. Then the root mean square distance (RMSD) between corresponding α-carbon atoms of the two structures is computed. Intermolecular alignments do not handle molecular flexibilities.
2. Intramolecular: These algorithms compare the relative positions or distances between α-carbon atoms in one protein against the relative positions or distances of the atoms in a second protein. These algorithms don't depend on the structural orientation of the atoms in 3D space but on 2D distance matrices.

Cohen and Sternberg (1998) point out that it is not always possible to draw equivalent conclusions using these approaches.

2 METHODOLOGY

In our research we look at the first approach and use a straightforward, geometrically sensible, and widely accepted comparison metric for deciding when regions from two proteins are similar, by calculating the distance between aligned α-carbon atoms from two proteins.

We follow a bottom up approach where superposition is done first. Then we compute fuzzy alignment of the α-carbon atoms in the two protein structures to further minimize the distance. The fuzzy measure gives an optimal alignment. Our method is summarized as follows:

1. Generate initial set of superpositions to evaluate by doing rigid body transformation of protein structures – translation and rotation of proteins and superimposition of one structure over another.
2. After superposition, identify the minimal RMSD (Euclidean) alignment.
3. Using a heuristic, reposition one of the proteins to compute a fuzzy alignment which is better than the original one and recompute the RMSD for this new alignment. This procedure can be iterated to improve the alignment.

The fuzzy alignment is very similar to the traditional sequence alignment algorithms, namely Needleman-Wunsch (Ewens and Grant, 2001) or Smith-Watermann (Liao et al., 2004), with some changes such as introducing a fuzzy alignment measure. The important steps involve formulation of a cost function and minimizing the cost function by using a heuristic in an iterative manner.

One of the advantages of this method is that it does not involve any complex calculations such as those needed in neural nets or genetic algorithms. This method is also very stable, as the distance is normalized with a fuzzy measure. It is also easy to extend this method for multiple structure alignment of proteins.

This method uses the geometric positions of the α-carbon atoms in the backbone of the protein structures as input. Our method is compared to popular methods such as DALI (Holm and Park, 2000), CE (Shindyalov and Bourne, 1998) and CHIMERA (Pettersen et al., 2004). The test data includes proteins of different lengths and different percentages of identity.

3 PROTEIN STRUCTURE ALIGNMENT

Consider two proteins P1 and P2 which are to be aligned by rigid body transformations. The alignment of the two proteins is done in two steps:

1. Rigid body superposition of P1 on P2 keeping P2 stationary. The rotations and translations are done to the entire structure and not to fragments. By doing this we get only parts of the structures that coincide and we need to calculate the distance between these atoms in the matched structure. So the structure comparison of the proteins is reduced to sub-structure comparison of proteins.
2. Calculation of fuzzy alignment and subsequent minimization of the distance between the aligned protein atoms.

Rigid body transformation is done to minimize the distance between the matched atoms of the two structures. If x'_i represents the coordinates of a translated and rotated atom originally at position x_i, then

$$x'_i = a + Rx_i \qquad (1)$$

where a is the translation vector and R is the rotation matrix (equation 1). We use CHIMERA to match atoms $X_1, \ldots, X_N$ in P1 with atoms $Y_1, \ldots Y_N$ in P2, where P2 is kept fixed and P1 is transformed. Now compute the Euclidean RMSD between the two aligned protein structures using equation 2, where ε is the minimized RMSD. Though this RMSD is supposed to be the minimum distance between the aligned structures, it can further be reduced by making a fuzzy alignment and we will show that to be a better alignment.

$$\varepsilon = \sqrt{\frac{1}{N}\sum_{i=1}^{N}(a + Rx_i - y_i)^2} \qquad (2)$$

3.1 Calculation of Fuzzy Measure

The calculation of the fuzzy measure is a three step process. It is described as follows.

Step 1: The important thing to be done in the calculation of the fuzzy measure is formulating a cost function and minimizing it. Let D_j denote the term of the cost function associated with Y_j, $1 \leq j \leq N$. This is in contrast to many other algorithms in the literature which use the scoring criterion that a maximum score is given if there is a match found between the amino acid residues, while ignoring the geometrical alignments of the atoms. The computation of D_j is a two step process which is described as:

1. Start with the superimposed structure as described above.
2. Now match Y_j with V_j where V_j is chosen from (X_{j-1}, X_j, X_{j+1}) to minimize D_j in equation 3 (with obvious modifications for j = 1 and j = N) and dst(A,B) is the Euclidean distance between 2 points A and B.

$$D_j = \min\{dst(X_{j-1}, Y_j), dst(X_j, Y_j), dst(X_{j+1}, Y_j)\} \qquad (3)$$

Step 2: This is followed by the computation of the fuzzy distance between each Y_j and V_j as shown in equation 4. The fuzziness is controlled by a temperature parameter T which is changed continuously and iteratively. Suppose V_j has co-ordinates (v_j^1, v_j^2, v_j^3) and Y_j has co-ordinates (y_j^1, y_j^2, y_j^3). For fixed T we compute all the values

$$d(1,j) = \left|v^1{}_j - y^1{}_j\right|,\ v^{1'}{}_j = \frac{e^{-d(1,j)T}}{\sum_{j=1}^{N} e^{-d(1,j)T}}$$

$$d(2,j) = \left|v^2{}_j - y^2{}_j\right|,\ v^{2'}{}_j = \frac{e^{-d(2,j)T}}{\sum_{j=1}^{N} e^{-d(2,j)T}} \qquad (4)$$

$$d(3,j) = \left|v^3{}_j - y^3{}_j\right|,\ v^{3'}{}_j = \frac{e^{-d(3,j)T}}{\sum_{j=1}^{N} e^{-d(3,j);T}}$$

T is the temperature parameter which is adapted as the heuristic, controlling the fuzziness. We start by setting the temperature at a high value (through experiment we chose T = 10). We reduce the temperature in steps of size dT until it reaches the minimum positive value dT. At one particular iteration I_{min} we find a minimum value for the sum of all the distances $v_j = v_j^{1'} + v_j^{2'} + v_j^{3'}$, $1 \le j \le N$, and consider that to be the optimal alignment. The fuzzy alignment of proteins may look like the one in Figure 2, for example, which is better than the normal alignment in Figure 1.

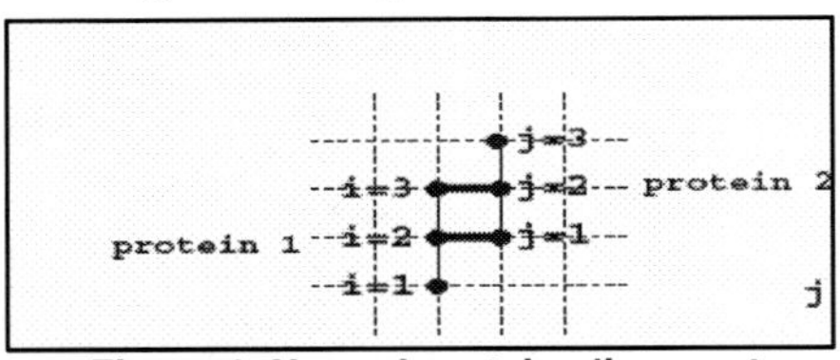

Figure 1. Normal protein alignment.

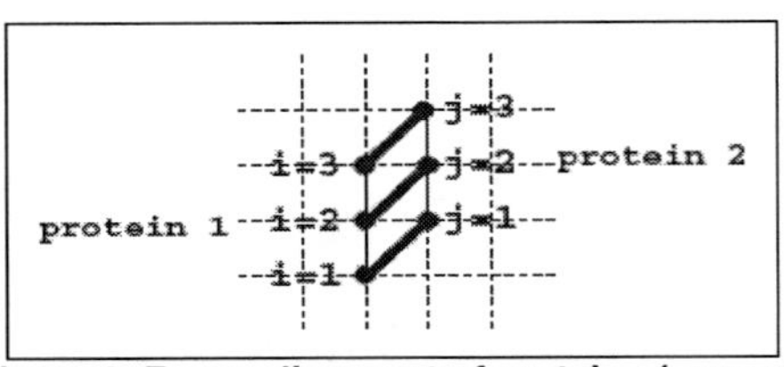

Figure 2. Fuzzy alignment of proteins (example).

For structural alignment the gap penalty is zero because we do not have any requirement that matched residue pairs should be contiguous. In the case of sequence alignment a gap penalty is defined when an atom at a particular position in P1 does not match with the atom in the same position in P2. Our method also gives the second best and the third best alignment, which can be used for further studies.

Step 3: The final step involves the calculation of the fuzzy distance. Let $(Y_1, V_1^{'})$, $(Y_2, V_2^{'})$.... $(Y_N, V_N^{'})$ denote the pairs of matched atoms. Let $v_j = v_j^{1'} + v_j^{2'} + v_j^{3'}$, $1 \leq j \leq N$ denote the minimal distances calculated at iteration I_{min} as described above. The fuzzy distance ε_f is computed finally using equation 5.

$$\varepsilon_f = \sqrt{\frac{1}{N}(v_1 + v_2 + ... + v_N)} \quad (5)$$

Our method uses the simple distance measure and uses a fuzzy alignment procedure to refine it further. This method is computationally simple compared to complex methods and prediction techniques such as genetic algorithms or neural nets.

4 RESULTS AND DISCUSSION

For the following comparisons, test pairs are taken from the Protein Data Bank (2007). Our method is compared against the following methods:

--DALI (Holm and Park, 2000), where the 3-dimensional coordinates are used to calculate residue-residue distance matrices and patterns are detected from that.

--CE (Shindyalov and Bourne, 1998), which involves a combinatorial extension of an alignment path defined by aligned fragment pairs.

--CHIMERA (Pettersen et al., 2004), an extensible 3-dimensional molecular modeling system where structure comparison is based on sequence alignment using algorithms such as Needleman-Wunsch.

--Protein 3D fit (Lessel and Schomburg, 1994), where similar protein fragments of a definable length are detected, combined in pairs and superimposed. Fits are improved until a maximum number of atom pairs show a distance below a given threshold value.

--FAST (Zhu and Weng, 2005), which uses a directionality-based scoring scheme to compare the intra-molecular residue-residue relationships in two structures.

In Table 1 we show the results for 5 representative pairs, each from a different protein family: cytochrome c (1YEA, 1YCC), cytochrome c3 (1WAD, 3CYR), ribosome inactivating proteins (1APA, 1FMP), glycosyl hydrolase family 7 (1CEL, 1EGL), and ATP synthase (1MAB, 1SKY).Our algorithm performs better in most of the cases compared to others and produces alignment with equal or smaller RMSD value.

Table 1. Distance value obtained using different alignment methods.
***No. is the number of atoms matched by CHIMERA.**

Structures & No.*	Fuzzy	CHIMERA (RMSD)	CE (RMSD)	DALI (RMSD)	Protein 3D fit (RMSD)	FAST (RMSD)
1YEA, 1YCC 106	0.302	.309	.38	0.7	.309	.381
1WAD, 3CYR 78	0.652	.789	1.59	1.8	.853	.983
1MAB, 1SKY 228	0.873	.904	1.1	1.8	1.01	1.365
1APA, 1FMP 188	0.915	.940	1.59	1.8	.966	.983
1CEL, 1EGL 98	1.297	1.475	4.8	-	1.023	1.031

Kolodny et al. (2005) have carried out a critical evaluation of different methods available for protein structure alignment. They note that CHIMERA is designed to overcome the defects of DALI and CE. From the results in Table 1 it is clear that our method produces results that are consistent with CHIMERA and gives better alignment. Note that the better the alignment, the smaller the distance. DALI and CE are also consistent with CHIMERA and our method. Another interesting observation is that the methods FAST and PROTEIN 3D fit are not consistent with the other methods considered.

From Table 1 we observe that the structures 1 YCC and 1 YEA are very closely related by their sequence. They have a sequence similarity of 80% and hence they get the minimum RMSD. When we look into structures 1 APA and 1 FMP where sequence identity is only 31%, the RMSD value is bigger. Another interesting result seen is that, in the case of the structures 1CEL and 1EGL, DALI does not calculate the RMSD because of the size complexity. Traditional algorithms do not even consider these structures because of the size complexity, but our algorithm focuses mainly on the structure alignment and hence we get a minimized distance compared to other methods.

So far we have tested our algorithm using protein structures taken from 15 different families of different percentages of identity and of different lengths. In these tests our method produces results which are on an average 8% better than CHIMERA, 45% better than CE and 80% better than DALI (For DALI, this comparison omits those cases where DALI returned no result).

Our future work involves extending the tool to measure the dissimilarity between protein structures taken from different families. Our method can be extended to study the multiple protein structure alignment and also to handle side-chain orientation.

REFERENCES

Cohen F.E., Sternberg M.J.E. (1998). "On the prediction of protein structure: The significance of the root mean squared deviation." *J.Mol.Biol.* 138, pp. 321-333.

Ewens W., Grant G. (2001). *Statistical Methods in Bioinformatics.* Statistics for Biology and Health, Springer-Verlag: New York.

Holm L., Park J. (2000). "DaliLite workbench for protein structure comparison." *Bioinformatics* 16 (6), June, pp. 566-567.

Kolodny R., Koehl P., Levitt M. (2005). "Comprehensive evaluation of protein structure alignment methods: scoring by geometric measures" *J.Mol.Biol* .346, pp 1173-1188.

Kolodny R., Linial N. (2004). "Approximate protein structural alignment in polynomial time." *PNAS* 101 (33), pp. 12201-12206.

Lathrop R.H. (1994). "The protein threading problem with sequence amino acid interaction preferences is NP-complete." *Protein Engineering* 7 (9), October, pp, 1059-1068.

Lessel U., Schomburg D. (1994). "Similarities between protein 3-D structures." *Protein Engineering* 7 (10), October, pp. 1175-1187.

Liao H.Y., Yin M.L., Cheng Y. (2004). "A parallel implementation of the Smith-Waterman algorithm for massive sequences searching." *Proceedings of the 26th Annual International Conference of the IEEE EMBS,* San Francisco, CA, September 1-5, 2004.

Pettersen E.F., Goddard T.D., Huang C.C., Couch G.S., Greenblatt D.M., Meng E.C., Ferrin T.E. (2004). "UCSF Chimera - A visualization system for exploratory research and analysis." J. *Comput. Chem.* 25, pp. 1605-1612.

Protein Databank. http://www.pdb.org/pdb/home/home.do. Accessed May 31, 2007.

Shindyalov I.N., Bourne P.E. (1998). "Protein structure alignment by incremental combinatorial extension (CE) of the optimal path." *Protein Engineering* 11 (9), September, pp. 739-747.

Zhu J., Weng Z. (2005). "FAST: A novel protein structure alignment algorithm." *Proteins* 58 (3), February, pp. 618-627.

CONDUCTANCE BASED NEURAL SIMULATOR: NEURAL EXCITABILITY, SPIKING, AND BURSTING

IREN VALOVA
Computer and Information Science,
University of Massachusetts
N. Dartmouth, MA 02747

GEORGE GEORGIEV
Computer Science, University of
Wisconsin, Oshkosh, WI 54901

NATACHA GUEORGUIEVA
Computer Science, City University
of New York
Staten Island, NY 10314

ABSTRACT
Neural networks can learn how to solve tasks based on the data given from training or initial experience, in real time and then reconfigure and carry out the computations in parallel with the received and processed new data. We present an extension of Integrate & Fire McGregor Spiking Neural Simulator (SNS) based on Hodgkin–Huxley model. The goal of this research is to study how the neuronal behavior depends on measurable physiological parameters, such as the maximal conductance, steady-state activation functions and time constants by altering the weight and delay factors of the synapses as well as changing the structure of the connections between the layers and the parameters of the NN. The simulation parameters are chosen to be close to those used by medical researchers in order to compare the results of our simulations with those reported in medical literature.

INTRODUCTION

A typical biological neuron consists [1] of dendrites (input device), axon (output device), soma (CPU), and synapses (chemical or electric, providing connection between neurons).

As neurons receive signals from nerve structures, they respond with spikes of electrical activity through the axon. The axon, splits into many branches which end in a synapse. The synapse converts the activity from the axon into electrical effects that restrict or accelerate activity from the axon into other neurons. Neurons communicate with each other through signals caused by the electrically-charged chemicals that operate inside the nervous systems such as potassium, sodium, calcium and chloride. As soon as a neuron receives a satisfactorily large acceleratory input, when compared to its threshold level, it sends a spike of electrical activity across the cell's membrane to the axon [2, 3].

The effects of an action potential arriving at a presynaptic terminal are transferred to the postsynaptic neuron through a chemically mediated process. If no input (spike) is received by the neuron, it maintains a rest potential which corresponds to a constant membrane potential. If some input was applied, the neuron potential changes immediately and then loses its strength gradually to its resting state. If the change is positive, the synapse is said to be excitatory. If the change is negative, the synapse is inhibitory. When the action potential arrives at the axon conclusion, it causes minuscule bubbles of chemicals called vesicles to discharge their contents into the synaptic fissure.

These chemicals are called neurotransmitters [4, 5]. These flow across the gap to the next neuron, where they find special spaces on the cell membrane of the next neuron called receptor sites. As soon as transmitter molecules have reached the postsynaptic side, they are detected by specialized receptors in the postsynaptic cell membrane and open specific channels so that ions from the extra-cellular fluid flow into the cell, which changes the membrane potential at the postsynaptic site. As a result, the chemical signal is translated into an electrical response, which is called the postsynaptic potential [6].

Currents for excitatory and inhibitory synaptic actions are generated in the postsynaptic neuron via changes in membrane conductance [7]. The biophysical processes involved in several aspects of synaptic transmission continue to inspire intense research. Our NN topology uses the MacGregor I&F neurons in order to model the behavior of the biological neuron [8].

MCGREGOR NEURON

I&F models simplify the spike generation mechanism but at the same time they provide a correct approximation of the membrane potential and other neuron properties. However, conductance based models have a higher degree of biological similarity and computational complexity, whereas McGregor neurons allow faster learning process, increased computational power and study of the NN from a wider perspective [10, 11].

The membrane potential is the potential difference between the interior and exterior of the cell. Before the input spike has arrived, the neuron i has membrane potential $u_i(t) = u_{rest}$, where u_{rest} is the rest potential. Assume that at time $t = 0$ the presynaptic neuron j fires its spike. For $t > 0$, neuron i changes its membrane potential immediately following the equation

$$u_i(t) - u_{rest} = \varepsilon_{ij}(t) \tag{1}$$

Each spike respectively evokes a postsynaptic potential. As long as there are only few input spikes, the total change of the potential is approximately the sum of the individual postsynaptic potentials (PSPs)

$$u_i(t) = \sum_j \sum_f \varepsilon_{ij}(t - t_j^{(f)}) + u_{rest} \tag{2}$$

When the membrane potential reaches a critical value, its trajectory shows a behavior that is quite different from a simple summation of PSPs: The membrane potential exhibits a pulse-like action potential with amplitude of about 100 mV. This action potential will propagate along the axon of neuron i to the synapses of other neurons. After the pulse, the membrane potential does not directly return to the resting potential, but passes through a phase of hyper-polarization below the resting value. This hyper-polarization is known as `spike-after-potential'. The generation of spike is described by the following equation

$$S = \begin{cases} 1, & \text{if } E \geq T_h \\ 0, & \text{if } E < T_h \end{cases} \tag{3}$$

where S is a flag showing a neuron spike, T_h is a threshold value and E is a transmembrane potential which depends on the membrane potential u and u_{rest}, i.e. $E = u - u_{rest}$.

The refractory properties are determined by

$$\frac{dG_k}{dt} = \frac{-G_k + B.S}{T_{G,k}} \tag{4}$$

where the transmembrane conductance $G_k = \frac{g_k}{G}$, g_k is potassium resting conductance, G is membrane resting conductance, B is postfiring potassium increment and $T_{G,k}$ is time constant for G_k decay.

$$\frac{dT_h}{dt} = \frac{-(T_h - T_{h0}) + cE}{T_{th}} \tag{5}$$

where T_{h0} is the resting value of threshold, T_{th} is the decay of threshold time constant and c determines the incrementation/decrementation of the threshold, $c \in [0,\ 1]$.

The transmembrane potential

$$\frac{dE}{dt} = \frac{-E + G_k(E_k - E) + G_e(E_e - E) + G_i(E_i - E) + SCN}{T_{mem}} \tag{6}$$

where transmembrane potentials E_k , E_e and E_i from potassium resting potential u_k , excitatory resting potential u_e and inhibitory resting potential u_i are:
$E_k = u_k - u_{rest}$; $E_e = u_e - u_{rest}$; $E_k = u_i - u_{rest}$, the membrane time constant T_{mem} is determined by the membrane capacitance C and the membrane resting conductance ($T_{mem} = C/G$) and SC is the current injected to the cell ($SCN = SC/G$).

G_k , G_e and G_i are transmembrane conductances which are calculated respectively from the following equations:

$$G_k = g_k/G; \quad G_e = g_{se}/G \quad G_i = g_{si}/G \tag{7}$$

where g_k , g_{se} and g_{si} are respectively potassium, excitatory and inhibitory resting conductances.

MODEL AND SIMULATIONS

The NN topology used for the simulation has two excitatory layers connected in a feed forward fashion and one inhibitory layer connected in a feedback manner to one of the excitatory layers – Fig. 1. The first excitatory layer represents the receptive/sensory region of the network. It is connected feed forward to the inhibitory layer. The second excitatory layer together with the inhibitory one represents the data processing section. The above model description corresponds to the CA3 region of hippocampus. Most of the neurons of CA3 are excitatory (pyramidal), while the remaining neurons are inhibitory (interneurons) [11]. In excitatory synapses signals are carried by glutamate while in inhibitory synapses they are carried by -amino-butyric acid (GABA). Glutamate then activates excitatory NMDA (long lasting) and AMPA (rapid) ionotropic receptors, while GABA activates inhibitory (fast) and (slow) receptors. Receptor activation changes postsynaptic conductance which in turn transforms the neuron's membrane potential. The synaptic connection strength and signal strength determine the magnitude and direction of this change.

In the proposed model, the arriving spikes change the postsynaptic conductance, thus changing the membrane potential. With each synaptic connection there is an associated synaptic strength (weight). Therefore, the weighted sum of all excitatory and the weighted sum of all inhibitory synaptic conductances determines the change in the membrane potential. The proposed model incorporates a membrane threshold which causes the neuron to require a stronger stimulus for the generation of consecutive spikes. For a short time immediately after the spike generation, the neuron is not capable of firing - absolute refractory period. In the interval after the absolute refractory period - relative refractory period - the neuron can only respond to a very strong stimulation.

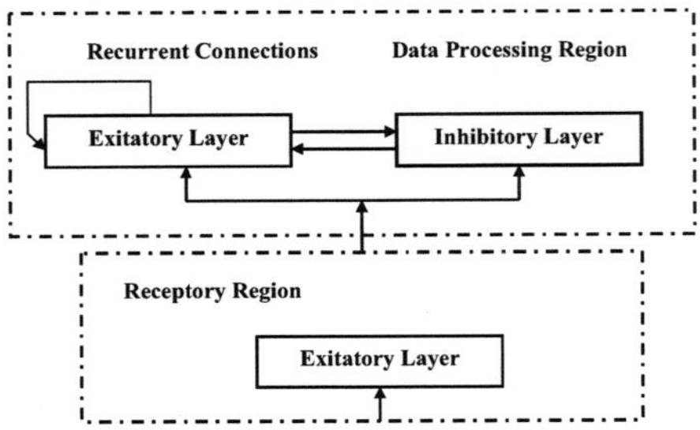

Figure 1. NN topology

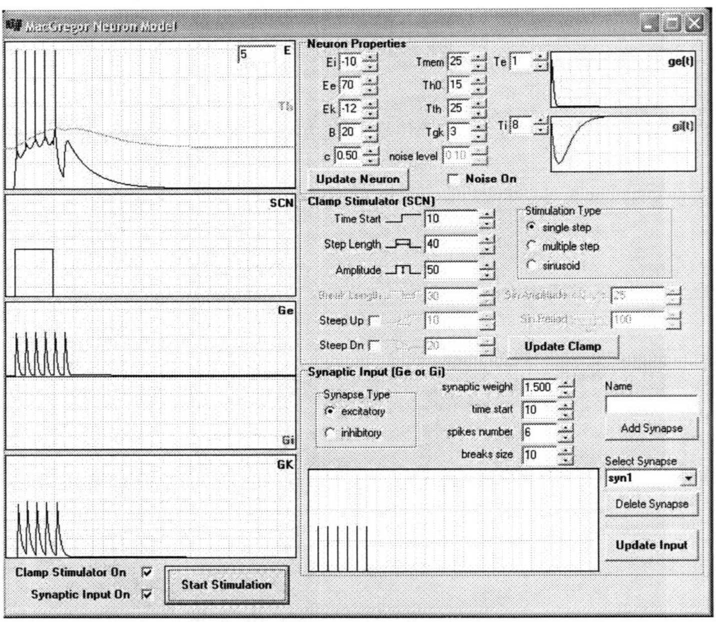

Figure 2. The NN performance with the increase of T_{h0}

The excitatory input to the i-th node in our topology is a simple linear summation of weighted inputs to that node;

$$g_{se} = \sum_{j} w_{ij} S_j \qquad (8)$$

where w_{ij} is the weight from node j to node i and S_j is the spiking variable of node j. For weight update we used the modified Hebbian-type synaptic plasticity with sigmoidal function as it produces a smooth learning curve [12].

$$w_{ij}(t+1) = \textbf{\textit{sigmoid}}\Big(w_{ij}(t) + \sum_{\substack{\textbf{\textit{excitatory}}\\ \textbf{\textit{nodes}}}} \mu.PSP_{ij}(t) - \sum_{\substack{\textbf{\textit{inhibitory}}\\ \textbf{\textit{nodes}}}} \eta.PSP_{ij}(t)\Big) \qquad (10)$$

where t is time, μ is the learning rate for excitatory nodes, η the learning rate for inhibitory nodes and $PSP_{ij}(t)$ is the postsynaptic value for connections between the neurons i and j.

We studied the influence of different parameters and group of parameters on the network performance. Fig.2 shows the NN output with the alteration of the resting value of threshold T_{h0}. When a certain threshold is reached as a result of neuron jumps, and an output spike is generated, the membrane potential goes back down to a given after-spike potential. After jumping, the neuron enters in a refractory period in which it recovers from the after-spike potential to the resting potential. Figs. 3,4 demonstrate the NN output with the alteration of the time constant $T_{G,k}$ for G_k decay.

CONCLUSIONS

In this research we investigated the importance of excitation, inhibition and various parameters on the proposed NN topology. As the area CA3 is a well-studied model of epileptic activity, we hope that the proposed NN will allow us to test specific hypotheses regarding the nature of real neural network activity, such as the idea that bursts of spontaneous activity are limited by the supply of releasable neurotransmitter at synaptic connection between neurons in the network.

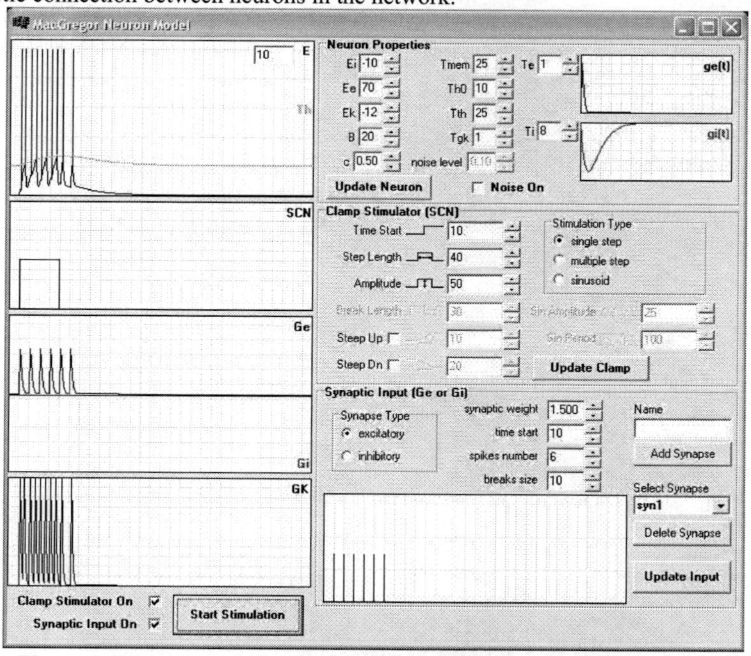

Figure 3. NN performance with the decrease of the time constant $T_{G,k}$

From the sets of experiments performed on the developed topology we may conclude that: Inputs have a greater effect when they arrive within a short time interval; Inhibitory inputs close to the soma can block the effect of stronger excitatory inputs that are applied at a greater distance; Increment of different variables which are dependant on membrane resting conductance G decreases the slope of each spike; Decrement of the above variables typically increases the numbers of spikes as well as the slope of each one.

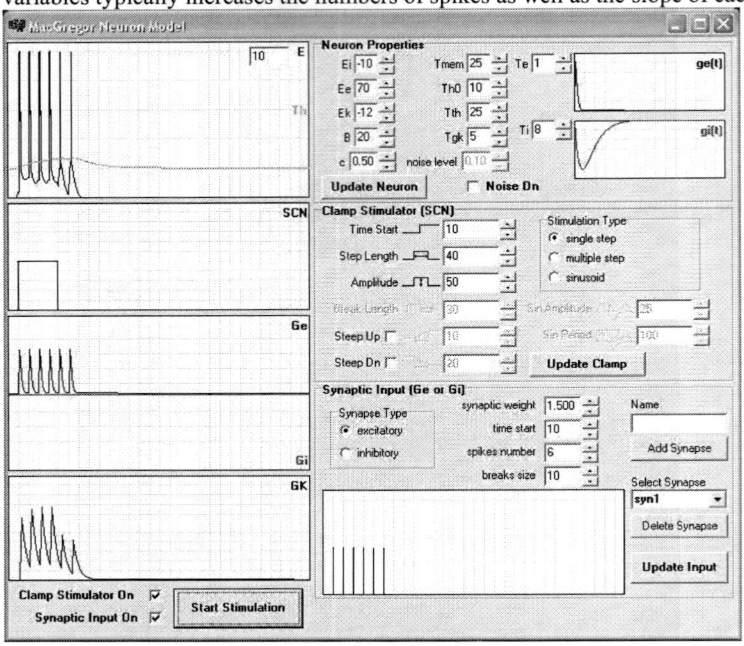

Figure 4. NN performance with the increase of the time constant $T_{G,k}$

The results based on the number of neuron spikes affected through these simulations suggest that it is possible to alter the number of neuron spikes through changing particular variables related to membrane resting conductance G .

REFERENCES

[1] Maas W., and Bishop C., Pulsed Neural Networks. MIT Press, Cambridge, MA, 1998.

[2] Gerstner W., Coding Properties of Spiking Neurons: Reverse and Cross Correlations. Neural Networks, 14:599-610, 2001.

[3] Maas W., Networks of spiking neurons: the third generation of neural network models" Neural Networks, 10(9):1659-1671, 1998.

[4] Gerstner W., Kistler W., Spiking Neuron Models, Cambridge, 2002.

[5] Cios K. J., Swiercz W., and Jackson W., Networks of spiking neurons in modeling and problem solving, Neurocomputing, vol. 61, pp. 99–119, 2004.

[6] Dayan P. and Abbot L. F., Theoretical Neuroscience. Cambridge, MA: The MIT Press, 2001.

[7] Gerstner W., Integrate and Fire Neurons and Networks. In The Handbook of Brain Theory and Neural Networks, 2/e, Cambridge, MA, 2002.

[8] MacGregor RJ, Oliver RM, A model for repetetive firing in neurons. Cybernetik 16: 53-64, 1974.

[9] Hodgkin A.L., and Huxley A. F., A quantitative description of ion currents and its application to conduction and excitation in nerve membranes. J Physiol 117: 500-544, 1952.

[10] MacGregor RJ, Theoretical Mechanics of Biological Neural Networks. Academic, 1993.

[11] Staley K., Longacher M., Bains J. S., and Yee A., Presynaptic modulation of CA3 network activity, Nature Neurosci., vol. 1, no. 3, pp. 201–209, 1998.

[12] Gerstner W. and Kistler W., Mathematical Formulations of Hebbian Learning, Biol. Cybern., vol. 87, no. 56, pp. 404–415, 2002.

DIGITAL NEURAL CORTEX

ALEXEI M. MIKHAILOV
FirstCortex, Inc, Potomac, MD

ABSTRACT
The ANNIE 2001 artificial neural cortex paper [1] formally introduced the direct/inverse columnar index pair. The power of direct/inverse index pair was clearly shown by the Google search engine, which is based on such a pair, where direct index is a {document}/word id and the inverse index is {word id}/document relationship [2]. An efficient way to achieving a generalization of information is in using a hierarchy of indexes. Litespeed, Ltd, Singapore, http://litespeed.com.sg, implemented such a system in 2002, which supports content-based marking of open-ended structured answers with a high level of accuracy. This paper introduces an indexing formalism and a set identification algorithm, which the implementation of the digital cortex is based upon. The paper also discusses an application of the suggested technology to the scene analyses, which is performed by the digital cortex without training sessions.

INTRODUCTION

The paper discusses a hierarchical indexing memory system, referred to as a digital neural cortex (DNC). A similar hierarchical memory was discussed in [3]. However, the work [3] provides only a verbal description of cortex-like memories and does not discuss their indexing properties.

DNC can be used for both indexing of large-scale volumes of textual information and indexing of bit-maps. DNC being an indexing system avoids any template matching, which becomes an intractable problem when the size of the application approaches billions of templates. The property of being a hierarchical system supports information generalizing capabilities. Generalization of lower-level features leads to stable higher-level features, so that noise-tolerance and invariance with respect to scaling and deformations of shapes is achieved. So far, DNC was tried as an engine for content-based marking of open-ended structured answers in schools and junior colleges in Singapore. We assumed that a search engine could work, within certain limitations, as a marking engine, which was proved to be true. This paper is largely devoted to the theory behind the DNC rather then to details of applications.

DNC does not belong to the class of traditional artificial neural networks. It rather follows weightless neural networks paradigm (WNN) that was suggested in [4] and was advanced in [5], [6], [7] and other works. The difference between DNC and WNN is as follows. (a)Unlike WNN, DNC features a hierarchical data structure. (b)Whereas WNN-data structure is a binary matrix, where the columns are accessed in a vertical fashion, each level of DNC-data structure is built as vertical and horizontal lists. (c) Unlike WNN, DNC is a self-leaning device.

DEFINITION, NOTATION AND SOME PROPERTIES OF COLUMNAR INDEXES

The term 'index' is used to denote several things: numerical scale, ratio, subscript, the finger next to the thumb and an alphabetical listing of names and/or topics along with page number where they are discussed. We are focusing on the last meaning, that is, on a book index. The book index contains a list of words, where each word is associated with a list of page numbers (Fig.1).

			277	115		11
170	222		206	90	25	34
90	170	153	191	72	23	24
Cycle	Levy	Lick	Light	Life	Log	Logic

Fig.1. Book Index

A formal definition of the book index (referred to as a columnar index) and a notation $\{y\}_x$ were suggested in [1]. Let X be the total number of columns. We denote the content of the column x as $\{y\}_x$ and call a set of columns $\{y\}_x$, where $x=1, 2,\ldots,X$, a columnar index. Unlike matrix, which is a table defined by two dimensions, that is, the number of columns and the number of rows, a columnar index is a table whose only dimension is defined by the number of columns. The vertical dimension is unavailable because the number of elements in the column may vary. A columnar index is a tool for memorizing of sets and it is subjected to the following limitation. Since each column is a set, no column can contain two or more equal elements. This is an implication of the fundamental definition of the set, according to which the set is a collection of distinct elements.

A columnar index can always be inverted. The inverse columnar index is also a table, where the number of columns is equal to the number of names that are stored in the columns of the direct index. The left-hand side of Fig.2 shows an example of the direct index I, whose inverse index I^{-1} is given on the right-hand side of Fig.2, where, for instance, the columns 1 and 3 of the inverse index I^{-1} contains (along with other elements) the element 1 because the column 1 of the direct index I contains the names 1 and 3.

	4	
3	3	
1	2	4
1	2	3

		2	3
1	2	1	2
1	2	3	4

Fig.2. Direct Index $\{y\}_x$ and its Inverse Index $\{x\}_y$

An inverse index is subject to the following limitation. Any two columns of the inverse index are supposed to be distinct. The subscripts of I^{-1}-columns are the patterns' names and the elements of I^{-1}-columns are the pattern components. Hence, the limitation imposed on the inverse index bans the situation where one pattern has more than one name. Theoretically, it is useful to represent columnar indexes with binary matrices. Figure 3 shows the indexes from Fig.1 represented by their binary matrices.

0	1	1
1	1	0
0	1	0
1	0	0
1	2	3

0	0	0	1
0	1	1	1
1	0	1	0
1	2	3	4

Fig.3. Representation of Indexes with Binary Matrices

If the index I is represented with the binary matrix A then the inverse index can be represented by a transpose matrix $I^{-1} = A^T$. But, practically, matrix-based representation of a columnar index leads to inefficient use of memory and excessive computations . Hence, practice, we implement the indexed data as a pair of lists: the vertical list supports the information stream within the direct index and the horizontal list supports the inverse information stream.

The notion of a function captures in a formal way the idea of a correspondence between two quantities. However, the notion of a columnar index formally captures the relation between one quantity and a set of quantities. Let us consider a function $y = f(x)$. In general case, the inverse function $x = u(y)$ does not exist unless it assigns to each y a unique element x, which is not the case even for the function $y = x^2$, where $x_1 = +\sqrt{y}$ and $x_2 = -\sqrt{y}$. That is, for each y there exist a set of two elements $\{x_1, x_2\}$. What does exist is the index $\{x\}_y$, $y=1,\ldots,Y$. The latter statement follows from the fact that, unlike functions, any columnar index can be inverted.

AVOIDING EXPONENTIAL MEMORY GROWTH BY LOSSLESS SPACE DECOMPOSITION

N-dimensional columnar index is denoted as $\{y\}_{(a, b,\ldots,z)}$. In this case, the address of any column of the index is specified by a vector $(a, b,\ldots, z)$. The space decomposition theorem states that any N-dimensional index can be decomposed into N one-dimensional indexes:

$$\{y\}_{(a, b,\ldots,z)}=\{y\}_a \cap \{y\}_b \cap \ldots \cap \{y\}_z$$

Note that, in general case, an N-dimensional function $y = f(x_1,\ldots,x_N)$, unlike the columnar index, can not be decomposed into N one-dimensional functions without a data loss. For instance, artificial neural networks collapse an N-dimensional space of input components $(x_1,\ldots,x_N)$ into N one-dimensional spaces by adding up the components: $W_n x_n$.

The space decomposition approach provides the solution to the problem of exponentially growing memory when the size of patterns grows linearly. For instance, a 2^{64} memory would hardly be ever achieved in one piece as the Universe contains less the 2^{100} atoms.

The computer screen is a good example of a space decomposition-based device. Clearly, the computer screen is not built with all different 256^3 = 16,777,216 color elements. But, it emits 256^3 colors by 3 light emitting diodes (red, green and blue) per pixel. An input pattern, say, (123, 22, 55), specifies the brightness of each diode, that is, the brightness of each color of the pixel. In the biological cortex, although each column responses to a single distinct feature, just 10 columns can distinguish up to 2^{10} feature combinations. Indexing in the digital cortex is different from RAM-type indexing, which uses a single address word to retrieve the data. Instead, the digital cortex uses a set of addresses, each address being a component of the input pattern, and retrieves the records by intersecting the active columns. For example, a book sentence can be used to activate a number of columns whose intersection could deliver a single page that contains this sentence. In the Fig. 1, the intersection of the columns "Life" and "Cycle" directly points to the page 90.

Clearly, the information space is enormously sparse as the number of incoming records will always be incomparably smaller than the number of potential records. This information space looks rather like the Universe, where the stars are light years away. Indeed, given a 100-component pattern, only few patterns will ever be recorded in the memory as the number 2^{100} already rivals the number of atoms in the Universe. The column, being a set of references, is a unit of the columnar indexing system. What mathematical representation of the column is computationally advantageous? We note that the use of matrices should be avoided for the reason that the encoding matrix will be sparsely populated and, thus, will lead to an inefficient use of memory and computational time.

SER EQUATION AS A SET IDENTIFICATION TOOL

Let $I=\{y\}_x$ be the columnar index and $I^{-1}=\{x\}_y$ its inverse index. An input to I is a subset $\{x\}$ of addresses of its columns. We want to find an inverse column y, such that $\{x\}=\{x\}_y$. In other words, we are looking for a solution to the following set equation.

$$\{x\}_y = \bigcap_{\{x\}} \{y\}_x = \{x\}$$

For example, given an unknown input $\{x\} = \{1, 2\}$, that is, the first two columns of I are active, find the column of I-1such that $\{x\}y=\{1,2\}$. Left-hand side of Fig.4 shows the columns that are activated by input $\{1, 2\}$, and right-hand side of Fig.4 shows the solution as the column 3 of the inverse index.

$\{Y\}_x =$

	5	
5	4	
3	3	5
1	2	4
1	2	3
^	^	

$\{x\}_y =$

				3
		2	3	2
1	2	1	2	1
1	2	3	4	5
		^		

Fig.4. Input Columns $\{1, 2\}$ and the Solution-Column $\{3\}$

The solution algorithm to this equation can be illustrated with the following block-diagram (Fig.5).

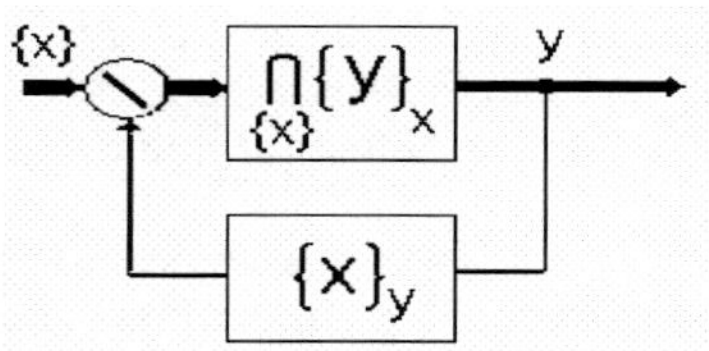

Fig.5. Set Equation Block-Diagram

The intersection of the columns $\{y\}_x$ in the direct path points to a subset of inverse columns y. The content $\{x\}_y$ of each inverse column from the subset is matched to the input $\{x\}$. No solution exactly fits with memory equation unless the unknown input set was previously stored in the indexing memory. In the latter case, there exists one and only one solution to this equation.

GENERIC TEMPLATE MATCHING PROBLEM AND AXIAL ROTATION

Generic template matching problem (TMP) can be described in terms of a two-variable function, where x is the number of sets or templates, Y is the number of each set components and z is the value of the component y of the set x. Each unknown set $z_1, z_2, \ldots, z_Y$ is to be matched to all X sets, so that a closest match can be found. If, for instance, we use a Manhattan distance measure, we have to calculate X sums of differences

$$|z_1-z_{1,1}| + |z_2-z_{1,2}| + \ldots + |z_Y-z_{1,Y}| \quad (1)$$

$$\ldots\ldots\ldots\ldots\ldots\ldots\ldots\ldots \quad \ldots$$

$$|z_1-z_{X,1}| + |z_2-z_{X,2}| + \ldots + |z_Y-z_{X,Y}| \quad (X)$$

This procedure amounts to XY operations. The beauty of solving the generic TMP by pattern indexing is that we only would need to perform XY/Z operations. A function of two variables $z = f(x, y)$ can be depicted with respect to a (x, y, z)-axes triplet, where z-axis is oriented vertically. Solving the TMP by pattern indexing is based on the index $\{x\}_{z,y}$, that is, not on the function $z = f(x, y)$. Creation of the index $\{x\}_{z,y}$ can be best visualized by rotation of the (x, y, z)-triplet such that z-axis is replaced by x-axis, which should be positioned vertically.

As an example, let us consider a problem of image retrieval by its colors. Let the number of images be $X = 2^{20}$ and the number of pixels in each image be $Y = 2^{16}$. Each pixel is represented by 3 bytes, which yield $Z = 2^{24}$ possible colors for each pixel. Firstly, we consider the template matching algorithm. Given an unknown image, we have to match it to all pre-stored template images, thus arriving at an overall matching time on the order of $T_M = XY = 2^{36}$ operations, that is, 64 billion operations. Secondly, we consider the indexing algorithm. The column $\{x\}_{y,z}$ contains names of those images, whose pixel y has the color z. Identification of an unknown image x amounts to reading Y columns of the index $\{x\}_{y,z}$ $(x=1,\ldots,X,\ y=1,\ldots Y)$ and creating a histogram of names found in these columns. The horizontal position of the histogram's maximum delivers the name of the best match image. We can work out the total number of operations for the index-based algorithm as the average columnar height across all columns (H) times the number of columns to be accessed (Y). The value H is equal to the total number of names in all columns, which is XY (each name X repeats itself Y times) over the total number of columns, which is YZ. Hence, $H = XY/YZ = X/Z$ and $T_I = HY = XY/Z = 2^{36}/2^{24} = 2^{12}$.In this case, the total of 4096 operations required by the indexing algorithm is much more attractive against 64 billion operations required by direct matching algorithm. For this example, the attained reduction of the computational load amounts to a factor of 16 millions.

Hence, even if the number of templates is real big, still there could be no point to go for hardware-based solution. For instance, large scale automated fingerprint identification systems store and search through hundreds of millions of feature patterns. There are many other large scale applications, where pattern indexing approach makes a lot of sense. It was shown that a program that implements a 1-Gbit indexed-based content accessible memory can store/retrieve 1-Kbit words on a 1.5GHz, 256M PC in a matter of 10 microseconds.

ACHIEVING FAULT-TOLERANCE WITH HELP OF GENERALIZATION

Noise and Fault-Tolerance issues are intrinsically important in the field of pattern recognition. The name histogram discussed in the previous paragraph provides a certain degree of noise-tolerance. Another degree is provided with the use of macro-columns, where not just a column specified by its address but this column's vicinity as well is accessed on the retrieval when the name histogram is created. However, the quantum leap in dealing with noise and deformation issues is achieved with the use of generalization, which is achieved by identifying common properties of objects and by using iterated index maps.

Let us suppose, for example, that we want a pattern recognition system to invariantly identify geometric figures on the plane with respect to their orientations, scaling and deformations that not change a figure topology. As a common property of pixels, a pixel's tangential line can be used. In this sense, an isolated pixel on the plane does not exist as it does not feature any orientation. Pixels inside a shadowed shape also do not show a prevailing orientation unless they are located on the circumference. Not only the tangential common property approach acts as an edge detector, it also acts as a generalizing property. In fact, all pixels on a line segment share the same angle. Hence, for all rectangles, however deformed they might be, four line segments will be identified, thus, the required invariance will be provided.

The pixels are placed in groups (lines or curves) with the help of an iterated indexing map denoted as $\{X\}_X$. The left-hand side of Fig.6 shows a spatial arrangement of nine pixels on the plane, and the right-hand side of Fig.6 shows the iterated index $\{X\}_X$ for the above arrangment.

```
1 2 3 4                                    9
      5   7 8 9              2 1 2   6 5 8 7 8
      6                      1 2 3 4 5 6 7 8 9
```

Fig.6. Nine Pixels Spatial Arrangement and its Iterated Index

The iterated index $\{x\}_x$ identifies three clusters of pixels, that is, 1-2-3, 5-6, 7-8-9. Each cluster is defined as an attractor. The pixel 4 is excluded from the index as it is featured by two different angles: 0 degrees and 90 degrees angles. The third cluster is not linked to the clusters 1 and 2 as the third cluster pixels' coordinates are far from coordinates of the pixels of the cluster 1 and 2. In the iterated indexing map, each column with the size greater than 1 represents a bifurcation, for instance, there exist two paths from the pixel 2, one leading to pixel 1 and the second leading to pixel 3.

Another example of an iterated indexing map, where 2 attractors were derived (left-hand side of Fig. 7) from the three rings that are shown on the right-hand side of Fig. 7. Note that despite the fact that two of these rings overlap on the plane, they were marked with 2 different colors.

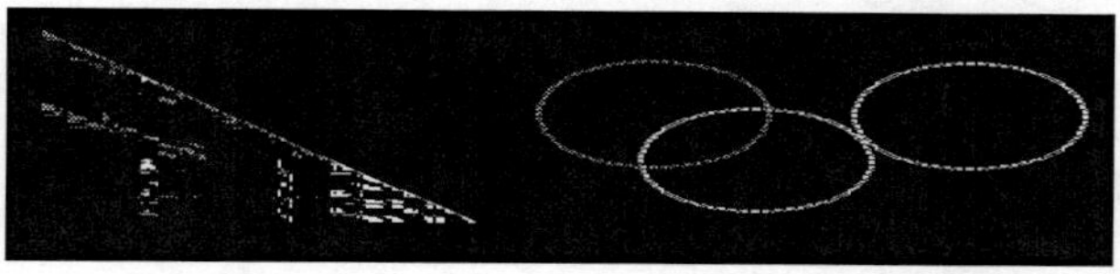

Fig.7. Three Rings Painted in Accordance with the Iterated Indexing Map

On the other hand, two out of three rings have the same color because not only these rings share common pixels, but they share a common tangential line at their touching point.

HIERARCHY AND INVARIANT REPRESENTATIONS

Generalization process is based on identification of common properties of objects and is supported be a hierarchy of indexes, where the outputs of a lower level index (pattern names) act as addresses (features) on higher level columns: the pixels are combined into lines and curves, the latter are combined into figures, which, in turn, are combined into scenes and so on.

In the scene shown on the left-hand side of Fig.8 the two shapes were identified and encoded as $2_6 4_6$, $1_1 2_5 3_1 4_2$, where first digit 2 indicates the number of neighbors (bonds) of a line segment in one of the two shapes and its subscript 6 indicates the number of occurrences of the 2-bond segment. The deformed shapes (right-hand side of Fig.8) were also encoded as $2_6 4_6$, $1_1 2_5 3_1 4_2$. In this way, higher indexing levels contain invariant representations of changeable real-world patterns.

Fig.8. Two Pairs of Topologically Equivalent Shapes

CONCLUSIONS

DNC is a hierarchical indexing memory system, which can be used in text- and image-based applications. It solves pattern recognition and clustering problems with the help of index-based search techniques, thus, providing dramatic speed and generalization advantages.

DNC employs a hierarchy of indexes because its lower levels outputs (pattern names) act as input components (features) on higher levels and so on. This process supports a generalization, which is an efficient tool in dealing with noise disturbances, rotation, scaling and deformation of shapes. A generalization itself is achieved by identifying common properties of objects.

The property of being an indexing system stems from DNC's columnar architecture, where the memory cells are linked in a book index-like fashion, such that all cells of a single column are activated by a single input address (feature). Each cortical level avoids template matching by using pattern indexing, which dramatically reduces the computational load from XY to XY/Z, where X is the number of templates, Y is the number of template components and Z is the dynamic range of the component.

REFERENCES

[1] Mikhailov, A. and Pok Y.M., 2001, "Artificial Neural Cortex", Proceedings of the Artificial Neural Networks in Engineering Conference (ANNIE 2001), St. Louis, Missouri.

[2] Brin, S. and Page, L.. The Anatomy of a Large-Scale Hypertextual Web Search Engine.

[3] Jeff Hawkins, J., 2004, "On Intelligence", Times Books.

[4] Bledsoe, W. and Browning, I., 1959, "Pattern Recognition and Reading by Machine", Proc. Eastern Joint Comp. Conf.

[5] Alexander, I. and Stonham, T., 1979, "A Guide to Pattern Recognition Using RAM", IEE J. Digital Systems and Computers, Vol.2., No.1.

[6] Morciniec, M., and Rohwer, R., 1995, "The n-Tuple Classifier: Too Good to Ignore", Intellix A/S Articles, Aston University, UK.

[7] Jorgenson, T., 1997, "Classification of Hand written Digits Using a RAM Neural Net Architecture", Special Issue on Neural Networks for Computer Vision Applications, World Scientific Publishing Company, pp.17-25.

PART II:

INFRASTRUCTURE SYSTEMS

MINIMIZING SPEED VARIATION AND SPEED NOISE USING DYNAMIC SPEED CONTROL IN SIGNALIZED NETWORKS

HUI CHEN
Department of Civil and
Environmental Engineering
Michigan State University

GHASSAN ABU-LEBDEH
Department of Civil Engineering
American University-Sharjah

YACOUB NAJJAR
Department of Civil Engineering
Kansas State University

ABSTRACT
Vehicle speed noise (fluctuation in individual vehicle speeds) and speed variations (among different vehicles) have negative impacts on road safety, fuel consumption and emissions. Dynamic-signal dynamic-speed (DSDS) control has the potential to reduce speed noise and speed variation in signalized networks. Vehicle speed profiles under DSDS control and dynamic-signal fixed-speed (DSFS) control were used to assess the advantages of control with dynamic speed. Speed noise and speed variations are compared under the two control schemes. The results show that DSDS control generates smoother flow profiles by reducing speed noise and speed variation. The results provide operational evidence that implementing DSDS control in signalized networks is an effective way to design safer and environment-friendly signalized network operations.

Keywords: Dynamic Speed, Signal Control, Fuel Consumption, Road Safety, VII, ITS

INTRODUCTION

Vehicle speed noise and speed variation between vehicles have direct impact on road safety, fuel consumption, and emissions. Vehicle speed noise refers to the deviation of vehicle speed from its desired travel speed, and vehicle variation refers to the speed difference between vehicles in a platoon. Thorton and Lyles (1996) concluded that a major factor leading to an accident is not speed itself but the variation of speed. Oh et al. (Oh et al, 2005) also proved that speed variation are the most distinguishing parameter between normal traffic conditions and traffic conditions right before accidents.

Results of previous studies of optimal speed profiles prove that fuel consumption in most conditions is minimized by operating at constant speed (schwarzkopf and Leipnik, 1977), and that the magnitude of fuel consumption increase as a result of deviations from the constant speed (Chang et al, 2005). Hence in order to operate safer and more fuel conservative signalized roadway systems, efforts should be directed at developing means to minimizing speed noise and speed variations. One possible approach is to use the dynamic speed approach to intelligently adjust the optimal speed according to changing road, weather, and traffic conditions. In signalized networks, a dynamic speed control algorithms could be effectively combined with dynamic signal control algorithms to form an integrated more flexible control system. This paper explores the merits and capabilities of the DSDS control developed in a previous study (Chen and Abu-Lebdeh, 2006) in reducing speed noise and speed variations by analyzing vehicle speed profiles obtained from microscopic simulated scenarios.

VEHICLE SPEED NOISES AND VEHICLE SPEED VARIATION IN SIGNALIZED NETWORKS

In signalized networks, vehicle speeds could be impacted by traffic signal operations, speed limits, and traffic flow conditions. For example, Figure 1 shows a typical vehicle speed profile along a signalized arterial with fixed speed control. The profile during t1 is typical of what happens when a downstream queue is encountered by the subject vehicle. The vehicle slowed down and then accelerated back to its desired speed after vehicles in the downstream queue accelerated to the desired speed and the queue dissipated. The profile during t2 shows what happens when the vehicle encounters a red signal. The vehicle slows down, stops for a while, and then accelerates back to the desired speed.

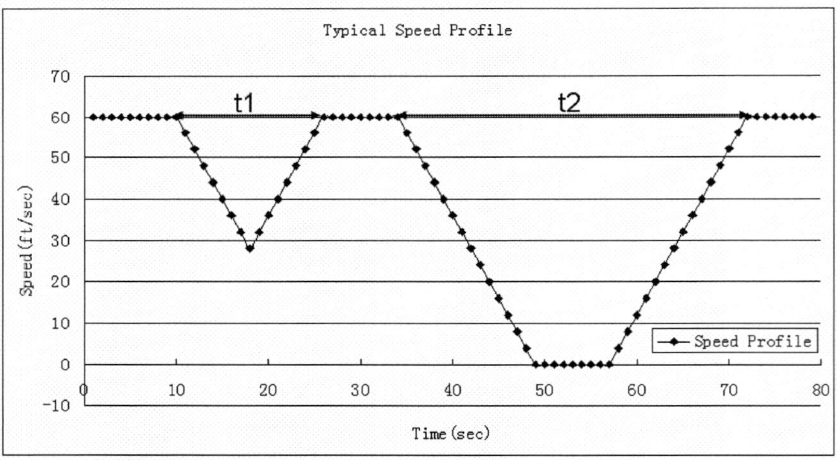

Figure 1: Typical Speed profile of a vehicle in signalized network.

Under both scenarios depicted in Figure 1, the vehicle speed deviated from its desired speed and thus speed noise occurred and, by implication, fuel consumption also increased. However these two scenarios have different impacts on the following vehicles: scenario 1 will cause the following vehicle to decelerate and then accelerate while in scenario 2 not only the subject vehicle decelerated and then accelerated, it also caused a more pronounced stopping and start-up shockwaves to travel upstream into the traffic platoon. It is known that the longer t1 and t2 are, the higher the number of vehicles that are impacted.

Also, during the acceleration and deceleration periods, there is more interaction between vehicles, and the speed variation is higher. These are also the periods when accidents are more likely to happen.

Using dynamic speed control with dynamic signal control in signalized networks can help minimize undesirable vehicle speed noise and speed variations by providing drivers with guidance on the optimal traveling speed, especially in over-saturated conditions where stop-and-go travel is more likely. A combined DSDS algorithm has been developed and presented in an earlier study where typical macroscopic measures of effectiveness (MOEs) were used for preliminary evaluations (Chen & Abu-Lebdeh, 2006). The results there showed that DSDS control can indeed improve throughput, decrease delay, and reduce the number of stops. For the benefit of the reader of this paper and to make the presentation of the material more tractable, the formulation of the DSDS control algorithm is briefly described in the following section. For more details, the reader is referred to the previous paper (Chen & Abu-Lebdeh, 2006).

DSDS CONTROL FORMULATION AND IMPLEMENTATION

Definitions and Variables

Let $N(ny, nx)$ denote a grid traffic network of ny intersections on all north-south arterials and nx intersections on all east-west arterials. Such a network has $ny*nx$ signals and $2*[(ny+1)*nx+(nx+1)*ny]$ one-way links. The variables used in the formulation are:

- Si: Saturation flow rate for approach i, veh/hg/lane;
- LV: Average effective vehicle length when stopped, ft;
- acc: Acceleration, ft/sec^2;
- $L_{(i,j),(l,m)}$: Length of link between signal (i,j) and signal (l,m), ft;
- h: Average time headway between two vehicles, sec/veh;
- vu: Speed of a start-up shockwave when a signal turns green, ft/s;
- vt: Speed of a stopping shockwave when a signal turns red, ft/s;
- $off_{((i,j),(l,m))k}$: Relative offset between signal (i, j) and (l, m) during cycle k, sec.

The following variables are used for all directions of flow. When used for a specific direction, an extra letter will precede the variable. For example $g_{(i,j)k}$ is the effective green time at signal (i,j) during cycle k; $eg_{(i,j)k}$ is the effective green time for the eastbound approach at signal (i,j) during cycle k.

- $g_{(i,j)k}$: Effective green time for an approach at intersection (i,j) during cycle k;
- $gs_{(i,j)k}$: Starting time of green for an approach at intersection (i,j) during cycle k;
- $ds_{(i,j)k}$: Dynamic speed on an approach to intersection (i,j) during cycle k;
- $dv_{(i,j)k}$: Number of discharged vehicles from an approach at intersection (i,j) during cycle k;
- $av_{(i,j)k}$: Volume of vehicles received on an approach at intersection (i,j) during cycle k;

$q_{(i,j)k}$: Length of queue on an approach at intersection*(i,j)* at the beginning of cycle *k*.

Formulation of DSDS Control

DSDS control is formulated as an optimization problem and solved by a Genetic Algorithm. In the sample 2-way grid network shown in Figure 2, the greens indicated by the arrows in the Figure and speeds of the independent streets are the decision variables. They are optimized directly by the GA. (Independent streets are those with offsets not bound by the network closure constraint). These decision variables are coded into a chromosome and their values that produced a near-optimal solution were obtained using the GA. The GA optimizes the values of these variables to maximize the following objective function. Traffic flow and variables are all represented at a macroscopic level.

$$throughput - \lambda * stop\ , \tag{1}$$

The first term is the system throughput and is computed as follows:

$$throughput = \sum_{k=1}^{n}\left(\sum_{i=1}^{ny} edv_{(i,nx)k} + \sum_{j=1}^{nx} ndv_{(ny,j)k}\right), \tag{2}$$

The number of stops is calculated as follows:

$$\sum_{k=1}^{n}\sum_{i=1}^{ny}\sum_{j=1}^{nx}\begin{pmatrix}\left[eq_{(i,j+1)k} + (eoff_{((i,j)(i,j+1))k} - c\_eoff_{((i,j),(i,j+1))k})\cdot vu/((vu-vt)\cdot h)\right] \\ +\left[nq_{(i+1,j)k} + (noff_{((i,j)(i+1,j))k} - c\_noff_{((i,j),(i+1,j))k})\cdot vu/((vu-vt)\cdot h)\right]\end{pmatrix} \tag{3}$$

λ in the objective function is a parameter (relative weight) that is determined based on the desired control strategy. The objective function in equation (1) is maximized subject to the following constraints:

1. $\min g \le g_{(i,j)k} \le \max g,\quad$ for $(i,j) \in N(ny,nx),\ k = 1,2,\ldots n$
2. $\min\ ds \le ds_{(i,j)k} \le \max\ ds,\quad$ for $(i,j) \in N(ny,nx),\ k = 1,2,\ldots n$
3. $q_{(i,j)1}$ are known;
4. $q_{(i,j)k} \le \max\ q\ ;\quad$ for $(i,j) \in N(ny,nx), k = 1,2,\ldots n$
5. $av_{(l,m)k} = dv_{(i,j)k},\quad$ for $(i,j),(l,m) \in N(ny,nx),\ k = 1,2,\ldots n$ and

(i, j) is the immediate upstream signal of (l, m)

6. $$dv_{(i,j)k} = \begin{cases} \min(q_{(i,j)k} + av_{(i,j)k}, \dfrac{g_{(i,j)k}}{h}), & c\_off_{((i,j),(l,m))k} \ge off_{((i,j),(l,m))k}; \\ \min(q_{(i,j)k} + av_{(i,j)k}, \dfrac{g_{(i,j)k} - (c\_off_{((i,j),(l,m))k} - off_{((i,j),(l,m))k})}{h}), & c\_off_{((i,j),(l,m))k} < off_{((i,j),(l,m))k} \end{cases}$$

7. $dv_{(i,j)k} \le q_{(i,j)k} + av_{(i,j)k};\quad$ for $(i,j) \in N(ny,nx),\ k = 1,2,\ldots n$

8.

$$q_{(i,j)k+1} = q_{(i,j)k} - dv_{(i,j)k} + av_{(i,j)k}, \qquad \text{for } (i,j) \in N(ny,nx),\ k = 1,2,...n$$

9. $g_{(i,j)k} \le g_{(l,m)k} + off_{((l,m),(i,j))k} + L_{(i,j),(l,m)} / vt, \qquad (i,j),(l,m) \in N(ny,nx),\ k = 1,2,...n$;

and *(i, j)* is the immediate upstream signal of *(l, m)*. This constraint is to prevent spillback.

10.

$$off_{((i,j)(l,m))k} = \begin{cases} L / ds_{(i,j)k} - q_{(l,m)k} \times h, & \textit{independent streets \& start from moving;} \\ (L + ds^2_{(i,j)k} / 2 \times acc) / ds_{(i,j)k} - q_{(l,m)k} \times h, & \textit{independent streets \& start from stoping;} \\ off_{((i,j),(i+1,j))k} = (ngs_{(i,j)k} + off_{((i,j)(l,m))k} + ng_{(l,m)k} + off_{((l,m),(l,m+1))k} + eg_{(l,m+1)k}) - (ngs_{(i,j)k} & \\ \quad + ng_{(i,j)k} + off_{((i,j),(i,j+1))k} + eg_{(i,j+1)k}), & \textit{dependent streets} \end{cases}$$

11. $gc_{(l,m)k} = cycle_{(i,j)k} + off_{((l,m),(i,j))k} + off_{((l,m),(i,j))k+1} - g_{(l,m)k}$;

12. $ngs_{(1,1)1} = 0$;

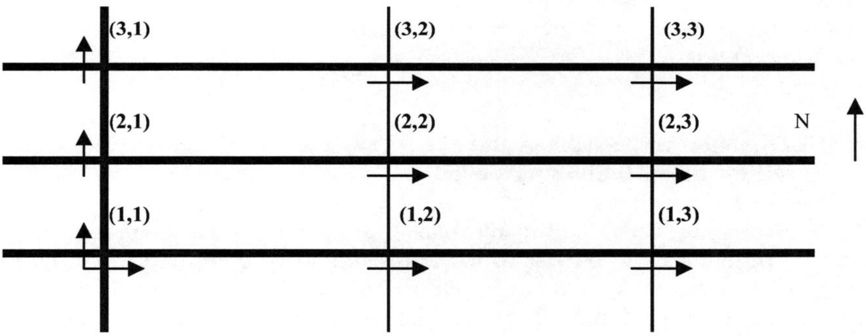

Figure 2: A sample grid network. (Thick lines indicate independent arterials; small arrows indicate optimized greens)

Implementation of DSDS Control

DSDS control is a futuristic one that is envisioned to be implemented in systems with vehicle integrated infrastructure (VII) setups with functionalities emulating those found in closed-loop control. Optimized speeds and signal control measures are communicated from the central computers of the authorized users (e.g., traffic operations centers) to either individual vehicles (for use with in-vehicle navigational aides) or to signal and speed controllers and displays in the field. To complete the loop, traffic flow information is continuously collected using sensors and communicated to the road side units (RSUs) or "hot posts" and from there on to the central control computers at the aggregation points.

EXPERIMENTAL SET-UP

Near optimal DSDS and DSFS control schemes were generated using the algorithm presented earlier for an over-saturated grid network, N(5, 4) for 40 cycles. Link lengths were 800 ft. For the major flow directions (eastbound and northbound), constant and higher than capacity demand was assumed at the entrance intersections, and initial queues were randomly chosen between 41 and 46 veh/lane (maximum is 53 veh/lane). For secondary directions, lower demands were assumed, and initial queues were set randomly between 5 and 10 vehicles. Minimum and maximum green times were set to 30 and 90 seconds, respectively. Speeds ranged between 15 mile/hr and 50 miles/hr for the DSDS and 40 mile/hr for the DSFS control. The GA solver developed by Goodman (Goodman, 1996) was used with a population of 1000 and 1000 generations to ensure near-optimal solution. All signals were assumed to have two phases. Turning movements were ignored. Once control solutions were obtained, signal timing and optimal speed information were built into a VISSIM (PTV, 2005) representation of the same network and used with the same network settings. Speed profiles of vehicles were obtained for vehicles under both dynamic-signal dynamic-speed (DSDS) control and dynamic-signal fixed-speed (DSFS) control. Speed profiles where then compared to assess the capability of dynamic speed control on speed noise and speed variations. Results are discussed in the following section.

RESULTS AND DISCUSSION

Speed Profile of Individual Vehicles

Figure 3 shows the speed profile of a lead vehicle arriving at a typical link in a typical cycle under both control schemes. Under DSDS control vehicles were given guidance to operate at lower desired speed compared to the DSFS control. The effect is that although both control schemes led to almost the same average speed, under DSDS there is less time when the actual speed deviated from desired speed to zero speed. Because of the minimum and maximum speed constraint, and also because of the randomness inherent in drivers' behavior which is explicitly accounted for in VISSIM but not accounted for in the macroscopic simulation models that are used to obtain the near-optimal solutions, speed noises are not completely eliminated (this reflected in the minor fluctuations of speed seen in the left parts of the graphs). However, the duration of time when actual speed deviated from the desired speed to zero speed is significantly shorter under the DSDS control. Similar results were observed for other vehicles.

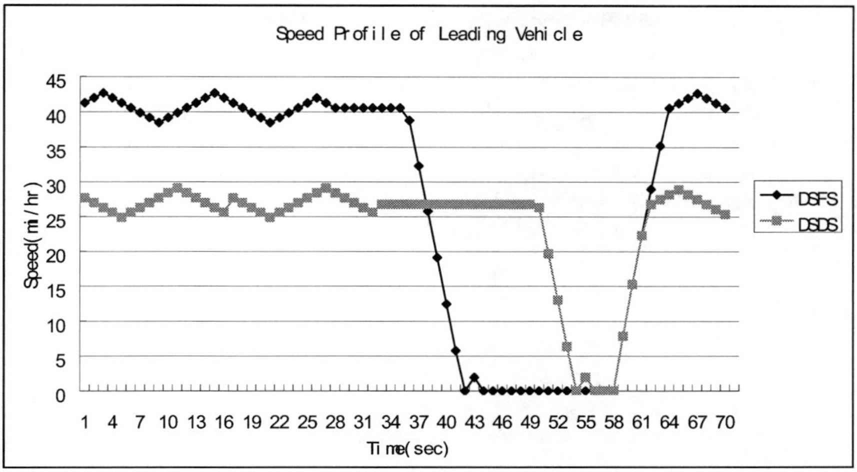

Figure 3: Speed profile of leading vehicle under DSDS and DSFS control

Speed Trajectory of Vehicles in a Platoon

Figure 4 shows the speed profiles of a platoon of vehicles arriving in a typical link during one typical cycle. Parts (a) and part (b) are for DSFS and DSDS, respectively. The Figure shows that the durations of time when speeds of individual vehicles deviated from the desired speed are shorter under the DSDS control. It is also clear from Figure 4 that fewer vehicles were impacted by the stopping shockwave and hence the resulting speed noise for the following vehicles is less. Because of the shorter acceleration and deceleration periods and the fewer vehicles involved, speed variation is smaller in the traffic stream under DSDS control. All this implies that traffic flow under DSDS control is safer and more fuel conservative.

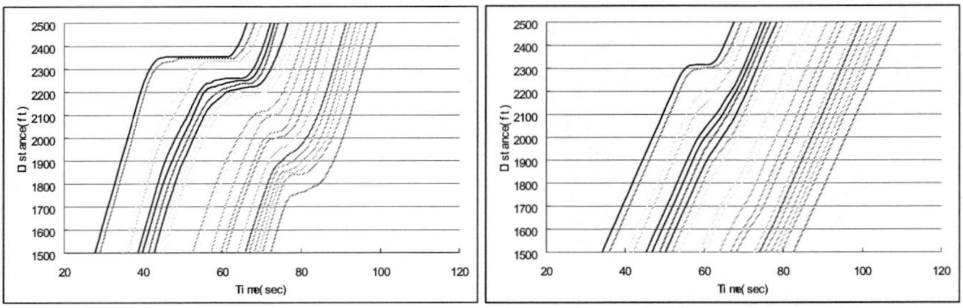

(a) DSFS control (b) DSDS control

Figure 4: Vehicle trajectories

CONCLUSIONS

This paper evaluated the potential capability of DSDS control to reduce speed noise and speed variations in signalized networks and hence enhance safety and promote more environment-friendly traffic operations in signalized urban networks. Simulated vehicle speed profiles obtained were analyzed and compared to those of DSFS control. The results show that DSDS control is able to promote smoother traffic flow by reducing speed noises and speed variations. Control with dynamic speed was shown to be an effective and necessary addition to signal control systems for signalized networks. DSDS control provides an untraditional but more flexible and effective control alternative to operate safer and more fuel conservative signalized network systems. DSDS control fits naturally in systems with VII capabilities.

REFERENCES

Chang, D. and Morlok, E. 2005. Vehicle Speed Profiles to minimize work and fuel consumption. Journal of Transportation Engineering, Vol. 131, No. 3, pp. 173-182.

Chen,H. and Abu-Lebdeh, G. 2006. Development of a Framework for an Integrated Dynamic Signal-Dynamic Speed Traffic Management Algorithm for Signalized Networks. Compendium of papers CD-ROM of 85th Transportation Research Board annual meeting. Washington DC. Jan, 2006

Goodman, E. 1996. An Introduction to GALOPPS, the "Genetic Algorithm Optimized for Portability and Parallelism," Release 3.2, (software and User's Guide), Technical Report #960801, MSU Genetic Algorithm Research and Applications Group and Case Center for Computer-Aided Engineering and Manufacturing, Michigan State University, 80pp., August, 1996

Oh, J., Oh, C., Ritchie, S. and Chang, M. 2005. Real-time Estimation of Accident Likelyhood for Safety Enhancement. Journal of Transportation Engineering. Volume 131, NO. 5, pp. 358-363

Thorton, M and Lyles, R. W. 1996. Freeway speed zones: Safety and compliance issues. Transportation Research Record 1635, pp 18-25.

Schwarzkopf, A. B. and Leipnik, R. B. 1977. Control of highway vehicles for minimum fuel comsumption over varying terrain. Transportation Research, 11(4), pp. 279-286.

VISSIM 4.10 User Manual(2005). PTV America, Inc.

HOT MIX ASPHALT DYNAMIC MODULUS PREDICTION MODELS USING NEURAL NETWORKS APPROACH

HALIL CEYLAN
Assistant Prof., Iowa State University, Ames, IA

KASTHURIRANGAN GOPALAKRISHNAN
Research Scientist, Iowa State University, Ames, IA

SUNGHWAN KIM
Postdoctoral Res. Assoc., Iowa State University, Ames, IA

ABSTRACT
The primary objective of this study is to develop a simplified Hot Mix Asphalt (HMA) dynamic modulus (|E*|) prediction model with fewer input variables compared to the existing regression based models without compromising prediction accuracy. ANN-based prediction models were developed using the latest comprehensive |E*| database that is available to the researchers containing 7,400 data points from 346 HMA mixtures. The ANN model predictions were compared with the existing regression-based prediction models which are included in the latest Mechanistic-Empirical Pavement Design Guide (MEPDG). The ANN based |E*| models show significantly higher prediction accuracy compared to the existing regression models although they require relatively fewer inputs. The findings of this study present a "paradigm shift" in the way the hot-mix asphalt material characterization has been handled by pavement materials engineers.

INTRODUCTION

The dynamic modulus (|E*|) is one of the Hot Mix Asphalt (HMA) stiffness measures that determines the strains and displacements in flexible pavement structure as it is loaded or unloaded. Many studies have been conducted over the last 50 years related to the development of HMA dynamic modulus (|E*|) test procedures and prediction models. Among these, the |E*| modulus test (also referred to as the Simple Performance Test) recommended by National Cooperative Highway Research Program (NCHRP) 9-19 study (Witczak et al., 2002) and the |E*| prediction equation developed by Witczak and his colleagues in 1999 (Andrei et al., 1999) has been incorporated into the new mechanistic empirical pavement design guide (MEPDG) (NCHRP, 2004). The input variables for the 1999 version |E*| model include aggregate gradation, mixture volumetrics, viscosity of the asphalt binder (η), and loading frequency (f). The aggregate gradation variables include percent passing #200 sieve ($\rho_{\#200}$), percent retained #4 sieve ($\rho_{\#4}$), percent retained 9.5 mm sieve ($\rho_{9.5mm}$), and percent retained 19 mm sieve (ρ_{19mm}). The mixture volumetrics includes air void (V_a) and effective binder content (V_{beff}).

A new revised version of the Witczak's |E*| predictive model has been developed to improve the characterization of asphalt mixtures in dynamic mode of loading (Bari and Witczak, 2006, Azari et al., 2007). Furthermore, the new |E*| prediction model adopting the asphalt binder shear modulus ($|G_b^*|$) and phase angle (δ) as variables instead of η and f would directly link to the Superpave Binder Performance Grading (PG) system and the associated binder testing (Bari and Witczak, 2006). In addition, the new |E*| model is based on a more comprehensive |E*| database containing 7,400 data points from 346 HMA mixtures covering a variety of binder stiffness as a result of testing lab-aged, plant mix, and field-aged cores. The new Witczak |E*| model was incorporated in a recent version of MEPDG (version 1.0) with the 1999 version |E*| model. Even though the new revised model predicts |E*| with higher accuracy compared to the 1999 model, the nonlinear regression technique was still employed in the model development so that there is still a significant scatter especially at the lower and higher |E*| modulus values (Bari and Witczak, 2006; Azari et al., 2007). By accurately characterizing the HMA dynamic modulus, it will be possible to predict the flexible pavement performance (rutting and cracking) with greater confidence.

Recently, researchers at Iowa State University (ISU) have developed a novel approach for predicting HMA dynamic modulus using the Artificial Neural Network (ANN) methodology. The comprehensive |E*| database containing 7,400 data records, which were used in the development of revised |E*| model (Bari and Witczak, 2006), were also used in developing the ANN models. This paper describes the ground-breaking work on the development of ANN-based |E*| prediction models, the comparison of ANN model predictions with the Witczak model predictions, and the feasibility of input variables optimization to ANN model predictions.

ANN DATABASE PREPARATION

Input variables for the |E*| ANN prediction model were retrieved from the NCHRP Report 567 CD-ROM (*CRP-CD-46*) "Simple Performance Tests: Summary of Recommended Methods and Database." (Witczak, 2005). The *CRP-CD-46* included as an appendix in the NCHRP report 567 contains not only |E*| new database but also all data and information collected and used during NCHRP 9-19 study. The eight input variables of the 1999 and 2006 version |E*| predictive equations were used in the two different types of ANN models (ANN 1999 and ANN 2006), respectively. The one output variable was the |E*| in both the ANN models. A total of 7,400 data records (which was also used in developing the new and revised |E*| model) was used in developing the ANN models.

The data were divided randomly into two different subsets: the training data subset containing 6,900 data points and the testing data subset which consisted of 500 data points. Both datasets were normalized within the range of -2 to 2 for input values and the range of 0.1 to 0.9 for output values to satisfy the transfer function (sigmoid) range and to prevent network saturation, which could impede the network's performance. The training data subset was used to train the backpropagation ANN |E*| prediction model and the testing data subset were used to examine the statistical accuracy of the developed ANN model. Backpropagation ANNs are very powerful and versatile networks that can be taught a mapping from one data space to another using a representative set of patterns/examples to be learned. The term "backpropagation network" actually refers to a multi-layered, feed-forward neural network trained using an error backpropagation algorithm. The learning process performed by this algorithm is called "backpropagation learning" which is mainly an "error minimization technique" (Haykin, 1999).

ANN |E*| PREDICTION MODEL DEVELOPMENT

A typical four-layered, i.e., one input- two hidden–one output layer, feed forward error-back propagation ANN architecture, as shown in Fig. 1, was used in this study. To ensure efficient convergence and the desired performance of the trained network, several parameters were incorporated in the training phase. These parameters included the training rate, the momentum term, and the number of learning cycles (epochs).

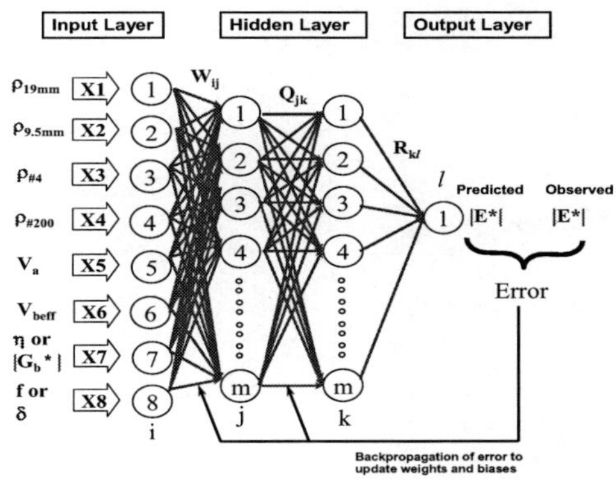

Figure 1. Four-layered neural network architecture

The training rate is a factor that proportions the amount of adjustment applied each time the weight is updated. A small training rate might result in slower convergence and dropping into the local minima conditions in the weight-error space. A large training rate often causes the convergence behavior of the network to oscillate and possibly never converge (Owusu-Aabio, 1998). The use of a momentum term could carry the weight change process through one or more local minima and get it into global minima. The training rate and the momentum coefficient used in the study were 0.4 and 0.6, respectively.

ANN 1999 |E*| prediction model has eight input parameters including the four aggregate gradation variables (ρ_{19mm}, $\rho_{9.5mm}$, $\rho_{\#4}$, $\rho_{\#200}$), two mixture volumetric variables (V_{beff}, V_a), one asphalt binder rheology property variable (η) and one loading frequency property (f). ANN 2006 model also has eight input parameters corresponding to the input variables of 2006 version Witczak |E*| prediction model, which adopted $|G_b^*|$ and δ as variables instead of η and f. Both ANN models have |E*| as one output neuron. Several network architectures with two hidden layers were examined to determine the optimum number of hidden layer nodes through a parametric study. Overall, the training and testing mean squared errors (MSEs) decreased as the networks grew in size with increasing number of neurons in the hidden layers. The error levels for both the training and testing sets matched closely when the number of hidden nodes approached 30 as in the case of 8-30-30-1 architecture (8 input, 30 and 30 hidden, and 1 output neurons, respectively). Figure 2 shows the training and testing MSE progress curves for the 8-30-30-1 network for 10,000 learning cycles or training epochs. The 8-30-30-1 architecture was chosen as the best architecture for both the ANN 1999 and 2006 model based on its lowest training and testing MSEs in the order of 2×10^{-4}. Both the training and testing curves for the output are in the same order of magnitude thus depicting proper training. The almost constant MSEs obtained for the last 2,000 epochs (see Fig. 2) also provided a good indication of adequate training for this network.

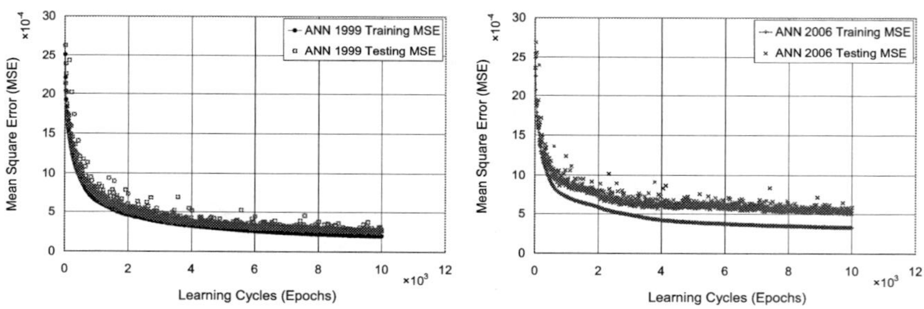

Figure 2. Training and testing progress curves of the ANN models

ANN |E*| PREDICTION MODEL RESULTS

The *goodness-of-fit* statistics for the ANN model predictions in arithmetic scale were performed using statistical parameters such as the correlation coefficient (R^2), the standard error of predicted values divided by the standard deviation of measured values (S_e/S_y) and the absolute average error (AAE). The R^2 is a measure of correlation between the predicted and the measured values and therefore, determines accuracy of the fitting model (higher R^2 equates to higher accuracy). The S_e/S_y and the AAE indicates the relative improvement in accuracy and thus a smaller value is indicative of better accuracy. A set of criteria originally developed by Pellinen (2001) were also adopted in this evaluation.

The results of statistical analysis are presented in Fig. 3 for the 500 testing data points. As mentioned previously, the 500 test vectors form an independent dataset which was not used in training the ANN and it was used to test the accuracy of the trained ANN. Clearly, the ANN 1999 and 2006 model predictions show "excellent" statistics compared to Witczak model predictions. Especially, the AAE obtained using ANN is almost half that of Witczak's model. It is also noticed that the 1999 and the 2006 Witczak predictions are more scattered below and above the line of equality with increasing |E*| values. Especially, the 1999 Witczak |E*| model seems to under-predict the actual measurement while the 2006 Witczak |E*| model tends to over–predict the actual measurements. In terms of performance, this prediction inaccuracy may translate into the risk of premature failure of the asphalt layer in rutting or fatigue. However, ANN model predictions are closely around the line of equality without bias and therefore there is a higher chance of preventing premature distress failure. It is also interesting to note that the ANN 1999 models show slightly better goodness of fit statistics than the ANN 2006 models.

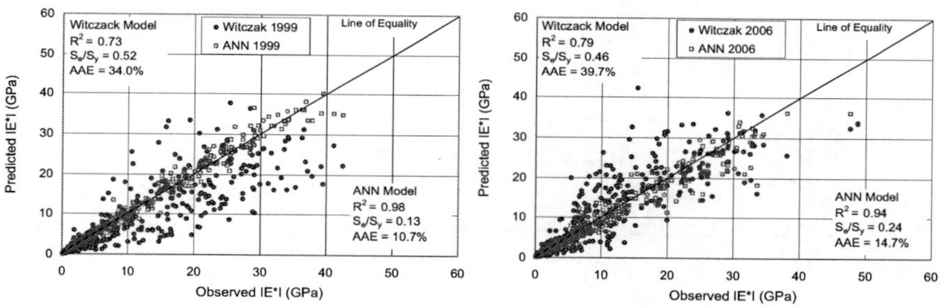

Figure 3. Predicted versus observed |E*| in Witczak and ANN models

OPTIMIZATION OF INPUT VARIABLES FOR ANN MODEL

Witczak equations include eight input variables. However, sometime, it is hard to obtain all of these variable values in actual situation. Preliminary investigations were carried out to obtain the optimum number of input variables for the ANN models compared to Witczak models without compromising prediction accuracy. The asphalt binder rheology properties (η, $|G_b^*|$, δ) were selected as the essential input variables and the effect of including/excluding other variables on ANN model predictions were examined. Among the developed ANN models, the ANN 1996-6 and ANN 2006-6 models using six input variables (ρ_{19mm}, $\rho_{\#4}$, V_a, V_{beff}, η or $|G_b^*|$, f or δ) and the ANN 1999-4 and ANN 2006-4 models using four input variables ($\rho_{\#4}$, V_{beff}, η or $|G_b^*|$, f or δ) are presented in Figs. 4 and 5 for illustration.

Even though the number of input variables were reduced in the development of models, ANN models shown in Figs. 4 and 5 still show better performance of |E*| predictions compared to Witczak models. Especially, the goodness-of-fit statistics for the ANN 1999-6 and the ANN 2006-6 models are very similar to the results obtained with the eight input variables.

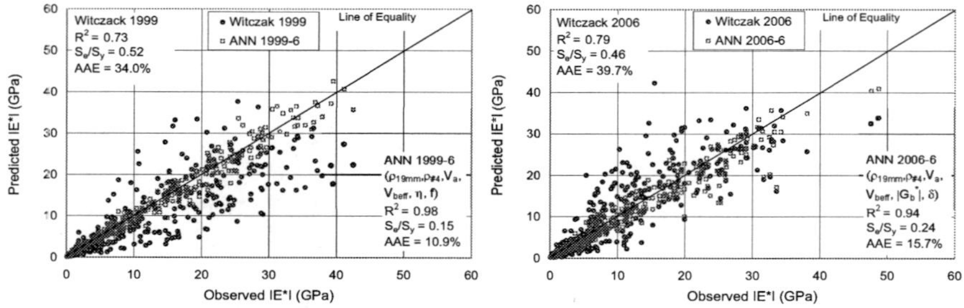

Figure 4. Comparison between Witczak and ANN-6 models

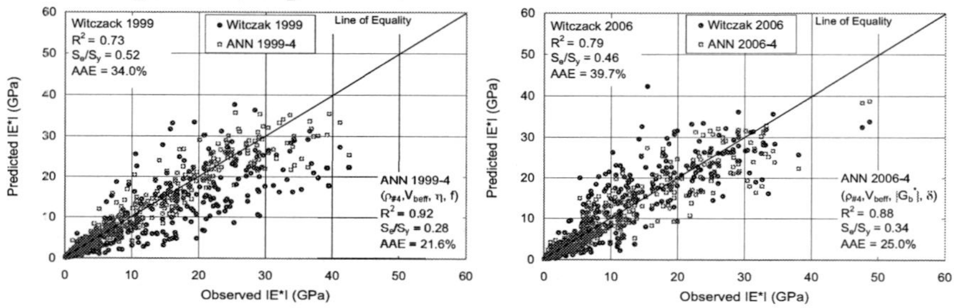

Figure 5. Comparison between Witczak and ANN-4 models

SUMMARY AND CONCLUSIONS

The primary objective of this study is to develop a simplified Hot Mix Asphalt (HMA) dynamic modulus (|E*|) prediction model with fewer input variables compared to the existing regression based models without compromising prediction accuracy. ANN-based prediction models were developed using the latest comprehensive |E*| database that is available to the researchers containing 7,400 data points from 346 HMA mixtures. It was found that by using less than eight input variables, ANN models still showed significantly better performance to Witczak models for |E*| prediction. Especially, the ANN models using six input variables ($\rho_{\#4}$, V_a, V_{beff}, η or $|G_b^*|$, f or δ) were very similar to the results obtained with the eight input variables.

The results of this study have significant implications in the context of advancing the state of the art in mechanistic-empirical pavement analysis and design. ANN models trained over comprehensive datasets could be successfully incorporated into MEPDG as surrogates for pavement materials characterization models and pavement performance prediction models. Because ANNs excel at mapping in higher-order spaces, such models can go beyond the existing univariate relationships between pavement structural responses and performance (such as the subgrade strain criteria for rutting). ANNs could be used to examine several variables at once and the interrelationships between them. ANNs could also be used to develop models for distress phenomena such as thermal cracking, block cracking, and rutting in HMA pavements, and faulting and D-cracking in concrete pavements. With the use of the computational intelligence tools, the findings of this study present a "paradigm shift" in the way the HMA material characterization has been handled by pavement materials engineers.

REFERENCES

Andrei, D., Witczak, M. W., and Mirza, M. W., 1999, "Development of a Revised Predictive Model for the Dynamic (Complex) Modulus of Asphalt Mixtures," National Cooperative Highway Research Program 1-37 A Inter Team Report, University of Maryland.

Azari, H., Al-Khateeb, G., Shenoy, A., and Gibson, N., 2007, "Comparison of Measured STP E* of ALF Mixtures with Predicted E* Using NCHRP 1-37 A and Witczak's new equations," CD-ROM, Presented at 86th Annual Transportation Research Board Meeting, Transportation Research Board, Washington, D.C.

Bari, J. and Witczak, M. W., 2006, "Development of a New Revised Version of the Witczak E* Predictive Model for Hot Mix Asphalt Mixtures," Journal of the Association of Asphalt Paving Technologists, Vol. 75.

Haykin, S., 1999, "Neural Networks: A Comprehensive Foundation," Prentice-Hall, Inc., NJ.

National Cooperative Highway Research Program (NCHRP), 2004, "Guide for Mechanistic-Empirical Design of New and Rehabilitated Pavement Structures," Available one line at: http://trb.org/mepdg, National Cooperative Highway Research Program 1-37 A, Transportation Research Board, Washington, D.C.

Owusu-Aabio, S., 1998, "Effect of Neural Network Topology on Flexible Pavement Cracking Prediction," Computer Aided Civil Infrastructure Engineering 13, pp. 349–355.

Pellinen, T. K., 2001, "Investigation of the Use of Dynamic Modulus as an Indicator of Hot–Mix Asphalt Performance," Ph.D. Dissertation, Arizona State University.

Witczak, M. W., Kaloush, K., Pellinen, T., El-Basyouny, M., and Quintus, H. V., 2002, "Simple Performance Test for Superpave Mix Design," National Cooperative Highway Research Program Report 465, Transportation Research Board, Washington, D.C.

Witczak, M. W., 2005, "Simple performance tests: Summary of Recommended Methods and Database," National Cooperative Highway Research Program Report 547, Transportation Research Board, Washington, D.C.

ESTIMATING RESILIENT MODULUS USING NEURAL NETWORK MODELS

MUSHARRAF ZAMAN
University of Oklahoma, Norman, Oklahoma

ALI EBRAHIMI
Burgess Engineering and Testing, Inc., Moore, Oklahoma

LUTHER WHITE
University of Oklahoma, Norman, Oklahoma

ABSTRACT
Resilient modulus (M_R) is an important material property in pavement design. Although generally determined by cyclic triaxial testing, the M_R values can be estimated from correlations with other material properties. Consequently, a combined laboratory and modeling study was undertaken to develop artificial neural network (ANN) models for subgrade soils. Sixty-three soil samples from different sites were tested for Atterberg limits, wet sieving, dry sieving, moisture-density, M_R, and unconfined compressive strength. The following parameters were used in the development of the ANN models: moisture content, dry density, plasticity index, percent passing No. 200 sieve, and compressive strength. Bulk stress and deviatoric stress were also used as model parameters. Six different ANN models were considered, namely Linear Network (LN), Generalized Regression Neural Network (GRNN), Radial Basis Function Network (RBFN), and Multi-Layer Perceptrons Network (MLPN) with one (MLPN-1), two (MLPN-2) and three (MLPN-3) hidden layers. Based on the R^2 values, the MLPN-2 model was found to be most effective. The proposed MLPN-2 model can be used in the design of pavements involving common subgrade soils in Oklahoma.

INTRODUCTION

Resilient modulus (M_R) is one of the fundamental material properties in the mechanistic design of roadway pavements. In the 2002 AASHTO design guide, three hierarchical design levels are suggested [1]. In Level 1, the required input values for M_R are obtained from laboratory or field tests. For Level 2 and Level 3 designs, however, the M_R values are either obtained from the agency database or estimated through correlations with other material properties.

To this end, a combined laboratory and modeling study was undertaken to develop a database for subgrade soils in Oklahoma and to develop relationships or models that could be used to estimate M_R from commonly used subgrade soil properties. Specifically, two categories of models, namely statistical models and Artificial Neural Network (ANN) models, are employed for the determination of M_R based on routine laboratory test results. In this paper, an overview of the ANN models is given, with an emphasis on the Multi-Layer Perceptrons Network (MLPN). Three different MLPN models are attempted with one (MLPN-1), two (MLPN-2), and three (MLPN-3) hidden layers. The MLPN-2 model was found to be most effective.

OVERVIEW OF SELECTED PREVIOUS STUDIES

As outlined in the 1993 AASHTO Guide for Design of Pavements, several researchers have attempted to develop empirical correlations to estimate resilient moduli in terms of other soil properties and stress states [1]. Different models have been developed to represent M_R as a function of stresses. One of the commonly used models is the power model [2-4]. In this model, the M_R is expressed as a function of the deviatoric stress (σ_d) and the bulk stress (θ) as $M_R = a(\sigma_d)^b$, for cohesive soils, and $M_R = k_1(\theta)^{k2}$, for granular soils, where a, b, k_1, and k_2 are regression constants. Another model proposed by Uzan [5] includes the bulk and deviatoric stresses together in the form $M_R = k_1(\theta/P_a)^{k2}(\sigma_d/P_a)^{k3}$, where k_1, k_2, and k_3 are model parameters and P_a represents atmospheric pressure. Some researchers have utilized other soil properties to estimate M_R. For example, Yau and Von Quintus [2] developed different equations for multiple soil types using a nonlinear regression technique. The equations proposed by these researchers could be generically expressed in the form: $M_R = k_1 P_a (\theta/P_a)^{k2} [(\tau_{oct}/P_a)+1]^{k3}$, where τ_{oct} is the octahedral shear stress. The model parameters k_1, k_2, and k_3 were expressed in terms of particle size (percent passing 9.5 mm (3/8″), 4.75 mm (No. 4), 0.425 mm (No. 40), and 0.075 mm (No. 200) sieves), percent clay, percent silt, liquid limit, in-situ moisture content, optimum moisture content, in-situ dry density, and optimum dry density.

Park et al. [6] used ANN models to predict the M_R values of subgrade and subbase materials using three basic parameters, namely maximum dry unit weight, uniformity coefficient, and percent passing sieve No. 200, and two stress parameters, namely confining stress and deviatoric stress. It was shown that the ANN models were capable of estimating M_R for subgrade and subbase materials for pavement design purposes. Compressive strength and plasticity index were not included as model parameters in their models.

The present study used the ANN models to develop correlations between M_R and routine subgrade soil properties. The routine soil properties used in the present study were unconfined compressive strength, dry density, moisture content, percent passing No. 200 sieve, and plasticity index. The deviatoric stress and bulk stress were used as the stress state to identify the condition of the specimens during the testing process.

NEURAL NETWORK MODELING

Sixty-three soil samples were collected from 14 different sites across Oklahoma. Data from these sites are collectively referred to as the *Development Dataset*. Additionally, 34 soil samples were collected from three sites for the evaluation of the developed models. These test data are collectively referred to as the *Evaluation Dataset*. Bulk samples were collected primarily from the B-horizon and the C-horizon according to the Oklahoma Department of Transportation (ODOT) standard sampling method, and the following routine tests were conducted: Atterberg limits, wet sieving (excluding hydrometer tests), dry sieving, and moisture-density. These tests are routinely used by ODOT to characterize subgrade soils. Resilient modulus (M_R) and unconfined compression tests were also conducted. Details of the laboratory testing program and the results are given by Ebrahimi [7].

A commercially available software, STATISTICA 7.1, developed by the StatSoft, Inc., was used in the development of the ANN models. It has four ANN modeling options to solve regression problems, namely Linear Networks (LN), Generalized Regression Neural Network (GRNN), Radial Basis Function Network (RBFN), and Multi-Layer Perceptrons Network (MLPN) [8]. The degree of complexity of these models increases from LN being the simplest model with no hidden layers to MLPN models with up to three hidden layers. STATISTICA 7.1 utilizes the feedforward structure, which is generally a simple network with stable behavior. In the models considered here, the input layer consists of seven nodes, one node for each of the independent variables, namely moisture content (w), dry density (γ_d), plasticity index (PI), percent passing sieve No. 200 (P_{200}), unconfined compressive strength (U_c), deviatoric stress (σ_d), and bulk stress (θ). The output layer consists of one node for the dependent variable, M_R. The selection of these parameters was partly based on the ODOT requirements for conducting pedagogical soil survey. In view of the limited scope and size of this paper, only the MLPN models are presented here.

The number of hidden layers in the MLPN models can range from one to three. In the present study, three MLPN models, henceforth referred to as MLPN-1, MPLN-2, and MLPN-3, were developed with differing numbers of hidden layers in each model. The number of nodes in each of the three hidden layers was set at six, based on the trial and error approach adopted. R^2 was used here as an indicator of model performance. Other parameters such as RMSE, MSE and AAE could also be used as performance indicators [9]. The R^2 values of the MLPN models were found to be 0.5733, 0.5744, and 0.5587 for one, two and three hidden layers, respectively. These R^2 values indicate that all three MLPN models are expected to better correlate the M_R/P_a values than the LN model (0.4323). However, the MLPN models were worse than the GRNN (0.6015) and the RBFN (0.6284) models. Figures 1-3 show comparisons between the experimental and predicted values of M_R/P_a for each of the three MLPN models. Increasing the number of hidden layers from one to two increased the R^2 value slightly for the development dataset. But, increasing the hidden layers from two to three, causes the R^2 and the predictive capability of the model to decline. Overall, the results did not show any significant differences among the three MLPN models. Figure 4 shows the back-prediction of the experimental M_R against the deviatoric stress for the MLPN-1 model for a selected specimen (MA-3B); similar comparisons for two other soils (NO-7A and OS-1B) are given by Ebrahimi [7]. The corresponding comparisons for two and three hidden layers are shown in Figures 5 and 6, respectively. Based on these figures, the predictive capacity of the MLPN-2 model is better, in an overall sense, than the MLPN-1 or the MLPN-3 models. The MLPN-3 might have reached over-learning or over-fitting [9]. A network with three hidden layers involves more weights and more complex functions, leading to possible over-fitting of the development dataset [10].

MODEL EVALUATION

Following the development of the ANN models, the evaluation dataset was introduced for evaluating the models. As noted before, the evaluation dataset was not included in the development of the models. Predictions of the M_R/P_a values were compared using the developed models with the evaluation dataset. Then, a comparison of the predictions and the experimental M_R/P_a values was made for each model using the R^2 values as the basis of comparison [6]. The R^2 values for the combined evaluation dataset for the MLPN-1, MLPN-2, and MLPN-3 models were found to be 0.5691, 0.5848 and 0.5500, respectively. Additional comparisons were made for two selected soils (WOE-4B and ROE-20B) from the evaluation dataset [7]. Although the R^2 values of the three models are similar, the MLPN-2 model shows the best prediction (see Figure 7).

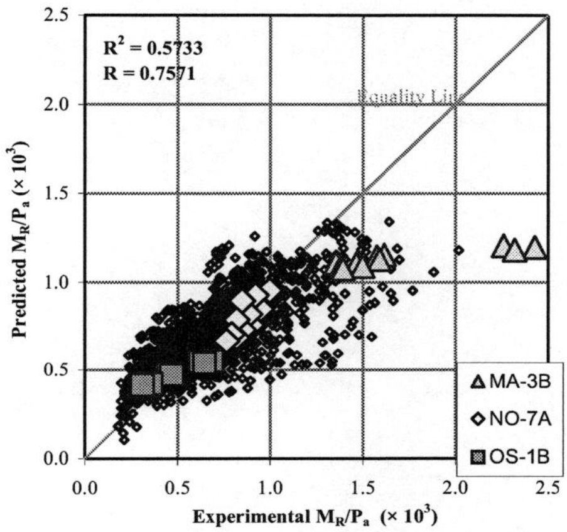

Figure 1 Comparison of Experimental and Predicted M_R/P_a for Development Dataset (MLPN-1) Model

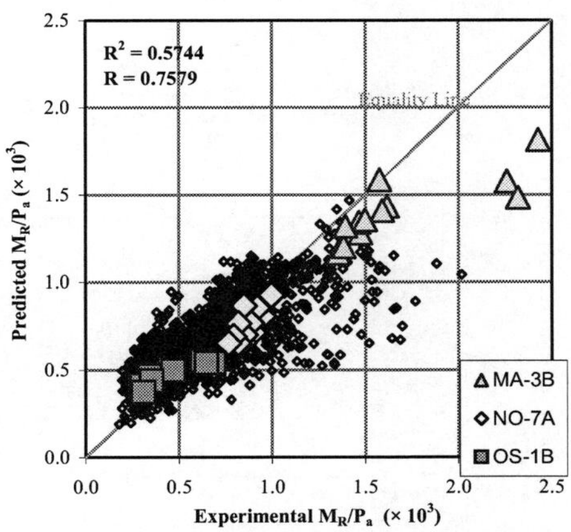

Figure 2 Comparison of Experimental and Predicted MR/Pa for Development Dataset (MLPN-2) Model

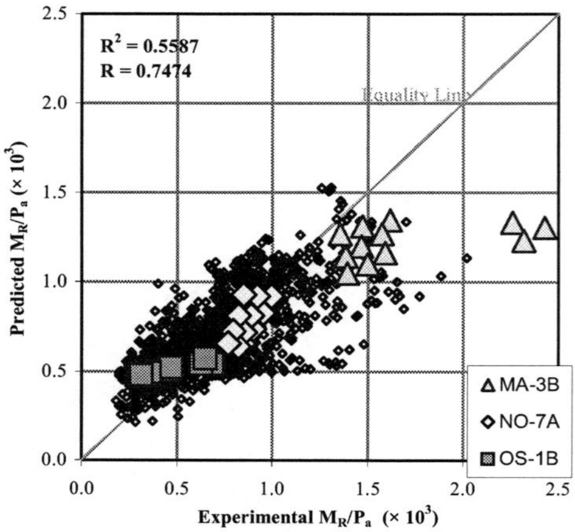

Figure 3 Comparison of Experimental & Predicted MR/Pa for Development Dataset (MLPN-3) Model

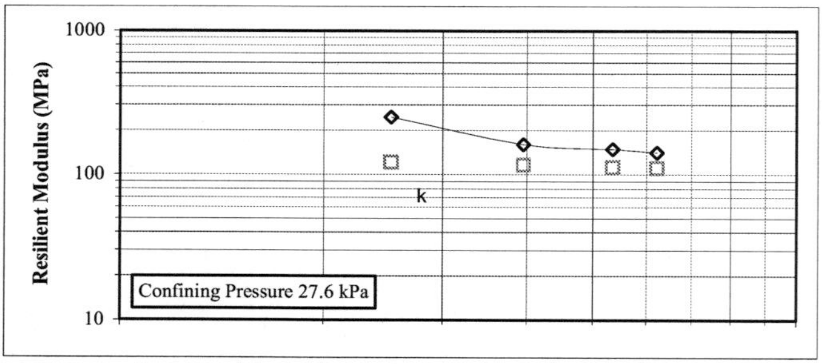

Figure 4 Comparison of Resilient Modulus from Experiment and Multilayer Perceptrons Network (MLPN-1) Model for Specimen MA-3B (1 kPa = 0.145 psi)

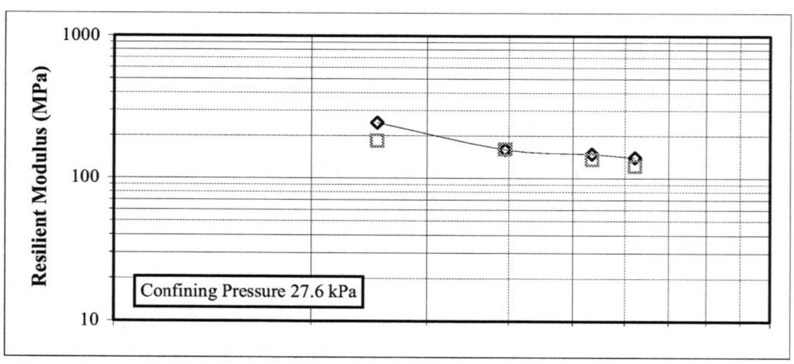

Figure 5 Comparison of Resilient Modulus from Experiment and Multilayer Perceptrons Network (MLPN-2) Model for Specimen MA-3B (1 kPa = 0.145 psi)

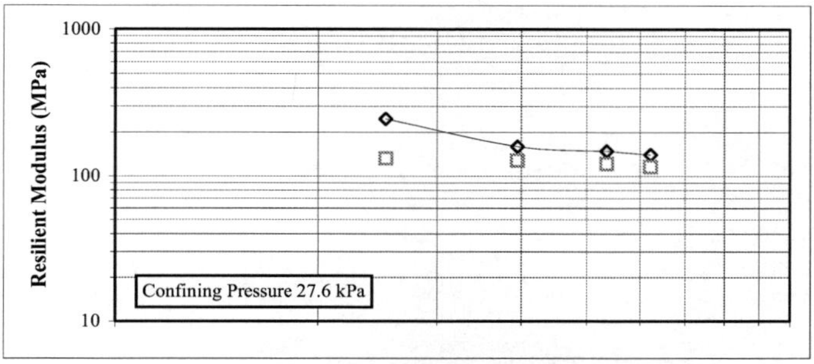

Figure 6 **Comparison of Resilient Modulus from Experiment and Multilayer Perceptrons Network (MLPN-3) Model for Specimen MA-3B (1 kPa = 0.145 psi)**

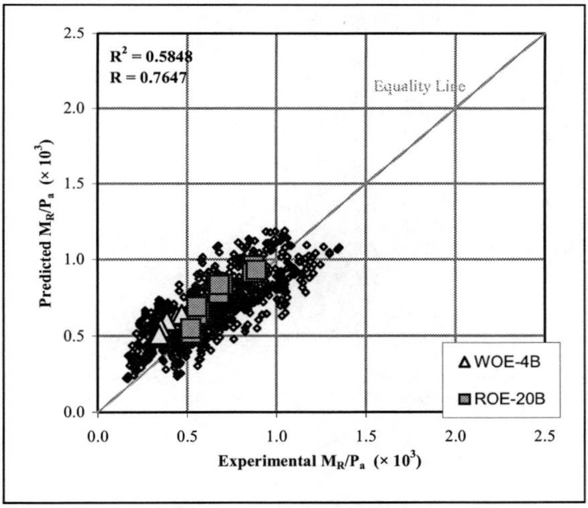

Figure 7 **Comparison of Experimental and Predicted MR/Pa for Combined Evaluation Dataset (MLPN-2 Model)**

CONCLUDING REMARKS

Artificial Neural Network (ANN) was employed in this study to develop correlations between resilient modulus, a commonly used pavement design parameter, with routinely used properties of subgrade soils obtained from laboratory testing. Six different models were considered, namely, Linear Network (LN), Generalized Regression Neural Network (GRNN), Radial Basis Function Network (RBFN), and Multi-Layer Perceptrons Network (MLPN) with one (MLPN-1), two (MLPN-2) and three (MLPN-3) hidden layers. The degree of complexity of these models increases from LN being the simplest model with no hidden layers to MLPN models with up to three hidden layers. Based on the R^2 values of both the development and the evaluation datasets, the MLPN-2 was found to be the most suitable model for estimating M_R in terms of moisture content (w), dry density (γ_d), plasticity index (PI), percent passing No. 200 sieve (P_{200}), unconfined compressive strength (U_c), bulk stress (θ) and deviatoric stress (σ_d). The selection of these parameters was partly based on ODOT requirements for conducting pedagogical soil survey. For some soils in the evaluation dataset, the MLPN-2 model under-predicted the M_R values. This may be a potential concern in pavement design applications where lower thicknesses may result from using the predicted M_R values in design. Also, the present study covered only limited soil types (alluvium and shale); additional soil types should be included in the future and the developed model(s) should be refined to enhance their applicability.

NOMENCLATURE

M_R : Resilient modulus
τ_{oct} : Octahedral shear stress
P_{200} : % passing # 200 sieve
GRNN : Generalized regression neural network
RBFN : Radial basic function network
MLPN : Multi-layer perceptrons network
U_c : Unconfined compressive strength
P_a : Atmospheric pressure
σ_d : Deviatoric stress
LN : Linear network
γ_d : Dry density
PI : Plasticity index
W : Moisture content
θ : Bulk stress.

ACKNOWLEDGEMENTS

The authors are thankful to Burgess Engineering and Testing, Inc. for assistance in conducting the laboratory tests. Thanks are extended to the Oklahoma Department of Transportation and the Oklahoma Transportation Center for technical assistance and support. Also, the authors are thankful to Sejuti Sylvia for her assistance.

REFERENCES

[1] American Association of State Highway and Transportation Officials (AASHTO) (2004), Washington, D. C.

[2] Yau, A. and Von Quintus, H.L. (2002). "Study of LTPP Laboratory Resilient Modulus Test Data and Response Characteristics," *Final Report October 2002*, FHWA-RD-02-051, USDOT, FHWA.

[3] NCHRP (1997). "Laboratory Determination of Resilient Modulus for Flexible Pavement Design," Final report, *Project 1-28*, NCHRP, TRB, Washington D.C.

[4] Hopkins, T.C., Beckham, T.L., Sun, L., and Pfalzer, B. (2004). "Kentucky Geotechnical Database," *Research Report KTC-03-06/SPR-177-98-1F*, University of Kentucky Transportation Center, College of Engineering, Lexington, Kentucky.

[5] Uzan, J. (1985). "Characterization of Granular Material," *Transportation Research Record 1022*, National Research Council, Washington, D.C., pp. 52-59.

[6] Park, H.I., Kweon, G.C., and Lee, S.R. (2006). "Prediction of Resilient Modulus of Subgrade Soils and Subbase Materials Based on Artificial Neural Network."

[7] Ebrahimi, A. (2006). "Regression and Neural Network Modeling of Resilient Modulus Based on Routine Soil Properties and Stress States." Ph.D. Dissertation, Univ. of Oklahoma, Norman.

[8] StatSoft, Inc. (2006). *Electronic Statistics Textbook*, Tulsa, Oklahoma.

[9] Bishop, C. (1995). *Neural Networks for Pattern Recognition*, University Press, Oxford, United Kingdom.

[10] Hill, T. and Lewicki, P. (2006). *STATISTICS Methods and Applications*, StatSoft, Tulsa, Oklahoma.

MRB SIGNATURES OF X-RAY CT IMAGES FOR ESTIMATING SOIL HYDRAULIC CONDUCTIVITY

S.H. ANDERSON
Dept. of Soil & Atmospheric Sci.
Univ. of Missouri
Columbia, MO 65211

Z. CHENG
Dept. of Soil & Atmospheric Sci.
Univ. of Missouri
Columbia, MO 65211

C.J. GANTZER
Dept. of Soil & Atmospheric Sci.
University of Missouri-Columbia

ABSTRACT

Recent advances in application of X-ray computed tomography (CT) for evaluation of soil properties have added new research tools for investigating structural features. Objectives of this study were to evaluate CT scan images using the Multiple Resolution Blanket (MRB) method and relate observed signature properties with soil saturated hydraulic conductivity. Forty intact samples taken at four depths were scanned at 7 mm intervals throughout each core to provide 8 scans per sample. Each scan was scaled so that CT numbers ranged from 0 to 255 relative attenuation, creating a 256 by 256 by 256 volume. The MRB method uses a theoretical blanket of varying thickness (k) to determine topological signatures representing variability. Signature graphs were divided into three regions (k < 0.8 mm, 0.8 < k < 3.5 mm, k > 3.5 mm) for analysis; these regions corresponded to relatively small, intermediate and large pore size characteristics. Regression slopes for the middle region (0.8 < k < 3.5 mm) and regression intercepts for the third region (k > 3.5 mm) were highly correlated with saturated hydraulic conductivity (r = 0.64). This study illustrates how X-ray CT images from soil cores can be evaluated using an image analysis technique and correlated with soil physical properties.

INTRODUCTION

Soil structure is of high interest to soil scientists. Scientists have worked on methods to relate hydraulic properties to soil structural properties such as porosity; however, the destructive nature of traditional methods do not allow collection of information on both temporal and spatial characteristics of soil macropores. The ability to describe and predict hydraulic properties is hence limited.

Image analysis of thin sections of soil has been used for direct study of soil structure. Thin sections of soil are one way to obtain images of intact soil structure (Ringrose-Voase, 1987); however, this method is time consuming. Moreover, even with the most delicate preparation, soil structure can change significantly during thin section preparation. Brakensiek et al. (1992) used thin sections to measure macropore sizes by manually counting them. The lower limit of resolution of this method was 0.4 mm. Hatano et al. (1992) used soil stained with methylene blue and digitized images measuring macropore area and perimeter to a resolution of 0.2 mm. Computed tomography (CT) scanning has been evaluated as a method for directly measuring the spatial distribution of soil macroporosity (Phogat and Aylmore, 1989; Zeng et al., 1996). Internal structure is reconstructed as a series across all image sections, which maybe used to analyze the 2- and 3-dimensional morphological parameters of soil pores. Medical CT scanners typically have a pixel resolution of 0.1 by 0.1 mm with a 1 mm slice thickness. This technique has the ability to characterize large macropores in intact soil with a better resolution than previous methods.

Fractal geometry is a useful analysis method for evaluating images and structural features of soils. These techniques, which are powerful topological methods for describing irregular objects, were developed two decades ago (Mandelbrot, 1983). Two important aspects of fractal geometry are (i) use of geometry and set theory in mathematics, and (ii) the graphical characteristization of intricate visual patterns and images (Jones 1991). Fractal methods have gained widespread attention from soil scientists over the past decade. Scientists use fractal geometry to relate difficult-to-measure soil hydraulic properties to other soil variables that are easier to measure, such as bulk density, soil aggregation, pore size distribution (PSD), and aggregate size distribution (Zeng et al., 1996, Perrier et al., 1995). Since the introduction of the concept of fractals, a set of water retention functions based upon fractal soil structure have been presented (Perrier et al. 1995). In addition, work by Bartoli et al. (1993) reports on soil structural data collected from thin sections that are consistent in a statistical sense with a fractal soil fabric.

The graphic aspects of fractals have not yet been used in soil structure studies. According to Pentland (1984), the image of a fractal surface is also itself fractal. Their work provides a theoretical underpinning to calculate fractal features of natural objects from their images. Among the fractal features that may be used to describe a surface, the fractal dimension and lacunarity have proven useful. Keller et al., (1989) successfully distinguished different image textures by using these two parameters. Work done by Albregtsen et al. (1992) showed that among the quantitative performance of three methods for estimation fractal dimension, the so-called Multiple Resolution Blanket (MRB) method developed by Peleg et al. (1984) gave the best measure of the true fractal dimension of the images.

Talibuddin and Runt (1994) examined several popular fractal techniques for accuracy, consistency and sensitivity by applying them to 257 by 257 pixel (picture element) self-affined surfaces of known fractal dimensions. He found that the MRB method and pyramid method were the most reliable and efficient for small images. Whichever method is chosen to measure the fractal dimension, fractal determination of real objects is based on measurements which are made across a limited range in scale. Scientists have found that this may conflict with the mathematical definitions of fractal dimension that assume an infinite scale. Ogawa (1998) used fractal methods to analyze soil images and suggested that the assumption of mathematical fractal was never truly met. Indeed, it is impossible to obtain a representation of natural systems with an infinite level of detail. Hence, any fractal model used to study soil structure should also account for the influence of structure at smaller and larger scales.

MRB signatures may be used to describe and classify differences in soil density distributions. Moreover, the MRB method can be applied on both fractal and non-fractal soil to obtain meaningful topological information. The MRB estimates will not be influenced by the mathematical limitation of fractal theory. Hence, the MRB method is potentially more useful for description of soil. The objective in this paper is to understand the physical meaning of the MRB signatures, and describe the relationships between MRB parameters and measured soil physical properties, with the goal of predicting soil properties from calculated MRB parameters.

MATERIALS AND METHODS

Sample site was Sanborn Field, University of Missouri-Columbia. Soil is classified as a Mexico silt loam (fine, smectitic, mesic, Aeric Vertic Epiaqualf). Soil cores were taken from four depths with 10 replications at each depth, total of 40 soil cores. Cores were collected from 80-156, 180-256, 280-356, and 380-456 mm depths with a core size of 76-mm-long by 76-mm-diam. Ten 0.1- by 0.1- by 2-mm scans at 7-mm spacings using a GE scanner set at a peak voltage of 120 keV were obtained from each core equilibrated to a water potential of –10 kPa. Eight scans were used for the calculation of MRB signatures discarding the top and bottom scans. A total of 320 images were used in the study. We cut off 128 pixels from each side of all CT images to exclude data outside

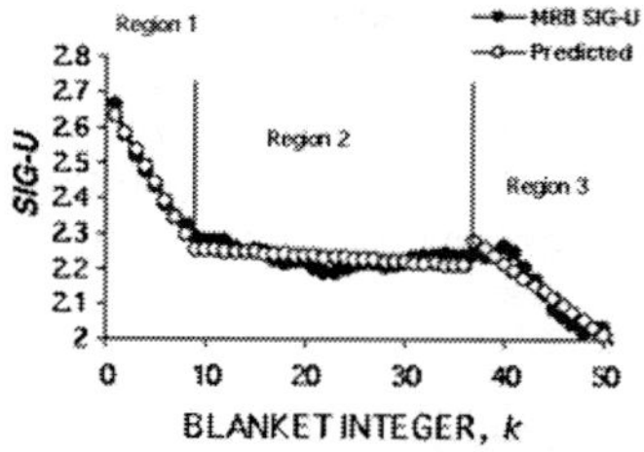

Fig. 1. Illustration of the three regions of the SIG_U curve.

the soil core area, and then assigned the remaining data to a 256 by 256 pixel array. The Hounsfield units or CT numbers (CTN) were scaled to a range of 0-255 gray levels.

To describe the relationship between the signature parameters and the macropore-scale properties, quantitative relationships between the macroporosity of synthetic images and MRB parameters were obtained using a C-language program. The upper MRB signature (SIG_U) value is related to the total area of pore size larger than k.

The MRB method is designed to allow the calculation of the gray-scale surface area (A) that is produced using a wide range of blanket sizes (k). The area of the pixel value surface measured at different blanket thicknesses ($k*\varepsilon$, where ε is a constant, and k is the blanket integer variable from 1 to an upper limit) will decrease at larger blanket thickness values. The gray-scale surface area is measured by calculating the volume between upper and lower blankets divided by 2ε. The SIG_C (composite signature) was calculated by finding the slope of the best fit straight line through the three points: $\{[\log(k-1), \log(A(k-1))], [\log(k), \log(A(k))], [\log(k+1), \log(A(k+1))]\}$. The upper signature identified as SIG_U is derived from the higher gray scale values using the maximum operator to detect the surface characteristic of the higher value pixels.

The CT images were not filtered nor processed in any way that would introduce potential artifacts. The MRB method consisted of developing a signature curve for each CT image. The MRB signature curve is correlated to variations in the amount and size of soil pores. To simplify the results, each MRB signature curve was divided into three regions. The first region included the region related to relatively smaller pores and was defined by a blanket integer less than 9. A blanket integer represents the thickness of the artificial blanket used to cover the image surface. An artificial blanket includes all points in three-dimensional space at a distance k from the original surface. The second region included the region related to relatively medium sized pores and was defined by blanket integers ranging from 9 to 35. The third region was related to relatively larger pores and was defined by blanket integers greater than 35. An example MRB signature curve is shown in Figure 1. Each region was best fit to a line, which is described by its slope and intercept. The three linear equations were represented by intercepts and slopes for the three regions: SIG_U(R1) and slope(k_1) for Region 1; SIG_U(R2) and slope(k_2) for Region 2; SIG_U (R3) and slope (k_3) for Region 3. The intercepts for the first and the third regions were used to develop a relationship between the MRB features and the laboratory measured soil properties. In essence, the intercept for the first region was the MRB signature value at a blanket integer equal to 0 and the intercept for the third region was the MRB signature value at a blanket integer equal to 35. The intercept value for the first region is related to the frequency of small pores and the intercept of the third region is related to the frequency of larger pores. Preliminary analysis indicated that these parameters were the best representations of the soil pore sizes or soil structure as defined by the CT images.

Field water contents of the soil samples were measured using the soil core method. A cylindrical metal sampler was driven into the soil and carefully removed to preserve a known volume of sample. The sample was dried to a constant weight at 105°C and recorded. These values were used to estimate the soil bulk density. Soil porosity was calculated from bulk density and particle density. Hydraulic conductivity of the saturated samples was measured by the constant head method with steady downward flow. Matric potentials ranging from 0 to –10 kPa were measured on the undisturbed soil cores using fine-pore glass funnels under pressure. The measured macroporosity was calculated by subtracting the water content at –10 kPa from the saturated water content.

RESULTS AND DISCUSSION

Figure 2 shows the composite signature (SIG_C) of the 3rd scan of soil cores representing the four depths of 80-156-, 180-256-, 280-356-, and 380-456-mm at transect position 4. These four SIG_C functions are not fractal across the whole range of blanket integers since the lines are not at a constant value. The slope decreases through Region 1 and through Region 3. The value has a bell-shaped curve through Region 2. Results show that these soil CT images are not fractal when analyzed by the MRB method. Shapes of the signatures are similar to those obtained by Talibuddin and Runt (1994). The dimension was greater than 2.0 for blanket integers less than 50. A dimension of ~2.0 occurred for blanket integers greater than *55*. Since the soil CT images were not found to be fractal using the MRB method, direct use of the fractal dimension does not appear to be useful for comparison of the soil treatments. However, following the suggestion of Peleg et al. (1984), only the MRB upper signature (SIG_U) was used to analyze the structure of non-fractal images.

Results of the SIG_U for one position are shown in Figure 3. From Fig. 3, we can see patterns in the data. In the 3rd region, the SIG_U values can be classified in the following order: depth 1 > depth 2 > depth 3 > depth 4. The order in the 1st region is reversed. There are differences between the four depths in the 2nd region but they are more complex. All the SIG_U values of the soil cores from the other nine positions have similar patterns, with the exception of the 2nd transect position.

The intercept for the third region $SIG_U(R3)$ was the MRB signature value at a blanket integer equal to 35. $SIG_U(R3)$ is related to the frequency of larger pores. Preliminary analysis indicated that this parameter was the best representation of the large pore sizes defined by CT images, suggesting that the major difference in soil structure among cores exists at the larger macropore scale.

Comparisons among the depths indicated that they were significantly different from each other at the <0.001% probability level. The surface samples were characterized with the highest values, with the second depth having slightly lower (not significant) SIG_U values. The depths of 280-356 mm and 380-456 mm had lower $SIG_U(R3)$ values. The first and second depths show few differences between each other, but they are different from the third and fourth depths. The third depth is also different from the fourth depth. This indicates that the macropores of the upper two depths are similar and have higher macroporosity, which is supported by the general knowledge that upper soil depths have more macropores than deeper soil.

Results indicate that the slope *k2*, and the $SIG_U(R3)$ are significantly related to other soil properties such as macroporosity, total porosity, hydraulic conductivity and bulk density (Table 1). The $SIG_U(R3)$ has the highest correlation with these soil physical properties, which shows that macroporosity is highly correlated with total porosity and bulk density. The third highest correlation with macroporosity is the intercept value of $SIG_U(R3)$ ($r = 0.84$), and supports the conclusion that SIG_U of Region 3 represents larger pores. In addition, the SIG_U may be an independent measure of macroporosity from the soil CT images. This measure might be developed as a comparison measure of soil macroporosity in the future to better describe the nature of macropores.

Results indicate that $SIG_U(R3)$ is the best predictor of macroporosity, which can be described by the following model:

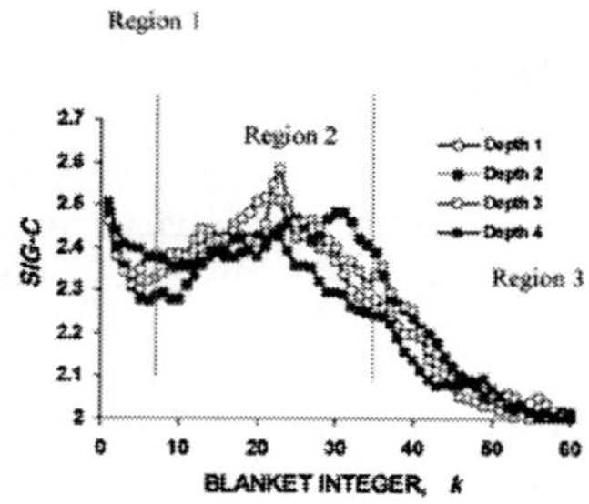

Fig. 2. The SIG_C of the 3rd scan representing depths of 80-156, 180-256, 280-356, and 380-456 mm within a soil core at transect position 4.

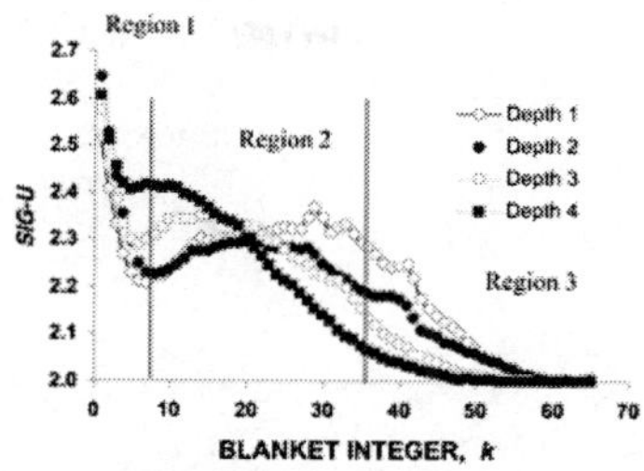

Fig. 3. Average SIG_U for the four depths at transect position 1 (n=8). The depths are 80-156, 180-256, 280-356, and 380-456 mm, respectively.

Table 1. Correlation matrix of physical parameters for samples from Sanborn Field.

	MRB Parameters						Measured Parameters			
	S-R1	*S-k1*	*S-R2*	*S-k2*	*S-R3*	*S-k3*	*M-P*	*L-K*	*f*	*ρ*
S-R1	1.000									
S-k1	-0.83	1.000								
S-R2	-0.44	0.73	1.000							
S-k2	0.068	-0.36	-0.82	1.000						
S-R3	-0.16	-0.03	-0.52	0.89	1.000					
S-k3	0.26	-0.08	0.40	-0.80	-0.94	1.000				
M-P	0.002	-0.10	-0.50	0.80	0.84	-0.74	1.000			
L-K	0.30	-0.29	-0.43	0.64	0.67	-0.55	0.75	1.000		
f	-0.11	-0.03	-0.49	0.80	0.81	-0.72	0.88	0.70	1.000	
ρ	0.013	0.093	0.58	-0.87	-0.88	0.80	-0.91	-0.70	-0.95	1.000

S-R1, S-R2, and S-R3 are the intercepts of 1st, 2nd and 3rd regions.
S-k1, S-k2, and S-k3 are the slopes of 1st, 2nd, and 3rd regions.
M-P is macroporosity, *L-K* is the logarithm of Ksat, *f* is porosity, and *ρ* is bulk density.

$$y = -0.45 + 0.239*x \qquad r^2=0.720 \qquad (1)$$

where y is macroporosity and x is $SIG_U(R3)$. $SIG_U(R3)$ is also the best predictor of hydraulic conductivity. However, this prediction only explains 45% of the variation. To improve the prediction, a stepwise multiple linear regression system was used to estimate which two soil properties would produce the best model. Macroporosity and $SIG_U(R1)$ using multiple linear regression were the best combined factors to predict the hydraulic conductivity. The prediction follows the linear model of

$$y = -38.26 + 108.23*x_1 + 11.84*x_2 \qquad r^2=0.657 \qquad (2)$$

where y is the value of logarithmic transformation of hydraulic conductivity [*ln(Ksat)*], x_1 is the macroporosity, and x_2 is $SIG_U(R1)$. The macroporosity is significantly related to *ln(Ksat)* with P<0.01, and the $SIG_U(R1)$ is significantly related to *ln(Ksat)* with P=0.026. The model predicts 65.7% of the variability of *ln(Ksat)* with these two factors. Since there exists a high correlation between $SIG_U(R3)$ with macroporosity, $SIG_U(R3)$ can be used instead of macroporosity to predict *ln(Ksat)*. This model is

$$y = -109.5 + 27.3*x_1 + 19.2*x_2 \qquad r^2=0.614 \qquad (3)$$

where y is the value of the logarithmic transformation of hydraulic conductivity *ln(Ksat)*, x_1 is *SIG*$_U$*(R3)*, and x_2 is *SIG*$_U$*(R1)*. Both the *SIG*$_U$*(R1)* and *SIG*$_U$*(R3)* are significantly related to *ln(Ksat)* with $P<0.01$. The model explains 61.4% of the variation in *ln(Ksat)* using these two factors. The basis for the prediction using *SIG*$_U$*(R1)* and *SIG*$_U$*(R3)* is that these parameters correlate with the porosity of the largest and smaller pore sizes, respectively. These two pore classes represent relatively more conductive and less conductive pores, respectively.

CONCLUSIONS

Results from this study found that parameters estimated from MRB signature curves determined from CT scan images are related to soil macroporosity and other soil physical properties. Due to the limitations in resolution of CT scanners, the MRB-measured parameters from CT images only represent the relative pores and solids of soil, which is one reason the blanket integer *k* does not exactly represent pore size. The value *k* can only characterize large differences in small and large macropore sizes; therefore, three regions of *SIG*$_U$ were used to represent the relative small, medium and large macropore sizes. The MRB parameters are useful for predicting soil physical properties, such as macroporosity and hydraulic conductivity. The *SIG*$_U$*(R3)* had the best correlation with macroporosity of the MRB signature parameters. The *SIG*$_U$*(R1)* and *SIG*$_U$*(R3)* combination was found to be a good predictor of saturated hydraulic conductivity. This study validates the potential for using the MRB method to search for relationships between parameters measured by image processing and soil physical properties.

REFERENCES

Albregtsen, F., B. Nielsen, and K. Yogesan. 1992. Fractal dimension, only a fraction of the truth? p. 733-736. Proc. 11th IAPR Int. Conf. on Pattern Recognition. Vol.III, IEEE Comput. Soc. Press. Los Alamitos, CA.

Bartoli, F., G. Burtin, R. Philippy, and F. Gras. 1993. Influence of fir root zone on soil structure in a 23 m forest transect: The fractal approach. Geoderma. 56:67-85.

Brakensiek, D.L., W.J. Rawls, S.D. Logsdon, and W.M. Edwards. 1992. Fractal description of macroporosity. Soil Sci. Soc. Am. J. 56:1721-1723.

Hatano, R., N. Kawamura, J. Ikeda, and T. Sakuma. 1992. Evaluation of the effect of morphological features of flow paths on solute transport by using fractal dimensions of methylene blue staining pattern. Geoderma. 53:31-44.

Jones, H. 1991. Fractals before Mandelbrot. A selective history. p. 7-33. *In* A.J. Crilly,
R.A. Earnshaw and H. Jones (eds). Fractal and chaos. Springer-Verlag, New York, NY.

Keller, J.M., S. Chen, and R.M. Crownover. 1989. Texture description and segmentation through fractal geometry. Vision, Graphics, and Image Processing. 45:150-166

Mandelbrot, B.B. 1983. Fractals: The fractal geometry of nature. W.H. Freeman. New York.

Ogawa, S. 1998. Fractal geometry of preferential flow pathways in field soils. Ph.D. Diss. Cornell Univ. (Diss. Abstr. 98-18499).

Peleg, S., J. Naor, R. Hartley, and D. Avnir. 1984. Multiple resolution texture analysis and classification. IEEE PAMI. 4:518-523.

Pentland, A. P. 1984. Fractal-based description of natural scenes. IEEE PAMI. 6: 661-674

Perrier, E., C. Mullon, and M. Rieu, 1995. Computer construction of fractal soil structures: Simulation of their hydraulic and shrinkage properties. Water Resour. Res. 31:2927-2943.

Phogat, V.K,. and L.A.G. Aylmore. 1989. Evaluation of soil structure by using computer assisted tomography. Aust. J. Soil Res. 27:313-323.

Ringrose-Voase, A.J. 1987. A scheme for the quantitative description of soil macrostructure by image analysis, J. Soil. Sci. 38:343-356.

Talibuddin, S., and J.P. Runt. 1994. Reliability test of popular fractral techniques applied to small two-deimensional self-affine data sets. J. Appl. Phys. 76: 5070-5078.

Zeng, Y., C. J. Gantzer, R.L. Peyton, and S.H. Anderson. 1996. Fractal dimension and lacunarity of bulk density determined with X-ray computed tomography. Soil Sci. Soc. Am. J. 60:1718-1724.

VOLUMETRIC SELF ORGANIZING FEATURE MAP FOR MODELING DEFORMABLE SOLIDS

GEORGE K. KNOPF AND PHILIP C. IGWE
Department of Mechanical & Materials Engineering,
Faculty of Engineering, The University of Western Ontario,
London, Ontario, Canada, N6A 5B9

ABSTRACT

Conceptual mechanical design requires geometric modeling tools that enhance user creativity by minimizing the need to follow precise formulas, rigorous procedures and rigid technical specifications. A mathematical representation of deformable solid objects that exhibit "realistic" elastic and plastic material behaviour with minimal details in a virtual reality (VR) environment is described in this paper. The physics-based deformation model consists of a set of point masses, connected through a series of springs and dampers, which undergo movement through the influence of external and internal forces. The location of each mass point corresponds to a node on a 3D mesh defined by the lattice of a volumetric self-organizing feature map (VSOFM). Material properties are introduced to the mesh by assigning a mass value to individual nodes and spring coefficients to the nodal connections. Once reshaped through external influences or haptic interactions, the mesh created by the VSOFM can be interactively modified during the conceptual design process undertaken in a VR environment.

INTRODUCTION

The conceptual design process has not benefited from highly precise computer-aided design (CAD) software tools to the same degree as detailed modeling and engineering analysis because the creative activities associated with developing "potential" solutions require minimal details and is far less formulaic in its implementation. At the early stages of product design the performance specifications and design constraints have not been fully established. Success at this stage can be enhanced if the industrial designers are provided the freedom to rapidly change and modify the product configuration and material behavior in order to explore a vast number of alternative solutions. Traditionally, product designers use their artistic talents to create 3D physical models from clay in an effort to explore competing concepts. Any CAD system that seeks to support and enhance conceptual design must, therefore, enable natural and haptic modes of human-computer interaction.

Recent advancements in high-speed computer hardware and virtual reality (VR) technology provide new opportunities to link the more fluid processes of creative conceptual design with the rigidly defined tasks of product detailing and engineering analysis. The need for a viable VR-based conceptual design tool comes from several case studies performed at Loughborough University (Sener, Wormald and Campbell, 2002). One key conclusion derived from this study is that the human-computer interface and related software tools for interacting with the virtual models must be intuitive to the user, provide sensory feedback during design, and mimic the natural way the consumer would interact with the product concepts that are being created.

The rendering software and haptic technologies used by a viable VR-based design system must provide the industrial designer with the means to interact with the virtual product during concept development but, on its own, it fails to satisfactorily reflect the subtleties of the designer's intent or enable the user to readily explore variations in part form and functionality. The realistic behaviour of a 3D model undergoing external forces depends on the underlying algorithms used to represent the shape and material behavior of the virtual object (Dani and Gadh, 1997) (Zhengyi, Lili and Yonghua, 2005). In addition to estimating the direction and position changes of model nodes over time, the mathematical algorithms must also compute the direction and magnitude of interactive forces which are reflected back to the user through the haptic tool.

The following sections summarize how a *volumetric self-organizing map* (VSOFM) can be used to create and represent a deformable virtual object that undergoes "realistic" shape change. During the early stages of concept design the lattice of the VSOFM is treated as a mesh of mass points that undergo direct influence from a haptic device or tool.

PHYSICS BASED MODELING

Shape modification of a virtual object can be simulated using either geometric- or physics-based algorithms. Geometric techniques only adjust the vertices of the underlying mesh model in response to external forces. Physics-based models (Hua and Qin, 2001), on the other hand, are able to both estimate the direction and magnitude of nodal movement based on realistic material properties and the external forces introduced to the model through the haptic tool. Elasticity and visco-elasticity properties have been modeled with some success but the computation times prevent these algorithms from being used effectively in real-time simulation.

The simplest and most popular physics-based approach are the *particle-based* deformation models where the volume enclosing the object consists of numerous particles connected together through a network of springs and dampers that move under the influence of both external and internal forces. Each particle or mass point in the network is defined by its own position, velocity and acceleration (Terzopoulos, Platt, Barr and Fleischer, 1987). The spring forces are often linear but nonlinear springs can be used to model elastic and viscoelastic objects such as human tissue or rubber. In general, real-time simulation has been restricted to objects with simple, non-complex geometries because these simple algorithms require a mathematical solver to calculate the inverse mass matrix and compute new stiffness and damping matrices at each time step. Although the stiffness and damping matrices are square symmetric matrices, the computation becomes inefficient when thousands of mass points are used to represent the complex object geometry.

DEFORMABLE VIRTUAL OBJECT

The deformable object is represented by a nodal mesh with a predefined topology called a *volumetric self-organizing feature map* (VSOFM). Each node, $\boldsymbol{x}_i$, in the underlying lattice, $\mathbf{X}$, is a mass point in the object space and connected to neighboring nodes through a spring. During the early stages of concept design the lattice is treated as a mesh of point nodes that undergo direct influence of the haptic device. The strength of the node connections are modified as the concept evolves in order to reflect different material properties. This permits the same concept shape to be analyzed as either a single homogeneous solid or a single part comprised of multiple materials.

Structure of the VSOFM

The VSOFM algorithm is divided into two parts (Figure 1): fitting the exterior nodes of the deformable lattice to representative surface points $\boldsymbol{p}^n$ (Knopf and Sangole, 2004), and uniformly distributing the interior nodes throughout the final closed surface. In essence, the algorithm is designed such that both the node $\boldsymbol{x}_i$ nearest the input data point $\boldsymbol{p}^n$ and all nodes residing in its neighborhood are simultaneously updated. On completion of the training process, the geometry mapped by the node locations represents a mesh that encloses the original object surface. The efficiency of the VSOFM algorithm for creating a particle-based solid is enhanced by uniformly positioning the interior nodes within the volume after the location of the exterior nodes have been identified.

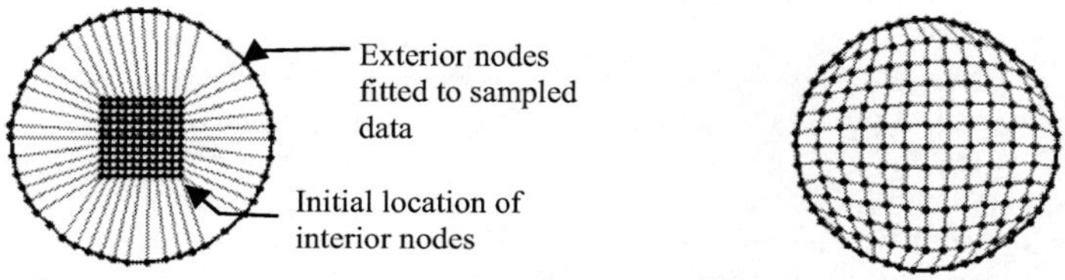

(a) Exterior nodes fitted to points sampled on a circle. **(b) Uniform distribution of interior nodes in (a).**

Figure 1. Two step process of creating a volumetric mesh.

Deformation Using Mass-Spring System

The topologically linked nodes of the initial reference mesh are defined as mass points linked together by a spring system, Figure 2. The external and internal forces directed at the mass points are calculated from haptic feedback and collision detection algorithms. The deformable modelling system must be able to compute the estimated position of the mass points at the next time step from the current forces being applied through the haptic tools. In this context, the spring located between nodes i and j has a natural or rest length of L_{ij}. The actual length of the interconnection spring when a force is applied, l_{ij}, is given by the vector separation between the two adjoining nodes; that is

$$l_{ij} = \| \mathbf{r}_{ij} \| \text{ where } \mathbf{r}_{ij} = \boldsymbol{x}_i - \boldsymbol{x}_j \tag{1}$$

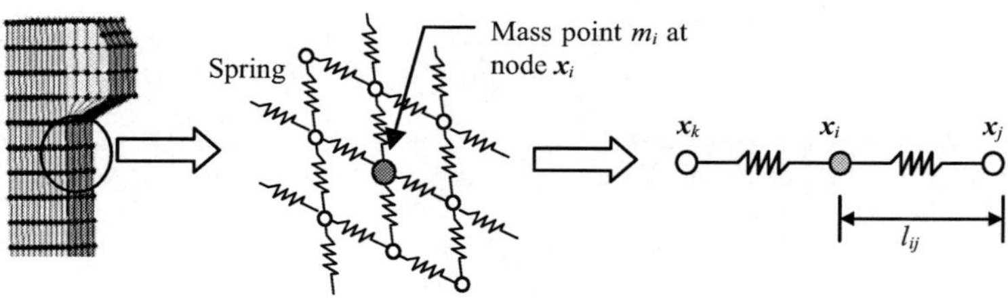

Figure 2. VSOFM mesh and the corresponding mass-spring representation.

The internal and external forces applied during shape deformation cause mechanical strain in the connecting spring. Hence, the total internal force applied to node i is due to all the springs that act on the node at time t and can be mathematically given as

$$\boldsymbol{g}_i(t) = \sum_{j \in \boldsymbol{ne}_i} c_{ij} \frac{(l_{ij} - L_{ij})}{L_{ij}} (\boldsymbol{x}_i - \boldsymbol{x}_j) \tag{2}$$

where the material stiffness is given by the coefficient c_{ij} and $\boldsymbol{ne}_i$ is the immediate neighborhood around node i. Although in this study c_{ij} is assumed constant, for some materials it is necessary to modify the stiffness value as the plastic strain changes (Hua and Qin, 2001).

Numerical simulation of the deformation process while the material experiences external forces from the haptic device are achieved using the discrete Lagrange equations of motion for a dynamic node-spring system. The system dynamics is given by the second-order differential equation

$$m_i \ddot{\boldsymbol{x}}_i(t) + \gamma_i \dot{\boldsymbol{x}}_i(t) + \boldsymbol{g}_i(t) = \boldsymbol{f}_i(t) \qquad (3)$$

where γ_i is the velocity-dependent damping coefficient which dissipates kinetic energy in the lattice through friction; m_i is the point mass of node i, $\boldsymbol{g}_i(t)$ is the total internal spring forces given by Eqn. (2), and $\boldsymbol{f}_i(t)$ is the external force vector applied to node i. For simplicity, the external force function is modeled as a time-varying 2D Gaussian operator. The magnitude of the external force $f_i(t)$ is simulated as an impulse, pulse, step or arbitrarily varying time function.

The rest length of the spring L_{ij} is determined at beginning of the deformation process. The spring is then allowed to vary in length l_{ij} with respect to the plastic deformations or other non-linear behaviour that is to be modelled by the system of equations. To simulate the dynamic behaviour of the adaptive mesh, it is necessary to provide initial positions $\boldsymbol{x}_i(0)$ and the initial velocities $\dot{\boldsymbol{x}}_i(0)$ for each node. The initial velocities are often assumed to be zero. It is then necessary to integrate the equation of motion forward through time until the mesh stabilizes; $\dot{\boldsymbol{x}}_i \approx \ddot{\boldsymbol{x}}_i \approx 0$.

DEVELOPING CONCEPTS USING A DEFORMABLE VSOFM

Concept Generation and Modification

The shape modeling tool for creating alternative design concepts requires a number of operational steps. It is first necessary to retrieve a reference VSOFM mesh from a library of primitive shapes or to create a new reference mesh with a priori knowledge about the desired surface. Once the reference mesh has been selected mass and damping coefficients are assigned to the individual nodes. If the virtual object is comprised of a single homogeneous material then all mass points, damping coefficients and spring properties in the mesh will have identical values. Conversely, if the material is non-homogeneous and the material property is assumed to vary throughout the solid then it is necessary to assign a unique mass value, damping coefficient and spring property to individual or clusters of nodes. External forces are then introduced to the mass-spring mesh through haptic tool or data glove.

In addition to applying different deformation operations it is possible to modify or replace the material properties in the deformed mesh in order to investigate mechanical behavior of the virtual object. Once the mass and spring coefficients have been altered then the designer applies forces from the haptic device to visualize how the concept would behave under different loading conditions.

The final step in the concept design process is to remesh the concept model to permit greater detail design and part analysis. Remeshing may also address data exchange issues between dissimilar modeling systems prior to transferring the concept model to a conventional CAD system.

Design and Analysis of a Telephone Handset

Consider the task of designing a functional, yet stylistic, telephone handset. Different shapes are possible with the exploitation of new plastic materials to reduce the overall weight and enhance features such as high friction gripping surfaces that prevent slippage during use. In this context, complex design parameters that cannot be effectively addressed using the conventional CAD design packages may be examined using concept design simulation tools in VR spaces, Figure 3.

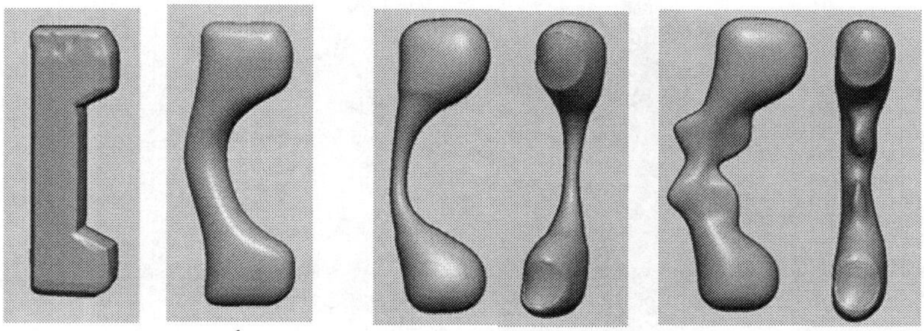

a b c d

Figure 3. Different concepts of a telephone handset for both form and function. Note that these concepts have been rendered using the Phantom® Omni™ system to demonstrate VR tools in design methodology.

In terms of the VSOFM technique for creating the underlying mesh of the original model, the lattice was first fitted to the prototype shape given in Figure 4a. The 17x15x5 reference mesh was built on top of VSOFM and the modeling material was assumed to be plasticine (Huang, Lucas, Adams, 2002). Nodal parameters for this freeform model are: Young's modulus of 130MPa, Poisson's ratio of 0.49, $m_i = 0.94$g, and $\Delta t = 0.001$sec. The damping coefficient is 18735 Ns/m.

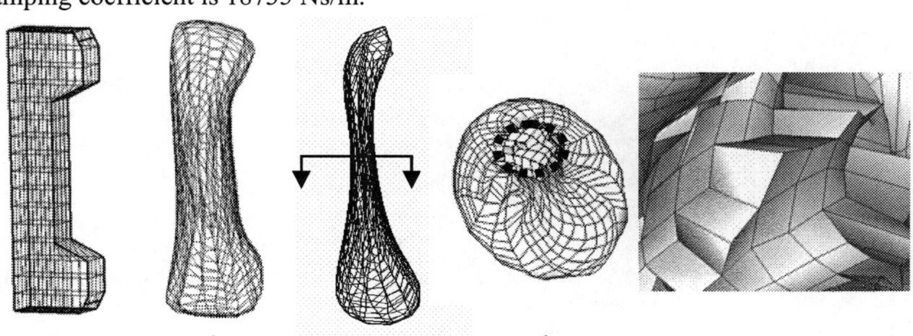

a. b c. d. e.

Figure 4. Modifications to the original VSOFM (a) as the design concepts varying from (b) to (c). Enlarged view of the interior nodes is given in (e).

Figure 5 shows the effect of external forces applied to the underlying VSOFM mesh. The applied external force of (0, 50, 0) N was centered at the node location (4, 4, 5). The forcing function was a step over 1 second. The external force was applied to a cluster of nodes having a neighborhood radius of 2 units. The result obtained from this algorithm shows that the degree and rate of shape deformation will vary if the values for the material parameters (Young's modulus, Poisson's ratio γ_i , m_i) are altered. The preliminary results show that our system is able to follow deformations due to external and internal forces acting on the model.

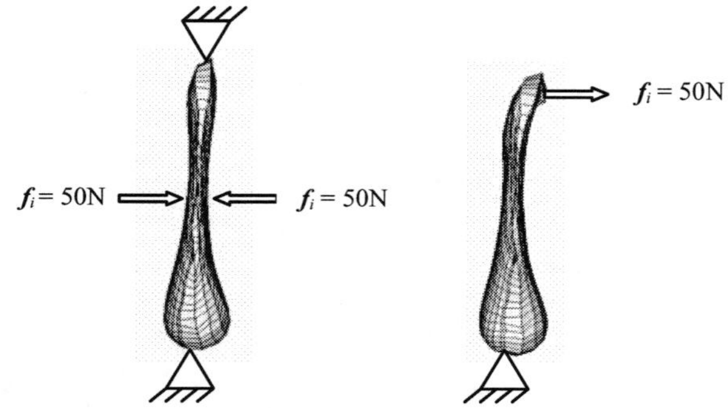

Figure 5. Deflection of virtual plastercine object created by VSOFM algorithm.

CONCLUSIONS

A volumetric self-organizing feature map (VSOFM) was presented as a computational framework for representing deformable solid objects. The physics-based deformation model consisted of a set of point masses, connected through a series of springs and dampers, which undergo movement through the influence of external and internal forces. To make the interactive modeling process more realistic and effective, additional research is required to introduce constraints that permit material cutting and removal processes. Material cutting can be simulated by assigning zero values to the effected spring constants, while material removal is achieved by assigning zero values to the corresponding mass and spring coefficients. Another area of investigation is the introduction of efficient collision prediction and detection algorithms for deformable objects. The algorithm exploits the limited number of nodes in the VSOFM lattice to improve computational efficiency without sacrificing graphical and haptic realism.

ACKNOWLEDGEMENTS

The authors wish to acknowledge the support from Integrated Manufacturing Technologies Institute – National Research Council of Canada, especially the extensive discussions with Dr. Roberto Canas and Mr. Dan Johnston.

REFERENCES

Dani, T.H., and Gadh, R., 1997, "Creation of Concept Shape Designs via a Virtual Reality Interface", *Journal of Computer-Aided Design (CAD)*, Vol. 29, pp. 555-563.

Hua, J., and Qin, H., 2001, "Haptic Sculpting of Volumetric Implicit Functions", *Proceeding of Ninth Pacific Conference on Computer Graphics and Applications*, pp. 254-264.

Knopf, G.K., and Igwe, P.C., 2004, "Sculpting Free-form Surfaces using Interactive CAD", *Flexible Automation and Intelligent Manufacturing,* pp. 146-153.

Knopf, G.K., and Sangole, A., 2004, "Interpolating Scattered Data using 2D Self-organizing Feature Maps", *Graphical Models*, Vol. 66, pp. 50-69.

Sener, B., Wormald, P., and Campbell, R.I., 2002, "Evaluating a Haptic Modeling System with Industrial Designers", *Proceedings of EuroHaptics*, pp. 165-169.

Terzopoulos, D., Platt, J., Barr, A., and Fleischer, A., 1987, "Elastically Deformable Models", *Siggraph Proceedings on Computer Graphics,* Vol. 21, No. 4, pp. 205-214.

Zhengyi, Y., Lili, L., and Yonghua, C., 2005, "Haptic Evaluation of Multi-material Part Design", *Journal of Computer Aided Design*, Vol. 37, pp. 727-736.

USING NEURAL NETWORKS TO MODEL INTERCITY MODE CHOICE

PRAVEEN EDARA
University of Missouri, Department of Civil and Environmental Engineering, Columbia, Missouri, USA

DUŠAN TEODOROVIĆ
University of Belgrade, Faculty of Transport and Traffic Engineering, Belgrade, Serbia

HOJONG BAIK
University of Missouri, Department of Civil, Architectural, and Environmental Engineering, Rolla, Missouri, USA

ABSTRACT Artificial intelligence techniques such as neural networks and fuzzy systems have been successfully applied in engineering, finance, and sociology, to model the decision making behavior of individuals when faced with choices. In transportation planning, predicting the mode chosen by users is essential to estimate the mode specific travel demand. Neural networks are able to capture complex relationships by learning from historical data, and produce good estimates of the dependent variables. In this paper, we present a neural network approach to predict the mode choice behavior of travelers making intercity trips. A case study of the intercity travel (>100 miles) in United States was conducted. Travelers choose from three modes: commercial airline, automobile, and general aviation, while making these trips. The results obtained were very satisfactory, meaning, there is good scope for the application of neural network techniques to the modal split problem in transportation.

Introduction

Artificial intelligence techniques such as neural networks and fuzzy systems have been successfully applied, in engineering, finance, and sociology, to model the decision making behavior of individuals when faced with choices. To demonstrate the neural network approach to mode choice problem in transportation, the intercity trips made in United States is studied in this paper. Intercity trips include all trips made between places separated by 100 miles or more. These trips are made primarily using two modes - the automobile (personal vehicle) mode and the airline mode (*1*). Although public transportation modes such as rail and bus exist, the number of trips made using them is pretty low. In addition, as the trip length increases the market share of airline increases and the automobile decreases. The data related to the inter-city trips made in United States was extracted from the American Travel Survey (ATS) database (*2*). ATS conducted by Bureau of Transportation Statistics (BTS) in 1995 is one of the most valuable data sources for analyzing long-distance trips. The survey contains data for 337,520 household trips and 556,026 person trips with more than 300 variables for each trip. The variables include: 1) trip characteristics such as trip purpose, trip origin and destination, trip distance, transportation mode, access/egress distance to/from airports, and 2) traveler's personal information such as income, family size, and education level. The survey, however, does not include information on travel cost and travel time

experienced by the trip maker for the trip. In order to overcome this limitation, the travel cost and travel time for each mode for a trip were synthesized using realistic assumptions (not shown in this paper). For example, travel speed of automobile is assumed to be 30 and 60 miles per hour on arterials and highways respectively. Total travel time for an air traveler is the sum of access/egress time to/from airport, processing time at airports and flying time. In order to consider the hub-and-spoke network structure in the commercial airline flights, a detouring factor was also applied.
The problem considered in this paper can be defined in the following way:

> *For known traveler income, distance traveled, travel times, and travel costs for three modes (commercial aircraft, automobile, and general aviation) and given the mode chosen by the set of trip makers, develop a model that will accurately predict the number of trips made by each mode for each zonal pair considered.*

The Neural Network Approach

After the synthesis of travel times and trip costs, the following input data for each trip maker is available – Income (I), Distance traveled in that trip (D), Travel Times (T) and Travel Costs (C) for three modes (commercial aircraft (CA), automobile (A) and general aviation (GA) and finally the Mode (M) chosen by trip maker for that trip.

In this section, artificial neural networks are used to model the mode choice problem. Three different networks were developed: one for each mode. For example, the following input and output variables are used for the commercial airline neural network.

Inputs
I - Income
D - Distance for the trip
ΔT - Time Differential (the difference between the travel time for the most competing mode (quickest among Automobile and General Aircraft) and the travel time for the Commercial aircraft)
ΔC - Cost Differential (ΔC) (the difference between the trip costs for the most competing mode (cheapest among Automobile and General Aircraft) and the trip cost for Commercial aircraft)

Output
P - Probability of a passenger choosing commercial airline mode for the trip.

The rationale behind developing such differential variables (ΔT and ΔC) is to include the marginal utility that a trip maker receives when he/she chooses a commercial airline. The chosen inputs include the socio-economic characteristics of the trip maker, the trip characteristics, and the service characteristics. Automobile and General Aircraft modes have the ΔT, ΔC, and P values specific to them.

The mode chosen by a trip maker is mentioned in the ATS database (1-A, 2-CA and 3-GA). This data needs to be converted to probabilities. The task of creating the probabilities is accomplished by grouping similar passengers into clusters, which results in the whole data set divided into a finite number of clusters. Each cluster represents a fictitious *artificial passenger* whose characteristics (income, distance, etc) are the average of all the passengers belonging to that cluster. Probability is obtained by dividing the number of passengers choosing a mode by the total number of passengers within that cluster. This procedure of creating probabilities is summarized in the algorithm below.

Step 1: Classify the whole data set using K-Means clustering. Form clusters. Each cluster represents one "artificial passenger".

Step 2: Calculate the trip characteristic variables for every "artificial passenger" as the average of all the passengers belonging to that cluster.

Step 3: For every cluster, count the number of passengers within the cluster choosing a particular mode. Using these counts, calculate for every "artificial passenger" the probability of "artificial passenger" choosing a particular mode.

In our case study, the initial data for business travelers consisting of 118,000 trips was grouped into 1,200 clusters using K-Means clustering algorithm. These clusters are used to train and validate the proposed neural network that consists of an input layer with four input variables and an output layer with one output. The number of hidden layers was decided based on trial and error. During training, when input data was fed to the network, initially, the weights (within a defined range) were arbitrarily assigned to the links and the output probability was calculated. Error was computed as the difference between the desired probability and the predicted probability. The objective was to minimize this error (for the whole data set) by adjusting the link weights. The weights were adjusted using the generalized delta rule. The artificial passenger data was divided into two parts - training data and testing data. The total number of artificial passengers was equal to 1,200. Four-fifth of this data (960) was used for training and one-fifth (240) for testing. At the end of training, a trained neural network that is able to predict the probability of a passenger choosing commercial airline was obtained. The validation of this network was done by feeding testing data, which was not used for training, and checking the accuracy of predicted probabilities (based on the error criteria described below).

Neuro Solutions tool was used for building, training, and testing the neural network. Upon several iterations - changing the number of hidden layers, number of neurons in each hidden layer, type of transfer function, and learning criteria, the following network configuration had the best performance: three hidden layers, tanhaxon transfer function, supervised learning and 1000 epochs. In this way, we developed one neural network model for each mode. The results obtained for all three models are shown in the next section.

Experimental Results

The following statistical indicators have been used for measuring the accuracy of predicted probability - Mean Square Error (MSE), Average Relative Variance (ARV), and Absolute Residuals (AR). (Nijkamp et al (1996)).

The Mean Squared Error is a very common indicator and can be represented as below:

$$MSE = \frac{\sum_{n=1}^{N} (y_n - \hat{y}_n)^2}{N} \quad n = 1, 2, \ldots N$$

where y_n denotes the expected values and $\hat{y}_n$ denotes the values calculated from the model by using a data sample with N observations.

Another indicator, often used in the literature on Neural Networks, is the Average Relative Variance (ARV (N)). It is defined as below:

$$ARV(N) = \frac{\sum_{n=1}^{N} (y_n - \hat{y}_n)^2}{\sum_{n=1}^{N} (y_n - \overline{y})^2} \quad n = 1,2,\ldots N$$

where y_n denotes the expected values and $\hat{y}_n$ denotes the values calculated from the model and $\overline{y}$ is the average of the expected values belonging to the set of data N.

By performing the residual analysis the errors of the test data set can be illustrated. The absolute residual can be defined as below:

$$AR = y_n - \hat{y}_n$$

The results of Mean Square Error (MSE) and Average Relative Variance (ARV) are shown in Table 1.

Table 1. Mean Square Error (MSE) and Average Relative Variance (ARV) for all three networks

Mode	Mean Square Error (MSE)	Average Relative Variance (ARV)
Commercial Airline (CA)	0.007	0.062
Automobile (Auto)	0.011	0.101
General Aviation (GA)	0.001	0.970

The performance of the neural network models was tested on a new data set. Model predictions were compared with the real values as shown in Figures 1, 2, and 3.

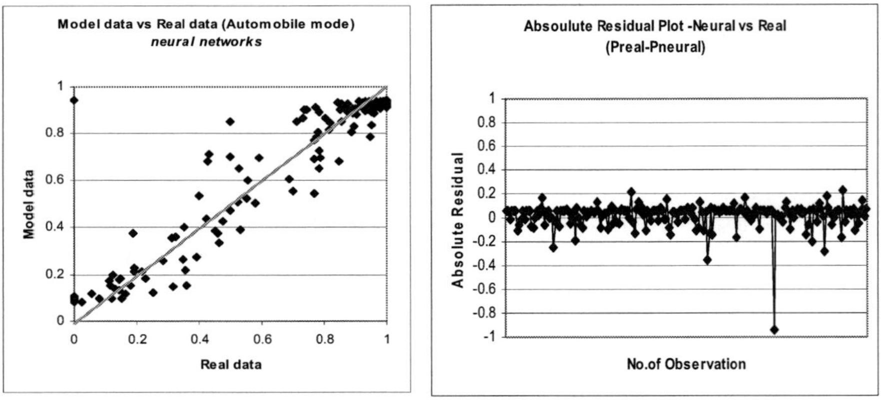

Figure 1. Comparison of the real data and the data obtained by the proposed model – Automobile Mode.

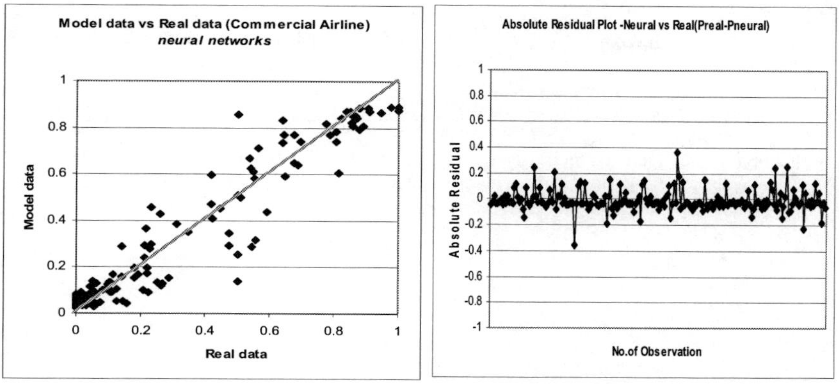

Figure 2. Comparison of the real data and the data obtained by the proposed model – Commercial Airline mode.

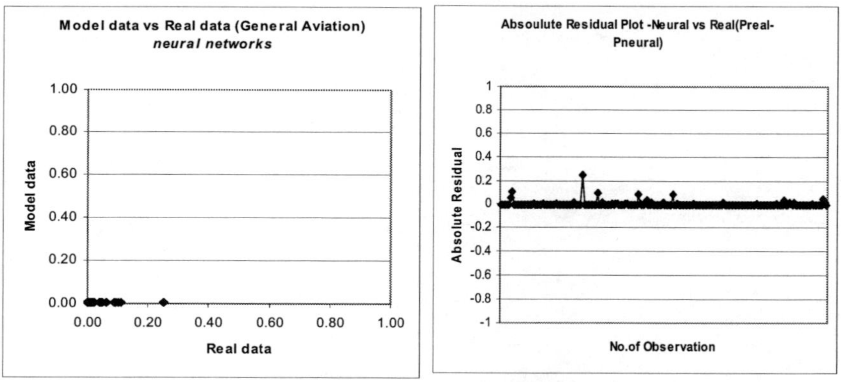

Figure 3. Comparison of the real data and the data obtained by the proposed model – General Aviation mode.

Conclusions

In this paper, we used neural networks to build mode choice models for intercity travel. A case study of intercity trips in United States was conducted. From the results of mean square error, average relative variance, and absolute residuals, it can be concluded that the neural network models produced highly satisfactory results for the automobile and the commercial airline mode. The results obtained for the general aircraft mode suggest that additional explanatory variables (e.g., aircraft ownership) should be included in the model specification depending on the data availability.

These promising results should encourage researchers to apply artificial intelligence methods (e.g.: neural networks) to the field of transportation when ever there is a need for forecasting, function approximations or predictions. Unlike the traditional logit and regression models, the neural network model does not have a physical equation that can be used to calculate the output for new data; instead it gives a trained network which is equivalent to a calibrated equation which can accurately predict the output for new data.

References

Market share of intercity trips, 1990. http://www.publicpurpose.com/ut-icshare.htm

American travel survey (ATS), 1995. http://www.transtats.bts.gov/Tables.asp?DB_ID=505&DB_Name=American%20Travel%20Survey%20%28ATS%29%201995&DB_Short_Name=ATS

Nijkamp, P., Reggiani, A. and Tritapepe, T., 1996, "Modeling inter-urban transport flows in Italy: A comparison between neural network analysis and logit analysis." *Transportation Research C,* 4, 323-338.

PREDICTION OF DAMAGE/REPAIR RATES IN WATER DISTRIBUTION SYSTEMS DUE TO SEISMIC EVENTS USING ARTIFICIAL NEURAL NETWORK

RONALDO LUNA*
Department of Civil, Arch. and Env. Engineering
University of Missouri Rolla
Rolla MO 65409
rluna@umr.edu
+1 (573) 341-4484
*Corresponding author

NANDINI K BALAKRISHNAN
CIHAN H DAGLI

Department of Engineering Management and Systems Engineering
University of Missouri Rolla
Rolla MO 65401

ABSTRACT

Water distribution systems are important lifelines that can be affected by natural disasters, such as earthquakes. It has been reported that it takes one to two weeks or more to restore the water distribution system to a fully functional level after a major earthquake strikes. So, it would be beneficial to society if one could predict the damage/repair rate experienced by the water distribution system. These predictions can be used by city planners or developers to identify the weak sections in the system and to improve its resilience. In this paper, feed forward back error propagating multilayer neural network is used for the prediction and it is trained using the damage statistics available for different earthquakes in twelve different locations. The training was done using earthquake magnitude, pipe diameter and material type, and peak ground velocity (PGV) as input parameters and the damage/repair rate per thousand ft as output parameter. The output of the neural network model was compared with two other empirical formulae and the results were found to be promising.

INTRODUCTION

Water delivery systems, particularly underground pipelines, can be severely damaged by strong ground motions during earthquakes. In addition to the direct impact on the system, such as interruption of water supply and inadequate supply for fire fighting, there will be other consequences such as pressure head losses, subsequent loss of water service, water pumping and hauling costs, and erosion of roadway subgrades due to leaks and spills, which can be detrimental to the local community. Other potential adverse consequences include flooding of roads and nearby basements and damage to adjacent utilities and foundations (Hong, 2006). In 2001, the US Environmental Protection Agency (EPA) released a national survey of drinking water infrastructure needs which concluded that approximately 151 billion dollars would be needed over twenty years to repair, replace, and upgrade the nation's 55,000 community drinking water systems to protect public health (ASCE, 2005). Upgrading the distribution system is a strategy adopted by authorities to reduce potential damage, and often the first step in the seismic upgrade of a pipeline system is the evaluation of the likely damage in the existing system due to earthquakes (O'Rourke and Deyoe, 2004).

During seismic events, multiple pipelines in a water distribution network are damaged simultaneously and, due to the complex nature of the physics affecting pipe damage under seismic waves, damage assessment is something that still remains challenging (Dong, 2004). This paper uses the Artificial Neural Network (ANN) method to predict the damage/repair rates in water distribution systems due to seismic events. Artificial neurons are typically nonlinear, and this is especially an important property when the physical process of interest is absolutely nonlinear (Ham, 2001). The term damage/repair rate is used herein since most databases available consist of a record of the repairs to restore the systems, and repair records are inherently associated with the damages caused by the earthquake event.

In many cases the data regarding damage to water distribution systems are expensive to collect, incomplete, or unavailable (Allouche and Bowman, 2006). However, ANNs can tolerate missing data and still make reasonable predictions (Brawley, 1993). The trained network can be used for different earthquake scenarios to minimize the damage to an existing network or to decide upon the pipe material type and diameter when laying out a new distribution network. Also, the results of the network can be used by government agencies while prioritizing the pipeline sections in the system for rehabilitation and can aid in the restoration activity. This method is envisioned for use in conjunction with other methods to give an accurate damage/repair rate for the scenario of interest.

DATA BASE DESCRIPTION

The data used for the ANN training was taken from the Part 2-Appendices of the report entitled "Seismic Fragility Formulations for Water Systems", which was published in April 2001 by the American Lifelines Alliance. (ALA, 2001) This Alliance is a public-private partnership between the Federal Emergency Management Agency (FEMA) and the American Society of Civil Engineers (ASCE). The various earthquakes that have been considered for this study and their respective magnitudes are given in Table 1. Four parameters that have a major impact on the distribution pipelines, pipe material type, pipe diameter, earthquake magnitude and peak ground velocity (PGV) were used to train the ANN. These parameters are not necessarily exhaustive, but are certainly significant factors based on review of relevant literature (PAHO, 2007; Pei Zongchang et al, 2005; O'Rourke and Jeon, 1999) that can influence the damage to the distribution system in a seismic event. The magnitude of the earthquakes (measured on the Richter scale) varied from a low of M=6.3 in the 1933 Long Beach earthquake to a high of M=7.9 in the 1923 Kanto and 1968 Tokachi-oki earthquakes. The pipe material type used in these distribution systems consisted of ductile iron, asbestos cement, steel, concrete pipe, cast iron, and in some cases a combination of these materials. The pipe diameters were divided into three categories namely *large* (> 12 in/~30.48 cm.), *small* (≤12 in.) and *Distribution Systems* (DS). The Distribution system consists of mostly small diameters, but may include some large diameters also. Peak Ground Velocity

(PGV) was recorded in in/sec at each of the sites considered. Data in which the wave propagation is measured using peak ground acceleration (PGA) and Modified Mercalli Intensity (MMI) were omitted to maintain consistency in data used in training. O'Rourke and Toprak (1997) carried out a study based on the 1994 Northridge earthquake and found that the zones of highest PGV show a spatial correlation with the damage/repair rate concentrations.

Table 1: Earthquakes whose data used for network training

Year	Place	Magnitude
1923	Kanto, Japan	7.9
1933	Long Beach, CA	6.3
1964	Niigata, Japan	7.5
1965	Puget Sound, WA	6.5
1968	Tokachi-oki, Japan	7.9
1971	San Fernando, CA	6.7
1979	Imperial Valley, CA	6.5
1983	Coalinga, CA	6.7
1989	Loma Prieta, CA	6.9
1989	Mexico	7.4
1994	Northridge, CA	6.7
1995	Hyogoken-nanbu, Japan	6.9

THE NEURAL NETWORK MODEL

A three layer back error propagating neural network model which has one input layer, one output layer and a hidden layer was used in this study Though not always practically feasible, Hornik et al. established that an Multilayer Perceptron Neural Network (MLP NN) that has only one hidden layer with a sufficient number of neurons can act as a universal approximator of nonlinear mappings (Ham and Kostanic, 2001). A supervised learning algorithm which was provided with a set of input and target values was used to train the network. Initially, the training was done using only two input parameters, the earthquake magnitude (M) and PGV, assuming they were the main physical variables affecting the phenomenon. However the training did not prove to be very successful, as the goal of 0.0001 MSE could not be achieved even after 10,000 epochs. Increasing the number of hidden neurons was also attempted, but it did not show significant progress in reducing the error and achieving the target.

Consequently, the pipe diameter and material type were included in the training and the number of hidden neurons was optimized to produce a minimum error. Including those two parameters significantly improved the training and the network achieved its goal in less than 100 epochs, which shows that those two parameters influence the damage/repair rates. From a data set of eighty, sixty four were used for training, the remaining 16 were used for testing. The Tansig transfer function was used in the hidden layer and Purelin in the output layer, both of which are commonly used transfer functions. Eighteen neurons in the hidden layer performed best, which resulted in an average prediction error of fifteen percent. Training results are shown in Fig. 1 while Fig 2 shows the testing results; the errors are reasonable and acceptable.

PERFORMANCE ANALYSIS

The developed network was analyzed by comparing the predicted value and actual damage/repair rates. The network was also analyzed for its performance by comparing with the empirical wave propagation fragility developed by O'Rourke and Jeon (1999) and also with the one developed by the American Lifeline Alliance (O'Rourke and Deyoe, 2004).

O'Rourke and Jeon (1999) developed an empirical relationship using data from the 1994 Northridge earthquake, whose formulation has the following form:

$$RR = \left(\frac{V_{max}}{266}\right)^{1.22} \qquad (1)$$

The American Life Alliance project developed the following wave propagation damage relation based on an analysis of 81 data points from 12 earthquakes,

$$RR = 0.0024\ Vmax \qquad (2)$$

where, RR = the repair rate in damage/repairs per kilometer,
Vmax = the peak horizontal velocity (PGV) in cm/sec

The same set of data was used in all the three cases to facilitate comparisons. Table 2 shows the percentage error noted in the estimation of the damage/repair rates using equations (1) and (2) and also the prediction error of the network model developed.

Table 2: Performance Comparison

Method	Percentage Error
Empirical Equation (O'Rourke)	13.01
Empirical Equation (American Lifeline Alliance)	13.96
Neural Network Model	14.99

The results show that the neural network prediction is comparable with other methods and can be used for the prediction of damage/repair rates with reasonable accuracy. As the empirical formula is based on only one of the variables, PGV, any error in recording that value might result in incorrect estimation of the damage rates, while the ANN model uses a number of contributing factors which will tolerate the missing and misrepresented data and still do a reasonable prediction. So, the ANN model can be effectively used in conjunction with other methods to get a more reliable picture of the potential damage scenario.

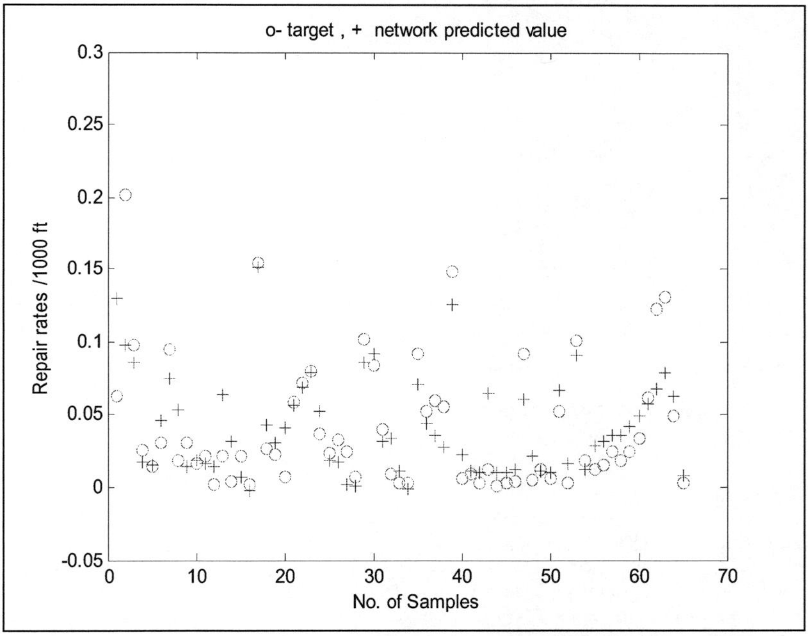

Figure 1: Training results of the ANN for water distribution systems

CONCLUSIONS

In this study a feed forward back error propagating multilayer neural network was used to predict the damage/repair rate of buried pipelines in water distribution systems for different earthquake scenarios. The observed error in predicting the damage/repair rates might be attributed to input parameters not included such as corrosion data. For example, corrosion is thought to have had an influence on the damage rate in the 1983 Coalinga earthquake. Similarly, sample size is also important in the wave propagation damage statistics, which means that, small damage/repair rates require data over a large length of pipe to enable good confidence in the data. If these factors are considered for and incorporated into the training, the neural network proves to be a promising methodology for the prediction of the rate of damage/repair of buried pipelines in water distribution systems.

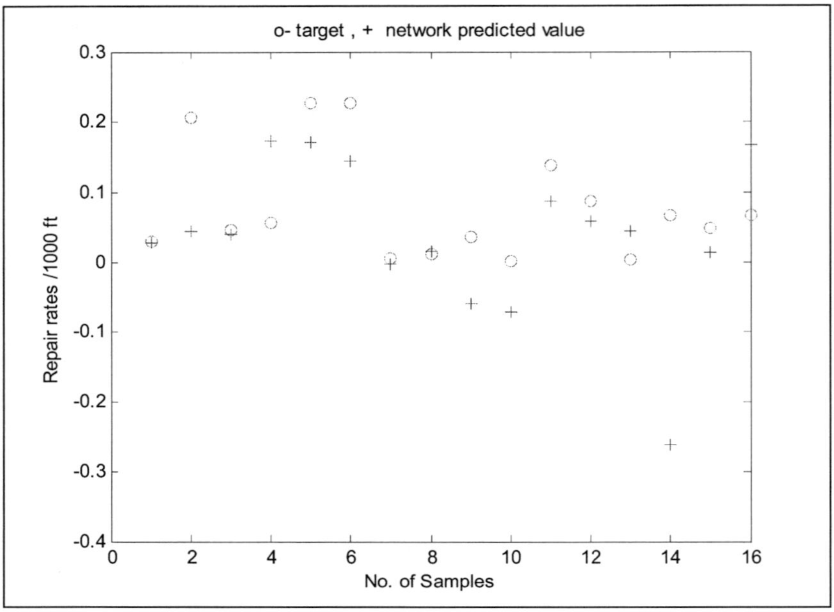

Figure 2: Testing results of the ANN for water distribution systems

REFERENCES

ALA., 2001, "*Seismic Fragility Formulations for Water Systems*", Part 2 – Appendices

Allouche, E.N., and Bowman, A.L., 2006, "Holistic Approach for Assessing the Vulnerability of Buried Pipelines to Earthquake Loads" *Natural Hazards Review.*, Vol. 7, Issue 1, pp. 12-18

ASCE.,2005."*Report Card for America's infrastructure*", <http://www.asce.org/reportcard/2005/index.cfm> (Aug.15, 2007)

Brawley, G. H., 1993, "Improved Characterization of Pipeline Defects using Neural Networks," *Proceedings of the Artificial Neural Networks in Engineering Conference*, November 14-17, St.Louis ,Missouri, USA.

Dong, X., and Shinozuka, M., 2004, "GIS Based Seismic Damage Localization for Water Supply Systems,"*Proceedings, 13th World Conference on Earthquake Engineering*, Aug 1-6 Vancouver, B.C, Canada

Ham, F.M., and Kostanic, I., 2001, "Principles of Neurocomputing for Science and Engineering," McGraw Hill, Inc, New York.

O'Rourke, M., and Deyoe, E., 2004, "Seismic Damage to Segmented Buried Pipe," *Earthquake Spectra,* Vol. 20, No. 4, pp.1167–1183

O'Rourke, T., and Jeon, S., 1999, "Factors Affecting the Earthquake Damage of Water Distribution, Monograph #16, ASCE, pp. 379–388.systems", *Proceedings, 5th U.S. Conference on Lifeline Earthquake Engineering*, TCLEE

O'Rourke, T. and Toprak, S., 1997, " *Using GIS to Assess Water Supply Damage from the Northridge Earthquake.*" <http://mceer.buffalo.edu/publications/bulletin/97/11-03/jul97n3.html >,(Mar 7, 2007).

PAHO.,2007, "*Natural Hazards and their Impact on Water Systems,*" pp.20-28 <http://www.paho.org/English/Ped/nd-chapter3.pdf>, (Feb 9, 2007)

Pei Z., Liu, Y., and Wang X., 2005,"Earthquake Damage Assessment Models of Water Supply Pipelines and Their Realization Based on GIS", IEEE, pp 5208-5210

PREDICTION OF STRAINING ACTIONS IN RIGID PAVEMENTS DOWEL BARS THROUGH ARTIFICIAL NEURAL NETWORKS

SAMIR N. SHOUKRY
Mechanical & Aerospace Eng.
Civil & Environmental Eng.
West Virginia University
Morgantown, West Virginia, USA

MOURAD Y. RIAD
Civil & Environmental Eng.
West Virginia University
Morgantown, West Virginia, USA

ADEL W. SADEK
Civil and Environmental Eng.
University of Vermont
Burlington, Vermont, USA

ABSTRACT

A heavily instrumented test section was constructed on Corridor 'H' in West Virginia (Robert C. Byrd's Highway near Elkins) consisting of 30 consecutive concrete slabs. The test section allows long term monitoring of the performance of full-scale slabs and joints. This investigation puts into perspective the mechanistic behavior of slabs on grade from the early age of construction. Various parameters such as strains, temperature profiles, joint openings provide key-performance and reliable data needed for analysis of rigid pavements response to environmental loading. In this paper time histories from the test section are used to verify the use of dynamic artificial neural networks (ANNs) in predicting straining actions of load transferring devices consisting of dowel bars. In this respect, the dowel bars deformations in response to temperature and moisture variations are predicted using Time Delay Neural Networks as well as Elman/Jordan Networks. The study shows that ANNs are capable of correlating the dowel bar bending moments to temperature gradients measured across the slab cross sections. In this study, comparing the results of ANNs and field measured bending moments indicate that Elman/Jordan networks provided slightly better match than Time Delay Neural Networks (TDNN).

INTRODUCTION

Climate and environmental variations are important factors that influence the functioning of concrete pavements. Ambient air temperature variations cause expansion and contraction of the slabs, while those will tend to curl when subjected to temperature gradients. The combined effect of those factors along with concrete shrinkage and mositure variations induce built in stresses that contribute to the early age cracking of concrete pavements. If the slabs were free to deform, those stresses would have been minimum. However, restrains such as slab-base friction, dowel-concrete interaction, tied shoulders, aggregate interlocks at joints and self weight resist the relief of such stresses. One of the main characteristic signs of the built in bending stresses in pavement joints is the dowel bar deformation due to slab curling.

Theoretical and experimental studies showed the effectiveness of circular dowel bars to transfer the load across rigid pavement joints through bending and shear. (Van Breemen 1955, Tabatabaie et al. 1978, Tayabji 1983, Vyce 1987, Titus et al. 1998, William et al. 2001). In this paper, recorded time history of measured Temperature Gradient Profile (TGP) through the thickness of concrete slabs is used to develop two different classes of dynamic neural networks that predict the Dowel-Bending Moment (DBM). The Measurements are provided from instrumented test section on Coridor H, (Robert C. Byrd's Highway, Route 33) near Elkins in West Virginia. Continuous measurements of DBM are recorded through post-processing strain time history of dowel bars using instrumentation technology. This work uses time history of 250 days where data are collected every 20 minutes starting from concrete pouring.

INSTRUMENTED SECTION

A heavily instrumented section was constructed on Corridor 'H' in West Virginia (Robert C. Byrd's Highway near Elkins), which allows long term monitoring of the performance of full-scale slabs and joints. The section provides data from several key performance parameters such as static and dynamic strains, joint openings, temperature profiles, dowel shear and bending strains. Those parameters help evaluate the slab response to environmental and seasonal changes, along with the effect of dynamic axle loads. Also a comprehensive highway test program is planned making use of the Falling Weight Deflectometer (FWD) to record the load transfer efficiency at transverse joints relative to seasonal changes. This investigation puts into perspective the mechanistic behavior of the slabs from the early age of construction and provides reliable data that would verify finite element models of rigid pavement structures.

The test site was constructed in August 2001, consisting of 30 slabs 15 x 12 ft and 11 inches thick with tied shoulders, resting on 4 inches free drainage asphalt stabilized base, that is placed on 6 inches of compacted gravel. Along the test section, six slabs were instrumented according to a system of sensors as shown in Figure 1. Before placement of concrete, a layer of course sand was placed on top of the base layer in order to fill the surface voids and provide a uniform and minimal slab-base friction. A detailed instrumentation plan as well as a description of the sensors and data acquisition systems could be found in Shoukry (2004).

Dowel bars of diameter 38.1 mm were instrumented to monitor their long-term response to environmental and dynamic loads, as well as their behavior during early age of construction. Strain data from dowels are collected since placement of concrete, and post processed to calculate the built-in bending moments dynamic shearing stresses and axial forces. Instrumented dowels are located at the shoulder corner, wheel path, mid-width of slab, corner with adjacent lane, and at the opposite corner of the adjacent lane. For collecting bending strains, two uniaxial vibrating wire strain gages were mounted on the top and bottom of the dowel in a slot within the bar center. Dynamic strains are collected by a 90 degrees wire resistance strain gage rosette mounted to the side of the bar at it's center. Each instrumented dowel was calibrated in the laboratory to check the accuracy of the collected readings and the calculated straining actions prior to installation in the field. Figure 2 illustrates a time history of the corner dowel bending moment through 300 days from placement of concrete.

Fig. 1 System of sensors in an instrumented slab

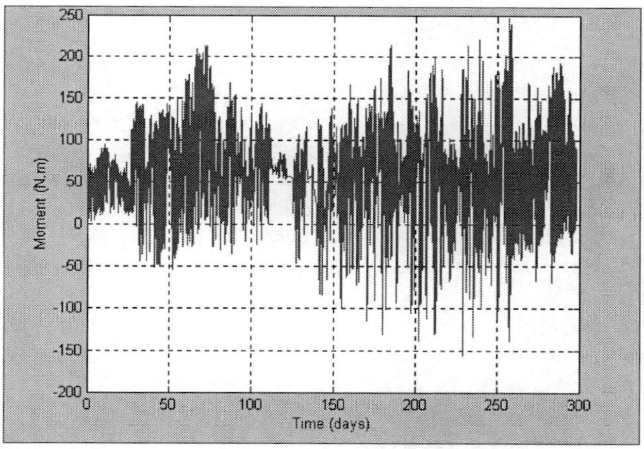

Fig. 2 Dowel bending for corner dowel through 300 days

ANN DEVELOPMENT

The data set for training the networks consisted of a total of 10,000 input-output exemplars. Each exemplar recorded the corresponding dowel bar bending moment in N.m. The sensor reading were recorded at a frequency of 20 minutes, which means that the 10,000-point data set corresponded to temperature and bending moment readings recorded over a period of about 139 days. For testing the network, a new data set, corresponding to data recorded during the 15 days following the end of the training set, was extracted. This test set thus consisted of a total of 1080 data points, which were used in testing all ANNs developed, as will be described later in this paper.

In this study, we experimented with two different dynamic ANN topologies (namely: (1) Time Lagged Neural Network (TLNN); and (2) Elman and Jordan networks) to predict dowel bending moment from the temperature gradient profile. Our decision to use dynamic ANNs as opposed to static networks was motivated by the fact that our data was indeed in the form of a time series, and from our belief that the TGP has a direct impact on the dowel bending moment.

Time Lagged Neural Networks

For the TLNN, the study used the Time Delay Neural Network (TDNN) architecture, whose memory structures are of the *tap delay line* type. A *focused* topology where the memory structures are only provided at the input layer was used to allow for using the standard back propagation algorithm in training the network. The tap delay was set at 10, meaning that the network would use the current temperature gradient, along with the temperature gradient for the previous 10 data points (i.e. last 200 minutes) in predicting the dowel moment. One hidden layer, consisting of 14 neurons, was used. The activation function for the hidden layer neurons was a Tanh function. The TDNN will be referred to as ANN1 in this paper.

Elman and Jordan Networks

Three different Elman and Jordan network topologies were tried in this study. The first topology was a variant of the Elman network, where the input was copied to the context PE layer (instead of feeding the hidden layer output), providing for an integrated past of the input (memory traces). This Elman-variant network will be referred to as ANN2 in this paper. The second topology was a traditional Elman network, where the hidden layer output was fed back to the context PE layer (ANN3). Finally, the third topology was a traditional Jordan network, connecting the network output to the context layer (ANN4). For all three models, one hidden layer was used consisting of "Tanh-type" neurons. The number of neurons in the hidden layer was 85 for ANN2, and 40 for ANN3 and ANN4. The use of the relatively large number of neurons in the hidden layer was possible since an adequate number of training exemplars was available (10,000 exemplars).

RESULTS FOR FOCUSED TDNN

During training, the TDNN (ANN1) had a Mean Square Error (MSE) of 0.0138 on the cross validation data set. The network performance on the test set (i.e. the new 1080 data points) was even slightly better, with a Mean Square Error of 0.0130. Figure 3 shows a plot of the bending moment predicted by ANN1 versus the actual moment as recorded by the sensors. As can be seen, the performance of the model appears quite good, with the predicted values very close to the recorded values, and exhibiting the same general trend. For this case, the average absolute difference between the predicted and recorded bending moment value was 14.21 N.m. Given that the average absolute value of the bending moment for the test set was 92.79N.m, the average absolute error represents only a value of 15.32% of the average absolute bending moment value.

RESULTS FOR AN ELMAN NETWORK

For the Elman/Jordan networks, the Elman-variant network (ANN2) gave the best performance. For ANN2, the MSE on the test set was only 0.010. This compares to a value of 0.026 for ANN3, and a value of 0.024 for ANN4. Figure 4 shows a plot of the.

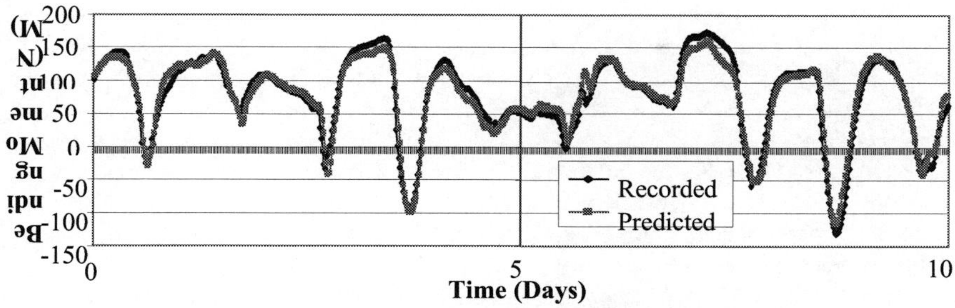

Fig. 3 Predicted versus Recorded Bending Moment for ANN1

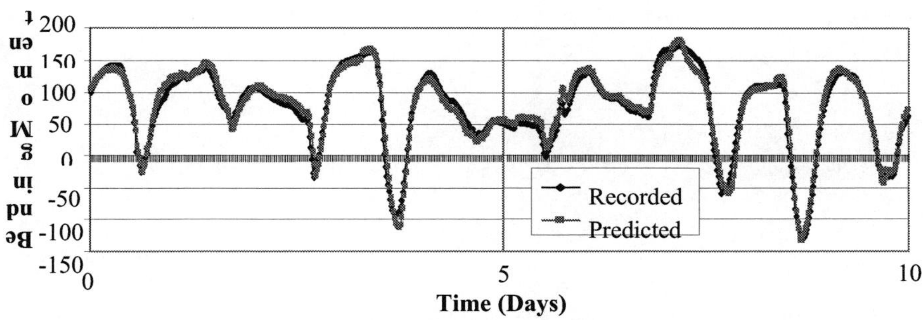

Fig. 4 Predicted versus Recorded Bending Moment for ANN2

predicted versus recorded bending moment for ANN2. Once again, the performance of the ANN appears quite good, with the predicted values very close to the recorded values. The average absolute difference between the predicted and recorded bending moment value in this case was 12.69 N.m, which represents a value of 13.68% of the average absolute bending moment value. Comparing the TDNN to the Elman-variant network, the Elman-variant network appears to slightly outperform the TDNN.

TUNING THE NETWORK PARAMETERS

Efforts were made to tune the network parameters. For the TDNN, we experimented with changing the tap delay value. Specifically, we experimented with values of 5, 10 and 15. We then experimented with changing the number of neurons in the hidden layer. We tried 40 neurons and 85 neurons. However, the original configuration of a tap delay value of 10, and 14 neurons in the hidden layer yielded the best results.

Similar tuning efforts were made to the Elman-variant network (ANN2). First, we experimented with changing the feedback parameter. Values of 0.70, 0.80, and 0.85 were tried. The number of neurons in the hidden layer was also changed to 40 neurons and to 14 neurons. The network with a feedback parameter of 0.80 and 85 neurons in the hidden layer yielded the best performance.

CONCLUSIONS

An instrumented section of rigid concrete pavements successfully provided continuous records of dowel strains as well as temperature gradients across the slab thickness. In this study, time history of 300 days was used to calculate straining actions of dowel bars in terms of bending moments. Dynamic ANNs were formulated for predicting the dowel bending moment as a function of the temperature gradient profile. Several topologies were examined, and their performance was compared on a test set. Dynamic ANN proved to be capable of capturing the relationship between the temperature gradient profile, and the dowel bending moment. While the performance of both the TDNN and Elman/Jordan type networks was quite good, the Elman-variant network appears to slightly outperform the TDNN. For the different Elman/Jordan network topologies examined in this study, the Elman-variant network, which fed the input to the context PE layer, appears to be the most appropriate topology for the problem at hand. For dynamic ANNs, a number of design parameters, such as the value of the tap delay for TDNNs, the value of the feedback parameter for Elman/Jordan networks, and the number of neurons in the hidden layer, could have an impact on the network performance. Given this, it is recommended that an analysis of the sensitivity of the network performance to the values of these parameters be conducted as a part of any ANN development effort, in an attempt to improve the network performance.

ACKNOWLEDGEMENT

The authors gratefully acknowledge the West Virginia Department of Transportation for sponsoring this research program, as well as for its continuous support and help through the various stages of construction.

REFERENCES

Hveem, F.N.,1949, "A Report of an Investigation to Determine Causes for Displacement and Faulting at Joints in Portland Cement Concrete Pavements on California Highways", Technical report, California Department of Transportation, CA.

Shoukry S. N., M Y. Riad and G. W. William, 2004, "Field Testing of Shok Bars", Final Report No. RP#170 Submitted to West Virginia Department of Transportation, Division of Highways.

Tabatabaie A.M. and E.J. Barenberg, 1978, "Finite Element Analysis of Jointed or Cracked Concrete Pavements", Proceeding of Transportation Research Board No 671.

Tayabji S.D. and B.E. Colley, 1983, "Improved Rigid Pavements Joints", Transportation Research Record 930, Transportation Research Board.

Titus-Glover L., E.B. Owusu-Antwi, T.Hoener, and M.I. Darter, 1998, "Design and Construction of PCC Pavements", Volume II: Design Features and Practices That Influence Performance of Pavements. Report No. FHWA-RD-98-127.

Van Breemen W., 1955, "Experimental Dowel Installations in New Jersey", Highway Research Board No. 34.

Vyce, J.M., 1987, "Performance of Load-Transfer Devices", Report No. 140, Engineering Research and Development Bureau, New York State Department of Transportation.

William, G.W., and S.N. Shoukry, 2001, "3D Finite Element Analysis of Temperature Induced Stresses in Dowel Jointed Concrete Pavements", The International Journal of Geomechanics, Vol. 1, No. (3), pp. 291-307.

Environmental Site Profiling: A Comparative Study

Yacoub Najjar, Professor
Kansas State University
2118 Fiedler Hall
Civil Engineering Department
Manhattan, KS 66506
USA
ea4146@ksu.edu

Sam Mryyan
Environmental Compliance Manager
Adjutant General's Department
2800 SW Topeka Blvd.
Topeka, KS 66611
USA
sam.mryyan@us.army.mil

ABSTRACT

Over the years, many methods have been developed to profile environmental contaminants in soil media and/or groundwater. These methods vary in their ability to make precise and accurate predictions. This paper highlights the differences between the following nine profiling methodologies: Inverse distance to a power, Kriging, Minimum Curvature, Modified Shepard's, Nearest Neighbor, Polynomial Regression, Radial Basis Function, Local Polynomial, and Artificial Neural Networks (ANNs). Because each method uses an individualized logic, the accuracy of the methods' predicted profiles is expected to vary. To illustrate this, a hypothetical data-rich contaminated site is used for this purpose. Accordingly, a small fraction of the available data (about 1%) is presented to each method for site profiling. A comparative study of the models' site profiling outcomes/predictions is then performed in order to assess the most accurate site profiling methodology. Overall, ANN-based profiling methodology outperformed the performance of all other profiling methodologies considered in this study.

INTRODUCTION

Environmental contaminants in geologic media such as soil and groundwater are of great concern in today's society. Millions of dollars are spent each year in efforts to clean up areas contaminated by pollutants from industrial and public waste such as solvents, fuels, and processing waste. Before cleanup efforts can begin, the area and extent of contamination must first be determined. This can be done by using one of several profiling methodologies. Typically, these methodologies take known data and use specific mathematical algorithms to predict the areas and levels of contamination. In this paper, the profiling performance of eight well-known profiling methodologies and Artificial Neural Networks (ANNs) are compared.

When trying to identify an area of contamination, the most accurate and precise means is to perform soil/groundwater sampling at regular designated intervals throughout the entire area in question. This, however, is not a practical method due to cost and time restraints. Instead, a limited number of samples are collected throughout the area in question. Many factors determine where and how samples are collected. Generally, sample locations are determined using the professional judgment of the site investigation team. Once known data is collected, one of several profiling methodologies can be used to predict areas of contamination. Eight of the most highly utilized methodologies (available in Surfer® 8.0 Software: http://www.goldensoftware.com/) for 2-D profiling are: Inverse distance to a power, Kriging, Minimum Curvature, Modified Shepard's, Nearest Neighbor, Polynomial Regression, Radial Basis Function, Local Polynomial. ANN-based methodology is investigated herein as an alternative efficient 2-D and 3-D profiling methodology. Although all profiling methodologies function in similar manner, some methodologies produce more accurate profiles than others.

In order to utilize any profiling methodology, a known set of data must be present. For the purpose of this paper, a hypothetical data-rich contaminated site scenario is utilized. Data was generated to compare 2-D pollution concentration profiles generated via different profiling methodologies.

CASE STUDY

Mathematical Equation

In order to determine the distribution of contaminates at the hypothetical site, the following mathematical equation (Equation 2) was developed in order to produce the pollutant concentration value at any given (x, y) location:

$$V = x^{0.5} + y^{0.6} + (x^{0.3} * y^{0.2}) + 2\sqrt[4]{(x*y)} + 3\ln\left(\frac{x^{1.1} + y^{1.5} + 20{,}000}{1000}\right) \qquad \textbf{(1)}$$

In Equation 1, V represents the contaminant concentration value. Note that, x and y coordinates used in this equation refer to the x and y distances (in feet) for the associated observation point measured from a reference point (i.e., x = 0 ft and y = 0 ft).

Databank

Two databases containing (x, y and V) values were generated for two 2-D cases at various locations across the site. The site size is 1000 ft in the X direction by 1000 ft in the Y direction. To achieve this objective, the hypothetical site was divided into two grid systems as follows:

1. 25-feet interval case: In this scenario, 25-ft interval (i.e., $\Delta x = \Delta y = 25$ ft) in both x (east) and y (north) directions was used to generate a total of 1,681 sampling points. The x, y coordinates and V values of selected 17 points (about 1% of the total sampling points) were provided for the eight profiling methodologies available in Surfer software. Each methodology was then used to predict the corresponding contamination value (V) for the 1,681 designated x and y coordinates representing the site. The resulting data banks were processed to construct 8 contamination distribution contour maps as well as to calculate the corresponding Root Mean Squared Error (RMSE) value (using Equation 2) associated with each profiling methodology.
2. 10-feet interval case: Utilizing a 10-feet interval (i.e., $\Delta x = \Delta y = 10$ ft) for both x (east) and y (north) directions it was possible to generate a total of 10,201 sampling points for the 1000 ft x 1000 ft site. Similar to the 25-ft interval case, x and y coordinates and V values of selected 103 points (about 1% of the 10,201 total sampling points) depicted in Figure 1 were provided for the eight profiling methodologies available in the Surfer software. Each methodology was then used to predict the corresponding contamination value (V) for all 10,201 designated x and y coordinates representing the site. The resulting eight data banks were processed to construct 8 contamination distribution contour maps and to calculate (using Equation 2) the corresponding RMSE value associated with each methodology.

ANN Model Development

Unlike Surfer® 8.0 profiling methodologies, ANN-based profiling model require the user to train or educate the network about the process it is to model. To train the network, a known set of input data along with the desired outcome is used [Dowla and Rogers (1995), Mryyan and Najjar (2005), Itani and Najjar (2000)]. The BPANN methodology/program [example, Najjar(1999)] using the supervised training approach can be used to train the desired ANN models to produce output values that are as close to the real values as possible via repeated modifications of the network's connection weights. This process, typically, continues until the error at the output layer is minimized. Once this training process is completed, the developed model can then be used for prediction tasks.

Neural Networks can reach a least-error structure by training using examples related to the problem under consideration. A least-error structure is the one responsible for producing outputs that are very close or equal to the real desired (target) values. Reasonable training input and output vectors should cover a wide range of the sampling domain. Deriving an appropriate and representative mapping between input and output vectors reflects the effectiveness of artificial neural networks. For proper modeling, a network should at least pass through two stages, namely; training and testing stages (Najjar & Basheer,1996). In the training stage, selected data with their input and output values are introduced to a network (having a certain number of hidden nodes and layers) so that the network trains itself to produce output values that are as close to the target values as possible. The training is achieved by modifying the values of the connection weights. The network stops learning when the weight adjustment process produces no improvement in the output values. The same network should be tested on data never used in training to verify its generalization capabilities. The procedure of training and testing should be repeated for various networks having different numbers of hidden layers and/or hidden nodes.

When developing any ANN model, it is important to determine what input and output values will be used. For the hypothetical data-rich contaminated site case considered herein, the x and y coordinates are used as the only input values to the model. The pollutant concentration value (V) was used as the output for their associated network model. In this case, x and y coordinates refer to the x and y distances (in feet), for the associated observation point, measured from a reference point (i.e., $x = 0$ ft, $y = 0$ ft).
For the 25-ft interval case (i.e., $\Delta x = \Delta y = 25$ ft), a network model was developed by using the same 17 points used by the eight Surfer Software methodologies. For the ANN case, 12 data sets were used for training and the remaining 5 data sets were used for testing purposes. The best performing network (BPN) was determined by carrying out a number of adaptive training and online testing trials [as indicated in Najjar(1999)] in order to arrive at the least error on the testing data sets. According to Najjar (1999), overall BPN is defined as the one yielding the least error [in terms of Average Squared Error (ASE) value] on the testing data sets from among all evaluated trial networks. In this case, overall BPN was achieved at ASE value of 0.010856 and a structure topology noted as 2-3-1 (i.e., 2 inputs representing x and y coordinates; 3 hidden nodes and one output denoting the associated value of the V variable). Once this network was established, it was then used to predict the V values at all 1,681 location points for the site. The predicted values were used to construct contamination distribution contour map and to calculate the corresponding RMSE value for this network model.

In the 10-ft interval case (i.e., $\Delta x= \Delta y=10$ ft), a network model was developed by using about 1% of the total 10,201 data points (i.e., the same 103 data points used by the Surfer Software methodologies). In this ANN development case, 75 data sets were used for training and the remaining 28 data sets were used for the online testing purposes

(Najjar, 1999). Figure 1 depicts the locations of the selected training and testing data points. Based on various ANN work by Najjar and his co-workers [Ali and Najjar (1998), Huang et al. (2006), Mandavilli et al. (2005), Mryyan and Najjar (2006), Najjar and Felker (2003)], it was pointed out that it is highly imperative that the training data sets contain all data sets that have the extreme attributes in terms of locations and values. Accordingly, the developed ANN model will always operate in an interpolation mode instead of an extrapolation mode. ANN-based prediction models are excellent when used in interpolation tasks, while may be unreliable when used in extrapolation tasks. Therefore, it is very important to appropriately select the distribution of the training and testing data sets. Following similar strategy, as the one used for the 25-ft interval case, the overall BPN was achieved at an ASE value (on the testing data sets) of 0.000228 and a 2-2-1 network topology structure. Once the 2-2-1 profiling network was established, it was then used to predict the V values at all 10,201 location points for the site. The predicted values were used to construct contamination distribution contour map and to calculate the corresponding RMSE value for this 10-ft interval case.

By comparing ASE values for the 10-ft and 25-ft interval cases, it can be observed that the ASE value for the 10-ft case was reduced by about 47 folds (i.e., 0.010856/0.000228) for the corresponding 6 fold increase in data richness (i.e., 103/17). This noted behavior is expected and logical. As more data become available, the profiling network should be able to characterize the site more accurately. Therefore, the more data is available the more accurate is the developed profiling network. Moreover, it can be observed that the 10-ft interval network only needed 2 hidden nodes to efficiently characterize the site in comparison with the 3 hidden nodes needed for the 25-ft interval network.

In order to compare (rank) the prediction accuracy of the profiling methodologies used herein, the following Root Mean Squared Error (RMSE) accuracy measure was used:

$$\text{Root Mean Squared Error (RMSE)}: = \frac{\sqrt{\sum_{1}^{n}(y'-y)^2}}{n} \qquad \textbf{(2)}$$

n = number of data sets used.
y′= the output generated by the model for the V variable
y= the actual value of the V variable
Accordingly, the best performing profiling methodology is the one yielding the least RMSE value.

RESULTS AND DISCUSSION

In order to compare the performance of all methodologies used herein RMSE values were utilized. RMSE values obtained for both 2-D profiling cases and the nine profiling methodologies (including ANN method) are listed in Table 1. By examining RMSE values listed in this table, it can be observed that for the 25-ft. interval case, all nine methodologies attain high RMSE values. The model achieving the least RMSE value is the ANN-based model. It attains about 19.17% error rate. The second most accurate methodology is the Local Polynomial with a 19.3% error rate. When compared to ANN performance, this represents less than 1% difference in prediction accuracy rate. The profiling methodology that produced the least accurate profile is the Inverse Distance to a Power. It has an RMSE value of about 42.4%. This represents more than double the RMSE value attained by the 2-3-1 ANN model.

For the 10-ft. interval case, all nine methodologies attained lower RMSE values than those obtained for the 25-ft interval case. This noted behavior is logical and consistent with our intuition. As more data become available, models will become more accurate. Again, the model with the least RMSE value is the ANN-based model. Its error rate is about 3.7%. The second most accurate methodology is the Radial Basis Function with about 4.8% error rate. When compared to ANN performance, this represents about 30% difference in the prediction accuracy rate. The profiling methodology that produced the least accurate profile is again the Inverse Distance to a Power, with an RMSE value of about 10.4%. This represents about 2.8 times the RMSE value attained via the ANN model. The only consistent thing in the RMSE comparison listed in Table 1 is that ANN-based profiling methodology is ranked best and the Inverse Distance to a Power methodology is ranked worst. The other seven methods seem to vary in terms of their ranking. Therefore, in order to assure that we are using the best profiling methodology, it is recommend to always use the ANN-based profiling methodology.

CONCLUSIONS

The use of ANN-based methodology for contaminates profiling, demonstrated in this study, provided the most reliable predictions about the location and extent of contamination for the hypothetical site. ANN-based models proved to yield the lowest RMSE values in the comparisons cases. ANN-based profiling models also produced the best contaminant distribution contour maps for profiling cases. This study clearly demonstrated that ANN-based methodology, when properly used, has the potential to provide the most accurate predictions and site profiling contour maps for a contaminated site.

REFERENCES

Ali, Hossam and Najjar, Yacoub (1998) "Neuronet-Based Approach for Assessing the Liquefaction Potential of Soils," Transportation Research Records, No. 1633, pp. 3-8.

Dowla, U. & Rogers, L. (1995) “Solving problems in Environmental Engineering and Geosciences with Artificial Neural Networks,” Cambridgem, MA: MIT Press.

Huang, C., Najjar, Y. and Romanoschi, S. (2006) “Characterizing the Fatigue Life of Asphalt Concrete,” Intelligent Engineering Systems through Artificial Neural Networks, Vol. 16, pp. 369-374.

Itani, Omar and Najjar, Yacoub (2000) "3-D modeling of Spatial Properties Via Artificial Neural Networks," Transportation Research Records, No. 1709, pp. 50-59.

Mandavilli, S., Najjar, Y. and Abu-Lebdeh, G. (2005) “Modeling Crash Rates for the Kansas Rural Expressway Network,” Intelligent Engineering Systems through Artificial Neural Networks, Vol. 15, pp. 721-730.

Mryyan, S.A. and Najjar, Y. M. (2006) “Using Neural Network to Investigate the Environmental Impact of an Abandon Landfill,” World Water and Environmental Resources Congress 2006, pp. 250-262. American Society of Civil Engineering.

Mryyan, S.A. and Najjar, Y. M. (2005) “Investigating the Environmental Impact of an Abandon Landfill,” Intelligent Engineering system through Artificial Neural Networks, Volume 15. pp. 751-761.

Najjar, Y. M. (1999) “Quick Manual for TR-SEQ1 ANN Training Program,” Department of Civil Engineering, Kansas State University, Manhattan, Kansas, 7 pages.

Najjar, Y. M. and Basheer, I. A. (1996) "A Neural Network Approach for Site Characterization and Uncertainty Prediction,” Uncertainty in the Geological Environment: From Theory to Practice, ASCE GSP No. 58, Vol. 1, pp. 134-148.

Najjar, Yacoub and Felker, Victoria (2003) “Modeling the Time-Dependent Roughness Performance of Kansas PCC Pavements,” Intelligent Engineering Systems through Artificial Neural Networks, Vol. 13, pp. 883-888.

Surfer User's Guide (2004). Golden Software, Inc.

Table 1: Profiling Methods and Their Corresponding Root Mean Squared Error (RMSE) Values for the 10 and 25- ft Interval Cases

	Method Used	RMSE (%) for 10-ft Interval case	RMS (%) for 25-ft Interval case
1	ANN	3.716%	19.169%
2	Radial Basis	4.795%	38.953%
3	Kriging	4.994%	23.663%
4	Modified Shepard's	4.999%	19.779%
5	Local Polynomial	5.153%	19.297%
6	Minimum Curvature	5.231%	20.396%
7	Nearest Neighbor	8.022%	38.953%
8	Polynomial Regression	8.710%	26.419%
9	Inverse Distance to a Power	10.427%	42.385%

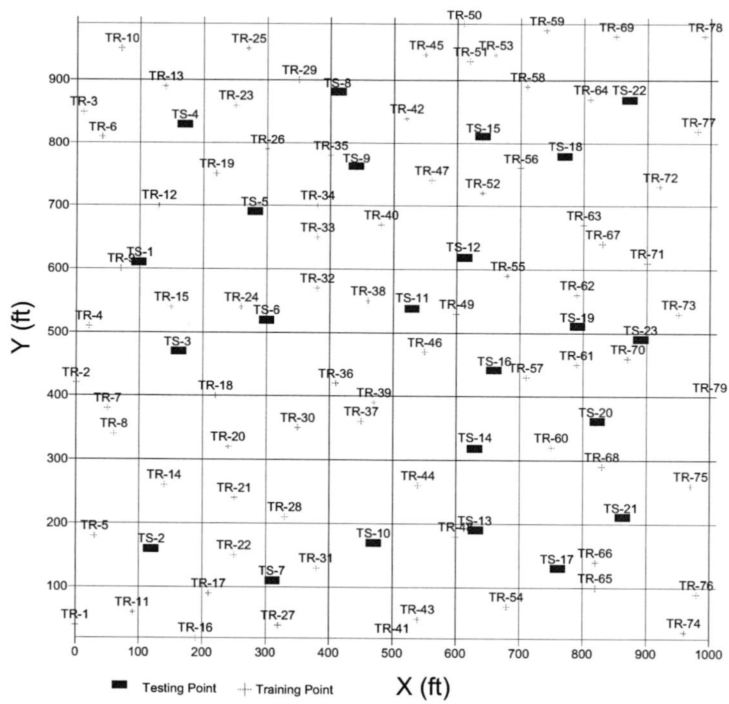

Figure 1 Location of testing and training points for the 10-ft interval case

UNSUPERVISED NEURAL NETWORKS FOR SITE PARAMETERS CHARACTERIZATION OF RECENT IRAN EARTHQUAKES

MOHSEN TEHRANIZADEH
Professor, Department of Civil Engineering, Amirkabir University of Technology

MOHAMMAD SAFI
Assistant Professor, Department of Civil Engineering, Power & Water University of Technology

ABSTRACT

Artificial neural networks have been used for determination of site parameters in the region lacking site investigation reports. This method benefits advantages of unsupervised neural network in classification problems. Using this method the ranges of shear wave velocity for determination of each soil type may be estimated for a large data base. This site classification has been performed using linear response spectra of the earthquake records. The local site conditions are an important factor in the recorded waveform of earthquake ground motions. This also becomes important in ground motion analysis and in earthquake resistant designs. Different site conditions can induce amplifications of different period ranges in the response spectra. The method has also been applied to 2003 Bam Earthquake in south of Iran. The effects of earthquake magnitude and duration on the classification results have also been discussed.

INTRODUCTION

The local site conditions are an important factor in the recorded waveform of earthquake ground motions. The objective of site classification is to classify a group of sites into several classes so that the conditions within the same site class are similar. A building may be more severely damaged if the natural periods of the structure happen to be close to the amplified periods of the ground. Therefore, the local site conditions become important in ground motion analysis and in earthquake resistant designs. Because quantitative subsurface soil properties are not commonly available for every site, the use of surface geology becomes important in understanding the subsurface geologic conditions. Empirical relationships between the surface geology and the subsurface shear-wave velocity have been developed and are used in ground motion amplification predictions (Joyner and Fumal, 1985; Boore et al., 1993; Borcherdt, 1994; Park and Elrick, 1998). Different researchers have applied the artificial intelligence phenomena in order to construct artificial acceleration time histories, estimating the fault motions and process earthquake data (Dai and MacBeth (1995), Savaidis et al. (2000)). To the authors best knowledge it is the first time that the neural networks are being used for site classification and determination of soil type and estimation of shear wave velocity in soil media.

In the first part of the paper the theory of self-organizing networks is briefly reviewed. Then Recent Iranian ground motion records are classified through a multi-step procedure. The results obtained from classification method have been verified comparing to the available experimental investigations (Zare (1999), Mirzaee and Ramazi (1999), Hamedi (2001)).

NEURAL NETWORK FOR CHARACTERIZATION OF SITE PARAMETERS

The most appropriate networks for classification problems are self-organizing networks that benefit the unsupervised learning method. In unsupervised learning, the designer does not explicitly specify mappings to be learned. The training sets are no longer pairs of inputs with desired output like back propagation networks, but simple patterns of data. Typically, such networks then cluster the patterns in certain ways. The designer in turn, can interpret the clusters as categories. A simple form of non-supervised networks is competitive schemes, which are a kind of self-organizing maps. The principal goal of the self-organizing map is to transform an incoming signal pattern of arbitrary dimension into one or two-dimensional discrete map, and to perform this transformation adaptively in a topologically ordered fashion (Boore et al. (1997)). In such network each neuron is fully connected to all the source nodes in the input layer. The algorithm of the self-organizing map proceeds first by initialization of the synaptic weights in the network, which can be done by assigning them small values, picked from random number generators. The most anonymous networks that can be formed in this manner are the Competitive Neural Networks (CNN).

Consider the architecture for a competitive network shown in fig. 1. The purpose of such a network is to categorize an input vector. A node in the output layer gives the category. The point is that the network has to cluster or categorize inputs such that similar inputs lead to the same categorization (Anderson and Rosenfeld (1988)). In the simplest case, the input nodes are connected to the output layer with excitatory connections or weights $w_{ij} \geq 0$. Assuming a network with inputs x_j, the winner is normally the unit with the largest input h_i defined as (Bishop (1995)),

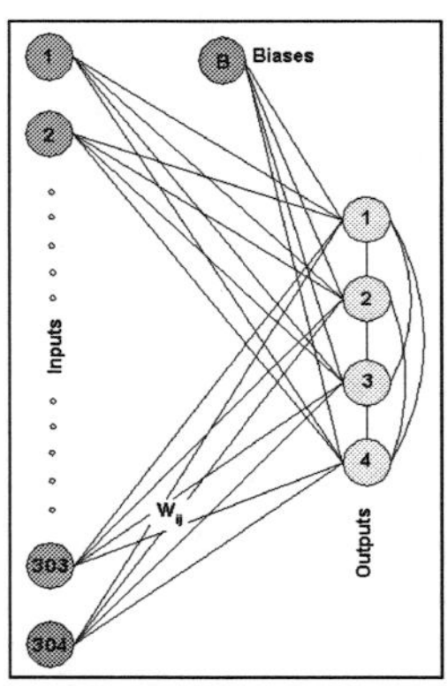

Figure 1– Neural Network Architecture Used For Data Classification

$$h_i = \sum_j w_{ij} . x_j = \boldsymbol{w}_i . \boldsymbol{x} \qquad (1)$$

If i^* is the winning unit, for any value of i we must have,

$$\boldsymbol{w}_{i^*} . \boldsymbol{x} \geq \boldsymbol{w}_i . \boldsymbol{x} \qquad (2)$$

If the weights for each output unit are normalized, $|w_i| = 1$, i.e. if the sum of the weights leading to output unit o_i sum to 1, for all i, this is equivalent to,

$$|\boldsymbol{w}_{i^*} - \boldsymbol{x}| \leq |\boldsymbol{w}_i - \boldsymbol{x}| \qquad (3)$$

In brief, the winner is the unit for which the normalized weight vector w is most similar to the input vector x. How the winner is determined is actually irrelevant. It can either be found by following the network dynamics, simply using the propagation rule for the activation. After a number of steps only one output unit will be active. Alternatively, simply h_i is calculated for all output units and the maximum is taken. If the network is winner-take-all, the node with the largest h_i is certain to win.

NUMERICAL STUDIES

For the purpose of efficiency analysis, the horizontal components of ground motion records of Iran, which have been selected base on the criteria in section two, were used for the classification process by the network. 304 records were finally selected and the acceleration response spectra for each were calculated. These vectors were classified in four categories using the unsupervised competitive neural network. The obtained network architecture has been shown in fig. 1. This network has 304 input and four output neurons indicating the soil types. The network outputs for all categories have been shown in fig. 2 for the learning ratio of 0.10.

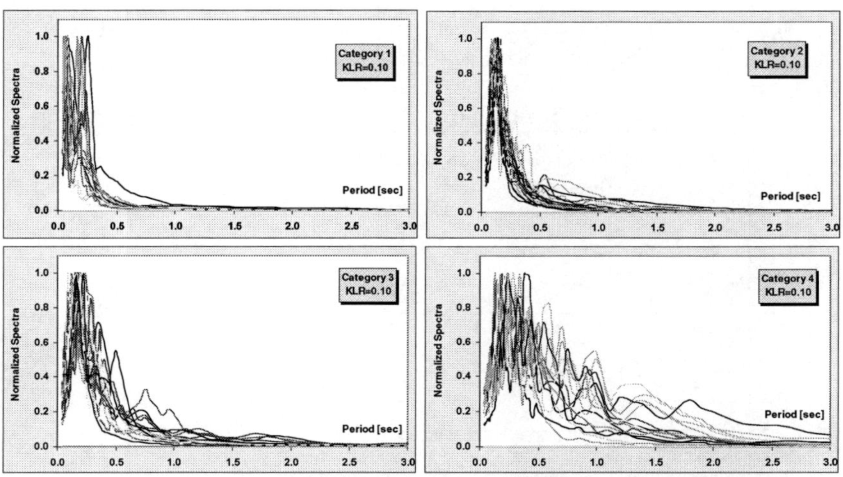

Figure 2- Classification results of S_a for learning rate of 0.10

Parametric studies were also performed for network parameters and it was seen that the most sensitive parameter was the learning rate. The increase in the network learning rate resulted in more errors in the classification process compared to the existing methods. After classification of spectra into four categories, those sites with known shear wave velocities (the sites where experimental shear wave velocities are available) have been used to obtain the ranges of shear wave velocities corresponding to each category. Artificial neural networks have been used to estimate the soil type of each station in recent Iran earthquakes. The results have been shown in figure 3 based on the Iranian seismic code and UBC code classification systems and for Bam earthquake in table 1. It can be seen that the correlation of the results for current and previous records is very good which demonstrate the stability of the method. The adaptation of the results with experimental results also presents the capability and efficiency of the proposed method for site characterization.

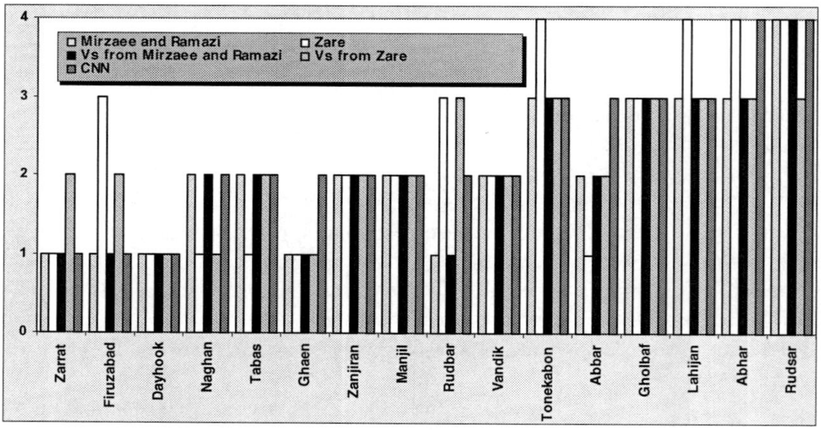

Figure 3- Comparison of Site Characterization Results for Recent Earthquake with Experimental Investigations

Table 1- Results Comparison for 2003 Bam Earthquake

No.	Station Name	Soil Type Iran 2800 (UBC 97)		
		Site Investigation	CNN Historical Data	CNN Bam Earthquake
1	**Zarand**	1 (B)	1 (B)	1 (B)
2	**Golbaf**	3 (D)	3 (D)	3 (D)
3	**Joshan**	1 (B)	1 (B)	2 (C)
4	**Kerman1**	2 (C)	2 (C)	2 (C)
5	**Mahan**	2 (C)	1 (B)	2 (C)
6	**Sirch**	1 (B)	1 (B)	1 (B)
7	**Bam**	2 (C)	1 (B)	2 (C)
8	**Shahdad**	1 (B)	1 (B)	1 (B)
9	**Kerman2**	2 (C)	2 (C)	2 (C)
10	**Andoohjerd**	1 (B)	1 (B)	2 (C)

The effects of earthquake magnitude and also effective duration of the records has also been investigated. Figure 4 and 5 showcase results of these sensitivity analyses. It can be seen that earthquake magnitude may affect the classification results if not appropriately considered. Table 2 also presents the Overall Assessment of Effect of Earthquake Parameters on Classification Results.

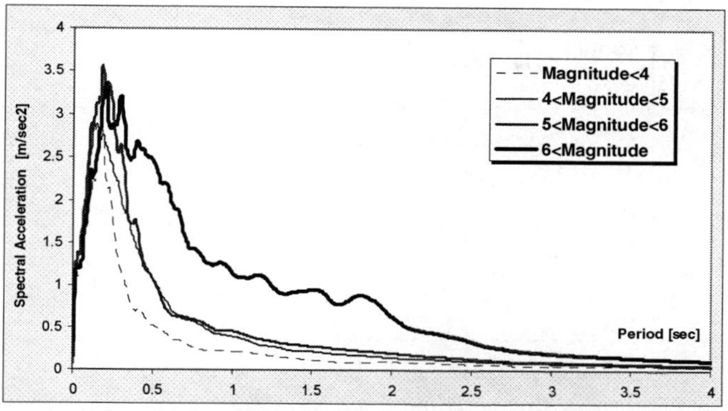

Figure 4- Effect of Earthquake Magnitude on Classification Results

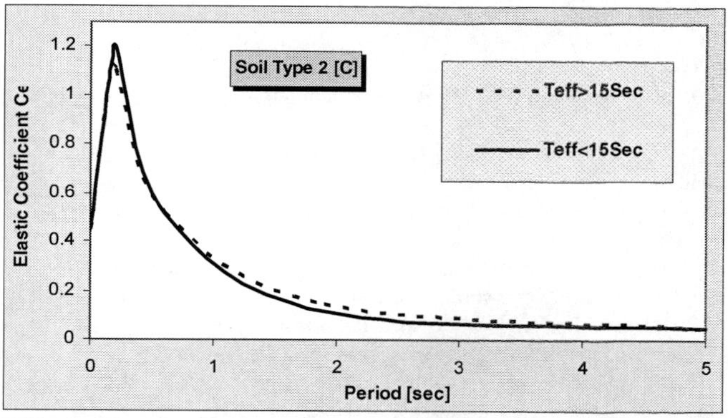

Figure 5- Effect of Earthquake effective duration on Classification Results

COMMENTS AND CONCLUSIONS

A new method for the classification of earthquake records was presented using a committee of competitive neural networks which uses unsupervised learning procedure and back error propagation neural networks. The classification was based on earthquake acceleration and displacement spectra rather than earthquake acceleration time histories. The results were used to determine the soil characteristics and range of variation of shear wave propagation velocity in the soil media. As this method is based on spectrum shape and soil characteristics, it should be considered together with the existing theoretical and experimental methods. The method was used for soil characterization of earthquakes in Iran as case study.

Table 2- Overall Assessment of Effect of Earthquake Parameters on Classification Results

Soil Type	Determination Method	PGA m/sec2	ED km	M Richter	T_t sec	T_{eff} sec
I	Experimental	1.54	36.84	4. 8	18.46	5.51
	Tentative	1.67	37.65	5.3	23.43	7.86
	CNN	1.63	32.70	4.9	22.18	6.14
II	Experimental	2.14	26.36	4.8	20.27	6.04
	Tentative	2.02	29.69	5.6	18.09	5.73
	CNN	1.00	42.43	4.7	16.67	7.06
III	Experimental	1.20	24.60	4.4	22.37	10.26
	Tentative	1.15	50.48	4. 6	28.53	15.06
	CNN	2.38	49.00	5.6	25.41	12.01
IV	Experimental					
	Tentative	0.98	65.91	6.4	34.55	18.31
	CNN	1.66	49.71	6.4	25.95	14.21

REFERENCES

Anderson, J.A, and Rosenfeld (ed.) (1988), Nero-computing: Foundations of research. Cambridge, MIT Press.

Bishop C.M (1995), Neural networks for pattern recognition, Oxford University Press, Oxford, U.K..

Boore, D. M., W. B. Joyner, and T. E. Fumal (1993). Estimation of response spectra and peak accelerations from Western North American earthquakes: an interim report, U.S. Geol. Surv. Open-File Rept. 93-509.

Dai, H., and MacBeth, C. (1995), Automatic picking of seismic arrivals in local earthquake data using an artificial neural network, Geophysical International Journal, No. 120, p758-774.

Hamedi F. (2001), Evaluation of earthquake records in Iran in order to determine design spectra, PhD Thesis, AmirKabir University of Technology, Tehran, Iran.

Haykin S. (1999), Neural networks; a comprehensive foundation, 2nd edition, Prentice Hall Inc., pp. 156-252.

Joyner, W. B. and T. E. Fumal (1985). Predictive mapping of earthquake ground motion, in Evaluating Earthquake Hazards in the Los Angeles Region – An Earth-Science Perspective, J. E. Ziony (Editor), U.S. Geol. Surv. Profess. Pap. 1360, 203-220.

Ministry of Building and Urban Planning, Standard 2800 (IS2800) Seismic Design of Building Structures, Tehran, Iran, 2000.

Mirzaee H. and Ramazi H (1999), Geological site investigation by geophysical methods for some regions in Iran, Technical report no. K-324 (In Farsi), BHRC publication.

Park, S. and S. Elrick (1998). Predictions of shear-wave velocities in southern California using surface geology, Bull. Seism. Soc. Am. 88, 677-685.

Savaidis A., Papazachos C., Soupios P., Grammalidis N., Saragiotis C., Hadjileontiadis L. and Panas, S.M. (2000), Rapid determination of earthquake parameters using a phase auto-detect system, Proceedings of European Seismological Committee, Lisbon, Portugal.

Uniform Building Code, UBC (1997), Minimum Design Loads for Buildings and Other Structures.

Zare M. (1999), Contribution a letude des mouvements forts eu Iran du catalogue aux loix dattenuation, PhD thesis, Universite Joseph Fourier, Grenoble, p237.

THREE DIMENSIONAL FINITE ELEMENT MODELING OF INTEGRAL BRIDGES SUBJECTED TO THERMAL LOADING

DUNJA PERIĆ
Department of Civil Engineering
Kansas State University
Manhattan, Kansas, USA

BHAVIK R. SHAH
Kansas State University
Manhattan, Kansas, USA

ASAD ESMAEILY
Kansas State University
Manhattan, Kansas, USA

ABSTRACT
Results of a full 3D finite element analysis of an integral abutment bridge subjected to a combined thermal and gravity loading are presented. The three-span, 45.7 m long integral bridge with steel deck over the Nashua River in Fitchburg, Massachusetts was slightly modified for the analysis and subjected to the temperature increase from -17.8° C to 37.8°C in the presence of a loose sand behind the abutment and dense sand behind the steel HP piles. The bridge responded with a restrained abutment motion, which is predominately translational, thus inducing significant bending stresses in piles. In the present analysis scenario the stresses in piles are larger than the compressive axial stresses in steel girders that develop due to a restrained bridge expansion.

INTRODUCTION

Due to the absence of expansion joints thermal loads in integral abutment bridges (IAB) are accommodated by abutments and piles, thus engaging the surrounding soil in a significant amount of a soil structure interaction. IABs have gained an increased popularity over their non-integral counterparts due to their multiple advantages, the most important of which is the increased durability that is achieved at lower costs. With increasingly aging infrastructure that requires repair and a replacement of many bridges as well as construction of new longer bridges there is a need to gain a deeper understanding of the soil structure interaction IAB. To this end we performed a full three-dimensional finite element analysis of an integral bridge subjected to a combined thermal and gravity loading, which accounts for the soil structure interaction. Results for a single loading scenario corresponding to a selected magnitude of thermal load and soil properties, which are described subsequently, are presented herein.

FINITE ELEMENT MODEL

A full 3D FE finite element model of the "Bemis Road Bridge: F-4-20" over the Nashua River in Fitchburg, Massachusetts (Ting and Faraji, 1998), was selected as a representative example of a typical integral bridge in Kansas. A slight modification of the bridge included adding a hinge connection between the superstructure and piers. The bridge was modified to prevent a moment transfer from the superstructure to the piers as requested by Kansas Department of Transportation. The schematic of the finite element model of the bridge including the soil substituting springs is shown in Fig. 1. Only half of the bridge is modeled due to the symmetry of its geometry and loading.

The bridge superstructure comprises a 45.7 m long three span integral bridge, with 18.3 m long central span, and two 13. 7 m long side spans. The bridge is 16.5 m wide. Its superstructure consists of: 1) 0.22 m thick, 45.7 m long and 16.5 m wide concrete slab, 2) seven W 36 x 135 (950 mm height, 1968 N/m weight) steel girders, which are spaced 2.7 m apart, and 3) seven transverse W 36 x 135 steel beams, 16.5 m long. On both sides of the central span three individual 0.9 m x 0.9 m x 1.1 m pier caps are used. The superstructure is connected to the piers by hinge connections. Concrete abutments are 2.44 m high, 0.76 m wide and 21 m long. Seven HP 12 x 74 (308 mm height, 1080 N/m weight) 8.84 m long piles are rigidly fixed into each abutment, allowing a full moment transfer.

Soil is modeled by nonlinear springs, which are attached to the nodes located on abutments and piles (Fig. 1). An iterative equivalent linear approach is adopted for determining the stiffnesses of these springs. More details about the soil model and properties are provided later. Finite element code ABAQUS Standard, release 6.5-1 was used for the analyses. The finite element model of the bridge consists of a total 191894 eight-node-coupled temperature-displacement elements (C3D8T), 277530 nodes, 12 connector elements (CONN3D2) modeling hinges, and 546 linear spring elements.

Thermal stresses are induced in the bridge due to the continuity of the deck and the presence of the soil behind abutments and piles, which prevent a free expansion. In the present analysis the temperature of the superstructure was increased by 55.6° C (from -17.8° C to 37.8°C), while the temperature of the portion of the substructure that is located under ground was held constant at 10° C. Consequently, there is a sharp temperature gradient across portions of abutments and piers. Loading was applied in two stages. First, the gravity acceleration of 9.81 m/s^2 was applied to the bridge. In the second stage the temperature increase was applied while the gravity remained in effect.

For the purpose of the present analysis concrete and steel are modeled as linear elastic materials. Their relevant material properties including mass densities, Young's moduli, Poisson's ratios, and coefficients of thermal expansion are given in Table 1.

In the analysis the action of the soil on the abutment and piles is modeled by using linear springs. The non-linearity of the soil is modeled by using an equivalent iterative linear approach. In this approach the abutment top displacement from a previous iteration is used to determine the stiffnesses of the springs for the current iteration. Similarly, the current stiffnesses of the springs attached to the piles are determined based on the pile top displacement from a previous iteration. Both types of iterations are continued until a prescribed convergence criterion is satisfied. The convergence criterion is met when the relative error becomes equal or smaller than 1%. Fig. 3 shows that the abutment rotation and translation exhibit the exponential convergence rate. No soil substituting springs were attached to the bridge in the very first iteration.

Design guidelines for retaining walls proposed by NCHRP (1991) were used to determine stiffnesses of the springs attached to the abutment. These guidelines contain graphical representation of the relationship between the coefficient of lateral earth pressure and horizontal displacement at the top of abutment for loose, medium dense and dense sands having friction angles of 30, 37 and 45 degrees respectively. Herein we present the results for the case in which a loose sand is present behind the abutment and dense sand is adjacent to the piles. A typical loose sand is represented by an actual dry unit weight of 17.7 kN/m^3 and maximum dry unit weight of 19.6 kN/m^3, thus corresponding to the relative compaction of 90%. According to Lee and Singh (1971) this relative compaction corresponds to relative density of 50%.

In accordance with the recommendations of Prakash and Kumar (1996) the stiffnesses of springs attached to piles vary linearly with depth. In addition, the stiffness of the spring attached at one meter depth below the pile head is a function of the shear strain in sand, which is given by Equation (1). The equation was proposed by Mwindo (1992) based on the analysis of 14 full scale laterally loaded pile tests.

$$\frac{k_h}{k_{h,\max}} = a\gamma^{-b} \qquad (1)$$

where k_h is the stiffness of the soil substituting spring at a given strain level, $k_{h,max}$ is the value of k_h at shear strain of 0.002 in sand, γ is the actual shear strain in sand, and a and b are empirical coefficients. Prakash and Kumar (1996) proposed to calculate the actual strain level (γ) as:

$$\gamma = \frac{1+\nu}{2.5B} y \qquad (2)$$

where ν is Poisson's ratio, which has a value of 0.3, B is the width of the pile, y is the lateral displacement of the pile head. Thus, Equations (1) and (2) provide the basis for the calculation of the stiffnesses of the springs adjacent to piles. Moreover, it is seen that the stiffness depends exclusively on the lateral displacement of the pile head, whereby $k_{h,max}$ is the input material parameter. In the present study its value is 70 MPa, which according to Prakash and Kumar (1996) represents medium dense to dense sand with the ground water table depth of 3 to 4.5 m. The remaining input parameters include the coefficients a, and b from Equation (1) whose values were selected to be equal to 0.05 and 0.5 as suggested by Prakash and Kumar (1996) for HP steel piles.

RESULTS AND DISCUSSION

Fig. 2 depicts the longitudinal displacement at the end of the bridge, along its centerline, from the abutment top to the pile bottom. A sudden change in the slope of this curve occurs at the location of the fixed connection between the abutment and the pile top. Fig. 2 indicates that the total abutment top displacement is equal to 1.35 cm, out of which the translational displacement is equal to 1.07 cm, while the rotational displacement is equal to 0.28 cm. Thus, the translation of the abutment is dominant over its rotation, whereby the later is equal to 0.066 degrees. A similar behavior was observed by monitoring a 66 m long instrumented pre-stressed concrete bridge near Rochester Minnesota. Lawver et. al (2000) reported that they measured a total abutment movement of 18 mm and the rotation of only 0.06 degrees. The response curve shown in Fig. 3 also demonstrates that in this case the abutment does not undergo only rigid body motion but also some bending. This again, can be attributed to the loose sand, which simply does not have enough stiffness to substantially suppress the bending of the abutment in the presence of an internal temperature gradient.

The dominant translational movement of the abutment, which occurs in loose sand, is likely to produce significant bending stresses in piles. A maximum bending stress is most likely to occur at the pile head and in this case it is equal to 326 MPa (Fig. 3), which is below the yield stress for most steels. However, it is important to note that piles in integral bridges are subjected to a cyclic loading due to daily and seasonal temperature changes, which requires that this stress be considered from the viewpoint of a fatigue.

In addition, the soil behind the abutment restricts a free expansion of the bridge, thus giving a rise to thermally induced axial stresses in the girders. The larger the resistance to the free expansion, the larger the axial stresses are. Thus, smaller axial stresses in girders will be produced by loose rather than by dense sand adjacent to the abutment. In the present case, maximum axial stress occurs in the bottom flange in the vicinity of the abutment and it is equal to 116 MPa, which is below the yield stress, but since it is a compressive stress it may be of concern for the assessment of buckling resistance.

CONCLUSIONS

A full 3D finite element analysis has been performed for a three-span, 45.7 m long integral abutment bridge whose superstructure is subjected to a combined thermal and gravity loading. The temperature of the superstructure was increased from -17.8° C to 37.8°C in the presence of a loose sand behind the abutment and dense sand behind the piles. The entire bridge-soil system was modeled with eight node brick elements, connector elements that model the hinges, and linear spring elements that represent the action of the soil on the bridge. The effect of a non-linear soil response has been modeled by using an equivalent iterative linear approach.

The results that were obtained for this loading scenario indicate that the surrounding soil has a significant effect on the response of the bridge. Presence of the loose sand behind abutment produces a dominant translational motion of the abutment in comparison with the rotation. This in turn causes a substantial pile head displacement, bending moment and axial stress. With an increased bridge length this stress may indeed reach the yield stress. While the presence of the loose sand behind the abutment still suppresses a free expansion of the bridge the resulting compressive stresses in girders are likely to be smaller than in the case of a dense soil. Both, the axial stresses in piles and compressive stresses in girders are the consequence of a significant soil structure interaction and they require considerations of safety against yielding of piles and buckling of girders in IAB.

The analysis enabled better understanding of the soil structure interaction induced by thermal expansion of IAB. In particular, the mechanism of a thermal load transfer to superstructure and substructure has been elucidated.

ACKNOWLEDGMENTS

All authors wish to acknowledge the support for this research, which was provided by Kansas Department of Transportation.

REFERENCES

Lawver, A., French, C., and Shield, C.K., 2000, "Field Performance of Integral Abutment Bridge", *Transportation Research Record*, Vol. 1740, pp. 108-117.

Lee, K.L. and Singh, A., 1971. "Relative Density and Relative Compaction", *ASCE Journal of the Soil Mechanics and Foundations Division*, Vol. 96, No. SM1, pp. 73-110.

Mwindo, J.M., 1992, "Strain dependent soil modulus of horizontal subgrade reaction", M.S. thesis, University of Missouri, Rolla, Missouri.

National Cooperative Highway Research Program, 1991, "Manuals for design of bridge foundations". Barker, R.M., Duncan, J.M., Rojiani, K.B., Ooi, P.S.K., Tan, C.K., and Kim, S.G., eds. Rep 343, *Transportation Research Board*, Washington, D.C.

Prakash, S. and Kumar, S., 1996, "Nonlinear lateral pile deflection prediction in sands", *ASCE Journal of Geotechnical Engineering,* Vol. 122 (2), pp. 130-138.

Ting, J.M., Faraji, S., 1998, "Streamlines analysis and design of Integral Abutment Bridges", Report UMTC 97-13, University of Massachusetts Transportation Center, Amherst, MA.

Table 1. Material properties

	Concrete	Steel
Coefficient of thermal expansion (per °C)	1.108 x 10^{-5}	1.2 x 10^{-5}
Young's Modulus (MPa)	30.015 x 10^{3}	207 x 10^{3}
Poisson's ratio	0.3	0.3
Thermal conductivity (W/m°C)	3.125	52.083
Mass Density (kg/m^3)	2400	7800

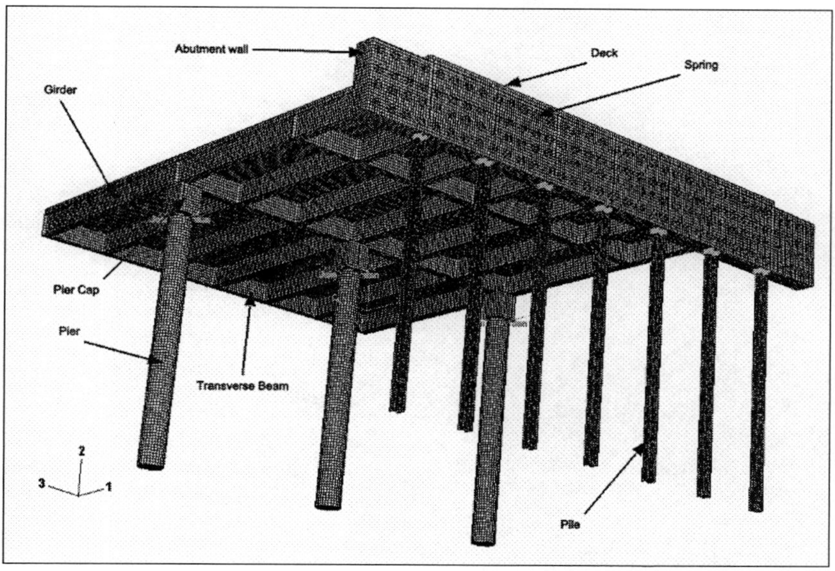

Figure 1. FE model of the bridge with springs

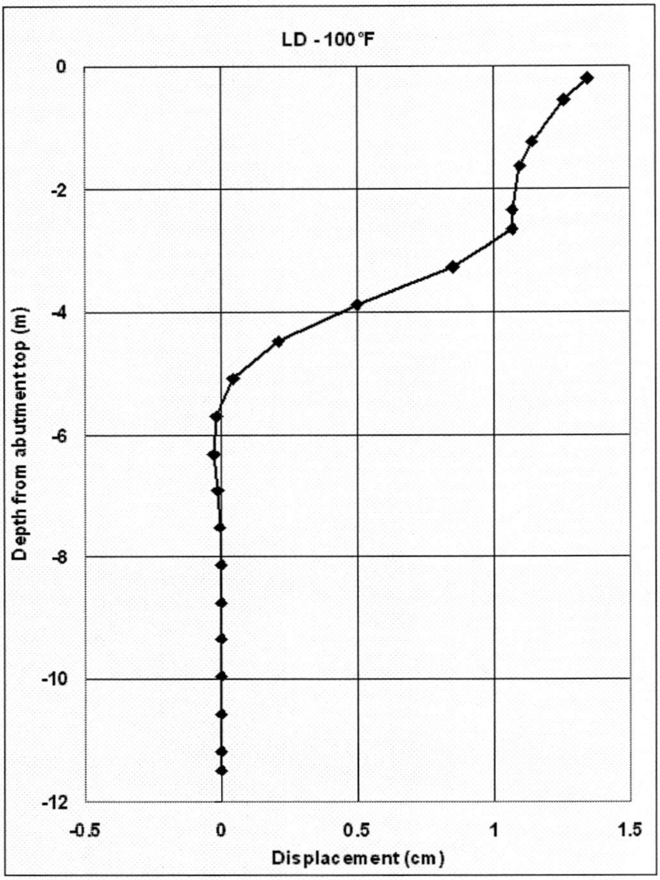

Figure 2. Longitudinal displacement U1 (cm) from the abutment top at the centerline

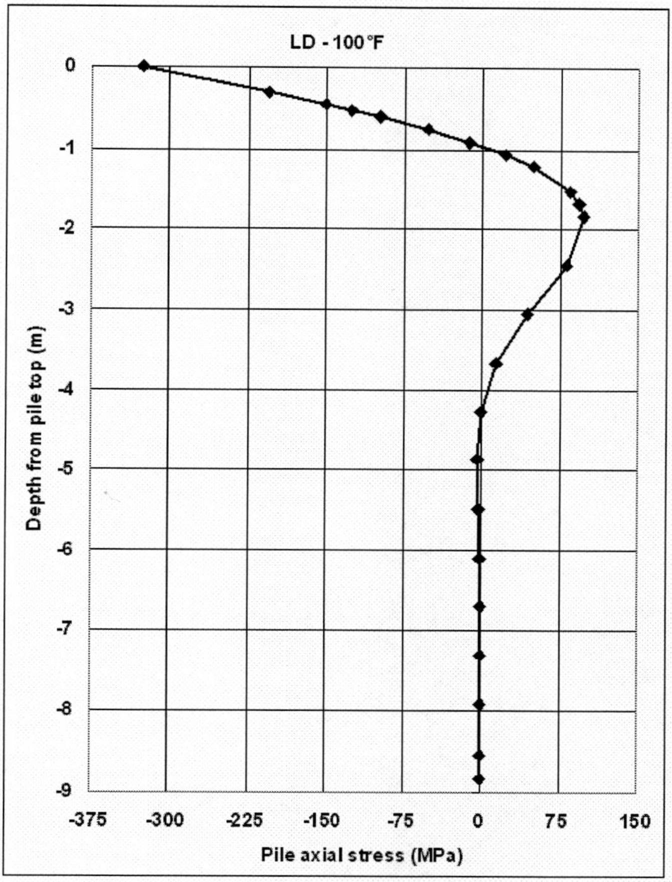

Figure 3. Bending stresses S11 (Mpa) in the central pile

PART III:

EVOLUTIONARY COMPUTATION

DISCOVERING BUILDING BLOCKS FOR HUMAN BASED GENETIC ALGORITHMS

TAKAOKI UEDA
University of Illinois at Urbana Champaign
Computer Science
Urbana, Illinois

XAVIER LLORÀ
University of Illinois at Urbana Champaign
National Center for Supercomputing Applications
Urbana, Illinois

DAVID E. GOLDBERG
University of Illinois at Urbana Champaign
Illinois Genetic Algorithms Laboratory
Urbana, Illinois

NORIKO IMAFUJI YASUI
University of Illinois at Urbana Champaign
Illinois Genetic Algorithms Laboratory
Urbana, Illinois

KUMARA SASTRY
University of Illinois at Urbana Champaign
Illinois Genetic Algorithms Laboratory
Urbana, Illinois

ABSTRACT
The push for rapid innovation and creativity in this Internet age places a premium on effective integration of both human and computer-generated knowledge. This paper takes the first step towards designing a *competent* HBGA, which can enable humans to innovate quickly, reliably, and accurately. Specifically, this paper proposes a methodology for discovering building blocks from text documents including reports, chat, transcripts and e-mail. The proposed method has been applied to simple test problems and to a news article set. The results show that the proposed BB-identification methodology is effective and enables humans to effectively exchange the BBs for rapid innovation.

1. INTRODUCTION

Due to the rapid growth of Internet, the amount of interactions among people has exponentially increased. Therefore, rapid visualization and representation of key points of discussions or any other form of documents is critical for successful brainstorming/innovation.

However, most documents and discussions are disorganized; the identification of key "building blocks" is difficult. Moreover, there is no absolute or relative quality measure of conclusions drawn from discussions or documents. Elsewhere, 4) has been pointed out that the primary steps towards a successful GA design involves Building Block(BB) identification, and HBGA(Kosorukoff, 2001) is no different. Therefore, the first step towards designing a *competent* HBGA is to develop a method that can automatically identify BBs which can be used to *innovation*, quickly, reliably, and accurately (Goldberg, 2002).

The purpose of this paper is to develop a method for discovering BBs from text-based discussions or documents. In this paper, BB discovery is analogous to discovering clusters of words in a sentence such that the set of word in each cluster indicate semantic of the sentence. Once the BBs are identified they can be used for both *innovation* (recombination) as well as *improvement*(mutation).

The remainder of this paper is organized as follows. We start with a brief description of related work in Section 2. In Section 3, we give an overview of our approach we propose for identifying BBs, followed by the details of each process of the proposed approach in Section 3-5. Section 6, presents experimental results obtained by applying our approach to simple text and a news article set. Followed by a note on future work, and key conclusions of the work is provided in Section 7.

2. RELATED RESEARCH

Finding building block is an important step toward designing competent GAs(Goldberg, 2002). Human-based genetic algorithms (HBGAs) (Kosorukoff, 2001) are first step toward modeling human innovation and creativity by means of computational embodiments. Despite HBGAs outsource GA evaluations and evolutionary operators to human beings, proper model building and linkage identification can help improving the scalability of such methods. Such situation usually leads to a need to be able to process streams of text—the normal encoding used by human communications. Therefore, we need to be able to identify building blocks that are defined based on the context and the semantics of the ongoing discourse under analysis.

GAs has been applied to a plethora of applications related to text processing, mining, and analysis. Those application include categorizing language independent multi-class text (Langdon, 2000), inducing relevant features for automatic word classification tasks (Johansson, 2003), and solving ambiguous labeling problems (Myers and Hancock, 1997) etc. Another group of applications have focused on methodology for document clustering, using some sort of evolutionary inspired approaches for instance. Some of these include collection clustering using divisive clustering (Dhillon and et al, 2003) and agglomerative clustering algorithm (1; 18). However, to the best of our knowledge few or null attempts have tried to use evolutionary approaches to discover context-aware semantic building blocks in streams of text. A key component of our work is the intensive usage of semantic networks. In this paper we have heavily relied on WordNet to get that semantic network. WordNet is a readable lexical database developed at Princeton(Miller et al, 1990). For further exploration about WordNet, see http://wordnet.princeton.edu/

3. APPROACH OVERVIEW

Our goal is to identify sets of words from given text stream that in such given particular context form semantic unity. In this section, we present a brief overview of the overall approach to facilitate its comprehension. Our approach consists of three processes:

1. Text processing of text streams
2. Semantic groups identification
3. Context-aware semantic building blocks discovery

For example, consider the following document, where each paragraph is identified by a unique number. All of these paragraphs have similar meaning when used in the same context.

1. This is a long story which cannot be described in one paragraph. However, we can describe this story by repeating paragraphs with same meaning many times.
2. This story is ambiguous and huge. Therefore, we cannot write it in one paragraph. One way to describe this story is to repeat paragraphs using similar words.
3. There is a story which cannot be described in one paragraph. However, we can write this paragraph repeatedly to describe it. Repeating is the only acceptable solution towards expression.
4. This is a story which cannot be expressed in one paragraph. There is one way to describe it and that is by repeating paragraphs with the same meaning as this paragraph many times.
5. One paragraph is not sufficient to express our story. By writing paragraphs with same meaning many times, it is possible to express our story.
6. Without repeating paragraphs with same meaning using similar words, we can not describe our story.

Text Processing: As a first step, we need to clean the document for enhancing efficiency of later processes. This step includes stop words elimination and word stemming. The former is to eliminate words irrelevant to document context, and the latter is to normalize each word to its base form. For example, this step converts the paragraph 1 in the sample document above to "story describe one paragraph. describe story repeat paragraph mean time".

Semantic Groups Identification: A semantic group indicates a cluster of words with similar meanings. Similarities between words are measured based on *hypernyms*. A *hypernym* is a word that is more generic than a given word. Our experiments use WordNet for obtaining hypernyms of each word. Note that every word in WordNet has at least one hypernym, that is, *entity*. For example, you can find semantic groups obtained from the sample document in the Section 7. Identified semantic groups are indexed by a word in the group.

Context-aware Semantic Building Block Identification: By the previous step, we have a set of words (including index words to semantic groups). Last step is to identify building blocks from the set of words. For example, as we report in the Section 7, obtained building blocks from the sample document were "one paragraph" and "describe repeating".

4. SEMANTIC GROUP IDENTIFICATION

A semantic group indicates a cluster of words with similar meanings. We use a design structure matrix (DSM) clustering method for identifying the semantic group. In this section, first we give a brief review of DSM clustering method. Next, the approach for identifying semantic groups by DSM clustering method is proposed. For elaborate exposition of DSM, please see MIT DSM web site: http://www.dsmweb.org/ and for DSM clustering, see http://www.illigal.uiuc.edu/pub/papers/IlliGALs/2003007.ps.Z

4.1 Semantic Groups Identification by DSM Clustering

Semantic groups are identified by DSM clustering method reviewed in the above. DSM clustering is mainly used because its ability to infer overlapping clusters—a common property of text where one word may have different meanings and, hence, belong to different groups. This subsection focuses on how to create DSM from a set of words. 5 Let *W* be a set of words contained in texts. Hypernym induced vectors *H*(*wi*) and *H*(*wj*) for *wi* ∈ *W* and *wj* ∈ *W* are obtained by hypernym sets of *wi* and *wj* . Let *Hwi* and *Hwj* be hypernym sets of word *wi* and word *wj* , respectively.

Let *H* = *Hwi* ∪*Hwj* = *{h*1, *h*2, . . ., *hn}* denote a combined set of hypernyms derived from *wi* and *wj* . *H*(*wi*) and *H*(*wj*) are defined by

$$H(w_i) = \{H_{w_i}(h_1), H_{w_i}(h_2), \ldots, H_{w_i}(h_n)\} \tag{4}$$

$$H(w_j) = \{H_{w_j}(h_1), H_{w_j}(h_2), \ldots, H_{w_j}(h_n)\}, \tag{5}$$

where

$$H_{w_i}(h_k) = \begin{cases} 0 & \text{if } h_k \notin H_{w_i} \\ 1 & \text{if } h_k \in H_{w_i} \end{cases} \tag{6}$$

Likewise, *Hwj* (*hk*) is 0 (if *hk* /∈ *Hwj*) or 1 (if *hk* ∈ *Hwj*). Let *θ*(*wi, wj*) be similarity between *wi*, and *wj* . *θ*(*wi, wj*) is measured by cosine similarity of hypernym induced vectors, as follows.

$$\left[\|H(w_i)\| \, \|H(w_j)\| \right]$$

Let *D* denote a DSM for identifying semantic BBs is a *n*-square matrix, where *n* = |*W*|. (*i, j*) entry (*i* _= *j*) of *D* is defined as follows.

$$D(i,j) = \begin{cases} 0 & \text{if } \theta(w_i, w_j) < \delta \\ 1 & \text{if } \theta(w_i, w_j) \geq \delta \end{cases} \tag{8}$$

where *wi, wj* ∈ *W* and *δ* is a threshold. Intuitively, the entry is 0, if the corresponding two words have similar meaning, and vice versa. In our experiment, *δ* = 1.10.

5. CONTEXT-AWARE SEMANTIC BUILDING BLOCKS IDENTIFICATION

A context-aware semantic building block indicates a cluster of words having close correlations, that is, similar semantic content of the given document. The BBs are identified by extended compact genetic algorithm (eCGA). In this section, first we give a brief review of eCGA, and then describe our approach to use eCGA on our problem.

5.1 Extended Compact Genetic Algorithm

The extended Compact Genetic Algorithm (eCGA) is probabilistic algorithm which is originated to learn linkage through correcting good probability distributions (Harik et al, 1999). The measure of desired distribution is quantified based on minimum description length (MDL) model. A class of probability model, marginal product models (MPMs), is used in eCGA for a probability distribution.

The key concept of MDL model is as follows. Simpler distributions are better than the more intricate one, and everything is equal. The MDL restriction penalizes both inaccurate and complex models, therefore conducting to an optimal probability distribution. MPMs are comprise of a product of marginal distributions on a partition of the genes. MPMs also make easier a direct linkage map with each partition separating tightly linked genes.

The eCGA has consecutive sequences as follows.

1. Generate a random population of size N.
2. Undergo tournament selection at rate S.
3. Model the population using a greedy MPM search.
4. If the model has converged, stop.
5. Generate a new population using the given model.
6. Return to step 2.

Two things need further explanation: (Baker and McCallum, 1998) the determination of MPM using MDL, and (Dhillon and et al, 2003) the generation of a new population based on MPM. A constrained optimization problem determines MPM in every generation. The determination of MPM in each generation can be represented as minimizing the sum of the model complexity (*Cm*) and the cost of using single model compare with complex one (*Cp*). Model complexity *Cm* and cost of using single model *Cp* can be represented as follows.

$$C_m = \log_2(n) \sum_{i=1}^{m} \left(2^{k_i} - 1\right) \tag{9}$$

$$C_p = \sum_{i=1}^{m} \sum_{j=1}^{2^{k_i}} N_{ij} \log_2 \left(\frac{n}{N_{ij}}\right) \tag{10}$$

where *n* is number of populations, *m* in the equations represent the number of BBs, *ki* is the length of BB $i \in [1, m]$, and *Nij* is the number of chromosomes in the current population possessing bit-sequence $j \in [1, 2ki]$ population size.

The greedy MPM search used in eCGA begin with amodel that all the variables are independent. Eventually, independent variables merge into subsets till the MDL model no longer progresses. The methodology of generating a new population is as follows, a new population of $n(1 - pc)$ where *pc* is the crossover probability are the best individuals in the current population. The others, $n \cdot pc$, are filled with randomly choosing subsets from the current individuals based on the probabilities of the subsets.

One of the most important parameters of eCGA is the population size. The analytical population sizing method is developed (Sastry and Goldberg, 2004). This analysis estimates the population size which enables to gain the solution as following equation.

$$n \propto 2^k \left(\frac{\sigma 2_{BB}}{d2}\right) m \log m, \tag{11}$$

where *m* is number of building blocks, which has size *k*, failure rate is $\alpha = 1/m$ and $\frac{\sigma 2_{BB}}{d2}$ is the noise-to-signal ratio (Goldberg et al, 1992).

5.2 Semantic BBs Identification by eCGA

Semantic BBs are identified by eCGA method. eCGA is mainly used because its ability to minimize complexity among BBs. eCGA requires vectors as an input. Suppose a given document consisting of paragraphs. Let *D* be the document which is a set of paragraphs, and denote each paragraph by *Pi*, that is, $D = \{P1, P2, \ldots Pn\}$. Each paragraph *Pi* is converted into a vector *Vi* for using eCGA as follows.

Let *W* be a set of semantic groups obtained from the document and other words (that is, not clustered words). Paragraph vector *Vi* is defined by

$$V_i = (P_i(w_1), P_i(w_2), \ldots, P_i(w_n)) \tag{12}$$

where $wk \in W$, $1 \leq k \leq n$ and

$$P_i(w_k) = \begin{cases} 0 & \text{if } w_k \text{ is not contained in } Pi \\ 1 & \text{if } w_k \text{ is contained in } Pi \end{cases} \tag{13}$$

This creates a population form from each paragraph. Each binary variable represents the existence of semantic group set in the paragraph. The final step is using eCGA to obtain context-aware semantic building blocks. The building blocks are identified by the MPM. MPM is provided by the model-building process of eCGA.

6 EXPERIMENTS

For examining our approach, we performed experiments using two documents: 1) a sample document consists of six paragraphs presented in Section 3, and 2) a news article set obtained from Google News on 29th January, 2007. In this section we report the experimental results from each document and show how our approach identify BBs from the documents.

6.1 Experiment 1

As a preliminary experiment, we examined our approach to a small sized sample document, which was shown in Section 3. The sample document consists of six paragraphs.

Table 3: Experiment 1: Semantic groups

Index word	group
describe	[describe write]
meaning	[possible meaning]
one	[one meaning]
solution	[solution meaning words]
words	[words way]
closer	[closer possible]

Table 4: # of BBs and its size

Size of BBs	1	2	3	4
# of BBs	69	69	24	1

After the first step of our approach (stop words elimination and stemming words), DSM clustering method identified six semantic groups (see Table 7.1). Each row indicates an index word (left column) of a word group (right column). Each pair (or triplet) has common hypernyms. For example, all of the words "possible", "meaning" and "one" have a hypernym "cognition" in common.

After clustering the semantic groups provided by DSM, we have a set of 36 semantic groups. Our approach identified two BBs having two words [one paragraph][describe repeating]. The example paragraphs uses either "describe" or "write". Only one paragraph uses both, but the semantic layer brought them together. This shows clustering semantic BBs raises efficiency and quality of BBs.

6.2 Experiment 2

To validate our approach, we performed an experiment using a larger document. The document consists of around hundred news articles about *a suicide bombing in Israel* obtained from Google News on Jan 29, 2007. The stop words elimination reduced the size of word set to half approximately (from 1224 words to 643 words). Then, DSM clustering identified various sizes of semantic groups. Figure 1 shows the number of identified semantic BBs and the sizes of groups (i.e., number of words contained in the group). DSM clustering identified 168 semantic groups. The average size of semantic groups was approximately 5.28. After DSM clustering step, we have a set of 283 semantic groups identified from the 608 paragraphs available. Table 7.2 shows number of identified BBs of each size. The average size of BBs was approximately 1.74.

Table 5 shows the top ranked BBs for each size. Each BBs in Table5 actually represents semantic context in document. In other words, we can have intuitive understanding of the article only by these BBs without reading through. These BBs are well suited for HBGA operation.

7 CONCLUSIONS

In this paper, we proposed an approach for identifying building blocks from text documents. Our approach used DSM clustering methods for finding semantic groups. By using eCGA, we could identify context-aware semantic building blocks. We report some experimental results that demonstrate how our approach effectively identifies the building blocks from text documents. This work is the first step towards designing a competent HBGA. As the future work, using this work, we will design a competent HBGA for obtaining innovative solutions.

Table 5: Examples of identified BBs

Size	BBs
4	[suicide israel april restaurant]
3	[agree occupation march], [gaza strip city], [violence undermine stage], [west bank military], [resort sea tip]
2	[brother thursday], [jihad joint], [iran february], [crackdown fire], [lead ramadan], [exile iranian]

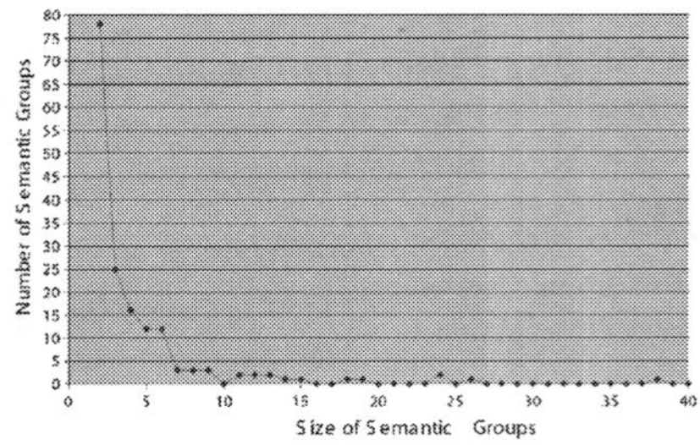

Figure 1: Distribution of Semantic Group Size

REFERENCES

L. D. Baker and A. K. McCallum., 1998 Distributional clustering of words for text classification. In W. B. Croft, A. Moffat, C. J. van Rijsbergen, R. Wilkinson, and J. Zobel, editors, *Proceedings of SIGIR-98, 21st ACM International Conference on Research andDevelopment in Information Retrieval*, pages 96–103, Melbourne, AU. ACM Press, New York, US.

Dhillon, I., Mallela, S., and Kumar, R., 2003 A divisive information-theoretic feature clustering algorithm for text classification. *Jurnal of Machine Larning Research*, 3:1265–1287.

D. Goldberg., 2002 *Design of innovation: Lessons from and for competent genetic algorithms*. Kluwer Acadamic Publishers, Boston, MA.

D. E. Goldberg, K. Deb, and J. H. Clark.,1992 Genetic algorithms, noise, and the sizing of populations. *Complex Systems*, 6:333–362. (Also IlliGAL Report No. 91010).

Harik, G.R., Lobo, F.G., and Sastry,K.,1999 Linkage learning via probabilistic modeling in the ecga. IlliGAL Report No. 99010, University of Illinois at Urbana-Champaign, Urbana, IL.

C. Johansson.,2003 Searching for features using a genetic algorithm, In Bjørnar Tessem, Pekka Ala-Siuru, Patrick Doherty, and Brian Mayoh (Eds.), *proceedings of the Eighth Scandinavian Conference on Artificial Intelligence (SCAI03)*, in the series *Frontiers in Artificial Intelligence and Applications*, IOS Press, Amsterdam. (pp.119-130)

A. Kosorukoff.,2001 Human based genetic algorithm. IlliGAL Report No. 2001004, University of Illinois at Urbana-Champaign, Urbana, IL, March.

W. B. Langdon.,2000 Natural language text classification and filtering with trigrams and evolutionary NN classifiers. In D. Whitley, editor, *Late Breaking Papers at the 2000 Genetic and Evolutionary Computation Conference*, pages 210–217, Las Vegas, Nevada, USA, 8.

Miller, G., Beckwith, R., Fellbaum, C., Gross, D., and Miller, K.,1990 Introduction to wordnet: An on-line lexical database. *International Journal of Lexicography*, 3(4):235–244.

R. Myers and E. R. Hancock.,1997 Genetic algorithms for ambiguous labelling problems. In *Energy Minimization Methods in Computer Vision and Pattern Recognition*, pages 345–360.

M. Porter.,1980 An algorithm for suffix stripping. *Program*, 14(3):130–137.

K. Sastry and D. E. Goldberg.,2004 Designing competent mutation operators via probabilistic model building of neighborhoods. *Proceedings of the Genetic and Evolutionary Computation Conference*, 2:114–125. Also IlliGAL Report No. 2004006.

S. M. Weiss, N. Indurkhya, T. Zhang, and F. J. Damerau.,2005 *Text Mining: Predictive methods for analyzing unstructured information*. Springer.

REVERSE LOGISTICS NETWORKS PROBLEM IN PRODUCT REMANUFACTURING SYSTEM BY PRIORITY-BASED GENETIC ALGORITHM

JEONG-EUN LEE
Graduate School of Information,
production & System,
Waseda university
leeje@toki.waseda.jp

MITSUO GEN
Graduate School of Information,
production & System,
Waseda university
gen@waseda.jp

KYONG-GU RHEE
Department of Accounting,
Dongeui university
rhee@deu.ac.kr

ABSTRACT

The interests about recovery of used products or materials have been increased. Therefore, reverse Logistics Network Problem (rLNP) will powerful and get a great potential for winning consumers in a more competitive context in the future.

We formulate a mathematical model of remanufacturing system as three-stage logistics network model for minimizing the total of costs of reverse logistics shipping cost and fixed opening cost of the disassembly centers and processing centers. And we consider a multi-stage, multi-product and some attached conditions for disassembly centers and processing centers respectively.

For solving this problem, we propose a Genetic Algorithm (GA) with priority-based encoding method consisting of 1st and 2nd stages combined a new crossover operator called as Weight Mapping Crossover (WMX). A heuristic approach is applied in the 3rd stage to transportation of parts from processing center to manufacturer.

1. INTRODUCTION

Today, consumer's demand has become more complicated in the global age that people can buy the products produced everywhere in the world. Therefore, the manufacturing companies have to produce excellent products in a short time and to offer it to consumers. As a result, the product's life-cycle is getting rapidly shorter and more and more products are being discarded.

Recycling problem can affect not only a cost performance of recycling of a company, but also whole strategy of the company including production strategy and purchase policy. To deal with this problem the government should treat it in one industry.

Researches related with the reverse logistics are going to progress in various fields and subjects.

In reuse logistics models, L. Kroon *et al.* (1995), they reported a case study concerning the design of a logistics system for reusable transportation packages. The authors proposed a mIP (mixed integer programming), closely related to a classical un-capacitated warehouse location model. In recycling models, R. K.

Pati *et al.* (2006), they developed an approach based on a mixed integer goal programming model to solve the problem. The model studies the inter-relationship between multiple objectives of a recycled paper distribution network. This study proposed reverse logistics network of remanufacturing process. In remanufacturing models, K. B. Kim *et al.* (2006), discussed a notion of remanufacturing systems in reverse logistics environment. They proposed a general framework in view of supply planning and developed a mathematical model to optimize the supply planning function.

J. E. Lee *et al.* (2007), proposed the minimizing of total shipping costs in reverse logistics.

Recently, GAs have received considerable attention regarding their potential as a novel approach to optimization problems and is often used to solve many real world problems, including the effective approaches on the SPR problem, capacity and flow assignment, and the dynamic routing problem. It is noted that all these problem can be formulated as some sort of a combinatorial optimization problem.

This paper propose multi-stage reverse Logistics Network Problem (m-rLNP) which consider the minimizing of total shipping cost and fixed opening costs of the disassembly centers and the processing centers in reverse logistics and a new genetic algorithm to solve this one. We developed priority-based Genetic Algorithm (priGA) and heuristic approach, and proposed a new crossover operator called as Weight Mapping Crossover (WMX).

This paper is organized as follows: in Section 1, the problem is described in detail and the previous researches are reviewed; in Section 2, the mathematical model of the reverse logistics network is introduced; in Section 3, the priGA approach and heuristic approach are explained in order to solve this problem; in Section 4, numerical experiments are presented to demonstrate the efficiency of the proposed method; finally, in Section 5, concluding remarks are outlined.

2. MATHEMATICAL FORMULATION

In this Section, we consider constituents, variables and assumptions for formulating a multi-stage logistics network model. In the remanufacturing process, parts which are disassembled of recovered products of be used are send to processing center. These parts become new products by combined with another parts of processed or new. In this paper, we considered two situations. First situation is that products are sent directly from the returning center to the processing center, according to the product type. And, second is that parts are sent recycling according to parts type from the disassembly center.

We make the following assumptions of this m-rLNP:

A1: If the quantity of provided parts from processing center is not enough for requirement of manufacturer, then manufacturer must buy parts from supplier.

A2: If the quantity of provided parts from processing center exceeds the requirement of manufacturer, then exceeded parts distribute in order of recycling and disposal.

A3: Not considering the inventories.

A4. Some products should be transported form returning center to processing center directly, not through disassembly center.
A5. Some products should be recycled or disposed form disassembly center directly, not through processing center.

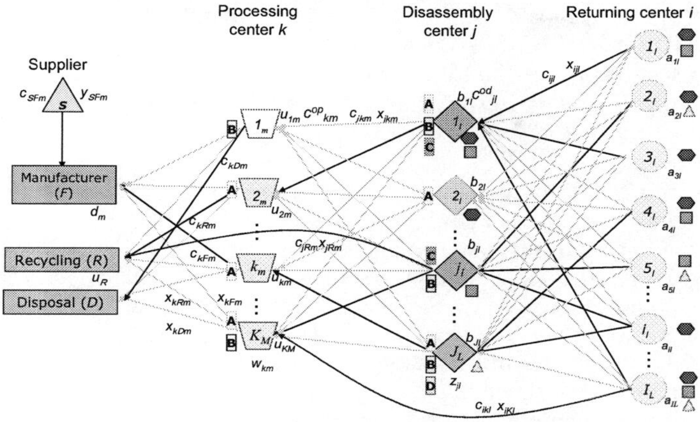

Fig. 1. Multi-stage reverse logistics network model

The mathematical model of the problem is:

$$\min \quad z = \sum_{j=1}^{J}\sum_{l=1}^{L} c_{jl}^{od} z_{jl} + \sum_{k=1}^{K}\sum_{m=1}^{M} c_{km}^{op} w_{km}$$
$$+ \sum_{i=1}^{I}\sum_{j=1}^{J}\sum_{l=1}^{L} c_{ijl}\, x_{ijl} + \sum_{i=1}^{I}\sum_{k=1}^{K}\sum_{l=1}^{L} c_{ikl}\, x_{ikl} + \sum_{j=1}^{J}\sum_{k=1}^{K}\sum_{m=1}^{M} c_{jkm}\, x_{jkm} - \sum_{j=1}^{J}\sum_{m=1}^{M} c_{jRm}\, x_{jRm}$$
$$+ \sum_{k=1}^{K}\sum_{m=1}^{M} c_{kFm}\, x_{kFm} - \sum_{k=1}^{K}\sum_{m=1}^{M} c_{kRm}\, x_{kRm} + \sum_{k=1}^{K}\sum_{m=1}^{M} c_{kDm}\, x_{kDm} + \sum_{m=1}^{M} c_{SFm}\, y_{SFm} \quad (1)$$

$$\text{s. t.} \quad \sum_{j=1}^{J} x_{ijl} + \sum_{k=1}^{K} x_{ikl} \geq a_{il}, \quad \forall\, i, l \quad (2)$$
$$\sum_{k=1}^{K} x_{jkm} + x_{jRm} \geq b_{jm}, \quad \forall\, j, m \quad (3)$$
$$\sum_{k=1}^{K} (x_{kFm} + x_{kRm} + x_{kDm}) \geq u_{km}, \quad \forall\, m \quad (4)$$
$$\sum_{j=1}^{J} z_{jl} \leq J, \quad \forall\, l \quad (5)$$
$$\sum_{k=1}^{K} w_{km} \leq K, \quad \forall\, m \quad (6)$$
$$y_{SFm} + \sum_{k=1}^{K} x_{kFm} \geq d_m, \quad \forall\, m \quad (7)$$
$$x_{ijl}, x_{ikl}, x_{jkm}, x_{jRm}, x_{kFm}, x_{kRm}, x_{kDm}, y_{SFm} \geq 0, \quad \forall\, i, j, k \quad (8)$$
$$z_{jl} = \{0, 1\} \quad \forall\, j, l \quad (9)$$
$$w_{km} = \{0, 1\} \quad \forall\, k, m \quad (10)$$

(1) is objective function for minimizing the total of costs of reverse logistics shipping cost and fixed opening cost of the disassembly centers and processing centers until become reusable parts from collection of used products. Constraints (2), (3) and (4) explain about capacities of returning, disassembly and processing center each other. Constraints (5), (6) limit the number of disassembly centers and processing centers that can be opened. Expression (5)

shows the constraint that demands of parts of manufacturer. Constraint (8) enforces the non-negativity restriction on the decision variables (9), (10) impose the integrality restriction on the decision variables used in this model.

3. HYBRID GENETIC ALGORITHM

In this section, we explain the priority-based genetic algorithm (priGA) for solving sub-problems of 1st stage and 2nd stage. And to solve the sub-problem of 3rd stage, we propose a heuristic approach.

3. 1. Representation

For m-rLNP, we use two priority-based encodings to represent the transportation trees on stages. This means that each chromosome in the population consists of two parts. While the first part (i.e., the first priority-based encoding) represents transportation tree between return centers and disassembly centers, the second part (i.e., the second priority-based encoding) represents transportation tree between disassembly centers and processing centers.

In the example for the 1st stage, we consider the case of that there are 5 returning centers, 3 disassembly centers and 1 processing center. When there are products of types more than two, the number of returning center is equal to the number of type of product. And it is excluded transport process form returning centers to processing centers directly, not through disassembly centers.

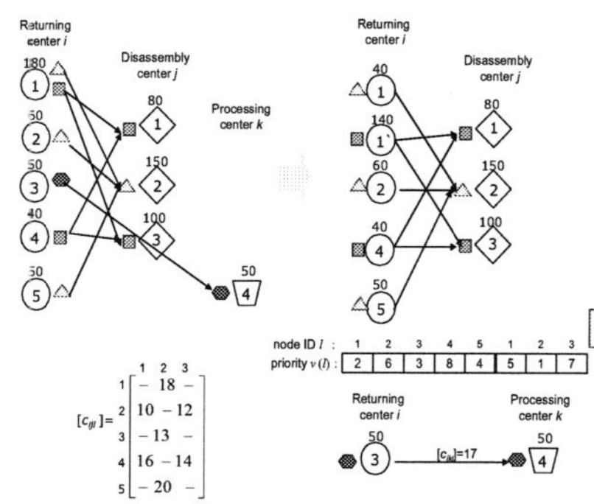

Fig. 2. An illustration of chromosome, transportation tree and costs for 1st stage

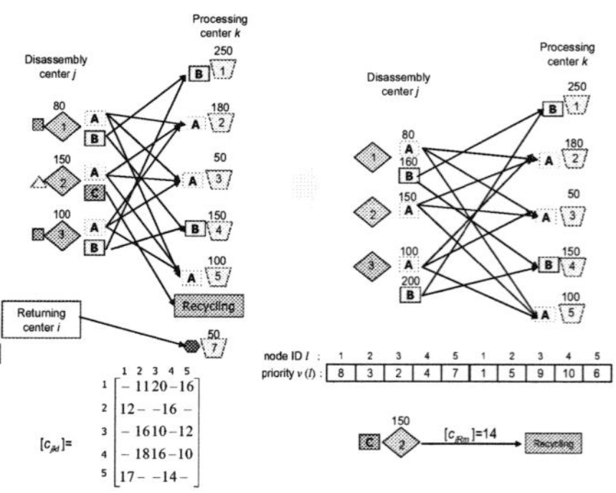

Fig. 3. An illustration of chromosome transportation tree and costs for 2nd stage

In the example for the 2nd stage, 3 disassembly centers and 5 processing centers and 1 recycling are considered. In the 2nd stage, we consider processable processing center according to type of part. It is excluded transport process recycling directly, not through processing centers.

The 3rd stage has three kinds of cases. Case 1 is the quantity of parts provided from processing center is not enough for requirement of manufacturer, the surplus demand of parts should be bought from supplier. Case 2 the quantity of provided parts from processing center is the same requirement of manufacturer and distribute the manufacturer provided parts from processing center. Case 3, the quantity of provided parts from processing center exceeds the requirement of manufacturer and then the exceeded quantity should be sent to orderly recycling and disposal.

3. 2. Genetic Operators
Genetic operators mimic the process of heredity of genes to create new orrspring at each generation. Using the different genetic operators has very large influence on GA performance(1993). Therefore it is important to exmined different genetic operators.
3. 2. 1 Crossover Operator: In this study, we proposed a new crossover operator, weight mapping crossover (WMX).
3. 2. 2 Mutation operator: in this study, insertion mutation has been adopted. Insertion mutation selects a gene at random and inserts it in a random position.

4. NUMERICAL EXPERIMENT

We used using pnGA (Prüfer number-based GA) proposed by Syarif and Gen (2003), to study the effectiveness of the developed GA with new encoding method (priGA).

As seen in Table 1, the number of returning centers changes between 5 and 15 on these problems, number of disassembly centers and number of processing centers change between 3 and 10, and 5 and 14, each other. The data such as transportation costs, demand of parts, capacities of returning centers, disassembly centers, processing centers, recycle and manufacturer were also randomly generated to provide realistic scenarios. And Table 1 also shows computational results for the pnGA and priGA.

Table 1. The size of test problems and computational results with pnGA and priGA

Problem No.	no. of *I*	no. of *J*	no. of *K*	no. of constraints	no. of shipping variables	no. of 0-1 variables	stage	pnGA	priGA
1	5	3	5	36	222	24	stage 1	5194	3150
							stage 2	10401	9850
							stage 3	12000	12000
							total	27595	25000
2	8	5	7	51	474	57	stage 1	12495	11711
							stage 2	35798	32345
							stage 3	34400	34400
							total	82693	78456
3	15	10	14	72	1659	93	stage 1	3812908	1013139
							stage 2	8316417	5917372
							stage 3	129820	129820
							total	12259145	7060331

In priGA, one-cut point crossover and insertion mutation operators were used as genetic operators and its rates were taken as 0.5. Each test problem is executed by 10 times using GA approaches. To solve large scale problem, we used priGA proposed. The results show that priGA better than pnGA according to solution quality. And because we use the heuristic approach, results of 3rd stage of each problem are same.

5. CONCLUSIONS

We formulated a mathematical model for the solution of rLNP by proposing priority-based genetic algorithm approach (priGA) and a heuristic approach. We also combined a new crossover operator, weight mapping crossover (WMX) with the hybrid priGA and adopted insertion mutation in this paper.

Numerical experiment demonstrated the efficiency and effectiveness of the hybrid GA approach for solving the m-rLNP problem.

Although memory requirement on new representation is greater than pnGA, i.e., Prüfer number-based GA, only 2 digits for each stage in transportation problem, this representation has got very important two advantages in real world applications. One of them is that its implementation is very easy. Another one is that after genetic operators, always feasible solutions are obtained. Based on this study, it was seen that the hybrid priGA with WMX demonstrated the best performance according to solution quality.

ACKNOWLEDGMENT

This research work is partly supported by Waseda University Grant for Special Research Projects 2006-2007 and the Ministry of Education, Science and Culture, the Japanese Government: Grant-in-Aid for Scientific Research (No.17510138).

REFERENCES

Ko, H. J., and Evans, G. W., 2007, "A genetic algorithm-based heuristic for the dynamic integrated forward/reverse logistics network for 3PLs", *Computers & Operations Research*, vol. 34, no.2, pp.346-366.

Lieckens, K., and Vandaele. N., 2007, "Reverse logistics network design with stochastic lead times", *Computers & Operations Research*, vol. 34, no.2, pp.395-416.

Biehl, M., Prater, E., and Realff, M. J., 2007, "Assessing performance and uncertainty in developing carpet reverse logistics systems", *Computers & Operations Res.*, vol. 34, no. 2, pp.443-463.

Kroon, L.,and Vrijens, G., 1995, "Returnable containers: an example of reverse logistics", *Inter. Journal of Physical Distribution & Logistics Management*, vol. 25, no. 2, pp. 56-68.

Pati, R. K., Vrat P., and Kumar, P., 2006, "A goal programming model for paper recycling system", *TheInternational Journal of Management Science, Omega, In press*.

Kim, K. B., Song, I. S., and Jeong, B. J., 2006, "Supply planning model for remanufacturing system in reverse logistics environment", *Computers & Industrial Eng.*, vol. 51, no.2, pp. 279-287.

Lee, J. E., Rhee, K. G., and Gen, M., 2007, "Designing a Reverse Logistics Network by Priority-based Genetic Algorithm", *Proc. of International Conference on Intelligent Manufacturing Logistics Systems*, pp. 158~163, Japan.

Shimamoto, N., Hiramatsu, A., and Yamasaki, K., 1993, "A Dynamic Routing Control Based on a Genetic Algorithm", *Proc. IEEE. Int. Conf. Neural Networks*, pp. 1123-1128.

Syarif, A., and Gen, M., 2003, "Double Spanning Tree-based Genetic algorithm For Two Stage Transportation Problem", *International Journal of Knowledge-Based Intelligent Engineering System*, vol. 7, no. 4, pp.388-389.

Gen, M., and Cheng, R. W., 2000, *Genetic Algorithm and Engineering Optimization*, Wiley, New York.

Gen, M., Altiparmak, F., and Lin, L., 2006, "A genetic algorithm for two-stage transportation problem using priority-based encoding", *OR Spectrum,* Vol.28, No.3, pp.337-354.

OPTIMIZING TARTARUS CONTROLLERS USING GRAPH BASED EVOLUTIONARY ALGORITHMS

STEVEN M. CORNS
Department of Mechanical Engineering
Iowa State University
Ames, Iowa

DANIEL A. ASHLOCK
Department of Mathematics and Statistics
University of Guelph
Guelph, Ontario

KENNETH MARK BRYDEN
Department of Mechanical Engineering
Iowa State University
Ames, Iowa

ABSTRACT

The Tartarus grid robot problem has been used for years to test different schemes to control grid robots. While there is a large amount of information available, there is no conclusive evidence as to whether there is a benefit to maintaining solution diversity in the population or to compare it to other test problems. This work investigates the affect of using Graph Based Evolutionary Algorithms to examine and classify the Tartarus problem using ISAc list controllers. By changing the underlying graph, it is possible to control the rate at information is spread, allowing the user to control the rate at which diversity is lost in the solution population. For the Tartarus problem diversity is not necessary to find intermediate fitness scores, but as the desired fitness level increases, there is more of a need for diversity preservation, making a diversity preserving graph desirable.

INTRODUCTION

Evolutionary algorithms manipulate a population of candidate solutions in ways that mimic evolution to achieve novel and hopefully superior solutions. One of the tasks EAs have been applied to is that of optimizing controllers for the Tartarus grid robot problem, an application in the area of artificial life. There is a large body of work on the Tartarus problem investigating both controller representation [Ashlock and Freeman, 2000; Ashlock, Willson, & Leahy, 2004] and sensor location [Corns et al., 2005] to better understand the problem. This knowledge can then be extended to similar real world problems such as controllers for UAVs, making it possible for them to more effectively navigate dynamic environments.

In this study we will investigate the use of graph-based evolutionary algorithms (GBEAs) on an evolving population of controllers for the Tartarus grid robot problem. GBEAs allow the user to control the amount of diversity in an evolving population by controlling the rate at which information is transferred [Bryden et al., 2005]. The goals of this research are to determine what amount of diversity is preferred in the Tartarus problem and to work towards establishing where the Tartarus problem fits into an on-going project to develop a taxonomy of evolutionary computation problems.

TARTARUS

The Tartarus problem is a standard evolutionary computation test problem. A 6 x 6 grid surrounded by impenetrable walls is first established. A Tartarus board is then constructed by placing six boxes onto the grid with none of the boxes touching the walls.

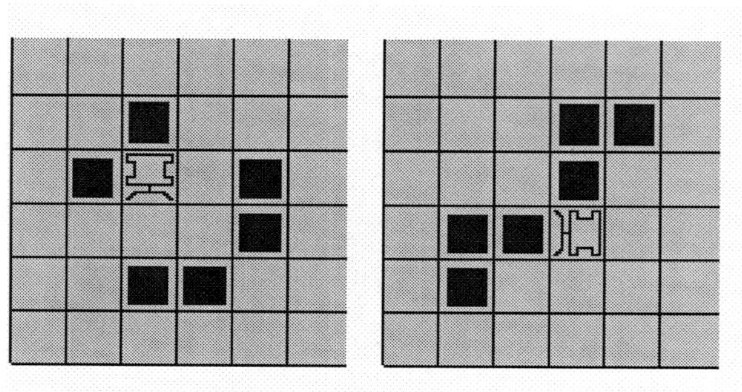

Figure 1, Valid Tartarus boards.

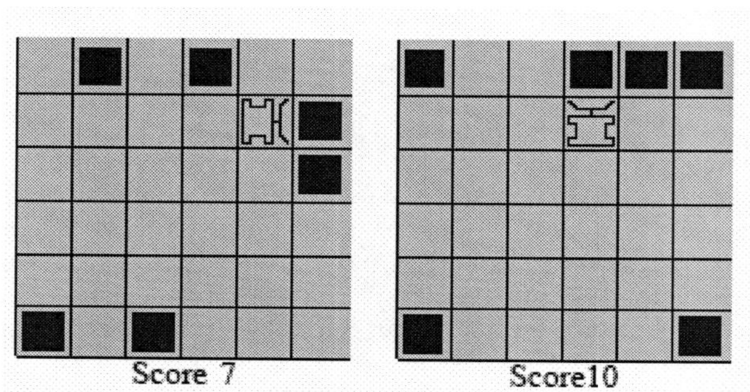

Figure 2, Final boards with scores.

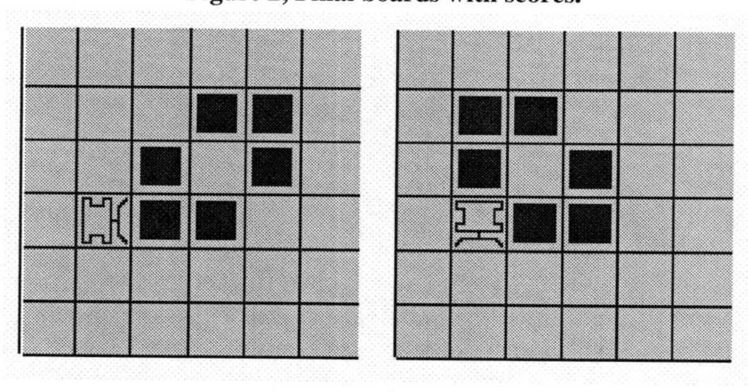

Figure 3, Willson configurations of Tartarus boards.

The only restriction on the configuration of the boxes in this study is that no boxes may form a closed group of four. An agent (or "bulldozer") is then placed on the grid, also away from the walls, and facing in a random direction. Examples of valid starting configurations are shown in Fig. 1.

Each of the bulldozers is then evaluated on 40 randomly generated Tartarus boards, with each bulldozer permitted 80 moves. There are three moves a bulldozer can take: turn right, turn left, or advance one grid space. The bulldozer can push one box ahead of it, but it cannot move two boxes, and it cannot push a box through a wall.

The object of the task is to push the sides of the boxes against the walls. A fitness point is awarded for each side of a box that is touching a wall at the end of the 80 moves, with a maximum fitness of 10 (corners are worth two points.) Fig. 2 shows some examples of final boards together with their fitness score. In the implementation used here, the bulldozer is given information about the eight adjacent grids.

The prohibition on having boxes in a close group of four deserves some discussion. Since the bulldozer cannot move any boxes in such a configuration, boards containing this configuration were judged to be of little worth in evaluating fitness. An additional configuration was discovered by Stephen Willson that also made it impossible for any bulldozer to clear the board. Examples of these configurations are shown in Fig. 3. Out of a desire to use the standard Tartarus problem, these "Wilson Configurations" were not excluded in this study. More information on possible layouts of Tartarus boards can be found in [Ashlock, Willson and Leahy, 2004].

GRAPH BASED EVOLUTIONARY ALGORITHMS

A *combinatorial graph* or *graph* (G), is a collection of vertices (V(G)) and edges (E(G)) where E(G) is a set of unordered pairs from V(G). Two vertices of the graph are *neighbors* if they are members of the same edge. GBEAs are evolutionary algorithms that impose a geography to restrict the number of possible mates a population member can interact with. Each vertex V(G) of the graph holds a population member, which can only interact with an individual whose vertex shares an edge E(G) with it. Evolution occurs one mating event at a time (steady state [Reynolds, 1992; Syswerda, 1991; Whitley, 1989]). The rate of information spread can be controlled by changing the graph type, thereby controlling diversity loss. For this study, the general strategy is to use the graph (population structure) to constrain mating by applying the graph as a geography on which the population lives, permitting mating only between neighbors and preserving diversity without sacrificing progress due to heterogeneous crossover.

LIST OF GRAPHS

In this section, the random graph used in this study is defined. All non-random graphs used in the study can be found in Graph Based Genetic Algorithms (Ashlock et al., 1999). The random toroidal graph was generated by randomly placing points on a unit torus and creating an edge between two points if they were within a certain distance of each other (for this case, 0.07 was used). Complete information on how the random graph was generated and descriptions of the other graphs can be found in Bryden et al. (2003). Table 1 lists the graphs used in the study.

Table 1. Graphs used, their regularity, whether they are random, and diameters.

Graph	Index Name	Index	Regularity	Random	Diameter
Cycle	C_{512}	0	2	No	256
Hypercube	H_9	1	$Log_2(n)$	No	9
Complete	K_{512}	2	n-1	No	1
Petersen, k=1	$P_{256,1}$	3	3	No	129
Petersen, k=17	$P_{256,17}$	4	3	No	18
Petersen, k=3	$P_{256,3}$	5	3	No	46
Petersen, k=7	$P_{256,7}$	6	3	No	22
Random Toroid	RTor_07	7	No	Yes	19
Torus, m=16	$T_{16,32}$	8	4	No	24
Torus, m=4	$T_{4,128}$	9	4	No	66
Torus, m=8	$T_{8,64}$	10	4	No	36
Simplexified	Z_{512}	11	4	No	19
Regular Tree, d=3	RTn512d3	12	3,1	No	16
Regular Tree, d=4	RTn512d4	13	4,1	No	11
Regular Tree, d=5	RTn510d5	14	5,1	No	9

EXPERIMENTAL DESIGN

In this study, a series of numerical experiments are performed to examine the effects of using combinatorial graphs on the Tartarus problem. The controllers for the Tartarus bots are represented using (I)f-(S)kip-(Ac)tion (ISAc) lists [Ashlock and Joenks, 1998]. ISAc lists store a linear array of simplified machine language instructions. ISAc lists can be used to simulate finite state machines, but can also hold programs that cannot be implemented by finite state machines. Of all the representations of Tartarus controllers, ISAc list have so far performed the best, with a best score of 8.2. Three experiments were conducted with scores of 5.0, 5.5 and 6.5 used as stopping criteria. Each of the experiments was run on all graphs listed and the mean number of generations to solution was calculated. 300 runs were performed for each experiment and a 95% confidence interval was calculated. In addition, 5 runs of 2500 generations were run for each graph to compare how well each graph performed when given a longer amount of run time. The highest scores achieved by these experiments using longer generations were averaged every 500 generations and compared to give an indication of which graph leads to a higher average score.

RESULTS

For the experiments with a fixed score as a stopping criteria, the mean number of generations to find that score and the associated confidence interval were calculated for each graph (Figs. 4, 5 and 6). While the required number of generations increased, there was little statistically significant change in graph performance for the three stopping criteria investigated. Graphs with a high level of connectivity (such as the complete and hypercube graphs) had populations that achieved the cut-off score in the fewest generations, while the sparser graphs (such as the cycle and the regular tree graphs) caused to population to take more generations to reach the desired score. This indicates that there is little need to preserve diversity to achieve scores below 6.5 for the Tartarus problem using an ISAc list encoding.

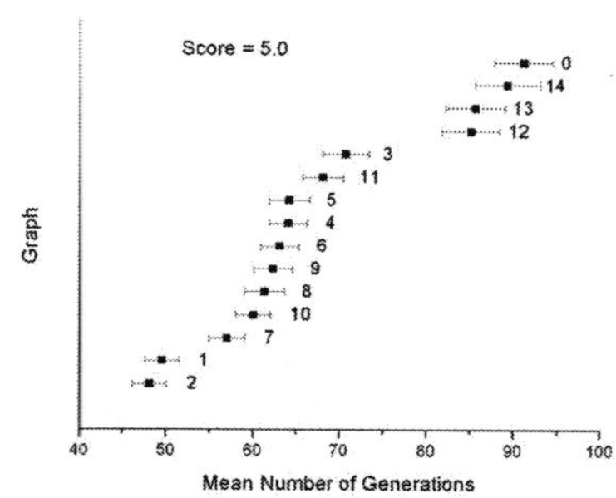

Fig. 4. Mean number of generations to solution by graph for a Tartarus score of 5.0.

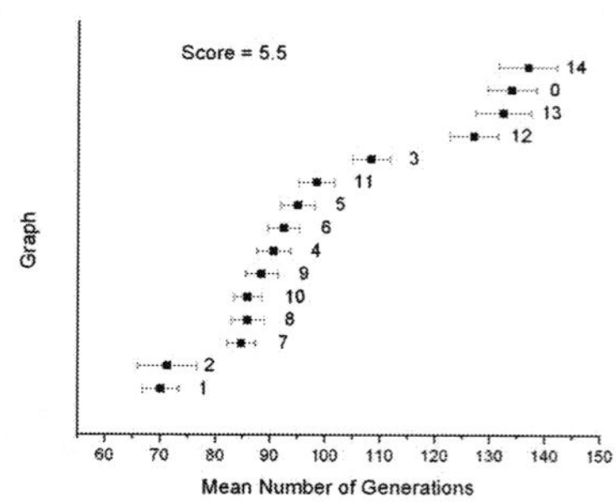

Fig. 5. Mean number of generations to solution by graph for a Tartarus score of 5.5.

When the experiments were allowed to run for 2500 generations (Fig. 7), the graph that gave the best results changed to the torus graphs. All three variations of the torus graph outperformed the other graphs, followed by Petersen graph an interior spacing

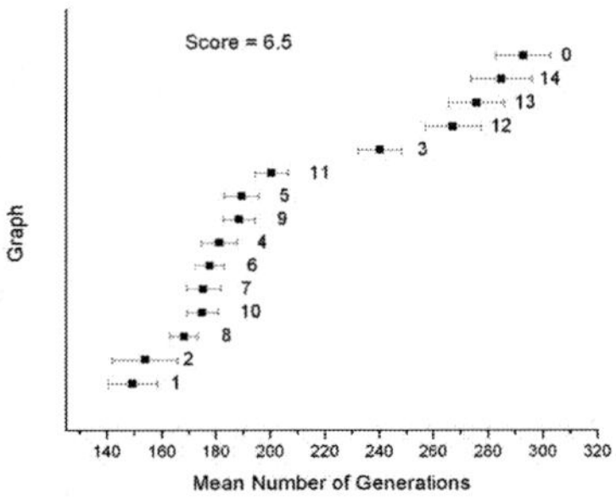

Fig. 6. Mean number of generations to solution by graph for a Tartarus score of 6.5.

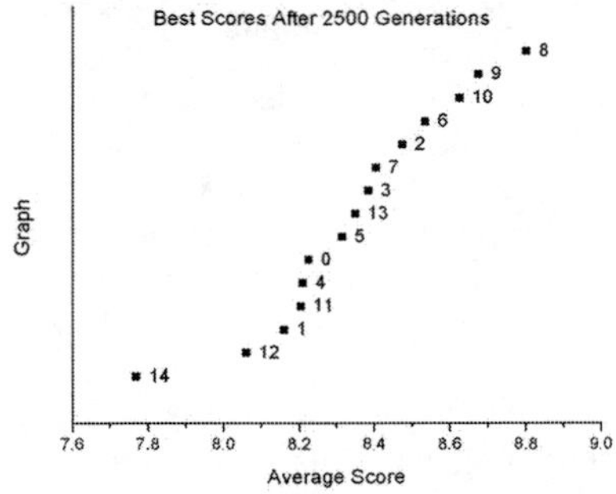

Fig. 7. Average scores on the Tartarus task by graph after 2500 generations.

value of 7, and then the complete graph. This was then followed by the random torus graph, the remainder of the Petersen graphs and then the sparser cycle and regular tree graphs. This order held true for roughly each measurement with the exception of the Hypercube graph, which showed little improvement as generation increased (Table 2).

Table 2, Average high score on Tartarus problem by graph at increments of 500 generations.

Graph (Index)	Number of Generations				
	500	1000	1500	2000	2500
C512 (0)	6.87	7.7	8.02	8.12	8.225
H9 (1)	7.78	7.98	8.08	8.15	8.16
K512 (2)	7.64	8.07	8.405	8.46	8.475
P256_1 (3)	7.46	7.885	8.005	8.305	8.385
P256_17 (4)	7.775	7.975	8.09	8.16	8.21
P256_3 (5)	7.355	7.825	8.07	8.235	8.315
P256_7 (6)	7.575	8.09	8.235	8.47	8.535
RTor07_3 (7)	7.625	8.05	8.3	8.37	8.405
T16_32 (8)	7.775	8.14	8.535	8.685	8.8
T4_128 (9)	7.865	8.345	8.46	8.565	8.675
T8_64 (10)	7.775	8.105	8.37	8.53	8.625
Z512 (11)	7.25	7.93	8.06	8.11	8.205
RT1n512d3 (12)	7.13	7.66	7.915	8.01	8.06
RT1n512d4 (13)	7.335	7.87	8.11	8.25	8.35
RT1n510d5 (14)	6.89	7.29	7.47	7.585	7.77

CONCLUSIONS AND FUTURE WORK

This experiment was designed to investigate the effect of imposing an arificial geography on an evolving population of Tartarus controllers. In this case, the recommeded type of graph would depend on the quality of solution desired. A relatively high score could be achieved quickly using the complete graph or, if sufficient computational power is avaiable, a better solution could be found using a less connected graph. Applying this philosophy to autonomous vehicles, a general strategy could be developed for the vehicle using a sparser graph, with an onboard processor using a more highly connected underlying graph could be used to respond to unexpected events that may arise, such as the introduction of an additional physical object into the vehicle's environment.

The Tartarus problem is a complex test problem, and so the run times involved in gathering the necessary data for a statistical anaylsis are substantial, especially for larger fitness scores. This work is continuing, so that the results may be incorporated into a the developing taxonomy of evolutionary computation problems. It should also be noted that with these large runtime constraints the number of tartarus boards against which the solutions were tested was limited. Because of this the best solutions found here cannot be fairly compared to those of previous works; they are presented here to show that the graph that leads to the best score changes as the number of generations increases. In future research a full set of tartarus boards will be tested to more fully understand the impact of using GBEAs to find higher fitness scores.

REFERENCES

Ackley, D.L. and Littman M.L., "A Case for Distributed Lamarckian Evolution," Working Paper, Cognitive Science Research Group, Bellcore, New Jersey, 1992.

Ashlock, D. and Freeman, J. "A Pure Finite State Baseline for Tartarus," *Proceedings of the 2000 Congress on Evolutionary Computation*, pp. 1223-1230. Addison-Wesley, 2000.

Ashlock, D. and Joenks, M. "ISAc lists, a Different Representation for Program Induction," *Genetic Programming 98, Proceedings of the Third Annual Genetic Programming Conference*, pages 3-10, Morgan Kaufmann, San Francisco, 1998.

Ashlock, D.A., Walker, J., and Smucker, M., "Graph Based Genetic Algorithms," *Proceedings of the 1999 Congress on Evolutionary Computation*, pp. 1362-1368, 1999.

Ashlock, D., Willson, S. and Leahy, N., "Coevolution and Tartarus," *Proceedings of the 2004 Congress on Evolutionary Computation*, pp. 1618-1624, Morgan Kaufmann, San Francisco, 2004.

Bryden, K.M., Ashlock, D.A., Corns, S.M. and Willson, S.J., "Graph Based Evolutionary Algorithms," *IEEE Transactions on Evolutionary Computations*, Vol. 10:5, October, 2006.

Bryden, K.M., Ashlock, D.A, McCorkle, D.S., and Urban, G.L., "Optimization of Heat Transfer Utilizing Graph Based Evolutionary Algorithms," *International Journal of Heat and Fluid Flow* Vol. 24, pp. 267-277, 2003.

Corns, S.M., Bryden, K.M., Ashlock, D., A., and Muth, D., *On the Effects of Representation on Evolving Grid Robot*, Proceedings of the 2005 IEEE Congress on Evolutionary Computation, pp. 1135-1140, 2005.

Goldberg, D.E. *Genetic Algorithms in Search, Optimization, and Machine Learning*. Addison-Wesley Publishing Company, Inc., Reading, MA, 1989.

Reynolds, C., "An Evolved, Vision-based Behavioral Model of Coordinated Group Motion," *From Animals to Animats 2*, pp. 384-392, 1992.

Syswerda, G., "A Study of Reproduction in Generational and Steady State Genetic Algorithms," *Foundations of Genetic Algorithms*, pp. 94-101, 1991.

Whitley, D., 1989, "The Genitor Algorithm and Selection Pressure: Why Rank Based Allocation of Reproductive Trials is Best," *Proceedings of the 3rd ICGA*, pp. 116-121.

AN EVOLUTIONARY ALGORITHM FOR IMPROVEMENT OF QOS OF NEXT GENERATION NETWORK IN DYNAMIC ENVIRONMENT

LIN LIN
Graduate School of Information,
Production & Systems,
Waseda University
lin@ruri.waseda.jp

MITSUO GEN
Graduate School of Information,
Production & Systems,
Waseda University
gen@waseda.jp

ABSTRACT
In this paper, we propose an effective evolutionary algorithm for solving the capacitated minimum spanning tree problem in dynamic environment. The purpose is to improve Quality of Service (QoS) of Next Generation Network (NGN) with considering the network provisioning capability and dynamic environment. We formulate this problem with minimizing the communication cost (as a kind of performance measures for NGN's QoS). As we known, this kind of models is NP-complete. We investigate the different encoding, crossover and mutation operators on the performance of EAs to design of MST problems. Based on the performance analysis of these encoding methods in EAs, we improve predecessor-based encoding, in which initialization depends on an underlying random spanning-tree algorithm. Numerical experiments with various scales of communication network problems show the effectiveness and the efficiency of our approach by comparing with the recent researches.

INTRODUCTION

As the network speed becomes faster and requirements about various services are increased, many researches are currently developing technologies aimed at evolving and enhancing the capabilities of existing network. A Next Generation Network (NGN) is a multi-service functional network, capable of differentiating between users and applications through policies of quality of service (QoS), access and security guidelines, providing Virtual Private Networks (VPNs). The NGN is characterized by the following main aspects (ITU-T Online): (1) Support for a wide range of services, applications and mechanisms based on service building blocks (including real time/ streaming/ non-real time services and multi-media); (2) Broadband capabilities with end-to-end QoS and transparency; (3) Unrestricted access by users to different service providers (4) A variety of identification schemes which can be resolved to IP addresses for the purposes of routing in IP networks, *etc*.

However, current Internet routing protocols such as Open Shortest Path First (OSPF), Routing Information Protocol (RIP), and Border Gateway Protocol (BGP) are called "best-effort" routing protocols, which means it will try its best to forward user traffic, but can provide no guarantees regarding loss rate, bandwidth, delay, delay jitter, *etc*. It's intolerable for NGN services, for example video-conferencing and video on-demand, which require high bandwidth, low delay, and low delay jitter. And provide the different types of network services at the same time is very difficult. Thus, the study of Quality-of-Service(QoS) is very important nowadays (Sun, 1999). To provide QoS in

NGN, many techniques have been proposed and studied, including Integrated Services (Braden *et al.*, 1994), Differential Services (Blake *et al.*, 1998), MultiProtocol Label Switching (MPLS)(Callon *et al.,* 1999), Traffic Engineering and QoS-based Routing (Sun, 1999). And most problems can be formulated as the optimization models, such as the network reliability optimization model, shortest path routing model and constrained minimum spanning tree model *etc.*

In this paper, we focus on the network topological design for providing NGN's QoS. We formulate the problem as a extended capacitated minimum spanning tree (cMST) problem, which the objective is minimizing the communication cost (defined as a kind of performance measures for NGN's QoS) with considering the following constraint. (1) Consider the capabilities of the network; (2) Define different priority for different types of services; (3) Dynamic environment. As we know, this cMST is NP-hard problem. In addition, the complex structures, complex constraints of this problem to be handled simultaneously, which make the problem intractable to traditional approaches. Recently, many evolutionary algorithms (EAs) have been successfully applied to solve constrained spanning tree problems of the real-life instances; and also have been used extensively in a wide variety of communication network design problems. In this paper, we proposed a new EA approach for this NGN's QoS problem, in which a PrimPred-based encoding, Prim-based crossover, LowestCost mutation, Immigration operators and a parameter auto-tuning strategy are adopted.

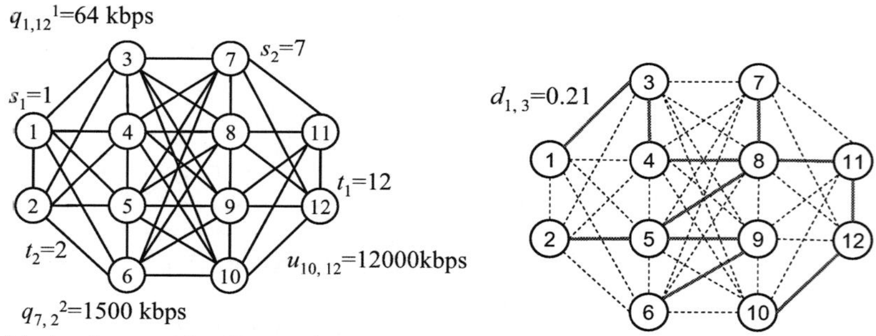

(a) simple example of network (b) a minimum spanning tree

Fig. 1 A simple network with 12 nodes and 40 edges[15]

NOMENCLATURE

Indices

i, j, k =1, 2, …, n, index of node

Parameters

n=|V|: number of nodes

m=|E|: number of edges

$q_{st}{}^l \in Q$: requirement of type l form source node s to sink node t

$u_{ij} \in U$: capacity of edge (i, j)

$w_l \in W$: weight (priority) of type l communication service

$d_{ij} \in D$: delay of edge (i, j) (or defined as a kind of performance measures for NGN's QoS), where $d_{ij} = \Sigma_l \, w_l \, \Gamma(q_{ij}{}^l \text{-} u_{ij})$

$\Gamma(q_{ij}{}^l \text{-} u_{ij})$: a function for delay defination of service type l

Decision variables

y_{ij}: the amount of requirement through arc (i, j)

x_{ij}: 0-1 decision variable

MATHEMATICAL FORMULATION

The communication network is modeled using an edge-weighted undirected graph $G=(V, E, Q, U)$ with n nodes and m edges. Fig. 1 presents a simple network with 12 nodes and 40 arcs. The objecitve is minimizing the total average delay. Mathematically, the problem is reformulated as a capacitated Minimal Spanning Tree (cMST) model is as follows:

$$\min \quad f(\boldsymbol{x}) = \sum_{(i,j)\in E}\left(\sum_{l=1}^{L} w_l \cdot \Gamma\left(\min\left\{0, \left|y_{ij} - u_{ij}\right|\right\}\right)\right) \tag{1}$$

$$\text{s.t.} \quad \sum_{i=1}^{n}\sum_{j=1}^{n} x_{ij} = n-1 \tag{2}$$

$$\sum_{i=1}^{n}\sum_{j=1}^{n} x_{ij} \le |S| - 1 \quad \text{for any set } S \text{ of nodes} \tag{3}$$

$$\sum_{j=1}^{n} y_{ij} - \sum_{k=1}^{n} y_{ki} = \begin{cases} q_{st}^{l}, & i = s \\ 0, & i' \in V - \{s,t\}, \\ -q_{st}^{l}, & i = t \end{cases} \quad \forall (s,t) \in \text{source node and sink node of } q_{st}^{l}, \forall l \tag{4}$$

$$y_{ij} \ge 0, \quad \forall i, j \tag{5}$$

$$x_{ij} \in \{0,1\}, \quad \forall i, j \tag{6}$$

In this formulation, the 0-1 variable x_{ij} indicates whether we select edge (i, j) as part of the chosen spanning tree (note that if $y_{ij}>0$ then $x_{ij}=1$, else $x_{ij}=0$). The constraint (2) is a cardinality constraint implying that we choose exactly n-1 edges, and the packing constraint (3) implies that the set of chosen edges contain no cycles (if the chosen solution contained a cycle, and S were the set of nodes on a chosen cycle, the solution would violate this constraint) constranint (4) implies a *flow conservation law* depended communication requirement on the is observed at each of the nodes other than s or t.

EVOLUTIONARY APPROACH

A. PrimPred-based Genetic Representation

We need to consider these critical issues carefully when designing an appropriate encoding method so as to build an effective EA. How to encode a spanning tree T in a graph G is critical for developing an EA to network design problems, it is not easy to find out a nature representation. We summarized the several kinds of classification of encoding methods as follows: (1) Characteristic Vectors-based Encoding (Davis *et al.* (1993), Piggott and Suraweera (1995), Bean (1994)); (2) Edge-based Encoding (Knowles and Corne (2000), Raidl and Julstrom (2003), Raidl (2000), Chou *et al.* (2001), and Julstrom (2003)); (3) Node-based Encoding (Cayley (1889), Zhou and Gen (1997, 2000))

As discussed in Chou *et al.* (2001)'s research paper, predecessor-based encoding generates some chromosomes that are illegal (*i.e.*, not a spanning tree). Combining the simple random initialization, most of the chromosomes will be illegal due to three reasons: missing node i, self-loop, or cycles. The complex repair process will be used at each generation (computational cost), and after repairing, the offspring of the crossover and mutation are difficult to represent solutions that combine substructures of their parental solutions (worst heritability and locality). In this paper, we proposed a PrimPred-based encoding, improved predecessor-based encoding. The initialization depends on an underlying random spanning-tree algorithm. Fig 2 shows a simple example of the PrimPred-based encoding and a decoded spanning tree with considering the network structure of Fig. 1. The detailed procedure of this PrimPred-based encoding and decoding is introduced in (Lin and Gen, 2006).

node id k:	2	3	4	5	6	7	8	9	10	11	12
chromosome $v(k)$:	5	1	3	8	9	8	4	5	12	8	11

T = {(1, 3), (2, 5), (3, 4), (4, 8), (5, 8), (5, 9), (6, 9), (7, 8), (8, 11), (10, 12), (11, 12)}

Fig. 2 An example of the PrimPred-based encoding

B. Genetic Operators

To adapt to the dynamic environment, we not only need to improve the search capability, but also shorten the computation time. So we need to propose effective genentic operators to improved both quality of solution and enhanced rate of convergence.

Crossover Operator is the main genetic operator. To provide good heritability, a crossover operator must build an offspring spanning tree that consists mostly or entirely of edges found in the parents. In this paper, we proposed Prim-based crossover which applying Prim's algorithm encoding process to the graph $G' = (V, T_1 \cup T_2)$, where T_1 and T_2 are the edge sets of the parental trees selected.

Mutation Operator is a background operator which produces spontaneous random changes in various chromosomes. For considering represent a feasible solutions after mutated, improving the heritability of offspring, we proposed a LowestCost mutation, first removing a randomly chosen edge, determining the separated components, then reconnecting them by the lowest cost edge.

In addition, designing a EA approach for dynamic environment, we heve to consider a important issue that how to improve the current population to adapt to changing service requirement dynamically.

Immigration Operator is modified to (1) include immigration routine, in each generation, (2) generate and (3) evaluate $popSize{\cdot}p_I$ random chromosomes, and (4) replace the $popSize{\cdot}p_I$ worst chromosomes of the current population (p_I, called the immigration probability).

C. Auto-tuning Strategy

As we know, how to set the genetic parameters is a key issue of EAs. In general, as system requirements increase, effectiveness of EA will be decrease. It is the other challenge how to give a exact settings for different system requirement in dynamic environment. In this paper, we propose an auto-tuning strategy by using fuzzy logic control. The main idea is adaptively regulation for taking the balance among the stochastic search and local search probabilities based on the change of the average fitness of parents and offspring which is occurred at each generation. The detailed procedure of this auto-tuning strategy is introduced in (Lin and Gen, 2007).

EXPERIMENTS AND DISCUSSION

In the experiment, proposed PrimPred-based EA is compared with various evolutionary algorithms (Zhou and Gen, 1997; Raidl and Julstrom, 2003) and traditional Prim's algorithm (without considering the capacity constraint). The 3 complete network structures (#of nodes n=20) with 3 kinds of service (Type 1: cable television; Type 2: IP phone; Type 3: data) are combined. The weight (priority) of these 3 types is 0.60, 0.30 and 0.10 respectively. The capacity of each edge (i, j) are represented as random variables depend on the uniform distribution runif(m, 100, 1000). The 30 time-period requirements of different service types from node s to node t are represented as random variables depend on the distribution functions (Type 1: exponential distribution rexp($|Q|$, 0.03); Type 2: lognormal distribution: 0.1*rlnorm($|Q|$, 0.1, 0.1); Type 3: normal distribution: rnorm($|Q|$, 0.01, 0.001)), where $|Q|$=100. The genetic parameter is initialized as follows: population size 20, crossover probability 0.50, mutation probability 0.50. There are two different termination conditions of per time-period are employed. One of them is computation time equal to 0.80 sec. (depend on the updating time of communication requirements). Another one is the objective function equal to 0 (all of requirements can be sufficed). The comparisons results are summarized in Fig. 3.

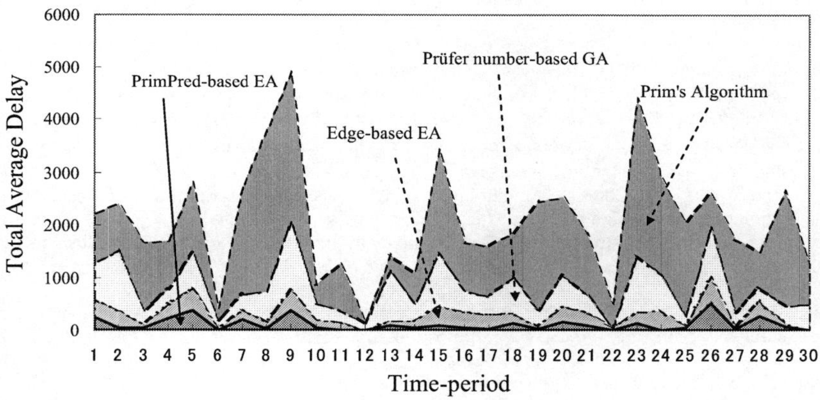

Fig. 3. Comparisons results of 3 different EA approaches and Prim's Algorithm

CONCLUSIONS

In this paper, we proposed an effective evolutionary algorithm for improvement of Quality of Service (QoS) on Next Generation Network (NGN). We formulated this problem as an extended capacitated minimum spanning tree (cMST) problem with considering capabilities of the network, different priority for different types of services and dynamic environment. In this evolutionary algorithm, we adopted a PrimPred-based encoding, Prim-based crossover, LowestCost mutation, Immigration operators and a parameter auto-tuning strategy. Numerical experiments with various scales of communication network problems show the effectiveness and the efficiency of our approach by comparing with the recent researches.

ACKNOWLEDGMENTS

This work is partly supported by the Ministry of Education, Science and Culture, the Japanese Government: Grant-in-Aid for Scientific Research (No.19700071, No.17510138).

REFERENCES

Ahuja, R. and Orlin, J., 2001, "Multi-exchange neighborhood structures for the capacitated minimum spanning tree problem," *Mathematical Programming*, vol. 91, pp. 71-97.

Blake, S. *et. al.*, 1998, "An Architecture for Differentiated Services", RFC 2475, 36 pages ftp://ftp.isi.edu/in-notes/rfc2475.txt

Braden, R., Clark, D. and S. Shenker, 1994, "Integrated Services in the Internet Architecture: an Overview", RFC 1633, 33 pages, http:/ /www.ietf.org/rfc/rfc1633.txt

Callon, R., *et. al.*, 1999, "A Framework for Multiprotocol Label Switching", 69 pages, draft-ietf-mpls-framework-05.txt

Chou, H., Premkumar, G. and Chu, C. 2001, "Genetic algorithms for communications network design – an empirical study of the factors that influence performance," *IEEE Trans. Evol. Comput.*, vol. 5, no. 3, pp. 236-249.

Gen, M., Zhou, G. and Takayama, M., 2000, "Matrix-based genetic algorithm approach on bicriteria minimum spanning tree problem with interval coefficients," *J. of Japan Society for Fuzzy Theory and Systems*, vol. 10, no.6, pp. 643-656.

ITU-T: http://www.itu.int/ITU-T/studygroups/com13/ngn2004/working_definition.html

Julstrom, B. and Raidl, G., 2001, "Weight-biased edge-crossover in evolutionary algorithms for two graph problems," *Proc. the 16th ACM Symposium on Applied Computing, Las Vegas, NV*, pp. 321-326.

Knowles, J. and Corne, D., 2000, "A new evolutionary approach to the degree-constrained minimum spanning tree problem," *IEEE Trans. Evol. Comput.*, vol.4, no.2, pp.125-134.

Lin, L. and Gen, M., 2006, "Node-based genetic algorithm for communication spanning tree problem", *IEICE Transactions on Communications*, vol.E89-B, no.4, pp.1091-1098, April 2006.

Lin, L and Gen, M., 2007, "Auto-tuning Strategy for Evolutionary Algorithms: Balancing Between Exploration and Exploitation", *Soft Computing*, Accepted.

Lo, C. and Chang, W., 2000, "A multiobjective hybrid genetic algorithm for the capacitated multipoint network design problem," *IEEE Trans. Sys., Man, Cyber.*, vol.30, no.3.

Neumann, F., 2004, "Expected runtimes of a simple evolutionary algorithm for the multi-objective minimum spanning tree problem," *Proc. 8th Inter. Conf. of Parallel Problem Solving from Nature*, pp. 80-89.

Raidl, G., 2000, "A weighted coding in a genetic algorithm for the degree-constrained minimum spanning tree problem," *Proc. SAC(1)*, pp. 440-445.

Raidl, G. and Drexe, C., 2000, "A predecessor coding in an evolutionary algorithm for the capacitated minimum spanning tree problem," *Proc. GECCO*, pp. 309-316.

Raidl, G. and Julstrom, B. A., 2003 "Edge Sets: An Effective Evolutionary Coding of Spanning Trees," *IEEE Trans. on Evol. Compu.*, vol.7, no.3, pp.225-239.

Rothlauf, F., Goldberg, D. and Heinz, A., 2002, "Network random keys a tree network representation scheme for genetic and evolutionary algorithms," *Evol Comput. Spring*, vol. 10, no. 1, pp. 75-97.

Rothlauf, F., Gerstacker, J. and Heinzl, A., 2003, "On the optimal communication spanning tree problem," *IlliGAL Technical Report*, Univ. of Illinois.

Schindler, B., Rothlauf, F. and Pesch, H., 2002, "Evolution strategies, network random keys, and the one-max tree problem," *Proc. Applic. of Evol. Computing on EvoWorkshops*, pp. 143-152.

Sun, W., 1999 "QoS/Policy/Constraint Based Routing", Technical Report, Ohio State University, http://www.cse.wustl.edu/~jain/cis788-99/qos_routing/index.html

Zhou, G. and Gen, M., 1997, "A note on genetic algorithm approach to the degree-constrained spanning tree problems," *Networks*, vol.30, pp. 105-109.

Zhou, G. and Gen, M., 1999, "Genetic algorithm approach on multi-criteria minimum spanning tree problem," *E. J. of Operational Research*, vol. 114, no. 1, pp. 141-152.

Zhou, G. and Gen, M., 2000, "Approach to degree-constrained minimum spanning tree problem using genetic algorithm," in M. Gen and R. Chen, *Genetic algorithms & Engineering Design*, New York: John Wiley.

USING A CIRCULAR LINDENMAYER-SYSTEM TO CREATE POLYPHONIC MUSIC COMPOSITIONS IN REAL TIME

KRIS A. BRYDEN
DEPARTMENT OF MUSIC
IOWA STATE UNIVERSITY
AMES, IOWA 50011 USA

ROBERT P. TAYLOR
DEPARTMENT OF MECHANICAL ENGINEERING
IOWA STATE UNIVERSITY
AMES, IOWA 50011 USA

ABSTRACT

Lindenmayer-systems (L-systems) are a specific type of fractal algorithm that incorporates a formal grammar of symbols and replacement rules. The authors have developed an L-system variant called a circular L-system (CL-system) that enables real-time mapping of a continuous data stream into polyphonic music. Polyphony in the system is created in a way similar to branching schemes used in graphical interpretations of bracketed L-systems. Most graphical interpretations of L-systems apply interpretations to only the top-most member of the stack, so the evolution of a branch occurs linearly, and no other portion of the system is evolved while developing that branch. In a sonic situation, this is equivalent to allowing only one voice at a time, but allowing many voices sequentially. To achieve polyphony in the current system, interpretations are applied in parallel to all members of the stack, thus allowing the evolution and rendering of multiple voices simultaneously.

1. INTRODUCTION

After a brief overview of sonification, music algorithms, and Lindenmayer systems (L-systems), this paper will discuss the authors' work on a sonification tool that uses circular Lindenmayer systems (CL-systems) to create polyphonic music in real time from data input. Sonification, that is, representing data with sound, can be an effective means other than or in addition to visualization to provide information about the data. Sonification is especially useful when the data is from a process that occurs over time. In the last few decades, researchers have developed numerous ways to map data into sound for various applications. An overview of research in sonification tool development and perception before 1997 is given in Kramer et al. 1997. and T. Hermann and H. Ritter, 1999.In 1999 Hermann and Ritter introduced a tool they call "model based sonification" that produces sound based on the physical properties of the data by applying "physical model sound-synthesis techniques." Recently Hermann and Ritter (2005) used this framework to develop the "Crystallization Model," which "provides a technique for browsing high-dimensional data." This is further explained in Hermann 2002.

2. MUSIC ALGORITHMS

The tools discussed above directly map aspects of sound to the data, which is common for most sonification projects. Indeed in the Concert Call for Sonification placed by the Listening to the Mind Listening (LML) research project, the criteria specified that the sonification must be a result of "explicit mapping" and that the "timeline of the data must map directly to the timeline of the sonification." Barrass et al. (2006) note that the most common strategy for mapping was to control sound generation parameters by the data values. They mention another common strategy was to matrix the data either by combining multiple data sources into a single sound generator or using a single data source to control various aspects of multiple sound generators (Barrass et al. 2006).

The approach discussed here, however, involves indirect mapping, using an algorithm to create music in response to the data. Our definition of algorithmic music composition, inspired by David Cope (1997), is that of a set of rules that solves the problem of what kinds of sounds to use and where these sounds will occur in a composition. Possible sources of algorithms listed by Cope include expert systems, neural networks, mathematical formulas, stochastics, genetic algorithms, markov chains, generative grammars, and other artificial intelligence techniques. Our system uses a unique fractal algorithm derived from a Lindenmayer-system (L-system) called a circular L-system (CL-system) to create polyphonic music in real time in response to a continuous data stream. After providing a brief overview of the system, this paper will discuss in greater detail its components.

3. SYSTEM DESCRIPTION

Figure 1 provides a diagram of the system, which creates musical output by using an incoming dataset to choose a L-system grammar from a table of grammars. The selected grammar is applied to a user-specified initial string, and the output string is sent to an interpreter that uses it to transform a set of user definable MIDI notes called the motive. The motive is a collection of MIDI notes specifying a short musical phrase or gesture. The user specifies a set of interpretive definitions that maps symbols in the L-system string to MIDI transforms. As each symbol is interpreted into a MIDI transform, it is deleted from the string, providing the basis for the CL-system discussed below. Once a string has undergone a complete L-system derivation, a new data point is sampled and the process continues. At set time intervals corresponding to the minimum allowed data sampling rate, CL-system productions are halted and the system behaves as though the current L-system derivation is complete. The output of the system is polyphonic (multiple voice) MIDI audio. The minimum data sampling rate can be set explicitly by the user or can be implied by the total duration of the initial motive. Changes in the data will then cause noticeable alteration of the sonic output at this rate.

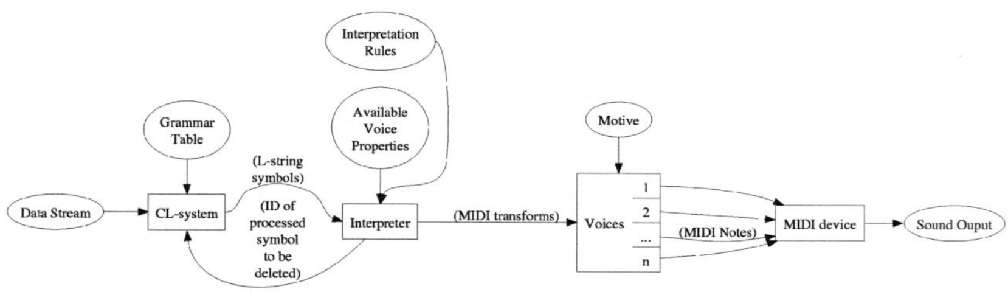

Figure 1. System diagram.

4. L-SYSTEMS

L-systems are a type of string re-writing system based on a formal grammar which defines replacement rules for symbols in the string. Replacement operations, called productions, are applied in parallel to a string to generate a new string. By iteratively applying productions to the string, the string evolves.

L-systems were originally developed in 1968 by Aristid Lindenmayer to model the growth of plants, but have found widespread application as a means of generating various types of computer graphics, particularly fractals (Prusinkiewicz, 1989). They have also been used in the automatic generation of music, typically through the musical interpretation of a graph generated by the L-system, e.g., the system used by G. L. Nelson (1996) to compose "Summer Song".

The system presented here is different in that it uses a direct musical interpretation of symbols in the resultant string from an L-system production via transforms of MIDI properties of a motive that is originally set by the user of the system. As new strings are generated via L-system productions, the interpreted transforms of these strings are applied sequentially to notes in the motive.

5. CIRCULAR LINDENMAYER SYSTEMS

Background

L-system productions can be of arbitrary length, for example, $F \rightarrow f$ and $F \rightarrow fFfFfFfFfFffffFFFFfFfFFfFfFf$ are both legitimate production rules in an L-system that uses the symbols f and F. An important result of this arbitrary length is that the output string of an L-system can potentially be orders of magnitude larger than the input string. In asynchronous or spatial interpretations, such as the graphical interpretation of a string produced by an L-system, this sort of string growth is of little consequence. However, in synchronous, nearly-synchronous, or fundamentally temporal interpretations, this sort of growth presents difficulties.

An example of an interpretation that is frequently both asynchronous and spatial is the standard graphical turtle, which interprets symbols as having spatial transforms such as "move forward", "turn left", etc. The turtle is an asynchronous interpretation because the entire history of the evolution of the turtle's path can be displayed in a single image. In fact the entire evolution is carried out in advance and then displayed all at once. Much information about the path is readily available from the image without the need to explicitly follow the path of the turtle in a time-like way. Due to the spatial nature of such a graphic, very long L-system strings can be meaningfully interpreted in a short time period.

Musical interpretations of an L-system string used for data sonification often need to be nearly synchronous, but by their nature they are time-based interpretations. They need to be nearly synchronous because it is desirable that a change in the data being sonified should manifest as a change in the sonic output within a reasonable time frame, usually a few seconds. Unlike the spatially based graphical interpretations, however, only a very limited amount of information can be usefully presented simultaneously, and these generally take at least a few seconds to be meaningful to a human listener. Due to the need to render output in a temporal way, the arbitrary growth of L-system strings becomes especially problematic. An L-system string with ten thousand symbols may correspond to a fairly small and simple graphical interpretation that can be rendered, seen, and understood in a matter of seconds, but a musical interpretation could easily require an hour just to play the required number of notes.

In order to address the issues of synchronicity of very long strings in a musical interpretation, Bryden, Bryden, and Ashlock (2004) developed a variation of the traditional L-system, which they termed a circular L-system (CL-system). In the CL-system, the parts of a string that have already been interpreted are discarded, so that when a set of productions begins, only some fraction of the resulting string from the previous set is operated on by the CL-system. Consider the grammar *a→ab*, *b→a* operating on the initial string *b*. A traditional L-system would lead to the set of productions shown in the left column below, whereas a CL-system might follow an evolution such as that in the right column.

b	*b*
a	*a*
ab	*ab*
aba	*aba*
abaab	*a* | *baab*
abaababa	*a* | *ababa*
abaababaabaab	*ab* | *aabaab*
abaababaabaababaababa	*a* | *babaababa*

For each column, the string on a given line is the predecessor of the string on the following line. The symbol | in the right column indicates that all preceding symbols on that line have been interpreted as output. On any line containing a |, the symbols preceding | are discarded, and only the symbols following it are used to generate the succeeding line. The position of | in a given string will depend on the speed at which the CL-system productions can be carried out and the time it takes to interpret and render output.

Implementation

Using this setup, the CL-system string grows more slowly; nevertheless it still may grow too quickly to allow musical interpretation to be accomplished within a reasonable time frame, and some additional means is needed to handle this growth. The means developed here is a variation of the above and treats the notes in the motive as though they reside in a ring buffer. While the notes from the previous cycle are playing on the audio device, CL-system productions are applied to the current string and that string is interpreted as transforms to notes in the ring buffer with symbols being discarded from the head of the string as they are interpreted. Once the audio from the previous cycle has ceased playing, the notes from the current cycle are sent to the audio device, and CL-system productions on the current string are halted. At this point the string may still be very long, so to bring it back to a more manageable size, the head is truncated, leaving only a user-definable number of symbols at the end of the string as the input into the next set of CL-system productions. Utilizing this system, thousands of symbols can be interpreted and rendered into audio within a few seconds, allowing the use of large and complex grammars.

Interpreter

The interpreter interprets the symbols from the CL-system productions into MIDI values that transform various parameters of individual notes found in the originating motive. If specified, it can also transform the entire motive by transposing, reversing or inverting it. The MIDI values are rendered using open source MidiShare software on both Windows and Mac OSX operating systems. MIDI is a control language that transmits instructions specifying the specific parameters of a note or group of notes to one or more devices that can render the information into sound.

6. ALGORITHM TO PRODUCE POLYPHONY

Recently we developed a procedure to produce polyphonic music; that is sonic output consisting of two or more independent musical lines. Polyphony in the current system is produced via a bracketed L-system. In bracketed systems, a pair of symbols in the grammar is reserved to indicate stack pushing and stack popping operations. When the pushing symbol is encountered by the interpreter, the current state of the interpreter is saved on a stack and interpretation continues on a branch. When the stack popping symbol is encountered, interpretation on the branch halts and the state of the system is restored to the previously saved state. A graphical interpretation of this bracketing is frequently used to draw trees and other plants that have branched structures.

Polyphony is created by adding a new voice (instrument) each time the stack is pushed and removing the most recently added voice each time the stack is popped. Typical graphical interpretations apply interpretations of the L-system string only to the top-most member of the stack, so the evolution of a branch occurs in a linear fashion and no other portion of the system is modified during the evolution of that particular branch. This is sensible because only one pixel at a time can be written to an image, and it doesn't matter in which order those pixels are written so long as they are placed in the proper spatial location. In a sonic situation this would be the equivalent of allowing only one voice at a time, leading to a monophonic rather than polyphonic output. To achieve polyphony, our algorithm applies interpretations in parallel to all members of the stack so that multiple voices can evolve and be rendered simultaneously. Figure 2 is a short sample of output from the program.

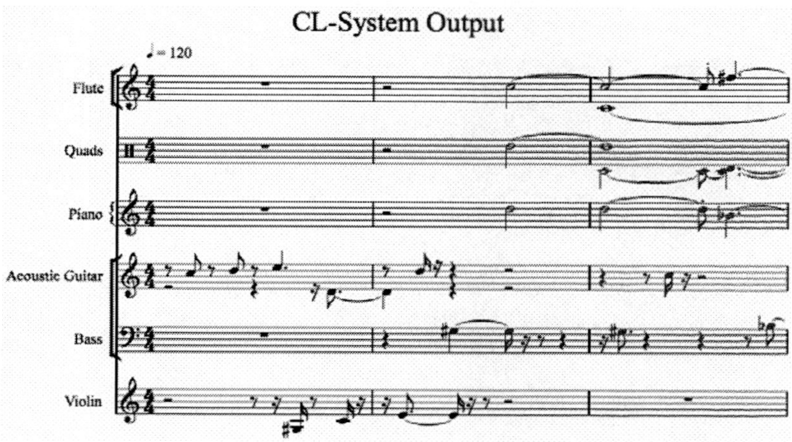

Figure 2. Output from the program.

7. FUTURE WORK

Our system currently handles a single stream of input data to be sonified. We plan to extend the system so that it can handle multiple streams with the capability to add and subtract data streams dynamically. It would be useful, for example, to concurrently sonify temperature data and windspeed data from a weather station. One way to accomplish this is to use a multi-threaded program that maintains a separate grammar for each stream.

The sonic output of our system is very sensitive to the grammar used, and a logical next step is to use various means of automatically evaluating the output music so that techniques such as genetic algorithms or evolutionary strategies can be used to generate grammar tables that meet some given sonic need, such as having a sense of forward motion or sounding like a particular genre of music. K. A. Bryden (2007) discusses a possible evolutionary strategy for encouraging a sense of forward motion in the sonic output.

In its current state, the output of our system is not reliably musical. We are exploring the use of directed graphs to indicate various voice hierarchies which can then be used along with harmonic and rhythmic constraints to make the individual voices alter their output in ways that provide more musicality. A familiar example of such a voice hierarchy is the way musicians in a jazz group listen to what is being played by various other members when deciding which notes are "allowed" in order to fit within the harmonic context of the moment.

ACKNOWLEDGMENTS

We gratefully acknowledge the Iowa state University Institute for Physical Research and technology (IPRT) for its generous support of this research.

REFERENCES

Barrass, S., Whitelaw, M. and Bailes, F., 2006, "Listening to the Mind Listening: an Analysis of Sonification Reviews, Designs and Correspondences." *Leonardo Music Journal* Vol. 16, pp. 13-19.

Bryden, A. N., Bryden, K. A. and Ashlock D. A, 2004, "Presenting a Large Audience Interactive Multimedia Performance Driven by a Circular Lindenmayer System" in *Smart Engineering System Design: Neural Networks, Fuzzy Logic, Evolutionary Programming, Complex Systems, and Artificial Life,* ed. by C. H. Dagli et al, ASME Press, Vol. 14, pp. 253-258.

Bryden, K. A., 2007, "Developing Metrics for Evolving a Musically Satisfying Result from a Population of Computer Generated Music Compositions" in the *Proceedings of the 2007 IEEE Symposium on Computational Intelligence in Image and Signal Processing*, Honolulu, Hawaii, pp. 171-176.

Cope, D., 1992, *Techniques of the Contemporary Composer,* U.S.: Schirmer.

Nelson, G. L., July 1996, "Real Time Transformation of Musical Material with Fractal Algorithms," *Computers and Mathematics with Applications*, Vol. 32, No. 1, pp. 109-116.

Hermann, T. and Ritter, H., 2005, "Crystallization Sonification of High-Dimensional Datasets," *ACM Transactions on Applied Perception (TAP),* Vol. 2 No. 4, pp. 550-558.

Hermann, T., 2002, *Sonification for Exploratory Data Analysis.* Dissertation. Universität Bielefeld.

Hermann, T. and Ritter, H. 1999, "Listen to your Data: Model-Based Sonification for Data analysis" in the Proceedings of the ISIMADE '99 Baden-Baden, Germany.

Kramer, G. et al., 1997,"Sonification Report: Status of the field and Research Agenda." Prepared for the National Science Foundation by members of the International Community for Auditory Display.

Prusinkiewicz, P. and Hanan, J., 1989, *Lindenmayer Systems, Fractals, and Plants,* Vol. 79 of *Lecture Notes in Biomathematics,* New York: Springer-Verlag.

TIME-DEPENDENT ALLOCATION OF DISPATCHING RULES IN JOB SHOP SCHEDULING USING GENETIC ALGORITHMS

GÜRSEL A. SÜER AND JING HUANG
Industrial and Systems Engineering
Ohio University
Athens, OH 45701

Abstract

A new solution methodology for job shop scheduling problem by time-dependent allocation of dispatching rules using Genetic Algorithms is proposed in this study. The objective is to minimize makespan. This problem is a well known NP-hard problem. Obviously, the chance to get a better solution increases when combinations of different rules are assigned to each machine and also changed as time goes on. Dispatching rules for each machine in different time intervals are defined as a rule matrix first. A scheduling model for job shop scheduling process is constructed to determine makespan. Then rule matrix is encoded and transferred to a chromosome to be used in Genetic Algorithms. Finally, the experimentation based on standard job shop scheduling data is performed where different parameters of Genetic Algorithm and dispatching rules are tested, and results show that the proposed methodology has a good potential.

1. Introduction

In a job shop, a finite set of jobs is processed on a finite set of machines. Each machine can only process one job at a time and each job is composed of a set of operations. The sequence of operations for each job is known. However, the sequence of operations varies for different jobs. Job shop scheduling problem is described as determining start time and completion time of each job on each machine without violating sequence of operations to achieve a specified performance measure. In this paper, the objective is to minimize makespan. We propose an approach to job shop scheduling problem by combining conventional heuristic procedures with genetic algorithm.

2. Literature Review

Genetic Algorithm has been widely researched in solving the job shop scheduling problem. Cheng et al. [4] provided a detailed review of study in job shop problem using genetic algorithm where 15 different gene representations are listed and encoding problem for chromosome feasibility is also emphasized. Various heuristic procedures and operators in each stage of genetic algorithm, and several hybrid methods are summarized by Cheng et al. [5]. In 1995, Dorndorf and Pesch [6] provided a new gene representation based on dispatching rules, and combined it with shift bottleneck procedure to improve the search performance of genetic algorithm. In the shifting bottleneck-based genetic algorithm, the selection of node in enumeration tree is determined by genetic algorithm. Cheung and Zhou [7] integrated heuristic rules to genetic algorithm and dealt with sequence dependent setup times in a job shop problem. In a dynamic complex job shop problem, Monch [8] made several modifications to shift bottleneck procedure to decompose overall scheduling into smaller scheduling problems, and then used genetic algorithm as a sub-problem solution procedure.

3. Problem Statement

Dispatching rules method is one of the earliest procedures proposed to solve scheduling problem. Over 100 different dispatching rules are summarized by Panwalker and Iskander [3]. However, no single rule that is suitable for every job shop scheduling problem has been suggested. The selection of dispatching rules will dramatically affect the quality of solutions. Obviously, the chance to get a better solution increases when different rules are assigned to each machine and also they are allowed to change during the course of scheduling period. In this paper, GA is applied to identify the best dynamic allocation of dispatching rules over a period of time to different machines where a delay strategy is also included to find the optimal solution.

4. Solution Methodology using Genetic Algorithm

In this section, the proposed GA methodology is described in detail.

4.1. Gene Representation

Different rules are dynamically allocated to machines when time advances. Both the time interval for reallocating a new rule and the number of alternative rules are parameters. For example, if there are 6 alternative rules for each machine and if there are m machines, Rule Matrix shown in Figure 1 is used.

$$\begin{pmatrix} R1_1 & R2_1 & R3_1 & R4_1 & R5_1 & R6_1 \\ R1_2 & R2_2 & R3_2 & R4_2 & R5_3 & R6_2 \\ \vdots & \ddots & & & & \vdots \\ \vdots & & \ddots & & & \vdots \\ R1_m & R2_m & R3_m & R4_m & R5_m & R6_m \end{pmatrix}$$

Figure 1. Rule matrix

Based on this rule matrix, a chromosome is set up. Each chromosome consists of (m) segments where each segment corresponds to a machine. Each segment is composed of (n) genes corresponding to a row in the above rule matrix. Each gene takes the value of rule index from 1 to 22 and it indicates which rule should be selected among those 21 available dispatching rules. If a gene takes the value of 22, a delay strategy will be applied. When a job is completed in machine j, none of the jobs in the queue waiting for machine j will be assigned if delay strategy is the current rule for machine j.

A sample chromosome representation for a 3-machine job shop scheduling problem is shown in Figure 2, where time interval is 10 minutes.

Figure 2. Gene representation for a sample job shop scheduling problem

Segment 1				Segment 2				Segment 3			
2	1	3	15	7	9	8	5	4	1	2	6
LPT Rule	SPT Rule	FOR Rule	NIN Q/SP T Rule	LWR Rule	LWR /OP Rule	MW R Rule	FIFS Rule	MOR Rule	SPT Rule	LPT Rule	LILS Rule
time: 0-9 40-	time: 10-19 50-59 ...	time: 20- 29 60-	time: 30-39 70-79	time: 0-9 40-49 ...	time: 10-19 50-59 ...	time: 20-29 60-69	time: 30-39 70-79 ...	time: 0-9 40-49 ...	time: 10-19 50-59 ...	time: 20-29 60-69	time: 30-39 70-79 ...

4.2. Reproduction

The performance measure used in this paper is makespan, and the lower the makespan, the better the schedule is. As a result, equations 1-2 are used to calculate reproduction probabilities for each chromosome. The chromosome with the lowest fitness value will have the highest reproduction probability.

$$TFF = \sum_{i=1}^{s} FF_i \qquad (1); \qquad p_i = \frac{TFF / FF_i}{\sum_{i=1}^{s} (TFF / FF_i)} \qquad (2)$$

where FFi is makespan and pi is reproduction probability for chromosome i.

4.3. Selection

Three different selection strategies are applied in this paper: 1) ($\mu+\lambda$) deterministic (Det), 2) ($\mu+\lambda$) stochastic (Stoc), and 3)top X% parents, top Y% offspring (X%).

4.4. Machine-Based Crossover

Pairs of chromosomes are randomly formed. A machine-based crossover is proposed (see Figure 3).

Figure 3. Machine-based crossover

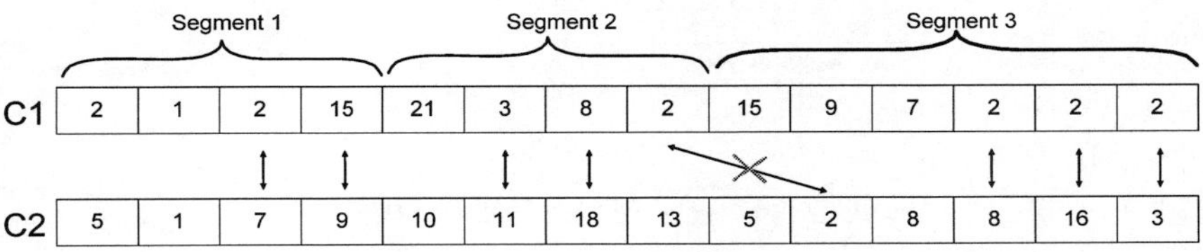

Two crossover operators are used in this paper: 1) Dual Crossover (D) (see Figure 4.a) and 2) Modified Single Crossover (M-S) (see Figure 4.b).

Figure 4. Crossover operators

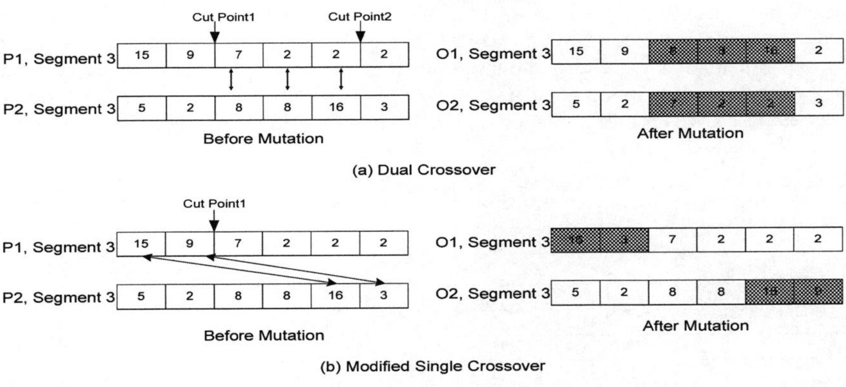

4.5. Machine-Based Mutation

Similar to the crossover strategy, the mutation also only occurs in the same segments. Three mutation operators are used in this paper; 1) Insertion Mutation (I) (see Figure 5.a), 2) Replace Mutation (R) (see Figure 5.b), and 3) Replace by Delay Mutation (RbyD) (see Figure 5.c).

4.6. Dynamic Crossover and Mutation

We propose a dynamic crossover and mutation approach. The number of machines (segments) being mutated increases while the number of machines being crossed over decreases as the number of generations increases.

Figure 5. Mutation operators

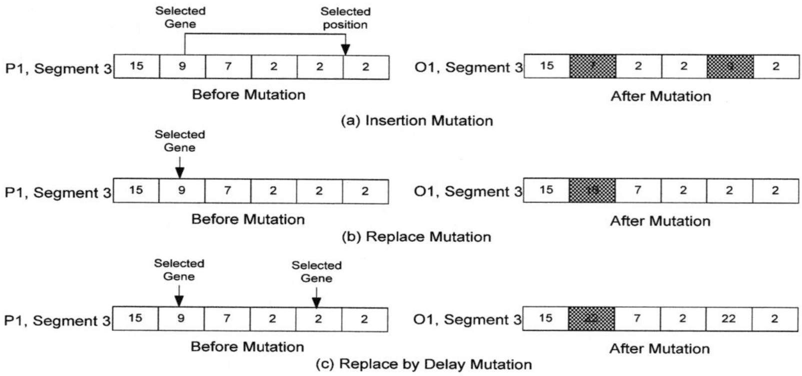

5. Results

In this section, the experimentations carried out are summarized.

5.1 Experimentation 1

In order to investigate the effect of different genetic operators, twelve cases are developed as shown in Table 1. Ten replications for each case have been run in the experimentation for a well known 6*6 job shop scheduling problem [1]. The regular parameters used in genetic algorithm are defined as; population size is 30, number of generations is 500. For gene representation, time interval is set as 2 units, and segment length for each machine is 30 thus having 180 genes in each chromosome. The results of experimentation 1 are also shown in table 1. The results indicate that the proposed delay strategy in the initial population improved the solution. On the other hand, when it was imposed as part of mutation operator, it didn't improve very much. Case 12 outperformed other cases followed by Case 4.

Table 1. Genetic operators used in each case

Case	In. Pop.	Mut.	C-O	Sel.	Best Value	Worst Value	Av. # Gen for Optimal Sol.	Freq
1	ND	I	M-S	Det.	57	69	>500	0
2	ND	I	D	50%	57	70	>500	0
3	ND	R(0.05)	D	50%	55	84	66.2	2
4	ND	R(0.3)	D	50%	55	89	97.9	7
5	ND	R(0.3)	D	Stoc.	55	134	234.3	3
6	ND	RByD(0.3)	D	50%	60	191	>500	0
7	ND	RByD(0.3)	D	Det.	58	104	>500	0
8	ND	RByD(0.03)	D	50%	57	99	>500	1
9	D	I	M-S	50%	55	87	237.5	4
10	D	I	D	50%	55	78	31	3
11	D	R(0.1)	D	50%	55	75	25	3
12	D	R(0.3)	D	50%	55	79	102	9

5.2 Experimentation 2

The following experimentation is constructed for analyzing the effect of dynamic mutation and crossover with the same data set (6*6 job shop scheduling problem [1]). Delay strategy is not included in the initial population, dual machine based crossover operator is adopted, replace mutation operator with 0.1 mutation probability is applied, and 50%-50% selection strategy is used. The population size is 10, and the number of generations is 100. Gene representation remains the same as the last experiment.

In case 2, the generations are divided into 3 stages, 1 to 25, 26 to 50 and 51 to 100. The number of machines being crossed over is 6, 4, 2 and the number of machines being mutated is 2, 4, 6 for stages 1, 2, and 3, respectively. All cases are summarized in Table 2 where frequency optimal is also reported. We run 50 replications for experiment 2. The results show that solution quality is improved when dynamic crossover and mutation approach is applied. Since case 3 outperforms other cases, it shows that decreasing exploration probability is more important than increasing exploitation probability as the generation number grows.

Table 2. Frequency of finding optimal solution for experimentation 2

	Case 1	Case 2 #gen 1-25; 26-50; 51-100	Case 3 #gen 1-25; 26-50; 51-100	Case 4 #gen 1-25; 26-50; 51-100
# Machines Crossed Over	6	6,4,2	*6,4,2*	6
# Machines Mutated	6	2,4,6	*6*	2,4,6
Frequency	3	14	*34*	26

5.3 Experimentation 3

Besides the 6*6 data set, the experimentation with other 10 data sets is also implemented using the proposed method in this paper. Those data sets are all from standard database and the optimal makespan values are known. Thus, deviation from optimal values can be calculated. Since the processing time significantly varies for each data set, thus, the time interval and chromosome representation might be changed for each data set. Relevant data in this experiment is summarized in Table 3.

Table 3. Parameters used for data sets

Data	1	2	3	4	5	6	7	8	9	10
Resource	FT10 [1]	FT20 [1]	ABZ05 [2]	ABZ06 [2]	LA01 [3]	LA02 [3]	LA03 [3]	LA04 [3]	LA05 [3]	LA06 [3]
# Machines	10	5	10	10	5	5	5	5	5	5
# Jobs	10	20	10	10	10	10	10	10	10	15
Time Interval	47	58	43	10	21	14	11	8	21	50
Segment Length	20	20	30	20	30	50	55	66	30	20
Pop. Size	100	100	100	100	100	100	100	100	100	100
# Gen.	1000	1000	1000	1000	1000	1000	1000	1000	1000	1000

Considering that parameters and operators might have an effect on data sets, a relatively generalized experimentation is developed. Delay strategy is included in the initial population, a replace operator with 0.3 mutation probability is adopted in mutation stage, and crossover stage uses machine-based dual crossover operator, in the selection stage, the top 50% percentage parents and top 50% offspring are selected to the next generation.

Table 6. Results for 10 data sets for experimentation 3

Data Sets	Best Solution	Optimal Value	Deviation	CPU time(second)
1	960	930	3.23%	30
2	1185	1165	1.72%	42
3	1239	1234	0.41%	236
4	948	943	0.53%	19
5	666	666	0.00%	<1
6	656	655	0.15%	78
7	605	597	1.34%	59
8	590	590	0.00%	127
9	593	593	0.00%	2
10	926	926	0.00%	<1

6. Conclusions

In this paper, a new approach is proposed using genetic algorithm to identify best configuration of dispatching rules which are allocated to machines every time interval. The experimentation results verify that this approach has a great potential to improve solution quality in a job shop scheduling problem. More importantly, this approach can be used not only to solve a particular instant of the problem but more importantly reveals more information about how the system behaves and trends can be observed relatively easily in terms of critical machines, rules, jobs, or combinations of all these factors.

References

1. Fisher, H., and Thompson, G.L., *Probabilistic learning combinations of local job-shop scheduling rules,* In: Muth, J.F., Thompson, G.L. (Eds.), Industrial Scheduling. Prentice-Hall, Englewood Cli.s, NJ, 1963, pp: 225-251.
2. Lawrence, S., *Resource constrained project scheduling: an experimental investigation of heuristic scheduling techniques (Supplement),* Graduate School of Industrial Administration, Carnegie-Mellon University, Pittsburgh, Pennsylvania, 1984.
3. Panwalker, S. S., and Iskander, W., *A survey of scheduling rules,* Operations Research, Vol. 25, 1977, pp: 45-61.
4. Cheng, R., Gen, M., and Tsujimura, Y., *A tutorial survey of job-shop scheduling problems using genetic algorithms, part I: Representation,* Computers & Industrial Engineering, Vol. 30, 1996, pp: 983-997.
5. Cheng, R., Gen, M., and Tsujimura, Y., *A tutorial survey of job-shop scheduling problems using genetic algorithms, part II: hybrid genetic search strategies,* Computers & Industrial Engineering, Vol. 36, 1999, pp: 343-364.
6. Dorndorf., U., and Pesch, E., *Evolution based learning in a job shop scheduling environment,* Complex System, Vol. 22, 1995, pp:25-40.
7. Cheung, W., and Zhou, H., *Using Genetic Algorithms and Heuristics for Job Shop Scheduling with Sequence-Dependent Setup Times,* Annals of Operations Research, Vol. 107, 2001, pp:65–81.
8. Monch, L., Schabacke, R., and Pabs, D., and Fowler, J., *Genetic algorithm-based subproblem solution procedures for a modified shifting bottleneck heuristic for complex job shops,* European Journal of Operational Research, Vol. 177,2007, pp: 2100–2118.

COMPARISON OF DIFFERENT GENETIC MIXER STRATEGIES FOR MULTI-OBJECTIVE MACHINE SCHEDULING

GÜRSEL A. SÜER & FATIH YARIMOGLU
Industrial and Systems Engineering
Ohio University
Athens, OH 45701

ABSTRACT
A deterministic multi objective single machine scheduling problem with non-zero ready times is solved in this paper. The objective is to obtain a sequence of jobs that will minimize total tardiness and number of tardy jobs. There is no preference made among objectives. A set of non-dominated solutions are provided at the end for the decision maker to make the final decision. A mixer approach, which is a new approach to deal with multi-objective genetic algorithms, is used to obtain high-quality solutions considering both objectives. The developed approach has two main components: 1) Individual Population Evolvers (IPEs), and 2) Population Mixer. Experiments showed that the proposed approach was effective in finding the optimal solution. Furthermore, mixing the populations that are coming from IPEs in the Mixer improved the set of non-dominated solutions. As expected, as the number of generations increased, better solutions are obtained. Among different mixer strategies considered, random-random-random worked very well when average values of replications are considered.

1. INTRODUCTION

Scheduling is one of the most challenging issues in manufacturing. Even in single machine scheduling which is simpler compared to the other scheduling problems, it is difficult to develop a high-quality sequence especially when multiple objectives are considered and there are non-zero ready times. We consider a deterministic scheduling problem and there are two objectives, namely Total Tardiness (TT) and the number of tardy jobs (n_T). No preference is established among the objectives and therefore, the aim is to find non-dominated solutions with respect to both TT and n_T. The final selection among non-dominated solutions is expected to be done by the decision maker.

Tardiness for job i is given by $T_i = \max(0, c_i\text{-}d_i)$ where, i is job index, T_i is tardiness, c_i is completion time and d_i is due date of job i. One of the objectives is minimizing total tardiness, TT= sum (T_i/i=1,2,..,n) where n is number of jobs. The other objective is to minimize the number of tardy jobs (n_T). A job is tardy when its completion time is higher than its due date ($c_i>d_i$). In this paper, setup times are assumed to be sequence-independent and they are included in the processing times of jobs.

2. LITERATURE REVIEW

The application of Genetic Algorithms to multi-objective optimization is very popular. It is possible to classify these methods as Pareto-based and Non-Pareto based approaches.

In 1989, Goldberg [3] studied Pareto ranking based approach in which Pareto ranking is used as the fitness function. In 1995, Fonseca and Fleming [2] improved that procedure and penalty is assigned according to the density of the related region. Non-dominated Sorting Genetic Algorithm (NSGA) was discussed by Srinivas and Deb [7] in 1993. Pareto Tournament Method was proposed by Horn, Nafpliotis and Goldberg [4] in which fitness is assigned according to the loneliness of the neighborhood of a certain solution.

Schaffer [6] proposed Vector Evaluated Genetic Algorithms (VEGA). Randomly Generated Weights and elitism approach is suggested by Murata, Ishibuchi and Tanaka [5] in 1996. Incremental Multiple Objective Genetic Algorithms (IMOGA) is one of the latest studies in the literature by Chen and Guan[1].

3. THE PROPOSED MIXER APPROACH

The main focus in this paper is the optimization of two performance measures at the same time. First, initial populations for Individual Population Evolvers (IPE) are generated randomly. Then, each IPE tries to improve the population considering their objective for a fixed number of generations. After that, chromosomes are selected from these populations and sent to the Mixer. The Mixer works on the populations for a given number of generations. Then chromosomes are selected from the Mixer and then fed back to IPEs. This whole process is called one "Run".

Figure 1. Proposed Mixer Approach

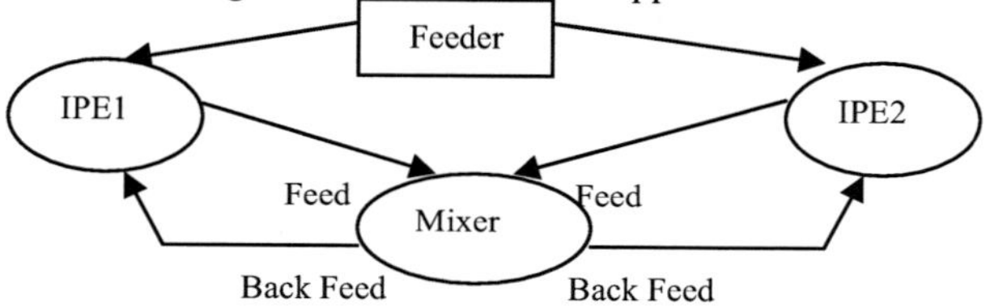

4. GENETIC OPERATORS

Genetic Operators for IPEs can be found in Suer [8]. The genetic operators used in the Mixer are discussed in this section. There are two strategies for the Feed operator, namely random feeding and adaptive feeding. In random feeding, each population evolver has a probability of 0.50 to be selected. Then, a random chromosome is selected from the population. This procedure continues until the population of the mixer is filled. In adaptive feeding, a feeding probability is assigned to each population evolver based on their respective percents of improvement. The lower the improvement percentage from one generation to the next, the higher the probability to select from that Individual Evolver. Different strategies are developed for the mate operator; purely random, Best-Best, and Best-Worst. In Purely Random, two chromosomes are selected randomly from the whole mixer population. In best-best strategy, the population of the mixer is divided into two subpopulations according to their identity (TT or n_T). Each chromosome gets "n_T" or "TT" identity based on better performance. Then, best chromosomes from each subpopulation are mated; and second best chromosomes are mated and so on. Finally in best-worst, the best chromosome of the first subpopulation is mated with the worst chromosome of the second subpopulation and so on. The strategies for back-feed operator are Random, Same Identity-Performance Based, Different Identity Performance-Based. In random, IPEs are fed randomly from the population of the mixer. Same Identity – Performance Based operator divides the mixer into two subpopulations according to their identity (TT or n_T). Then, they are sent to the Individual Population Evolver of their identity based their reproduction probability. In Different Identity – Performance Based operator, they are sent to the Individual Population Evolver of their opposite identity based their reproduction probability.

5. EXPERIMENTATION AND RESULTS

The results of the experimentation are given in four subsections.

5.1 Comparison with the Optimal Solution

There is no known optimizing procedure to the best knowledge of the authors for these two objectives with non-zero ready times. However, a data set with known optimal solution with respect to both measures can be generated where $n_T=0$ and $TT=0$. Then, we can check if the proposed approach can find the optimal solution. The mutation probability used for IPEs and Mixer is 0.05. The population size is 100. The number of runs is 6 and the number of replications is 20. Feeding, mating, back feeding strategies are kept random. The results for this experiment are given in Table 1. As seen in that table, the optimal result is found with great success as the number of IPE generations is increased.

	#IPE gen / # MIXER gen						#IPE gen / # MIXER gen			
Total number of Jobs	300/ 100	400/ 100	500/ 100	600/ 100		Total Number of jobs	300/ 100	400/ 100	400/ 100	600/ 100
10	20	20	20	20		**35**	4	20	20	20
15	20	20	20	20		**40**	0	17	20	20
20	20	20	20	20		**45**	0	2	19	20
25	20	20	20	20		**50**	0	0	3	8
30	18	20	20	20						

Table 1. Number of Generations versus Frequency of Optimal Solutions

5.2 Effect of Number of Runs

In this experiment, the total number of generations was kept constant at 2520, and the number of runs is increased from 3 to 10. The calculation of the numbers of generations for each case is illustrated in Table 2. Thirty replications are carried out for each case. Other parameters of the experiment are kept the same as in the previous experiment.

The non-dominated solutions for this experiment are given in Table 3. Three out 5 overall non-dominated results were found when number of runs was set to 3. Minimum difference (Min diff), maximum difference (Max Diff) and average difference (Av Diff) between the individual solutions and the overall non-dominated solutions are also summarized in the same table. We have also observed that non-dominated solutions continued to improve with 4-run and 9-run cases.

# runs	#gen for IPE1 and IPE2	#gen for Mixer		# runs	#gen for IPE1 and IPE2	#gen for Mixer
3	630	210		**7**	270	90
4	473	157		**8**	236	79
5	378	126		**9**	210	70
6	315	105		**10**	189	63

Table 2. The calculation of number of generations for different number of runs

nT		4	5	6	7	8	9	10	**Final**

	3 Runs	Runs	Runs	Runs	Runs	Runs	Runs	Runs	**Sol.**
2	***199***								**199**
3	***130***	171	185		174	166	183	202	**130**
4	117	154	131	144	162	151	***116***	152	**116**
5		***96***	119	131	111	130		139	**96**
6	106		101	110	105	101	99	115	
7	***81***	94	97	101	102			99	**81**
8		90	91	93	97				
9			90		90	89	95	94	
10					88				
Freq	3	1	0	0	0	0	1	0	
Min	0	0	15	20	15	34	0	18	
Max	1	41	55	35	46	36	53	72	
Av	0.25	23.00	27.25	27.67	31.50	35.00	26.50	42.25	

Table 3. The non-dominated solutions of 30 replications with 2520 generations

5.3 Effect of Number of Runs and Generations

In this case, the number of generations is raised to 10080 and the results are summarized in Table 4. The number of generations for IPE1, IPE2 and mixer will be 4 times higher than those of used in experiment 2. In this case, as expected the quality of the solutions improved significantly. The most of the best results were obtained with 3, 9 and 10 runs.

nT	**3 Runs**	**4 Runs**	**5 Runs**	**6 Runs**	**7 Runs**	**8 Runs**	**9 Runs**	**10 Runs**	Final Sol.
2									
3	173	166	***118***	175	122	134	142	197	**118**
4	135	***105***	113	134			122	125	**105**
5	114		102	106	110	115	***92***	102	**92**
6	102	94	101	93	97	101		***90***	**90**
7	94	91	92	92	94	97		***88***	**88**
8	***87***	90	91	91	89	95			**87**
9	87	89	86	***85***	87	90	***85***	86	**85**
10	***62***	87	72	83	84			78	**62**
11		84							
12						88			
13					81				
Freq	2	1	1	1	0	0	2	2	
Min Diff	0	0	0	0	2	5	0	0	
Max Diff	55	48	11	57	22	23	24	79	
Av Diff	15.87	12.43	6.00	16.50	8.71	12.00	10.25	18.00	

Table 4. The non-dominated solutions of 30 replications with 10080 generations

5.4 Effect of Strategies

In the previous three experiments, random strategy was used in feed operator, mate operator and back feed operator. In this section we wanted to test the effect of different strategies. The strategies used are summarized below:

Mate Operator: Random, Best-Best (BB), Best-Worst (BW)
Feed Operator: Random (R), Adaptive (A)
Back – Feed Operator: Same Identity-Performance Based (SIPB), Different Identity-Performance Based (DIPB)

There are nine combinations of strategies included in this experiment and their results are summarized in Tables 5 and 6. The highest frequency strategy was S1 (Best-Best, Adaptive, Same Identity Performance Based) and the second highest frequency strategy was S9 (Random, Random, Random). It is also worthy to note that S9 also produced the lowest average difference. Table 6 shows that S9 dominated all other strategies with respect to average best solutions.

	Total Tardiness (TT)									***Overall Non-D Sol.***
nT	**S1** BB: A: SIPB	**S2** BB: A: DIPB	**S3** BW: A: SIPB	**S4** BW: A: DIPB	**S5** BB: R: SIPB	**S6** BB: R: DIPB	**S7** BW: R: SIPB	**S8** BW: R: DIPB	**S9** Random: Random: Random	
2										
3		196	156	177	171	212	157	238	***142***	***142***
4	132	141	137	131	143	138	128	***125***	134	***125***
5	122	120	110	113	121	111	119	123	***96***	***96***
6	***90***	109	106	101	105	103	105	91		***90***
7		93	93	98	93	***86***	101		94	***86***
8	***82***	86		91			97		90	***82***
9			89							
10			86							
11	***76***									***76***
Freq	3	0	0	0	0	1	0	1	2	
Min Diff	0	4	7	6	7	0	3	0	0	
Max Diff	26	54	16	35	29	70	23	96	9	
Av Diff	6.60	20.66	12.60	15.00	18.80	22.20	14.33	31.00	5.00	

Table 5. Comparison of Non-Dominated Solutions with Different Strategies

6. CONCLUSIONS

The results of Experiment I show that the proposed approach was successful in finding the optimal solution for n_T and *TT*. In terms of the best non-dominated solution values, there was no clear cut pattern and different number of runs produced good results as shown in Experiments II and III. Another important observation was that, as the number of generations increased better solutions found both with respect to the best values. As a result, differences between individual observations and the best observations were lower. Finally, despite the efforts to create new effective Mixer strategies, Random-Random-Random combination proved to be a robust strategy.

nT	**S1** BB: A: SIPB	**S2** BB: A: DIPB	**S3** BW: A: SIPB	**S4** BW: A: DIPB	**S5** BB: R: SIPB	**S6** BB: R: DIPB	**S7** BW: R: SIPB	**S8** BW: R: DIPB	**S9** Random: Random: Random	***Overall Non-D Sol.***
2										
3		205.66	205.50	208.33	201.25	212.00	186.66	238.00	***175.25***	**175.25**
4	198.64	200.05	201.97	202.76	***194.73***	205.45	203.51	200.93	206.88	**194.73**
5	157.09	156.47	158.08	159.37	160.51	158.06	157.34	161.10	***155.68***	**155.68**
6	135.22	136.02	133.06	137.40	135.31	135.84	138.35	139.13	***132.77***	**132.77**
7	125.05	120.56	123.31	122.64	126.93	123.16	124.12	128.18	***118.29***	**118.29**
8	114.77	116.51	118.03	114.34	117.87	117.42	114.46	117.20	***114.10***	**114.10**
9	111.61	112.41	111.07	***110.63***	118.00	113.25	116.50	115.04	111.20	**110.63**
10	111.08	108.23	***107.57***	109.91	113.07	115.33	115.35	112.75	114.90	**107.57**
11	***102.25***	108.62	117.40	106.33	116.50	111.33	115.25	113.00	110.70	**102.25**
12	120.50	107.75	108.83	***103.00***	109.66	112.00	110.50	110.50	***103.00***	***103.00***
13						***114.00***			111.50	***114.00***
14					***111.00***					***111.00***
Freq	1	0	1	1	1	1	0	0	6	

Table 6. Comparison of Averages of with Different Strategies

References.

1. Chen, Q. and Guan, S. U., *Incremental Multiple Objective Genetic Algorithms*. IEEE Transactions on Systems, Man, and Cybernetics—Part B: Cybernetics, vol. 34, no. 3, 2004.
2. Fonseca, C. and Fleming, P., *An Overview Of Evolutionary Algorithms In Multi-Objective Optimization*, Evolutionary Computation, vol.3, no. 1, pp. 1-16,1995
3. Goldberg, D., *Genetic Algorithms in Search, Optimization and Machine Learning*, Addison-Wesley, Reading, MA, 1989.
4. Horn, J., Nafpliotis, N. and Goldberg, D., A Niched Pareto Genetic Algorithm For Multi-Objective Optimization, in Fogel, D., editor, Pro*ceedings Of The First Ieee Conference On Evolutionary Computation*, pp. 82-87, IEEE Press, Piscataway, NJ, 1994.
5. Murata, T., Ishibuchi, H., and Tanaka, H., *Multi-Objective Genetic Algorithms and Its Application to Flowshop Scheduling*, Computers and Industrial Engineering, vol. 30, no. 4, pp. 957-968, 1996.
6. Schaffer, J., *Multiple Objective Optimization with Vector Evaluated Genetic Algorithms*, in Grefenstete, J. J., editor, Proceedings of the First International Conference on Genetic Algorithms, pp. 93-100, Lawrence Erlbaum Associates, Hillsdale, NJ, 1987.
7. Srinivas, N. & Deb, K. *Multi-objective Optimization Using Non-dominated Sorting in Genetic Algorithms,* Tech. Rep., Indian Institute of Technology, Delhi, India, 1993.
8. Süer, G., A., Vázquez, R., and Santos, J., *Evolutionary Programming For Minimizing The Average Flow Time In The Presence Of Non-Zero Ready Times*, Computers and Industrial Engineering, Vol. 45, pp.331-344, 2003.

HYBRID EVOLUTIONARY CODE GENERATION OPTIMIZING BOTH FUNCTIONAL FORM AND PARAMETER VALUES

DALE E. COURTE
University of Dayton
Computer Science
Dayton, Ohio

ABSTRACT
Evolutionary computation (EC) is an effective tool in the optimization of complex systems. It is desirable to model such a system with appropriate computer commands and parameter settings. Automated determination of both commands and settings, based on observed system behavior, is a desirable goal.

Of the many forms of evolutionary computation, one recently developed discipline is that of *grammatical evolution* (GE). This approach can evolve executable functions in any computer language that can be represented in BNF form. The ability to synthesize arbitrary functions from a formal grammar is an attractive alternative to the expression tree generation of the more common *genetic programming* (GP) approach. However, the GE approach may not be ideal for the optimization of any real-valued parameters of the functions generated. This work combines the use of grammatical evolution for function synthesis with the use of *evolutionary programming* (EP) to optimize the parameters (constants) required by the synthesized functions. These two evolutionary processes combine to explore a rich and complex search space of functional forms and floating point values. A prototype system is implemented and applied to the problem of function approximation.

INTRODUCTION

The term *evolutionary computation* (EC) refers to a collection of related programming disciplines inspired by biological theories of evolution (Bäck and Fogel, eds., 2000). Each of these disciplines represents a specific population-based approach to complex optimization problems. Many have a long history, such as *evolution strategies* (Rechenberg, 1964) (Schwefel, 1968), *genetic algorithms* (Holland, 1975), and *evolutionary programming* (Fogel, et al., 1965) (Fogel, et al., 1990). Others, such as *genetic programming* (Koza, 1990) and *grammatical evolution* (O'Neil and Ryan, 2001) are relative newcomers.

These population-based approaches have much in common, and all borrow terminology from biology. All manage a *population* of candidate solutions to the problem at hand. Each candidate solution is encoded in a linear structure, often referred to as a *chromosome*. Positions within a chromosome are known as *genes* or *codons*, and the permitted values are called *alleles*. The nature of these encodings varies by both evolutionary discipline chosen and the specifics of the problem at hand. In fact, determining an appropriate encoding for potential solutions is the most difficult design decision faced by the evolutionary programmer. Once an encoding is chosen, a *fitness function* must be defined to estimate the validity of candidate solutions. This enables chromosomes to be ranked based on how close they come to solving a problem, and that ranking affects the evolutionary process by allowing solutions to compete for survival in the population.

The basic evolutionary algorithm randomly creates an initial population, evaluates and ranks the members of that population, then repeats an evolutionary cycle through a number of *generations*. During each of these cycles, members of the population *reproduce* to create new members, using the chromosomes of existing members. Those new members then undergo random changes known as *mutations*. A competition known as *selection* then determines which members survive until the next cycle, based on the relative fitness of the members. This process has been shown to successfully navigate very large problem spaces, driving the population to near-optimal solutions.

Of the EC disciplines mentioned above, this paper seeks to combine two of them, forming a hybrid system that leverages the strength of each. The disciplines chosen are grammatical evolution and evolutionary programming, for reasons described in the following section.

MOTIVATION

It has long been a goal of computational science to somehow enable computers to program themselves. This lofty goal was first considered very early in the computer age (Friedberg, 1958) (Friedberg, 1959), and has been met with moderate success over the years. Two EC disciplines generate executable code as their output, those being genetic programming (GP) and grammatical evolution (GE). It is a very attractive idea, that a computer program can not only discover a solution to a problem, but also provide as output, a computer program to solve it. Of these two, GE is chosen here for its flexibility in choice of target language. A GE system can produce executable code in any computer language that can be defined by a BNF grammar (Naur, Ed., 1960).

GE systems use non-negative integer values as alleles, and a chromosome is simply a vector of such values. BNF grammars consist of production rules that define syntactically correct strings in a given language. One can begin at the *start symbol* for the language, and, through successive application of appropriate production rules, build a string that is guaranteed to be syntactically correct. Stated briefly, a given GE chromosome determines a specific sequence of productions to be applied, and therefore uniquely defines a valid string. Given an appropriate grammar, that string represents a valid computer program that can be executed. Discussion of the process that maps a chromosome and a grammar to a string can be found in (O'Neil et al., 2001), and online tutorials are available[1].

Optimization of Constants in GE

Even the simplest computer programs contain numeric constants. In fact, one aspect of a computer program that makes it correct is that the appropriate constant parameter values are used in the computation. These constants may be from either discrete or continuous domains, and a computer program may contain any number of them. In fact, given a computer program, the process of determining the appropriate setting for its constants is itself a considerable optimization problem.

There has been research into how constant generation and optimization can be integrated into the GE paradigm. All approaches extend the language grammar to facilitate the production of, and incremental adjustment (mutation) of numeric values. Some approaches are *digit concatenation* grammars, *ephemeral* constants (grammars to produce expressions used to compute constants arithmetically from a fixed set of values), and *meta-grammars* which allow the actual language grammar, specifically the portion that generates constants, to evolve along with the problem solution (Dempsey et al., 2005) (Dempsey et al., 2007). While effective, these approaches seem to complicate the process of defining the target language grammar. This work defines an alternative to these grammar-based solutions to constant generation that allows for continuous parameter optimization without manipulating the language grammar.

EP Constant Optimization

The evolutionary programming (EP) discipline has long been applied to problems involving the optimization of real-valued continuous parameters (Fogel et al., 1990). Chromosomes in EP systems contain floating-point values. During each generation of the evolutionary process, mutation causes every allele in a chromosome to be perturbed a small amount by adding a random value drawn from a normal probability distribution having zero mean and a given standard deviation. The standard deviation used can be either static or dynamic. In the dynamic case, it is encoded as a mutable part of the chromosome. Using such Gaussian perturbations, EP systems have proven value in the optimization of continuous parameters. Given that fact, it seems desirable to incorporate aspects the EP approach into a grammar-based, hybrid evolutionary system.

PROTOTYPE SYSTEM FOR FUNCTION SYNTHESIS

This initial investigation into GE/EP hybrid system is intended to demonstrate that elements of these two disciplines can operate in parallel to solve the problem of function synthesis. Given

[1] http://www.grammaticalevolution.org/

sufficient x values and corresponding values $y=f(x)$, for some unknown function f, the system presented attempts to synthesize a function $g(x)$ which approximates the unknown $f(x)$. The target language is a subset of MATLAB[2] allowing a full complement of algebraic expressions in x, and including the trigonometric functions sine and cosine. This choice was driven by the ability of MATLAB programs to build strings representing valid commands and execute them at runtime, as well as by the rapid prototyping capability the environment provides.

A chromosome for a given solution consists of two parts: 1) A vector of integers values to be used along with the target grammar to produce executable expressions. 2) A vector of floating point values, one for each constant that appears in the expression. References to constants in the generated code are simply encoded to access this vector (i.e. c(1), c(2), etc.).

The Grammar

The grammar in shown in Figure 1 is a subset of the executable algebraic expressions in x. In MATLAB, all arithmetic operators (+, -, etc.) have an alternative functional form. The functional forms are used here to simplify the grammar without sacrificing expressivity. As an example, the simple expression "x+5x" would be coded as follows:

```
plus(x,times(5,x))
```

Recall that the grammar is used in a GE system to generate an executable expression from a chromosome of integer alleles. The grammar shown makes possible the generation of nested expressions of any length, making the search space of possible expressions infinite. Space does not permit a discussion of this process here, but one can be found in (O'Neil and Ryan, 2001),

```
start-> <exp> ";"

exp     ->      <op2> "(" <term> "," <term> ")"
        |       <op1> "(" <term> ")" | <term>

term    ->      "x" | <exp> | {constant}

op2     ->      "plus" | "minus" | "times" | "rdividep"  "power"

op1     ->      "sin"| "cos" | "sqrtp" | "uminus"
```

Figure 1 : BNF grammar for MATLAB expressions

There are many sources of information on BNF grammars, the seminal discussion being in the Algol 60 report (Naur, ed., 1960). Of note here is the string "{constant}" in the *term* productions. This is an *action symbol*, and each time this term production is applied a new reference to the constant parameter vector will be generated ("c(1)", "c(2)", etc.). For example, were the expression "$5x+2^x$" generated, it would appear as:

`plus(times(c(1),x),power(c(2),x));` where c = [5.0 2.0]

Also of note, the functions "rdividep" and "sqrtp" were defined by the author, to avoid division by zero and complex values respectively. At execution, divide by zero defaults to the identity function, as does the square root of a negative value. In short, these violations become null operations.

The Evolutionary System

At the heart of every evolutionary system is the representation chosen for the chromosomes. Being a hybrid system, a chromosome here is defined in two parts. There is an integer vector of length 128 representing the GE portion of the chromosome. Given this portion of the chromosome and the grammar provided, an expression can be built through a process that will be referred to as *incubation*.

---

[2] Mathworks, Inc.. - http://www.mathworks.com

There is also an EP portion of each chromosome which is a floating-point vector holding all parameter (constant) values for the expression generated. This vector is initialized as empty when the initial chromosomes are created, and is afterwards managed by the incubation process. Every time the incubation process generates a constant reference, the vector is examined to see if a constant value exists for that reference. If not, the vector is extended and a new value is added, drawn randomly from a user-defined range. After incubation, if the vector is too long, its size is truncated. This is necessary as the number of constants associated with a chromosome may change when mutations occur.

During each evolutionary cycle, mutations are applied to both the GE and EP portions of all chromosomes, and a novel, two-step selection technique is applied. A mutation of the GE portion of a chromosome replaces a randomly determined number of alleles with new randomly chosen values. The values in the EP portion undergo Gaussian perturbations using a fixed standard deviation. Note that recombination (combining chromosomes of two or more parents) was not explored in this work.

Supervised learning is used to determine fitness. Recall that a collection of *x* and *y=f(x)* values are known. Given a function approximation *g(x)* encoded in a chromosome *c*, the fitness of that chromosome is computed (given n data points) as follows:

$$fitness(c) = \left(\sum_{i=1}^{n} \left| f(x_i) - g(x_i) \right| \right) + b(c) + p(c),$$

where *b(c)* and *p(c)* are optional heuristic bonuses and penalties that may be assigned based on qualities of the chromosome or the function it represents. The system will attempt to minimize this function.

GE mutations are carried out first. Re-incubation and evaluation occur after the mutations. Mutations that result in fitness greater than that assigned to the non-mutated chromosome are added to the population where they will compete for survival. The most fit members of the population will then be retained for the EP mutations, maintaining a constant population size. EP mutations compete only against themselves. In other words, if the mutation results in improved fitness, the original chromosome is replaced in the population by its mutation. There is no competition against the rest of the population. It was determined experimentally that this two-step selection process both encouraged sufficient exploration and avoided convergence to local minima.

Results and Observations

Experiments have been conducted using a variety of functions, and the results have been promising. Results of two of these experiments are highlighted here. For each, graphs are provided to highlight the evolutionary process, and the synthesized function *g(x)* is shown in MATLAB functional form,

Experiment 1: $f(x) = x^2 + 9.9x - 25\sin(x)$

best evolved g(x)= minus(times(x,plus(c(1),x)),times(sin(x),c(2)));

c=[9.8651 24.7562]

In multiple runs, the system successfully converged to a near-optimal solution. In fact, every run honed in on some form of the correct function, lacking only in complete accuracy of the constants. Below, graphs show the accuracy of the most fit member of the population after 200 and 800 generations in Figure 2.

Experiment 2: $f(x) = 0.25x^5 - 4.24x^3 + 6x$

best evolved g(x)= minus(times(cos(x),times(x,c(1))),times(x,
times(cos(rdividep(x,plus(x,rdividep(c(2),x)))),c(3))));

c=[10.1208 -18.5145 3.1334]

Here, while the evolved function bares little resemblance to *f(x)*, the graphs in Figure 3 clearly show that the function *g(x)* is a reasonable approximation within the given range of *x* values. It was found this function needed more generations to converge to a reasonable fit. 1600 generations were used in the run pictured.

The evolutionary progress of experiment 2 can be seen in Figure 4. At left, the best fitness in the population is displayed over generations. The large decreases most likely represent the introduction of new expressions via GE mutations, while the smaller decreases are more likely due to constant adjustment provided by the GE mutations. At right, the success rate of both types of mutation in the population is shown over generations. The figure shows that both types of mutation continue to contribute throughout the evolutionary process. The exponentially decreasing rate of success is expected as the system converges.

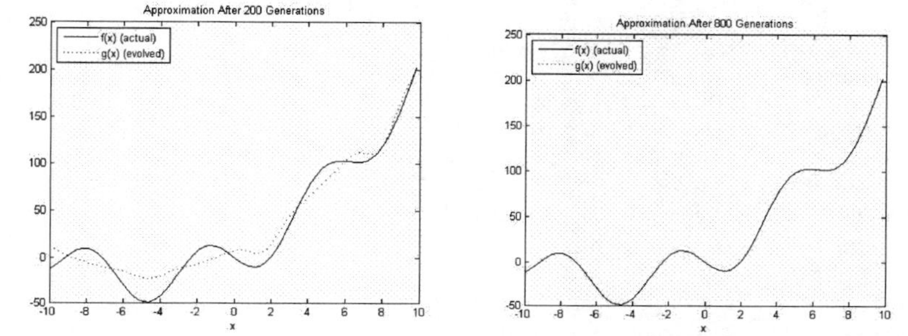

Figure 2: Experiment 1 - Comparison of approximation after 200 and 800 generation

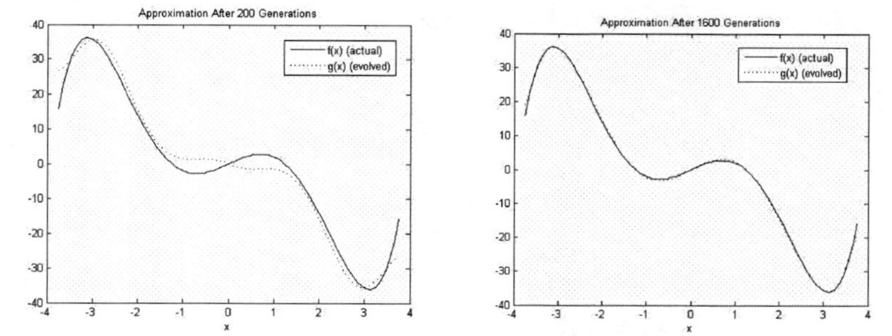

Figure 3: Experiment 2 - Comparison of approximation after 200 and 1600 generation

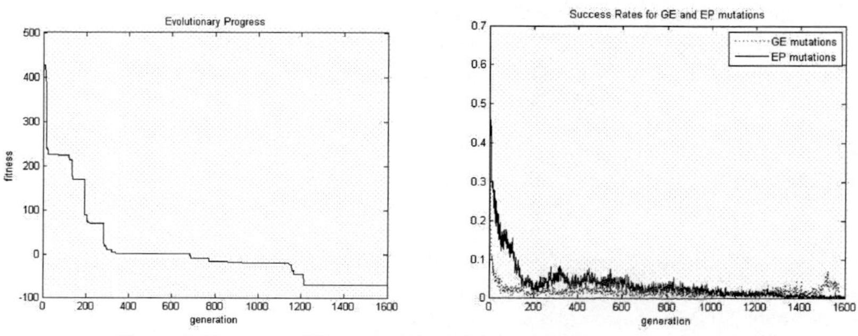

Figure 4: Progress of fitness and mutation success rates over generations

The above experiments were conducted with a population of 800 chromosomes. Constant values for the EP portion of the chromosomes were initially generated in the range [-20.0,+20.0], but were permitted to mutate outside of that range. EP mutations were carried out using a fixed standard deviation of 3.0. Fitness values below zero seen in Figure 4 were made possible by inclusion of a bonus function that rewarded members of the population that generated expressions containing constants. (After all, it is hard to optimize constants in an expression that does not contain constants.)

CONCLUSIONS AND FUTURE WORK

This work has been conducted with an eye towards application to pattern recognition problems. It is hoped that complex feature extractors can be synthesized to identify discriminating information in data of high dimension such as digital images, audio or video. The system presented here demonstrates that this hybrid system architecture can optimize functional forms and constant parameters in parallel, which is a promising result.

It should be noted that many improvements remain to be made to this system. Certainly, the lack of recombination in the evolutionary process can be addressed, especially given that the value of crossover in GE systems is documented in (O'Neil and Ryan, 2001). In addition, dynamic management of standard deviation in the EP portion of the evolutionary process would improve constant optimization. In short, this is a limited prototype system, but one that is nonetheless providing encouraging results.

REFERENCES

Bäck, T. Fogel, D.B., and Michalewicz, T., eds., *Evolutionary Computation I*, Institute of Physics, 2000.

Dempsey, I., O'Neill, M., and Brabazon, A. "Constant creation in grammatical evolution", *International Journal of Innovative Computing and Applications 2007 - Vol. 1, No.1 pp. 23 – 38*, 2007.

Dempsey, I., O'Neill, M., and Brabazon, A. 2005. "Meta-grammar constant creation with grammatical evolution by grammatical evolution", *Proceedings of the 2005 Conference on Genetic and Evolutionary Computation* (Washington DC, USA, June 25 - 29, 2005). H. Beyer, Ed. GECCO '05. ACM Press, New York, NY, 1665-1671.

Fogel, L.J., Owens, A.J., and Walsh, M.J., "Artificial Intelligence through a Simulation of Evolution", *Biophysics and Cybernetic Systems: Proceedings of the 2nd Cybernetic Sciences Symposium*, Spartan Books, pp. 131-155, 1965,

Fogel, L.J., Owens, A.J., and Walsh, M.J., "On the Evolution of Artificial Intelligence", *Proceedings of the 5th National Symposium on Human factors in Electronics*, pp. 63-76, 1990.

Friedberg, R.M., "A Learning Machine: Part I", *IBM Journal of Research and Development*, Vol. 2:1, pp. 2-12, 1958.

Friedberg, R.M., Dunham, B., and North, J.H., "A Learning Machine: Part II", *IBM Journal of Research and Development*, Vol. 3, pp. 282-287, 1959

Holland, J.H., *Adaptation in Natural and Artificial Systems*, University of Michigan Press, 1975

Koza, J.R. *Genetic Programming, on the Programming of Computers by Means of Natural Selection*, MIT Press, 1992.

Naur, P., Ed., "Report on the Algorithmic Language Algol 60", *Communications of the Association for Computing Machinery*, May 1960

O'Neill, M. and Ryan, C., "Grammatical Evolution", *IEEE Transactions on Evolutionary* Computation, Vol. 5 No.4, August 2001.

O'Neill M. and Ryan C., "Crossover in Grammatical Evolution: A Smooth Operator?", *Lecture Notes in Computer Science 1802, Proceedings of the European Conference on Genetic Programming*, pages 149-162. Springer-Verlag, 2000.

Rechenberg, I., (Toms, B.F., transl.), "Cybernetic Solution Path of an Experimental Problem (Kybernetische Lösungsansteurerung Einer Experimentellen Forschungsaufgabe)", Annual Conf. Of the WGLR, Berlin, 1964.

Schwefel, *Experimentelle Optimierung Einer Zweiphasendüse Tiel I*, AED Research Institute Project MHD-Staustrahlrohr 11034/68 Technical Report 35, 1968.

MULTIPLE FEATURE SELECTION USING POLYMODAL EVOLUTIONARY SEARCH

SERGEY A. SUBBOTIN
Zaporozhye National Technical University, Ph.D., Assoc. Prof., department of program tools, Zhukovskiy str., 64. Zaporozhye, 69063, Ukraine

ANDREY A. OLEYNIK
Zaporozhye National Technical University, M.Sc., department of program tools, Zhukovskiy str., 64. Zaporozhye, 69063, Ukraine

ABSTRACT
The evolutionary approach for a decision of feature selection problem is considered in this paper. The new method of polymodal evolutionary search with a chromosome clusterization is offered. The developed method is based on idea of simultaneous search of several optimums, thus informative feature sets (chromosomes) are grouped in clusters on their arrangement in informative feature space that results in more uniform covering of feature search space. So stable subpopulations in different clusters are formed, diversity of search is provided, and convergence to different local minima is reached that allows to find more optimal combinations of features.

INTRODUCTION

At the construction of diagnostic models of complex objects and processes there is a necessity of their description reduction based on feature selection [1–2]. There are different ways of feature selection, among them are exhaustive search, randomized search and heuristic search. These approaches are connected with the necessity of the combinatory problems decision.

Evolutionary methods are the group of stochastic heuristic optimization methods based on the idea of evolution with the help of natural selection.

Evolutionary methods as the information for optimization use only fitness function. Therefore they do not require calculations neither derivative, nor any other characteristics so such methods are applicable to a wide class of optimization problems [3]. These methods allow to solve different combinatory optimization problems, including feature selection problem [6]. However classical methods of evolutionary search are highly iterative.

Classical evolutionary methods [3–7] directly search for a unique combination of informative features. Thus the optimum feature set can be not found because classical evolutionary methods can non-uniformly cover the search space and significant in the variable space can appear non investigated for the limited amount of iterations.

The purpose of this work is the development of a polymodal evolutionary search method based on idea of simultaneous search of several optimum which groups informative feature sets (chromosomes) in clusters on their arrangement in informative feature space that results in more uniform covering of feature space.

POLYMODAL EVOLUTIONARY SEARCH WITH THE CHROMOSOME CLUSTERIZATION

The suggested method during evolutionary search defines the groups of similar chromosomes and raises a variety of a population by reducing the of fitness function values of chromosomes depending on a closeness to the center of their group.

The developed polymodal evolutionary search with the chromosome clusterization assumes the execution of the following steps.

Step 1. Set: the quantity of optimums (optimum sets of informative features) k which is required to be found during evolutionary search; N represents the quantity of chromosomes in a population, $N >> k$.

Step 2. Set the counter of iterations: $t = 1$.

Step 3. Set the quantity of elite chromosomes: $k_e = k$.

Step 4. To initialize an initial population with chromosomes H_j ($j = 1, 2, \ldots, N$;) with length L (the quantity of features).

The model error E_j is used as the fitness function corresponding to chromosome H_j, Thus bits of chromosome that equal one correspond to the features included in model, and zero to non informative features. The model error E_j is calculated using the formula:

$$E = \frac{1}{m}\sum_{i=1}^{m}\left(y^i - y^i_{calc}\right)^2 ,$$

where y^i_{calc} is a value of a predicted parameter calculated on a model with selected features, y^i is an actual value of a predicted parameter, m is a number of exemplars in a training set.

Step 5. To group chromosomes in k clusters based on their fitness function values and an arrangement in an informative feature space [2].

Step 5.1. For each chromosome H_j to calculate Hamming distance to all other chromosomes in a population. Hamming distance d between chromosomes H_j and H_l is calculated under the formula: $d = \sum_{u=1}^{L} o_u$, where L represents the size of chromosomes; $o_u = 1$ if values of u-th bits of chromosomes H_j and H_l do not equal, $o_u = 0$ – otherwise.

Step 5.2. Set the counter of generated clasters: $m = 1$.

Step 5.3. Choose a chromosome with the best fitness function value as the center of m-th cluster. Thus the chromosomes which yet have been not grouped in clusters are considered.

Step 5.4. Add in cluster (N/k – 1) chromosomes nearest on Hamming distance to a chromosome, being the center of current m-th cluster.

Step 5.5. If $m = k$ then go to a step 6.

Step 5.6. Set: $m = m + 1$. Go to a step 5.3.

Step 6. Reduce fitness function values of the chromosomes which are not being the best in cluster using the formula:

$$f_{n,j} = \left(\frac{d_j}{d_{\max,j}}\right)^s f_j ,$$

where f_j – the fitness function value before j-th chromosome changing; $f_{n,j}$ – new fitness function value of j-th chromosome; d_j – Hamming distance from j-th chromosome to the center of its group; $d_{max,j}$ – the maximal Hamming distance in the cluster of j-th chromosome; s – the parameter determining a degree of fitness function reduction of chromosomes, not being the centers of cluster, $s \geq 1$.

Step 7. Apply crossover and mutation operators.

Step 8. Generate new generation. Thus the best (elite) chromosomes in every cluster are guaranteed pass in the new generation.

Step 9. If $t = T$ (T is the maximum possible quantity of iterations), then go to a step 12.

Step 10. Set: $t = t + 1$.

Step 11. Go to a step 5.

Step 12. For each of k chromosomes being the centers of clusters to construct mathematical model. As such model it is possible to use neural networks.

Step 13. Choose the best from k models constructed on the previous step. The feature set corresponding to such model, makes as the decision.

Step 14. Halt.

SOFTWARE IMPLEMENTATION OF EVOLUTIONARY SEARCH

The suggested method of polymodal evolutionary search with the chromosome clusterization has been implemented with the help of a language of Matlab. The choice of Matlab is connected by that it contains powerful libraries of mathematical functions for construction of numerical models of dependences.

The software is consist of functions:

– ModelError makes the model corresponding to chromosome H_j, and calculates its error E_j. Thus features which correspond to zero bits of a chromosome are not included in model.

– FeatureSelection carries out polymodal evolutionary search with the chromosome clusterization in order to defining of an optimum combination of informative features. As the fitness function it is used ModelError.

The interaction of the developed functions among themselves and with Matlab environment and Matlab components is represented on fig. 1.

EXPERIMENTS AND RESULTS

Experimental research of the offered method of feature selection was carried out based on the decision of a vehicles classification problem under bidimentional images.

Initial sample contained the transformed graphic representations of various vehicles received from video cameras. Sample consisted of 1062 copies, each of which was characterized by 4096 features representing normalized values of the image points intensity projected on a sensor matrix of points of dimension 64×64. Using these 4096 features there were calculated 26 generalizing the graphic information about the. Vehicles were classified on cars, minibuses, motorcycles, lorries and buses. For each class of transport the model was constructed and the most informative features were selected. Thus the problem has consisted in synthesis of four classification models of each type of vehicles based on the most informative of 26 generalizing features.

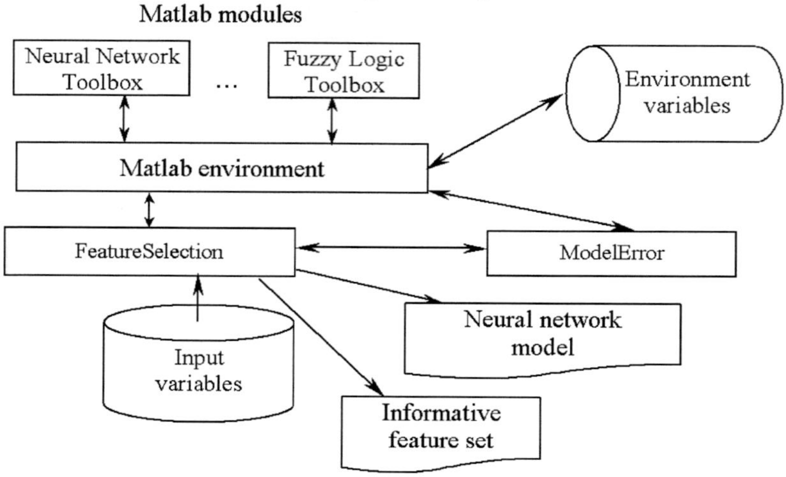

Fig. 1. The interaction of the developed functions with Matlab environment

Initial parameters of all evolutionary methods were established by the following: population size – 100; crossover probability – 0,8; mutation probability – 0,1. Stopping criteria: maximum number of iterations – 20; achievement of comprehensible value of the fitness function equal 0,002; absence of changes in the best value of fitness - function during ten iterations.

The purpose of the experiments was to find out the combination of the most informative features and synthesis the neural network model based on this combination. The error of classification on this model must be not more then 0,2 %. Results of experiments for different quantity of clusters are presented in the table 1 (ε_{ls} represents minimum error of classification for learning sample, ε_{ts} represents minimum error of classification for test sample, n represents the quantity of the obtained models providing sufficient accuracy).

Table 1

Results of experiments

Number of clusters	ε_{ls}, %	ε_{ts}, %	n
1 (classical evolutionary search)	0,18	1,84	0
2	0,19	0,72	0
3	0,12	0,18	1
4	0,15	0,19	2
5	0,14	0,14	2

The experiments have shown that as a result of application of polymodal evolutionary search with the chromosome clusterization stable subpopulations are formed in different clusters, heterogeneity of search is provided, and also convergence to different local minima is reached.

CONCLUSIONS

The new method of polymodal evolutionary search with a chromosome clusterization is offered in this paper. The developed method is based on idea of simultaneous search of several optimums, thus informative feature sets (chromosomes) are grouped in clusters on their arrangement in informative feature space that results in more uniform covering of feature search space.

Comparison of the results obtained with the help of the developed method with results of application of classical evolutionary methods shows that the offered method allows to find more optimum feature set for the account of more uniform covering of search space. Thus, the suggested method can be recommended for application in practice at the decision of technical and biomedical problems.

NOMENCLATURE

d_j: Hamming distance from j-th chromosome to the center of its group
$d_{max,\,j}$: the maximal Hamming distance in the cluster of j-th chromosome
E_j: model error
f_j: the fitness function value before j-th chromosome changing
$f_{n,j}$: new fitness function value of j-th chromosome
H_j: j-th chromosome
k: quantity (number) of optimums
k_e: quantity of elite chromosomes
L: chromosome length (the quantity of features)
m: number of exemplars in a training set
n: the quantity of the obtained models providing sufficient accuracy
N: population size
s: the parameter determining a degree of fitness function reduction of chromosomes, not being the centers of cluster
t: iteration counter
y^i_{calc}: value of a predicted parameter calculated on a model with selected features
y^i: actual value of a predicted parameter
ε_{ls}: minimum error of classification for learning sample
ε_{ts}: minimum error of classification for test sample

REFERENCES

Birger I.A. Technical diagnostics. – M.: Machinebuilding, 1978. – 240 p. (In Russian).

Dubrovin V.I., Subbotin S.A., Boguslayev A.V., Yacenko V.K. Intelligent means of diagnostics and reliability prediction of airengines: Monograph. – Zaporozhye: Motor Sich JSC, 2003. – 279 p. (In Russian).

Holland J. H. Adaptation in natural and artificial systems. Ann Arbor. The University of Michigan Press. – 1975. – 97 p.

Haupt R., Haupt S. Practical Genetic Algorithms. - New Jersey: John Wiley and Sons, 2004. – 261 p.

The Practical Handbook of Genetic Algorithms. Volume I. Applications / Ed. by L.D. Chambers. – Florida: CRC Press, 2000. – 520 p.

The Practical Handbook of Genetic Algorithms. Volume II. New Frontiers / Ed. by L.D. Chambers. – Florida: CRC Press, 2000. – 421 p.

Xin Yao. An Overview of Evolutionary Computation. – Canberra: Australian Defence Force Academy, ACT, 1996. – 19 p.

STATISTICAL COMPARISON OF EVOLUTIONARY ALGORITHMS

MATHIEU BARRETTE
Electrical engineering department
École de technologie supérieure

TONY WONG
Automated manufacturing
École de technologie supérieure

BRUNO DE KELPER
Electrical engineering department
École de technologie supérieure

ABSTRACT
Benchmarking evolutionary algorithm performance is an important task in computational intelligence research. However evolutionary algorithms are stochastic processes and to obtain significant results proper statistical tools must be used. This paper proposes a generic comparison procedure to assess the relative performance of several evolutionary algorithms. The proposed procedure is based on a simple ranking scheme and is statistically relevant under a control risk of error. A useful feature of the procedure is the ability to visualize the dynamical behavior of the algorithms along with the confidence intervals of their performance. It also has the advantage of eliminating the use of arbitrary weighting coefficients when several comparison criteria are involved.

1. INTRODUCTION

Comparing the performance of different evolutionary optimization algorithms (EAs) is an important task especially when developing new approaches using the computational intelligence paradigm. Usually a benchmarking process is performed in order to show the effectiveness of the newly devised evolutionary algorithm against some other well-known EAs. There exists several performance criteria commonly used to the literature. For example, the mean speed of convergence excluding failed trial runs [1], the number of trial runs that reached global optimum [2], mean solution fitness [3] and the best solution fitness [4, 5, 6]. All these criteria are valid but they should be used within some proper comparison procedure so that it is possible to determine if an EA's performance is statistically significant. Benchmarking EAs involves several trial runs and sample statistics can not reveal their underlying distribution. Worst, misinterpretation of the benchmarking results may lead to erroneous conclusions. Thus, to perform comparison analysis, it is also necessary to determine if the observed differences are truly significant.

This paper is organized as follows: section 2 describes the basic evolutionary optimization approach. Section 3 describes the proposed comparison procedure followed by a discussion of the complete experimental protocol. The last section shows the comparison results for five well-known evolutionary algorithms.

2. EVOLUTIONARY OPTIMIZATION

An evolutionary algorithm (EA) is and iterative stochastic process often used to optimized difficult numerical problems such as $F(\mathbf{x}) : \Re^N \rightarrow \Re$ where $\Re$ is the set of real numbers, N is the problem dimension and $\mathbf{x} = \{x_1, x_2, ..., x_N\}$ the set of decision variables bounded by $l_i \leq x_i \leq h_i$, $i = 1, 2, ..., N$. Discussions in this work are oriented towards minimization problems but they are equality applicable to maximization problems since $\min\{F(\mathbf{x})\} = -\max\{-F(\mathbf{x})\}$. Let $\mathbf{x}_n$ be the best solution found by an instance of an evolutionary algorithm after n iterations and $f(n) \equiv F(\mathbf{x}_n)$ be the corresponding objective function value or more succinctly the cost value. We shall consider that an iteration is synonymous to one evaluation of $F(\cdot)$, thus $f(n)$ is the cost value obtained after n evaluations of the objective function. In the context of evolutionary optimization, the number of iterations n^* needed to reach a global optimum is not known a priori and is considered as a random variable with unknown distribution. Since n^* is not known beforehand, it is often desirable to install a stopping criterion which limits the execution time of an EA.

2.1 STOPPING CRITERION

Optimum solutions are usually known in a benchmarking context. It is thus possible to stop the evolutionary process once it reaches a small zone Z enclosing the global optimums. We can define this zone by adding a small threshold value ε (usually in the order of 10^{-6}) to the optimal solution f^* of a given problem. In other words, the evolutionary process can be stopped as soon as $f(n) \leq Z$ where $Z \equiv f^* + \varepsilon$.

However, evolutionary optimizers are not immune to local attractors. They may be imprisoned in a local optimum and not able to reach Z. It is also possible that the EAs converge too slowly and must be stopped before reaching Z. A maximum allowable number of iterations $n_{\max}$ is recommended to confine the optimization process within a reasonable processing time. So, there are usually two stopping criteria in play while evaluating EA performances: *i*) reaching the optimum zone Z; *ii*) exhausting the maximum number of iterations $n_{\max}$.

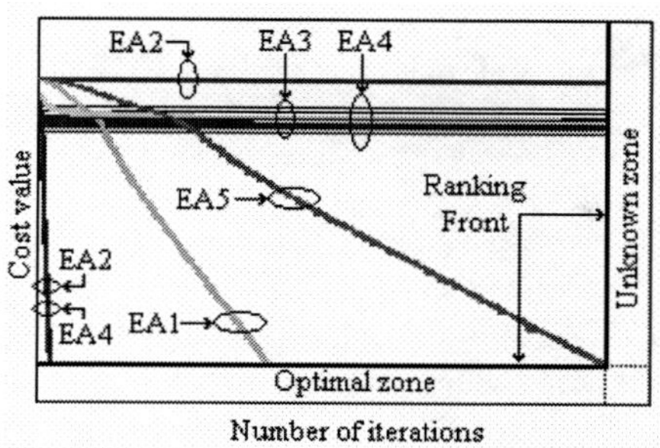

Fig.1 Typical performance results for five different EAs.

3. PERFORMANCE COMPARISON

Since EAs are stochastic processes, it is necessary to execute several trial runs in order to obtain a fair comparison. In practice, not all EAs can reach the optimum zone for all trials. This fact is shown in Fig. 1 where five different evolutionary optimizers were used to optimize a numerical function. In this experiment only EA1 and EA5 were able to reach Z for all trial. For EA2 and EA4 they failed to reach Z for some trials but succeeded for some others. Finally EA3 always misses the optimum zone. Thus, the finite confinement provided by Z and $n_{\max}$ give rise to two different types of optimization result for each trial run: *i*) the EA instance reached the optimal zone; *ii*) the EA instance failed to reach the optimal zone. Therefore, a fair performance comparison procedure must also take into account the number of successful (failed) trial runs and the number of iterations needed to reach the optimal zone.

3.1 RANKING SCHEME

As mentioned in the introduction, EA performance evaluation can be performed using different criteria. In this work, we propose the integration of several comparison criteria by ranking the trial runs according to their relative performance. Consider M trial runs within which $I \leq M$ trials reached the optimal zone Z while the remaining $M - I$ trials failed to reach Z. We assign ranks to the trial runs according to the following scheme: assign ranking numbers, in increasing order, for the I successful trial runs with the best trial results receiving the smallest rank number. For example, the rank number may be based on the number of iterations n needed to reach the optimal zone Z. For the $M - I$ failed trial runs, $n = n_{\max}$ and the ranking should be based on another criterion. To illustrate the proposed ranking scheme, we will use the cost value $f(n)$ to rank the failed trial runs. This ranking scheme can be carried out graphically by using the results shown in Fig. 1 as follows: starting from the leftmost of the optimal zone which is parallel to the x-axis, move along this axis and assign a ranking number to the trails intersecting the optimal zone. When we reach the $n_{\max}$ value continue ranking the trials with the cost value $f(n_{\max})$ given by the y-

axis. This ranking scheme is given by Eqs. (1) and (2) where q denotes de 'quality' of a trial run and R its rank.

$$q_j = n_j + f_j \quad j = 1,2,\ldots,M \; ; \tag{1}$$

$$R_j = r(q_j) = 0.5 + \sum_{k=1}^{M} s(q_j - q_k); \; s(u) = \begin{cases} 0 & \text{if } u < 0, \\ 0.5 & \text{if } u = 0, \\ 1 & \text{if } u > 0. \end{cases} \tag{2}$$

Extending the ranking procedure to classify multiple EA performance is straightforward. Consider the case of k different EAs using m_i trial runs for a total of $M = \sum_{i=1}^{K} m_i$ trials. The resulting ranks can be arranged in a $m_k \times k$ matrix as shown by Eq. (3)

$$\begin{matrix} R_{1,1} & R_{2,1} & \cdots & R_{k,1} \\ R_{1,2} & R_{2,2} & \cdots & R_{k,2} \\ \vdots & \vdots & \cdots & \vdots \\ R_{1,m_1} & R_{2,m_2} & \cdots & R_{k,m_k} \end{matrix} \tag{3}$$

3.2 HYPOTHESIS TESTING

Performance comparison can be initiated after the ranking assignment is completed. Statistical hypothesis testing (SHT) is a fair and recommend approach to determine the significance of the observed differences. In SHT the probability of obtaining a result at least as extreme as that observed, based on the truthfulness of the null hypothesis, is called the p-value. If the p-value is smaller than a threshold α then the difference is statistically significant. In this work, the null hypothesis H_0 and the alternate hypothesis H_1 can be defined as

$$\begin{aligned} &H_0\text{: The } k \text{ distributions are equivalent} \\ &H_1\text{: At least two distributions differ} \end{aligned} \tag{4}$$

Here the data are ranks, and one nonparametric test between k different and independent samples based on their ranks is the well-known Kruskall-Wallis test defined by the following statistic [11, 12]:

$$H = \frac{12}{M(M+1)} \sum_{i=1}^{k} \frac{\left(\overline{R}_i\right)^2}{m_i} - 3(M+1) \;, \tag{5}$$

where $\overline{R}_i$ is the mean rank of ith EA. It is possible to compute the exact distribution of H, but that would require complex tables because of the multiple combinations of M, k and m_i. It is possible to approximate H by using the χ^2 distribution with $k - 1$ degrees of freedom. This approximation is accurate except when $k = 3$ and $m_i \leq 5$, $i = 1,2,\ldots,k$. Also, in the case of tied ranks, as it change the variance; an adjustment should be made by dividing H with

$$1 - \frac{\sum t^3 - t}{M^3 - M} \tag{6}$$

where t represents the number of tied rank over all samples and the sum is computed for the complete set of tied ranks. If the SHT rejects the null hypothesis then at least one of the k EAs performs differently from at least one other EA. However, SHT will not indicate which of the k EAs is involved. To get this information multiple comparisons must be used. This is explained in the next subsection.

3.3 MULTIPLE COMPARISON PROCEDURE

Standard SHT works well in evaluating one null hypothesis but it cannot be iterated to account for more hypotheses. The reason is that multiple SHT would increase the α threshold value and thus lower the significance level of the tests. For more than one test, it is also necessary to carefully control the so-called familywise error rate which is the probability of making one or more type I error (accepting the null hypothesis which it is in fact false) over all the pairwise tests. In this work, multiple comparison procedure (MCP) is used as a post-hoc comparison procedure. It is only executed when the null hypothesis is rejected by the Kruskall-Wallis test. For a ranking analysis, Tukey's honestly significant difference criterion (HSD) MCP defined by Eqs. (7) and (8) is recommended [7, 13].

$$\left|\bar{R}_i - \bar{R}_j\right| \le \frac{Q_{k,\infty}^{(\alpha)}}{\sqrt{2}} \cdot \sqrt{d_{ij}} \quad i,j < k \,, \tag{7}$$

$$d_{ij} = \left(\frac{M(M+1)}{12} - \frac{\sum t^3 - t}{12(M-1)} \right) \cdot \left(\frac{1}{m_i} + \frac{1}{m_j} \right), \tag{8}$$

Where $Q_{k,v}^{(\alpha)}$ is the upper α point of the studentized range distribution with parameter k and degrees of freedom v. Only if Eq. (7) is true then EA i and j can be assumed to have significant differences.

4. EXPERIMENTAL PROTOCOL AND RESULT ANALYSIS

According to the above discussion, an experimental protocol incorporating result ranking, statistical hypothesis testing and multiple comparison features can be described in eight simple steps.

1. Define a fair testbench comprising several well-known optimization problems.
2. Define the maximum number of evaluations n_{max}.
3. Define the threshold for the optimum zone Z.
4. Select the set of EAs for comparison.
5. Run a predefined number of trials for each EA.
6. Use Eqs. (1) to (2) to compute the rank of all trials.
7. Execute the Kruskall-Wallis test with α = 0.05 (for example).
8. If the null hypothesis is rejected, do the HSD MCP.

Following the above protocol, five well-known EAs, in their canonical forms, were chosen for comparison. They are: differential evolution (DE) [1], evolution strategy (ES) [8], particle swarm optimizer (PSO) [9, 10], harmony search (HS) [4, 5, 6] and the stochastic hill climbing with learning by vector of normal distribution (SHCLVND or

HCL) [2]. The meaning and role of the EA controlling parameters are explained in the cited references. The testbench also includes seven synthetic optimization problems with known optimum and they are detailed in Table 1.

Table 1. Testbench functions

Function		$\lvert H_i - L_i \rvert$	Optimum
Ackley	$F_1(\vec{x}) = -20 \cdot \exp(-0.2\sqrt{\frac{1}{n}\sum_{i=1}^{n} x_i^2}) - \exp(\frac{1}{n}\sum_{i=1}^{n}\cos(2\pi \cdot x_i)) + 20 + \exp(1)$	60	$F_1(\vec{0}) = 0$
Corana parabola	$F_2(\vec{x}) = \sum_{i=1}^{n} \begin{cases} c_r\,(sgn\,(z_i)\,t_i + z_i)^2\,d_i & \text{if } (\lvert x_i - z_j \rvert < \lvert t_i \rvert) \wedge (\lvert x_i \rvert > \lvert t_i \rvert) \\ d_i \cdot x_i^2 & \text{otherwise} \end{cases}$ $c_r = 0.15 \quad s_i = 0.2 \quad t_i = 0.05 \quad z_i = \left\lfloor \frac{x_i}{s_i} + 0.49999 \right\rfloor \cdot s_i \cdot sgn(x_i)$ $n = 2 : d = [1, 1000] \quad n = 30 : d = [1, 1000, 10, 10, 1, 10, 100, 1000, 1, 10, ...]$	2000	$F_2(\vec{0}) = 0$
Griewangk	$F_3(\vec{x}) = \sum_{i=1}^{n}(\frac{x_i^2}{4000}) - \prod_{i=1}^{n}(\cos(\frac{x_i}{\sqrt{i}})) + 1$	1200	$F_3(\vec{0}) = 0$
Hyper-Ellipsoid	$F_4(\vec{x}) = \sum_{i=1}^{n} i \cdot x_i^2$	10.24	$F_4(\vec{0}) = 0$
Michalewicz	$F_5(\vec{x}) = \sum_{i=1}^{n} \sin(x_i) \cdot (\sin(\frac{i \cdot x_i^2}{\pi}))^{20}$	π	-
Rastrigin	$F_6(\vec{x}) = 10 \cdot n + \sum_{i=1}^{n}(x_i^2 - 10 \cdot \cos(2\pi \cdot x_i))$	1200	$F_6(\vec{0}) = 0$
De Jong's #1 sphere	$F_7(\vec{x}) = \sum_{i=1}^{n} x_i^2$	10.24	$F_7(\vec{0}) = 0$

For each optimization problem, 50 trials are to be conducted using all five EAs while the optimum zone Z is defined by $\varepsilon = 10^{-6}$. Finally, two versions of the synthetic problems (two-variable and 30-variable) will be used in order to verify the effects of problem scaling on the EAs. In all, 14 problems are to be solved by five evolutionary optimization algorithms using this experimental protocol.

Table 2 details the benchmarking results. In this table column D gives the problem dimension, column n_{max} denotes the maximum allowable number of iterations, column F is the mean objective value found by the EAs, column CR (convergence rate) is the percentage of trial that have reached the optimum in the sample, CS (convergence speed) is the mean number of iteration done by the trials that reached the optimum and thus the ones that haven't reach Z are not include in that statistic. The last column is the most important one. It is the graphical results of the proposed comparison procedure. The result analysis of a MCP can be quite difficult to investigate numerically. That is the reason why they are generally supported by a graphical representation that expose simultaneously all the relation between the k EAs on one explicit graphical representation. The horizontal lines are the confidence intervals of the mean ranks. When two EAs exhibit overlapping confidence intervals, this means that, under the α error probability, there is no significant difference between the two EAs. Since the ranking is given in increasing order (good results have small ranking numbers – see section 3.1), the best performing EAs are the leftmost ones in the last column of Table 2.

Table 2. Benchmarking results

F	D	N_{max}	EA	F	CR	CS	MCP
F1	2	9.96E5	DE	1E-6	100	1630	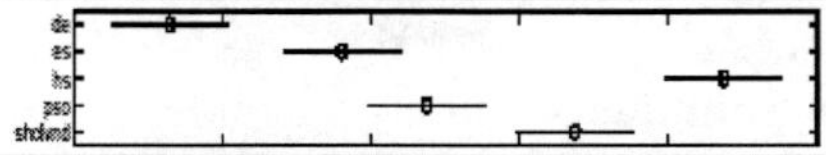
			ES	2.8	86	1763	
			HS	2.6E-5	0	-	
			PSO	1E-6	100	5016	
			HCL	1E-6	100	943 770	
F1	30	1.311E6	DE	1E-6	100	531 750	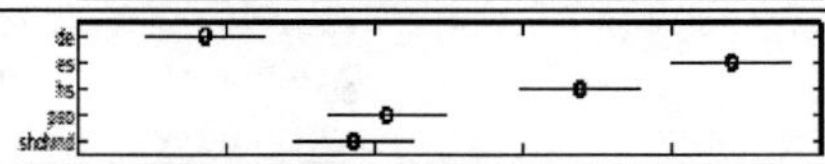
			ES	18.8	2	27 649	
			HS	1.9	0	-	
			PSO	1.1	32	30 593	
			HCL	1E-6	98	1 305 200	
F2	2	8.5E5	DE	1E-6	100	1456	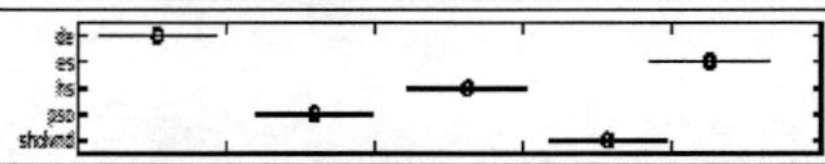
			ES	59 841	8	1644	
			HS	1E-6	100	179 050	
			PSO	1E-6	100	4182	
			HCL	1E-6	100	747 830	
F2	30	1.5E6	DE	1E-6	100	640 710	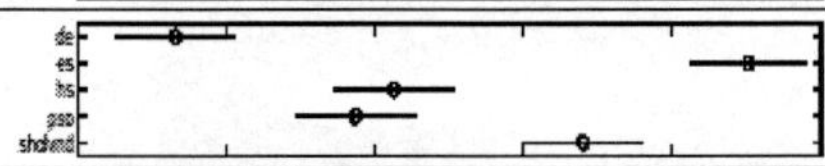
			ES	1.69E8	0	-	
			HS	4372	0	-	
			PSO	112 726	16	373 730	
			HCL	198 088	0	-	
F3	2	6.33E5	DE	2E-3	72	3557	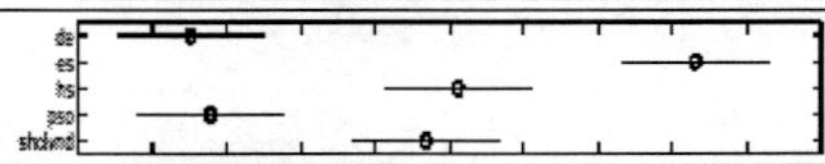
			ES	3.2E-2	6	1160	
			HS	3.7E-3	50	62 256	
			PSO	8.9E-4	88	17 330	
			HCL	1E-6	100	553 590	
F3	30	9.41E5	DE	1E-6	100	409 130	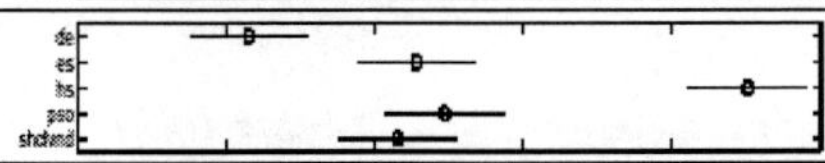
			ES	1E-2	32	20 520	
			HS	10E-2	0	-	
			PSO	1.7E-2	32	20 932	
			HCL	1E-6	100	934 580	
F4	2	3.128E5	DE	1E-6	100	676	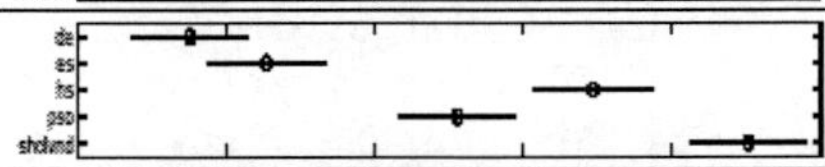
			ES	1E-6	100	756	
			HS	1E-6	100	10 328	
			PSO	1E-6	100	1602	
			HCL	1E-6	100	251 280	
F4	30	8.1E5	DE	1E-6	100	315 690	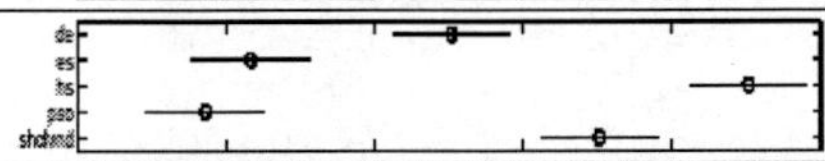
			ES	1E-6	100	18 251	
			HS	20.4	0	-	
			PSO	1E-6	100	17 888	
			HCL	1E-6	100	795 660	
F5	2	3.5E5	DE	1E-6	100	707	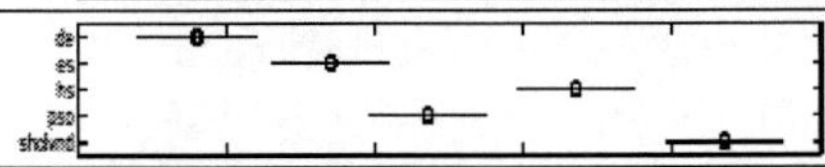
			ES	0.11	84	760	
			HS	1E-6	100	45 703	
			PSO	1E-6	100	1715	
			HCL	1E-6	100	268 050	
F5	30	3E6	DE	1E-6	100	1 249 900	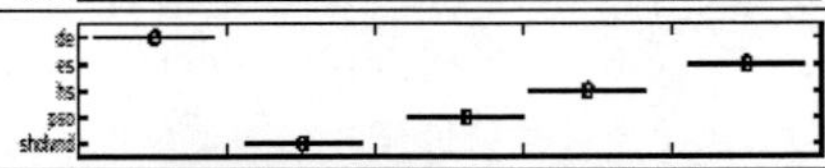
			ES	11.4	0	-	
			HS	6.9	0	-	
			PSO	5.1	0	-	
			HCL	0.41	0	-	
F6	2	8.48E5	DE	1E-6	100	1636	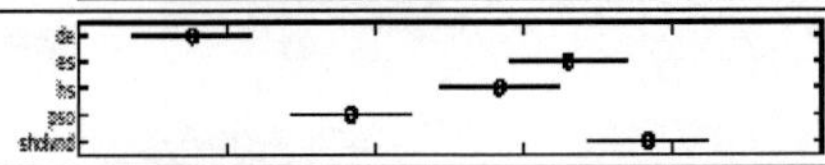
			ES	0.975	32	1520	
			HS	1E-6	100	119 790	
			PSO	1E-6	100	4731	
			HCL	1E-6	100	776 820	
F6	30	1.3E6	DE	1E-6	100	821 740	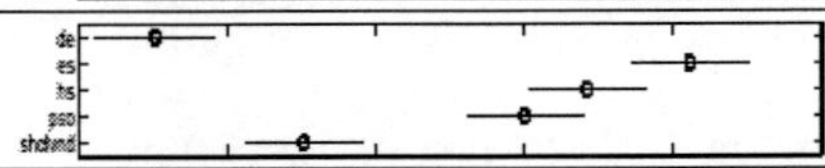
			ES	155.7	0	-	
			HS	118.4	0	-	
			PSO	107.2	0	-	
			HCL	1.4	24	1 235 200	
F7	2	2.972E5	DE	1E-6	100	649	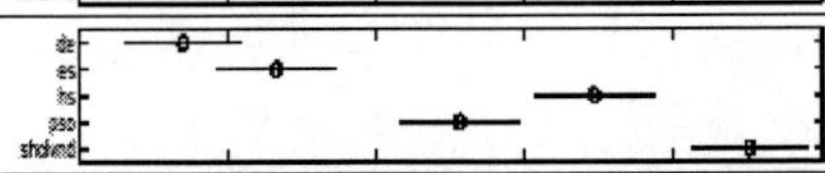
			ES	1E-6	100	732	
			HS	1E-6	100	6907	
			PSO	1E-6	100	1637	
			HCL	1E-6	100	234 310	
F7	30	7.4E5	DE	1E-6	100	276 690	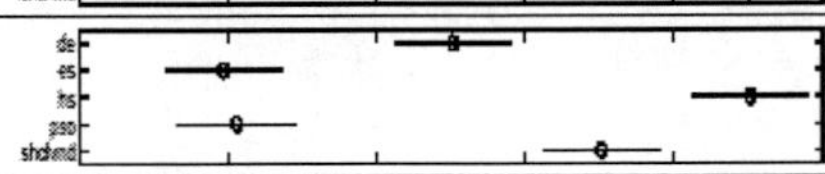
			ES	1E-6	100	15 442	
			HS	1.73	0	-	
			PSO	1E-6	100	15 692	
			HCL	1E-6	100	701 230	

We observed that the solution quality and the trial ranks are dependent on $n_{\max}$ and by extension it is also true for the convergence rate, the convergence speed and the fitness. Since $n_{\max}$ is decided beforehand it is often desirable to assess the dynamical behavior of the algorithms during their execution. By computing the rank at different temporal states and executing a MCP each time a rank change occurs, it is now possible to visualize the algorithm's performance for different numbers of iteration. In Fig. 2, the y-axis is the mean rank of each EA and each horizontal band represents the mean rank of an EA for different temporal range. The band's height is the confidence interval obtained with the MCP. As shown in Fig. 2, rank changes may occur at any point during the comparison process and if n_{max} is small the outcome could be quite different.

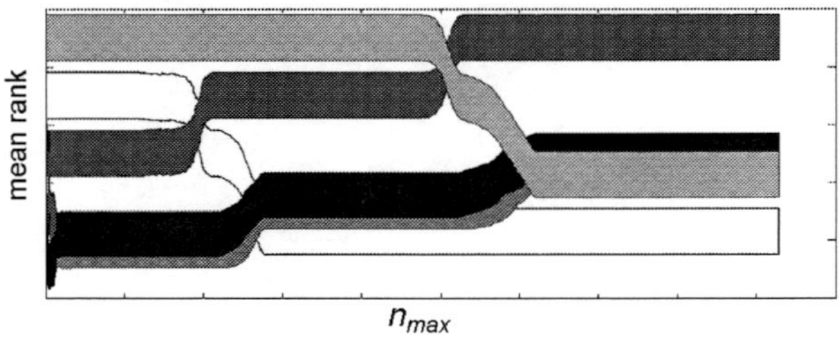

Fig. 2 MCP progress plot of EA performances.

Interestingly, the benchmarking performed in the work revealed that the differential evolution algorithm is efficient for all seven test problems. DE's performance for the Ackley, Corona's parabola, Griewangk, Michalewicz and Rastrigin functions is significantly different compared to all other EAs. However, this benchmarking also exposed influences by the problem dimension on the EA performances. For instance, DE and ES performed equally well for the 2D hyper-ellipsoid problem but DE is outperformed by PSO and ES when the problem is scaled to 30 dimensions. The same phenomenon is observed for the De Jong sphere problem. In this benchmarking, HS and SHCLVND are outperformed by DE on all test functions.

6. CONCLUSIONS

The proposed comparison procedure has the advantage to be statistically consistent and adequate to evaluate performance analysis of the EAs. With the ranking scheme, it also allows a unification of the three major criteria under a unique ranking number without the use of arbitrary artefacts such as weighting coefficients or function objective transformations. For these reasons, the proposed comparison procedure is more rigorous and more robust than the usual single-value objective numerical comparisons.

REFERENCES

[1] R. Storn and K. Price, "Differential Evolution A Simple and Efficient Heuristic for Global Optimization over Continuous Spaces", Journal of Global Optimization, vol. 11, Dordrecht, pp. 341–359, 1997.

[2] S. Rudlof and M. Koeppen, "Stochastic Hill Climbing with Learning by Vectors of Normal Distributions", Fraunhofer-Institut for Production Systems and Design Technology , 1996.

[3] Y. Shi and R.C. Eberhart, "Empirical Study of Particle Swarm Optimization", Proceedings of the 1999 Congress on Evolutionary Computation, 1945–1950. Piscataway, NJ: IEEE Service Center.

[4] Z. W. Geem and C.-L. Tseng, "New Methodology, Harmony Search and Its Robustness", Late-Breaking Papers of GECCO-2002, pp. 174–178, New York City, USA, July. 2002.

[5] Z. W. Geem and C.-L. Tseng, "Engineering Applications of Harmony Search", Late-Breaking Papers of GECCO-2002, pp. 169-173, New York City, USA, July 2002.

[6] K. S. Lee and Z. W. Geem, "A new structural optimization method based on the harmony search algorithm", Computers and Structures vol. 82 pp781–798.

[7] Y. Hochberg and A.C. Tamhane, "Multiple Comparison Procedures", Wiley, USA, 1987.

[8] H. G. Beyer and H. P. Schwefel, "Evolution strategies A comprehensive introduction", Netherlands: Kluwer Academic Publishers, 2002. vol. 1,no. 1 pp. 3–52.

[9] M. Clerc and J. Kennedy, "The Particle Swarm Explosion, Stability, and Convergence in a Multidimensional Complex Space", IEEE Trans. on Evolutionary Computation, vol. 6 no. 1 pp. 58–73, february. 2002.

[10] J. Kennedy and R. Eberhart, "Particle Swarm Optimization", Proceedings of the 1995 IEEE International Conference on Neural Networks, pp. 1942–1948.

[11] J.D. Gibbons, "Nonparametric Methods for Quantitative Analysis" 2nd ed., Ohio: American Sciences Press Inc., 1985.

[12] W.J. Conover, "Practical Nonparametric Statistics" 3rd ed., John Wiley & Sons, Inc., 1998

[13] J.D. Gibbons and S. Chakraborti, "Nonparametric statistical inference" 3rd ed., New York: Marcel Dekker, 1992.

STUDY FOR MULTI-PRODUCT LOGISTICS NETWORK DESIGN PROBLEM WITH CONSTRAINTS OF DELIVERY COURSE

SHINICHIRO ATAKA
Waseda University
ataka@fuji.waseda.jp

MITSUO GEN
Waseda University
gen@ waseda.jp

ABSTRACT
Logistics network design is one of the most important fields in supply chain management (SCM). For some real-world applications, it is often that the Transportation Planning (TP) model is extended to satisfy other additional constraints or performed in several stages. In addition, a product produced in a factory may be several kinds of products. Even if it is a factory of the same company, delivery is different by a kind of a produced product. The restrictions which this model has are deeply concerned with TP model in the real world. In this paper, we formulate a multi-stage logistics network model with considering the constraints of delivery rute. To solve the problem, we also propose an effective Genetic Algorithms (GA) approach by using a priority-based encoding method with special decoding process for a delivery rute. Finally, numerical experiments with various scales of logistics network problems show the effectiveness and the efficiency of our approach by comparing with the recent researches.

1. INTRODUCTION

A logistics network design is an important element when constructing Supply Chain Management (SCM), and it is the problem that should be optimized for long-term promotion of efficiency of the whole supply chain. In late years, by globalization of a market or the spread of Internet, the form diversifies, too. Transportation Planning (TP) was formulated by Hitchcock and the purpose is to design a delivery rute of the smallest cost. However, there are the following problems in general TP model.

(1) A general TP model considers it only about one phase of delivery plans almost DC (Delivery center) and customer.
(2) In addition, a delivered product is unitary.
(3) A problem paid its attention to in delivery section is only a delivery cost

As well as these, a consideration is not simply weather and a traffic condition, physical condition of a driver as reality. For some real-world applications, it is often that the TP is extended to satisfy several other additional constraints or performed in several stages.

By the way, Genetic Algorithm (GA) is metaheuristics solution search technique copied from evolution of a creature. It is applied to various fields since it was suggested, and application to TP is one of example. GA is the technique that is effective for a non-linear problems, it can support the problem that is impossible of a calculation by general LP software. However GA has some problems. These are great computational complexity, precocious convergence to a limited part solution, complexity of parameter setting.

In this study, we suggest multi-product logistics network with constraints of delivery rute. This is a more realistic TP model, and a product produced in a plant is having many kinds, and this model owns limitation on the delivery rute. A reason give constrain of delivery rute is to express various delivery forms showing for figure 1. Therefore, we combine hybrid GA (hGA) for solving this problem and priority-based encoding was used as encoding method. It was improvement type of GA, and hGA was developed to make up for a weak point of GA which we mentioned above.

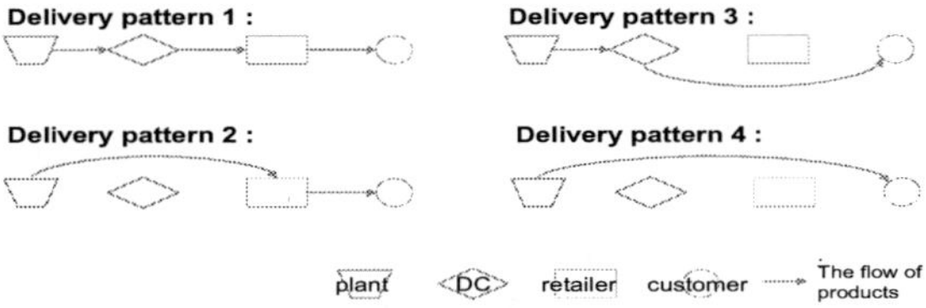

Figure 1. Various circulation forms

2. MATHEMATICAL FORMULATION

As expansion of TP, we suggest a network model to show for figure 2. All nodes are connected in this model and have a flexible delivery form. A product to be considered has many kinds, and the delivery costs are different each, too. Furthermore, products gives limitation each so that delivery rutes are different. From these, it is thought that this network model is the thing which is near to reality.

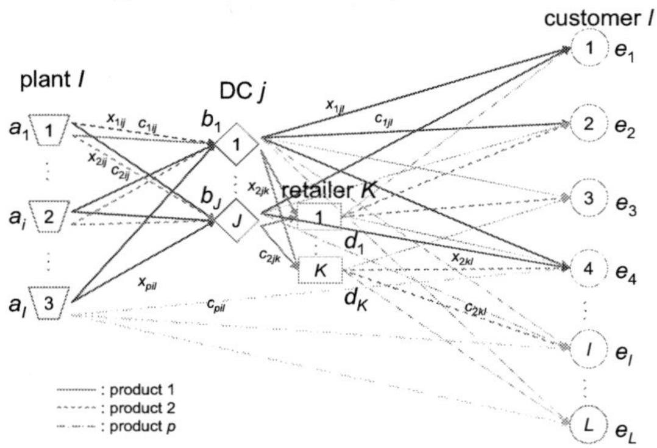

Figure 2. Illustration of a network model

The purpose of in this study is a delivery rute design of the smallest cost to satisfy customer demand subject to the plant and DCs capacity and also the minimum number of DCs to be opened. The mathematical model of the problem can be formulated as follows.

$$\min \quad z=\sum_{p=1}^{P}\left(\sum_{i=1}^{I}\sum_{j=1}^{J}t_{pij}c_{pij}x_{pij}+\sum_{j=1}^{J}\sum_{k=1}^{K}t_{pjk}c_{pjk}x_{pjk}+\sum_{k=1}^{K}\sum_{l=1}^{L}t_{pkl}c_{pkl}x_{pkl}+\sum_{i=1}^{I}\sum_{k=1}^{K}t_{pik}c_{pik}x_{pik}\right.$$
$$\left.+\sum_{j=1}^{J}\sum_{l=1}^{L}t_{pjl}c_{pjl}x_{pjl}+\sum_{i=1}^{I}\sum_{l=1}^{L}t_{pil}c_{pil}x_{pil}\right)+\sum_{j=1}^{J}g_j z_j \tag{1}$$

$$\text{s. t.} \quad \sum_{j=1}^{J}t_{pij}x_{pij}+\sum_{k=1}^{K}t_{pik}x_{pik}+\sum_{l=1}^{L}t_{pil}x_{pil}\le a_i,\ \forall i,p \tag{2}$$

$$\sum_{p=1}^{P}\sum_{i=1}^{I}t_{pij}x_{pij}\le b_j,\ \forall j \tag{3}$$

$$\sum_{p=1}^{P}\sum_{j=1}^{J} t_{pjk} x_{pjk} \le d_k,\ \forall k \tag{4}$$

$$\sum_{i=1}^{I} t_{pil} x_{pil} + \sum_{j=1}^{J} t_{pjl} x_{pjl} + \sum_{k=1}^{K} t_{pkl} x_{pkl} \ge e_{pl},\ \forall l, p \tag{5}$$

$$x_{pij}, x_{pik}, x_{pil}, x_{pjk}, x_{pjl}, x_{pkl} \ge 0,\ \forall i, j, k, l, p \tag{6}$$

$$z_j = \{0,1\},\ \forall j \tag{7}$$

where:

I: number of plants (i = 1,2,…,I), J : number of DCs (j = 1,2,…,J), K : number of retailers (k=1,2,…,K), l : number of customers (l=1,2,…,L), p : a kind of a product, a_i : capacity of plant i, b_j : capacity of DC j, d_k : capacity of retailer k, e_{pl} : demand of customer l, c_{pijj} : unit cost of delivery from plant i in product p to DC j, c_{pjk} : unit cost of delivery from DC j in product p to retailer k, c_{pkl} : unit cost of delivery from retailer k in product p to customer l, c_{pik} : unit cost of delivery from plant i in product p to retailer k, c_{pjl} : unit cost of delivery from DC j in product p to customer l, c_{pil} : unit cost of delivery from plant i in product p to customer l, x_{pij} : the amount of shipment from plant i in product p to DC j, x_{pjk} : the amount of shipment from DC j in product p to retailer k, x_{pkl} : the amount of shipment from retailer k in product p to customer l, x_{pik} : the amount of shipment from plant i in product p to retailer k, x_{pjl} : the amount of shipment from DC j in product p to customer l, x_{pil} : the amount of shipment from plant i in product p to customer l, g_j : fixed cost for operating distribution center j, z_j : 0-1 variable that takes on the value 1 if DC j is opened, t : If product p goes along a route of a subscript, otherwise, 1, it is 0. While constraints (2) and (3) ensure that the plant-capacity constraint and the DC-capacity constraint, respectively, constraint(4) is the retailer-capacity constraint, and using the constraint (5), it is guaranteed that all demands are met.

3. GENETIC APPROACH

Here, we explain elucidation by GA for this model. A network model suggesting it this paper has a flexible delivery rute, and chromosome representaion becomes particularly important from the complexity.

3.1 REPRESENTATION

In this study, we use a priority-based encoding developed by Gen & Cheng (2000). As for the feature a gene in a chromosome is characterized by two factors: locus, the position of the gene within the structure of chromosome, and allele, the value the gene takes. In priority-based encoding, its length is equal to total number of plants and DCs, and the position of a gene is used to represent a node (plant/DC in transportation network), and the value is used to represent the priority of corresponding node for constructing a tree among candidates.

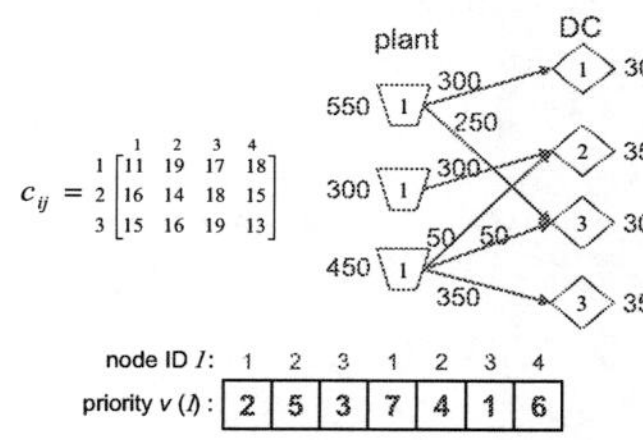

Figure 3. Sample of priGA

For a TP, a chromosome consists of priorities of plants and DCs to obtain transportation tree and its length is equal to total number of plants (|*I*|) and DCs (|*J*|), *i.e.* |*I*|+|*J*|. The transportation tree corresponding with a given chromosome is generated by sequential arc appending between plants and DCs. At each step, only one arc is added to tree selecting a plant (DC) with the highest priority and connecting it to a DC (plant) considering minimum cost. Figure 2 represents a transportation tree with 3 plants and 4 DCs, its cost matrix and priority based encoding.

The encoding algorithm of priority- based encoding, From 1 to the total of a nodes (|*I*|+|*J*|), The number for a total of nodes is put into the gene seat, and operation exchanged at random after that is performed. At the first step of decoding procedure, an arc between DC 1 and plant 1 is added to transportation tree since DC 1 has highest priority in the chromosome and the lowest cost is between Plant 1 and DC 1. After determining the amount of shipment that is g_{11} = min{550, 300} = 300, capacity of plant and demand of DC are updated as a_1 = 550 - 300 = 250, b_1 = 300 - 300 = 0, respectively. Since b_1 = 0, the priority of DC 1 is set to 0, and DD 4 with next highest priority is selected. After adding arc between DC 4 and Plant 3, the amount of shipment between them is determined and their capacity and demand are updated as it is explained above, and this process repeats until demands of all DCs are met.

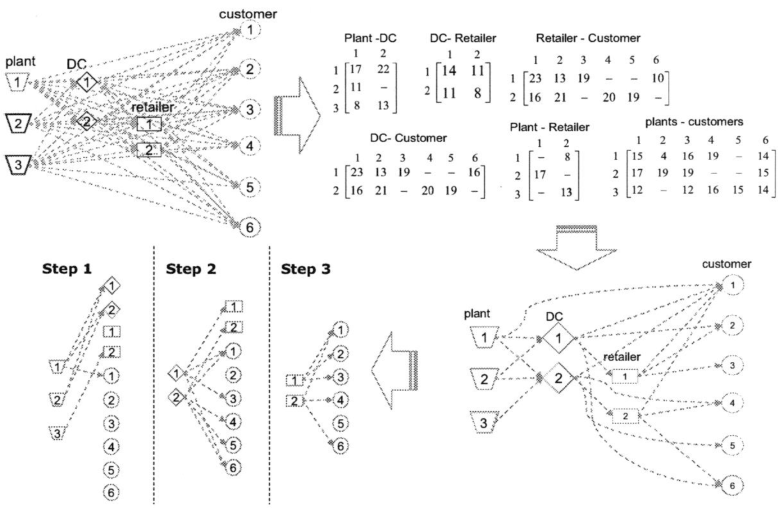

Figure 4. A solution procedure of this problem

On the basis of this, we show the example which applied the priority-based encodings model in figure 4. In this study, we consider a multi-product model. But here, I explain solution method for delivery of a kind of product as example. When we apply priority-based encoding to this model, at first we consider the three-stage TP model which all nodes are connecte. Next, cost matrix which encompass limitation on a delivery rute is given. A part of a hyphen of a cost matrix of figure 3 is it. When we really calculate in GA, a vast cost is given this part. Because priority-based encoding is based on priority in decode processing, and it chooses the rute where a cost is low. It is rare that the rute where a vast cost is chosen. In addition, if the rute include vast valu the corresponding chromosome will not be selected in genetic selection operator. In other words, limitation of a delivery rute is possible without preparing for a special routine by this way.

If a delivery rute is settled, we divide a network model into three steps. It become the form that was suitable for chromosome expression of priority-based encoding by doing it this way. This model has a complicated delivery rute, but can untie it in the same way as three-stage TP. An example of a chromosome is shown in Figure 5.

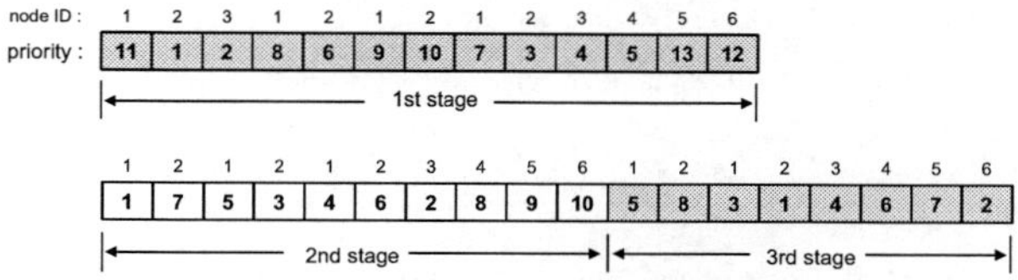

Figure 5. A sample of transportation tree and chromosome

3.2 GENETIC OPERATOR

The crossover is done to explore new solution space and crossover operator corresponds to exchanging parts of strings between selected parents. In this study, we use weight mapping crossover (WMX). It can be seen as extension of one-point intersection. This technique does not need to perform Repair processing after crossover. Swap mutation was used for this study. Two digits are randomly selected and their positions are exchanged. The fitness function has been taken as inverse of objective function. And we employ roulette wheel selection as a selection mechanism. The roulette wheel selection is one of proportional selection mechanisms.

4. NUMERICAL EXPERIMENTS

We perform a numerical value experiment here. But there are problems such as great computational complexity or complexity of parameter setting in GA. Usually, GA use fixed parameter valu, and evolution is based on this value. For these problems, we adopt hybrid GA (hGA) to solve these problems. This is GA which did it to improve effectiveness ability for problems of GA. In hGA, fuzzy logic controller (flc) is incorporated, and it performs self-regulation of a parameter by a process of evolution. In this study, we perform a comparison experiment between priority- based (priGA) and hGA which carried flc in priGA.

Table 1 is the test data that we used for an experiment, and these data were generated randomly. The cost generated from a uniform distribution between 30 and 50. The rute that cannot use for delivery was made at 15% of the number of all nodes at random, and it was given a vast cost. In addition, the number of products set it to 2. An experiment is performed by 20 times each and shows the result for table 2.

Table 1. Experimental data

Problem No.	No. of Plants (i)	No. of DCs (j)	No. of retailers (k)	No. of customers (l)	Population size	Crossover rate	Mutation rate
1	2	5	8	15	100	0.3	0.3
2	3	7	12	30	100	0.3	0.3
3	4	5	8	45	100	0.5	0.3
4	6	8	10	60	100	0.5	0.3

Table 2 gives best, average, average, standard deviation and average computation time (ACT) of 20 runs.

Table 2. As a result of comparison experiment

Problem No.	priGA Best	AVG	SD	ACT	hGA Best	AVG	SD	ACT	Improvement Rate (%)
1	9937.0	9941.6	2.8	3.1	9912.0	9912.7	1.3	2.8	0.25
2	10610.0	10625.0	19.1	4.8	10481.0	10509.0	15.7	3.2	1.22
3	11833.0	11890.0	32.0	15.7	11742.0	11782.0	23.2	6.3	0.77
4	11943.0	12021.0	46.2	17.6	11828.0	11870.0	34.6	7.0	0.96
5	13039.0	13141.0	45.9	33.6	12925.0	12994.0	33.7	10.6	0.87

As a result, hGA give a superior experiment result in all problems and its quality of a solution is better than before method. Improvement rate is not a big value, but should pay attention to a difference of opening calculation time in particular as problem size becomes big.

CONCLUSION

In this study, we suggest multi-product logistics network with constraints of delivery rute. This is a more realistic TP model, and a product produced in a plant is having many kinds, and this model owns limitation on the delivery rute. I used an encode method of a priority base for chromosome expression on untying this model and introduced effective technique to limit a delivery rute. In a numerical value experiment about the conventional Priority type (priGA: priority-based Genetic Algorithm) and hybrid model (hGA: Hybrid Genetic Algorithm) we performed a comparison experiment by and clarified utility of hGA in quality of a solution and two points of the shortness of calculation time from an experiment result. However, today's distribution channel becomes a flexible form, and a product may not be delivered by order of a plant - DC - customer like the model that we always used this time. A study of the model that can be equivalent to such a flexible circulation form or the shortness of product life cycle like these days is a future problem.

REFERENCES

Gen, M. and R. Cheng : Genetic Algorithms and Engineering Design, John Wiley & Sons. (1997)

Gen, M and R, Cheng : Genetic Algorithms and Engineering Optimization. New York: John Wiley & Sons (2000)

Gen, M. Altiparmak, F. & Lin, L. : A genetic algorithm for two-stage transportation problem using priority-based encoding. OR Spectrum, vol.28, no.3, pp.337-354 (2006)

Gen, M. & Li, Y. Z. Solving : Multi-objective Transportation Problem by Spanning Tree-base Genetic Algorithm. Adaptive Computing in Design and Manufactured, pp.98-108, Springer-Verlag (1998)

Altiparmak, F., Gen, M., & Lin, L. : A genetic algorithm approach for multi-objective optimization of supply chain networks. Computers and Industrial Engineering, vol.51(1), pp.196-215 (2006)

Sun, M. : A. Tabu search heuristic procedure for solving the transportation problem with exclusionary side constraints, Journal of Heuristic, 3, pp.305-326, (1998)

Sun, M. : The transportation problem with exclusionary side constraints and two branch-and-bound algorithms. European Journal of Operational Research, vol.140, no.3, pp. 629-647(19) (2002)

Syarif, A. & Gen, M. : Solving exclusionary side constrained transportation problem by using a hybrid spanning tree-based genetic algorithm. Journal of Intelligent Manufacturing, vol.14, no.3-4, pp.389-399 (2003)

PART IV:

FINANCIAL ENGINEERING AND DATA MINING

STOCK TRADING BASED ON NEURAL NETWORK MODELING AND FUZZY-TECHNICAL INDICATORS

YU MENG
Department of Engineering Management and Systems Engineering
University of Missouri-Rolla
DAVID ENKE
The University of Tulsa
Department of Finance and Operations Management

ABSTRACT

Directional technical indicators are often considered a useful tool for forecasting stock market trends. In this study two feed-forward neural networks with backpropagation learning are trained to forecast the directional technical indicators +DI, –DI and ADX. By combining technical analysis and fuzzy logic, a new hybrid trading system has been developed with the goal of generating early buy/sell signals, and hence better returns. The historical price data of the S&P 500 index from the year 2001 to the year 2005 are used in this study. The results obtained by the hybrid trading system are compared against those obtained using a buy-and-hold strategy, a technical indicator "rule-of-thumb" trading strategy, and a trading system based only on the neural network model without the aid of fuzzy logic.

INTRODUCTION

When making strategic decisions under uncertainty, it is necessary to make forecasts. Accurate forecasting is especially necessary when taking responsive actions in a dynamic stock market due to the fact that "time is money" in financial activities. Technical analysis is an approach that helps traders to identify market trends in their early stages of development, allowing for trades in the direction of the existing trends through the study of market action (Murphy, 1999). Technicians believe that market action discounts everything, which implies that the study of the price action is all that is needed since the price itself contains all the information about the factors that can possibly affect it. Technicians also believe that prices move in trends, therefore, various techniques have been created to discover trending patterns in the stock market (Stevens, 2002). Specific trend-following technical indicators are beginning to be widely used by traders for forecasting market trends. Furthermore, traders that use technical analysis often develop trading strategies according to the trading signals generated by technical indicators. However, the lagging problem of some technical indicators leaves room for improvement in the decision making process. Often, the optimal time to trade has passed before the signals are generated by the technical indicators. As such, finding a way to forecast technical indicator signals, as opposed to just reacting to past data, should prove useful and help traders overcome the lagging characteristics of many technical indicators.

Neural networks are a form of computational intelligence that may help improve the forecasting ability of technical indicators. Among all the useful tools and techniques for forecasting, neural networks have their own advantages for discovering complex non-linear relationships without too many prior assumptions. There have been many successful applications of neural networks in the financial area (Chenoweth, 1996). In addition to neural networks, fuzzy logic is another form of computational intelligence that may potentially help one to better interpret the trading signals generated by a certain indicator by giving various degrees of truthfulness to the linguistic variables (Bogullu and Enke, 2002). In this study, neural networks and fuzzy logic will be incorporated into a trading model based on the rules of technical analysis. It is expected that this new hybrid model will allow traders to take positions in the market when a developing trend is at its early stage, and therefore help them gain additional returns.

MODEL DEVELOPMENT

Technical Indicators

Various technical indicators are used to help traders spot the presence of a market trend. One such indicator is the Average Directional Index (ADX), developed by J. Welles Wilder to evaluate the strength of the current trend, whether it is up or down (Chande, 2001). ADX is derived from two other indicators, called the Positive Directional Indicator (+DI) and the Negative Directional Indicator (–DI), calculated as:

$$+DI_t = \frac{100 wSum_t(+DMI, n)}{wSum_t(TR, n)} \tag{1}$$

$$-DI_t = \frac{100 wSum_t(-DMI, n)}{wSum_t(TR, n)} \tag{2}$$

$$DX_t = \frac{100\,|(+DI_t) - (-DI_t)|}{(+DI_t) + (-DI_t)} \tag{3}$$

$$ADX_t = \frac{(n-1)ADX_{t-1} + DX_t}{n} \tag{4}$$

The Wells Summation (wSum) and the True Range (TR) measures that are used in the +DI and –DI formulas are determined as follows:

$$wSum_t = Value_t + \frac{(n-1)wSum_{t-1}}{n} \tag{5}$$

$$TR_t = max\ \{|HI_t\text{-}LO_t|,\ |HI_t\text{-}CL_{t-1}|,\ |CL_{t-1}\text{-}LO_t|\} \tag{6}$$

where n = number of periods, HI_t = current high price, LO_t = current low price, and CL_{t-1} = previous close price. The following rules are used to calculate the DMI:

IF $HiDiff_t < 0$ and $LoDiff_t < 0$ or $HiDiff_t = LoDiff_t$ **THEN** $+DMI_t = 0$ and $-DMI_t = 0$
IF $HiDiff_t > LoDiff_t$ **THEN** $+DMI_t = HiDiff_t$ and $-DMI_t = 0$
IF $HiDiff_t < LoDiff_t$ **THEN** $+DMI_t = 0$ and $-DMI_t = LoDiff_t$

where $HiDiff_t$ = HI_t – HI_{t-1}, $LoDiff_t$ = LO_{t-1} – LO_t, HI_{t-1} = previous high price, and LO_{t-1} = previous low price.

+DI measures the force of the up moves and –DI measures the force of the down moves over a set period *n*. The default setting for *n* is 14 periods. ADX combines +DI with –DI and then smoothes the data with a moving average to provide a measurement of trend strength. Since it uses both +DI and –DI, ADX does not offer any indication of trend direction, just trend strength.

ADX oscillates between 0 and 100, with readings above 40 indicating strong trends and readings below 20 indicating weak trends. Readings above 60 are very rare. To catch a trend in its early stage, investors look for an ADX that goes above 20.

When using the DI indicators, trading signals are generated during crossovers, such that a buy signals occur when +DI moves above –DI from below, while sell signals occur when -DI moves above +DI from below. However, in a trading range (the market is flat), this system may produce many non-profitable signals, hence +DI/–DI should be used in conjunction with other technical analysis patterns and indicators.

One thing that needs to be pointed out is that since previous daily prices rather than real time prices are used to calculate +DI and –DI in this study, it is not possible to know the exact time during a day when the trading signals occur. Therefore, it must be concluded that there is a buy signal whenever the sign of the difference between +DI and –DI (DDI) changes from negative to positive. A sell signal occurs when the opposite is true.

Neural Network Modeling

After ADX and DDI are calculated based on the historical prices, an appropriate neural network structure needs to be designed to forecast these two indicators. Initially, proper attributes should be selected for the task. When studying the stock market trend as one type of time series pattern, modelers must assume there is some inherit relationship between the successive values in the data set. One of the goals is to explore this relationship. Therefore, for each indicator, the successive values in a time period of 7 days are used to predict the next-day value. For instance, seven variables, ADX_{t-6}, ADX_{t-5}, ADX_{t-4}, ADX_{t-3}, ADX_{t-2}, ADX_{t-1}, and ADX_t are used to forecast $ADX_{t+1.}$

Historical data from January 2, 2001 to December 30, 2004 are used as the training data set for the selected neural network. Data from January 3, 2005 to December 30, 2005 are used for testing.

Two feed-forward neural networks with backpropagation learning (BPFF) are used to train the networks to forecast the indicators. For training the DDI network, the transig and the logsig transfer functions are used in the two hidden layers and the purelin transfer function is used in the output layer. There are 15 neurons in the first hidden layer and 10 neurons in the second hidden layer. For training the ADX network, the transig transfer function is used in a single hidden layer, while the purelin transfer function is used in the output layer. There are 15 neurons in the hidden layer.

Fuzzy-Technical Analysis

According to the rules of technical analysis, traders should be in a buy position when both the –DI signal line crosses the +DI line from above and the ADX is large enough to suggest a strong ascending trend. Likewise, traders should be in a sell position when both the +DI signal line crosses the –DI line from above and the ADX is large enough to suggest a strong descending trend. Yet, in order to take advantage of an early signal and gain additional profit, it is possible to buy or sell when both the distance between the +DI/-DI signal lines becomes small enough and the ADX becomes large enough, even while other traders wait for the trading signals that are actually generated at the end of the day. The problem is that there is some linguistic uncertainty in precisely interpreting the words "small enough" and "large enough". As such, it is reasonable to introduce fuzzy logic as a solution to this problem. The uncertainty regarding the level of the technical indicators can therefore be accounted for by calculating the degrees of membership for linguistic variables such as "small", "around middle", and "large".

For the following, the distance between the +DI line and –DI line (DDI) will be labeled as *d*, with the value of ADX labeled as *v*. These two variables will be fuzzified using the membership functions illustrated in Figure 1 and defined in Table 1.

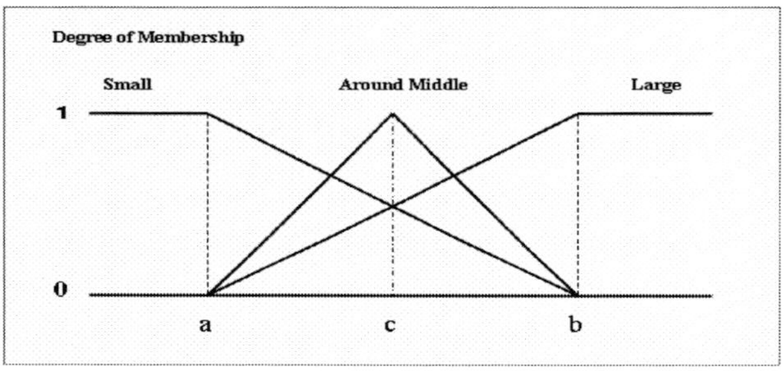

Figure 1. Membership Function for *d* and *v*

Conditions	Degree of Membership for " Small"	Degree of Membership for "Around Middle"	Degree of Membership for "Large"
m≤a	1	0	0
a<m≤c	1-(m-a)/(b-a)	(m-a)/(c-a)	(m-a)/(b-a)
c≤m<b	1-(m-a)/(b-a)	(m-c)/(b-c)	(m-a)/(b-a)
m≥b	0	0	1

Table 1. Degree of Membership for *d* and *v*

In Table 1, *m* represents the values of *d* or *v* depending on whether the purpose is to decide the degree of membership for DDI or the degree of membership for ADX. In the case of calculating the degree of membership of *d*, the values of a, c, and b are 5, 7.5 and 10, respectively. As for the case of calculating the degree of membership of *v*, the values of a, c, and b are 16, 18 and 20, respectively.

After the neural networks have been trained to forecast DDI and ADX, the degrees of membership for *d* and *v* can be calculated using the membership functions defined above. The fuzzy-technical analysis is performed according to the following general rules:

1. If d is small and v is large, then there is a strong sell signal.
2. If d is large and v is large, then there is a weak sell signal.
3. If d is small and v is small, then there is a weak sell signal.
4. If d is large and v is small, then there is a weak sell signal.
5. If -d is small and v is large, then there is a strong buy signal.

6. If -d is large and v is large, then there is a weak buy signal.
7. If -d is small and v is small, then there is a weak buy signal.
8. If -d is large and v is small, then there is a weak buy signal.

RESULTS

The returns in dollar amounts of the four trading models are compared against each other in Table 2. All the comparisons between the different trading systems are made based on the following assumptions to ensure consistency in the results:

1. The trader purchases 100 shares of stock on the first full day of the trading period. Therefore, for the BPFF model, although the first buy signal actually occurred on January 14, 2005 as forecasted by the neural network, this signal is ignored since it is presumed that the trader had already bought the shares on January 03, 2005 - the first day of the trading period.
2. The number of shares in each transaction remains the same, i.e., 100 shares.
3. All the shares held by the trader must be sold by the last day of the trading period.
4. All the transactions are made at the open prices.
5. There is a transaction fee of $10 involved in each trade of 100 shares.

From Table 2 it can be seen that the BPFF model and the hybrid trading system with fuzzy-technical analysis based on neural network modeling outperform the rule-based model and the buy-and-hold strategy. The reason why the BPFF model generates better results compared to the rule-based model is that if traders can forecast the technical indicators with greater accuracy, they will not have to wait until the trading signals occur in order to take the correct buy or sell trading actions. It is important to get into a trade before the majority of other traders spot the move. Table 2 also shows that the hybrid trading system outperforms all the other models and gives superior results. This system allows traders to buy or sell a little earlier than when the trading signals are generated by the system, which is potentially beneficial in that traders might be able to buy at a lower price or sell at a higher price when the market trend appears more certain to them, or at least gives them earlier signals of trend reversal.

Trading System	**Buy and Hold**	**Rule of Thumb**	**BPFF NN**	**Fuzzy-TA Based NN**
# of Trades	2	12	12	12
$ Return (per 100 shares)	4250	3203	9944	16695
Fees ($)	20	120	120	120
Profits ($)	4230	3083	9824	16575

Table 2. Comparisons between Different Trading Models

CONCLUSIONS AND FUTURE WORK

Computational intelligence tools, such as neural networks and fuzzy logic, have been shown to be helpful during a technical analysis-based stock market decision-making. Incorporating neural network modeling and fuzzy logic into a trading model based on rules of technical analysis can improve the prediction ability of the traditional rules by offering earlier, timely, and more accurate trading decisions.

However, the current modeling leaves some room for improvement in data selection to increase the forecasting accuracy of this study. For instance, when one looks at the historical price chart of the S&P 500 Index, they notice that the data set used for training is from a down-trend period, while the data set used for testing is from an up-trend period. Since +DI and –DI are indicators used to measure the force of trends in different directions, the accuracy in forecasting their difference DDI may be affected by the trend patterns in the time period from where the data sets are selected. To avoid problems of this kind, it is necessary for future work to increase the size of data set over an extended time period and try to include different trend patterns. Nonetheless, the current modeling, which provides a hybrid model trained on data from one trend, and tested on data from another, shows good results and leads us to believe that the model will be robust. In addition, although fuzzy logic seems to be a good method to account for the uncertainty of linguistic variables, difficulties still exist in the subjectivity of designing proper fuzzy rules and membership functions to achieve desirable results. Automated techniques for generating membership functions, such as using neural networks and/or genetic algorithms, should also be considered.

In summary, future research should work towards improving the forecasting ability of the trading system for different market trends over different time frames. Also, since each indicator may work differently for individual stocks, various indicators should be examined, and an intelligent system should be developed to combine the effects of different indicators to achieve optimal and more robust results.

REFERENCES

Bogullu, V., and D. Enke (2002) "Using Neural Networks and Technical Indicators for Generating Stock Trading Signals", *Intelligent Engineering Systems through Artificial Neural Networks*, Volume 12, pages 721-726.

Dong, Y., and D. Enke (2005) "A Modified Trading Strategy Model Combining Neural Networks with the Bollinger Bands Technical Indicator", *Intelligent Engineering Systems through Artificial Neural Networks*, Volume 15, pages 661-669.

Ham, F.M, and I. Kostanic (2001) *Principles of Neurocomputing for Science & Engineering*, McGraw-Hill, Inc. New York, NY.

Headly, P. (2002) *Big Trends in Trading: Strategies to Master Major Market Moves*, John Wiley & Sons, Inc., New York.

Klir, G., and B. Yuan (1995) *Fuzzy Sets and Fuzzy Logic: Theory and Applications,* Upper Saddle River, N.J.

Murphy, J. (1999) *Technical Analysis of the Financial Market: A Comprehensive Guide to Trading Methods and Applications,* NYIF, NY.

Stevens, L. (2002) *Essential Technical Analysis: Tools and Techniques to Spot Market Trends*, John Wiley & Sons, Inc., New York.

Tirasirichai, C., and D. Enke (2005) "Using Neural Networks to Identify The Type of Market Trend for Choosing the Proper Technical Analysis Indicator", *Intelligent Engineering Systems through Artificial Neural Networks*, Volume 15, pages 651-659.

Chande, T. (2001) *Beyond Technical Analysis: How to develop and Implement a Winning Trading system*, John Wiley & Sons, Inc., New York.

NEURAL BASED TECHNICAL ANALYSIS IN STOCK MARKET FORECASTING

A. MURAT ÖZBAYOĞLU
TOBB University of Economics and Technology
Computer Engineering Department
Ankara, Turkey
mozbayoglu@etu.edu.tr

ABSTRACT

In this study, a Jordan/Elman neural network model is proposed for stock market forecasting using technical trading indicators in order to predict the next day closing price of a particular stock value. The technical indicators chosen to represent the stock characteristics are sampled from independent indicator types, such as trend, mode, momentum and volatility. The training data is obtained from NYSE historical data of TKC stock and QQQQ ETF between 1999 and 2005. With the trained network, 2006 data has been tested and promising results are achieved. For performance evaluation, different buy-sell rules are created using the most common technical indicators and the rate of returns are compared with the neural network forecaster's performance. The results indicate that the neural network prediction model outperformed all of the other models. This system can be used for the management of a combination of stocks and ETFs providing short term buy-sell signals.

INTRODUCTION

Stock market, being one of the core components of a market economy is among the most fundamental indicators of the health of an economy. Predicting the stock market behavior becomes an important issue for the individuals or financial analysts for maximizing their investments. There have been a lot of economic models or studies for the stock market forecasting problem in order to identify the general market trend, seasonal swings, or predicting the next day's close value.

In order to predict the index or stock values, financial experts use two main methods: fundamental analysis and technical analysis. Fundamental analysis emphasizes on the financial health of the stock or index by analyzing certain indicators like PE ratio, profit margin, free cash flow, book value, debt/asset ratio, etc. Technical analysis takes an opposite path, pays minimum attention to the company financial status, instead the recent stock market action and the charts are the main concerns for this analysis [1].

Automated stock market prediction systems mostly use technical analysis. There are two main approaches in automated stock market forecasting systems, using the stock data directly or using filtered data. When the data is used directly, the model has the advantage of using the real information, no loss or delay occurs, but the problem of extracting the core information from the noisy signal data is difficult. There is no dominant frequency range for applying frequency domain techniques to extract information. Using filtered data results in some information loss, also the filtered version of the signal usually lags the original one, but it is easier to interpret the results and predictions are generally acceptable.

Most of the automated stock market forecasting models base their predictions on technical indicators, chart patterns, function approximation methods, linear and nonlinear regression techniques. However, the vast amount of technical indicators that are used in the prediction models might provide different, sometimes conflicting results for the near term stock behavior. This is mostly due to the fact that a technical indicator is nothing but a filter applied to a time series, in this particular case the time series belong to the price or volume. As a result, each indicator might provide a different perspective of the same signal. Hence, selecting the necessary technical indicators and interpreting the results is a challenging problem itself.

As most of the technical indicators are not independent, the problem is transferred into identifying the most important indicators which are either not related to each other or has minimum correlation. Selecting the most appropriate features for the prediction model would reduce the noise and increase the system performance. The technical indicators can be grouped into the following four categories: mode, trend, cycle and volatility of the market [2]. In order to achieve independence in feature selection, it is preferable to include at least one technical indicator from each category.

The market mode indicators (advance/decline ratio, new highs/new lows ratios, John Ehlers's MESA filter [3], etc.) indicate the general behavior type of the market, it can be in a trend mode, cycle mode or sideways mode.

The market trend indicators show the general direction the market is following recently; most used indicators in that category are the Moving Averages (SMA, EMA, etc.). SMA and EMA are calculated as

$$SMA^N(t) = \frac{\sum_{t=1}^{N} p(t)}{N} \tag{1}$$

$$EMA^N(t) = \frac{2}{N+1} \cdot p(t-1) + (1 - \frac{2}{N+1}) \cdot EMA(t-1) \tag{2}$$

$$EMA^N \approx \frac{\sum_{t=N-1}^{0} p(t) \cdot \left(1 - \frac{2}{N+1}\right)^{N-1-t}}{\sum_{t=N-1}^{0} \left(1 - \frac{2}{N+1}\right)^{N-1-t}} \tag{3}$$

Cycle indicators might catch the cycles in the market behavior especially at times when there is no identifiable trend present. Stochastic, momentum, RSI or MACD can be considered one of the cycle indicators. MACD and RSI can be calculated as follows:

$$MACD = \left(0.15 \cdot p(t) + 0.85 \cdot EMA^{12}(t-1)\right) - \left(0.075 \cdot p(t) + 0.925 \cdot EMA^{26}(t-1)\right)$$

$$RSI^N = 100 \cdot \frac{up\ EMA^N}{up\ EMA^N + down\ EMA^N} \tag{4}$$

The volatility indicators provide information about how fast and how much the market might move up and down in the near future. Standard deviation and Bollinger bands are among the techniques used as volatility indicators.

$$BB_U{}^N = SMA^N(t) + F \cdot \sigma(N)$$
$$BB_L{}^N = SMA^N(t) - F \cdot \sigma(N) \qquad (5)$$

Automated stock market forecasting models might also use the chart patterns by identifying resistance and support levels. Several attempts have been also made in financial forecasting applications that use neural networks in the recent years [4]. Some developed models that use fuzzy logic [5], neuro-fuzzy systems [6], genetic algorithms [7] or Hidden Markov Models [8] also exist.

MODEL DESIGN

In this study a Jordan/Elman neural network model [9,10] is proposed for stock market prediction. The technical indicators are analyzed and the ones which provided statistically better returns are chosen. In order to have feature independence, at least one indicator from each technical group is included in the feature vector which had 6 elements and consisted of the difference between the stock value and MACD, 50-SMA, 20-SMA, MESA, position in the Bollinger Band and RSI values. In order to achieve scale invariance among different stocks, the relative percentage positions of the stock value compared to the technical indicator results are used. The input vector is also normalized between 0 and 1. There are 6, 16 and 1 neurons at the input, hidden and output layers respectively, with each neuron using a hyperbolic tangent activation function. The output indicates the normalized difference between next day's predicted value and 50-SMA. The context units chosen for the Jordan/Elman network use hyperbolic tangent as the transfer function and are fed with the input vector with a "remembering factor through time" of 0.8, where 0 indicates only present vector being factored in, 1 indicates all of the past being factored in. Also a momentum value of 0.7 is used in the learning rule at each layer.

The neural network application developed in Java calculated the technical indicator values and combined these results with historical stock data in order to train the network. When the combined error dropped below a threshold value, training stage was finished and the weights were stored for each trained stock.

The system was trained using the 1150 closing prices of TKC stock in NYSE during the period of 6/1/2001 to 12/30/2005. The training process took 50,000 iterations, at the end of the training, the MSE went down to 0.012 the corresponding weight matrices were stored. In the second training set QQQQ closing values were trained using the data taken from 5/19/1999 to 12/30/2005 which consisted of 1665 data points. The training process took 50,000 iterations, bringing the MSE down to 0.0016. The weights were also stored separately.

TEST RESULTS

Both training sets were tested using the recent data for both TKC and QQQQ. The testing dataset (138 data points) had closing prices from 1/3/2006 to 7/20/2006 for both TKC and QQQQ. The neural network prediction results compared with the actual closing prices for TKC data set are shown in Figure1.

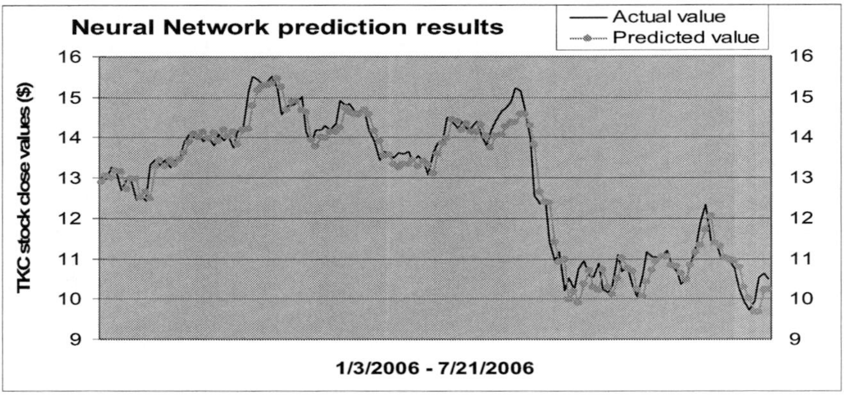

Figure 1 NN prediction values compared to actual closing prices of TKC

In order to evaluate the effectiveness and performance of the system, the results were compared to the performance of the most commonly used technical analysis indicators. The comparison was done individually and by a hybrid system which combined the signals provided by technical indicators. A virtual investment account with an amount of money equivalent to the value of 1000 TKC shares on 1/3/2006 was created for each compared model. The closing value of TKC on 1/3/2006 was $12.80, so all accounts started with $12,800. On 7/21/2006 all accounts were closed and the results were compared. On 7/21/2006, TKC had a closing value of $10.49 which indicated a %18.17 loss of the stock value. The combined results are tabulated in Table 1.

Virtual investment account management style	7/21/06 Value	Net % change
Keep in cash	12,800	0
Buy and Hold	10,490	-18.17
Buy when MACD > 0, otherwise sell	15,737	22.75
Buy above 50 days SMA, otherwise sell	12,560	-2.18
Buy above 20 days SMA, otherwise sell	10,771	-15.98
Buy above MESA, otherwise sell	13,706	6.91
Buy below %10 BB, sell above % 90 BB	13,082	2.04
RSI Indicator, buy below 30, sell above 70	12,504	-2.31
6 of 6 technical indicators show same action (100%)	12,333	-3.80
5 of 6 technical indicators show same action (>80%)	15,469	20.66
4 of 6 technical indicators show same action (>70%)	13,385	4.57
Buy if neural network predicts increase, sell if decrease	19,209	49.84

Table 1 Performances of different accounts starting with 1000 TKC stocks

Each account was maintained by buy-sell commands triggered by the corresponding indicator's rules. For example a 50 day MA buy-sell rule was "When the stock value goes above the 50 day line, buy the stock, if the stock value goes below the 50 day line sell the stock". The neural network buy sell rule checked the predicted next day value for the stock. The buy-sell rule was formed as "If the predicted value is higher than the present day's stock close value by a margin of %0.5 buy the stock, if it is less than the present day's stock close value by a margin of %0.5 sell the stock, if it is within that -%0.5 to +%0.5 band of the stock price, take no action". At the same time in order to make the account management realistic, a $20 transaction cost was charged for each buy/sell command, regardless of the number of stocks traded.

Analyzing the results indicate that the neural network was able to learn from the previous data better than the most common technical indicators. Similar results have been obtained by testing 2006 QQQQ data which was trained with 1999-2005 QQQQ data. The net returns were dropped slightly when TKC data was tested by QQQQ training data and when QQQQ data was tested by TKC training data, but in both cases the neural network still provided the best returns compared with the other methods. While the neural network predictions resulted in a profitable outcome, the global stock indices during the same period either had a downturn or mostly stayed flat.

RESULTS AND CONCLUSION

In this study, a stock market forecasting system is proposed using a Jordan/Elman network. The proposed model uses a combination of stock values with the technical indicators as inputs and it tries to predict the closing price of the next day. The feature vector components are chosen through independent technical analysis indicators and their relative positions based on the stock or index value. With this approach, the signal characteristics are obtained and can be used on different stocks or indexes regardless of their relative value.

The training data had a span of 5 years which included not only increasing and decreasing periods, but also covered times where no obvious trend was observed. Also the training data was long enough to include most, if not all the chart patterns and these were implicitly used in the training process.

The developed model used TKC and QQQQ stock values between 2001 and 2006 for training and testing. TKC was chosen for analysis, because the purpose of this study was to create a neural based stock trading application program for Turkish stock market ISE100, TKC was the only stock that could be traded in NYSE and ISE100 at the same time, also it was easier to access historical data from NYSE. Later on, to validate the results, QQQQ was added for further analysis. The results indicated that the model was able to predict most of the next day trend successfully. When the model was used as part of a trading expert to manage a virtual investment account, it provided better results than any of the most commonly used technical indicators by the analysts. These encouring results indicate the proposed model might be used as a decision model for buy-sell activities on an actively participated portfolio.

In the near future, the ETFs in NYSE will be included in the model. The results will be tested and analyzed similarly. Since ETFs are a basket of stocks, their performance is more dependent on the market dynamics and technical analysis rather than company specific issues. Only one type of neural network has been tried in this initial study. In the next phase of the work, other type of neural networks might be introduced to the model for performance comparison. The final goal is to provide a good alternative model for trading compared to traditional systems.

NOMENCLATURE

σ(N) : N-point Standard Deviation
BB_L^N, BB_U^N : Lower and Upper Bollinger Bands of N data points
Down EMA^N : EMA of N data points of which only down days are considered.
EMA^N : Exponential Moving Average of N data points
F : Number of Standard Deviations
MACD : Moving Average Convergence Divergence
MSE : Mean Square Error
N: number of data points sent to technical indicator
p(t) : Value of the stock price at time t
QQQQ : Nasdaq-100 Trust Fund
RSI^N : Relative Strength Index of the Stock for the last N data points
SMA^N : Simple Moving Average of N data points
TKC : Turkcell GSM Mobile Operator Company
Up EMA^N : EMA of N data points of which only up days are considered.

REFERENCES

[1] Murphy,J.: Technical Analysis of the Financial Markets: A Comprehensive Guide to Trading Methods and Applications, Prentice Hall Pres, (1999).

[2] Pan, H.: A Joint Review of Technical and Quantitative Analysis of Financial Markets Towards a Unified Science of Intelligent Finance, 2003 Hawaii International Conference on Statistics and Related Fields (2003).

[3] Ehlers, J.: MESA and Trading Market Cycles, 2nd Edition, John Wiley and Sons (2002).

[4] Atiya, A., Talaat, N. and Shaheen, S.: An efficient stock market forecasting model using neural networks, International Conference on Neural Networks, 1997, Vol.4, 2112-2115 (1997).

[5] Kourouliotis, D.E., Diakoulakis I.E. and Emiris,D.M.: A fuzzy cognitive map based stock market model: synthesis, analysis and experimental results, The 10th IEEE International Conference on Fuzzy Systems, Vol.1, No.8 465-468 (2001).

[6] Li, R.J. and Xiong, Z.B.: Forecasting stock market with fuzzy neural networks, Proceedings of 2005 International Conference on Machine Learning and Cybernetics, Vol.6, 3475-3479 (2005).

[7] Rui, J. and Szeto, K.Y.: Discovering investment strategies in portfolio management: a genetic algorithm approach, ICONIP'02 Proceedings of the 9th International Conference on Neural Information Processing, Vol.3, 1206-1210 (2002).

[8] Hassan, M.R. and Nath,B.: Stock market forecasting using hidden markov model: a new approach, ISDA'05 Proceedings of the 5th International Conference on Intelligent Systems Design and Applications, 192-196 (2005).

[9] Jordan, M.I., Attractor dynamics and parallelism in a sequential connectionist machine, Proceedings of 9th Annual Conference of Cognitive Science Society, pp.531-546, (1986).

[10] Elman, J.L., Finding Structure in time, Cognitive Science, vol. 14, pp.179-211 (1990).

Fuzzy Logic-Based Japanese Candlestick Pattern Recognition and Financial Forecast

TAKENORI KAMO
Smart Engineering Systems Lab
University of Missouri – Rolla
Rolla, MO 65401, USA
Phone (573) 341-6556
Fax (573) 341-6567
takenori@umr.edu

CIHAN DAGLI
Smart Engineering Systems Lab
University of Missouri – Rolla
Rolla, MO 65401, USA
Phone (573) 341-7211
Fax (573) 341-7238
dagli@umr.edu

ABSTRACT

This paper discusses fuzzy-logic based Japanese candlestick pattern recognition and financial forecasting. Fuzzy logic is used to recognize candlestick patterns and is applied to financial forecasting. Candlestick patterns recognized by fuzzy logic give impact to the new model presented in this paper, and the results of stock market forecasting. In this paper, the daily stock quotes of Exxon Mobile, General Electric, General Motors, Google, Microsoft, and Wells Fargo are used as input data sets. The output of the model is the forecast of the next day's closing price. For the new model, IF-THEN rule-based fuzzy logic is used for recognizing basic candlestick patterns. The performance of the previous study (Kamo and Dagli, 2006) is used for comparison of the data. The results of the models are evaluated on the basis of mean squared error. Graphs and charts visually illustrate the trend generated by the new model and the actual stock market movement.

Introduction

Many financial forecasting models developed by researchers and scholars worldwide exist. Those models use financial tools and investment instruments as key features, and the Japanese candlestick method is one of them. The Japanese candlestick method has several unique features, and some basic features were introduced in the paper *Hybrid Approach: Neutral Networks and Japanese candlestick method for financial forecast* (Kamo and Dagli, 2006). This paper is the continuation of that study. In the above paper, the Japanese candlestick method was incorporated into the gating networks in the committee machine. In the gating networks, some basic candlestick patterns were identified and used as a static weight generator that is assigned to the experts of the committee machine. After the author discovered that the Japanese candlestick method can improve financial forecasting, it was found that a better way to apply the candlestick methods into a model is needed. In reality, the Japanese candlestick method is a perception-based technique, and precision is not necessary. Therefore, the advantage of the Japanese candlestick method was not appreciated in the previous model. For improving the forecasting results, there is a need for handling imprecise knowledge of candlestick patterns. Because fuzzy logic provides a wide variety of concepts and techniques to represent knowledge that is uncertain or imprecise (Zadeh, 1992), fuzzy logic is introduced for experimenting the new model. For the purpose of comparison for this study, the model from the previous experiment is shown. The results of its forecasts are evaluated on the basis of the mean squared error.

Previous Study

In the previous study, the effect of the Japanese candlestick method was investigated. The model with basic candlestick techniques were compared to a simple generalized regression neural network (GRNN) based model. The results showed the candlestick technique-based model improved the overall performance of the forecasting. Japanese candlestick method was developed for trading commodities in the 1700s using the historical data of exchanging rice from the 1600s (Nison, 2001). The patterns of candlesticks are believed to describe the market condition (Nison, 2001). The Japanese candlestick chart requires open, high, low, and close data to draw the graphic chart. The figures below represent two basic candlesticks. The real body shows the difference between the opening price and closing price; a white candlestick has a white real body that shows the closing price is higher than the opening price, and a black candlestick has a black real body that shows the closing price is lower than the opening price. The upper shadow is the line above the real body. The lower shadow is the line below the real body. For a white candlestick, the difference between the high price and the closing price becomes the upper shadow, and the difference between the opening price and low price becomes the lower shadow. For a black candlestick, the difference between the high price and the opening price becomes the upper shadow, and the difference between the closing price and low price becomes the lower shadow.

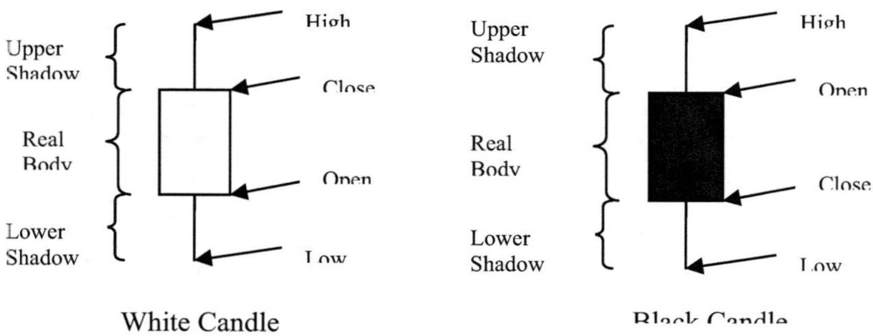

Figure 1. Candlestick Parts and Names

The results showed that the model with candlestick techniques performed better than a simple GRNN-based model. Also, the simulation occasionally showed that the model with candlestick techniques reacted faster to the market trend change than the simple GRNN-based model.

Fuzzy Logic

Lotfi Zadeh's *Fuzzy Sets* (1965) contributed to the growth of various application of fuzzy logic (Lin and Lee, 1996). Fuzzy logic acts like the brain concerning exact information and structured numerical estimators in that it controls highly formalized insights found in the real world and combines fuzzy IF-THEN rules for expert knowledge (Lin and Lee, 1996). A fuzzy set does not have a sharp boundary that divides numbers from non-members in the group ("Fuzzy" 2006). A fuzzy set Ã is expressed as below.

$$\widetilde{A} = \{(x, \mu_{\widetilde{A}}(x)) \mid x \in U\} \tag{1}$$

$\mu_{\tilde{A}}(x)$ is a degree of membership function of x in $\tilde{A}$, and U is a universe of discourse (Lin and Lee, 1996). Fuzzy set can be expressed through the support of a fuzzy set.

$$Supp(A) = \{x \in U \mid \mu_{\tilde{A}}(x) > 0\} \tag{2}$$

α-cut is one of important properties of fuzzy sets. A crisp set is expressed in (3).

$$A_\alpha = \{x \in U \mid \mu_A(x) \geq \alpha\}, \qquad \alpha \in (0,1] \tag{3}$$

When the condition of (3) becomes $\mu_A(x) > \alpha$, A_α is a strong α-cut. When all sets meet all levels of $\alpha \in (0,1]$, this fuzzy set A with a distinct α-cut becomes a level set of A, and Lin and Lee (1996) expressed this set as,

$$\Lambda_A = \{\alpha \mid \mu_A(x) = \alpha, \qquad for\ some\ x \in U\}. \tag{4}$$

Model

Figure 2 is the structure of the model for experimenting fuzzy logic for candlestick techniques. The experts are GRNN-based, and the gating network becomes a fuzzy logic-based weight generator.

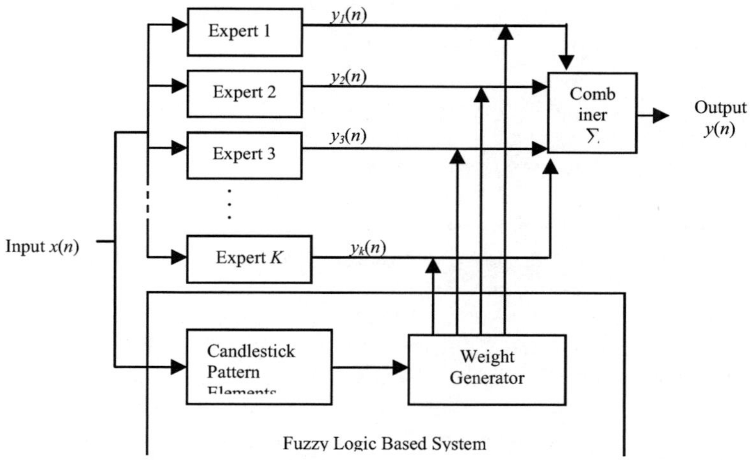

Figure 2. The structure of the model

Each element is identified with a fuzzy expression; for example, the real body A is white and big, the real body B is white and small, the upper shadow A is very short, and so on. Each element is now assigned to the membership function with a specific range; IF-THEN rules evaluate those elements. The results generated by the rules are combined and produce a crisp number. The number indicates a strength level of the market, and is assigned to each expert as a weight.

Experiment

This is an experiment to observe the effect of fuzzy logic on the previous study of Japanese candlestick method-based financial forecasting. Figure 2 shows the structure of this experimental model. The results of the model are compared with a simple GRNN-based forecasting model. The model forecasts the next-day stock price for the following companies' quotes: Exxon Mobil (XOM),

General Electric (GE), General Motors (GM), Google (GOOG), Microsoft (MSFT), and Wells Fargo (WFC). The period of historical data used for the experiment is between 4/15/2002 and 3/31/2006. This period has 1000 trading days. Because Google's IPO date was 8/19/2004, the data between 8/19/2004 and 3/31/2006 were used for forecasting its stock price. There are 410 trading days of data for this period. The membership function type used for this experiment is a triangular membership function. This membership function is simple and suitable for this experimental study. An example of membership function retrieved from the Fuzzy Logic Toolbox of MATLAB is shown in the figure below.

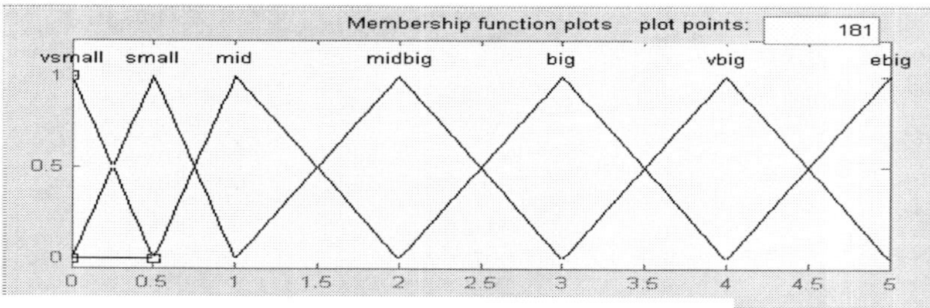

Figure 3. Membership Function Plots

IF-THEN rules are used to formulate the condition of elements. Examples are stated below.

If the real body size is very small,

→ *then*, the stock strength is weak.

When more than one element is being considered:

If the real body size is very small and the upper shadow line is long and the lower shadow line is long,

→ *then*, the stock strength is neutral.

Inputs are ranged between 0 to the universe of discourse. In Figure 3, the input range is between 0 and 5. The linguistic values big, small, long, short, and so on, are based on the previous 5 days' candlestick pattern observations. Based on the results from the fuzzy logic process, the weight is calculated and assigned to the experts. The output of the model is compared with the actual data, and the performance of the model is evaluated based on the mean squared error.

Results

Table 1 shows the results of two models used for this experiment. The statistics imply that a model with fuzzy logic produces better forecasting results.

	A model from the previous study	A model with fuzzy logic
Exxon Mobile (XOM)	0.3920	0.2522
General Electric (GE)	0.1949	0.1363
General Motors (GM)	0.6136	0.4913
Google (GOOG)	52.08	39.57
Microsoft (MSFT)	0.1272	0.0925
Wells Fargo (WFC)	0.2700	0.1994

Table 1. Results of the experimental models in the mean squared error

The line chart shows the forecasted data generated by two models and the actual stock price. The data is from the Exxon Mobile experiment. Figure 4 shows that a fuzzy logic-based model produces smoother curves than the previous model. The previous model could detect the candlestick patterns, but there was no fuzzy membership function to generate the results. In the previous study, the real body size of candlesticks is identified as big or not big and the length of shadow lines of candlesticks are identified as long or not long in the model. Therefore, when a large candlestick appears in the chart, the strength of the stock significantly affects the final forecasting results. Figure 5 shows the candlestick chart of the same period as Figure 4.

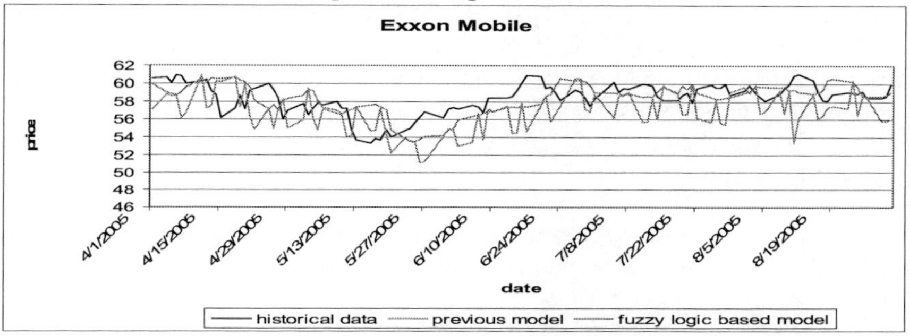

Figure 4. Result in line chart (Exxon Mobile)

In Figure 5, at x = 25, five large candlesticks appear. In the previous model, the large white candlestick followed by the large black candlestick became a very strong bullish sign. On the other hand, the large black candlestick after the large white candlestick became a very strong bearish sign. Figure 5 shows that the results of the model of the previous study move up and down during this period. However, with the fuzzy logic-based model, the size of these candlesticks is not considered as large because the relative size of them is not significantly large; fuzzy logic considers the size of these candlesticks the same or average. Therefore, the result of the fuzzy logic-based model in Figure 4 shows a smooth curve.

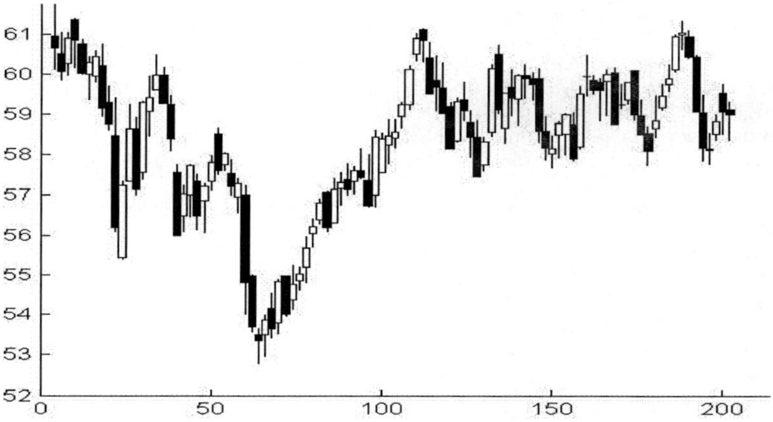

Figure 5. Candlestick Chart of Exxon Mobile

Conclusions and Future Works

This experiment shows that fuzzy logic improves the performance of the forecasting results, and fills the void that is hard to achieve in a classical way or a well-defined boundary crisp membership method. The candlestick method is very flexible and tolerant of imprecise data. Also, the reading of candlestick method is built on hundreds of years of experience of traders. Therefore, the use of fuzzy logic technique is good for describing the definition of candlesticks which is sometimes vague. This model is the first fuzzy logic-based model, and it is an experimental study; therefore, not much study has done with adjusting the number of membership functions, changing the types of membership functions, and so on. For this experiment, all fuzzy logic membership functions used for are triangle type. IF-THEN rules are simple, and the number of rules may not be enough. The logical operation used for fuzzy logic is mainly AND for more than two fuzzy sets.

The number of candlestick pattern elements is still small. However, increasing the number of candlestick pattern elements could smooth out the significant impact toward the market movement.

The further study of candlestick patter techniques and fuzzy logic is needed to express the market condition with candlestick patterns.

References

Fuzzy Logic Toolbox User's Guide, 1995-2006, Math Works, Inc.

Demuth, H., and Beale, M., *2001, Neural Network Toolbox; for use with MATLAB version 4*. The Mathworks, Inc. 2001. Chapter 7.

Hagan, M. T., Demuth, H. B., and Beale, M., 1996, *Neural Network Design*. PWS Publishing Co.

Haykin, S., 1999, *Neural Networks.* , Second edition. Prentice-Hall.

Kamo, T., Dagli, C., 2006, *Hybrid Approach: Neural Networks and Japanese Candlestick Method for Financial Forecast*, ASME 2006, pg. 795-800.

Kim, K, Youn, H., and Kang, Y. 2000, *Short-Term Load Forecasting for Special Days in Anomalous Load Conditions Using Neural Networks and Fuzzy Inference Method,* IEEE Transactions on Power Systems, Vol. 15, NO.2, May 2000.

Klir, G.J., 1995, *Fuzzy Logic,* Potentials, IEEE, Oct/Nov 1995, Volulme 14, Issue 4, pg 10-15.

Lin, C.T., Lee, C.S.G, 1995, *Neural-Fuzzy Systems: a neuro-fuzzy synergism to intelligent systems.*, 1996, Prentice Hall.

Min S., Mitra B., 2001, *Gating Improves Neural Network Performance,* Neural Networks, 2001. Proceedings. IJCNN '01. International Joint Conferrence on, Volume 3, pg. 2159 - 2164 vol.3.

Morris, G. L., 1992, *Candle Power: Advanced Candlestick Pattern Recognition and Filtering Techniques for Trading Stocks and Futures.*, Probus Publishing Company.

Nisson, S., 2001 *Japanese candlestick charting techniques: a contemporary guide to the ancient investment techniques of the Far East.,* Second Edition. New York Institute of Finance.

Ross, T. J. 1995, *Fuzzy logic with engineering applications.*, 1995, McGraw-Hill, Inc.

Schwarts, D.G., Klir, G.J., 1992, *Fuzzy Logic flowers in Japan*, Spectrum, IEEE, July 1992. Volume 29, Issue 7, pg. 32-35.

Zadeh, L.A., 1983, *Commonsense Knowledge Representation Based on Fuzzy Logic*, Computer, October 1983 /volume 16, Issue 10. pg 61-65.

Zadeh, L.A., 1983, *Commonsense Knowledge Representation Based on Fuzzy Logic*, Computer, October 1983 /volume 16, Issue 10. pg 61-65.

Zadeh, L.A., 1988, *Fuzzy Logic*, Computer, April 1988, Volume 21, Issue 4, pg 83-93.

Zadeh, L.A., 1994, *Soft Computing and Fuzzy Logic*, Software, IEEE, November 1994, Volume 11, Issue 6, pg. 48-56.

Zadeh, L.A., 1996, *Fuzzy Logic = Computing with Words*, Fuzzy Systems, IEEE Transaction on, May 1996 Volume 4, Issue 2, Pg. 103-111.

Zhang, B., Joung J., 1999, *Time series prediction using committee machines of evolutionary neural trees,* Evolutionary Computation, 1999. CEC 99 Proceedings for the 1999 Congress on Volume 1.

FORECASTING THE JORDANIAN STOCK PRICES USING ARTIFICIAL NEURAL NETWORK

AYMAN A. ABU HAMMAD
Civil Engineering Department
Applied Science University
P. O. Box: 926296, Amman 11931 Jordan
SOUMA M. ALHAJ ALI
Industrial engineering department
The Hashemite university
B. O. Box: 330127, Zarka 13115 Jordan
ERNEST L. HALL
Center for Robotics Research
University of Cincinnati
Cincinnati, OH 45221-0072 USA

ABSTRACT
Artificial Neural Network (ANN) technique was used in forecasting the Jordanian stock prices. The algorithm was developed using a feedforward multi layer neural network; the network was trained using backpropagation algorithm. Software was developed by using MATLAB to simulate the performance and efficiency of the algorithm. Simulation was conducted for seven Jordanian companies from service and manufacturing sectors. The companies were sampled from different categories which vary according to the degree of stock stability. The results were accurate and acceptable for Jordanian brokers. The use of ANN provides fast convergence, high precision and strong forecasting ability of real stock prices.

INTRODUCTION

Stock market is an exchange where secured trading is conducted by professional stockbrokers; it is one of the main indicators to the financial performance. In addition, stock market represents an essential part of the economy in the developing countries. Apparently, it is significant for shareholders and investors to estimate the stock price and select the best trading opportunity accurately in advance. This brings high return and reduces potential loss to investors. Jordanian stock market faces continuous fluctuating values due to the political, economical, and psychological factors. Furthermore, this market witnesses a noticeable sharp growth in the last few years. Thus, it is imperative for the Jordanian stockbrokers to employ new analysis tools.

Traditional methods for stock price forecasting are based on the statistical methods, intuition, or on experts' judgment. Time series analysis, Arema / Arma model are usually used for forecasting the stock prices. However, their performance depends on the stability of the prices, as more political, economical, and psychological impact-factors get into the picture, the problem becomes non linear, and need a more heuristic or nonlinear methods like ANN, Fuzzy logic, or Genetic Algorithms (Greene, 2003, InvestorWords, 2005).

Hassoun (1995) defines ANN as " parallel computational models comprised of densely interconnected adaptive processing units, they are viable computational models for a wide range of problems including pattern classification, speech synthesis and recognition, adaptive interfaces, function approximation, image compression, associative memory, clustering, forecasting and prediction, combinatorial optimization, nonlinear system modeling, and control". ANN can outperform other methods of forecasting due to its remarkable ability to derive meaning from complicated or imprecise data, it had been used successfully to extract complex patterns and trends (Stergiou and Siganos, 1996). Literature shows that ANN can be used in prediction, classification, data association, data conceptualization, and data filtration (Anderson and McNeil, 1992).

A lot of research had been conducted for using ANN in stock prices forecasting, as Steiner and Wittkemper (1997) had developed a portfolio optimization model that was built based on ANN embedded in the nonlinear dynamic capital market model. An economic approach to the analysis of highly integrated financial markets and econometric methods had been developed by Poddig and Rehkugler (1996). Donaldson and Kamstra (2000) proposed a methodology for forecasting future stock prices and return volatilities for fundamentally valuing assets such as stocks and stock options. Sheng proposed a neural network-driven fuzzy reasoning system for stock price forecast, his experimental result shows that the fuzzy neural network has fast convergence, a high precision, and strong function approximation ability which make it suitable for real stock price prediction.

ANN MATHEMATICAL MODEL:

One layer ANN that has L cells, all fed by the same input signals $x_j(t)$, and producing one output per neuron $y_l(t)$ can be modeled as (Lewis, et al., 1999):

$$y_l = f(\sum_{j=1}^{n} v_{lj} x_j + v_{l0}); \; l = 1,2,..., \; L. \tag{1}$$

The summing function can be replaced by a function that finds the average, the largest, the smallest, the ORed values, or the ANDed values (Anderson and McNeil, 1992).

Most ANN's consist of more than one layer, where the second layer input is the first layer output and so on. For the forecasting model of stock prices, the network consists of two or three layers depending on the degree of stock prices stability of the case study. There were 13 inputs to the network which are the stock prices for the first 13th working days of the month while the network output was the price for the 14th working day of the month.

SIMULATION RESULTS

To test the efficiency and effectiveness of the model a software program was developed using MATLAB. Seven Jordanian companies from different sectors were used as case studies. For each company, a full year was used for training the network; each month was used as a different pattern. The data starting from February 2002 and ending with January 2003 was used for the training, the validation was done by using another year which starts with February 2003 and ends with January 2004.

Different training functions, activation functions, number of layer, and number of neurons were tried till the error converged to the set value which is 10^{-6}. The performance function used was the mean square error (MSE). MSE is the average squared error between the network outputs and the target. The weights and biases of the network were automatically initialized to small random numbers by the software.

Case 1: Arab Engineering Industry

The training was done using one step secant backpropagation, a two layers network was used with Hyperbolic tangent sigmoid activation function for the first layer and hard limit activation function for the second layers, the first layer consists of 14 neurons and the second layer consists of one neuron. The network was trained using the stock prices for this company during the year starting February 2002 and ending January 2003, the network was able to train the data with a MSE of $9.7245*10^{-7}$ in only 11 epochs. To put things into perspective, the output of the network was plotted against the target as shown in Fig. 1, after the network passed the validation stage; the network was used to forecast the prices for the year starting from February 2003 until January 2004. Fig. 2 reveals the forecasted prices against the actual prices, as shown in the figure the forecasted price is very close to the actual one.

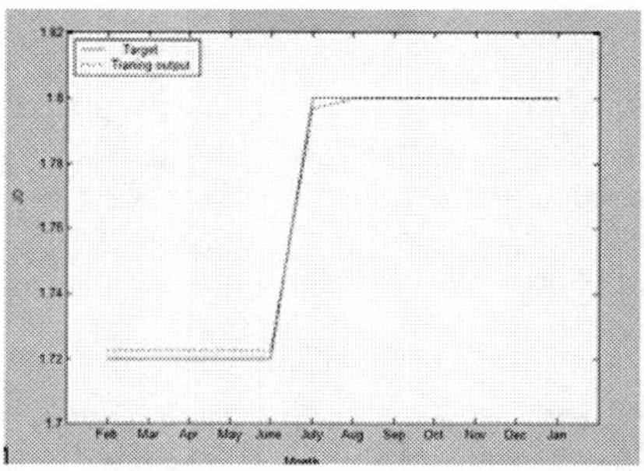

Figure 1: Training output against the target for Arab Engineering Industry Company.

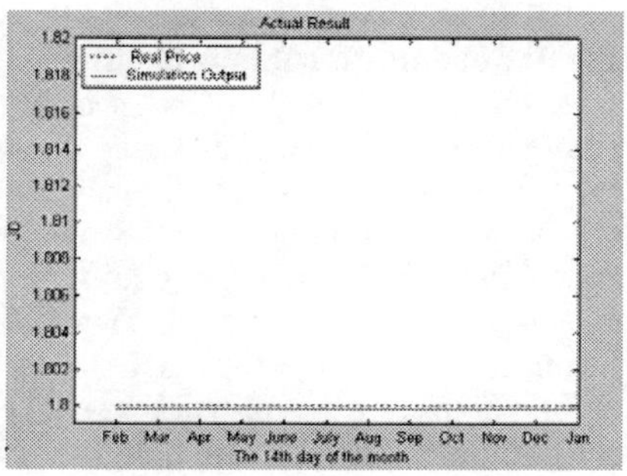

Figure 2: Forecasted prices against the actual prices for the Arab Engineering Industry Company.

Case 2: NUTRIADAR (Jordanian Drug Company)

The same training function was used, however, this time three layers contain (14,10,1) neurons respectively was needed to converge to a small training error, the first and second layers used positive linear transfer activation function, while the third layer used a hard limit transfer activation function. The company stock prices exhibits a noticeable variation between the days of each month, which make the forecasting job more difficult. The network was able to train the data in 1000 epochs that took only 30 seconds. The output of the network is plotted against the target as shown in Fig. 3, the figure prove that the network output matches the actual prices, after the network passed the validation stage, the network was used to forecast the prices for the year starting from February 2003 until January 2004. Fig. 4 reveals the forecasted prices against the actual prices. This time the forecasted price was slightly different than the actual prices, however, the gap did not exceed .08 JD (1 JD=$1.40).

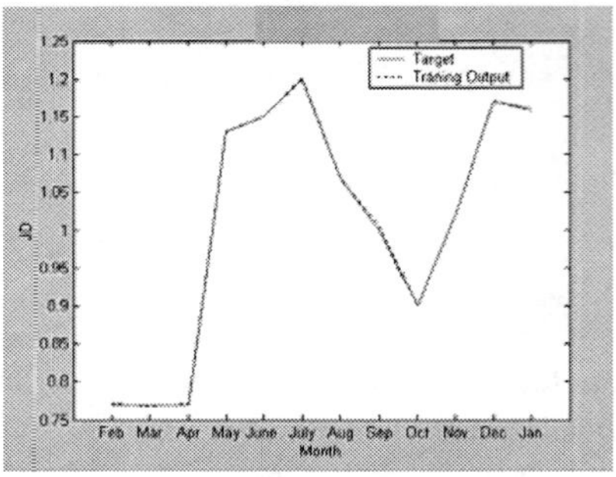

Figure 3: Training output against the target for the Nutriadar Company.

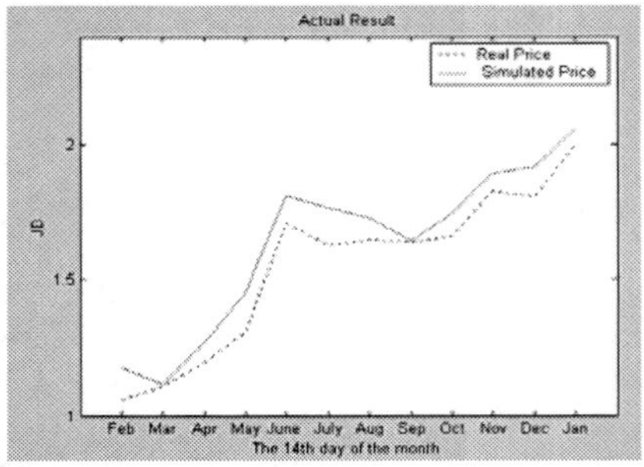

Figure 4: Forecasted prices against the actual prices for the Nutriadar Company.

The rest of the seven case studies were very close to the ones presented before where the network was able to train the data very quickly and produce a very good forecast, except for one case that will be presented next.

Case 3: Jordan Petroleum Refinery

A three layers ANN was used consist of (14,7,1) neurons respectively, the training was done using one step secant backpropagation, the first and second layers used positive linear activation function, while the third layer used hard limit activation function. The stock prices for this company during the year starting February 2002 and ending January 2003 are very volatile and varies substantially from one month to month and even during the same month. The training mean square error reaches 10^{-3} within 90 second in 1000 epochs.

The output of the network against the target is shown in Fig. 5. As shown in the figure the network output slightly differs from the target, although it exhibits the same pattern. The network was used to forecast the prices of the year starting from February 2003 until January 2004. The forecasted prices against the actual prices are shown in Fig. 6. The network was able to produce a very good forecast for the first four months (February-June), after June the actual prices fall from 15 JD's to 4 JD's. As shown in the figure the forecasted prices fall too at the same time but with different amplitude (from 14 to 1 JD) then it follow the same pattern of the actual price but with a difference of less than 2 JD's. Further investigation into this case reveals that the company broke the shares which caused the prices to drop.

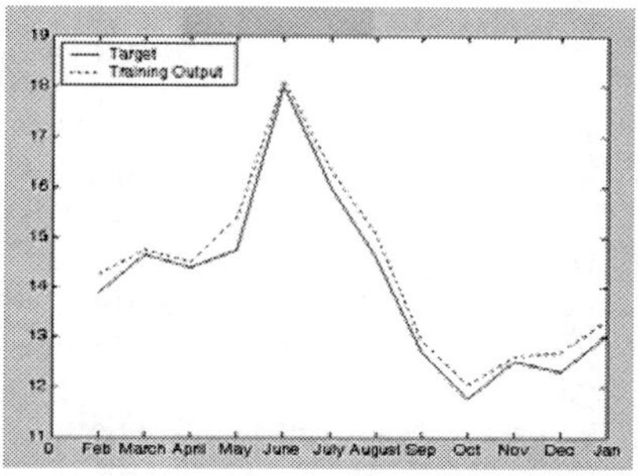

Figure 5: Training output against the target for Jordan Petroleum Refinery Company.

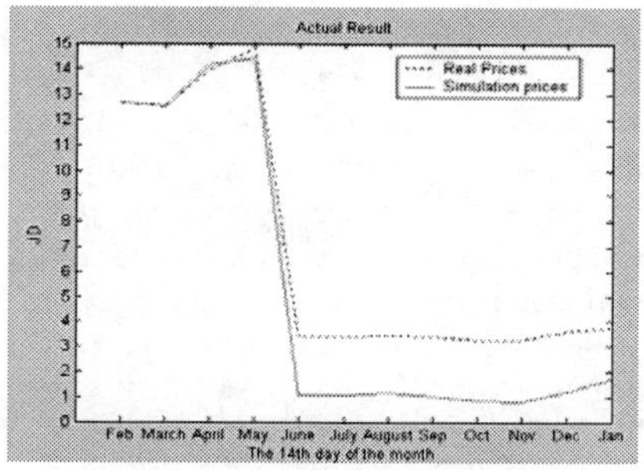

Figure 6: Forecasted prices against the actual prices for Jordan Petroleum Refinery Company.

DISCUSSION

The results obtained from the software were accurate for six out of seven cases, thus, ANN can be used for forecasting stock prices.

In the last case, the network did not give a good forecast but it was able to detect the pattern of the change in prices. Even when the actual stock prices change dramatically for assignable causes, the network was able to detect that and the forecast change at the same month and even with the same pattern as the actual data. Therefore, ANN can give a good indication about the trends of stock prices.

MODEL EVALUATION BY STOCKBROKERS

The model was evaluated by stockbrokers through the use of a questionnaire that was distributed in Amman Stock Market. The questionnaire was designed so that it can be filled in no more than two minutes. The questionnaire first presents the forecasting results obtained from the network and then asked the participant to answer seven multiple choice questions, which focus on the techniques currently used by the broker, and whether he/she would be willing to use the network in the future, only seven responses were returned.

In response to a question: “Would you depend on ANN technique, and by how much percent?”, four out of the seven participants stated that they will depend on ANN technique by 75%, two of them stated that they will depend on ANN technique by only 25%, and one participant stated that he will depend on ANN technique by 50 %.

In response to another question: “Do you believe that this technique will be applicable to any company regardless of the range of stability of its stock prices?”, four participants decided that the ANN technique will be applicable 100% for any company. Two of the participants believed that ANN will be applicable 50% for any company, and one participant believed that it is only 25% applicable for any company.

CONCLUSIONS

A forecasting model for stock prices was developed using ANN. The model was developed using a feedforward neural network with two to three layers. The network was trained using one step secant backpropagation. The activation functions used were hyperbolic tangent sigmoid, positive linear and hard limit transfer function. Simulation software developed by MATLAB was used to evaluate the network performance on seven Jordanian companies sampled from service and manufacturing sectors. The companies selected have different degree of stock prices stability. The network was trained on the data of year 2002. The network was able to produce the output within a MSE of 0.00228871-$9.75237*10^{-8}$ from the target. The network performance was evaluated by using the stock prices of the year 2003. The network output was very close to the actual data, except for one case, for which the company broke it's shares in the middle of the year, however, even in that case the network output drops dramatically to values close but not exactly the same as the ones of the actual data. The results of the network were further evaluated via a questionnaire sent to stockbrokers from Amman stock market. The returned responses indicated an average reliability measure of 75% in using the model for forcasting. Additionally, the responses indicated that this technique is 100% applicable for any company.

The model significance stems from the fact that stock market represents an essential part of the economy in the developing countries. The ANN model can help stockbrokers to forecast the stock prices and select the trading chance that will maximizes their profits more accurate than the available methods.

REFERENCES

Anderson, D. and McNeil, G., 1992, "Artificial Neural Networks Technology," NY, Rome NY, Rome Laboratory.

Donaldson, R. G. and Kamstra, M., 2000, "Forecasting Fundamental Stock Price Distributions", University of British Columbia, Canada.

Greene, W. H., 2003, "Econometric Analysis," 5th ed., Prentice Hall.

InvestorWords, InvestorGuide.com, Inc., 2005, http://www.investorwords.com.

Hassoun, M. H., 1995, "Fundamentals of Artificial Neural Networks," USA: Massachusetts Institute of Technology Press, pp. 57-134.

Lewis, F. L., Jagannathan, S., and Yesildirek, A., 1999, "Neural Network Control of Robot Manipulators and Nonlinear Systems," Padstow, UK: Taylor and Francis Ltd, T. J. International Ltd, pp. 147-167.

Poddig, T., Rehkugler, H., 1996, "A 'world' model of integrated financial markets using artificial neural network," Neurocomputing, Vol. 10, pp. 251-273.

Sheng, L., "A Fuzzy Neural Network Model for Forecasting Stock Price," Zhejiang University, Hangzhou, China.

Steiner, M., Wittkemper, H. G., 1997, "Theory and Methodology: Portfolio optimization with a neural network implementation of the coherent market hypothesis," European Journal of Operational Research, Vo. 100, pp. 27-40.

Stergiou, C. and Siganos, D., 1996, "Neural networks," Surprise Journal, Vol. 14, http://www.doc.ic.ac.uk/~nd/surprise_96/journal/vol4/cs11/report.html.

FORECASTING AGGREGATE SALES WITH INTEREST RATES USING MULTIPLE NEURAL NETWORK ARCHITECTURES

ANTHONY JOSEPH, PH.D.
Department of Computer Science
Pace University
163 William Street, 2nd Floor
New York, New York 10038, USA

MAURICE LARRAIN, PH.D.
Department of Finance
Pace University
One Pace Plaza
New York, New York 10038, USA

ESHWAR SINGH, M.S.
Department of Computer Science
Pace University
163 William Street, 2nd Floor
New York, New York 10038, USA

ABSTRACT

This paper examines out-of-sample forecasts of aggregate sales using 3-month treasury bills interest rate in NeuroSolutions environment referenced against forecasts of linear regression models. Two types of dynamic neural network models trained with the Levenberg-Marquardt backpropagation algorithm under supervised learning were used to investigate the hypothesis that aggregate manufacturing and trade sales are positively related to inverted interest rates and that inverted interest rates are an efficient predictor of the future values of aggregate sales. Interest rates and aggregate sales are important macroeconomic variables used as business cycle indicators. Interest rates are a lagging economic indicator while inverted interest rates are a leading one. Manufacturing and trade sales are a coincident indicator. Seven performance metrics were used to measure the relative efficacy of the predictions with the neural network models out-performing the linear regression models and the Elman models producing smoother forecasts and better performance on the sectionalized data.

INTRODUCTION

The many types of interest rates (Hall and Lieberman, 2008; King and Watson, 1996) essentially tend to rise and fall together. They do so across national boundaries less they cause destabilization in national financial markets because of investors' willingness to seek higher rates of returns on their investments in global financial markets (Hall and Lieberman, 2008). Interest rates are inversely related to both stock and bond prices (Hall and Lieberman, 2008). Changes in interest rates are a consequence of market forces in the economy or governmental monetary policy designed to regulate the economy. The government influences interest rate changes indirectly through the prudent purchase (or sale) of bonds, which ultimately leads to decreases (or increases) in interest rates (Hall and Lieberman, 2008). A decrease in interest rates causes increased spending in planned investments on plant, equipment, software, new housing, and consumer durables that result in the rise of aggregate expenditure and the total output. An increase in interest rates produces the opposite effect. Interest rates are a lagging economic indicator in the average prime rate, but they are "'inverted leading indicators' of real economic activity"

(King and Watson, 1996). Larrain (2002) used spectral analysis to show that changes in inverted treasury bills interest rate are correlated to changes in sales and leads them by about 20 months.

Aggregate expenditure and final sales are equivalent. They are both equal to gross domestic product minus inventory investment (Hall and Lieberman, 2008; Larrain, 2002). Final sales include the lesser volatile service sector sales and manufacturing and trade sales. Manufacturing and trade sales are a coincident economic indicator. Aggregate expenditure is capable of explaining fluctuations in the economy and is inversely related to interest rates (Hall and Lieberman, 2008). The same can be assumed about sales.

The main tools used in this study to perform the experiments are dynamic neural network models in NeuroSolutions environment. They are the focused time-delay neural network (TDNN) and the Elman network, which are simpler examples of the time-lagged feedforward network (TLFN) and the recurrent neural network architectures respectively (Haykin, 1999; Principe et al, 2000). They are trained with the Levenberg-Marquardt backpropagation algorithm under supervised learning. Gencay and Liu (1997) reported that the backpropagation algorithm "asymptotically converges to the estimation which locally minimizes the expected squared error loss" while Lawrence et al (1997) discovered that this algorithm may only find suboptimal solutions to a problem. Both TLFNs and recurrent neural networks are extensions of the multilayer perceptron structure (Haykin, 1999; Principe et al, 2000). Multilayer perceptrons are universal nonlinear approximators.

This study examines out-of-sample forecasts of aggregate sales using inverted 3-month treasury bills interest rate. The forecasting was done with the focused TDNN and the Elman network relative to the linear regression models of the Excel *forecast* command. The investigated hypothesis was that aggregate sales are positively related to inverted interest rates, and that inverted interest rates are an efficient predictor of the future values of aggregate sales. Prior to the forecasting regimens, the interest rate and sales monthly data of 575 samples were preprocessed and tested for normality and persistence. Additionally, the inverted interest rate relationship to and lead over sales were established. Seven performance metrics and visual analysis were used to determine the effectiveness of the focused TDNN and Elman recurrent models' forecasts relative to themselves, to each other, and to the linear regression models. The metrics were correlation (r), root mean square error (RMSE), percentage of correct directions (POCD), Theil inequality coefficient, mean absolute percentage error (MAPE), Durbin-Watson statistic DW), and power spectrum density test in the standardized cumulative periodogram (SCP). Overall, a focused TDNN model produced the best forecast while the Elman network models yielded seemingly smoother forecasts, and the TDNN and Elman network produced better forecasts than the linear regression models.

MATERIALS AND METHODS

The datasets used in the experiments were obtained from Global Insight Database. They consisted of U.S. 3-month treasury bills (T-bills) not seasonally adjusted and real manufacturing and trade sales chained to 1996 dollars. Each dataset comprised 575 samples of monthly data ranging from January 1959 to November 2006. The data were preprocessed to obtain relevant knowledge of their individual characteristics, relationships to each other, and to put them in a convenient form to facilitate the experimentations. The additive inverse of 3-month T-bills interest rate levels was taken to invert it. The 12-month backward first difference of both inverted T-bills and sales data was taken to remove possible serial correlation and to put the data in a more robust form for analysis than would be allowable with their levels. It reduced each dataset size to 562 samples: February 1960 to November 2006. Furthermore, the T-bills and sales data means were removed and they were scaled to within ±1 to bring their magnitudes within the range of values suitable for neural network processing. Because the scatter plots of T-bills and sales showed that their levels and differences to be nonlinearly related, T-bills and sales were nonlinearly filtered using the running median filter (Arce, 2005) with a window size of 5. The nonlinear T-bills and sales relationship provided evidence to support the weak correlations of 0.15 for the levels and about -0.40 for unfiltered and filtered differences. Tests for normality and persistence revealed that T-bills and sales were not normal but were persistent. Their kurtosis and the skewness values were other than 3 and 0 respectively for each instance of their levels and differences (filtered and unfiltered); their Hurst exponents (Peters, 1994) were in the interval (0.5, 1] indicative of long-term correlations between widely separated T-bills and sales samples. Moreover, T-bills lead sales by 15, 22, and 21 months for their levels and unfiltered and filtered differences respectively that correspond to maximum correlation values of 0.18, 0.53, and 0.57.

The focused TDNN and Elman neural networks models used in the forecasting experiments were either of the single or multiple type depending on whether the out-of-sample set was expected to be forecasted in a single run or not. These network architectures were similarly design to the extent possible. They had two inputs, a 4-node hidden layer and a single node output layer with tanh activation functions. The supervised learning training regimen was used with the Levenberg-Marquardt backpropagation learning algorithm in batch mode for weight updates. The training was done in the 1-step ahead prediction mode; the error termination criterion was mean square error with increments threshold of 0.0001 and maximum allowable epochs of 1000. The best weights obtained during training were used for the testing procedures. Overall, the Focused TDNN used less weights and biases than the Elman network: 25 compared to 37. More specifically, the choice of the 2-input and 1-output nonlinear neural network predictor models was motivated, in part, by the NeuroSolutions software design strategy of employing the targeted variable (sales) to be predicted as a delayed input into the network to improve its predictive capability. In the multiple model scenarios, the network models differed only in the randomly assigned initial values of their weights and biases.

The T-bills and sales datasets of 562 samples each were subdivided in a training set of the first 64% (360 samples) and a testing set of the remaining 202 samples. When more than one model of the focused TDNN and Elman architectures was used to forecast the 202 samples, this out-of-sample testing set was partitioned into three subsets of 50 and one subset of 52 samples and the sliding window technique was used to maintain the training set size of 360 samples. However, the linear regression models generated by Excel *forecast* command was only employed in the single model forecasts regime. No model was provided with prior information of T-bills lead on sales.

RESULTS

The relative qualities of the sales forecasts produced by the focused TDNN, Elman network, and linear regression models were determined by evaluating the prediction error using visual examination and seven different performance metrics -- correlation, root mean square error, percentage of correct direction, power spectrum density test, Theil inequality coefficient, mean absolute percentage error, and Durbin-Watson statistic. The performance metrics values move in different directions – some (RMSE, Theil, and MAPE) move towards 0.00; some (r and POCD) move towards 1.00 or 100%; and the others (DW and SCP) move towards a structureless prediction error (or white noise). Since the performance metrics producing similar types of results may have values that do not move equally well towards the desired optimum values, the forecasts with the higher number of relatively more favorable performance metric values are graded more highly.

The focused TDNN models performed the best in single model forecasting while the Elman network performed better than TDNN in the multiple model forecasting. The single model linear regression models performed the worst (see Table I). The focused TDNN forecasted values came closest to the desired sales but the Elman network produced smoothest results as evidenced by the more gradual changes in transitions in its graphs, and the linear regression models' forecasts follows desired sales poorly compared to the neural network models (see Fig. 1). The focused TDNN single model forecasts of the out-of-sample dataset are better than its multiple model forecasts of about 50 samples per model per run for both the filtered and unfiltered differences of sales data. For example, the single model forecast of the filtered sales was better than the multiple model forecasts on five performance metrics (r = 0.98, POCD = 52%, Theil = 0.22, DW = 2.08; and the power spectrum density passes); they have equal RMSE of 0.08. A similar type of outcome was found for the forecasts of the unfiltered differences of sales (see Table I). The Elman network single model forecast was better than its multiple model forecast for the filtered sales out-of-sample data; the supporting performance metrics values are r = 0.97, POCD = 49%, Theil = 0.21, MAPE = 0.22, and DW = 1.78 .However, the two forecasting methods performed equally well on the unfiltered sales data in that there was a tie on one performance metric, RMSE of 0.15, and otherwise three performance metrics were more favorable for each type of forecast. In general, the forecasts of the filtered sales data were better than those of the unfiltered with the best forecast produced by the focused TDNN and the worst with linear regression models (see Table I and Fig. 1).

DISCUSSION

Forecasting is very important in economics and finance. It guides many decisions. Therefore, its accuracy is important to decision makers and forecasters; decisions based on inaccurate forecasts can be quite costly (Diebold and Mariano, 1995).

From the results of the experiments, the neural network architecture single model forecasts were better than its multiple model forecasts in every instance. However, across architectures, the Elman network produced better multiple model forecasts. This finding suggests that for some time series data using a single model to forecast the entire out-of-sample dataset should be preferred since multiple models would not perform as well. A further consequence of this finding is not to use two network models where one would suffice. While focused TDNN forecasted sales were generally the closest to the actual sales values, the Elman models produced smoothest results. This is consistent with the finding of Gencay and Liu (1997) who stated "that recurrent networks filter noise more successfully than feedforward networks," but no evidence was found to confirm that recurrent models necessarily produced better out-of-sample forecasts of nonlinear "noisy time series." Moreover, the relative inferior forecasting capability of the linear regression models compared to neural networks is consistent with previous findings of other researchers as reported by Kingdon (1997) who provided evidence from the Makridakis and the Santa Fe Institute competitions to show that neural networks performed at least as good as, but generally better, than linear and other nonlinear methods.

The criticality of interest rates regulatory property in the economic and its importance in financial markets should never be underestimated. The 22 months lead of inverted T-bills over sales is consistent with Larrain's (2002) finding of 19 months using spectral analysis. The models used in the experiments were expected to learn the 21 or 22 months lead of inverted T-bills over sales. That is, they were to learn the relationship between the current sample of inverted T-bills and the 21st sample of filtered sales differences and 22nd sample of unfiltered sales differences. With adequate training of the neural network models this would be possible since it was established that both inverted T-bills and sales were persistent with respective Hurst values of 0.8314 and 0.8646 for their levels, 0.7243 and 0.7215 for their unfiltered differences, and 0.7306 and 0.7306 for their filtered differences. Therefore, under these conditions where the network models had knowledge of the information stored in the inverted T-bills and sales data samples, 1-step ahead prediction in the NeuroSolutions environment produced the effect that past and current T-bills samples were forecasting unfiltered sales' 23rd month sample and filtered sales' 22nd month sample in the future.

CONCLUSION

The focused TDNN produced the best forecast in terms of closeness to the actual out-of-sample sales data, but the Elman network produced the smoothest forecasts and the better multiple model forecasts. The POCD and MAPE values were generally insufficiently high and low respectively. If POCD and MAPE values were affected by the nonlinear filtering method applied to the data, it would be instructive in a further study to use a different nonlinear filtering technique such as appropriate Volterra filters to determine whether generally higher values of POCD and lower values of MAPE are achievable. Regardless of the nonlinear filtering method used on the data, could better forecasts be yielded in a different neural network environment? Another question to be explored is whether larger networks would produce significantly better results.

TABLES AND FIGURES

Table I

Summary of Performance Statistics

Architecture/Model	Test Data	r	RMSE	POCD	Theil	MAPE	DW	SCP
FTDNN	**Sectionalized filtered dataset***	0.94	0.01	41.00	0.27	0.32	2.41	
		0.94	0.07	35.00	0.37	0.58	2.09	
		0.95	0.17	55.00	0.16	0.23	1.04	
		0.93	0.08	53.00	0.28	0.29	1.98	
	Mean of sectionalized filtered dataset*	**0.94**	**0.08**	**46.00**	**0.27**	**0.35**	**1.88**	**Failed**
	Entire filtered dataset	**0.98**	**0.08**	**52.00**	**0.22**	**0.41**	**2.08**	**Passed**
	Sectionalized unfiltered dataset*	0.66	0.17	37.00	0.48	0.59	2.47	
		0.72	0.15	33.00	0.34	0.51	2.58	
		0.93	0.14	35.00	0.34	0.41	2.39	
		0.73	0.14	35.00	0.54	0.68	2.34	
	Mean of sectionalized unfiltered dataset*	**0.76**	**0.15**	**35.00**	**0.43**	**0.55**	**2.44**	**Passed**
	Entire unfiltered dataset	**0.88**	**0.14**	**38.00**	**0.46**	**0.66**	**2.41**	**Passed**
Elman	**Sectionalized filtered dataset***	0.94	0.63	51.00	0.28	0.32	1.09	
		0.93	0.77	43.00	0.25	0.36	1.46	
		0.98	0.06	56.00	0.15	0.22	1.50	
		0.94	0.08	42.00	0.28	0.30	1.87	
	Mean of sectionalized filtered dataset*	**0.95**	**0.39**	**48.00**	**0.24**	**0.30**	**1.48**	**Failed**
	Entire filtered dataset	**0.97**	**0.71**	**49.00**	**0.21**	**0.22**	**1.78**	**Failed**
	Sectionalized unfiltered dataset*	0.67	0.17	41.00	0.48	0.40	2.31	
		0.82	0.15	39.00	0.53	0.50	2.06	
		0.94	0.14	39.00	0.35	0.40	2.21	
		0.72	0.14	37.00	0.55	0.62	2.06	
	Mean of sectionalized unfiltered dataset*	**0.79**	**0.15**	**39.00**	**0.48**	**0.48**	**2.16**	**Failed**
	Entire unfiltered dataset	**0.87**	**0.15**	**35.00**	**0.55**	**0.43**	**2.30**	**Passed**
Linear Regression**	**Entire filtered dataset**	**0.66**	**0.31**	**38.00**	**0.69**	**0.47**	**0.05**	**Failed**
	Entire unfiltered dataset	**0.62**	**0.30**	**49.00**	**0.99**	**0.34**	**0.23**	**Failed**

*The sectionalized datasets were partitioned into the subsets of 50, 50, 50, and 52 samples respectively.

**The linear regression models were produced with the Excel Forecast command.

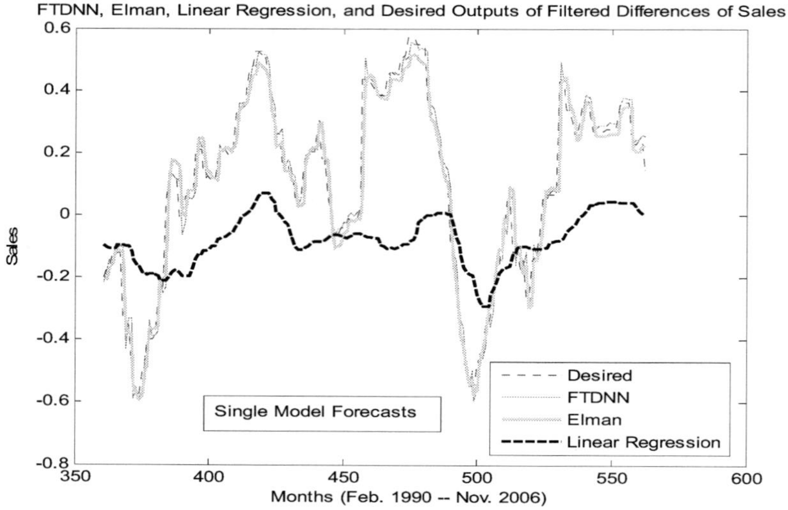

Figure 1: Focused TDNN, Elman network, and linear regression single model forecasts of filtered differences of sales given inverted 3-month T-bills as input.

REFERENCES

Arce, G., 2005, *Nonlinear Signal Processing: A statistical Approach*, John Wiley & Sons, Hoboken, N.J., pp. 81-250.

Diebold, F. and Mariano, R., 1995, "Comparing Predictive Accuracy," *Journal of Business and Statistics*, Vol. 20, January, pp. 134-144.

Gencay, R. and Liu, T., 1997, "Nonlinear Modelling and Prediction with Feedforward and Recurrent Networks," *Physica D,* Vol. 108, pp. 119-134.

Hall, R. and Lieberman, M., 2008, *Economics: Principles and Applications*, 4th Edition, Thomson South-Western, Mason, OH, pp. 406-440, 595-661, 673-712.

Haykin, S., 1999, *Neural Networks: A Comprehensive Foundation*, 2nd Edition, Prentice Hall, Upper Saddle River, pp. 1-49, 635-663, 732-789.

King, R. and Watson, M, 1996, "Money, Prices, Interest Rates, and Business Cycle," *The Review of Economics and Statistics*, Vol. 78, February, pp. 35-53.

Kingdon, J., 1997, *Intelligent Systems and Financial Forecasting*, Springer-Verlag, London, pp. 29-53.

Konur, U. and Okatan, A., 2004, "Time Series Prediction using Recurrent Neural Network Architectures and Time Delay Neural Networks," *Transactions on Engineering, Computing, and Technology*, Vol. 3, December, pp. 5-7.

Kuan, C. and White, H., 1994, "Artificial Neural Networks: An Econometric Perspective," *Econometric Reviews*, Vol. 13, pp. 1-91.

Larrain, M., 2002, "Do Interest Rates Lead Real Sales and Inventories? A Spectral Analysis Approach," *Business Economics*, pp. 33-43.

Lawrence, S., Giles, C., and Tsoi, A., 1997, "Lessons in Neural Network Training: Overfitting may be Harder than Expected," *Proceedings of the 14th National Conference on Artificial Intelligence*, AAAI-97, AAAI Press, Menlo Park, CA, pp. 540-545.

Peters, E., 1994, *Fractal Market Analysis: Applying Chaos Theory to Investment and Economics*, John Wiley & Sons, New York, N.Y., pp. 53-64.

Principe, J., Euliano, N., and Lefebvre, W., 2000, *Neural and Adaptive Systems: Fundamentals Through Simulations*, John Wiley & Sons, New York, pp. 473-587.

ENFORCED KNOWLEDGE EXTRACTION WITH BP-NETWORKS

IVETA MRÁZOVÁ
Faculty of Mathematics and Physics, Charles University
Prague, Czech Republic

ZUZANA REITERMANOVÁ
Faculty of Mathematics and Physics, Charles University
Prague, Czech Republic

ABSTRACT
We introduce a new method for training BP-networks inspired by the rapid method of scaled conjugate gradients and enhanced with the enforcement of a transparent internal knowledge representation.

INTRODUCTION

To clarify the role of hidden neurons in BP-networks, we want to group the outputs of the hidden neurons around three possible values: +1 (active states), −1 (passive states) and 0 (silent states). Mrázová and Wang (2007) call such a kind of internal knowledge representation condensed. The criterion for developing a condensed internal representation can be formulated as an additional term of the objective function to be minimized during training. BP-networks trained in this way provide better generalization, robustness and a transparent network structure. Our goal is to develop a new faster method of enforcing a transparent network structure based on the scaled conjugate gradients technique (SCG).

RELATED WORKS

For the BP-training algorithm proposed by Rumelhart, Hinton and Williams (1986), the training set T is a finite non-empty set of P ordered pairs of input/output patterns: $T = \{[\vec{x}_1, \vec{d}_1], \ldots, [\vec{x}_P, \vec{d}_P]\}$. A neuron with the weights $(w_1, \ldots, w_n)$, the threshold ϑ and the input vector $\vec{z} = (z_1, \ldots, z_n)$, computes its potential value ξ as $\xi = \sum_{i=1}^{n} z_i w_i + \vartheta$. The thresholds are represented by weights coming from fictive neurons with a constant output value equal 1. For the output neurons, we will consider the linear transfer function $y = f(\xi) = \xi$. For the hidden neurons, we will use the hyperbolic tangent transfer function:

$$y = f(\xi) = \frac{1 - e^{-2\xi}}{1 + e^{-2\xi}} \quad \text{, with derivative} \quad f'(\xi) = (1 + y)(1 - y) . \quad (1)$$

Standard BP-networks use the objective function E:

$$E = \frac{1}{2} \sum_p \sum_j (y_{j,p} - d_{j,p})^2 , \quad (2)$$

where p is an index over all training patterns, j is an index over all output neurons, y is their actual and d is their desired output value. Omitting the index p, the weights of the network will be adjusted iteratively by:

$$w_{ij}(t+1) = w_{ij}(t) + \alpha \, \delta_j \, y_i \quad (3)$$

$$\text{with} \quad \delta_j = \begin{cases} d_j - y_j & \text{for an output neuron} \\ f'(\xi_j) \sum_k \delta_k \, w_{jk} & \text{for a hidden neuron} \end{cases} \quad (4)$$

w stands for weights and thresholds, respectively, i and k index neurons in the layers below and above the neuron j, respectively. d_j is the desired and y_j the actual output value of the neuron j. $f'(\xi_j)$ corresponds to (1), $t+1$ and t index next and present weights, respectively, α is a positive constant representing the learning rate. Möller (1993) proposed an extremely efficient variant to standard BP-training – the method of scaled conjugate gradients (SCG). Below we will combine this approach also with learning from hints introduced by Abu-Mostafa (1993). Its variant described by Gallmo and Carlstrom (1995) consists in supplying the network with additional target outputs which express some knowledge about the problem. After training, the additional outputs can be removed.

LEARNING INTERNAL REPRESENTATION

We will restrict our observations w.l.o.g. to the last hidden layer. The criterion used to clarify the role of hidden neurons already during training should have its local minima corresponding to active (+1), passive (-1) and silent (0) states. This can be achieved by formulating the wanted representation criterion F as:

$$F = \sum_p \sum_j (1 + y_{j,p})^s (1 - y_{j,p})^s y_{j,p}^2 , \qquad (5)$$

where p indexes the training patterns and j the hidden neurons. y represents their output value. In our experiments we used $s = 4$. The new objective function G has the form:

$$G = E + c_F F \qquad (6)$$

where E corresponds to (2) and F to (5). The constant $c_F \geq 0$ reflects the trade-off between the influence of E and F in G. Omitting the index p, we get the following term weight adjustment:

$$w_{ij}(t+1) = w_{ij}(t) + \alpha\delta_j y_i + \alpha_r \varrho_j y_i \qquad (7)$$

In this formula, δ_j is defined as for the standard BP-training algorithm (4) and

$$\varrho_j = \begin{cases} 0 & \text{for output neurons} \\ 2\left[(s+1)y_j^2 - 1\right]\left[f'(\xi_j)\right]^s y_j & \text{for neurons from } l\text{-th hidden layer} \\ f'(\xi_j) \sum_k \varrho_k w_{jk} & \text{for other neurons} \end{cases} \qquad (8)$$

w stands for weights and thresholds, respectively. i and i' index neurons connected with neuron j via the weights w_{ij} and $w_{i'j}$, respectively. k indexes the neurons from the layer above the neuron j. $f'(\xi_j)$ corresponds to (1), s is the parameter for tuning the shape of the representation error function. $t+1$ and t index next and present weights, respectively. α and α_r from Equation (7) are constants representing the particular learning rates.

The entire process of training BP-networks by means of scaled conjugate gradients with enforced internal representation (SCGIR) is summarized in the algorithm box below. For a given network, $G(w_k)$ can be evaluated according to:

$$G(w) = \frac{1}{2}\sum_p \sum_j (y_{j,p} - d_{j,p})^2 + c_F \sum_p \sum_h (1 + y_{h,p})^s (1 - y_{h,p})^s y_{h,p}^2 \qquad (9)$$

where p goes over all training patterns, j and h index the output and hidden neurons, respectively. y denotes the actual output of a neuron while d is its desired output value. s is a parameter for tuning the shape of the representation criterion and c_F is a non-negative constant. Omitting the index p, the elements $\partial G/\partial w_{ij}$ of $G'(w) = (\partial G/\partial w_{11}, \ldots, \partial G/\partial w_{m_{last}m})$ correspond to:

$$\frac{\partial G}{\partial w_{ij}} = \begin{cases} -\,\delta_j\, y_i & \text{for an output neuron} \\ -\left\{\sum_k \delta_k w_{jk} + c_F\left[2(s+1)y_j^2 - 1\right]\left[f'(\xi_j)\right]^{s-1} y_j\right\} f'(\xi_j)\, y_i & \\ & \text{for neurons from the } l\text{-th hidden layer} \\ -\,f'(\xi_j)\left[\sum_k (\delta_k + c_F\varrho_k)\, w_{jk}\right] & \text{for other hidden neurons} \end{cases} \qquad (10)$$

w stands for the weights. i and i' index neurons connected with neuron j via w_{ij} and $w_{i'j}$, respectively. k indexes the neurons from the layer above the neuron j, m and m_{last} denotes the number of output neurons and the number of hidden neurons placed in the last hidden layer, respectively. y_j, s and c_F have the same meaning like in 9). $f'(\xi_j)$ corresponds to (1), δ_k and ϱ_k can be determined according to (4) and (8), respectively.

SUPPORTING EXPERIMENTS

The experiments compare the SCGIR-method with pure SCG, standard BP (GD) and BP with an enforced internal representation (GDIR). The data provided by the World Bank (2001) consists of 972 records comprising 25 world development indicators to be used as input attributes for 162 countries over the years 2001–2006. As the output attribute, we used Gross domestic income according to purchasing power parity. The data was normalized such that in every dimension the mean was 0 and standard deviation was 1. Missing values were either estimated from known values (linear interpolation) or they were replaced by mean (0). As output hints we used the attribute Income group (*hintIG*) comprising 5 classes. Alternatively we have used clustering of the input data into k groups as a hint. In such a case, the k-means algorithm with Euclidean distance measure has been applied – *hint*9 (for $k = 9$) and *hint*6 (for $k = 6$). Testing was done by means of 10-fold cross-validation. The tested topology was 25-37-1 with randomly initialized weights from $[-1; 1]$. Learning rates were set to 0.3. c_F was experimentally chosen 0.00004 for *GDIR* and 0.0005 for *SCGIR*. To check for early stopping we used increased mean squared error (MSE) on validation set during 10 immediately following training cycles.

In Table 1, we follow this notation: E_{tr} is the average MSE on the training set, E_{v} denotes average MSE on the validation set, E_{t} corresponds to average MSE on the test set. $n_{\mathrm{s}}(0.3)$ represents the average number of hidden neurons with outputs out of $-0.85, 0, 0.85 \pm 0.15$, while $n_{\mathrm{s}}(0.1)$ indicates the average number of hidden neurons with outputs out of $-0.9, 0, 0.9 \pm 0.1$. n_{epochs} states the average number of training cycles, n_{neur} is the number of hidden neurons with a too small weight to the output neuron, n_{neur1} indicates the average number of hidden neurons, that have formed always the same internal representation, identical or complementary one to another hidden neuron.

In this test, *SCGIR* clearly outperformed *GDIR*. *SCGIR* is much quicker and with a smaller error E_{t}. When learning with hints, it yields a lower number of non-transparent hidden neurons. Both, the *SCG* and *SCGIR* provide a much

Training with scaled conjugate gradients and an enforced internal representation:

1. Initialize the weight vector $\vec{w_1}$ with small random values, set scalars such that $0 < \sigma \leq 10^{-4}$, $0 < \lambda_1 \leq 10^{-8}$ and $\overline{\lambda_k} = 0$. Set $\vec{r_1} = \vec{g_1} = -G'(\vec{w_1})$, $success = true$ and the discrete time variable k equal to 1.
2. If $success = true$, then calculate the second-order information:

$$\sigma_k = \frac{\sigma}{\| \vec{g_k} \|}, \qquad \vec{s_k}^{SCG} = \frac{G'(\vec{w_k} + \sigma_k \vec{g_k}) - G'(\vec{w_k})}{\sigma_k}$$

$$\text{and} \quad \delta_k^{SCG} = \vec{g_k}^T \vec{s_k}^{SCG}$$

3. Scale $\vec{s_k}^{SCG}$, δ_k^{SCG}:

$$\vec{s_k}^{SCG} = \vec{s_k}^{SCG} + (\lambda_k - \overline{\lambda_k}) \vec{g_k}$$
$$\delta_k^{SCG} = \delta_k^{SCG} + (\lambda_k - \overline{\lambda_k}) \| \vec{g_k} \|^2$$

4. If $\delta_k^{SCG} < 0$ then make the Hessian matrix positive definite:

$$\vec{s_k}^{SCG} = \vec{s_k}^{SCG} + \left(\lambda_k - 2\frac{\delta_k^{SCG}}{\| \vec{g_k} \|^2}\right) \vec{g_k}, \quad \overline{\lambda_k} = 2\left(\lambda_k - \frac{\delta_k^{SCG}}{\| \vec{g_k} \|^2}\right),$$

$$\delta_k^{SCG} = -\delta_k^{SCG} + \lambda_k \| \vec{g_k} \|^2 \quad \text{and} \quad \lambda_k = \overline{\lambda_k}$$

5. Calculate step size α_k^{SCG}: $\mu_k = \vec{g_k}^T \vec{r_k}$; $\alpha_k^{SCG} = \dfrac{\mu_k}{\delta_k^{SCG}}$
6. Calculate the comparison parameter Δ_k^{SCG}:

$$\Delta_k^{SCG} = \frac{2\,\delta_k^{SCG} \left[G(\vec{w_k}) - G(\vec{w_k} + \alpha_k^{SCG} \vec{g_k}) \right]}{\mu_k^2}$$

7. If $\Delta_k^{SCG} \geq 0$ then a successful reduction in the value of the objective function G can be made:

$$\vec{w}_{k+1} = \vec{w_k} + \alpha_k^{SCG} \vec{g_k}, \qquad \vec{r}_{k+1} = -G'(\vec{w}_{k+1}), \qquad \overline{\lambda_k} = 0$$

and $success = true$

8. (a) If $k \bmod N = 0$ then restart the algorithm: $\vec{g}_{k+1} = \vec{r}_{k+1}$.
   else create a new conjugate direction:

$$\beta_k = \frac{\vec{g_k}^T \cdot (\vec{g_k} - \vec{g}_{k-1})}{\| \vec{g}_{k-1} \|^2} \quad \text{and} \quad \vec{g}_{k+1} = \vec{r}_{k+1} + \beta_k \vec{g_k}$$

   (b) If $\Delta_k^{SCG} > 0.75$ then reduce the scale parameter: $\lambda_k = \dfrac{1}{4}\lambda_k$.

   else reduction in error not possible: $\overline{\lambda_k} = \lambda_k$; $success = false$.
9. If $\Delta_k^{SCG} < 0.25$ then increase the scale parameter:

$$\lambda_k = \lambda_k + \frac{\delta_k^{SCG} (1 - \Delta_k^{SCG})}{\| \vec{g_k} \|^2}$$

10. If steepest descent direction $\vec{r_k} \neq 0$, set $k = k+1$ and go to 2. Otherwise, terminate and return the weight vector $\vec{w_k}$ as the desired minimum for the objective function G.

Table 1: The performance of the methods tested for networks with the topology 25-37-1

method	n_{epochs}	E_{tr}	E_{v}	E_{t}	n_{neur}	n_{neur1}	$n_{\mathrm{s}}(0.3)$	$n_{\mathrm{s}}(0.1)$
GD	1709.1±706.1	0.046±0.009	0.021±0.003	0.060±0.033	3.8±1.9	4.0±2.4	8992.5±998.7	21133.0±1879.0
$GD - hint9$	2498.4±11.4	0.050±0.005	0.025±0.003	0.068±0.037	5.0±1.4	3.5±1.6	9040.8±762.9	21217.0±1214.8
$GD - hint6$	2430.5±123.9	0.051±0.007	0.024±0.002	0.067±0.041	3.9±1.2	5.3±2.6	8170.9±1252.2	19463.0±2052.6
$GD - hintIG$	2480.3±68.6	0.047±0.005	0.026±0.004	0.066±0.029	3.4±1.5	3.6±1.9	8477.6±877.7	19900.0±1632.5
GDIR	1182.2±975.6	0.076±0.018	0.043±0.010	0.107±0.067	3.4±1.3	9.0±3.2	2776.9±2092.3	9914.2±5468.4
$GDIR - hint9$	2484.6±54.7	0.072±0.009	0.043±0.008	0.091±0.048	4.7±2.2	12.1±2.3	1131.0±228.8	4947.1±814.7
$GDIR - hint6$	2441.8±147.9	0.070±0.005	0.043±0.008	0.092±0.047	3.6±1.5	9.0±2.4	1405.4±223.7	6001.9±890.9
$GDIR - hintIG$	2453.2±117.9	0.076±0.008	0.044±0.007	0.090±0.044	4.9±1.9	8.6±2.5	1330.9±233.9	5610.0±977.2
SCG	62.7±11.6	0.042±0.009	0.028±0.006	0.056±0.026	3.5±2.0	4.7±1.5	8670.1±1441.9	20324.0±2232.6
$SCG - hint9$	251.4±85.9	0.034±0.004	0.030±0.006	0.064±0.034	7.5±3.1	5.4±2.8	2767.5±1196.4	7283.0±2683.5
$SCG - hint6$	154.0±53.2	0.040±0.006	0.030±0.004	0.071±0.047	5.9±2.3	3.8±3.0	4984.9±1937.7	12874.0±4199.0
$SCG - hintIG$	188.0±13.2	0.030±0.006	0.027±0.005	0.053±0.025	14.0±2.9	0.8±1.0	3885.4±871.5	9886.8±1829.7
SCGIR	226.5±127.1	0.067±0.005	0.033±0.008	0.087±0.045	4.0±1.9	8.6±1.6	3820.8±583.5	13469.0±1547.8
$SCGIR - hint9$	233.3±62.9	0.056±0.012	0.043±0.014	0.075±0.024	8.9±2.5	6.9±2.2	729.6±346.5	3329.6±1298.9
$SCGIR - hint6$	238.2±64.0	0.050±0.009	0.036±0.008	0.071±0.026	8.8±2.8	8.6±1.6	786.0±252.5	3338.9±994.9
$SCGIR - hintIG$	180.6±51.7	0.050±0.010	0.037±0.009	0.084±0.040	10.4±2.7	8.1±2.8	689.2±143.8	2696.9±507.7

Table 2: The performance of the methods tested for networks with the topology 25-37-1 and noisy data

method	E_{tr}	E_{noisy}	*difference*	n_{neur}	n_{neur1}	n_{epochs}	$n_{\mathrm{s}}(0.3)$	$n_{\mathrm{s}}(0.1)$
GD	0.0202	0.0233	0.0097±0.0047	0.8	1.6	10002	7268.6	17147
GDIR	0.0427	0.0450	0.0056±0.0082	6.8	4.3	10002	756.0	3712
SCG	0.0070	0.0132	0.0100±0.0141	10.1	2.1	985.2	2142.9	5494.5
SCGIR	0.0070	0.0136	0.0056±0.0165	16.8	3.9	1292	404.9	1632

higher number n_{neur} and n_{neur1} when used with hints. This means, that they need less hidden neurons to learn the task.

At last we tested the so far best performing methods on noisy data – namely *GD*, *GDIR*, *SCG* and *SCGIR*, all trained with the hint *hintIG*. We used the data from the entire training set with 1–5 % random noise added to each data element. Both the training and testing has been repeated 10-times. Table 2 presents the errors achieved for the training and (noisy) test sets as well as the difference *difference* between internal representations of the original and of the noisy data (mean over all activities of hidden neurons over all input patterns). Obviously, the mean *difference* for both the *GDIR* and *SCGIR* is half of the difference achieved for the *GD* and *SCG* methods. *SCG* is, however, quicker than *SCGIR* for comparable errors. But when considering the other parameters (n_{neur}, n_{s} and *difference*), *SCGIR* appears to be much better. *SCGIR* seems to be better also in comparison with *GDIR* due to its speed and improved performance – n_s is two times smaller and the same holds for *difference*, too.

CONCLUSIONS

The new SCGIR-method supports very fast training of BP-networks with a simultaneous enforcement of a transparent internal representation. This strategy reflects the concept of learning with hints of Abu-Mostafa (1993) that is expected to improve generalization. Experimental results obtained so far confirm that SCGIR can provide much faster convergence than the classical BP-training algorithm with enforced internal representation (GDIR). The performance of SCGIR improves significantly when enhancing it with hints (SCGIR-hint). On the other hand, SCGIR seems to be less powerful than GDIR when forcing transparent internal representations, yet it yields comparable results also for noisy data.

ACKNOWLEDGEMENTS

This research was supported by the grant No. 358/2006/A INF/MFF of GA UK and by the program 'Information Society' under project 1ET100300517.

REFERENCES

Abu-Mostafa, Y. S., 1993, "Hints and the VC Dimension," *Neural Computation*, 5, pp. 278–288.

Gallmo, O. and Carlstrom, J., 1995, "Some Exoperiments Using Extra Output Learning to Hint Multi-Layer Perceptrons," *Niklasson, J. and Boden, M. (eds): Current Trends in Connectionism*, pp. 179–190.

Möller, M., 1993, "A scaled conjugate gradient algorithm for fast supervised learning," *Neural Networks*, 6, pp. 525–533.

Mrázová, I. and Wang, D., 2007, "Improved generalization of neural classifiers with enforced internal representation," *Neurocomputing*, (available also online via the DOI link: `http://dx.doi.org/10.1016/j.neucom.2006.10.143`).

Rumelhart, D. E., Hinton, G. E. and Williams, R. J., 1986, "Learning representations by back-propagating errors," *Nature*, (323), pp. 533–536.

World Bank, 2001, *World Development Report 2000/2001: Attacking Poverty*, The World Bank Group, Washington, D.C.

PRINCIPAL COMPONENT ANALYSIS CONSIDERING WEIGHTS BASED ON DISSIMILARITY OF OBJECTS IN HIGH DIMENSIONAL SPACE

MIKA SATO-ILIC
Department of Risk Engineering
School of Systems and
Information Engineering
University of Tsukuba
Tsukuba, Ibaraki
305-8573, Japan

SHOTA ITO
Graduate School of Systems
and Information Engineering
University of Tsukuba
Tsukuba, Ibaraki
305-8573, Japan

ABSTRACT
A weighted principal component analysis based on the result of fuzzy clustering has been proposed. The merit of this analysis is to obtain the result of the principal component analysis considered the dissimilarity structure of objects in the high dimensional space. The dissimilarity structure is represented through weights of objects which show how each object contributes to the classification structure based on the dissimilarity of objects in the high dimensional space. Since the degree of contribution is defined by a function, according to the change of the function, the weights will be changed. This paper proposes weighted principal component analyses based on several weight functions. Moreover, we investigate the features of the weights by using numerical examples.

INTRODUCTION

Since ordinal principal component analysis (Anderson, 1984) is based on the distance between objects and the orthogonal projection of objects into a low dimensional space, the distance between two objects in high dimensional space (observation space) is larger than the distance between the projected objects obtained in low dimensional space (projected space). However, since only the order of values of the distance is satisfied, even if we obtain a very small value of distance between two objects in the projected space the real value of distance between the objects in the observation space may be very large.

In order to solve this problem, a principal component analysis considering the dissimilarity of objects in the high dimensional observation space (Sato-Ilic, 2002) and a modified analysis for the interval-valued data (Sato-Ilic, 2003) have been proposed. The essential difference between this principal component analysis for interval-valued data and conventional principal component analyses for interval-valued data (Lauro and Palumbo, 2000) (Gioia and Lauro, 2006) is whether the property of the dissimilarity structure in high dimensional space is considered or not. Structure of the dissimilarity of objects in the high dimensional space is implemented by using a function of the weights which can show a classification structure obtained as a fuzzy clustering result. This exploits the merit of fuzzy clustering whose result is obtained as continuous values. In the case of conventional hard clustering, the result is obtained as discrete values. So, the idea that the degree of the contribution of each object for classification structure is represented by a clustering result can not be used.

According to the definition of the function which transforms a fuzzy clustering result to the degree of contribution of each object for the classification structure, the values of degree or weights will be changed. This paper proposes several functions to define the weights and investigate features of the weighted principal component analyses based on the change of the weight functions.

This paper consists of the following sections. In the next section, we briefly explain the fuzzy clustering method used in this paper. In the third section, we mention a conventional weighted principal component analysis for interval-valued data (Sato-Ilic, 2003). The fourth section proposes several weight functions. In the fifth section, several numerical examples are described. Finally, in the sixth section we conclude the paper.

FUZZY CLUSTERING

The state of fuzzy clustering is represented by a partition matrix $U = (u_{ik})$ whose elements show the degree of belongingness of the objects to the clusters, u_{ik}, $i = 1, \cdots, n$, $k = 1, \cdots, K$, where n is the number of objects and K is the number of clusters. In general, u_{ik} satisfies the following conditions:

$$u_{ik} \in [0,1], \forall i, k; \ \sum_{k=1}^{K} u_{ik} = 1, \forall i. \tag{1}$$

Fuzzy c-means (FCM) (Bezdek et al., 1999) is one of the methods of fuzzy clustering. FCM is the method which minimizes the weighted within-class sum of squares:

$$J(U, \boldsymbol{v}_1, \cdots, \boldsymbol{v}_K) = \sum_{i=1}^{n} \sum_{k=1}^{K} u_{ik}^m d^2(\boldsymbol{x}_i, \boldsymbol{v}_k), \tag{2}$$

where $\boldsymbol{v}_k = (v_{ka})$, $k = 1, \cdots, K$, $a = 1, \cdots, p$ denotes the value of the centroid of a cluster k, $\boldsymbol{x}_i = (x_{ia})$, $i = 1, \cdots, n$, $a = 1, \cdots, p$ is i-th object, and $d^2(\boldsymbol{x}_i, \boldsymbol{v}_k)$ is the square Euclidean distance between $\boldsymbol{x}_i$ and $\boldsymbol{v}_k$. p is the number of variables. The exponent m which determines the degree of fuzziness of the clustering is chosen from $[1, \infty)$ in advance. The purpose is to obtain the solutions U and $\boldsymbol{v}_1, \cdots, \boldsymbol{v}_K$ which minimize equation (2). From conditions shown in equation (1), the local extrema of equation (2) can be obtained as follows:

$$u_{ik} = 1 / \sum_{l=1}^{K} \{d(\boldsymbol{x}_i, \boldsymbol{v}_k) / d(\boldsymbol{x}_i, \boldsymbol{v}_l)\}^{\frac{2}{m-1}}, \quad \boldsymbol{v}_k = \sum_{i=1}^{n} (u_{ik})^m \boldsymbol{x}_i / \sum_{i=1}^{n} (u_{ik})^m, \ \forall i, k. \tag{3}$$

WEIGHTED PRINCIPAL COMPONENT ANALYSIS FOR INTERVAL-VALUED DATA

Suppose that the observed interval-valued data which are values of n objects with respect to p variables are denoted by the following:

$$X = ([\underline{x}_{ia}, \overline{x}_{ia}]), \quad i = 1, \cdots, n, \ a = 1, \cdots, p,$$

where $[\underline{x}_{ia}, \overline{x}_{ia}]$ shows the interval-valued data of the i-th object with respect to a variable a which has the minimum value $\underline{x}_{ia}$ and the maximum value $\overline{x}_{ia}$.

The data X is decomposed into two data sets $\underline{X}$ and $\overline{X}$ for the minimum values and the maximum values as follows:

$$\underline{X} = (\underline{x}_{ia}), \ \overline{X} = (\overline{x}_{ia}), \quad i = 1, \cdots, n, \ a = 1, \cdots, p. \tag{4}$$

We assume the following linear combination of $\underline{X}$ and $\overline{X}$:

$$\hat{X} \equiv \underline{WX} + \overline{WX}. \tag{5}$$

$\underline{W}$ and $\overline{W}$ are diagonal matrixes defined as follows:

$$\underline{W} = \text{diag}(\underline{w}_1, \cdots, \underline{w}_n),\ \overline{W} = \text{diag}(\overline{w}_1, \cdots, \overline{w}_n),\ \underline{w}_i = \prod_{k=1}^{K} \underline{u}_{ik}^{-1},\ \overline{w}_i = \prod_{k'=1}^{K'} \overline{u}_{ik'}^{-1},\ \forall i, \tag{6}$$

where $\text{diag}(a_1, \cdots, a_n)$ shows a $n \times n$ diagonal matrix whose diagnal elements are $a_1, \cdots, a_n$. $\underline{U} = (\underline{u}_{ik})$ and $\overline{U} = (\overline{u}_{ik'})$, $i = 1, \cdots, n$, $k = 1, \cdots, K$, $k' = 1, \cdots, K'$ are the results of fuzzy clustering obtained for $\underline{X}$ and $\overline{X}$, respectively. That is, $\underline{u}_{ik}$ shows the degree of belongingness of an object i to a cluster k for $\underline{X}$ and $\overline{u}_{ik'}$ shows the degree of belongingness of an object i to a cluster k' for $\overline{X}$. Here $\underline{u}_{ik}$ and $\overline{u}_{ik'}$ are assumed to satisfy the following conditions:

$$\underline{u}_{ik},\ \overline{u}_{ik'} \in (0, 1),\ \forall i, k, k';\ \sum_{k=1}^{K} \underline{u}_{ik} = \sum_{k'=1}^{K'} \overline{u}_{ik'} = 1,\ \forall i, \tag{7}$$

where K is the number of clusters for $\underline{X}$ and K' is the number of clusters for $\overline{X}$ and they are assumed to be given in advance so as to obtain the clustering result satisfactory. Using a property of the algebraic product and the condition of the degree of belongingness shown in equation (7), these weights show the clustering situations in the minimum and maximum parts of the interval-valued data. That is, the weight $\underline{w}_i$ shows that if the belonging status of the object i to the clusters in the minimum part is clearer, then the weight becomes larger. If the belonging status of the object i is less obvious in the minimum part, the classification structure of the data which consists of minimum values is close to uniformity, and then the weight becomes small. Therefore, the following equation

$$\underline{WX} = \text{diag}(\underline{w}_1, \cdots, \underline{w}_n)\underline{X} = (\underline{w}_i \underline{x}_{ia}),\ i = 1, \cdots, n,\ a = 1, \cdots, p$$

shows that clearer objects in the part of minimum values under the classification structure have larger values and that the objects which are vaguely situated under the classification structure are avoided. That is, those which do not have any significant relation to the classification structure for example objects that are treated as noise are avoided. For $\overline{WX}$ shown in equation (5), we can see the same relationships as that obtained above from condition (7).

Suppose that $\hat{X}$ is centered with respect to p variables and $\hat{\boldsymbol{z}}_1$ is the first principal component of the transformed data $\hat{X}$. $\hat{\boldsymbol{z}}_1$ is defined as:

$$\hat{\boldsymbol{z}}_1 = \hat{X}\hat{\boldsymbol{l}}_1, \tag{8}$$

where $\hat{\boldsymbol{l}}_1 = (\hat{l}_{11}, \cdots, \hat{l}_{1p})'$. The purpose of principal component analysis for interval-valued data based on fuzzy clustering is to estimate $\hat{\boldsymbol{l}}_1$ which maximizes the variance of $\hat{\boldsymbol{z}}_1$ under the condition of $\hat{\boldsymbol{l}}_1'\hat{\boldsymbol{l}}_1 = 1$. The estimate of $\hat{\boldsymbol{l}}_1$ is obtained as the corresponding eigen-vector for the maximum eigen-value of $\hat{\Sigma} = \hat{X}'\hat{X}$. In order to obtain the second principal component $\hat{\boldsymbol{z}}_2$ for $\hat{X}$, we define the linear combination, $\hat{\boldsymbol{z}}_2 = \hat{X}\hat{\boldsymbol{l}}_2$, where $\hat{\boldsymbol{l}}_2 = (\hat{l}_{21}, \cdots, \hat{l}_{2p})'$. $\hat{\boldsymbol{l}}_2$ is obtained as the eigen-vector corresponding to the second largest eigen-value of $\hat{\Sigma}$. If we get s ($s \leq p$)

different eigen-values of $\hat{\Sigma}$, according to the above, we find the s principal components for $\hat{X}$. The proportion of the α-th principal component for principal component analysis for interval-valued data based on fuzzy clustering, $\hat{P}_\alpha$, is as $\hat{P}_\alpha = \hat{\lambda}_\alpha/\mathrm{tr}(\hat{\Sigma})$, where $V\{\hat{\boldsymbol{z}}_\alpha\} = \hat{\lambda}_\alpha$ and $\sum_{\alpha=1}^{p} \hat{\lambda}_\alpha = \mathrm{tr}(\hat{\Sigma})$. The accumulated proportion up to the s-th principal component for principal component analysis of the interval-valued data based on fuzzy clustering is defined as $\hat{P} = \sum_{\alpha=1}^{s} \hat{P}_\alpha$.

WEIGHT FUNCTIONS

We propose two new weight functions that will replace equation (6) in the following way.

$$\underline{w}_i = \left\{ -\sum_{k=1}^{K} \underline{u}_{ik} \log \underline{u}_{ik} \right\}^{-1}, \quad \overline{w}_i = \left\{ -\sum_{k'=1}^{K'} \overline{u}_{ik'} \log \overline{u}_{ik'} \right\}^{-1}, \quad i = 1, \cdots, n, \tag{9}$$

$$\underline{w}_i = \sum_{k=1}^{K} \underline{z}_{ik}, \quad \overline{w}_i = \sum_{k=1}^{K} \overline{z}_{ik}, \quad \underline{z}_{ik} \equiv \exp\left[-\frac{1}{2} \left(\frac{\underline{d}_{ik}}{\underline{b}_k} \right)^2 \right], \quad \overline{z}_{ik} \equiv \exp\left[-\frac{1}{2} \left(\frac{\overline{d}_{ik}}{\overline{b}_k} \right)^2 \right], \tag{10}$$
$$i = 1, \cdots, n.$$

In equation (10), $\underline{d}_{ik} \equiv d(\underline{\boldsymbol{x}}_i, \underline{\boldsymbol{v}}_k)$, $\overline{d}_{ik} \equiv d(\overline{\boldsymbol{x}}_i, \overline{\boldsymbol{v}}_k)$, where $d(\underline{\boldsymbol{x}}_i, \underline{\boldsymbol{v}}_k)$ is Euclidean distance between $\underline{\boldsymbol{x}}_i$ and $\underline{\boldsymbol{v}}_k$ and $d(\overline{\boldsymbol{x}}_i, \overline{\boldsymbol{v}}_k)$ is Euclidean distance between $\overline{\boldsymbol{x}}_i$ and $\overline{\boldsymbol{v}}_k$. $\underline{\boldsymbol{x}}_i$ shows i-th object of $\underline{X}$ and $\overline{\boldsymbol{x}}_i$ shows i-th object of $\overline{X}$ which are shown in equation (4). $\underline{\boldsymbol{v}}_k$ and $\overline{\boldsymbol{v}}_k$ are k-th cluster centers obtained as results of fuzzy c-means shown in equation (3) for minimum and maximum data, respectively. $\underline{b}_k$ and $\overline{b}_k$ are parameters which show the estrangement of the cluster k from the average of all objects for minimum and maximum data. Here we define as

$$\underline{b}_k = \frac{1}{n} \sum_{i=1}^{n} \underline{d}_{ik}, \quad \overline{b}_k = \frac{1}{n} \sum_{i=1}^{n} \overline{d}_{ik}, \quad k = 1, \cdots, K.$$

$\underline{z}_{ik}$ and $\overline{z}_{ik}$ are degree of belongingness of an object i to a cluster k for minimum and maximum data, respectively. Weights of n objects for each cluster in minimum and maximum values of data are denoted as follows:

$$\underline{Z}_k = \mathrm{diag}(\underline{z}_{1k}, \cdots, \underline{z}_{nk}), \quad \overline{Z}_k = \mathrm{diag}(\overline{z}_{1k}, \cdots, \overline{z}_{nk}), \quad k = 1, \cdots, K. \tag{11}$$

Using equation (11), we define the following matrix replaced by equation (5).

$$\tilde{X} \equiv \sum_{k=1}^{K} (\underline{Z}_k \underline{X} + \overline{Z}_k \overline{X}) = \sum_{k=1}^{K} \underline{Z}_k \underline{X} + \sum_{k=1}^{K} \overline{Z}_k \overline{X}. \tag{12}$$

Since $\sum_{k=1}^{K} \underline{Z}_k = \mathrm{diag}(\sum_{k=1}^{K} \underline{z}_{1k}, \cdots, \sum_{k=1}^{K} \underline{z}_{nk})$, $\sum_{k=1}^{K} \overline{Z}_k = \mathrm{diag}(\sum_{k=1}^{K} \overline{z}_{1k}, \cdots, \sum_{k=1}^{K} \overline{z}_{nk})$, from equations (5) and (12), we obtain the weights shown in equation (10). Here, we assume the same number of clusters for minimum and maximum data in order to define equation (12).

Weights shown in equations (6) and (9) have similar features, that is, if the belonging status of each object to the clusters is clearer, then the weight becomes larger. If the belonging status of each object is less obvious, then the weight becomes small. So, the weights work to avoid objects which are unclearly situated to the classification structure. However, if $\underline{u}_{ik}$ or $\overline{u}_{ik}$ are very close to 0, then the weights shown in equation (6) become very large. In order to avoid this extreme case and relax the sharp fluctuations, we propose the weight function shown in equation (9). Weights shown in equation (10) use a clustering result directly, that is, the weights show degree of belongingness of each object to the clusters.

NUMERICAL EXAMPLES

Landsat data observed over the Kushiro marsh-land is used. The values of the data show the amount of reflected light from the ground with respect to 7 kinds of light for 1048576 pixels. We use data from July and October of 1993 and treat this data as an interval-valued data by using the minimum and the maximum values.

Figures 1-3 show the results of the first principal component shown in equation (8). Figure 1 shows the result when we use the weight function shown in equation (6) and figure 2 is the case when we use a weight function shown in equation (9) and figure 3 is the case when we use the weight function shown in equation (10). The number of clusters is assumed as three. The values of $\underline{b}_1$, $\underline{b}_2$, $\underline{b}_3$, $\overline{b}_1$, $\overline{b}_2$, and $\overline{b}_3$ are given as 32.5, 26.6, 32.5, 43.6, 45.5, and 38.3, respectively. They are obtained as the average distances from the objects to each cluster in minimum and maximum data. From a comparison of these figures, it can be seen that the weight functions shown in equations (9) and (10) can avoid extreme values as the result of PCA, however the weight shown in equation (6) makes the extreme values.

Figures 4-6 show the results shown in figures 1-3 as the form of images. From figures 4-6, we can see that figures 4 and 5 are similar to each other. These figures are the results of using weights which change according to situation of classification structure and show the lakes and marsh-land area. Figure 6 shows that only the lake area reflected the clustering result because this weight directly uses the clustering result. From these results, we can verify the features of weight functions mentioned in the previous section.

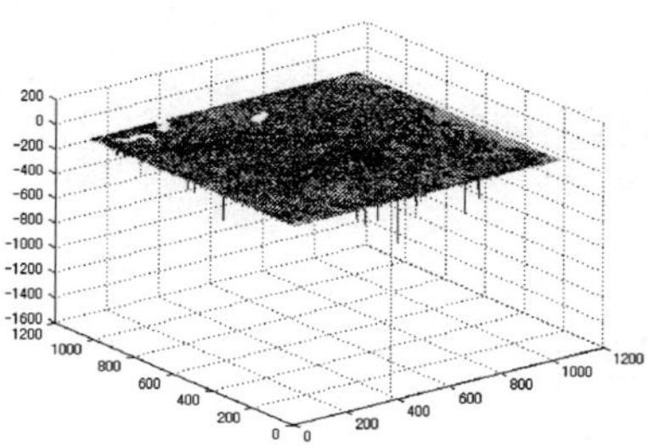

Figure 1 Result of PCA using Weights shown in (6)

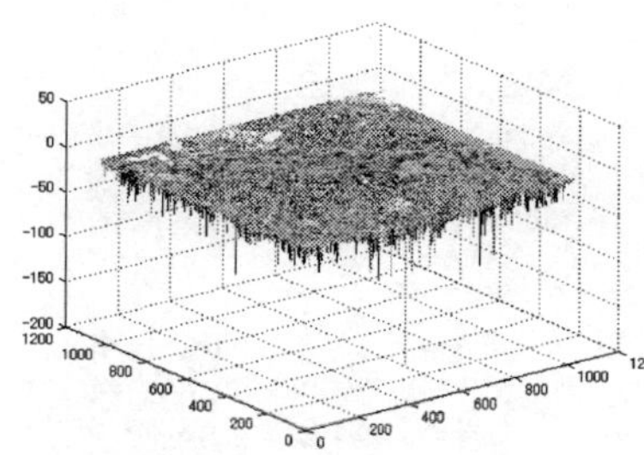

Figure 2 Result of PCA using Weights shown in (9)

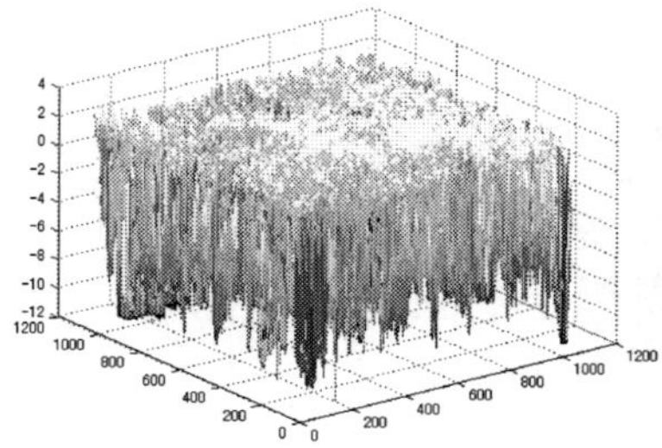

Figure 3 Result of PCA using Weights shown in (10)

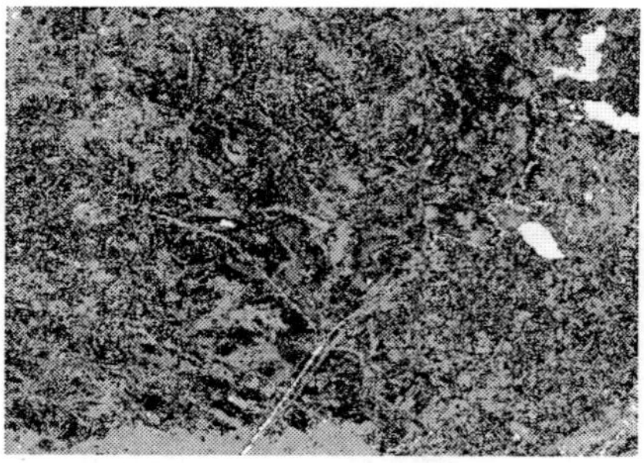

Figure 4 Result of Image for PCA using Weights shown in (6)

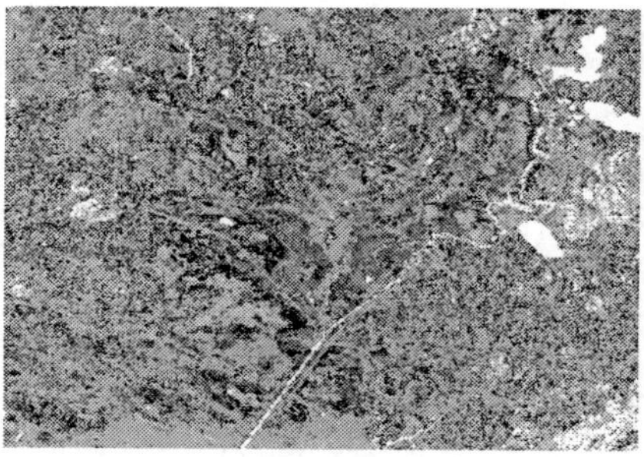

Figure 5 Result of Image for PCA using Weights shown in (9)

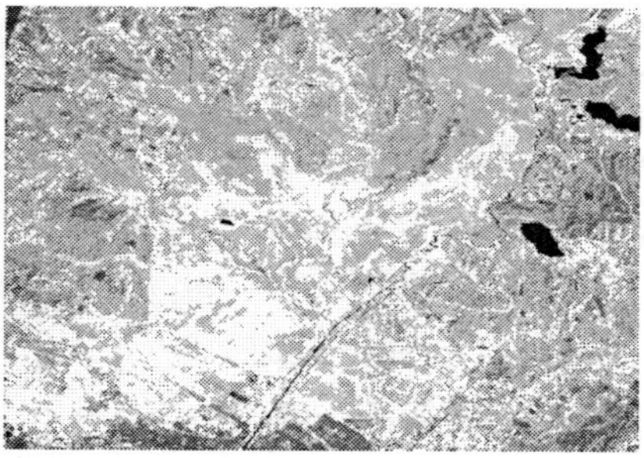

Figure 6 Result of Image for PCA using Weights shown in (10)

CONCLUSION

This paper proposes several weighted principal component analyses. The difference is caused by the different weight function which reflect the dissimilarity structure in observation space. Since real data have various structures and the dissimilarity structure is represented by a classification structure in which the weight is defined as degree of contribution of each object to the classification structure, various functions which transform from the clustering result to the weight are needed. Therefore, this paper proposes several weight functions and discusses the features.

REFERENCES

Anderson, T. W., 1984, "*An Introduction to Multivariate Statistical Analysis*", Second ed., John Wiley & Sons.

Bezdek, J. C., Keller, J., Krisnapuram, R. and Pal, N. R., 1999, "*Fuzzy Models and Algorithms for Pattern Recognition and Image Processing*", Kluwer Academic Publishers.

Gioia, F. and C. N. Lauro, 2006, "Principal Component Analysis on Interval Data", *Computational Statistics*, vol. 21, pp. 343-363.

Lauro, C. N. and F. Palumbo, 2000, "Principal Component Analysis of Interval Data: A Symbolic Data Analysis Approach", *Computational Statistics*, vol. 15, pp. 73-87.

Sato-Ilic, M., 2002, "Fuzzy Cluster Structure based Weighted Principal Component Analysis", *5th Czech-Japan Seminar on Data Analysis and Decision Making under Uncertainty*, pp. 97-102.

Sato-Ilic, M., 2003, "Weighted Principal Component Analysis for Interval-Valued Data based on Fuzzy Clustering", *IEEE International Conference on Systems, Man and Cybernetics*, pp. 4476-4482.

VISUAL EXPLORATION OF NUMERIC DATA USING 3D SELF ORGANIZING FEATURE MAPS

GEORGE K. KNOPF AND HARISH PUNGOTRA
Department of Mechanical & Materials Engineering,
Faculty of Engineering, The University of Western Ontario,
London, Ontario, Canada, N6A 5B9

ABSTRACT
Scientific data visualization techniques are used to transform large quantities of numeric data vectors into colorized patterns and geometric shapes that enable scientists and engineers to observe the underlying behaviour of complex systems in new informative ways. In this paper a three-dimensional graphical form, called a "glyph", is created by a self-organizing feature map (SOFM) and used to visually display patterns inherent in large volume *N*-dimensional data sets. The level of similarity amongst the input vectors assigned to a common lattice node, or neighboring nodes, will determine the local shape of the glyph. Colour coding is used to show prior vector classification by an expert. Benign and malignant breast cancer data are used to illustrate how this approach can automatically create colorized glyphs for assisting the data analyst to explore the database for meaningful patterns.

INTRODUCTION

Recent advances in computer hardware have enabled scientists and engineers to automatically gather and store huge amounts of numeric data captured by a variety of sensors. Furthermore, numerous parameters are recorded simultaneously leading to large multivariate high-dimensional data sets. All data acquired by these sensors represent useful and valuable information for future analysis and developing a thorough understanding of the underlying phenomena that created the data. Unfortunately, many data management systems display the information by means of text lists or simple 2D graphs thereby limiting the user's ability to only viewing small amounts of numeric data at a particular instant in time with little opportunity to adequately explore the entire database without first establishing a well-defined *search objective*. As a consequence, the amount of data often far exceeds the capability of the data analyst to efficiently explore and effectively interpret the information content of the stored data. In many applications the technical capability of gathering data has far outpaces the ability to process and decipher the data, resulting in massive and ever expanding "data dumps" (Keim, 2001).

Visual data exploration techniques have become important tools for searching through strings and lists of numeric data in order to provide the user with some insight into the global behavior of complex physical systems. These techniques are often developed to directly integrate human beings into the data exploration process by enabling the analyst to use visual perception, skilled expertise and long-term experience to better interpret large data sets and draw meaningful conclusions. Data visualization is particularly useful in applications where little is known about the relationship between the data vectors or when the exploration goals are poorly defined. In these situations the user must be an active participant in the exploratory process and possess the ability to dynamically modify the search goals (Keim, 2002). The process of visual data exploration may further assist the user in developing a new hypothesis or introducing a unique interpretation of the data. Verification of the presented hypothesis or subjective interpretation will require additional exploration of the data using quantitative measures derived from statistical or AI data analysis techniques.

Interactive data exploration depends upon the visualization tools used to assist the investigator in discovering and analyzing patterns. Simplified graphical and pictorial forms, called *glyphs*, are often preferred for exploration purposes than endless streams of numbers and tables, Figure 1. Visual cues such as shape, color, size, orientation, or relative location of symbolic forms (Tufte, 1983) (Sangole and Knopf, 2003) can provide scientists and engineers with a high-level perceptual tool for quickly identifying coherent patterns in seemingly random data sets. The visual attributes displayed on the glyph are determined and bound by the multivariate data set being represented by the visual form.

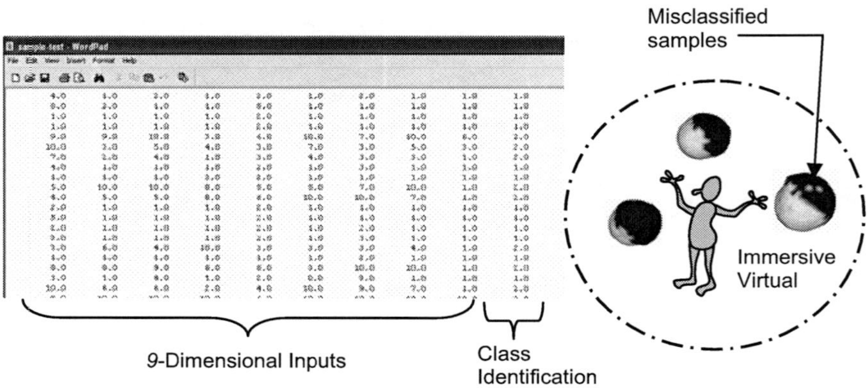

Figure 1. Visual comparison between table (list) of 9-dimensional data vectors and colorized 3D pictorial form.

In this paper a self-organizing feature map (SOFM) with a three-dimensional closed lattice is used to transform *N*-dimensional vectors into reproducible glyph where the shape and color reflect local content and classification (Sangole and Knopf, 2003) (Knopf and Sangole, 2006). The SOFM adaptation algorithm is able to cluster high-dimensional data vectors and assign similar clusters to neighboring nodes on the 3D lattice structure. The level of similarity amongst the input vectors assigned to a node, or cluster unit, will determine the local shape and color-code of the glyph. These colorized graphical shapes can be displayed and manipulated within an immersive virtual reality environment for interactive data analysis. The following sections briefly describe how the SOFM can be used for exploratory clustering and summarizes how the algorithm is used to create 3D glyphs. An illustration on classifying benign and malignant breast cancer data is presented.

EXPLORATORY CLUSTERING USING THE SOFM

Exploratory clustering methods have been developed to assist the analyst in searching for meaningful patterns within large databases by reducing both the volume and dimensionality of the original data to a more manageable size. These techniques are often called clumping, data grouping, unsupervised pattern recognition, and cluster analysis. Most clustering algorithms attempt to find a balance between the "ideal" number of clusters and the number of data vectors assigned to each cluster. A small number of clusters will enable fast computational processing and real-time navigation through the data set. However, if the number of clusters is too small then the level of visible information content will become too low. In some situations it is difficult to interpret the patterns formed by clustering in a lower dimensional space and, therefore, the vectors assigned to the clusters are mapped back into the original *N*-dimensional space for detailed study.

Two common exploratory clustering methods used to identify groups of data vectors that form patterns are the *K*-means clustering algorithm and Kohonen's self-organizing feature map. *K*-means is a clustering algorithm that finds groups of similar vectors that are represented by their respective cluster centre, which is computed as the mean of the data vectors assigned to it. The self-organizing feature map (SOFM) is an unsupervised clustering algorithm that searches for hidden patterns or relationships by correlating data vectors and establishing an ordered topology for the cluster centers in the feature space. Associations among the cluster centers and consequently those among groups of similar data vectors are reflected in the relative position of the nodes in the lattice of the SOFM (Gross and Seibert, 1993) (Ritter, 1999) (Sangole and Knopf, 2003). The relative *similarity, dissimilarity,* or *variation* amongst the vectors assigned to a cluster, or between cluster units in a common map, can be graphically displayed on the SOFM using the appropriate metric or measure-of-information. The selection of this metric, $\mathcal{M}^{d}_{i,j,k}$, is a non-trivial task. It is often application dependent and defined by the observer such that it quantifies a desired characteristic of the data set to be displayed on the 3D glyph.

Figure 2 illustrates how the node positions can be modified based on the measure of information for a 2D circular map. The locations of the nodes in the lattice are shifted by scaling the radial distance of the nodes in proportion to the magnitude of the similarity measure. An alternative approach is shown in Figure 2c where a unique circle is assigned to each node (Ebert, Rohrer, Shaw, Panda, Kukla and Roberts, 2000). The combination of spheres with various radii results in an implicit surface. The radii of the constituent circles and spheres correspond to a metric such as the variance in the vectors assigned to the node.

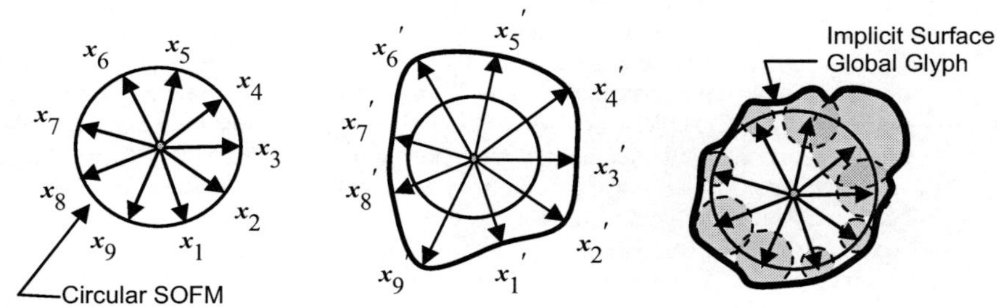

a. Original SOFM. **b. Local measure displayed as vector length.** **c. Local measure displayed as circle radius.**

Figure 2. Different approaches to displaying information on a circular SOFM.

CREATING A GLYPH USING A 3D SOFM

The individual merits of both data self-organization and shape-based visualization have been combined into a single mechanism. The *N*-dimensional data vectors are initially mapped onto a closed surface that is allowed to deform based on the desired measure of information. The self-organizing feature map establishes a topological order for the original data by forming cluster units and automatically rearranging the weights of the units on the lattice such that neighboring units correspond to data vectors that lie in close proximity in the original N-dimensional space. Local surface distortions and colors are introduced to the lattice after weight adaptation based on local metrics and expert class assignment.

The numeric data vectors are mapped into a SOFM where each lattice node corresponds to a particular cluster unit. An input $\boldsymbol{X}^{p}$ is randomly sampled from P training data vectors and compared to the weight vector, $\boldsymbol{w}_{ijk}$, associated with the cluster unit at node location (i,j,k). The cluster at the $(i,j,k)^{\text{th}}$ node has a variable neighborhood ($\boldsymbol{NE}_{ijk}$) with a decreasing radius. Initially, the weights assigned to each unit are set to small random values distributed throughout the input space. During the self-organization process, the cluster unit with a weight vector that is closest to the applied input, and its associated neighborhood, are simultaneously updated to minimize error. At the end of the adaptation process, the N-dimensional weights for the cluster unit at each node on the lattice form prototypical representations of the inputs assigned to them and reflect some global physical characteristic of the numeric data set (Knopf and Sangole, 2006).

EXPERIMENTAL STUDY

To illustrate the process of visualizing N-dimensional data consider the Wisconsin breast cancer data (Blake and Merz, 1998) that is available obtained from the machine learning repository maintained by the Computer Science Department at the University of California, Irvine. The data set consists of 683 fine needle aspirate (FNA) tissue samples provided by Dr. William H. Wolberg, University of Wisconsin Hospitals, Madison, and is pre-classified into two categories: malignant and benign tissue samples. Every FNA tissue sample is a vector consisting of nine attributes: clump thickness, uniformity of cell size and shape, marginal adhesion, single epithelial cell size, bare nuclei, bland chromatin, normal nucleoli and mitoses, all of which need to be considered in order to make a diagnosis. Each attribute is assigned a value on a scale of 1-10 by the pathologist, 10 indicating a very high probability of malignancy (Wolberg and Mangasarian, 1990).

The data vectors are randomly selected and presented to a SOFM network consisting of 600 nodes, or cluster units, distributed on the surface of a sphere. This lattice is created by adaptively fitting a 20x20x20 SOFM to 2562 coordinate points sampled from the surface of a unit sphere. In this manner the neighboring nodes on the SOFM lattice form quadrilateral elements. Furthermore, the interior nodes of the 3D SOFM are ignored for this application because the goal is to create a colorized surface glyph that reflects the information content of the original data set. An initial neighborhood of 5 unit radius is selected and the network is trained for approximately 300 cycles.

On completion of the weight adaptation process, the spherical lattice is color-coded and deformed to reflect information about the underlying physical phenomenon. The deformed "spherical" lattice is colour-coded using prior information on the two categories of samples, blue indicating malignant tissue and red representing benign tissue samples. Distortions are introduced using a metric representing the distance of a lattice node from the center of all nodes; that is,

$$\mathcal{M}_{ijk}^{d} = scale \bullet \left\| \boldsymbol{w}_{ijk} \right\| \tag{1}$$

where $\boldsymbol{w}_{ijk}$ is the weight vector of the cluster unit located at (i,j,k) and *scale* is used to amplify or reduce the magnitude of the measure.

A second glyph is created from the same data set and spherical lattice structure for the SOFM. However, in this case an implicit surface model is created by assigning a *similarity measure* to each cluster and representing it as a sphere at the node location. The radius of the sphere corresponds to how similar the length of the cluster center vector (ie. SOFM weight vector) is to its immediate neighboring clusters. The radius is given by the measure

$$\mathcal{M}_{ijk}^{d} = \sum_{l,m,n=-1}^{1} \left\| \mathbf{w}_{ijk} - \mathbf{w}_{i-l,j-m,k-n} \right\| \qquad (2)$$

Figure 3 presents several views of two graphical objects created from the Wisconsin breast cancer data. The colorized 3D shape glyph shown on the top row, Figure 3a, was created using Eqn. (1) and represents the magnitude of the feature vectors assigned to the corresponding node in the SOFM. Multiple views of the second type of glyph are shown in the bottom row, Figure 3b. In this illustration the spheres at the node locations correspond to the level of similarity, or dissimilarity, between the cluster center and neighboring clusters. The metric used to determine the radii of the spheres is given in Eqn. (2). A small sphere indicates that the feature vector for the cluster is similar to immediate neighbors in the lattice whereas a large sphere shows a high degree of dissimilarity.

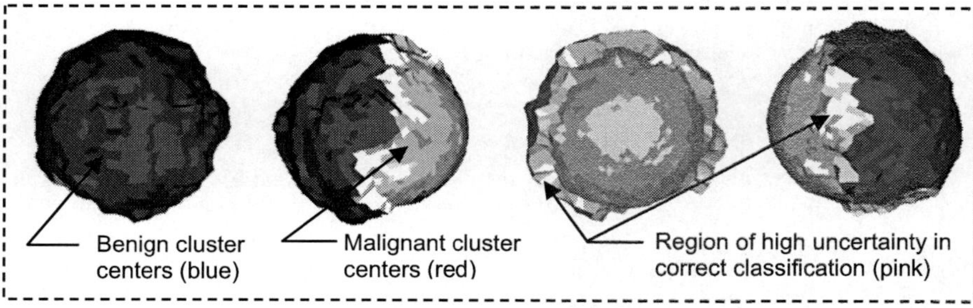

a. Colorized shape glyph composed of triangular facets (multiple views).

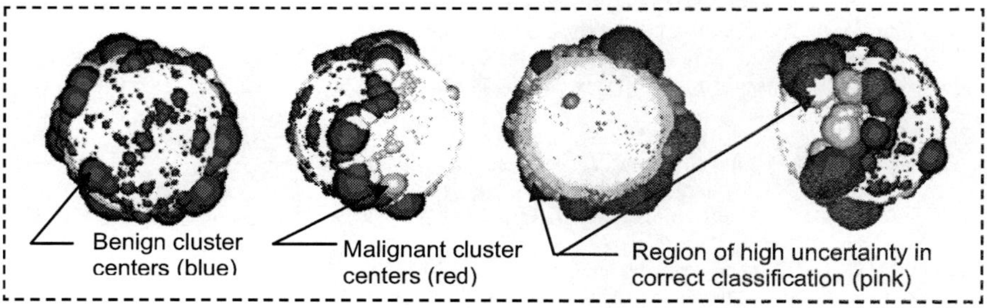

b. Colorized glyph composed of localized spheres (multiple views). A small sphere radius corresponds to high degree of neighborhood similarity.

Figure 3. Two colored glyphs used to cluster 683 nine-dimensional data vectors into two classes (benign and malignant). Note that in the greyscale representation above the blue clusters appear dark and red are shown as medium grey (pink is very light).

CONCLUSIONS

Scientists and engineers are relying more and more on graphical tools to analyse and interpret large volumes of complex, high-dimensional data. This paper described how a

SOFM with a predefined lattice could generate a closed geometric representation of the information embedded in arbitrary *N*-dimensional numeric data. The geometric shapes were created by grouping, or clustering, input vectors and assigning the cluster centers to nodes on a unit sphere. Cluster and class attributes were then displayed by introducing surface deformations to the spherical lattice and color-coding the facets according to expert classification.

Two types of glyphs were illustrated in this work. The first was a deformed sphere where the surface geometry reflected the magnitude of the 9-dimensional vectors assigned to each node (cluster center) on the lattice. A sharp elevated ridge formed between the lattice regions associated with benign and malignant cancer data indicating clear separation between the classes. The colorized visual cues also showed the regions of the surface with greatest uncertainty in correct classification. The size of the individual spheres in the second type of glyph corresponded to the "degree of similarity" where a small spherical icon represented minimal differences whereas a larger sphere corresponded to a high-degree of dissimilarity. As expected, the greatest differences were found along the boundaries between the benign and malignant classes. However, several relatively large spheres are observed in the interior of the class regions and these results suggest that some of the original data vectors require further study by an expert. Information about specific input vectors can be recovered by maintaining a record of which input vectors were assigned to each node.

ACKNOWLEDGEMENTS

This work has been supported, in part, by the Natural Sciences and Engineering Research Council of Canada.

REFERENCES

Blake, C.L., and Merz, C.J., 1998, "UCI Repository of Machine Learning", Department of Information and Computer Science, University of California, Irvine, CA. [http://www.ics.uci.edu/~mlearn /MLRepository.html]

Ebert, D.S., Rohrer, R.M., Shaw, C.D., Panda, P., Kukla, J.M., and Roberts, D.A., 2000, "Procedural Shape Generation for Multi-dimensional Data Visualization", *Computers and Graphics*, Vol. 24, pp. 375-384.

Gross, M., and Seibert, F., 1993, "Visualization of Multidimensional Image Data Sets using a Neural Network", *The Visual Computer, International Journal of Computer Graphics*, Vol. 10, No. 1, pp. 145-159.

Keim, D.A., 2001, "Visual Exploration of Large Data Sets", *Communications of the ACM*, Vol. 44, No.8, pp. 38-44.

Keim, D.A., 2002, "Information Visualization and Visual Data Mining", *IEEE Transactions on Visualization and Computer Graphics*, Vol. 8, No. 1, pp. 1-8.

Knopf G.K., and Sangole, A., 2006, "Data Visualization", *Wiley Encyclopedia of Biomedical Engineering*, Metin Akay, ed., John Wiley [13 pages – electronic access through Wiley].

Ritter, H., 1999, "Self-organizing Maps on Non-Euclidean Spaces", *Kohonen Maps*, E. Oja and S. Kaski, ed., Elsevier Science B. V., pp. 97-109.

Sangole, A., and Knopf, G.K., 2003, "Visualization of Randomly Ordered Numeric Data Sets using Spherical Self-organizing Feature Maps", *Journal of Computers & Graphics*, Vol. 27, pp. 963-976.

Tufte, E., 1983, *The Visual Display of Quantitative Information*, Graphics Press, USA.

Wolberg, W.H., and Mangasarian, O.L., 1990, "Multisurface Method of Pattern Separation for Medical Diagnosis Applied to Breast Cytology", *Proceedings of the National Academy of Sciences*, Vol. 87, pp. 9193-9196.

Discovery of Useful Concepts Using The Hierarchy of Attributes and Concepts

Sang C. Suh, Sam I. Saffer, Viral Masarani
Department of Computer Science
Texas A&M University – Commerce
Commerce, TX 75429-3011

Abstract
Clustering is the subject of active research in several fields such as statistics, pattern recognition, and machine learning. This research focuses on clustering in data mining. Data mining adds to clustering the complications of very large datasets with very many attributes of different types. This imposes unique computational requirements on relevant clustering algorithms. A variety of algorithms have recently emerged that meet these requirements and were successfully applied to real-life data mining problems. Clustering, in data mining, is useful to discover distribution patterns in the underlying data. Clustering algorithms usually employ a distance metric based (e.g., Euclidean) similarity measure in order to partition the database such that data points in the same partition are more similar than points in different partitions. In this paper, we study clustering algorithms for data with categorical attributes. Instead of using traditional clustering algorithms that use distances between points for clustering which is not an appropriate concept for Boolean and categorical attributes, we propose a novel concept of HAC (Hierarchy of Attributes and Concepts) to measure the similarity/proximity between a pair of data points. We present a robust clustering algorithm HAC that employs hierarchy of concepts and not distances when merging clusters. Our methods naturally extend to non-metric similarity measures that are relevant in situations where a domain expert/similarity table is the only source of knowledge. For data with categorical attributes, our findings indicate that HAC not only generates better quality clusters than traditional algorithms, but it also exhibits good scalability properties.

Introduction
One of the numerous definitions of Data Mining, or Knowledge Discovery in Databases is [2, 7]:

Data Mining, or Knowledge Discovery in Databases (KDD) as it is also known, is the nontrivial extraction of implicit, previously unknown, and potentially useful information from data. This encompasses a number of different technical approaches, such as clustering, data summarization, learning classification rules, finding dependency net works, analyzing changes, and detecting anomalies.

Generally, much of this ever increasing data contains valuable knowledge that is not easily retrievable. The increasing speed and capacity of computer technology has made feasible the utilization of various data mining techniques to automatically extract knowledge from this information. One of the techniques is clustering. Clustering is a division of data into groups of similar objects. Representing the data by fewer clusters necessarily loses certain fine details, but achieves simplification. It models data by its clusters.

1) Parametric Clustering: The goal of this kind of algorithm is to solve an optimization problem to satisfy the optimality criterion imposed by the model, which is often means minimizing the cost function. K-Means (Median) Algorithm is an example of parametric clustering algorithms [1, 5].
2) Non-Parametric Clustering: In non-parametric clustering methods, a cluster is loosely defined as a region surrounding a local maximum of the probability density function. Given a sufficiently large sample, nonparametric clustering methods are capable of detecting clusters of unequal size and dispersion and with highly irregular shapes [6, 7].

Hierarchical Clustering
Clustering builds a cluster hierarchy or, in other words, a tree of clusters, also known as a dendrogram. Every cluster node contains child clusters; sibling clusters partition the points covered by their common parent. Such an approach allows exploring Hierarchical clustering is one of the most frequently used methods in unsupervised learning. Given a set of data points, the output is a binary tree (dendrogram) whose leaves are the data points and whose internal nodes represent nested clusters of various sizes. The tree organizes these clusters hierarchically, where the hope is that this hierarchy agrees with the intuitive organization of real-world data. Hierarchical structures are ubiquitous in the natural world [4].

Data on different levels of granularity hierarchical clustering methods are categorized into agglomerative (bottom-up) and divisive (top-down) [4]. An agglomerative clustering starts with one-point (singleton) clusters and recursively merges two or more most appropriate clusters. A divisive clustering starts with one cluster of all data points and recursively splits the most appropriate cluster. The process continues until a stopping criterion (frequently, the requested number k of clusters) is achieved [8].

Conceptual Clustering

Conceptual clustering is closely related to formal concept analysis (FCA), decision tree learning, and mixture model learning. Conceptual clustering is a technique that forms concepts out of data incrementally by subdividing groups into subclasses iteratively; thus building a hierarchy of concepts. Conceptual clustering builds a structure out of the data incrementally by trying to subdivide a group of observations into subclasses. The result is a hierarchical structure known as the concept hierarchy. Each node in the hierarchy subsumes all the nodes underneath it, with the whole data set at the root of the hierarchy tree [5]. The representation of each concept takes the form of a probabilistic distribution of each possible attribute value calculated over the set of instances associated with this concept [3].

Conceptual clustering enhances the value of existing databases by revealing patterns in the data. These patterns may be useful for understanding trends, for making predictions of future occurrences from historical evidence, or for synthesizing data records into meaningful clusters. A conceptual clustering system accepts a set of object descriptions (events, observations, facts) and produces a classification scheme over the observations. These systems use an evaluation function to discover classes with "good" conceptual descriptions [12].

HAC (Hierarchy of Attributes and Concepts) Method

HAC is both a hierarchical and conceptual clustering system that organizes data so as to maximize inference ability. This algorithm implements clustering by discovering substructures in database which compress the original data and represent structural concepts in the data. Once a substructure is discovered, the substructure is used to simplify the data by replacing instances of the substructure with a pointer to the substructure definition. The discovered substructures allow abstraction over detailed structures in the original data. Iteration of the substructure discovery process constructs a hierarchical description of the structural data in terms of the discovered substructures. This hierarchy provides varying levels of interpretation that can be accessed based on the specific data analysis goals [11].

HAC accepts database of structured data (concepts) as input. This type of data is naturally represented using a graph. The graph representation includes labeled vertices with vertex id numbers, attributes on X-axis and downward directed edges. Each vertex represents a concept and the value of that concept is given by the directed edges (usually map to the attribute's value on X-axis or to some other concept). With this graph we can make a concept table for every attribute in the database and this table will have all the concepts and their value which are relevant according to the attribute. Ultimately with the help of these concept tables we can get a table in which each record will have an attribute name, its value and relationship with different concepts. Each attribute in this table represents a cluster. In the following we give a brief summary of how HAC works [9, 13].

Algorithm:

1. Make concept tables for different attributes using HAC. In this basically we have to make a hierarchy graph such that each level or concept is meaningful. Example in fig.1 concept c1 has name "skin diseases" which is perfectly meaningful since, it has got two attributes (skin allergy, skin cancer) under it. In the example explained below we used two concept tables for attributes 'diseases' and 'counties'.
2. Now, make a "Cluster Point Table" which will have attribute name, attribute value and relationship with different concepts. Attribute name is basically the name of cluster; Attribute value will give the value of cluster which is a combination of different concepts. Attribute values can be generated by using following steps:

   Let total_num= Total number of attributes in cluster point table;

   Concept_1=total number of concepts in first concept table;

   Concept_2=total number of concepts in second concept table;

   total_num= Concept_1 * Concept_2;

   For i=0 to total_num

   V_i = total number of records under different combination of $C1_{(j, j+1---m)}$ and $C2_{(k, k+1----n)}$

   Here, $V_{(i, i+1----total\_num)}$ are different clusters or attributes in Cluster Point Table.

   $C1_{(j, j+1---m)}$ are different concepts of concept table1

   $C2_{(k, k+1---m)}$ are different concepts of concept table2
3. From this Cluster Point table and Concept tables (made through HAC) we can generate rules. In the following example, in Figure 7, we have provided a hypothetical section of Cluster Point Table. The rule generated is "there are 51 people with "skin diseases (Cluster C1)" in the state of "Arkansas (Cluster D1)", in other words, there are 51 people with "skin allergy" and "skin cancer" in the counties "Jefferson" and "Washington".

Example: Clustering Using the HAC Method

In the following example, we have used a database which contains information about various diseases found in the various counties in the USA. Following is an attribute table for the "diseases" followed by its concept table and HAC graph. Similar way, second entry in the concept table for C2 comprises A4 and A13, which are eye and ear diseases. So C2 is appropriately given the name of "Eye – Ear Diseases". In the same way, all different clusters are created and concept tables are built. (See Figure 1).

Attribute names	Attribute values
A1	Skin Allergy
A2	Skin Cancer
A3	Asthma
A4	Cataracts
A5	Heart – Muscle Damage
A6	Lung Cancer
A7	Pulmonary Fibrosis
A8	Hemoglobin Deficiency
A9	Bladder Cancer
A10	Pre-pregnancy malaise
A11	Brain Damage
A12	Hepatitis
A13	Acoustic Neuronal
A14	Coronary Heart Disease

Figure 1. Attribute Table for Diseases

Concept names	Concept values	Concept attributes
C1	Skin Diseases	A1, A2
C2	Eye – Ear Diseases	A4, A13
C3	Acute Visceral Diseases	A3, A8, A9
C4	Acute Diseases	C1, C2, C3
C5	Heart Diseases	A5, A14
C6	Lung – Liver Diseases	A6, A7, A12
C7	Chronic Visceral Diseases	C6, A10, A11
C8	Chronic Diseases	C5, C7
C9	General Diseases	C4, C8

Figure 2. Concept Table for Diseases

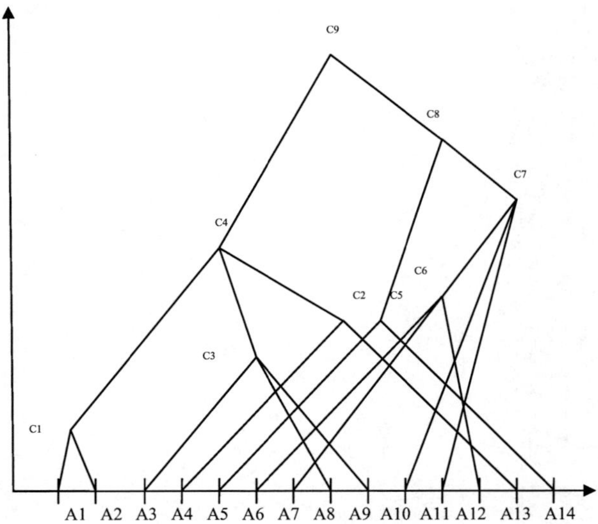

Figure 3. HAC for Diseases

In Figure 2, all the appropriate attributes are clustered together in different concept values and the concept table is created. For example, cluster C3, "acute visceral diseases" is comprised of "Asthma", "Hemoglobin Deficiency" and "Bladder Cancer". HAC graph has been created for the Diseases attribute, and various clusters are shown as depicted on the graph with attributes they are comprised of (See Figure 3). In the same way, Figure 4 shows all the values of the attribute Counties and in Figure 5, they have been clumped together to form up some clusters. For example, cluster name "Georgia" is comprised of "Clay", "Jackson" and "Lincoln" counties. HAC for Counties has been created much as the same way as the HAC for Diseases, as can be seen in Figure 6.

Attribute names	Attribute values
A1	Jefferson
A2	Washington
A3	Union
A4	Franklin
A5	Madison
A6	Jackson
A7	Clay
A8	Lincoln
A9	Montgomery

Figure 4. Attribute Table for Counties

Concept names	Concept values	Concept attributes
D1	Arkansas	A2, A1
D2	Georgia	A7, A6, A8
D3	Indiana	A5, A4, A3,A6
D4	Oregon	A1, A8
D5	States	A9, D1, D2, D3, D4

Figure 5. Concept Table for Counties

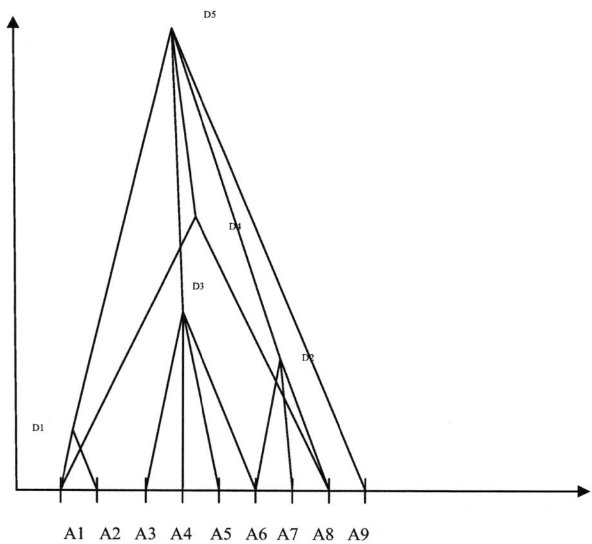

Figure 6. HAC for Counties

Conceptual Points	Support	Concepts (Diseases)	Concepts (Counties)
V1	51	C1	D1
V2	22	C2	D2
V3	23	C3	D3
V4	12	C5	D2
V5	65	C6	D1

Figure 7. Hypothetical Section of Cluster Point Table

Figure 7 shows a section of Cluster Point Table. This table will be created by counting up the total number of lines from the database which contains attribute values from the both relative clusters, and later this table will be used to create Concept Cluster Point Graph as shown in Figure 8. Using the attribute and concept tables for the selected attributes, diseases and counties in this case, a graph can be constructed to represent clusters at concept levels. Each point in the graph represents a concept that is formed by a combination of attribute values. This point in the graph is called a Concept Cluster Point (CCP). Since the cluster point represents a high level concept, it naturally converts to a rule that matches with a concept.

Furthermore, the support of each rule can be given by calculating the number of contributing entries in the original relational table to form the concept. In the table shown in Figure 8 below, we have used hypothetically simulated database to create Cluster Pointer Table.

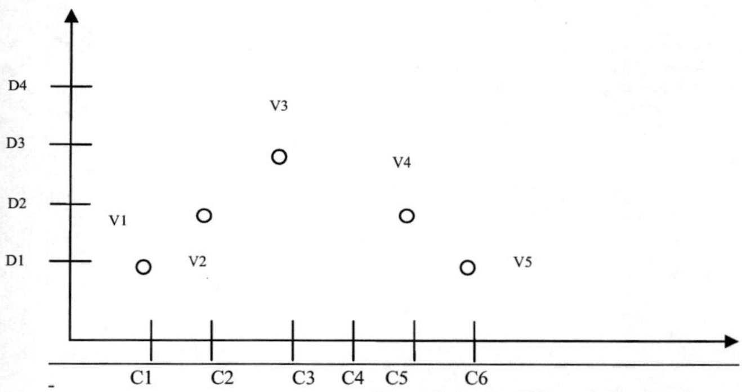

Figure 8. Concept Cluster Points Graph

Results

This HAC-based algorithm basically generates different types of rules based upon clusters depending upon the number of concept tables used to make a cluster. Each cluster is a combination of two or more different concepts. We generated the rules using two different sets of concept tables. The two concept tables for the attributes 'diseases' and 'counties' were used respectively. In general case, the HAC algorithm generates the rules carrying the following facts on the "relationship among various attributes in different concepts." Hence, in this case, each point in the generated CCP graph showed the rules representing the relationship among specific diseases, regions, and eventually the number of people. So, the general prototype of the rules generated from the CCP graph will take the following form:

Rule ⇒ Total number of people with specific diseases in specific counties.

For example, from the Cluster Point Table given above, the second entry shows 22 as the support value for the combination of two concepts, C2 and D2, which means "there are 22 people with 'eye–ear disease' in the counties of Jackson, Clay and Lincoln." Similarly, each point in the table can be interpreted as a rule representing a relationship between diseases and counties. In this way, different cluster point entries can be used to generalize different rules with different values.

Acknowledgement

This research has been partially supported by L-3 Communication Corporation, ComCept Division under Project Corvus.

References

[1]Usama Fayyad, Gregory Piatetsky-Shapiro, Padhraic Smyth, and Ramasamy Uthurasamy, "Advances in Knowledge Discovery and Data Mining", AAAI Press/ The MIT Press, 1996.
[2]J. Ross Quinlan, "C4.5: Programs for Machine Learning", Morgan Kaufmann Publishers, 1993.
[3]Alex Freitas and Simon Lavington, "Mining Very Large Databases with Parallel Processing", Kluwer Academic Publishers, 1998.
[4]A. K. Jain and R. C. Dubes, "Algorithms for Clustering Data", Prentice Hall, 1988.
[5]V. Cherkassky and F. Mulier, "Learning From Data", John Wiley & Sons, 1998.
[6]Leonardo Angelini, Mario Pellicoro and Sebastiano Stramaglia. "Cost functions for pair wise data clustering", Universit`a di Bari and I.N.F.N., Bari, Italy, Physics Letters A, Volume 285, Issue 5-6, p. 279-285,2001.
[7]Graham Cormode, "Some Key Concepts in Data Mining – Clustering", DIMACS Series in Discrete Mathematics and Theoretical Computer Science Volume 70, 2006.
[8]Martin H. C. Law, Mario A. T. Figueiredo, and Anil K. Jain. "Simultaneous Feature Selection and Clustering Using Mixture Models", IEEE TRANSACTIONS ON PATTERN ANALYSIS AND MACHINE INTELLIGENCE, VOL 26, NO 9, 2004.
[9]Rakesh Agrawal and Tomasz Imielinski, "Database Mining: A Performance Perspective", IEEE Transactions on Knowledge and Data Engineering, 5(6):914-925, December 1993.
[10]Usama Fayyad, David Haussler, and Paul Stolorz, "Mining Scientific Data", Communications of the ACM, vol. 39, no. 11, pp. 51-57, November 1996.
[11]David J. Hand, "Data Mining: Statistics and more?", The American Statistician, vol. 52, no. 2, pp 112-118, May 1998.
[12]Istvan Jonyer, Diane J. Cook and Lawrence B. Holder, "Graph-Based Hierarchical Conceptual Clustering", journal "International Journal on Artificial Intelligence Tools", volume 10, number 1-2, pages 107-135, 2001.
[13]Sang C. Suh, Nikhil Goel, and Y.S.Kwon, "Generating Meaningful Rules Using Attribute Concept Hierarchy", Intelligent Engineering Systems Through Artificial Neural Networks Vol. 16, pp., ASME Press, New York, November, 2006, ISBN 0-7918-2222-0.

DETERMINING THE NUMBER OF DATA CLUSTERS IN ANY DATASET IN THE PRESENCE OF CONSIDERABLE NOISE

S. EASWARAN
Dept. of Computer Science & Computer Engineering
Pacific Lutheran University
Tacoma, Washington

ABSTRACT
In this paper, a technique to determine the number of data clusters in a dataset in the presence of considerable extraneous noise data is proposed. The technique is based on intelligently combining the seed-growing algorithm for number-of-clusters determination with the weighted voting-scheme for analyzing data clusters. The generally satisfactory performance of this technique is seen through a variety of artificially created datasets intentionally corrupted with considerable noise data. Some issues with this technique and situations where this algorithm may not perform too well are also discussed.

1.0 INTRODUCTION

Accurately determining the number of clusters in any dataset with no a-priori information about their grouping structure is an essential aspect of cluster analysis. It is a difficult problem [1-8] and none of the existing methods have been established as satisfactory. In addition, there has been limited progress in this direction especially for arbitrary datasets, and more so for data in the presence of considerable noise.

In this paper, a technique to determine the number of data clusters in an arbitrary dataset especially in the presence of considerable extraneous noise data is proposed and discussed. The technique proposed is based on suitably combining the seed-growing algorithm for number-of-clusters determination of reasonably-separated data clusters [1][2] with the weighted voting-scheme to obtain a measure of cluster core-ness of data points in a dataset [3], and thereby eliminating non-core data. The generally satisfactory performance of this technique to determine the number of data clusters in an arbitrary dataset in the presence of considerable noise data is demonstrated using a variety of artificially created datasets corrupted with extraneous noise data.

2.0 THE SEED-GROWING ALGORITHM FOR NUMBER OF CLUSTERS DETERMINATION

A seed-growing algorithm was recently proposed [1][2] to accurately determine the number of data clusters in any arbitrary dataset that consists of reasonably separated clusters and to partition the data into these clusters. The objective of the seed-growing algorithm was, "given any arbitrary dataset of data points that are formed into reasonably separated data clusters, … to correctly classify these data points into classes that make up the different data clusters and to correctly determine the number of such data clusters in the given dataset". In that paper, the term "reasonably separated" was explained in context, with representative illustrations. The paper mentioned two issues with the algorithm – namely, the issue of island formation and data-bridging between clusters. It mentioned and demonstrated that the algorithm performs poorly under these situations, though perfect in most (if not all) other situations. Implicit in the paper was the fact that the dataset does not contain much noise data as it would then miserably fail.

In the original paper [1] and in a follow up paper [2], it was mentioned that "island formation" is an issue where data-point islands are formed because of stray data points, or certain data points within a cluster being somewhat separated from other data points within the same cluster. These result in errors being made by the algorithm in determining the total number of classes in the dataset. On the other hand, it was mentioned that "cluster bridging" is due to coalescing of two or more separate data clusters by the algorithm into a single cluster because of unfavorably situated stray data points. This too results in errors being made by the algorithm in determining the total number of classes in the dataset.

The original paper [1] argued that "island formation" can be somewhat controlled by suitably choosing the value of the clustering parameter "p" of the algorithm. The paper also suggested that "data bridging" can be overcome by "stray data" removal (as it was the source of the problem) using some yet to be determined technique prior to the application of this algorithm. However, none of these were investigated in the paper, nor was any suitable technique proposed.

In a follow up paper [2], the original seed-growing algorithm was augmented by a pre-processing and a post-processing step to greatly overcome these two issues suffered by the original algorithm. As a pre-processing step, a singleton-isolation algorithm was proposed to overcome the issue of data bridging and was experimentally verified to perform satisfactorily. This pre-processing step was not intended to eliminate the problem of island formation, however. In fact, in a number of situations, though the problem of data bridging was overcome by this pre-processing step, it inadvertently contributed further towards the issue of island formation. However, the proposed post-processing step suitably handled this problem of island formation (even those which became a problem due to the pre-processing step). This was accomplished by augmenting the original algorithm again. The modification incorporated the number-of-data-points in each cluster as part of the outcome of the original algorithm during classification, and then discarded extremely (relatively) small-sized clusters during the post-processing step.

Representative experimental evaluations were provided [1][2] to confirm that the pre and post-processing steps when used together with the original algorithm (the enhanced seed-growing algorithm), considerably improved data classification and number of clusters determination (even in difficult datasets). It can be said that with this enhanced seed-growing algorithm (and in the absence of any significant noise data), the number of data clusters in a dataset with reasonably-separated clusters can be accurately determined.

3.0 THE WEIGHTED VOTING-SCHEME FOR CLUSTER ANALYSIS

A weighted voting-scheme [3] had been proposed and experimentally verified in the literature as an excellent technique for analyzing data points with the objective of obtaining quantified information on the degree of "cluster core-ness" of data points in the dataset – that is, to answer the question, how much part of a cluster's core is a given data point in a dataset? This quantitative measure enables one to determine if a given data point in a dataset is deep within a cluster (close to or part of a cluster's core), inside a cluster but some what removed from a cluster's core, peripheral to a cluster, close to a cluster, or away from any cluster based on its relative score.

4.0. USING THE WEIGHTED VOTING-SCHEME FOR NOISE/ NON-CORE DATA REMOVAL

It is proposed in this paper that an excellent application of the weighted voting-scheme for cluster analysis is in progressively removing extraneous noise data from a dataset, or in removing non-core data points from a dataset.

Since the overall vote for a particular data point in the weighted voting scheme is a quantitative measure of the degree to which the data point is a part of a cluster's core, by choosing a suitable (noise separation) threshold, data points can be separated into noise and regular data. And, by (progressively) choosing higher thresholds, one could separate "core" ("central") regions of clusters from the rest of the data in the dataset (Figure 1).

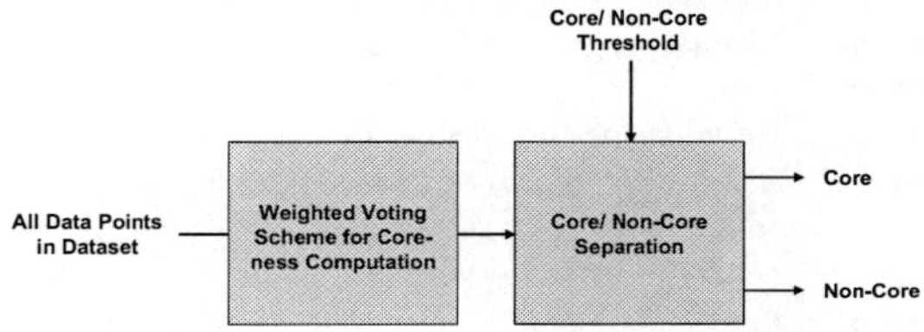

Figure 1. Separating Noise, or Core and Non-Core Regions in a Dataset

As a representative illustration, Plate 1(a) graphically shows a dataset corrupted with considerable extraneous noise data, Plate 1(b and c) shows its corresponding (logarithmic) voting scores – graphically and through a score-to-color mapping, and Plate 1(d and e) shows core separation at different threshold levels (the retained cores are showed in yellow – unfortunately, the print is in black and white – nonetheless, the dark core regions seen are what was retained though the rest too are shown).

5.0. USING THE SEED-GROWING ALGORITHM FOR CLUSTER CORE COUNTING

Once the non-core (or noise – which is non-core) regions in a dataset are removed (using the weighted voting scheme for degree of core-ness scoring and subsequently applying a threshold for cluster core separation/ noise data removal), what remains is a (sub-) dataset of well-separated data clusters (cluster cores). Thus, at this stage, it is straight-forward for the seed-growing algorithm (basic or enhanced) to classify the data points in this (non-core removed) dataset into clusters, and provide an accurate count of the number of cores (clusters) in this dataset. This is because it has been well established that the seed-growing algorithm (basic or enhanced) is very successful in accurately determining the number of clusters when the clusters are reasonably-separated and there are no stray data points [1][2] (which is now the case) in the dataset.

It is therefore proposed in this paper that a count of the number of clusters in this (noise removed, or core/ non-core separated) dataset can be used as the actual count of the clusters in the original dataset corrupted with considerable noise.

6.0 A VISUAL SOFTWARE TOOL FOR INVESTIGATIONS

A visual software tool was implemented, initially for experimentally evaluating how well the noise removal and cluster core/non-core separation function, and for experimentally evaluating how well these algorithms (seed-growing, weighted voting-scheme, and noise/ core/non-core separation) perform when operating together to determine of the number of clusters in a dataset corrupted with considerable noise data.

A dataset creation section was incorporated for visually generating, placing, and displaying differently shaped and distributed clusters – two-dimensional uniform and Gaussian distributions with the possibility of controlling the position and various physical shape-related parameters such as the width, height, radius, elongation, and rotation. The predetermined physical shapes of the data clusters could be rectangles, ellipses (including circles), annular shapes, circular/ annular sectors and arcs, and spirals. Also, a feature to point-by-point, manually generate and visually place any number of arbitrarily-shaped and distributed data points was incorporated. Any particular distribution cluster could contain any number of data points, and any number of distribution clusters could be generated to form a complex dataset which could be saved and reloaded.

The weighted voting scheme was implemented for cluster data scoring, and a coloring scheme where colors progressively change (through intuitive color changes) as the overall scores change from the highest (red) to the lowest (blue) was also implemented for visually evaluating the success of the technique.

A noise-removal section was incorporated based on the weighted voting scheme and a threshold to remove data points whose cluster core-ness score was below a user definable threshold. This allows for observing the data points that become isolated as the threshold is progressively increased from the minimum to maximum. Also, a "number-of-clusters" determination section was incorporated based on the seed-growing algorithm.

7.0 EXPERIMENTAL INVESTIGATIONS AND OUTCOMES

Using this software tool, a variety of representative datasets (corrupted with uniformly distributed or other noise data) were generated for evaluating the proposed technique(s) at different levels. It was important to first evaluate the successfulness of the weighted voting-scheme scores along with different threshold parameters for progressively separating out (increasingly more) noise data, and for progressively separating out core/non-core data. In this regard, difficult datasets too were generated and used in evaluating the proposed technique under various trying conditions. It was then experimentally evaluated if the seed-growing algorithm would correctly determine the number of clusters in the original dataset (which contains noise data) by determining the number of "cluster cores" in the modified (noise/ non-core removed) dataset.

Representative outcomes for a single dataset (shown for illustration) are as follows. Plate 1 (a through f) illustrates the outcomes of the investigations on a particular dataset with three data clusters – the first being spiral shaped (2500 data points), the second being somewhat annular shaped (2500 data points), and the third being elliptical shaped (600 data points). The dataset was corrupted with 1000 data points of (uniformly distributed) noise data. Plate 1(a) shows the dataset (the same color is used for all the data points). Plate 1(b and c) shows the corresponding (logarithmic) voting scores – illustrated graphically and through a score-to-color mapping, and Plate 1(d and e) shows core separation/ noise removal at different threshold levels (the retained cores are showed in yellow while those shown in red are the data points removed from the dataset – unfortunately, the print is in black and white – nonetheless, the dark core regions seen are what was retained though the rest too are shown). The retained "cluster cores" are used in determining the number of clusters using the seed-growing algorithm. Plate 1(f) shows the result of classification after noise removal. It shows the voting scores (linear) (blue) obtained by the data points classified into the different classes (three in this case), and their corresponding class id's (red). As can be seen, the technique correctly determined the number of clusters (Plate 1f) in the original dataset corrupted with noise.

One of the biggest challenges in all this is in automatically determining a suitable threshold for core separation. Although it was shown that progressively increasing the threshold will progressively remove noise or non-core data starting from the most outliers and progressing towards the cores, automatically determining a suitable threshold range may be a challenge at times. This is being further investigated as future work.

Although only "representative" outcomes were given in order to keep this presentation manageable, the investigations have brought to light the following. That this is indeed a promising technique for determining the number of data clusters even in the presence of considerable noise data. Determining a suitable separation threshold is sometimes a challenge although it would work satisfactorily (correctly) over a very wide range of threshold values as could be seen from Plate 1 (d and f). It is obvious that as long as the core of all the clusters have been retained (while removing non-core data), the seed-growing algorithm will successfully determining the number of cores and hence the number of clusters correctly (since the cores are well separated).

Thus, the challenge is in providing a suitable threshold (range of values), and ensuring that all the cores are retained. Another point to be noted is that if the clusters are of considerably different sizes, their (voted score) "strengths" may be widely different causing some of them to become separated out from the dataset (if weak). This would require careful investigations to draw stronger conclusions. Another point to note is that the underlying nature of the data distribution or the shape of the data clusters, or the nature of the noise data do not matter (the seed-growing algorithm and the weighted voting scheme are both independent of these). All that matters is in coming up with a suitable threshold (range) to retain the cluster cores while removing the rest of the data from the dataset prior to determining the number of clusters. The seed-growing algorithm will indeed do an excellent job (as established previously) if it is presented with the well separated cores. The weighted voting scheme too will do an excellent job in rightly scoring the data points with regard to their cluster core-ness.

7.0 SUMMARY AND CONCLUSIONS

A technique was proposed in this paper for accurately determining the number of clusters in a dataset in the presence of considerable noise data. It was by judiciously combining the seed-growing algorithm for number-of-clusters determination and the weighted voting scheme for cluster core-ness scoring. The proposed technique was analyzed and discussed, and representative outcomes were provided. It was shown to be a very satisfactory and promising though determining a suitable threshold could become challenging at times. Some additional investigations and developments that need to be made were also explained and discussed.

REFERENCES

[1] S. Easwaran, "Determining the Number of Clusters in any Arbitrary Dataset containing Reasonably-Separated Data Clusters", ANNIE 2006 Annual Conference, (November 5-8, 2006). ASME Press, New York 2005, ISBN: 0-7918-0204-3.

[2] S. Easwaran, "Enhancements to the Seed-Growing Algorithm for Determining the Number of Data Clusters", ANNIE 2007 Annual Conference, Missouri (November 12-14, 2007).

[3] S. Easwaran, "A Weighted Voting-Scheme Based Technique for Investigating Data-Clusters in any Dataset", ANNIE 2007 Annual Conference, Missouri (November 12-14, 2007).

[4] Sandrine Dudoit, Jane Fridlyand, "A Prediction-Based Re-Sampling Method to Estimate the Number of Clusters in a Dataset", Genome Biology 2002, 3, 0036.1-0036.21

[5] Chris Fraley, Adrian E. Raftery, "How Many Clusters? Which Clustering Method? Answers via Model-Based Cluster Analysis", The Computer Journal, Vol. 41, No. 8, 1998

[6] Trevor Hastie, et al, "Estimating the Number of Clusters in a Dataset via the Gap Statistic", Journal of the Royal Statistical Society, B, 63, 411-423, 2001

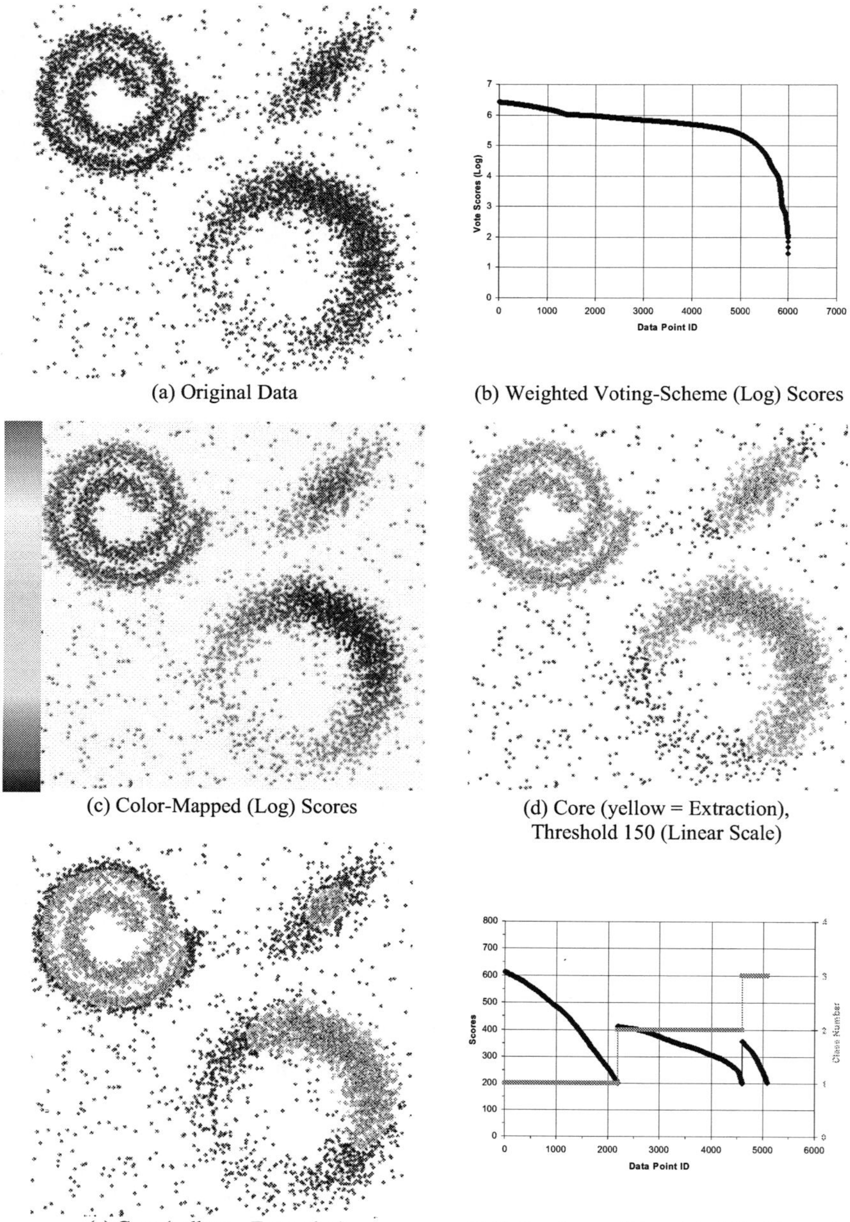

(a) Original Data

(b) Weighted Voting-Scheme (Log) Scores

(c) Color-Mapped (Log) Scores

(d) Core (yellow = Extraction), Threshold 150 (Linear Scale)

(e) Core (yellow = Extraction), Threshold 300 (Linear Scale)

(f) Voting Scores (Linear) (within Clusters), and Classification

Plate 1: Outcomes on Three Clusters with Extensive Noise Data
(Original Data, Weighted Voting Scheme Scoring, Thresholding, and Classification)

A WEIGHTED VOTING-SCHEME BASED TECHNIQUE FOR INVESTIGATING DATA CLUSTERS IN ANY DATASET

S. EASWARAN
Dept. of Computer Science & Computer Engineering
Pacific Lutheran University
Tacoma, Washington

ABSTRACT
The ability to automatically identify core regions of clusters and obtain useful placement information about each data point with respect to these core regions are important in analyzing clustered data. They provide additional perspectives on our understanding of the underlying data distribution, and serves as a springboard for further analysis in innovative ways. In this paper, a voting-scheme based technique is proposed to quantify the degree of cluster core-ness of data points in a dataset. It is demonstrated that this technique performs extremely well with respect to the above objectives.

1.0 INTRODUCTION

Automated cluster-analysis techniques are used in numerous fields for a variety of purposes [1-7]. They are becoming increasingly important due to rapid growth in size and variety of datasets collected and stored electronically. They have brought about a need for extracting and analyzing valuable information from this large collection of data.

2.0 A QUANTITATIVE MEASURE FOR DEGREE OF CLUSTER CORE-NESS

Data points and data clusters are important components of a dataset. It is proposed in this paper that an important aspect when analyzing data points for clusters is the ability to determine the degree to which each data point is a part of a cluster's core (the "central" portion of a cluster). This is because it is not only important to know which cluster a data point belongs to, but it is equally important to know the degree to which it belongs to a cluster's core in order to understand how the data points in a dataset are clustered. It will enable one to quantitatively determine whether a given data point is deep within a cluster (close to or part of a cluster's core), inside a cluster but some what removed from a cluster's core, peripheral to a cluster, close to a cluster, or away from any cluster (Figure 1). It is also proposed in this paper that a quantitative measure of the degree to which a data point belongs to a cluster's core can be obtained based on a weighted voting-scheme.

3.0. THE WEIGHTED VOTING-SCHEME FOR QUANTIFYING CLUSTER CORE-NESS

The Voting Function: It is proposed in this paper that a quantitative measure of the degree to which a data point belongs to a cluster's core can be obtained based on a weighted voting-scheme. In this scheme, each data point in the dataset votes for every other data point as to the degree each of the other data points is part of a cluster's core (from the voting data points perspective only, and not on how other data points perceive it). On the whole, how much a given data point is actually part of a cluster's core is determined by summing all votes the data point under consideration receives from all other data points to qualify it as a core point within a cluster. This is done for all data points in the dataset to get a quantitative feel for the "distribution" of the data points.

The proposed voting scheme can be mathematically written as,

$$V(i) = \sum_{j=1}^{M} w_{i,j} \; (j \neq i) \; : \; i = 1,2,3,...,M.$$

Here, w(i,j) is the vote given by the voting data point "j" to the voted data point "i", "M" is the total number of data points in the dataset, and V(i) is the total votes (score) received by the voted data point "i" from all data points other than itself.

A desirable feature that should be incorporated in this voting scheme is that the vote given by a data point (weighting function for the votes) should be a full vote (voting weight = 1.0) if the voting point is exactly on (overlays) the voted point, and zero if the voting point is as far away from the voted point as possible (infinitely far). Every other voting point should vote in some meaningful manner with a value between these limits (0 and 1.0) depending on how far they are from the voted point.

Thus, each data point voting for a data point under consideration should in some way base its vote on how far the voted point under consideration is from it. If it is close to it, the voting-point should vote by indicating that the point under consideration is a strong cluster core point as far as the voting point perceives it (by providing a higher-weight vote), and the opposite should be the case if the voting point is far away from the voted point. How much a given data point is actually part of a cluster's core is then determined by summing all the votes received by it to qualify it as a cluster's core point.

Another important feature to be satisfied by the vote-weighting function (w_{ij}) is that the voting weights should monotonically and non-linearly (exponentially or similarly) decrease with distance from the voted point. This is because, it is important that the cumulative effect of many data points far away from a given voted point (the combined small-valued votes from a very large number of voting points) should not falsely indicate that the voted point as a strong cluster core. This could otherwise undo the effect of an actual cluster core point getting very high votes (but) from a smaller number of very close by points (thus not becoming high enough in comparison to the former case).

The Weighting Function: Based on these important and desirable features, the function $w_{i,j} = \exp\left[\ln(W).\left({r_{i,j}}/{R}\right)^{n}\right]$ is proposed in this paper as a suitable weighting function for use with the above voting scheme. Here "r" is the distance of the voted point from the voting point, and "R", "W", and "n" are predetermined constants. These constants are for initial adjustments to vary the voting weight at a given distance to obtain desirable characteristics of the weighting equation for analyzing the problem at hand. It should be noted that if one chooses the constants "W" such that ln(W) = -0.5 (i.e., W = 0.606) and n = 2, this weighting function essentially becomes a Gauss function.

This weighting function satisfies a desirable feature that (i) if the voting point is exactly at the voted point (r = 0), the voting weight is one (the voting point perceives this voted point as a cluster's core in the strongest possible terms), (ii) if it is infinitely far (r = infinity) the weight is zero, and (iii) the weight decreases monotonically and exponentially as the voting point moves progressively away from the voted point. Also, when r = R, then w = W. Thus, "R" can be chosen such that when the voting point ("r") is at this predetermined distance "R", a suitable predetermined vote ("W") is given.

In order to describe this weighting function, the variation of the weight "w" of this weighting function with "r/R" (the "normalized" distance from the voted point) is shown in Figure 2. This variation is shown for different values of the constant "n" with fixed "W" (Figure 2a), and for different values of the constant "W" with fixed "n" (Figure 2b).

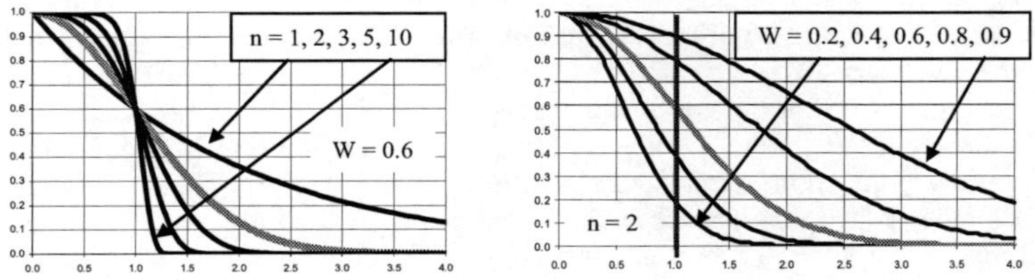

(a) "W = 0.6" fixed, "n" Varying

(b) "n = 2" Fixed, "W" Varying

Figure 2. The Plot of Voting Weight ("w") Vs. Normalized Distance ("r/R")

In the weighting function, the values of "R" and "W" are suitably predetermined based on what weight we want to assign when the voting point is at what distance. For a given "n", this fixes the voting weights assigned by voting points closer than "R", and by voting points further away than "R". Therefore, for a given "n", the values of "R" and "W" are chosen based on this consideration.

n = 2		
W = 0.6		
Normalized Distance (r/R)	w (weight)	Number of Data Points Needed to Match Single Point at r = 0
0.0	1.000000000000000	1.0
0.5	0.880111736793393	1.2
1.0	0.600000000000000	1.7
1.5	0.316840225245622	3.2
2.0	0.129600000000000	7.8
2.5	0.041062493191833	24.4
3.0	0.010077696000000	99.3
3.5	0.001915811682358	522.0
4.0	0.000282110990746	3,544.8
4.5	0.000032178279547	31,076.9
5.0	0.000002843028803	351,737.6
5.5	0.000000194569751	5,139,545.1
6.0	0.000000010314425	96,951,601.3
6.5	0.000000000423536	2,361,074,394.1
7.0	0.000000000013471	74,231,584,564.5
7.5	0.000000000000332	3,012,951,674,158.3
8.0	0.000000000000006	157,877,403,574,268.0
8.5	0.000000000000000	10,680,022,306,806,300.0
9.0	0.000000000000000	932,714,479,721,904,000.0
9.5	0.000000000000000	105,159,776,127,866,000,000.0
10.0	0.000000000000000	15,306,467,074,865,100,000,000.0

Table 1. The Number of Data Points at Given "r/R" to Match Single Data Point at r=0

Thus, by suitably choosing "R" and a corresponding "W", we can assign a desirable weight for a predetermined "r=R". It is also evident that the parameter "n" controls how slow the weights change for r <R, and how fast the weights die with distance for r > R. The value of "n" can thus be chosen based on how we want the votes to be for distances of the voting point less than "R", and how we want it to be (die) for distances greater than "R". Therefore, the constants "R", "W", and "n" can be chosen a-priori to control the votes from data points as they move away from the voted points, to ensure that high overall votes to any data point under consideration can be confidently interpreted as representing a strong cluster core and vice versa.

As mentioned earlier, an important aspect that is needed to ensure meaningful overall votes for any data point is that a large number of data points with small votes alone should not even collectively provide a high overall vote to the data point under consideration. This can indeed be achieved by this weighting function by reasonable choice of the constants "R", "W", and "n". Table 1, shows (for a typical value of the constant n = 2, and W = 0.6) the voting weights ("w") provided by data points at different normalized distances (r/R) from the voted point under consideration. It also shows how many data points at different normalized distances are needed to provide the same voting effect as a single data point located at the voted point (r=0) itself. The table clearly shows that meaningful overall scores can indeed be achieved with this weight function.

What is also desired in the weighting function is that "R" and "W" are chosen to provide a predetermined weight (w=W) at a predetermined distance (r=R), and "n" (along with "R" and "W") is chosen such that high weights are provided for data points at distances less than "R" and low votes are provided by data points at distances larger than "R" – the transition being made in some desired manner. The value of the constants "R", "W", and "n" can be suitably chosen to accomplish this as can be seen in Figure 2.

4.0 A VISUAL SOFTWARE TOOL FOR EVALUATIONS

A visual software tool was implemented for evaluating the performance of this scheme in providing a measure of the degree of cluster core-ness of data points.

A dataset creation section was incorporated for visually generating, placing, and displaying differently shaped and distributed clusters – two-dimensional uniform and Gaussian distributions with the possibility of controlling the position and various physical shape-related parameters such as the width, height, radius, elongation, and rotation. The predetermined physical shapes of the data clusters could be rectangles, ellipses (including circles), annular shapes, circular/ annular sectors and arcs, and spirals. Also, a feature to point-by-point, manually generate and visually place any number of arbitrarily-shaped and distributed data points was incorporated. Any particular distribution cluster could contain any number of data points, and any number of distribution clusters could be generated to form a complex dataset which could be saved and reloaded. The proposed voting scheme with the weighting function was implemented with simple means for varying the voting parameter constants "R", "W", and "n".

For visually evaluating the successfulness of the technique, a coloring scheme where colors progressively change (through intuitive color changes) as the overall scores change from the highest (red) to the lowest (blue) was implemented. However, the colors are lost in the printed (black and white) copy of this paper (Plate 1 (a), (b), (c), and (d)).

5.0 EXPERIMENTAL INVESTIGATIONS AND THEIR OUTCOMES

Using this software tool, a variety of different datasets were generated and used for evaluating the proposed technique for its usefulness in data scoring towards a satisfactory quantitative measure of the degree of cluster core-ness of data points in a dataset. Different and difficult datasets too were generated and used to evaluate the proposed algorithm under various trying conditions. Plate 1 (a), (b), (c), and (d) (different cluster shapes/ data distributions) graphically illustrate representative outcomes of the investigations, and its usefulness in showing the degree of cluster core-ness of data points. A score-to-color mapping was incorporated and used to simplify the presentation of the numerical scores, and is given alongside the plates (unfortunately, the printed copy is in black and white). Further as an illustration, Figure 3 shows a plot of the sorted (maximum to minimum) voted scores for the dataset of Plate 1(c) (both belong to the same dataset). Both linear (regular) and logarithm of the voting scores are shown in the plots of Figure 3. As can be seen from Figure 3 and Plate 1 (a), (b), (c), and (d), the logarithm based score is sometimes useful in better resolving regions of low scores. As can be seen, the scores correctly indicate the cluster core-ness of the data points.

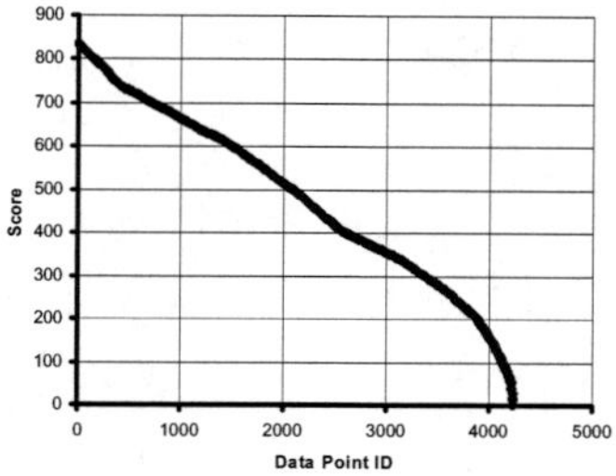

Sorted Linear Score

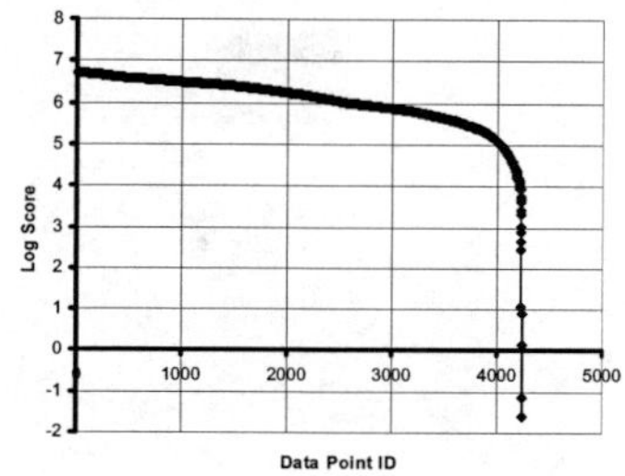

Sorted Log Score

Figure 3. Sorted Data Point Scores for Dataset in Plate 1(c)

7.0 SUMMARY AND CONCLUSION

The need for introducing and defining "degree of core-ness" and to quantitatively describe these were proposed as useful tools in analyzing clustered data. A voting-scheme based technique was proposed as a means to quantitatively ascertain the degree of cluster core-ness of data points. A suitable weighting function with desirable properties too was proposed for use with this voting scheme, resulting in the weighted voting-scheme based technique which was theoretically analyzed and justified. Experimental evaluations were conducted, and the outcomes clearly indicated the validity and usefulness of all the propositions. The outcomes showed that the proposed weighted voting-scheme based technique could indeed be used to analyze data points in a dataset with respect to knowing the degree of cluster core-ness of each data point in any dataset.

REFERENCES

[1] S. Easwaran, "Determining the Number of Clusters in any Arbitrary Dataset containing Reasonably-Separated Data Clusters", ANNIE 2006 Annual Conference, Missouri (November 5-8, 2006). Published in "Intelligent Engineering Systems through Artificial Neural Networks, Volume 15, ASME Press, New York 2005, ISBN: 0-7918-0204-3.

[2] Katherine S. Pollard, et al, "A Method to Identify Significant Clusters in Gene Expression Data", Collection of Biostatistics Research Archive (COBRA), Berkeley Electronic

[3] Nadia Bolshakova, et al, "An Integrated Tool for Microarray Data Clustering and Cluster Validity Assessment", Proceeding of Neural Networks and Computational Intelligence, 2003.

[4] Sandrine Dudoit, Jane Fridlyand, "A Prediction-Based Re-Sampling Method to Estimate the Number of Clusters in a Dataset", Genome Biology 2002, 3, 0036.1-0036.21

[5] Chris Fraley, Adrian E. Raftery, "How Many Clusters? Which Clustering Method? Answers via Model-Based Cluster Analysis", The Computer Journal, Vol. 41, No. 8, 1998

[6] Trevor Hastie, et al, "Estimating the Number of Clusters in a Dataset via the Gap Statistic", Journal of the Royal Statistical Society, B, 63, 411-423, 2001.

[7] H. Ressom, et al, "Computational methods for Identifying Number of Clusters in Gene Expression Data", Proceeding of Neural Networks and Computational Intelligence, ACTA '03

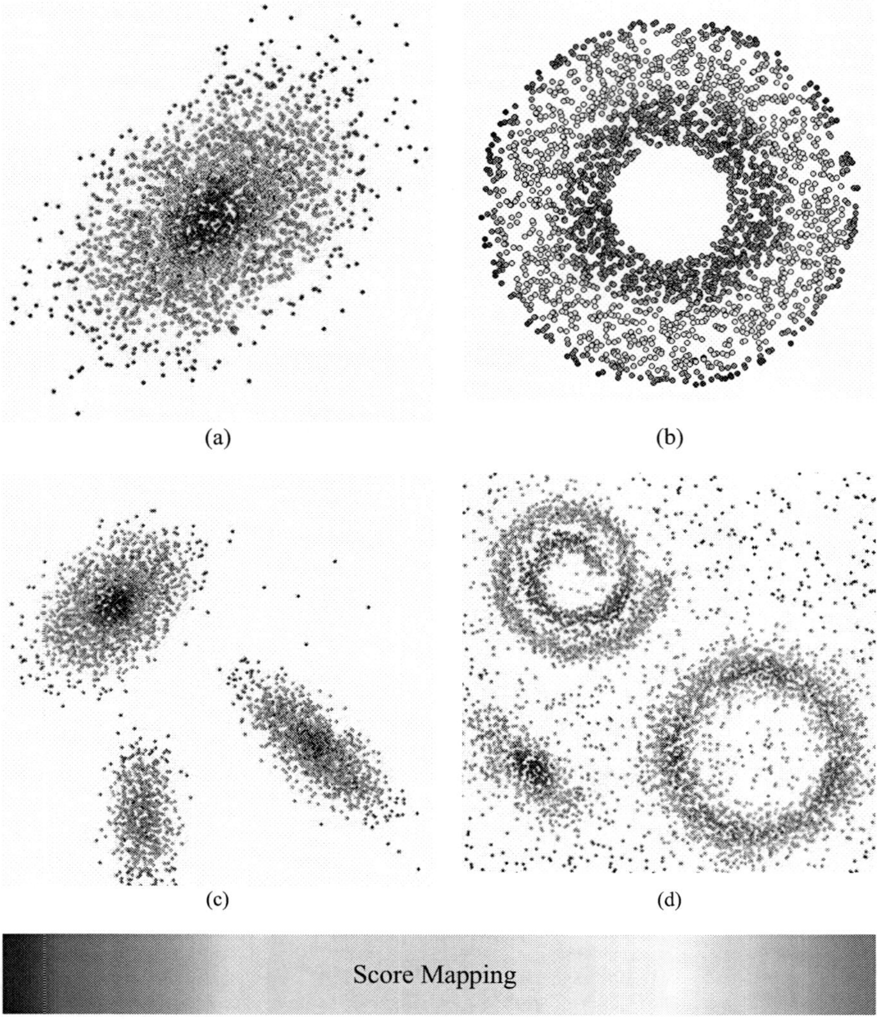

Plate 1. Weighted Voting Scheme (Color-Mapped Scores)

ENHANCEMENTS TO THE SEED-GROWING ALGORITHM FOR DETERMINING THE NUMBER OF DATA CLUSTERS

S. EASWARAN
Dept. of Computer Science & Computer Engineering
Pacific Lutheran University
Tacoma, Washington

ABSTRACT

A seed-growing algorithm was proposed in the literature [1] to accurately determine the number of clusters in any arbitrary dataset with reasonably separated clusters and accurately classify the data points into these clusters. Island formation and data bridging were demonstrated and discussed as situations under which the algorithm performs poorly although extremely well otherwise. In this paper, singleton isolation as a pre-processing step in combination with determining and eliminating extremely small-sized clusters during post-processing is proposed as an enhancement to the algorithm. With this enhancement, it is demonstrated that the seed-growing algorithm performs extremely well even in situations where it otherwise performs poorly.

1.0 INTRODUCTION

An essential aspect of cluster analysis is accurately determining the number of clusters in any arbitrary dataset with no a-priori information about its grouping structure [1-6]. It is a difficult problem with limited progress made, and none of the existing methods have been established as satisfactory, especially for arbitrary datasets.

A seed-growing algorithm was proposed in the literature [1] to accurately determine the number of clusters in any arbitrary dataset with reasonably separated clusters, and to accurately classify the data points into these clusters – the term "reasonably separated" was explained in context with representative illustrations. The algorithm was verified by demonstrating its performance on a wide and representative variety of artificially created datasets. The paper mentioned island formation and data-bridging between clusters as two main issues with the algorithm. It mentioned that the algorithm performs poorly under these situations, though perfect in all other situations.

"Island Formation" is where data islands are formed because of stray data points or because of data points within a cluster being somewhat separated from other data points within the same cluster. These result in errors being made by the algorithm in determining the total number of classes in the dataset. "Cluster Bridging" is due to coalescing of two or more data clusters by the algorithm into a single cluster because of some unfavorably situated stray data point. This too results in errors by the algorithm.

2.0. THE ORIGINAL "SEED-GROWING" ALGORITHM

The original seed-growing algorithm [1] is summarized as follows. Starting from an arbitrary data point as reference, distances from it to all other data points in the dataset are computed, rank ordered, and tabulated in ascending order. The pth minimum distance is obtained from this table ("p" is a pre-determined number based on a clustering criteria

[1] – it is set to one in most cases). This is repeated with each data point as reference. Using another table formed by this collection of pth minimum distances, the largest of the pth minimum distances is obtained (Max-p^{th}MinDistance).

Thereafter, starting from any arbitrary data point as seed for a new class, all data points within Max-p^{th}MinDistance from this data point are assigned to this class. This class assignment process is repeated from all data points assigned to this class, growing this cluster. When this cluster no longer grows in size, the seed growing process is repeated from another arbitrarily selected seed point not previously assigned to any class, and the entire cluster growing process is repeated until all data points in the dataset have been clustered. The number of classes in the dataset it is the number of seed points used.

3.0. ENHANCEMENTS TO THE "SEED-GROWING" ALGORITHM

The main cause for the two issues with the original algorithm is the presence of isolated stray data-points. These could occur as one or more isolated single stray data points or as one or more groups of isolated stray data points of varying numbers. When using the algorithm, the presence of such data points results in these data-points forming data islands or bridging (coalescing) two or more regular data clusters through them to make these regular clusters appear as a single cluster to the seed-growing algorithm [1].

It was shown [1] that in the seed-growing algorithm, the pth-MinDistance parameter "p" for seed-growing serves in controlling the minimum number of "isolated" data points in a group that would be recognized as a separate cluster and not be lumped with another cluster. For a chosen "p", the value "p+1" becomes the minimum number of "isolated" data points in a group that would be recognized as a separate cluster if they are otherwise reasonably-separated from other data points in the dataset. This implies that groups of "p" or less isolated data points will always be lumped with another cluster.

Thus, for example, when using the seed-growing algorithm with p=1, an isolated single data point will never get recognized as a single (one point) cluster but would become lumped with some other cluster (however far it may be from that cluster) upsetting the cluster growing process (by Max-p^{th}MinDistance becoming a very large value) if such isolated single data points are present in the dataset. In this situation, a group of two or more isolated data points would however form separate data clusters (even though it just may be a cluster of "stray" data points) if they are present in the dataset. This would result in the number of clusters determined by this algorithm being erroneous – the result being more than the actual number of clusters. Also, this stray data point could form a bridge that coalesce two or more regular clusters into one. This again would result in the number of clusters determined by this algorithm becoming erroneous – the result being less than the actual number of clusters.

From the above, it is evident that by automatically identifying and removing all isolated single data-points (singletons) from the dataset prior to the application of the seed-growing algorithm, the issue of cluster-bridging (cluster coalescing) can be overcome. This would make in the original seed-growing algorithm correctly classify and determine the number of clusters in the original dataset.

Despite this, it is obvious that the presence of two or more isolated data points existing as groups (stray or otherwise) in the original dataset will not get removed by this singleton removal operation. These isolated groups of data points will contribute towards errors in the total number of clusters determined by the algorithm (due to island(s) formation) – the error manifesting as the number of clusters determined being higher than the actual number of clusters in the original dataset. These were stray data points occurring as a group, and should not have been counted as a cluster. However, a technique is proposed and investigated in this paper to overcome this type of error too, in order to obtain the correct number of clusters in the original dataset. Hence, a definite improvement to the original seed-growing algorithm can be expected by isolating singletons (in a preprocessing step) prior to the application the original algorithm, and using post-processing to not count (what could be considered) stray clusters.

4.0. PRE-PROCESSING FOR SINGLETON REMOVAL

The following pre-processing operation is proposed for singleton isolation. It is based on the fact that distances of singletons from its closest data point is considerably larger than minimum distances between any other two data points (stray or otherwise).

The Singleton-Isolation Pre-Processing: For each data point in the dataset, the distance of that data point to all other data points is computed and the distance from it to its closest data point is determined. For any data point in the dataset, if this distance is greater than some predetermined minimum, it is declared as a singleton. All singletons are removed from the dataset prior to applying the basic seed-growing algorithm. Figure 2 illustrates the distribution of the minimum inter data-point distances for the example dataset shown in Figure 1. Although no rigorous method is presented to automatically determine this threshold distance, it is provided as proof-of-concept and is being further investigated for a more rigorous and automated method.

5.0. POST-PROCESSING FOR DISCARDING STRAY CLASSES

In the number of clusters determination step of the seed-growing algorithm, the following adjustment is proposed based on arguments made in Section 3.0.

Once cluster growing for a particular cluster has come to a natural end (Section 2.0 – The Seed-Growing Algorithm, Step 6), in addition to labeling those data points as belonging to a particular cluster, the number of data points assigned to that cluster too is automatically recorded. Thus, at the end of cluster growing for the entire dataset, there would be a table that contains data points and their cluster ids. This is equivalent to having a record of cluster labels and data points corresponding to that data label, along with the number of data points belonging to that cluster. Therefore, by discarding classes that have less than a predetermined number of data points (declaring them as merely stray classes formed by stray data points), we can automatically get an accurate determination of the actual number of clusters in the dataset. The minimum number of data points that can be considered as forming a valid class is based on the problem at hand.

6.0. EXPERIMENTAL SETUP AND INVESTIGATIONS

A visual software tool was implemented, and various complex and representative datasets were created using this for investigations. For this paper, the following datasets were created. The number of clusters created was one through four (a different number of clusters for different trials). The number of data points in each of these clusters was set at 2000, 1500, 1000, and 500 respectively. The number of isolated stray data points was set at one through four, occurring as single isolated stray data points, or as isolated groups of single, dual, triple, and quad data points. The left portion of Table 1 shows the dataset composition in each of these situations. For illustrative purposes, graphical display of a typical dataset so created is shown in Figure 1. This figure corresponds to a dataset of two Gaussian clusters of 2000 and 1500 data points each, along with four isolated stray data points occurring as a single isolated data point and a single triple data point group (dataset number 21 in Table 1).

A variety of other datasets with different number of differently shaped clusters with varying number of data points, along with a varying numbers of stray data points too were used in the investigations. However, as the outcomes were along expected lines and similar to that reported in this paper, these results are not included in this paper.

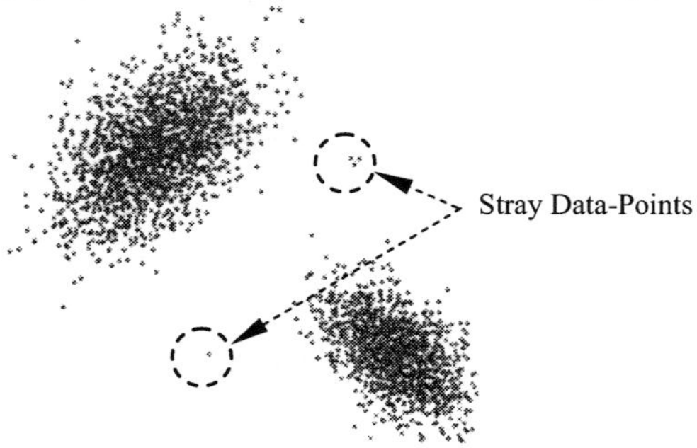

Figure 1. Graphical Display Corresponding to Dataset # 21 of Table 1

It should be noted that the dataset was intentionally created as a dataset that would be very problematic for the basic seed-growing algorithm to correctly determine the number of classes in the dataset. This was done to show the power of the enhanced algorithm even under trying conditions. Thus, the results for the basic seed-growing algorithm reported in this paper are in no way indicative of its regular performance (the illustrative cases were originally declared as problematic areas for the basic algorithm – island formation, and data bridging). The original paper [1] should be consulted for fully understanding the overall performance of the basic algorithm.

In this research, the number of clusters in the dataset was then determined without and with the use of singleton isolation prior to the application of the basic algorithm. The number of data points within each of the so determined clusters (class composition) too was determined (by simply counting the data points within each cluster during the classification process) and recorded (Table 1).

7.0. OUTCOMES OF INVESTIGATIONS AND DISCUSSIONS

Figure 2 illustrates a typical plot of the ordered (maximum to minimum) minimum inter data-point distance for each of the data points in Dataset # 21 of Table 1 and displayed in Figure 1. This plot is shown to illustrate the utility of the method used for singleton isolation described in Section 4.0. The plot shows the much larger then usual minimum inter data-point distance for singletons. It shows that in this case there is only one ("strong") singleton present in the dataset. The plot also shows that there are possibly a few (between 2 and 6) "marginal" singletons in the dataset in total. It shows that all other data points have at least one close neighbor.

Complete outcomes of the representative investigations are given in Table 1. The left portion of the table shows the composition of the dataset that was created for the investigations, while the right portion of the table shows the outcomes of the investigations without and with singleton isolation prior to the application of the seed-growing algorithm. The composition (number of data points) of each of the so determined classes (with singleton isolation) too is given in the same table. The table further shows the implied number of classes (with singleton isolation) based on discarding extremely small-sized classes (post-processing) by interpreting them as classes formed by stray data points. As can be seen, the table clearly shows the positive effect of singleton isolation in determining of the number of classes in a dataset.

In the case of single cluster, since the value of "p" used is "one" (no other value for "p" is needed with singleton isolation), although there was the issue of data bridging, the number of clusters determined was "one" even without the use of singleton isolation. Whether the stray data points occurred as singles, duals, trios, or quads did not matter in this case. This is not because the data bridging problem was absent for single clusters, but because there was no other cluster to bridge with, and with "p=1", single data points will never be determined as separate cluster but be lumped with another cluster [1].

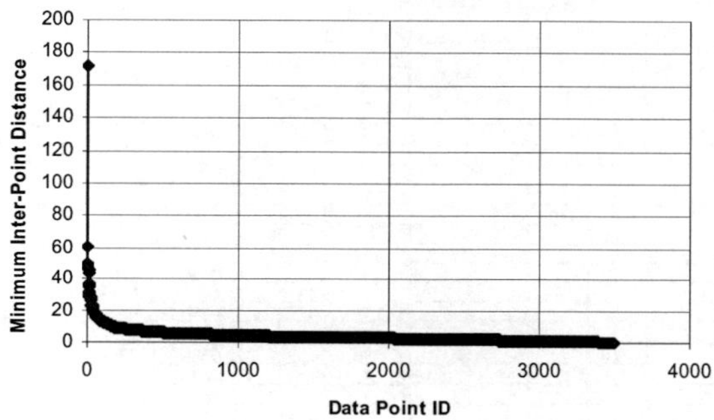

Figure 2. Minimum Inter Data-Point Distance Vs. Data-Point ID (Dataset # 21)

With singleton isolation, only singletons became isolated. Thus, when stray data points occur as groups of more than one data point, they do not get isolated but continue to remain as isolated groups of stray data points and became considered as classes (islands) of their own (also seen in Table 1). This is not a problem, however, because this can be overcome during post-processing (Section 5.0) when class composition is determined and extremely small-sized classes are discarded. This was always the case with singleton isolation in any of the cases (Table 1). As expected (also in Table 1), these stray data points became problematic only when singleton isolation was not used (a known issue with the basic seed-growing algorithm with unfavorably situated stray data).

Thus, the problem of cluster bridging is eliminated when singleton isolation is used with the basic seed-growing algorithm. Although these points now form data islands, they are easily detected and discarded during post processing by evaluating class composition. It should be noted that although the issues with problematic areas have been considerably reduced by the methods discussed in this paper, problematic situations are still present in very unfavorable situations. They are being further investigated.

8.0. SUMMARY AND CONCLUSION

In this paper, causes for the unfavorable situations faced by the original algorithm were investigated, and a combination of pre and post-processing was proposed as an enhancement to this algorithm to improve its performance even under unfavorable situations. Experimental evaluations were conducted using a variety of problematic datasets to verify the proposed enhancement. It was shown that with singleton isolation as pre-processing to eliminate data bridging and post-processing to discard stray classes, the number of classes in any arbitrary dataset with reasonably separated clusters can be accurately determined using the "Seed-Growing" algorithm.

As future work, a more rigorous technique for determining the threshold for singleton isolation will be investigated based on the fact that there is usually a much larger gap between the minimum inter data point distance for a singleton and than for other data points. Although this basis was used in the research reported in this paper to determine the threshold for singleton isolation, it is being further investigated.

Dataset #	Dataset Structure						Results								
							Determined # of Clusters		With Singleton Isolation						
	# of Clusters	# of Stray Points	Organization of Stray Points						Determined Cluster Composition With Singleton Isolation (Cluster # and # of Points)						Implied # of Clusters
			single	dual	trio	quad	Without Singleton Isolation	With Singleton Isolation	1	2	3	4	5	6	
1	1 Cluster	1	1				1	1	2000						1
2		2	2				1	1	2000						1
3		2		1			2	2	2000	2					1
4		3	3				1	1	2000						1
5		3	1	1			1	2	2000	2					1
6		3			1		2	2	2000	3					1
7		4	4				1	1	2000						1
8		4	2	1			1	2	2000	2					1
9		4		2			3	3	2000	2	2				1
10		4	1		1		1	2	2000	3					1
11		4				1	2	2	2000	4					1
12	2 Clusters	1	1				1	2	2000	1500					2
13		2	2				1	2	2000	1500					2
14		2		1			3	3	2000	1500	2				2
15		3	3				1	2	2000	1500					2
16		3	1	1			1	3	2000	1500	2				2
17		3			1		3	3	2000	1500	3				2
18		4	4				1	2	2000	1500					2
19		4	2	1			1	3	2000	1500	2				2
20		4		2			4	4	2000	1500	2	2			2
21		4	1		1		1	3	2000	1500	3				2
22		4				1	3	3	2000	1500	4				2
23	3 Clusters	1	1				1	3	2000	1500	1000				3
24		2	2				1	3	2000	1500	1000				3
25		2		1			4	4	2000	1500	1000	2			3
26		3	3				1	3	2000	1500	1000				3
27		3	1	1			1	4	2000	1500	1000	2			3
28		3			1		4	4	2000	1500	1000	3			3
29		4	4				1	3	2000	1500	1000				3
30		4	2	1			1	4	2000	1500	1000	2			3
31		4		2			5	5	2000	1500	1000	2	2		3
32		4	1		1		1	4	2000	1500	1000	3			3
33		4				1	4	4	2000	1500	1000	4			3

Table 1. Comparison Without and With Singleton Isolation
(33 Different Problematic Datasets)

REFERENCES

[1] S. Easwaran, "Determining the Number of Clusters in any Arbitrary Dataset containing Reasonably-Separated Data Clusters", ANNIE 2006 Annual Conference, Missouri (November 5-8, 2006). Published in "Intelligent Engineering Systems through Artificial Neural Networks" (Volume 15), ASME Press, New York 2005, ISBN: 0-7918-0204-3.

[2] Sandrine Dudoit, Jane Fridlyand, "A Prediction-Based Re-Sampling Method to Estimate the Number of Clusters in a Dataset", Genome Biology 2002, 3, 0036.1-0036.21

[3] Chris Fraley, Adrian E. Raftery, "How Many Clusters? Which Clustering Method? Answers via Model-Based Cluster Analysis", The Computer Journal, Vol. 41, No. 8, 1998

[4] Trevor Hastie, Robert Tibshirani and Guenther Walther, "Estimating the Number of Clusters in a Dataset via the Gap Statistic", Journal of the Royal Statistical Society, B, 63, 411-423, 2001

[5] Stan Salvador, Philip Chan, "Determining the Number of Clusters/ Segments in Hierarchical Clustering/ Segmentation Algorithms", (http://www.cs.fit.edu/~pkc/papers/ictai04salvador.pdf)

[6] H. Ressom, et al. "Computational methods for Identifying Number of Clusters in Gene Expression Data", Proc. of Neural Networks and Computational Intelligence, 2003, ACTA

Towards A More Human-Oriented Data Mining Application

Sang C. Suh
Reno Alex
Shree Kumar Menon
Department of Computer Science
Texas A&M University Commerce
Commerce, TX 75 429

Abstract:
Information technology has been in the market for a long time now. Organizations have implemented one or many forms of database applications and have accumulated data to provide historical backgrounds. But these data merely represent transactional elements and are of no use to a user unless some meaningful information can be generated from such vast volumes. Data mining is the next step in the process of analyzing data. It tries to find hidden nuggets of information from large databases containing raw data. It equips a user with the ability to identify useful trends and generate valuable information from the database which wouldn't be possible with traditional SQL query processing.

One of the important problems arisen in this area is the user's inability to find the easy and faster way to formulate queries to request the generation of valuable information from the database. SQL queries normally generate answers based on the facts stored in the database. They do not support the generation of information from data. Our proposed system addresses this problem by allowing the users to use natural language based queries. The proposed system internally performs data analysis processing and generates the analysis result in response to the natural language based queries which are pre-formulated by the system for the users to choose. Furthermore the system also addresses the problem of choosing the most suitable algorithm for the database in focus and the problem to be solved. Hence, the system allows any non-technical users to request the information discovery process on a database of interest.

Key Words: data mining, data mining application, intelligent systems

1.0 Introduction

"Data mining is the process of discovering meaningful new correlations, patterns, and trends by sifting through large amounts of data stored in repositories, using pattern recognition technologies as well as statistical and mathematical techniques (SPSS)."[2] Even though the term "data mining" has been used relatively sparingly, the technology that goes behind it has been around for long time now. Technically data mining is about discovering patterns from data that is hidden inside database. Till now the data mining concept was focused on large databases and development aimed at large organizations. How the data mining tool analyses the data, and the type of information it provides, depends on the techniques that are used.

Knowledge discovery process consists of following key steps ([10]):

- learning the application domain.
- data cleaning and transformation.
- choosing functions of data mining (summarization, association, clustering ...)
- analyze the data through a series of datamining algorithms.
- pattern evaluation and knowledge presentation.

2.0 The data mining application

Our proposed application is implementation of common and proven algorithms in the field of datamining using latest technologies available. It is basically a very light weight application that will provide for a framework for building better enhanced datamining utilities. Since our team consisted of both Java and C# developers, this project led to parallel development of the application in two different technologies. In order not to constrain our self with specific technology resource, the two most used technology in the present era, C# and Java, were used. Moreover resource constraints in one technology area will not lead to stagnation of the project. The general architecture of the project is as shown below in Figure 1.

The front end exists as both desktop application as well as web application. The middle tier consists of classes being developed as libraries. These can be accessed in the form of components so that it can be utilized by any front end system. The system is designed to be highly maintainable. This is achieved through the use of XML configuration files (See Figure 2) that describe the new libraries added to the middle tier. This way any ne implementations can be added on the fly and corresponding user interface can be developed for the same.

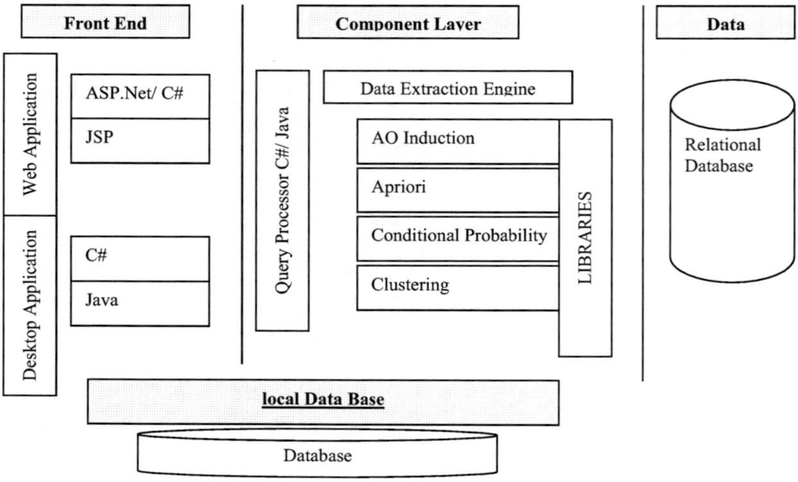

Figure 1. Application Architecture

The user interface is developed based on the interface provided by the middle tier components. The user select the queries interactively from given examples (See Figure 3). The query analyzer will then analyze the quer and selects the algorithm that is most suitable for the present query. It then gives the user the choice of database that can be used by the algorithms that are going to be used. The user then selects the placeholder fields base on the initial query example. The user is then given a range of values, which are obtained from the databas directly using normal SQL query, in case of fields that needs values to be entered manually. This will make sur that the user is not entering any out of range or invalid values.


```
<?xml version="1.0" encoding="utf-8" ?>
<Modules xmlns="http://www.tamu-commerce.edu/istg">
    <Module ShortName="MiningLibrary" LongName="Datamining Library Management">
        <Components Name="Apriori"  RootFolder="C:\Inetpub\wwwroot\Datamining\">
            <Component ScreenName="Apriori_likelihood" AssemblyName="AprioriLK.dll"  AssemblyP
                <CustomProperties>
                    <Property Name="Support" Type="double" Source="Entry"/>
                    <Property Name="Confidence" Type="double" Source="Entry"/>
                </CustomProperties>
            </Component>
            <Component ScreenName="Apriori_Sequence" AssemblyName="AprioriSQ.dll"  AssemblyPat
                <CustomProperties>
                    <Property Name="Support" Type="double" Source="Entry"/>
                    <Property Name="Confidence" Type="double" Source="Entry"/>
                </CustomProperties>
            </Component>
        </Components>
```


Figure 2. XML Configuration file

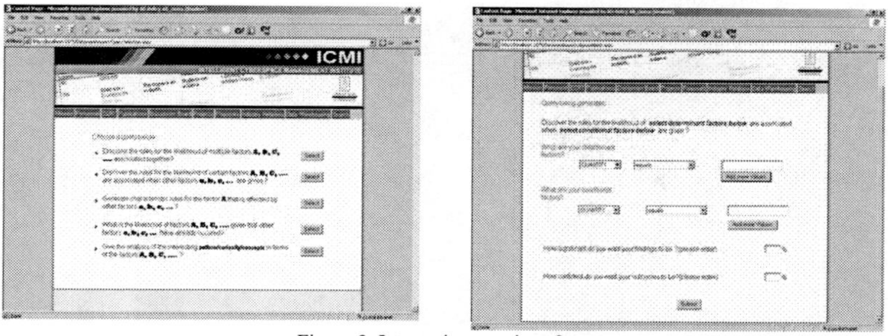

Figure 3. Interactive user interface

The controller for the user interface invokes the algorithm library chosen and marshals the input data to it. The data is used by the library to find relationships as asked and the result is stored in a local database. Then the result will be passed back to the calling program. The calling program then fetches the data result and displays it in a predefined order to the end user as the result of query execution. With the present system the user is presented with the results which can be either predictive or non-predictive (i.e. analytic summary).

3.0 Brief review of generic algorithms used in the application

Four sets of basic data mining algorithms are implemented in the system. They have predefined interfaces in the system as to what the input and output are to be. An effort was made to keep the interface same so that it can support the varying implementation of the same algorithm through the life of the application. The algorithms used for the system are as follows:

- Apriori
- Conditional probability
- AO induction
- Clustering

Apriori:

It is also known as market basket analysis. It is used to generate association rules that show the relationship among attributes of databases. Two variances of Apriori algorithm are used in the present application that

- gives support and confidence for a particular likelihood.
- gives a set of associations whose likelihood satisfy given confidence and/or support.

General Definitions

- **Itemset:** Set of items that occur together
- **Association Rule:** Probability that particular items are purchased together.
    - $X \Rightarrow Y$ where $X \cap Y = 0$
- **Support**: supp(X) of an itemset X is the ratio of transactions in which an itemset appears to the total number of transactions.
- **Confidence** of rule $X \Rightarrow Y$, denoted as conf($X \Rightarrow Y$):
    - conf($X \Rightarrow Y$) = supp($X \Rightarrow Y$) / supp(X)
    - Confidence can also be defined in terms of the conditional probability. conf($X \Rightarrow Y$) = P($Y \mid X$) = P($X \Rightarrow Y$) / P(X)
- **Transaction Database** stores transaction data.

Association rule mining algorithms typically produce too many rules, their operation is not intuitive to many users and interpreting the results can also be hard ([2], [12]). An algorithm for Apriori is summarized below in Figure 4 [1]:

```
L1 = {large 1-itemsets};
for (k=2; Lk-1≠0; k++) do
          begin
           Ck=apriori-gen(Lk-1); // new candidates
           forall transactions t∈D do
                  begin
                   Ct=subset(Ck, t);
                   forall candidates c∈Ct do
                          c.count++;
                  end
           Lk={c∈Ck| c.count ≥ minsup}
          end
answer = ∪kLk;
```

Figure 4. Algorithm Apriori

Conditional Probability:

The present system is designed to generate conditional probabilities between two values. Work on generating conditional probabilities with more than 2 attributes is underway. The algorithm finds the probability of occurrence of an attribute given that another one has occurred. The very basic equation of conditional probability is:

$$P(A|B) = \frac{P(A \cap B)}{P(B)}$$

Note that $P(B) > 0$. The line between A and B is read "given." This means "the probability of A given that B has happened." The event on the right side of the line is the event that has already happened.

The conditional probability is used in mining to detect dependency. As pointed out in "Dependency detection in Mobi Mines", the dependency is a very relative term and can be expressed as a correlation between two random variables [11]. Suppose there is a frequently occurring pattern P1 in data A that is followed by pattern P2 in data B, a rule can be formed between these two data that tells us that P1 triggers P2 with some confidence. The confidence of the rule is the conditional probability of pattern P2 occurring if P1 occurs. The more the number of rules with higher confidence, the greater the dependency among the data.

AO Induction:

The attribute oriented induction is performed based on hierarchical relationships among the attributes which are specified in a table called concept table. The algorithm used data as described in the concept table to generalize the database according to the concept hierarchy of the attributes. The generalization process is continued until no further generalization is possible from the result set. The concept table, having a tree structure like the one shown in Figure 5, is to be created by someone with expert knowledge in the content of that particular database.

Basic Principles of AO Induction are [8]:

- Attribute-removal: remove attribute *A* if there is a large set of distinct values for *A* so there is no generalization operator on *A*, or *A*'s higher level concepts are expressed in terms of other attributes.
- Attribute-generalization: If there is a large set of distinct values for *A*, and there exists a set of generalization operators on *A*, then select an operator and generalize *A*.
- Attribute-threshold control: typical 2-8, specified by default.
- Generalized relation threshold control: control the final relation/rule size

One of the main issues that arise is which fields should be selected for dimensioning and to what level you should generalize. The AO induction mainly leverages on learning by example technique set oriented database approach, and generalization of data from database [7].

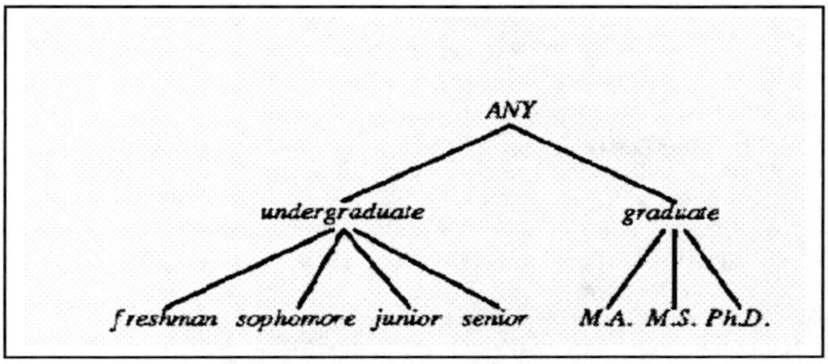

Figure 5. Concept Tree Hierarchy

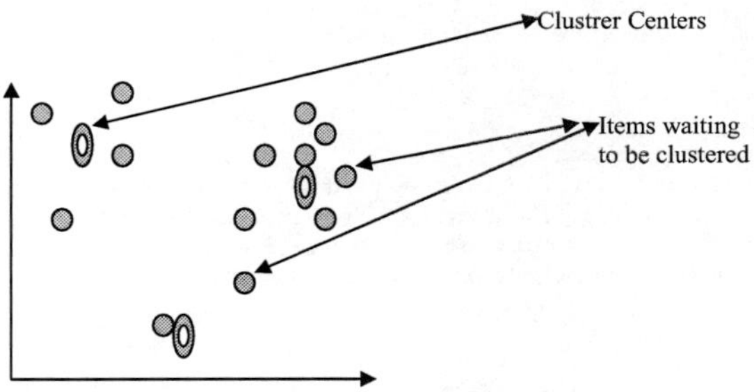

Figure 6. Cluster Formation

<u>Clustering:</u>
Clustering is an example of unsupervised operation. It is used where you wish to find groupings of similar records in your data without any preconditions as to what that similarity may involve. Clustering is usually achieved using statistical methods. In our application we have used k-means algorithm implementation. Whichever method is used, the basic operation is the same. Each record is compared with a set of existing clusters, which are defined by their 'centre' as can be seen as circles with double lines in Figure 6. A record is assigned to the cluster that is nearest to, and this in turn changes the value that defines that cluster. Multiple passes are made through a data set to re-assign records and adjust the cluster centers until an optimum solution is found.

Clustering is often undertaken as an exploratory exercise before performing further data mining using a classification technique [6]. For this reason, good visualization support is a helpful adjunct to clustering since it enables you to play with your clusters in order to see if the clusters identified make sense and are useful in a business context [10].

4.0 Results

The system prototype that was built was able to achieve the following benefits:

- It assists a non technical user who has no knowledge about data in the database to find co-relations and patterns based on general queries.
- It implements a dynamic architecture in which new implementation or update of implementation of present algorithm can easily replace the older version without much effort.
- More user-friendly query generation and answering are achieved through the use of pre-defined natural language queries that make full use of the result of data mining.

5.0 Advantages and discussion

The main advantage of this application is its simplicity and ease of use. A user can generate desired output for his query without much understanding of domain data. The system utilizes expert knowledge to find which algorithms are suitable for which type of queries and appropriately guides the user towards the results. The developed prototype has been proved to be of great benefits but there are still many areas that have room for improvement before practical value of the application can be fully accessed. The research is going to now work towards

- Make sure the user have greater degree of flexibility for query selection by a query engine based on artificial intelligence to select the best possible fit of algorithm.
- Compatibility for all types of data storages. Create new libraries that will be used to help the system utilize not only relational databases but also flat files and spreadsheets.
- Ability to dynamically load a database and have real-time ability to data mine the target database.
- Generate helper tables like concept table in AO induction on the fly.
- Increasing the accuracy of the results with implementation of improved algorithms.
- Automatic indexing of specific fields in the database to increase query performance.
- Output to be presented in general layman language.

6.0. Future directions of research

The research will be used to add values to already existing products in small and medium scale industries. The system can be extended to a specialized version for various industries. Further research can be performed to make user interface fully natural language based. More algorithms for different applications can be added to the system. The effectiveness of multiple algorithm integration should be conducted.

7.0. Acknowledgement

This research has been partially supported by L-3 Communication Corporation, ComCept Division under Project Corvus.

8.0 References

1. Agrawal R., R. Srikant: "Fast Algorithms for Mining Association Rules", Proceedings of the 20th VLDB Conference, Santiago, Chile, 1994
2. Agrawal C. and P. Yu: "A New Approach to Online Generation of Association Rules", IEEE Trans. On Knowledge and Data Engineering, Vol. 13, No. 4, pp. 527-540, 2001.
3. Yonatan Aumann and Yehuda Lindell. "A statistical theory for quantitative association rules", *Proceedings of the 5th ACM SIGKDD international conference on Knowledge discovery and data mining*, August, 1999.
4. P. Palmerini, S. Orlando and R. Perego. "Data mining (DM): Statistical properties of transactional databases", *Proceedings of the 2004 ACM symposium on Applied computing*, March 2004.
5. Dubes, Richard C. and Anil K. Jain. *Algorithms for Clustering Data,* New Jersey: Prentice Hall, 1988.
6. Bhatia, S.K. and Deogun, J.S. "Conceptual clustering in information retrieval", *IEEE Transactions on Systems, Man and Cybernetics*, Vol. 28, No. 3, pp. 427-436, 1998.
7. J. Han, Y. Cai, and N. Cercone, "Knowledge Discovery in Databases: An Attribute-Oriented Approach", Proceedings of the 18th VLDB Conference, Vancouver, British Columbia, Canada 1992.
8. Jiawei Han, Yongjian Fu, "Mining Multiple-Level Association Rules in Large Databases", IEEE Transactions on Knowledge and Data Engineering, Vol. 11, No. 5, pp. 1-8, 1999.
9. Dana Avram Lups, "Unsupervised Single Link Hierarchial Clustering", BOLYAI, INFORMATICA, Volume L, Number 2, 2005
10. Data Mining: Concepts and Techniques, 2nd ed., Jiawei Han and Micheline Kamber Morgan Kaufmann Publishers, (ISBN 1-55860-901-6) March 2006.
11. Sweta Pittie, Hillol Kargupta , Byung-Hoon Park, "Dependency detection in MobiMine: a systems perspective", Information Sciences—Informatics and Computer Science: An International Journal, Volume 155 , Issue 3-4, October 2003.
12. Claudio H. Yamamoto1, Maria Cristina F. de Oliveira,Visualization to assist users in association rule mining tasks, Volume 3379/2005, Integrated Publication and Information Systems to Information and Knowledge Environments, 2005.

OCEAN SURFACE WIND VECTOR FORECASTING USING SUPPORT VECTOR REGRESSION

HICHAM MANSOURI
School of Industrial Engineering
The University of Oklahoma
Norman, Oklahoma
hmansouri@ou.edu
ROBIN GILBERT
School of Industrial Engineering
The University of Oklahoma
Norman, Oklahoma
dioelaclaire@ou.edu
THEODORE B. TRAFALIS
School of Industrial Engineering
The University of Oklahoma
Norman, Oklahoma
ttrafalis@ou.edu

LANCE M. LESLIE
School of Meteorology
The University of Oklahoma
Norman, Oklahoma
lmleslie@ou.edu
MICHAEL B. RICHMAN
School of Meteorology
The University of Oklahoma
Norman, Oklahoma
mrichman@ou.edu

ABSTRACT

The ocean surface wind vector field is a key element for short range weather forecasting and for wind driven ocean wave prediction. Such forecasts are needed to issue timely and accurate ocean weather warnings and avoid major catastrophes. Many recent research attempts to measure and forecast ocean surface wind speed using polarimetric microwave radiometry provided by satellites. However, the amount of data provided by a satellite is very large and has many redundant data. Consequently, any type of analysis of this data should be done on a subset of the data, that is, by some form of data thinning. In this paper we use support vector machines to extract a subset that is composed of support vectors. The size of the subset is fewer than eight percent of the total data for this data type. The results obtained show that the support vectors allow reconstruction of the ocean surface wind speed vectors with high accuracy.

Keywords: Data Mining Applications, Data Thinning, Supervised Training, Regression Analysis.

1. INTRODUCTION

Support vector machines (SVMs) are a family of learning algorithms used in supervised learning tasks such as statistical classification and regression analysis. SVMs were introduced by Vapnik (1982) decades ago but they only have been investigated by the machine learning community since 1995. Kernel methods and SVMs were heavily studied these past ten years as alternatives to artificial neural networks, decision trees, or Bayesian networks for classification tasks (Taylor and Cristianini, 2004) and (Scholkopf and Smola, 2002).

To use a SVM method, we need to solve a quadratic programming problems with linear constraints. Therefore, the number of data points used during the training period is critical for the speed of the algorithm. Many methods were proposed to cope with this problem (Platt, 1999) and (Musicant and O. L. Mangasarian, 2000). In this work, we use Voronoi tessellation to deal with large data sets (Gilbert, 2006).

The main idea of the method used will be presented in Section 2. Then, the results will be described in Section 3. Finally, Section 4 concludes the paper.

2. METHOD USED

2.1 Support Vector Regression

We assume that there is a source that provides a sequence of l observations according to an unknown probability distribution function. Each observation (data point) is represented as a vector with a finite number of continuous and/or discrete attributes that can be represented as a point in the Euclidean space $\Re^n$. Each data point will be represented by $x_i \in \Re^n, \forall\, i \in [1,l]$.

In regression problems, each observation x_i is related to a unique real valued scalar denoted by y_i. The couplets (x_i, y_i) in $\Re^{n+1}$ then form a set of points that has a continuous-unknown shape. The objective of SVR is to find a prediction function, denoted by *f*, such that the differences between $f(x_i)$ and the target values y_i are minimized. By introducing for each observation x_i a set of positive slack variables ξ_i which have to be minimized, we can formulate, the following set of constraints for the regression problems:

$$\begin{cases} |f(x_i) - y_i| \le \xi_i & \forall i \in [1,l] \\ \xi_i \ge 0 & \forall i \in [1,l] \end{cases} \tag{1}$$

In the SVM literature, the prediction function *f* belongs, for the linear regressions, to a class of functions denoted by *F* such that:

$$F := \{x \in \Re^n \mapsto \langle w \cdot x \rangle + b : \|w\| \le B\} \tag{2}$$

where $B>0$, $w = \sum_{j=1}^{l} \alpha_j x_j$, and $\alpha_j \in \Re\ \forall j \in [1,l]$.

In the case of nonlinear regression, the class of functions *F* needs to be changed in a way that it will still allow for a linear regression, but in a different Hilbert space where the observations x_i will be mapped. To do so, we introduce a nonnegative definite kernel $k : \Re^n \times \Re^n \to \Re$, we induce a new Hilbert space *H* and a map $\phi : \Re^n \to H$ such that $k(x,y) = \langle \phi(x), \phi(y) \rangle_H$ for any *x* and *y* in $\Re^n$. Hence, *F* becomes:

$$F := \{x \in \Re^n \mapsto \langle w \cdot \phi(x) \rangle_H + b : \|w\|_H \le B\} \tag{3}$$

where $w = \sum_{j=1}^{l} \alpha_j \phi(x_j)$, and $\alpha_j \in \Re\ \forall j \in [1,l]$. The explicit knowledge of *H* and ϕ is not even required. Therefore, the set of constraints (1) becomes:

$$\begin{cases} \left| \sum_{j=1}^{l} \alpha_j k(x_j, x_i) + b - y_i \right| \le \xi_i & \forall i \in [1,l] \\ \xi_i \ge 0 & \forall i \in [1,l] \end{cases} \tag{4}$$

The SVM literature proposes an objective function that reduces the slack variables and the expected value of $\left|f(x_i)-y_i\right|$. To achieve that objective we need to minimize the quantities b, ξ_i, and $\|w\|_H$.

We have $\|w\|_H^2=\left\langle\sum_{j=1}^{l}\alpha_j\phi(x_j)\cdot\sum_{j=1}^{l}\alpha_j\phi(x_j)\right\rangle_H=\sum_{i=1}^{l}\sum_{j=1}^{l}\alpha_i\alpha_j\phi(x_i)\phi(x_j)$.

Hence, $\|w\|_H^2=\alpha^T K\alpha$, where $(K)_{ij}=k(x_i,x_j)$. Finally, the quadratic programming that needs to be solved is:

$$\min_{\alpha,\xi,b}\ \alpha^T K\alpha+C\xi^T\xi+b^2$$
$$subject\ to:\left|K\alpha+b1-y\right|\le\xi \tag{5}$$

where $C>0$ is a positive trade-off constant that penalizes the nonzero values of the ξ_i, **1** is a $l\times 1$ vector of ones, and y is the vector which entries are the y_i.

The optimal solution $\left(\alpha^*,b^*\right)$ of this problem gives the following prediction function: $x\mapsto\sum_{i=1}^{l}\alpha_i^* k(x_i,x)+b^*$ (6)

The vectors x_i for which the values α_i are nonzero are called support vectors.

Experiments show that over a few thousand data points the algorithm tends to be very slow and requires a lot of core memory. Moreover, the ocean surface wind speed sets contain tens of thousands of data points. Consequently, we cannot apply support vector regression directly. Hence, we have opted to apply a Voronoi tessellation on the data set and then apply the SVR in each cell.

2.2 Voronoi Tessellation

For any discrete set S of points in Euclidean space and for almost any point *x*, there is one point of S closest to *x*. The word "almost" is occasioned by the fact that a point *x* may be equally close to two or more points of S.

In general, the set of all points closer to a point *c* of S than to any other point of S is the interior of a (in some cases unbounded) convex polytope called the Dirichlet domain or Voronoi cell for *c*. The set of such polytopes tessellates the whole space, and is the Voronoi tessellation corresponding to the set S. If the dimension of the space is only 2, then it is easy to draw pictures of Voronoi tessellations (figure -1-).

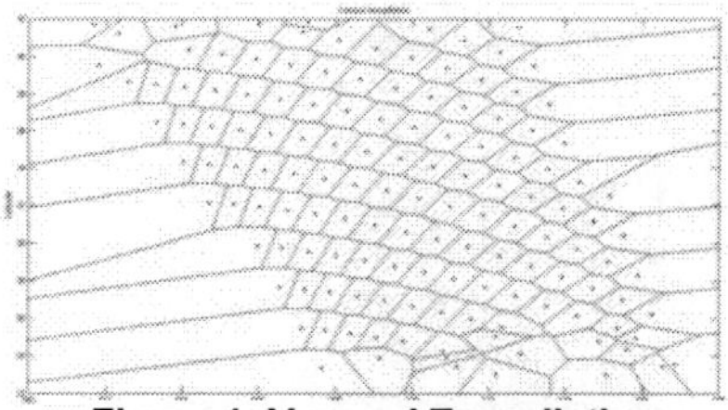

Figure 1: Voronoi Tessellation

3. EXPERIMENTS AND RESULTS

3.1 Data Set

WindSat is designed to demonstrate the viability of using polarimetric microwave radiometry to measure the ocean surface wind vector from space. The objective of the satellite data obtained from WindSat is to measure the ocean surface wind vector and consequently to generate ocean surface wind data for numerous practical applications. In general terms these data are essential for short-term weather and sea state analysis, prediction and verification, the issuing of timely weather warnings, and the generation of climatological data sets. Particular applications include marine weather warning that affect ship movement and positioning, improve the predictions of storms at sea and to help understand better the mechanisms of global climatic change.

Satellite derived winds have been used for data analysis and assimilation in meteorology and oceanography for about 30 years. For satellite derived winds it is necessary to account not only for random errors but also spatially correlated errors. In such cases it was found long ago that high density wind data can degrade model forecasts compared with low density winds. For single tracking methods, like those aplied to WindSat, spatially correlated errors are inevitable and ther have been many attempts to reduce the correlated errors by a range of data thinning, or selection, methods. The basis of all operational data thinning methods is to divide the atmosphere into three dimensional boxes and to produce one observation per box, a procedure also called "superobbing" in meteorology and oceanography because one "super" observation is generated from all those present in the analysis box. The thinning stategies applied to these boxes are numerous and here we mention five commonly used approaches. The simplest approach is to randomly select one observation per box. Alternatively, all the observations in a box can be averaged. Methods three and four are more sophisticated and use quality marks to thin the data. These are the Recursive Filter Flag (RFF) (Hayden and R. J. Purser, 1995) and the Quality Indicator (QI) (Holmlund, 1998) which select data by a range of methods such as comparison of measurements from the satellite with colocated rawindsondes or with analyses and very short ran\ge model predictions. RMS wind differences between the satellite observed winds and the rawindsondes or model generated winds are employed to estimate the quality of the observations. Finally, the RFF and the QI can be combined to produce a fifth and even more selective data thinning technique.

In the present study, the primary data provided by WindSat are Sea Surface Wind Speed and Sea Surface Wind Direction. In addition to ocean surface wind vector, the WindSat system will provide a host of secondary ocean-scene environmental data products. These products include Cloud Liquid Water, column integrated Precipitable Water, and Sea Surface Temperature (SST).

3.2 Experiments

During the experiment, we used data from January 01, 2005, over the domain spanning the region defined by (127W, 145E) and (23N, 42N). After we removed all data points that have some missing information, we end up with 13540 points. As the SVR allows only 1 output, we need to find 2 prediction functions; one for the u-component and one for the v-component of the wind speed vector. The input vector contains longitude, latitude, SST, water vapor, cloud liquid water, rain rate. The target is either the u-component or the v-component. We will test the effect of using different kernels.

3.3 Results

The main objective is to thin the data sent by the satellite. Hence, after applying the SVR, we only keep the support vectors. To show the quality of those support vectors, we reconstruct the wind field and then compare it to the observed values. Table -1- shows the MAE, MSE, and the correlation coefficient for different kernels. "RBF 10" is the Radial Basis Function with parameter 10. "Poly 1 2" is the polynomial kernel with parameters 1 and 2. "Sig 1 1" is the Sigmoid kernel with parameters 1 and 1. As we can see, the best results were obtained for the Radial

Basis Function with parameter 1. The MAE was around 1 m/s for both wind vector components. The correlation coefficients are greater than +0.90 for both components. Therefore, we will use the RBF kernel with parameter 1 for any subsequent applications.

	U-component			V-component		
	MAE	MSE	Correlation	MAE	MSE	Correlation
RBF .5	1.25	8.27	0.90	0.88	3.07	0.98
RBF 1	1.15	5.99	0.91	0.72	2.71	0.98
RBF 5	1.30	6.72	0.90	0.99	3.29	0.98
RBF 10	1.48	8.32	0.87	1.09	3.71	0.98
RBF 20	1.60	9.30	0.85	1.22	4.25	0.98
RBF 50	1.86	10.95	0.82	1.62	5.94	0.97
RBF 100	2.03	12.41	0.80	1.89	7.34	0.96
Linear	2.06	12.84	0.79	1.96	7.82	0.95
Poly 1 2	2.06	12.82	0.79	1.96	7.81	0.95
Poly 1 3	2.05	12.80	0.79	1.96	7.80	0.95
Sig 1 1	2.06	12.85	0.79	1.96	7.82	0.95

Table 1: U and V prediction's errors for different kernels

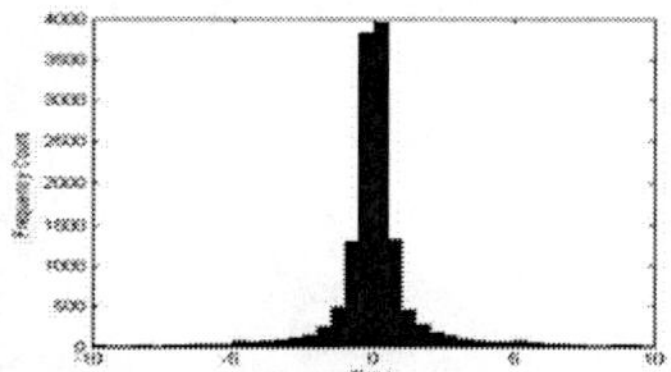

Figure 2 (a): Error distribution for the U-component

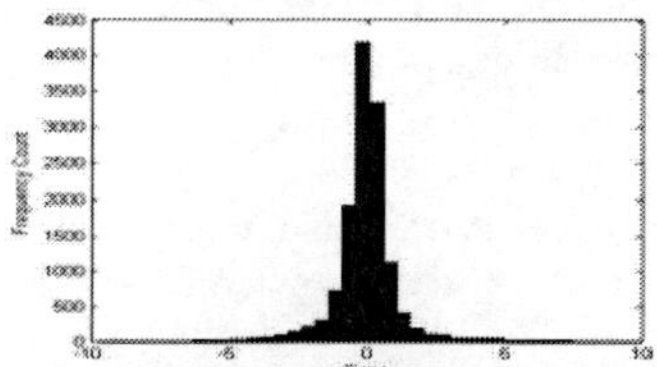

Figure 2 (b): Error distribution for the V-component

Figure 2a and 2b show the error distributions for the U-component and the V-component respectively. In figure 3a, 3b, 3c, and 3d we can see the reconstructed and the observed wind field contours for the U-component and the V-component respectively.

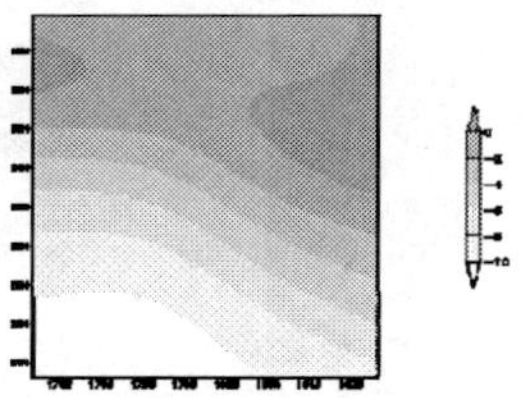

(a): the reconstructed wind field for the U-component

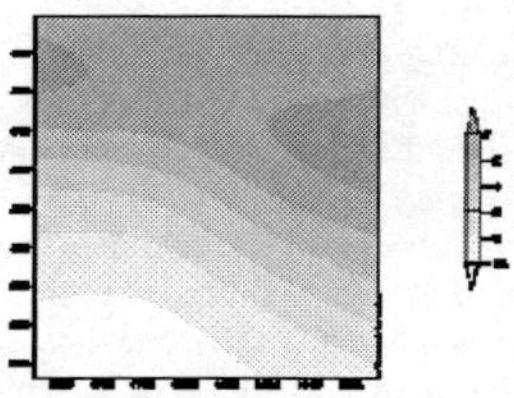

Figure 3 (b): the observed wind field for the U-component

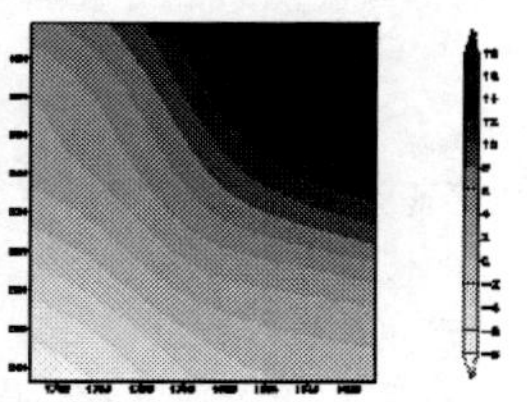

(c): the reconstructed wind field for the V-component

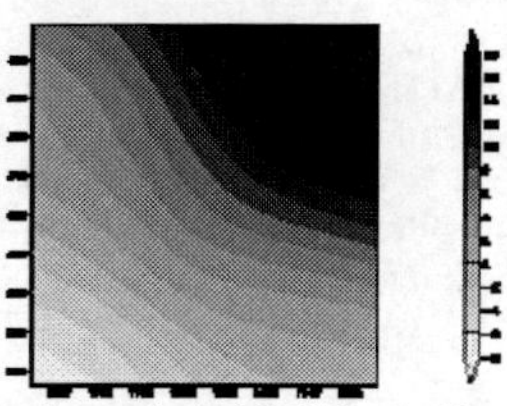

Figure 3 (d): the observed wind field for the V-component

Figure 4 shows that out of the available 13540 data points we need only around 1000 support vectors to reconstruct the wind vector field. As we can see, most of the support vectors have an alpha value near zero. Hence, they have an insignificant contribution to the final solution. Hence, we can thin further the data by removing more support vectors with very small alpha values.

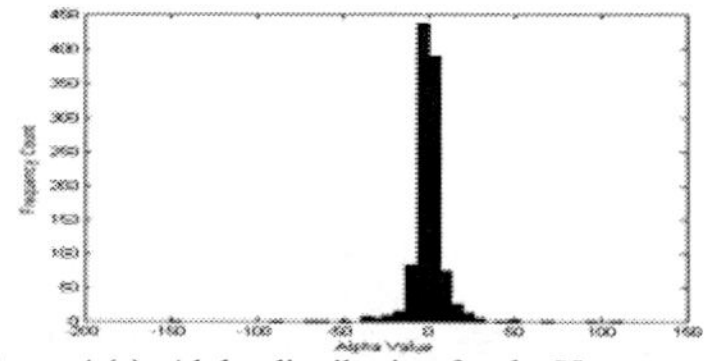

Figure 4 (a): Alpha distribution for the U-component

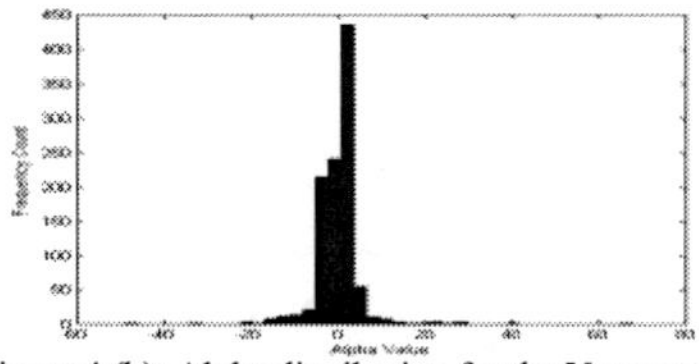

Figure 4 (b): Alpha distribution for the V-component

4. CONCLUSIONS

Recently, the Support Vector Machines have been widely studied and used for their ability to search for complex prediction patterns, such as in classification and regression problems. In this paper we used Support Vector Regression to thin important meteorological satellite data. The results obtained show that fewer than 8% of the data are needed to reconstruct the wind field with high accuracy (the correlation coefficient is greater than +0.9). We have conducted two separate analysis; one for the U-component and one for the V-component. In future work, we will try to formulate a new Support Vector Regression that allows for more than 1 output for each datapoint.

ACKNOWLEDGEMENTS

The authors wish to acknowledge NOAA under grant NA17RJ1227 and NSF under grant EIA-0205628 for providing financial support for this work. The opinions expressed herein are those of the authors and not necessarily those of NOAA or NSF. The authors wish also to thank Andrew Mercer for his help with the contour plots during this work.

REFERENCES

R. Gilbert. *Learning with kernels and applications. Masters thesis*. The University of Oklahoma, Norman, USA. 2006.

C. M. Hayden and R. J. Purser. *Recursive filter objective analysis of meteorological fields, applications to NESDIS operational processing*. J. Appl. Meteor. 13: 3-15, 1995

K. Holmlund. The utilization of satellite properties of satellite-derived atmospheric motion vectors to derive quality indicators. Wea. Forecasting. 13: 1093-1104, 1998.

D. R. Musicant and O. L. Mangasarian. *Large scale kernel regression via linear programming*. Machine Learning, 46(1-3):255–269, 2000.

J. Platt. *Using sparseness and analytic QP to speed training of support vector machines*. In M. S. Kearns, S. A. Solla, and D. A. Cohn, editors, Advances in Neural Information Processing Systems 11. MIT Press, 1999.

B. Schölkopf and A. J. Smola. *Learning with Kernels*. MIT Press, 2002.

J. Shawe-Taylor and N. Cristianini. *Kernel Methods for Pattern Analysis*. Cambridge University Press, 2004.

V. Vapnik. *Estimation of Dependences Based on Empirical Data*. Springer, 1982.

TREPAN-PLUS: AN EXTENSION OF A DECISION TREE EXTRACTION ALGORITHM UTILIZING ARTIFICIAL NEURAL NETWORKS

MAIMUNA H. RANGWALA
Department of Industrial and
Systems Engineering
Ohio University

GARY R. WECKMAN
Department of Industrial and
Systems Engineering
Ohio University

JON MARVEL
Department of Management
Gettysburg College

WILLIAM A. YOUNG II
Department of Industrial and
Systems Engineering
Ohio University

ABSTRACT

The purpose of this paper is to present heuristics for employing TREPAN, an algorithm for extracting decision trees from neural networks. Typically, several parameters need to be chosen to obtain a satisfactory performance of the algorithm. The current understanding of the various interactions between these is not well understood. By empirically evaluating the performance of the algorithm on a test set of databases chosen from benchmark machine learning and real world problems, several heuristics are proposed to explain and improve the performance of the algorithm. The experimentation is further validated by performance statistic measures. The algorithm is extended to work for multi-class regression problems and its ability to comprehend generalized feed-forward networks is investigated.

INTRODUCTION

One of the more pronounced drawbacks of Neural Networks is their lack of explanation capability. They act like 'black boxes' and do not give any reasoning behind the conclusion of a learning system. However, it is essential to understand the basis of decisions of this type of computer support systems for various reasons. Thus, classification decisions are extremely useful when they are in human comprehensible form. Representing the extracted knowledge in the form of 'if-then' rules is most likely the best method of explaining the output of a Neural Network model. Decision trees can be easily represented in the form of 'if-then' rules and hence extracting decision trees is probably one of the best methods of explaining a Neural Network. Decision trees are fast, simple to implement and can convert the learned hypothesis to a set of easily interpreted rules. The basic idea is to mimic human reasoning in a way that gives an insight into the decision process.

Schmitz et al (1999) have proposed an algorithm ANN-DT that extracts decision trees from trained neural networks. Craven and Shavlik (1996) have developed the TREPAN algorithm that mimics the behavior of a neural network in the form of decision trees. One of the research objectives of this paper was to enhance TREPAN to be able to handle not only multi-class classification type but also multi-class regression type problems. Another objective is to investigate if TREPAN can understand and analyze generalized feed-forward networks (GFF). TREPAN is tested on different datasets and best settings for TREPAN algorithm are explored based on database type to generate heuristics for various problem domains.

ARTIFICIAL NEURAL NETWORKS

Neural networks generate their own implicit rules by learning from examples. Neural networks are one of the most widely used algorithms for classification problems. The output layer is indicative of the decision of the classifier. The cross-entropy error function is most commonly used in classification problems. In a classification problem, each output node of a neural network represents a different hypothesis and the node activations represent the probability that each hypothesis may be true. Each output node represents a probability distribution and the cross entropy measure calculates the difference between the network distribution and the actual distribution (Plunkett and Elman, 1997).

Although neural networks are known to be robust classifiers, they have found limited use in decision-critical applications such as medical systems. Trained neural networks act like black boxes and are often difficult to interpret (Craven and Shavlik, 1995). The availability of a system that would provide an explanation of the input/output mappings of a neural network in the form of rules would thus be very useful. Rule extraction is one such system that tries to elucidate to the user, how the neural network arrived at its decision in the form of if-then rules.

Two explicit approaches have been defined to date for transforming the knowledge and weights contained in a neural network into a set of symbolic rules: decompositional and pedagogical (Thurn, 1995). The pedagogical approach treats neural networks like black boxes and aims to extract rules that map inputs directly to its outputs. The Validity Interval Analysis (VIA) proposed by Thrun and TREPAN is an example of one such technique.

DECISION TREES

Decision trees are probably the most widely used symbolic learning algorithms, as are neural networks in the non-symbolic category. Decision trees classify data through recursive partitioning of the data set into mutually exclusive subsets which best explain the variation in the dependent variable under observation. Decision trees classify instances (data points) by sorting them down the tree from the root node to some leaf node. This leaf node gives the classification of the instance. Each branch of the decision tree represents a possible scenario of decision and its outcome.

Decision trees have been proved useful in their applications to various real world problems. Leech (1986) applied a decision tree induction approach to a chemical nuclear power plant process. Michie (1989) used an induction algorithm to produce a decision tree for making decisions whether to grant credit to a loan applicant or not. Evans and Fisher (1994) applied decision tree induction to the problem of banding in Rotogravure printing. Decision trees continue to be an active research area, the current focus being on improving methods for building, controlling and executing the decision tree algorithms to achieve maximum efficiency.

TREPAN ALGORITHM

The TREPAN algorithm, developed by Craven, is a novel rule-extraction algorithm that mimics the behavior of a neural network. Given a trained Neural Network, TREPAN extracts decision trees that provide a close approximation to the function represented by the network. TREPAN uses a concept of recursive partitioning similar to other decision tree induction algorithms. In contrast to the depth-first growth used by other decision tree algorithms, TREPAN expands using the best first principle. That node which increases the fidelity of the tree when expanded is deemed the best.

In conventional decision tree induction algorithms the amount of training data decreases as one traverses down the tree by selecting splitting tests. Thus, there is not enough data at the bottom of the tree to determine class labels and is hence poorly chosen. In contrast, TREPAN uses an 'Oracle' to answer queries, in addition to the training samples during the inductive learning process. Since the target here is the function represented by the Neural Network, the network itself is used as the 'Oracle'. This learning from larger samples can avoid the lack of examples for the splitting tests at lower levels of the tree, which is usually a problem with conventional decision tree learning algorithms. It ensures that there is a minimum sample of instances (min. sample) available at a node before choosing a splitting test for that node where min. sample is one of the user specified parameters. If the number of instances at the node; say m is less than min. sample then TREPAN will make membership queries equal to (min. sample-m) from the 'Oracle' and then make a decision at the node.

A disadvantage of TREPAN is that the trees generated have normalized values. Denormalizing these values to comprehensible form is a cumbersome task. An algorithm can be incorporated into TREPAN that denormalizes the values and provides them to the user.

METHODOLOGY

The first step in this research is to modify and enhance the TREPAN software in order to be able to handle a wider variety of problems. TREPAN currently works with multi-class classification type problems and two-class regression problems. The next step was to run experiments to investigate if TREPAN can analyze the weights associated with generalized feed forward networks. A set of instructions in C language were written to enable the user to input the number of classes pertaining to the problem. The final step was to identify potential databases for TREPAN analysis. The six applications studied here are the Iris plant, body fat percentage, level of chlorophyll, inhibition for corrosion, student matriculation nad survivability of wireless networks

NeuroSolutions (1995) software is used to develop different neural network models for the datasets described earlier. Various networks types such as multi-layer perceptron and generalized feed forward networks were experimented with to obtain the best neural network model. Except for the corrosion dataset, all networks use 60%, 15%, and 25% of the data for training, cross-validation, and testing respectively. Due to its limited size, corrosion is split as 70%, 15%, and

15% for training, cross-validation, and testing respectively in order to provide for more training data. The network configurations for each problem domain are summarized in Table 1.

Table 1: Network Configurations

Database	No. of hidden layers	No. of processing elements each layer	Transfer function in hidden layers	Transfer function in output layer	Training,Cross-validation, Testing %
Iris	1	4	Tanh	Tanh	(60,15,25)%
Body fat	1	4	Tanh	Tanh	(60,15,25)%
Saginaw Bay	2	8,4	Tanh	Tanh	(60,15,25)%
Corrosion	2	5,3	Tanh	Tanh	(70,15,15)%
Admissions	2	15,10	Tanh	Tanh	(60,15,25)%
Outages	1	3	Tanh	Bias	(60,15,25)%

The classification accuracy attained by TREPAN is studied by varying user specific parameters with the objective of identifying possible heuristics. The TREPAN tool developed by Craven consists of three available decision tree induction algorithms: Single test TREPAN, TREPAN and Disjunctive TREPAN. A total of 66 experiments having different combinations of the above-mentioned parameters were run on each of the six datasets. Beam width values of 2, 3, 5, 7, 10 and minimum sample (min. sample) sizes of 1, 10, 50, 100, 500 and 1000 were used. The tree size that resulted in best classification accuracy was obtained by a plotting graph of tree node against the classification accuracy obtained at that node. These steps were repeated for each algorithm with the objective of first identifying the best set of parameters among each algorithm and then the best model among the three algorithms.

Assessing a classifier's performance is a very important aspect of machine learning. It is also essential for comparing different classifiers. It is assessed using the confusion matrix. A confusion matrix is a primary tool in visualizing the performance of a classifier. However, it does not take into account the fact that some misclassifications are worse than others are. To overcome this problem we use a measure called the Kappa Statistic (Cohen, 1960), which considers the fact that correct values in a confusion matrix are due to chance agreement.

The comprehensibility of the tree structure decreases with the increase in tree size and complexity. The principle of Occam's Razor states, "when you have two competing theories which make exactly the same predictions, the one that is simpler is the better" (Thorburn, 1915). Thus, between two trees induced by different classification algorithms and having the same prediction accuracy, the one with fewer leaves, or in other words the simpler tree is usually preferred.

RESULTS

The bold values show the maximum accuracy obtained in the respective single test TREPAN run. Table 2 shows the accuracies as percentage of correct classifications obtained. The results were compared to a baseline program for decision trees called C4.5. Accuracy and Comprehensibility. It is observed that excluding the iris database, TREPAN performs considerably better at classifying instances in terms of accuracy. For iris, the classification accuracies in both cases are equal. This may be because iris is a comparatively simple database and can more easily identify the underlying relationships among the attributes. It is interesting to note that in case of iris TREPAN surpassed the accuracy of the neural network itself from which it learned. This can be attributed to the ability of TREPAN to generate more data during the induction process.

The most significant difference in classification accuracy can be seen in the corrosion dataset. TREPAN was able to approximate the neural network averaging fidelity of over 90%. The kappa statistic, which is a better indicator of a classifier's performance than classification accuracy, showed higher values for all models of TREPAN than C4.5. The accuracies attained by both algorithms in the admissions case are nearly equal but there is a considerable difference in the tree size, the TREPAN tree being smaller than C4.5. The TREPAN confusion matrix is also more balanced in predicting the two classes than C4.5. Although C4.5 achieved equal or nearly equal accuracies to TREPAN in two cases (iris and admissions respectively), it generated larger incomprehensible trees compared to TREPAN.

Table 2: Summary of results

Problem Domain	Method	Classifier accuracy	No. of Nodes	No. of leaves	Kappa Statistic
Iris	TREPAN	94.70%	2	3	0.815
	C4.5	94.70%	2	3	0.801
Body fat	TREPAN	96.80%	3	4	0.956
	C4.5	92.06%	3	4	0.892
Corrosion	TREPAN	85.70%	12	13	0.808
	C4.5	57.14%	11	12	0.44
Outages	TREPAN	85.33%	6	7	0.689
	C4.5	76.00%	17	18	0.513
Saginaw	TREPAN	89.30%	12	13	0.651
	C4.5	84.67%	23	24	0.629
Admissions	TREPAN	72.10%	23	41	0.416
	C4.5	71.90%	78	157	0.404

The complexity of the database is defined here by comparing the multiple regression (R2) values for the datasets analyzed. A regression database with an R2 of 0.85 or higher is considered simple; R2 between 0.70 and 0.85 is considered complex and that with R2 of 0.70 and lower is considered as a highly complex database. The datasets fall into three categories based on size. A database with 200 instances or less is taken as small; number of instances between 200 and 1000 is considered medium and that with number of instances larger than 1000 is considered a large database.

CONCLUSIONS

Various experiments were run to investigate its compatibility with generalized feed forward networks. The weights and network files were restructured to present GFF networks in a format recognizable by TREPAN. The empirical investigation of TREPAN on these datasets reveals the following heuristics.

- A minimum sample (min. sample) size of 50 or less should be used for small databases. A min. sample size higher than 50 results in low model accuracy
- Best models are obtained by setting beam width values in a range from 2 to 7.
- If a high min. sample value is selected, a high beam width value should be selected for better model accuracy.
- Beam width values higher than 7 do not result in good model accuracy.
- For complex and highly complex databases, best model accuracy is obtained within a tree size range of (number. of inputs ± 5). Inputs refers to the total number of processing elements in the input layer after symbolic attributes have been translated.
- TREPAN generalizes better at lower min. sample sizes for highly complex databases having little data. The same does not hold true for highly complex databases having large amounts of data. For such datasets, a higher min. sample size should be used.
- It is observed that single test TREPAN and TREPAN algorithm perform better than disjunctive TREPAN on most occasions.

REFERENCES

Cohen, J. (1960). A coefficient of agreement for nominal scales. Educational and Psychological Measurement, 20, 37-46.

Craven M. W, Shavlik J. W. (1995). Using sampling and queries to extract rules from trained neural networks, Machine Learning. Proceedings of the Eleventh Inter-national Conference, Cohen W. W & Hirsh H. (Eds.), San Francisco, CA: Morgan Kaufmann.

Craven, M. W., and Shavlik, J. W., (1996) Extracting tree-structured representations of trained networks. In Advances in Neural Information Processing Systems, volume 8, pages 24–30, Denver, CO, MIT Press.

Evans, B. and Fisher, D. (1994). Overcoming process delays with decision tree induction, IEEE Expert, 9: 60-66.

Leech, W. J. (1986) A rule based process control method with feedback. Advances in Instrumentation, 41: 169-175.

Michie D. (1989) Problems of computer-aided concept formation, In Quinlan, J.R., (Ed). Applications of Expert Systems Volume 2. Wokingham, UK: Addison-Wesley, 310-333.

NeuroSolutions, (1995). Software developed and distributed by Neurodimension Incorporated [http://www.neurosolutions.com/products/ns/].

Plunkett, K., Elman, J., L. (1997). Exercises in Rethinking Innateness: A handbook for connectionist simulations, MIT Press, pp. 166.

Schmitz, G. P.J., Aldrich, C., & Gouws, F. S. (1999). ANN-DT: An algorithm for extraction of decision trees from artificial Neural Networks. IEEE Transactions on Neural Networks, 10(6), 1392-1401.

Thorburn, W. M. (1915). "Occam's razor", Mind, 24, pp. 287-288. available at http://math.ucr.edu/home/baez/physics/General/occam.html.

Thrun S. (1995). Extracting rules from artificial neural networks with distributed representations, In Tesauro G.,Touretzky D., and Leen T. eds. Advances in Neural Information Processing Systems 7, Cambridge, MA:MIT Press, pp.505- 512.

KNOWLEDGE CONSOLIDATION IN SOCIAL NETWORK DATA MINING

SUSEELA T. SARASAMMA
Northrop Grumman Mission Systems,
Bellevue, NE 68123, USA.
suseela_ts@yahoo.com

PREMCHAND S. NAIR
Computer Science Department,
Creighton University,
Omaha, NE 68178, USA.
psnair@creighton.edu

ABSTRACT

The information revolution has enabled us to acquire vast amount of data on any topic in any domain, hidden in email messages, web logs, technical papers, analyst's reports and so on. Exploitation of such data in harnessing relevant information for analysis and decision making needs to be automated. The facts and their inter-relationships can be represented as a graph. Techniques from Social Network Analysis (SNA) are used in this context. SNA models relations that exist among entities such as people, events, and organizations. The underlying graph can be quite complex even for simple applications. We propose a fuzzy theoretical approach to reduce this complexity through elimination of parallel edges. It results in a simpler graph that consolidates and derives hidden relationships between entities. A new measure of centrality called influence of an entity on the network is introduced and used to show the efficacy of consolidation operation.

I. INTRODUCTION

In this paper, we apply fuzzy theory to perform Social Network Analysis (SNA). Social Network Analysis models meaningful relations that exist between entities. These entities may be people, events, organizations, symbols in text, sounds in verbalizations, nations of the world and so on. The primary focus is on the relations among actors and not on the actors themselves or their attributes. A social network can be modeled as a fuzzy graph with the entities as the nodes or actors and the relations among them as the edges or links. The actors in a fuzzy social network may be embedded in fuzzy networks which are themselves embedded in other fuzzy networks. For instance the employees (persons) within a department have a certain structural fuzzy relationship or fuzzy network. Now the departments will be interrelated by another type of relation. Such structure where entities are embedded in fuzzy networks that are embedded in another fuzzy network is termed a multi-modal fuzzy network. In this example since there are two types of actors, people and departments, it is a two-modal fuzzy network.

Research in Social Network Analysis has provided several techniques for detecting structural patterns in the network. A key contribution is the determination of an actor's centrality in the network. The measures of centrality, defined for SNA are degree centrality, closeness centrality, graph centrality, stress centrality, Bonacich centrality and betweenness centrality. More information on social network analysis techniques can be found in [9]. These centrality measures have application in tasks such as discovering the emergent leaders in organizations, identifying key players in criminal investigation, discovering cross-border knowledge flows in scientific communities, and study of how diseases spread in communities of people or animals. Many of these centrality measures from SNA can be straight away applied to Fuzzy Social Network Analysis (FSNA).

The aim of this paper is two-fold. First, we propose a method to consolidate the information content in a multi-modal fuzzy graph. Second, we introduce a new centrality measure that is more consistent with works done on fuzzy graph. We study the effect of consolidation on the new centrality measure as well.

II. FUZZY GRAPHS

A fuzzy graph G is a (V, E, s, d, μ, ρ) where V is a non-empty set, E a set that can be empty, $s:E\rightarrow V$, $d:E\rightarrow V$, $\mu:V\rightarrow[0,1]$, $\rho:E\rightarrow[0,1]$ such that $\forall e \in E$, $\rho(e) \leq \mu(s(e)) \wedge \mu(d(e))$.

We call s(e) the source of e and d(e) the destination of e. Observe that there can be many elements in E with identical source and destination. We call μ the fuzzy vertex set of G and ρ the fuzzy edge set of G, respectively. We will assume that the underlying set V is finite. Note that in the most general case, both vertices and edges have membership value. However, in the special case where $\mu(x)=1$, for all $x \in V$, edges alone have fuzzy membership. So, in this case, we use the abbreviated notation G=(V, E, s, d, ρ). More information on fuzzy graphs can be found in [1-8].

A fuzzy graph G = (V, E, s, d, ρ) is called multi-modal if each of its edges is labeled from a finite set of labels. Let $G_i = (V, E_i, s_i, d_i, \rho_i)$, i = 1, 2 be two fuzzy graphs. Define $G_1 \cup G_2$ as the fuzzy graph (V, E, s, d, ρ) such that $E = E_1 \cup E_2$, s, d, and ρ restricted to E_i are s_i, d_i and ρ_i (i = 1, 2) respectively.

Theorem 1. Let G be a k-modal fuzzy graph. Then there exist k unique uni-modal fuzzy graphs G_i, i=1, 2,..., k such that $G = G_1 \cup G_2 \cup \ldots \cup G_k$.

III. CONSOLIDATION OPERATOR

As an example consider the fragment of a fuzzy social network as shown in Figure 1. The Intelligence Analysis Agency (IAA) frequently receives many reports from its field agents on activities that might be of significance. In this example, for simplicity, we consider a maximum of two reports on each entity. The solid line indicates an older report and the dashed line indicates a newer report. There are four types of entities: event shown as a shaded square, person as a shaded triangle, location as open circle and organization as a shaded circle. Thus it was reported that Person_1 is believed to be the leader of organization Org_0 with a confidence of 0.5. Later a new report asserts the same with a confidence of 0.8. Person_2 is reported as friend of Person_1 with a confidence of 0.7 and so on.

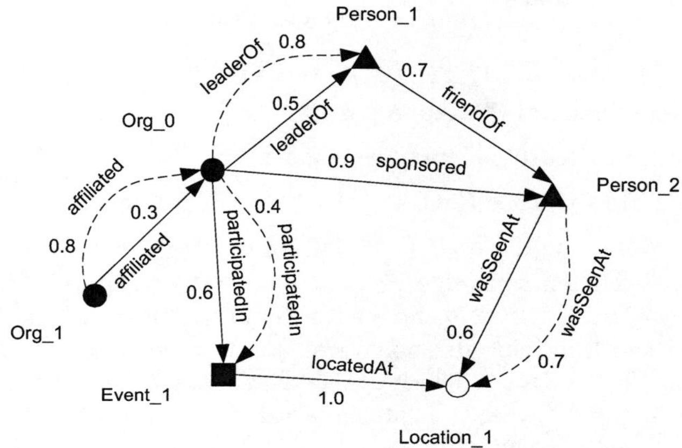

Figure 1. Partial SNA of IAA.

The Intelligence Analysis Agency needs to consolidate these reports by replacing parallel edges by a single edge. If we follow the traditional approach, we need to replace two parallel edges with one edge having maximum confidence, see Figure 2 (left). Since vertex labels are identical to one on Figure 1, for the sake of clarity, they are omitted in Figure 2.

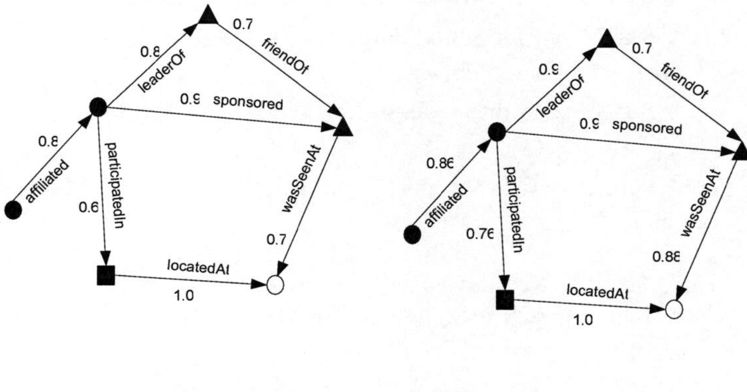

Figure 2. Modified SNA.

However from our day to day life experience, we know that this does not reflect reality. Selecting the maximum of the confidence values of parallel edges does not capture the cumulative effect of multiple reports. If α and β are the confidence values of two parallel edges the consolidation of those two parallel edges must yield a value, say $\alpha\otimes\beta$, with following properties:

1. [**Closure**]If $\alpha, \beta \in [0,1]$ then $\alpha\otimes\beta \in [0,1]$.
2. [**Commutativity**]If $\alpha,\beta\in [0,1]$ then $\alpha\otimes\beta = \beta\otimes\alpha$
3. [**Associativity**]If $\alpha,\beta,\gamma \in [0,1]$ then $(\alpha\otimes\beta)\otimes\gamma = \alpha\otimes(\beta\otimes\gamma)$.
4. [**Incremental Effect**]If $\alpha,\beta \in (0,1)$, $\alpha, \beta < \alpha\otimes\beta$.
5. [**Zero Gain**]If $\beta \in [0,1]$ then $0\otimes\beta = \beta$.
6. [**Solid Evidence**]If $\alpha,\beta \in [0,1]\alpha\otimes\beta = 1$ if and only if $\alpha = 1$ or $\beta = 1$.
7. [**Monotone**]If $\alpha,\beta,\gamma \in [0,1]$, If $\beta \leq \gamma$ then $\alpha\otimes\beta \leq \alpha\otimes\gamma$.

An operation satisfying properties 1, 2, 3, 5 and 7 is known as a t-conorm. Properties 1, 2, 3, 4 and 7 needs no justification. If both the values are between 0 and 1, the consolidation operator must produce a result that is strictly greater than both of them. If one of the values is zero, there should not be any gain and that is the reason for property 5. Finally, property 6 states that solid evidence or value 1 can be produced only if one of the values is 1. Once the value 1 is attained, no more incremental effect is possible.

Definition 2. If $\alpha, \beta \in [0,1]$, then $\otimes:[0,1]\times[0,1]\rightarrow[0,1]$ is defined by $\alpha\otimes\beta = \alpha + \beta - \alpha\beta$ and is called the consolidation operation on [0,1].

Theorem 3. The consolidation operator satisfies all seven properties.

Definition 4. Let G = (V, E, s, d, ρ) be a uni-modal fuzzy graph. Then the consolidated fuzzy graph corresponding to G, denoted by $G^{\otimes}$, is the fuzzy graph (V, E′, s, d, ρ′) such that

1. $\forall x,y \in V$, there exists at most one $e' \in E'$ such that $s(e')=x$ and $d(e')=y$.
2. $\forall e' \in E'$, $\rho(e')>0$ if and only if $C(e') =\{e \in E| \, s(e) = s(e'), d(e) = d(e'), \rho(e) > 0\}$ is non-empty.
3. $\forall e' \in E'$ such that $\rho(e')>0$,let $C(e') =\{e_i \mid i = 1,2,\ldots,n\}$. Then $\rho(e') = e_1\otimes e_2\otimes\ldots\otimes e_n$.

See Figure 2 (right) for the consolidated fuzzy graph corresponding to Figure 1.

In this section, we showed how to consolidate a uni-modal fuzzy graph. Now, let G be a k-modal fuzzy graph. Then by Theorem 1, there exist k unique uni-modal fuzzy graphs G_i, i = 1,2,…,k such that $G = G_1\cup G_2\cup \ldots\cup G_k$. Now the consolidated fuzzy graph of G, denoted by $G^{\otimes}$, is defined as $G^{\otimes} = G_1^{\otimes}\cup G_2^{\otimes}\cup \ldots\cup G_k^{\otimes}$.

IV MEASURE OF INFLUENCE

A path P in a fuzzy graph G is an alternating sequence of vertices and edges $v_0,e_1,v_1,\ldots,v_{n-1},e_n,v_n$, where $v_0,v_i \in V$, $e_i \in E$, $s(e_i) = v_{i-1}$, $d(e_i) = v_{i-1}$, $i=1,\ldots,n$ and all the vertices are distinct, except v_0 may be the same as v_n such that $\rho(e_i) > 0$, $1 \le i \le n$. Here $n \ge 1$ is called the length of the path P. The strength of P is defined as minimum of all $\rho(e_i)$, $1 \le i \le n$. In other words, the strength of a path is defined to be the weight of the weakest edge of the path. A single vertex v may also be considered as a path. In this case, the path is of length 0. If the path has length 0, it is convenient to define its strength to be $\mu(v)$.

Let v, u be two vertices of a fuzzy graph G. If P is the set of all paths from v to u, then influence of v over u, denoted by influence(v,u), is defined as maximum$\{\alpha|\alpha$ is the strength of path P from v to u$\}$. Now, the influence of v, denoted by influence(v), in the social network can be defined as $\sum_{u\in V, u \neq v}$ influence(v, u).

Consider the social network modeled as a fuzzy graph and the corresponding consolidated fuzzy graph in Figure 3. The edge labels and relationship names have been omitted for clarity of presentation.

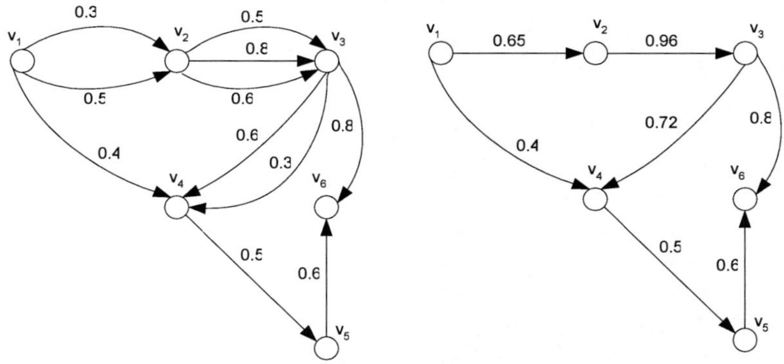

Figure 3 A fuzzy social network.

The influence of vertices in Figures can be summarized as follows:

vertex	**social network**	**consolidated social network**
v_0	2.5	3.1
v_1	2.7	2.98
v_2	1.9	2.02
v_3	1.0	1.0
v_4	0.6	0.6
v_5	0	0

It is worth noticing that if the influence of a vertex was zero, it remained zero even after the consolidation. The influence of each vertex either continues to be the same or went up in value. Note that consolidation operator may alter the ordering of vertices based on their influence. For example, influence(v_1) < influence(v_2) before applying the consolidation operator and influence(v_2) >

influence(v_1) after the consolidation operator has been applied. Thus, while it is perfectly legitimate to use maximum operator to "union" two serial edges of a graph, we need to use consolidation operator to "union" two parallel edges of a graph.

CONCLUSION

In this paper, we proposed a technique to model multi-modal social networks as fuzzy social networks. We defined a uni-modal fuzzy graph with parallel directed edges. Using union operation on fuzzy graphs, we showed that a k-modal fuzzy graph is in fact the union of k uni-modal fuzzy graphs. Then we presented a technique to consolidate the parallel edges in a uni-modal fuzzy graph using a new operator called consolidation operator. We achieved the consolidation of k-modal fuzzy graphs by using the graph union operation on the consolidated uni-modal fuzzy graphs.

REFERENCES

[1] Cerruti, U., “Graphs and fuzzy graphs”, *Fuzzy Information and Decision Processes* 123--131, North-Holland, Amsterdam-New York, 1982.

[2] Chen, Q. J., “Matrix representations of fuzzy graphs (Chinese)”, *Math. Practice Theory* **1**:41--46, 1990.

[3] Dodson, C.T.J., “A new generalization of graph theory”, *Fuzzy Sets and Systems* **6**:293--308, 1981.

[4] Dubois, D. and H. Prade, H., *Fuzzy Sets and Systems, Theory and Applications*, Academic Press, New York,1980.

[5] El-Ghoul, M., “Folding of fuzzy graphs and fuzzy spheres”, *Fuzzy Sets and Systems* **58**:355--363, 1993.

[6] Mordeson, J.N. and Nair, P.S, “Cycles and cocycles of fuzzy graphs”, *Inform. Sci.* **90**:39--49, 1996.

[7] Mordeson, J.N. and Nair, P.S, *Fuzzy Graphs and Fuzzy Hypergraphs*, Physica-Verlag, New York,2000.

[8] Rosenfeld, A., “Fuzzy graphs”, In: L. A. Zadeh, K. S. Fu, M. Shimura, Eds., *Fuzzy Sets and Their Applications*, 77--95, Academic Press, 1975.

[9] Wasserman, S., and Faust, K., *Social Network Analysis:methods and applications,* Cambridge University Press, 1994.

PART V:

ARCHITECTING AND ENGINEERING COMPLEX SYSTEMS

A NOVEL PARTICLE SWARM OPTIMIZER WITH KRIGING MODELS

HEPING LIU ALICE E. SMITH
Department of Industrial and Systems Engineering
Auburn University
Auburn, Alabama 36849-5346

ABSTRACT
In Particle Swarm Optimization (PSO), fitness functions play a very important role. However, if we only have some observation points and have no explicit fitness function of independent and dependent variables of these points, the applications of PSO will meet some obvious difficulties. In this paper, we used kriging models to address this problem. We used kriging models to construct fitness functions based on some known observation points and then according to the constructed fitness functions, PSO was adopted to perform the optimization operation. A novel algorithm combining PSO with kriging models was developed and three simulated cases were investigated. The results from these cases showed our algorithm successfully found the optimal points using only some known observation points.

1. INTRODUCTION

Particle Swarm Optimization (PSO) is a population-based stochastic optimization method, and is developed by Eberhart and Kennedy (1995). PSO uses a fitness function to calculate the fitness values of particles. According to the fitness values, PSO performs evolution operations to update their position vectors. PSO continues this loop until the stopping criteria are satisfied.

However, when we use PSO to optimize problems which only have some observation points and have no specific function relationship between the independent and dependent variables, the optimization process of PSO will encounter some essential difficulties because PSO loses the evaluation foundation for newly updated position vectors. Obviously, the evolutionary optimization process of PSO will have to be stopped.

In the research of engineering and science, we could obtain the needed data by doing experiments. Suppose we conduct an experiment to obtain the actual values of dependent variables in lieu of using an actual fitness function to evaluate a solution. The evolutionary optimization operation of PSO could continue when we continuously adopt an experiment to evaluate a solution. However, any experiment involved incurs some economic cost. The economic cost incurred by experiments may be an important constraint when we keep doing experiments to evaluate solutions.

One possible way to solve these problems is to use known observation points to fit an approximate fitness function and use this function to conduct the evaluation operation so that the optimization search process could be continued without incuring additional economic or compuational costs. The paper used kriging models to predict the function values of unknown points according to the data of known points. Kriging models are named after Krige, a mining engineer in South Africa (Krige, 1951; Journel and Huijbregts, 1981). Essentially, kriging models can be simply explained to be a best linear unbiased estimator (BLUE) of various functions of spatial processes.

The paper is organized as follows. In the second part, the methodologies of PSO and kriging modeling are introduced. Next, the paper does some case analyses and compares the results from the PSO with kriging models with those obtained by Simulated Annealing (SA), Evolutionary Strategy (ES), and PSO. Finally the paper gives some concluding remarks and discusses future work.

2. METHODOLOGY

2.1 BASIC PRINCIPLES OF PSO

In PSO, each particle has several common attributes: *xVector, pVector, vVector, xFitness* and *pFitness. xVector* records the current position of particle. If the investigated space is *D* dimensions, *xVector* can be denoted by $X_i=(x_{i1},x_{i2},\dots,x_{iD})$. *pVector* is the best position found so far by the particle and can be represented by $P_i=(p_{i1},p_{i2},\dots,p_{iD})$. *vVector* is a velocity vector, represented as $V_i=(v_{i1},v_{i2},\dots,v_{iD})$. *xFitness* and *pFitness* record the fitness values of *xVector* and *pVector*, respectively.

In PSO, the iteration process of each particle mainly includes two steps, which can be described as follows (Kennedy and Eberhart, 1995):

- **First step:** The algorithm finds particle *g* with the best *pFitness* in the neighborhood of particle *i,* and then combines the *pVector* of particle *g* and the *pVector* of current particle *i* to generate the new *vVector* for particle *i.* Meanwhile, the algorithm introduces the random disturbance *rand()* to increase the search pressure. The following is the specific formula updating the dth dimension of *vVector*:

$$v_{id} = v_{id} + \varphi_1 rand()(p_{id} - x_{id}) + \varphi_2 rand()(p_{gd} - x_{id}) \qquad (1)$$

- **Second step:** The algorithm uses the updated *vVector* to obtain a new *xVector* for current particle *i*. The related formula is as follows:

$$x_{id} = x_{id} + v_{id} \qquad (2)$$

Two constants φ_1 and φ_2 are the cognition and social components, respectively. Maurice Clerc (1999) introduced a constriction factor *K* to control the growth speed of *vVector.*

2.2 INTRODUCTION TO KRIGING MODELS

Considering a stochastic process with the form below:

$$Z(X) = \mu(X) + \varepsilon(X) \qquad (3)$$

where *X* is a position vector, $\mu(X)$ is a drift function showing the average behavior of $Z(X)$, and $\varepsilon(X)$ is an error term with $E[\varepsilon(X)]=0$. Suppose *n* observed values are $Z(X_1)$, $Z(X_2)$, ……, $Z(X_n)$. Kriging models use a linear combination of the observed values to estimate the value at an unobserved point *X*.

$$\hat{Z}(X) = \sum_{i=1}^{n} \lambda_i Z(X_i) \qquad (4)$$

where the coefficients λ_i need to satisfy two constraints: the estimator need be unbiased and the variance of the estimation errors need be minimized. Based on the unbiased condition, we get

$$E[\hat{Z}(X)] = E[\sum_{i=1}^{n} \lambda_i Z(X_i)] = \sum_{i=1}^{n} \lambda_i E[Z(X_i)] \quad (5)$$

which can be expressed as

$$\mu(X) = \sum_{i=1}^{n} \lambda_i \mu(X_i) \quad (6)$$

Suppose that the drift function $\mu(X)$ consists of M basis functions $f_l(X)$, $l=1, 2, \ldots$ ..., M, that is,

$$\mu(X) = \sum_{l=1}^{M} a_l f_l(X) \quad (7)$$

Then the above unbiased condition becomes

$$f_l(X) = \sum_{i=1}^{n} \lambda_i f_l(X_i), \qquad l = 1, 2, \ldots\ldots, M \quad (8)$$

And the variance minimization of estimation error becomes the following optimization problem with constraints:

$$\begin{aligned} \text{Minimize} \quad & Var[\hat{Z}(X) - Z(X)] = \sum_{i=1}^{n}\sum_{j=1}^{n} \lambda_i \lambda_j Cov[Z(X_i), Z(X_j)] \\ & -2\sum_{i=1}^{n} \lambda_i Cov[Z(X_i), Z(X)] + Var[Z(X)] \quad (9) \\ \text{Subject to} \quad & f_l(X) = \sum_{i=1}^{n} \lambda_i f_l(X_i) \qquad l = 1, 2, \cdots, M \end{aligned}$$

By introducing Lagrangian multipliers u_l where $l=1, 2, \ldots\ldots, M$ to solve this optimization problem, we can obtain the following "kriging system":

$$\begin{bmatrix} C_{11} & \cdots & C_{1n} & f_1(X_1) & \cdots & f_M(X_1) \\ \vdots & \ddots & \vdots & \vdots & \ddots & \vdots \\ C_{n1} & \cdots & C_{nn} & f_1(X_n) & \cdots & f_M(X_n) \\ f_1(X_1) & \cdots & f_1(X_n) & 0 & \cdots & 0 \\ \vdots & \ddots & \vdots & \vdots & \ddots & \vdots \\ f_M(X_1) & \cdots & f_M(X_n) & 0 & \cdots & 0 \end{bmatrix} \bullet \begin{bmatrix} \lambda_1 \\ \vdots \\ \lambda_n \\ u_1 \\ \vdots \\ u_M \end{bmatrix} = \begin{bmatrix} C_1 \\ \vdots \\ C_n \\ f_1(X) \\ \vdots \\ f_M(X) \end{bmatrix} \quad (10)$$

where C_{ij} represents covariance $Cov(Z(X_i), Z(X_j))$ between sample points X_i and X_j, and C_i means covariance $Cov(Z(X_i), Z(X))$.

In kriging models, the covariance functions we are using are the pure nugget effect, the triangle model, the cubic covariance, the spherical covariance and the exponential covariance (Mathern, 1973; Repley, 1981). In covariance functions, the influence ranges of covariance greatly influence the performance of kriging models. This paper uses the sample standard deviation as the metric of the influence ranges of covariance. It is calculated as follows:

$$\delta = \sqrt{\frac{1}{n-1}\sum_{i=1}^{n}\{Z(X_i) - [\frac{1}{n}\sum_{i=1}^{n} Z(X_i)]\}^2} \quad (11)$$

Choosing sample standard deviation as the metric of the influence ranges of covariance ensures that different problems have a consistent unit.

2.3 DEVELOPMENT OF THE PSO WITH KRIGING MODELS

In this part, the paper develops an algorithm of the PSO with kriging models. The specific details are defined in Fig. 1.

```
Initialize a kriging model by doing experiments to obtain a set of M observation points;
Remember the particle bestParticle with the best fitness value among M observation particles;
For each particle i {
    For each dimension d {
            Initialize x_id, p_id and v_id;}
    Calculate xFitness_i based on the predicted value of kriging model;
    Set pFitness_i to xFitness_i and set g to i if pfitness_i is better than pfitness_g;}
t=0;
While (t<(NFE-SwarmSize)/SwarmSize) {
      For each particle i {
        Find gParticle with the best fitness in the neighbors of particle i;
        For each dimension d {
                Update v_id and x_id;}
                Calculate xFitness_i based on the predicted value of the kriging model;
                Update p_id and pFitness_i if xFitness_i is better than pFitness_i}
      If (correction condition of fitness) {
         Correct xFitness_i and pFitness_i by actual experiment values;
         Add the corrected points into the set of known points and update the kriging model;
         Update bestParticle;}
      t++;}
Do an experiment to obtain the actual pFitness value of gParticle;
Set bestParticle to gParticle if pfitness of gParticle is better than pfitness of bestParticle
Report results.
```

Figure 1. The Algorithm of the PSO with Kriging Models

In the algorithm above, first we need to obtain some initial observation points to fit an initial kriging model. Before this process, the specific type and structure of kriging models could be chosen by prior knowledge that we have.

Fitting kriging models is a dynamic process. By doing experiments to correct the predicted values of kriging models, the algorithms will correspondingly update the fitted kriging models to make them more precise. However, the update uses the same structure as the fitted kriging models.

In Fig. 1, NFE means the number of function evaluations which is calculated according to the evaluation times of kriging models. In this paper, it is the focus of our concerns about how to decrease the experiment cost, that is, how to decrease the actual function evaluations. The idea behind the algorithm above is to integrate the optimization function of PSO with the prediction feature of kriging models to search for optimal points based on limited observation points.

3. COMPUTATONAL EXPERIENCES

We used three simulated cases to test the developed algorithm. The testing process was divided into two steps. First, we randomly generated 200 solutions. The first 150 solutions were used to construct different kriging models and the remaining 50 solutions were used to validate the fitted kriging models. Based on the absolute average error percentage of these 50 validation solutions, we chose the best fitted kriging model. Next, for each case we randomly generated 10 different initial solution sets by simulation. For each solution set, we used the PSO with kriging models to optimize it. In order to see the robustness of the algorithm, we randomly generated 10 different initial populations for each initial solution set and ran the algorithm for 10 times.

Case 1 is the six hump camelback function (SHC) (Branin, 1972) where x lies between ±3 and y lies between ±2. The SHC function has six local minima and two global minima: $(x_1, x_2)^* = (0.089, -0.713)$ and $(-0.089, 0.713)$, $f^* = -1.03$.

$$f(x_1,x_2)=(4-2.1x_1^2+\frac{x_1^4}{3})x_1^2+x_1x_2+(-4+4x_2^2)x_2^2 \quad (12)$$

Cases 2 and 3 are the B2 function and the Goldstein and Price (GP) function (Fan, et al, 2006). The search domain of B2 is $100 < x_j < 100, i = 1, 2$, and it has several local minima and one global minimum ($B2((0,0)^*) = 0$). The search domain of GP is $-2 <= x_j <= 2, i = 1, 2$ and it has four local minima and one global minimum ($GP((0, -1)^*) = 3$).

$$B2(x_1,x_2)=x_1^2+2x_2^2-0.3\cos(3\pi x_1)-0.4\cos(4\pi x_2)+0.7 \quad (13)$$

$$\begin{aligned}GP(x_1,x_2)=[1+(1+x_1+x_2)^2\times(19-14x_1+3x_1^2-14x_2+6x_1x_2+3x_2^2)]\times[30+\\(2x_1-3x_2)^2\times(18-32x_1+12x_1^2+48x_2-36x_1x_2+27x_2^2)]\end{aligned} \quad (14)$$

In order to see the effectiveness of our algorithm, we compared the results from our algorithm with those obtained by SA, ES and PSO. For SA, ES and PSO, we used the actual fitness functions. We want to see how many actual function evaluations are needed if we use those directly for optimization, rather than the combined actual function evaluations with kriging model. The detailed results are listed in Tab. 1. which shows that our algorithm used the least number of actual function evaluations. For three test cases, the fitted kriging models are ordinary kriging with cubic covariance and the influence distance of covariance equal to five standard deviations (for SHC), ordinary kriging with cubic covariance and the influence distance of covariance equal to one standard deviation (B2) and ordinary kriging with exponential covariance and the influence distance of covariance equal to four standard deviations (GP).

Table 1. Performance Comparison Table of Different Methods for Three Cases

Test Case	Actual Function Evaluations				Successful Times (per 10 times)			
	Krig-PSO	SA	ES	PSO	Krig-PSO	SA	ES	PSO
SHC	160	1600	610	380	8	8	9	9
B2	160	2000	1810	910	9	8	9	9
GP	460	4800	970	610	9	7	8	9

Of course, as the number of function evaluations increases, SA, ES, PSO and our algorithm can all reach the optimal solutions with a 100% success rate. Therefore, it is difficult to judge the proper comparison point in evolution. Thus, this paper chose to use the numbers of function evaluations called as the specific comparison criterion when about 8-9 times among the 10 runs for each initial solution reach an optimal solution. Using this criterion, the numbers of function evaluations considered are more meaningful. On the one hand, the success rates at this stage are acceptable; on the other hand, the exact numbers of function evaluations needed can be easily recognized.

4. CONCLUSIONS AND FUTURE WORK

The paper combined PSO with kriging models to give a novel algorithm which could optimize problems without explicit fitness functions. In comparing with PSO without kriging models and two other well known metaheuristics (SA and ES), the PSO with kriging models used far fewer actual function evaluations to achieve comparable optimization performance. Of course, the PSO with kriging has the additional computational burden of building kriging models frequently. This is a preliminary study but results indicate some promise, and further research is merited.

REFERENCES

Branin, F.H., 1972, "Widely Convergent Methods for Finding Multiple Solutions of Simultaneous Nonlinear Equations", IBM Journal of Research Developments 16, 504-522.

Clerc, M.., 1999, "The Swarm and the Queen: Towards a Deterministic and Adaptive Particle Swarm Optimization," Proceedings of the IEEE Congress on Evolutionary Computation (1999CEC), Washington, DC, 1951-1957.

Journel, A. G. and Huijbregts Ch. J., 1981, "Mining Geostatistics", Academic Press, New York.

Kennedy, J. and Eberhart, R., 1995, "Particle Swarm Optimization", Proceedings of the 1995 IEEE International Conference on Neural Networks, 1942-1948.

Kennedy, J., 1997, "The Particle Swarm: Social Adaptation of Knowledge", Proceedings of the 1997 International Conference on Evolutionary Computation, 303-308, IEEE Press.

Krige, D.G., 1951, "A Statistical Approach to Some Basic Mine Valuation Problems on the Witwatersrand", J. Chem. Metall. Min. Soc. S. Afr., 52, 119-139.

Matheron, G., 1973, "The Intrinsic Random Functions and Their Applications", Adv. in appl. prob. 5, 439-468.

Reynolds, R.G., 1994, "An Introduction to Cultural Algorithms", Proceedings of Evolutionary Program. (EP-94), 131-139, San Diego, CA.

Ripley, B., 1981, "Spatial Statistics", J. Wiley and Sons, New York.

Shu-Kai S. Fan, Yun-Chia Liang, Erwie Zahara, 2006, "A Genetic Algorithm and a Particle Swarm Optimizer Hybridized with Nelder-Mead Simplex Search", Computers & Industrial Engineering, 50, 401–425.

CONTRACTUAL AGENT-BASED WORKFLOW DESIGN PATTERNS IN WEB-ENABLED E-BUSINESS

V. K. MURTHY
School of Business Information Technology,
RMIT University, Melbourne, Victoria 3000, Australia
kris.murthy@rmit.edu.au

ABSTRACT
An agent-based contractual workflow paradigm is described for E-business. The various types of task patterns that arise in E-business (commerce, control and manufacturing, e.g. Purchasing, manufacturing and negotiation) require a *"what if"* programming approach consisting of intention and actions for trial-error design, before an actual commitment is made. Such an approach enables us to take care of the unpredictable nature of connectivity among the devices and the networks and also provide for the trial and error program design required. We also describe several types of task patterns that arise in E-business and how the contract-based workflow becomes a useful vehicle for modeling and realising web-enabled services.

INTRODUCTION

This paper describes an agent-based contractual workflow paradigm to support long and short duration transactions in commerce, control and manufacturing. environment. In this environment, the traditional transaction model needs to be replaced by a more realistic model, called a "workflow model" between the agents (peers) that interact, compete and cooperate, to realise a distributed program . The various types of task patterns that arise in commerce, control and manufacturing, e.g. Purchasing, manufacturing and negotiation require a "what if" programming approach consisting of intention and actions for trial-error design, before an actual commitment is made. Such an approach enables us to connect business partners anywhere and anytime and take care of the unpredictable nature of connectivity among the devices and the networks . It also provides for seamless integration of differing applications and communications and the trial and error program design required, de Bruijn et al [2005].

AGENT-BASED WEB-SERVICE

An agent-based software system is a collection of agents interacting through messages among themselves and the environment. Each agent maintains its own share of data and has its own program piece to manipulate it. The agents are active and behave like actors in a movie, each following its own script and interacting with other agents. Accordingly agents can model the changing world by perceiving the changes themselves at any specific time. Thus agents can adequately provide web-service for a given request in a particular scenario.

We define an agent-based web-service thus: Given an input set I= (I1,I2,I3.....) the agent -based web service execution W (I1,I2,..In) is an execution of a sequence of valid agent-based states (s0,s1,s2,...sn) that represent a meaningful state transition from s0 to achieve a desired final state sn which satisfies the context specified by the invoker, subject to the condition that all the elements of the input set I are valid objects that exist during the state transformation.

In web-service a task submitted is called a workflow. It is a distributed task that can be executed partly within that agent as an internal workflow (Inflow) and partly in other agents as external Workflow (Exflow). Usually, transactions are assumed Bacon [1993], Krishnamurthy and Murthy [1992] to satisfy the ACID properties, namely :

Atomicity : All or none of transaction happens;
Consistency: A transaction preserves the consistency in database before and after its execution;
Isolation: Intermediate results are not externally made visible until commitment;
Durability: The effects are made permanent when a transaction succeeds and recovers under failure.

The ACID properties are restrictive for agent-based web-service tasks and need to be relaxed as illustrated by the following example on flight reservation. Here one needs to have the following tasks: select a suitable airline that offers cheaper fares, ensure there is vacancy, and make an advanced booking (that is to be confirmed later). These individual steps are really not traditional transactions, but well-defined program pieces that can be carried out concurrently and may need to follow a predefined partial order to satisfy certain predicates (e.g., seat availability) invariant criteria (number of seats is determined by the size of aircraft) and synchronization (one step requires the completion of other step). Therefore, the steps need not be atomic and need not be immediately consistent, and need not satisfy isolation property since intermediate non-commit states are to be made visible(called externalization). Further such steps are liable to be cancelled eventually and require a suitable rollback preserving some local states. Also, in a cooperative agent-based environment it may be necessary to allow data exchange between some transactions during their execution thereby necessitating the relaxation of the Isolation property.

Thus in an agent-based web- service environment, we need a new model where:
(i) The isolation property is removed making the intermediate results visible.
(ii) The precedence order in execution and other dependencies are taken care by modifying the atomicity requirement.

This model is called "a workflow paradigm" between the agents. The workflow model, its contract based formulation and the E-business patterns arise from the basic work on contract model of Wachter and Reuter [1995], the design by contract by Meyer [1992] and the design patterns and contracts by Jezequel et al. [2000].

WORKFLOWS : EXTERNAL AND INTERNAL

The main thrust of this paper is to describe the workflow paradigm for applications in agent-based web-service for E-business (commerce, control and manufacturing). Unlike the transactional paradigm, which is severely restricted by the ACID properties, the workflow permits more general properties that are suitable for these applications. We illustrate some of the important applications and the design of suitable protocols. We also classify some workflow patterns ariing in E-business and the language support needed.

A global workflow (we call it an External workflow or *Exflow*) T(ij) is defined as a workflow between two agents A(i) and A(j) ; this consists of a message sent from A(i) to execute a desired workflow in A(j); this message is received by A(j). A(j) has a behavior specified by: Pre(T(ij)), G(j), C(j), Post (T(ij))S(j), where Pre() and Post() are respectively the pre and post states of the world that are active before and after the workflow T(ij). G(j) is a guard of A(j) to signal when the required precondition is met, and C(j) is the command function ; S(j) signals when the post condition is achieved. Here the script specifies what message A(j) can accept and what actions it performs when it receives the message while in state Pre(T(ij)) to satisfy the post condition post(T(ij)). The Exflow T(ij) can trigger in A(j) numeric, symbolic or database computations.

Each **Exflow** T(ij) from agent i to agent j triggers a set of serializable computations in A(j) either in a total order or in a partial order depending upon whether parallelism, concurrency and interleaving are possible locally within A(j). If the agent A(j) is "made up" of sub-agentsts, we may have to execute a workflow consisting of several local workflows (called internal workflow - **Inflow**). After executing Inflow, each agent reaches a new state from an old state using its internal command set C(j); before executing the commands, the required precondition is met, and after completion of the command set , the post condition is ensured in the new state. This is the **design by contract approach**, Meyer [1992] and widely used in the language Eiffel. The precondition is specified by "require" and post condition by "ensure", Meyer[1992.].

CONTRACTUAL WORKFLOW PARADIGM

The principal aim of the contractual paradigm is to ensure that every action is evaluated in terms of the requirements on the state of the world before the execution of the service (i.e., a given context ,e.g.user, time, location, type of data, ordering actions) so that the collaboration, communication and action framework take place infallibly. This paradigm was first suggested by Smith [1980] and later by Wachter and Reuter [1995] for long duration transactions and by Parunak [1989] in decentralized AI in manufacturing and control.

The concept of a contractual workflow paradigm is quite simple. A contract is a consistent and fault tolerant execution of an arbitrary sequence of predefined actions carried out according to an explicitly specified control flow description., called a script. The script has a condition event structure that describes a stereotyped sequence of event in a particular context . Events form a causal chain and during the execution of a workflow, a contractual obligation should take the program from one consistent state of the world to another, de Bruijn et al [2005].This means the precondition and post condition of a contract holds at every elementary contextual step ensuring a consistent and fault tolerant execution of any E-business task.

Exflow and Inflow have more general properties than the ACID properties mentioned earlier; these are given below.

CHEMICAL REACTIVITY -LIKE PROPERTIES

The Exflow and Inflow have more general properties called "chemical Reactivity properties", since they resemble chemical reactions: Molecularity, Contractual obligation, Opacity during a molecular action, and retry or rescue through a recovery protocol bringing the system back into the invariant state. These are defined as below:

(i) Molecularity : If there is a crash during a composite operation all the effects of the sub-operation are lost. If there is no crash the composite or molecular operation is complete. That is a molecule is synthesised fully or not at all.

(ii) Contractual obligation: Invocation of a single composite operation takes the program from one consistent state to another. This means precondition and post condition of a contract holds. Thus conventional consistency is replaced by contractual obligation.

(iii) Opacity: The results of the sub-operations of composite operation should not be revealed until the composite operation is complete.

(iv) Durability: If a crash occurs and contract fails and a component cannot meet its obligation and fails in its contract, an exception is raised. Then we have three possibilities:

a. Exception is not justified: it is a false alarm; we may ignore.

b. If we have anticipated the exception when we wrote the routine and provided an alternative way to fulfil the contract, then the system will try that alternative. This is called resumption.

c. If, however, we are still unable to fulfil the contract we go into graceful degradation or surrender with honour. Then bring all the objects to an acceptable state (pre-committed- state) and signal failure. This is called organized panic. This should restore the invariant. At this point we initiate retry. The effect of retry is to execute the body of the routine again.

Remark: In Eiffel Jezequel et al.[2000] Meyer[1992], the rescue clause does all the above (this is essentially RESTART after recovery in Transaction processing)

Also the chemical reactivity properties are generalized further: Since a number of remote objects are invoked, the agents should ensure all the remote actions and the local actions are complete. If any one fails the whole program has to be abandoned, and we need to retry, rescue and bring the system to its invariant state. Contractual obligation is extended to all agents under concurrent invocation and partial failures. No results are revealed from any agents until all the actions are complete and committed.

(v) Retry /Rescue and Reset :If false alarm then retry; else rescue and restart so that all the invariants in all objects are rest to their pre-action state.

Recall that we have two types of transactions - Exflow between peers, and inflow within each peer. We split the Exflow into Intention and Action transactions , Winograd [1996] where the intention transactions have the ACID properties, and the Action transactions are long duration transactions supported by a recovery protocol. The intention transactions are again local to each agent and based on the decision in this phase, the action transaction takes place through a protocol (called Intention - Action Protocol -IAP) provided with a time-out strategy and recovery to cope up with failures of disconnection. We now illustrate the applications of the intention-action protocol.

HOTEL RESERVATION PROBLEM

Consider the Hotel reservation problem. The hotel database has a set of rooms and each room has a status available or unavailable on specified dates. The object Alloc stores the confirmed allocation of rooms to named guests and Int-alloc holds the intentional reservations intention-committed. We assume that each guest has a unique ID and is not permitted to hold multiple reservations in the same hotel.

The hotel database has two integrity constraints (preconditions)

1. The number of confirmed + intentional reservations do not exceed the number of rooms on any date.
2. The set of rooms that are unavailable on any date are those already allocated to the guests in Alloc plus those in Int-Alloc on those dates.

The following agent-based transactions act on the hotel database.

1. Reservation request from guest within a deadline (Exflow)
2. Reservation availability from hotel and intention commit (Inflow)
3. Send negative intent or (positive intent + deadline) to guest to confirm (Exflow).
4. Reservation acceptance or cancellation by the guest (Extflow).
5. After cancellation by the guest deadline the intent reservations are cancelled and rooms are made available and the Int-Alloc is suitably modified and committed (Inflow)
6. After the confirmation by guest within deadline the Int -Alloc rooms are transferred to Alloc and committed (Inflow)-post condition fulfilled.(Note that transaction 2, 5, 6 are inflows with ACID properties while 1,3,4 are exflows using IA-protocol).

E-CHECKING

The Electronic check (E-check) was introduced by Financial Services Technology Consortium (FSTC) [www.fstc.org]. This system has functions, such as: deposit and clear, cash and transfer. It has entities: payee, payer, payer bank, payee bank and clearing house. E-check can be an effective system over integrated networks. E-check can be used for high value transactions, Business to business (B2B) payments etc. At every stage of this system the Intention -Action protocol and time-tolerant and time-critical transactions arise, Dani and Radhakrishna, [2000]. Using intention-action, log and timestamps we can incorporate fairness, non-repudiation and timeliness of financial transactions.

When a check is presented for withdrawal, the bank-agent verifies the preconditions: adequate balance in the account , signature and validates the check and security codes ; the post condition is acceptance or rejection of a check. Also agents maintain logs so that the participants cannot deny a particular intention, event or action.

We now describe the workflow patterns that arise in E-business.

WORKFLOW PATTERNS IN E-BUSINESS

Certain common behavioral patterns, Gamma et al., [1995],Jezequel et al.[2000], Shalloway, & Trott [2002] , Parunak [1989]) of workflows occur in Commerce and control applications. A Pattern enforces a problem solving discipline for a design architecture. It consists of three parts: a context, a problem and a solution. The context is a description of the environment before the pattern is applied. It outlines the preconditions under which the problem and its solution appear. The context shows where and when the pattern will work. It is usually introduced with the help of a scenario. Problem describes the goals and objectives of a situation. Solution describes the means to achieve the goals and objectives through a suitable protocol.

In commerce, control and manufacturing we encounter the following patterns; there can be other workflow patterns which are combinations of these patterns.

a. ***Conventional transactions with ACID properties***: These occur at the ***lowest level of a workflow***.

b. ***Supply- chain pattern***: Here the context is ***a sequence of conventional transactions between a customer and a seller through a set of intermediary agents***.

This pattern can arise in manufacturing problem too where a workstation can request different other workstations for parts and the movement of these parts are transactions. This workflow is successful, if and only if each individual transaction in a chain is successful. An abort in any transaction will be considered unsuccessful, restoring the consistency of the initial state.

c. ***Mutually exclusive- collectively exhaustive pattern***:

Here, the context is ***a single buyer trying to acquire goods from several suppliers to complete a given task through several mutually exclusive transactions*** and collectively exhaustive transactions. This pattern is also required in control and manufacturing where a total assembly of parts are required from several component parts. This workflow is considered successful only if all the mutually exclusive transactions are all committed. If any one is aborted this workflow is unsuccessful and the initial states are restored.

d. ***Negotiated choice pattern***:

Here, the context is that ***a customer bargains and negotiates with several suppliers*** simultaneously to obtain a particular product at a bargain price. In manufacturing sector, this pattern arises, when a workstation is negotiating with other workstations for parts at different levels of completion. This workflow is successful, if and only if one of the transactions is committed ;soon after that the rest of the transactions are aborted respecting the contract and the states are properly restored.

e. ***Auction Pattern***

Auction process is ***a controlled competition among a set of clients and auctioneer*** coordinated by the auctioneer. In the auction pattern, the transactions happen between an auctioneer and several clients through communication and these are successively revised. The rules of the English auction pattern are as follows:

1. The auctioneer-agent begins the process and opens the auction.
2. At every step, decided by a time stamp, only one of the client-agent is permitted to bid ; the auctioneer relays this information. The bidding client agent is called active and it does not bid more than once and this client becomes inactive until a new round begins.

3.After the auctioneer relays the information a new client becomes active and bids a value strictly greater than a finite fixed amount of the earlier bid. (This is English auction; it can be modified for other auctions) .
4.If within a time-out period no client-agent responds, the last bid is chosen for the sale of the goods and the auction is closed.

LANGUAGE SUPPORT

Eiffel, Java and UML are powerful languages to implement Mobile Object Programming Sytems. They provide for software contract that captures mutual obligations through program constructs such as : " require [else]" for precondition and "ensure [then]"for post condition , assertions, invariants. Eiffel provides for a good semantics for exception handling through a "rescue" clause and "retry" clause for dealing with the recovery and resumption .The tool called "iContract", Kramer [1998], provides developers with supports for design by contract using Java. UML (Unified Modeling Language) and OCL (Object constraint Language), Warmer and Kleppe [1999] has also been gaining importance.

CONCLUSION

We described the mobile-object based workflow paradigm for mobile E-business by splitting the workflows into external (exflow) and internal workflows (inflow). Also we described the intention-action protocol and how to implement them by contractual design using Eiffel, i -contract tool of Java and UML. We also described several kinds of behavioral patterns that arise in E-business and how the contract based workflow scheme becomes a useful vehicle for implementing these patterns through software.

REFERENCES

Bacon, J.(1993).*Concurrent systems,* Reading: Addison Wesley .
Dani,A.R and Radha Krishna, P . 2001,An E-check framework for E-payment systems in Web based environment, 91-100, in *E-commerce and web technologies, Lecture notes in Computer Science*, 2115 , New York: Springer Verlag.
de Bruijn ,J . et al. (2005), Using th web service modeling ontology to enable semantic E-business,Comm.ACM, Vol. 48, No.12, pp.43-47.
Gamma,E et al (1995). *Design Patterns*, Boston:Addison Wesley.
Jezequel, M., Train,M., and Mingins,C.(2000). *Design Patterns and contracts,* Reading: Addison Wesley.
Kramer,R (1998). "iContract-The java Design by contract tool,, *26 th Conference on Technology of object oriented Systems,(TOOLS USA'98)* Santa Barbara.
Krishnamurthy , E.V and Murthy,V.K. (1992). *Transaction Processing Systems*,Sydney: Prentice Hall.
Krishnamurthy, E.V and Murthy,V.K(2006),Distributed Agent paradigm for soft and hard computation, Journal .of Network and Computer Applications, Vol.29,pp.124-146.
Meyer, B. (1992). Applying design by contracts, *IEEEComputer* 25(10), 40-52.
Parunak,H.Van Dyke(1989), Distributed AI and manufacturing control: Some issues and Insights,pp.81-104, in *Decentralized AI,* Eds. Y.Demazeau and J.-P.Muller, North Holland, Amsterdam.
Shalloway, S and Trott,J.R. (2002).Design patterns Explained, New York: Addison Wesley.
Smith,R.G.(1980), The Contract net Protocol: High level communication and Control in a distributed Problem Solver, *IEEE Transactions on Computers, Vol.C29*, 1104-1113.
Thomas, P and Weedon, R. (1998),*Object-Oriented Programming in Eiffel* ,Reading, Mass., Addison Wesley.
Wachter,H and Reuter, A.(1995). The ConTract Model, in (Ed A.K.Elmagarmid) *Database Transaction Models* , San Mateo,California: Morgan Kaufmann, 219-264.
Warmer, J and Kleppe,A.(1999).*The Object Constraint language*,Reading, Mass., Addison Wesley.
Winograd, T.(1996), *Bringing Design to Software*. Addison Wesley. Reading, Mass..

ULTRA HIGH-SPEED MICROBRIDGE CHAOS DOMAIN

DAVOUD ARASTEH
Department of Electronic Engineering Technology
Southern University and A & M College, Baton Rouge, Louisiana, USA

ABSTRACT

We solve the tetrode microbridge dynamic system problem and its variational equations to determine the largest Lyapunov exponent. We have characterized the dynamic behavior of Josephson tetrode, which consists of five weak-link Josephson junctions for different values of one of the control parameters. We determined the region of control space where noisy oscillation at frequencies of hundreds of gigahertz occurs. Other regions show periodic oscillations. Stewart-McCumber resistive shunt model for NbN electrodes is used in the numerical analysis. We calculate the largest Lyapunov exponent through Gram-Schmitt reorthonormalization algorithm using the dynamics of electrical voltages across the junctions when one of the normal resistances is varied. Result confirms the existence of chaotic oscillation in microbridge device in specified control parameter range.

INTRODUCTION

Josephson-junction devices have nonlinear dynamical characteristics and display a wide variety of nonlinear phenomena. The dynamics of tunnel junctions can be described by resistively and capacitively shunted Josephson junction (RCSJ) model [1],[2]. A Josephson junction circuit with an inductor is an autonomous system that has a model circuit consists of RLC components with nonlinearity. Chaotic oscillations are observed in tunnel junctions with an inductor in autonomous systems and in non-autonomous systems due to the mixing of the external and internal frequencies like when an external microwave oscillation is applied to a Josephson junction at the characteristic frequency of the Josephson junction. Various nonlinear dynamics have been reported in SQUID, which consists of many Josephson junctions and inductors [3]-[7]. There have been some studies on the use of different coupling configurations in order to obtain chaotic oscillations, such as one-dimensional linear coupling and two-dimensional coupling configurations [8]-[9].

In this paper, we simulate the Josephson tetrode device model where instead of tunnel junctions; a microbridge configuration of Josephson junctions with a negligible capacitance is used [10]. The model device consists of five Josephson junctions in which a four-terminal superconductive device is made of five Josephson weak-link junctions through microbridge configurations (Fig. 1). Two junctions are connected in series and three junctions are connected in parallel with three independent variables to satisfy the necessary condition for the generation of chaos. To describe the dynamics of Josephson junctions, the resistively shunted Josephson junction RSJ model is used for our numerical calculation.

The tetrode microbridge dynamic system equations and its variational equations are solved for determining the largest Lyapunov exponent and the dynamics of electrical voltages across the junctions when one of the normal resistances is varied. The generation of chaotic, quasi-periodic and periodic oscillations has been demonstrated by Lyapunov characteristic exponent and phase space attractors portrait.

SYSTEM MODEL

Josephson tetrode equivalent circuit consists of five Josephson junctions, is shown in Fig. 1. Four superconductive devices, 1, 2, 3, and 4 are weakly connected through five Josephson junctions. Two of the junctions are series-connected (1-2 and 2-3) and the other three junctions are parallel connected (1-4, 2-4, and 3-4) in the Josephson tetrode. In microbridge configurations for the Josephson junctions the capacitance between the microbridges is negligible. We use a conventional RSJ model to describe the dynamics of the Josephson tetrode. Three independent differential equations are derived for the phase difference φ_{12} , φ_{23}, φ_{14} of wave functions. The other variances can be described as dependent variances because by assumption there is no magnetic flux in the loop of the Josephson junctions whose length is much smaller than the coherent length. Therefore, the sum of the phase differences of wave functions in one loop is zero. The equations of the phase difference in terms of normalized explicit functions in the Josephson tetrode and its coefficients are given below. $R_{n,kl}$ is the normal resistance of the Josephson junction, and all the normal resistances are normalized by $R_{n,12}$, hence $r_{kl}=R_{n,kl}/R_{n12}$. $I_{c,kl}$ is the critical current in the five Josephson junctions, and I_i is the dc drive current. The critical currents and the dc drive currents are also normalized by $I_{c,12}$. Since $I_{c,kl} R_{n,kl} = V_{c,kl}$ is a constant value for microbridge junctions, hence we derive the following relationship $i_{c,kl.}r_{kl} = 1$, $i_1=I_1/I_{c,12}$; $i_2=I_2/I_{c,12}$; $i_3=I_3/I_{c,12}$. The critical currents are defined as $i_{c,kl}=I_{c,kl}/I_{c,12}$, with values $i_{c,12}=1/r_{12}$; $i_{c,23}=1/r_{23}$; $i_{c,24}=1/r_{24}$; $i_{c34}=1/r_{34}$.

$$\frac{d\varphi_{12}}{d\tau}=\frac{1}{\eta}(\alpha_{12}+\beta_{12}\sin\varphi_{12}+\gamma_{12}\sin\varphi_{23}+\delta_{12}\sin\varphi_{14}+\sigma_{12}\sin(\varphi_{14}-\varphi_{12})+\rho_{12}\sin(\varphi_{14}-\varphi_{23}-\varphi_{12}))$$

$$\frac{d\varphi_{23}}{d\tau}=\frac{1}{\eta}(\alpha_{23}+\beta_{23}\sin\varphi_{12}+\gamma_{23}\sin\varphi_{23}+\delta_{23}\sin\varphi_{14}+\sigma_{23}\sin(\varphi_{14}-\varphi_{12})+\rho_{23}\sin(\varphi_{14}-\varphi_{23}-\varphi_{12}))$$

$$\frac{d\varphi_{14}}{d\tau}=\frac{1}{\eta}(\alpha_{14}+\beta_{14}\sin\varphi_{12}+\gamma_{14}\sin\varphi_{23}+\delta_{23}\sin\varphi_{14}+\sigma_{14}\sin(\varphi_{14}-\varphi_{12})+\rho_{14}\sin(\varphi_{14}-\varphi_{23}-\varphi_{12}))$$

Where $\eta=h\omega/4\pi eI_cR_n$, $\tau=\omega t$ and α_{kl}, β_{kl}, γ_{kl}, δ_{kl}, σ_{kl}, ρ_{kl} are coefficients defined as follows:

$\xi = \{(r_{n34}+r_{n23})(1+(1+r_{n14})/r_{n24})+r_{n14}+1\}^{-1}$

$\alpha_{12} = \xi\ \{r_{n14}+r_{n14}(r_{n34}+r_{n23})/r_{n24}\}i_1-\xi\ (r_{n34}+r_{n23})i_2-\ \xi r_{n34}i_3$

$\beta_{12} = -\xi\{r_{n14}+(r_{n34}+r_{n23})+(r_{n14}(r_{n34}+r_{n23})/r_{n24})\}$; $\gamma_{12} = \rho_{12} = \xi$

$\delta_{12} = -\xi\{1+(r_{n34}+r_{n23})/r_{n24}\}$

$\sigma_{12} = \xi(r_{n34}+r_{n23})/r_{n24}$

$\alpha_{23}=[\xi r_{n23}(1+(1+r_{n14})/r_{n24})(r_{n14}+r_{n14}(r_{n34}+r_{n23})/r_{n24})-(r_{n14}r_{n23})/r_{n24}]i_1+$
$+[r_{n23}-\xi r_{n23}(1+(1+r_{n14})/r_{n24})(r_{n34}+r_{n23})]i_2-\xi r_{n34}r_{n23}(1+(1+r_{n14})/r_{n24})i_3$

$\beta_{23} = -\xi r_{n23}(1+(1+r_{n14})/r_{n24})(r_{n14}+(r_{n34}+r_{n23})+r_{n14}(r_{n34}+r_{n23})/r_{n24})+(r_{n14}r_{n23})/r_{n24}+r_{n23}$

$\gamma_{23} = \xi r_{n23}(1+(1+r_{n14})/r_{n24})-1\}$

$\delta_{23} = -\xi r_{n23}(1+(1+r_{n14})/r_{n24})(1+(r_{n34}+r_{n23})/r_{n24})+r_{n23}/r_{n24}$

$\sigma_{23}=\xi r_{n23}(1+(1+r_{n14})/r_{n24})((r_{n34}+r_{n23})/r_{n24})-r_{n23}/r_{n24}$

$\rho_{23} = \xi r_{n23}[1+(1+r_{n14})/r_{n24}]$

$\alpha_{14} = [r_{n14}-\ \xi r_{n14}(r_{n14}+r_{n14}(r_{n34}+r_{n23})/r_{n24})]i_1+\xi r_{n14}(r_{n34}+r_{n23})i_2+\xi r_{n14}r_{n34}i_3$

$\beta_{14} = -r_{n14}+\xi r_{n14}[r_{n14}+(r_{n34}+r_{n23})+(r_{n14}(r_{n34}+r_{n23})/r_{n24})]$

$\gamma_{14} = -\xi r_{n14}$

$\delta_{14} = \xi r_{n14}[1+(r_{n34}+r_{n23})/r_{n24}]-1$

$\sigma_{14} = -\xi r_{n14}(r_{n34}+r_{n23})/r_{n24}$

$\rho_{14}= -\xi r_{n14}$

RESULTS

We have implemented the Gramm-Schmitt Reorthonormalization (GSR) algorithm [12] to calculate the Lyapunov exponent spectra when $R_{n,14}$ varies between zero to 3, with step size 0.01Ω. In three-dimensional systems like microbridge tetrode, the stretching and contraction along the principal axis in the phase space produce a spectrum of Lyapunov exponents, (λ_1, λ_2, λ_3). The rate of expansion of the initial volume under the dynamic flow is the sum of the Lyapunov exponents, and this sum will necessarily be negative for an attractor of any kind. Lyapunov exponents are ordered from the largest (most positive) to the smallest (most negative). Stable periodic attractors have only zero and negative Lyapunov exponents. In three-dimension systems the spectrum of the Lyapunov exponents is (+, 0, -), where the zero corresponds to the chaotic trajectory itself, with some trajectories expanding, while others are contracting. A necessary condition for the dynamics in the region, to be chaotic is that, $\lambda > 0$. In GSR method, the fiducial trajectory generated using a RK4 integrator applied to dynamic equations with time step $\varepsilon = 0.001$. Error and convergence analysis was carried out to the fiducial trajectory sampled over time intervals $t_s \geq \varepsilon$. As a function of t_s, the GSR method were convergent as expected, possessing a smaller coefficient for the positive Lyapunov exponent and a larger one for the negative exponent.

In Fig. 3 we have shown the bifurcation diagram of V_{12} and first Lyapunov exponent for tetrode system versus junction inductance $R_{n,14}$. The results indicate three distinct regions corresponding to periodic, quasi-periodic and chaotic behaviors. When Josephson tetrode microbridge system has quasi-periodic behaviors, the Lyapunov exponent has small positive and close to zero values. This corresponds to the intermittent chaotic trajectories (quasi-periodic) attractors. On the other hand chaotic attractors with some trajectories expanding, while others are contracting, have larger Lyapunov exponent. This is mirrored by the larger positive values for first Lyapunov exponent. We observe chaotic dynamics between the two different quasi-periodic attractors when $R_{n,14}$ is changed. The chaotic domain is located in the range of 0.4 to 1.4 for $R_{n,14}$.

The plot of three-dimensional phase portraits of a set attractors in the phase space of the three normalized voltages V_{12}, V_{23} and V_{14} at different $R_{n,14}$ are shown in Fig. 2. The Normalized voltages fluctuate chaotically in the range 0.4 to 1.4. The phase portraits are also chaotic when $R_{n,14}$ is between 0.4 and 1.4 Ω. For example, chaotic oscillations phase portraits are obtained at $R_{n,14}$=0.7, 1.25 and 1.399. As it is decreased further from 0.4, the phase portraits become quasi-periodic oscillations, and the intermittent chaotic oscillatory output is obtained. When $R_{n,14}$ is increased above 1.4 the quasi-periodic oscillations returns. In this range the shape of the quasi-periodic attractor is slightly changed, and breaks down at $R_{n,14}$ = 1.4 Ω. These are located when normalized resistance, $R_{n,14}$ takes values from 1.5 to 2.8. Therefore, chaotic oscillations are obtained at the transition between two different quasi-periodic attractors as is varied. The transition from quasi-periodic to chaotic attractors indicates that the intermittent breakdown route to chaos is occurred in the Josephson microbridge tetrode. Periodic responses are also distinguishable through a sharp drop in to negative values for first Lyapunov exponent. For example, in the region around $R_{n,14}$=3Ω behavior is completely 1-cycle periodic oscillation. Furthermore, in quasi-periodic regions, there are some tiny regions of periodic windows with negative first Lyapunov exponent, where limit-cycle oscillation occurs. The oscillation frequencies of the chaotic temporal waveforms are around hundreds of gigahertz, which could be useful for cryptosystem key generation by generating ultra fast random-like signals and chaotic modulation in secure communication systems.

CONCLUSION

The dynamics of electrical voltages across the junctions of microbridge tetrode, which consists of five weak-links Josephson junction, is studied when one of the normal resistances is varied. We confirm the generation of chaos by calculating the largest Lyapunov exponent. By changing the normal resistance, regions of quasi-periodic, chaotic and periodic attractors can be distinguished by exploring the Lyapunov exponent values and 3-D phase space portraits. The chaotic output from this device can be used in applications like ultra fast noise generators, or random secret key generator for secure communication. Behavior of this system under the microwave radiation is proposed for future work. Preliminary experiments suggest a strong chaotic behavior in a wide range of its control parameters.

REFERENCES

[1] W. C. Stewart, "Current-voltage characteristics of Josephson junctions," *Appl. Phys. Lett.*, vol. 12, pp. 277–280, Apr. 1968.

[2] D. E. McCumber, "Effect of ac impedance on dc voltage-current characteristics of superconductor weak-link junctions," *J. Appl. Phys.*, vol. 39, pp. 3113–3113, June 1968.

[3] T. Ueta and H. Kawakami, "Bifurcation of an inductively coupled Josephson junction circuit," *IEICE Trans. Fundamentals*, vol. E77-A, pp. 1758–1763, Nov. 1994.

[4] K. Fukushima, T. Yamada, and H. Shibata, "Chaotic fluxon-motion in a Josephson junction transmission line," *Physica B*, vol. 239, pp. 134–136, 1997.

[5] D. Dominguez and H. A. Cerdeira, "Spatiotemporal chaos in rf-driven Josephson junction series arrays," *Phys. Rev. B*, vol. 52, pp. 513–526, July 1995.

[6] D. Dominguez and H. A. Cerdeira, "Turbulence, chaos and noise in globally coupled Josephson junction arrays," *Phys. Lett. A*, vol. 200, pp. 43–50, Apr. 1995.

[7] K. Wiesenfeld, P. Colet, and S. H. Strogatz, "Synchronization transitions in a disordered Josephson series array," *Phys. Rev. Lett*, vol. 76, pp. 404–407, Jan. 1996.

[8] T. D. Clark, J. F. Ralph, R. J. Prance, H. Prance, J. Diggins, and R. Whiteman, "Bifurcations and chaos in a coupled superconducting-quantum-interference-device ring-resonator system," *Phys.Rev. E*, vol. 57, pp. 4035–4040, Apr. 1998.

[9] R. Rangel, A. Gimenez, and M. Octavio, "Dynamics and chaos of a current driven two-dimensional Josephson junctions array under 1/3 Phi magnetic field," *Physica A*, vol. 261, pp. 409–416, 1998.

[10] A. Uchida et al. "Chaotic Oscillations in Josephson Tetrode", *IEEE Transaction on Applied Superconductivity*, Vol. 14, No. 4, 2004

[11] A. Uchida, P. Davis, and S. Itaya, "Generation of information theoretic secure keys using a chaotic semiconductor laser," *Appl. Phys. Lett.*, vol. 83, pp. 3213–3215, Oct. 2003.

[12] *A. Wolf, J. B. Swift, H. L. Swinney, and J. A. Vastano*, "Determining Lyapunov exponents from a time series", Physica D, vol. 16, pp. 285–317, 1985.

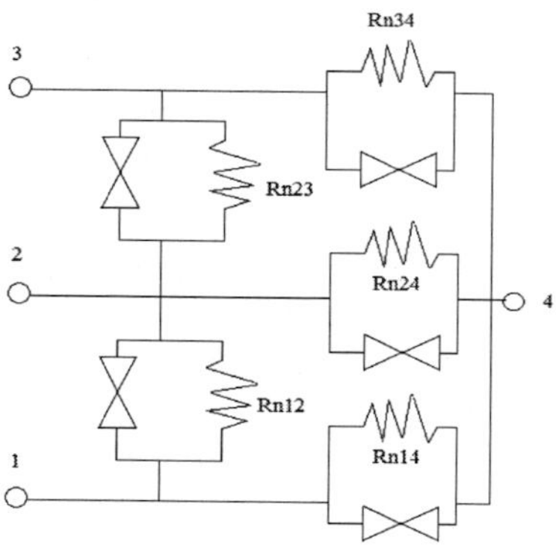

Figure 1 Equivalent Circuit of the Josephson Tetrode and Tetrode Microbridge device configuration.

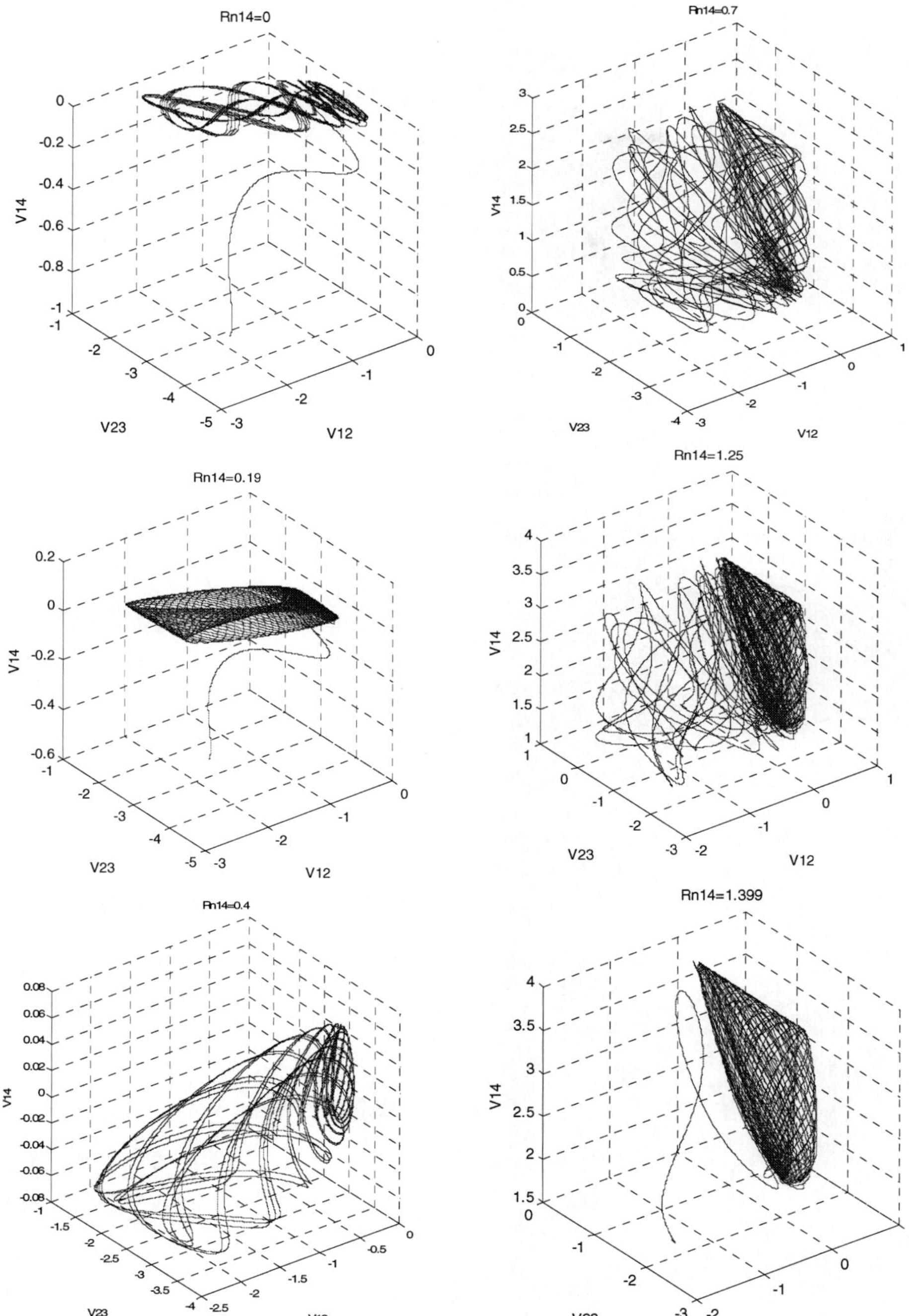
Rn14=0
V14
V23
V12
Rn14=0.7
V14
V23
V12
Rn14=0.19
V14
V23
V12
Rn14=1.25
V14
V23
V12
Rn14=0.4
V14
V23
V12
Rn14=1.399
V14
V23
V12

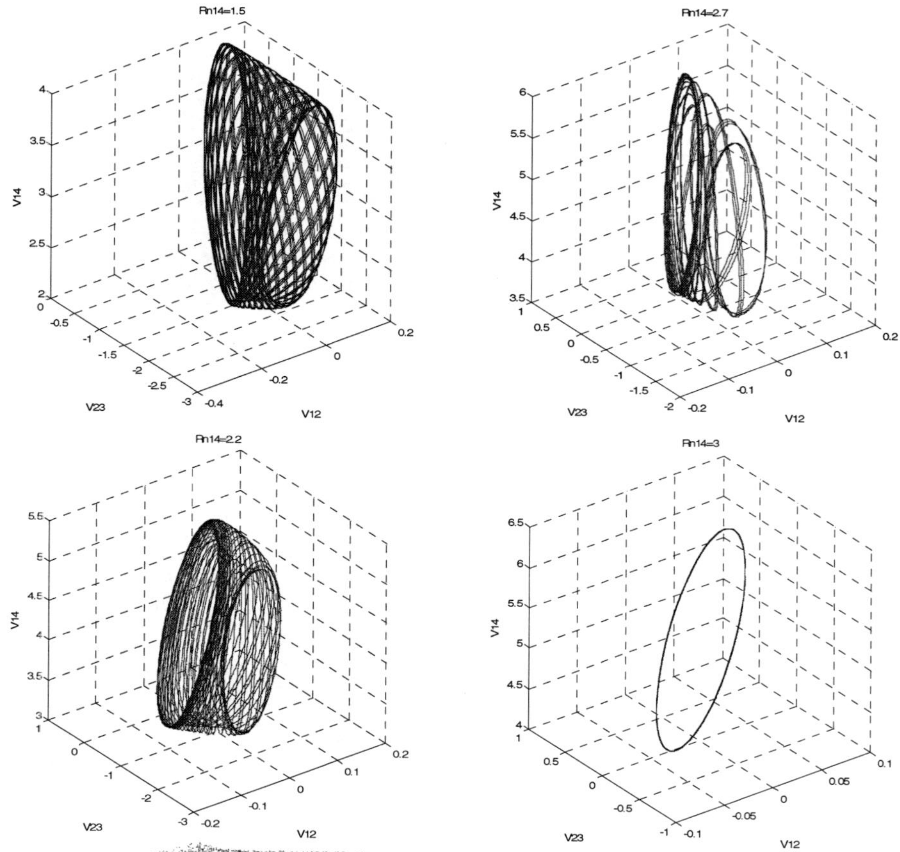

Fig. 2 Phase portraits corresponding to $R_{n,14} = 0$, $R_{n,14} = 0.19$, $R_{n,14} = 0.7$, $R_{n,14} = 1.25$, $R_{n,14} = 1.399$, $R_{n,14} = 1.5$, $R_{n,14} = 2.2$, $R_{n,14} = 2.7$, and 1-cycle periodic $R_{n,14} = 3$.

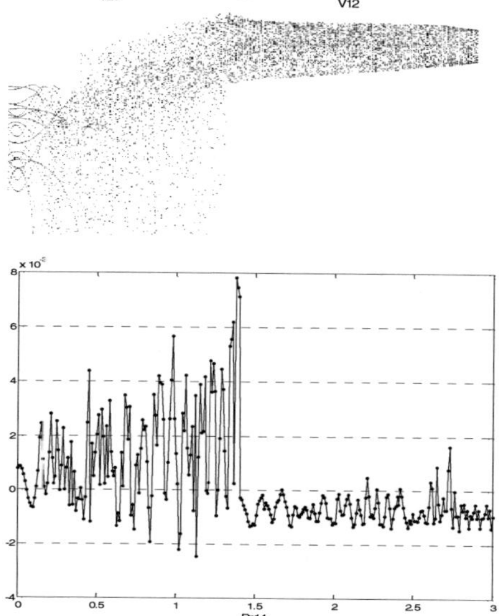

Fig. 3 Tetrode microbridge bifurcation diagram for voltage V_{12} and first Lyapunov exponent vs. $R_{n,14}$ with step size 0.01

Controlling Security API Attacks: An Artificial Neural Network Approach

Richard T. Gordon and Allison Gehrke

ABSTRACT:
This paper describes a new approach to controlling vulnerabilities in Security APIs. Since 2001, research has documented attacks that work by executing valid API commands but in unexpected sequences that the designers could not have intended. The attacks are often not classified as flaws by the API vendors since the command set also serves desired functionality. Consequently, vendor response places the burden on end-users to rigorously enforce security policy that is not well understood and not implemented in practice. Fixing the APIs themselves is not a trivial task since they are installed in millions of computers and a fix in one module can introduce additional un-foreseen logic flaws. APIs are, by design, supposed to support a broad set of commands that can be issued in any order. The API Security research field is predominantly focused on automated identification of logic flaws in algorithms and describing vulnerabilities. We propose a different approach independent of the inherent complexities of API design and security. We describe a way to control the threats in Security APIs by front-ending them with a neural network trained to detect attack command sequences. The neural network implements a simple policy of allowing valid sequences to proceed, tagging questionable sequences for proper analysis, and denying sequences that break security.

INTRODUCTION

This paper will analyze and report on attacks against security application programming interfaces (security APIs) with the intent to propose a system designed to defend and control such attacks. A security API attack is a breach of a target API's security goals with respect to the API boundary [2]. These attacks are characterized by legal sequences of calls to the API that are chained together in an unexpected way to cause a security failure. This paper focuses on defending "pure" API attacks that result from a logical flaw in the API's design. Security APIs are used in Automated Teller Machines (ATMs) and Point of Sale (POS) devices, to name a few. Their application in these devices is considered the "killer application" of modern cryptography [1].

Hardware security modules (HSM) are an instantiation of a security API. The IBM 4758 Common Cryptographic Architecture (CCA) API is a key-management API used with HSMs. The CCA provides such services as: commands to generate, process, and verify PINs; securing the master key (MK) in the register;

key-wrapping (encrypting a key under the device's MK) for storing keys externally on host computers; and key-typing to determine the subset of operations from the API a given key has access to [2]. The 4758 is used as the basis of attack analysis and as the security API model in this paper but many of the principles extend to other comparable HSMs such as the Visa Security Module (VSM).

A properly defined security API provides two high level services: a set of function calls and a security policy [2]. The security policy is made of security invariants meant to be guaranteed by the API; Security API vulnerability is defined as a sequence of legal function calls (and possibly offline computation) which leads to the violation of the security policy [2].

In 2000, Anderson gave a talk where he described an HSM failure based on a single bad API command [3]. He pondered a question that led to the discovery of many vulnerabilities and how we understand Security API attacks today: "How can you be sure that there isn't some chain of 17 transactions which will leak a clear key?" Anderson looked into the API of the VSM and discovered just such an attack.

We trained the neural network against five published attacks in total (described below) and tested its ability to generalize to unseen attacks by including an additional four attacks in the just the test set. The network achieves at least 97% accuracy even when presented with untrained attack sequences in the test set (98% without untrained sequences in test).

API Attack Sequences

All of the attacks described in this section are good examples of a class of attacks that can be mitigated using a neural network.

1. XOR to Null Key Attack: described in [1, 6]. A common method for importing top-level keys in many HSMs was from clear components using secret-sharing (i.e., n-key parts required n-people to collude to obtain the full key). In attack operations, the same key is submitted twice to the API command which was designed to combine two different keys. The keys are xor'ed together to create a key of all zeros which is a known key in the system that is then exploitable.

The XOR to Null Key attack affected the 4758, the VSM, and compatible devices. HSM vendors addressed this vulnerability by making the combine key parts command privileged. This approach is ineffectual given the known focus of hackers to first obtain necessary access permissions in order to exploit a system's vulnerabilities.

2. Poor Type System Attack, described in [2, 5]: Terminal Communications (TC) keys are allowed to be encrypted with a Terminal Master Key (TMK) for transfer to an ATM, but TMKs and PIN keys share the same type in the VSM so the TC key can be encrypted under the PIN key in the same way leading to known PINs for any target account.

3. Check Value Attack, described in [1, 6]: this is a flaw in

the implementation of check value using a command to test keys. There are no restrictions on the type of key supplied and no control vector checking is done internally at all. The left half of a 3DES key can be supplied twice (how the attack is modeled in the network), and the check value retrieved as a test pattern [6]. This can then be input to the meet-in-the-middle attack to establish a known DES key in the system.

4. Decimalization Table Attack, described in [1, 6]: exploits IBM's method for calculating PINs from primary account numbers. The first stage of the attack requires determining the digits of the PIN. The PIN verification command returns a "yes/no" depending on if the correct trial PIN is supplied. The decimalization table is parameterized. So, send the PIN verification command a trial PIN and a chosen decimalization table with a 1 set for a given digit. Try this for each digit. Any matches confirm the PIN doesn't contain that digit.

5. Key conjuring, described in [5, 6]: part of an attempt to harvest data that are indications of attack. Generate a large number of unknown DES keys using Key_Part_Import command, or generate related keys with known differences. Multiple generations of DES keys of the same type is modeled as the attack for this stage.

The following attacks (#6-#9) were only included in the test set for the neural network to evaluate its ability to generalize.

6. Encrypt multiple test patterns for keys of the same type to create a test vector, described in [5]. This is modeled as multiple calls to the key_test command with keys of the same type. #5 above and this attack (#6) set the stage for a meet-in-the-middle attack. The security API literature promote preventing test vector generation and imposing restrictions on key generation for secure API design heuristics [5], so we modeled each as attacks.

7. Import/Export Loop Attack, described in [5, 6]: import and export keys without transmission just to do a type conversion on the keys for unauthorized use.

8. Code Book Attack, described in [1]: Build a 'code book' containing every PIN and the result of that PIN encrypted under a given key. It should not be possible for the attacker to build up such a codebook (repeated call to PIN encrypt is the modeled attack).

9. PIN Format Attack, described in [1, 6]: Manipulate parameters to functions like the PIN block format. For example, submit an ANSI X9.8 formatted PIN to the reformat function but specify it as a PIN block of another format (e.g. a VISA-3 format) [1]. The translation exposes the PIN through trivial analysis (following the formatting rules).

Controlling Security API Attacks

Automating formal methods for analyzing APIs for logic flaws is an active area of research [2, 8]. Formal methods have a long way to go before they can "guarantee" security; independent evaluation can protect against these attacks, but it must cover the whole system [6]. Designing effective security APIs is difficult due to the nature of APIs and the difficulty in securing systems in general; that is, it's very difficult to plan for the infinite number of valid but unintended ways the system can be used.

The potential attack paths in an average API are quite numerous. Youn et al [2] defined the depth of an attack as the number of consecutive API and offline function calls required. Search through all attacks up to a depth of n takes time and space slightly worse than exponential [2]. Therefore, exhaustive analysis even with automated methods is currently intractable.

Even if one could prove a given API is secure, the fix would require an upgrade to installations in millions of computers and would likely not support backward-compatibility. Two business reasons make such an undertaking highly unlikely in practice: backward-compatibility is a driving concern for most HSM vendors (as demanded by their clients) and attacks may not be classified as flaws by the API vendors since the command set also serves desired functionality. The IBM practice of "Pre-exclusive-OR technique" is one example of the latter which allows the type-casting that the "Type System Attack" described in [1, 6] depends on. IBM considers this technique a valuable feature that facilitates key interchange between CCA and non-CCA cryptographic systems. IBM recognizes its potential for abuse and offers restricted access as a solution. IBM claims the system was designed so these functions can be restricted to authorized systems personnel [9]. The language implies this is optional and not enforced by the security API itself. Consequently, the responsibility of attack mitigation falls on end-users to enforce tight procedural controls, with no detail in IBM's user guide as to what these controls should be [1].

Research today is focused on identifying logic flaws in APIs and on identifying threats. No research has been done that takes a practical approach to mitigating API security threats independent of the complexities of API design or possible attack scenarios. The next section describes a proof of concept on how API threats might be mitigated with little cost and without changing security APIs or their infrastructure.

Artificial Neural Network Approach

Here we propose a different approach to controlling vulnerabilities present in security APIs. The idea is to front-end security APIs with a neural network trained to detect invalid command sequences. The neural network implements a simple policy of allowing valid sequences to proceed, tagging questionable sequences for proper analysis, and denying sequences that compromise security.

We utilized a back-propagation neural network architecture consisting of 570 input nodes (95 characters encoded with 6 bits each), 48 hidden nodes and 12 output nodes (logically two nodes since each output character is represented by six bits). An activation of .8 or less on each node designates an inconclusive validity confirmation and the call would be dispatched for further examination and validation. The characters represent the sequence of API commands and the inputs each command requires. For example, the first ten characters in XOR to Null sequence is the name of the first command used in the procedure followed by input parameters. The remaining characters are the name of the second command and its associated input parameters. For the XOR to Null attack sequence the command was passed the same key-part twice; in valid sequences, different key-parts are used.

Training Data

This section describes how we handled the representation of the data to accurately train the network to recognize attack and valid sequences with generalized and robust training data. All training and test sets were automatically generated programmatically. We ultimately used a dataset of 2,000 training examples with randomly generated attack and normal operation sequences from each attack vector to train the network. Training does not take long (less than one minute) and the use of more training data for the inclusion of additional attack sequences is a reasonable approach for future enhancements. The neural network achieves 98% accuracy or better (0-2% were not classified and would be subject to an additional layer of analysis) against the five attack vectors it is trained for. The network in all cases is tested with multiple iterations of independent 100 example test sets.

The process to generate training and test data underwent several iterative improvements as we trained the network. We originally noticed a pattern bias in experimental training data that was affecting network performance. The network must be able to recognize arbitrary key names (that programmers may use to reference their keys) in several ways to recognize attack sequences. Careful attention to the random generation of key names ensured the network learned the attack pattern over any arbitrary key naming convention.

The vast majority of erroneous predictions throughout our tests are not classified due to activation ambiguity. The network never lets an attack sequence through. Increasing the training data to 2,000 examples gave the network more information to clearly recognize valid sequences. However, it still has trouble with valid sequences where the key names are different but very similar to each other (only one character difference, for example). This is characteristic of not enough data to learn the arbitrary pattern of any two different keys. We modified the training data to randomly generate sequences with more examples of similar key names and achieve 98% accuracy on average. This demonstrates that with enough data, the network can be trained to recognize very general patterns with high accuracy.

We also used test sets that included 10% of untrained attacks to test the network's ability to generalize. The network maintained 97% accuracy and the errors were always from the untrained set with ambiguous classifications; the network never allows an attack to proceed as a valid sequence. We propose a second layer to route the ambiguous call sequences to a second network that has been specifically trained for the characteristics inherent in the sequences with inconclusive activation levels.

CONCLUSIONS

We can confidently conclude that the network was trained to recognize the intended patterns and can be used to mitigate API attacks. We are continuing with research to generalize to additional known and unknown attack sequences as defined by the API (attacks can be data entry typo mistakes, or input that varies in subtle ways from the expected input). This test demonstrates that it is possible to use this approach to design an adaptive security model for security APIs. The network is trained to learn generalized valid sequences across the API command set, and is able to recognize as yet undiscovered attacks and does not require previous knowledge of attack details to mitigate new threats.

Based on the promising results achieved so far, we plan to continue this research and focus on each section of the 4758 command set. For example, we will generate training and test data for commands in the following services provided by the security API: DES Key Processing and Key Storage, Data Confidentiality and Data Integrity, Coprocessor Control, RSA Key Administration and Key Storage, and Financial Services Support [10].

REFERENCES

1. Clulow, J. S.: 2003, 'The Design and Analysis of Cryptographic Application Programming Interfaces for Devices'. Master's thesis, University of Natal, Durban.
2. Youn, P., Adida, B., Bond, M., Clulow, J., Herzog, J., Lin, A., Rivest, R., Anderson, R., (2005) "Robbing the Bank with A Theorem Prover," *Technical Report, Number 644, UCAM-CL-TR-644, ISSN 1476-2986*, University of Cambridge.
3. Anderson, R. (1994) "Why Cryptosystems Fail" in *Communications of the ACM* vol 37 no 11 (November 1994) pp 32-40; earlier version at http://www.cl.cam.ac.uk/users/rja14/wcf.html
4. Bond, M. (2000) "A Chosen Key Difference Attack on Control Vectors," Retrieved from http://www.cl.cam.ac.uk/~mkb23/research.html on 4/20/2006.
5. Bond, M.: (2001), "Attacks on Cryptoprocessor Transaction Sets". *In: CHES*. Berlin, pp. 220-234.
6. Bond, M.: 2004, 'Understanding Security APIs'. Ph.D. thesis, University of Cambridge.
7. Bond, M. and R. Anderson: 2001, 'API-Level Attacks on Embedded Systems'. IEEE Computer Magazine 34(10), 67-75.
8. Ganaphathy, V., Seshia, S., Jha, S., Reps, T., Bryant, R., (2004) "Automatic Discovery of API-Level Vulnerabilities," UW-Madison Computer Sciences Technical Report: UW-CS-TR-1512, July 2004.
9. "IBM Comment on 'A Chosen Key Difference Attack on Control Vectors'" (Jan 2001) Retrieved from: http://www.cl.cam.ac.uk/~mkb23/research.html on 4/15/2006
10. IBM 4758 PCI Cryptographic Coprocessor, CCA Basic Services Reference And Guide, *Release 1.31 for the IBM 4758-001*

Modular Architecting for Effects Based Operations

Emel Meteoglu
University of Missouri-Rolla
223 Eng. Mngmt. & Sys. Engineering
1870 Miner Circle
Rolla, MO 65409-037
emtyf@umr.edu

Cihan H. Dagli
University of Missouri-Rolla
223 Eng. Mngmt. & Sys. Engineering
1870 Miner Circle
Rolla, MO 65409-037
dagli@umr.edu

ABSTRACT

Effects Based Operations (EBO) is a way of thinking for planning, executing, and assessing any operations for the effects they produce, rather than dealing with actions, targets or even objectives. The literature on EBO has been growing day by day; however, there is still a need for modeling techniques and tools that provide more efficient and effective effects based assessment, planning and analysis in order to further develop the capabilities of the operations. In this context, this paper presents an introduction to EBO by focusing on its methodology and its challenges. Moreover, this paper illustrates the importance of modular architecting in effects based planning stage. Modular architecting provides synchronization of the right actions and decisions, makes strategic aim consideration easier and provides efficiency in the cases where there are multiple strategic aims. The most important benefit of this study is its ability to facilitate the achievement of economy of national power for military EBO and economy of action sources for other systems. Approaches presented in this paper utilize clustering of effects and actions by using a Neural Network tool which is called Adaptive Resonance Theory. The application of the approach is illustrated with an example of the defense industry on the development of a modular EBO architecture, which can also be applied to other industries.

INTRODUCTION

The literature on effects based operations (EBO) has been growing each day. There is no agreed definition of EBO; however, it can be described as a way of thinking for planning, executing and assessing military operations for the effects they produce; rather than dealing with actions, targets or even objectives (Smith, 2002). The most critical issue in EBO is to understand the adversary's system of systems and design architecture for the execution of EBO in order to reach victory. However, cause and effect chains are complicated and are often beyond the capacity of an expert to comprehend and provide practical advice. There is an ongoing argument about past wars and there is no completely objective method for defining the most suitable actions to create the same effect or even better effects. Finally, there is also the risk of over reliance on experts' opinions such as a field commander and his staff.

In this paper, the aim is to overcome these limitations and make an improvement for EBO methodology. The study considers adversary systems as a system of systems (SoS) which includes political, military, economic, social, informational and infrastructural (PMESII) systems of adversary. The definition of the EBO and its methodologies are given in the fooling section. Also, the challenges for the EBO are explained in order to point the proposed solution for one of those challenges. The second section of the paper illustrates the importance of modular architecting as a modeling methodology for the study. Modular architecting was chosen as the modeling methodology because it provides the EBO synchronization of the right actions and decisions; makes strategic aim consideration easier, and provides efficiency in the cases where there are multiple strategic aims. A major benefit of modularity is creating modules which provide the ability to achieve multiple effects on different systems. This means that modules provide the ability to achieve strategic aim variability or effects variability through the combination and standardization of actions. Therefore, the most important advantage of this research is the ability to facilitate the achievement of economy of action resources which are national power (diplomatic, informational, military and economic (DIME)) for military EBO. In this paper, new approaches are presented that allow effects based planners to develop a modular architecture that can be shared across different systems and different strategic aims. The model of a hypothetical example of EBO for a terrorist country based on literature review is also developed. One of the neural network architectures; namely, Adaptive Resonance Theory I, is used in order to cluster effect-action matrix.

1. EFFECTS-BASED OPERATIONS

Effects based operations are operations to produce influences on the long-term or short-term state of a system in order to attain desired effects-with different degree of probability utilizing the integrated application of all applicable instruments of hard/soft power (Davis, 2002). These desired effects are concurrently planned, executed, assessed within a complex adaptive system view.

Complexity is one of the main problems in traditional warfare. Lack of understanding about effects based operations (EBO) from a system of systems perspective could be attributed as one of the reasons for the perception of complexity in any military operation. Complexity in the EBO can be grouped around three main challenges (Smith, 2002).

The first complexity is determining and estimating direct and indirect effects of planned actions. The second one is determining actual conclusions of executed actions. Finally, the last challenge which is the reason and motivation of this study is synchronization of the right actions to generate desired effects. Right actions are chosen after course of actions (COA) development and analysis step in the EBP. The tools to choose right actions are presented by some researchers (Davis, 2001 and Cohen et al., 2005). However, the aim of this paper is to orchestrate these selected right actions with modular architecting modeling methodology. The most beneficial contribution of this paper is in the process between action selection and converting actions to concept of operations. In the following section, the methodology used to overcome this challenge is described and in the methodology development is conducted.

2. MODULAR ARCHITECTING

In section 1, the steps EBO and the challenges for EBO process are described. In this section modular architecting will be explained as an alternative method to overcome these challenges. Modular architecting can be applied to entire systems engineering field. It can be used in order to overcome system design complexities and to create variety in design of systems. Modular architecting is achieved throughout the systems engineering process. The importance of system modeling in the development of modular vehicle design is demonstrated in Rushton and Zakarian's study (2000). Their approach creates modular systems and optimized integration and interactions of the system elements. For the effects based operations (EBO), their approach which composes a requirement analysis, integration analysis, and design analysis is adapted in this research with a few changes. The modeling methodology for this paper is shown in Figure 1.

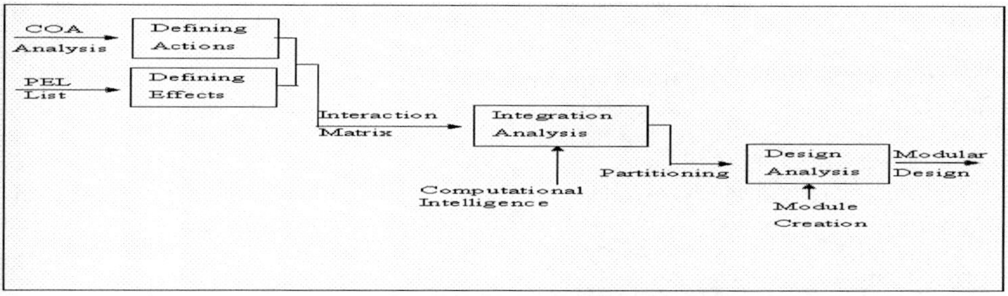

Figure 1. Modular Design Methodology for Synchronization of the Right Actions

2.1. Defining Actions

In this study, actions associated with hypothetic example of EBO for a terrorist adversary system are stated in the below. It is assumed that actions are already chosen to create desired effects as a result of COA analysis and development phase. Actions are grouped according to their source of national power. Since defining each variable for each action is beyond of this study, actions have limited scalable attributes.

Diplomatic Actions:

1. Mild contribution of NATO
2. Long-term over-flight authority from many countries
3. Receiving grants for landing rights from majority of coalition
4. Having granted bed-down and basing authority from some part of coalition
5. Agreement with some part of coalition on hosting forces on their land
6. Employing hundreds of officers for investigations
7. Mild supply of new investigative tools to law enforcement and national security agencies
8. Severe strengthening legislation to combat terrorism
9. Anti-terrorism acts among many countries
10. Aeronautics act among some countries in order to maximize effectiveness of aviation security system
11. Severe cooperation between the police and the justice departments of the coalition countries
12. Creation of an adequate infrastructure for supply and exchange info among some countries
13. Severe border monitoring

Informational Actions:

14. Frequent radio broadcasting
15. Gaining deep knowledge about enemy people, culture, religion and language
16. Training several number of officers about enemy system in psychological operations
17. Dropping thousands of leaflets to AOR
18. Knocking out huge percentage of enemy's radio capability
19. High level usage of graphics on leaflets
20. Applying higher level disinformation rarely
21. Making propagandas daily
22. Aggressive advertising and PR campaign

23. Releasing a few number of video cassettes about enemy leader

Military Actions:

24. Providing a few number of C-130s for strategic and tactical airlift
25. Designating thousands of troops which includes naval, air and land personnel to AOR
26. Employing plenty of aircraft for long flight hours for hundreds of missions with airbus and helicopters
27. Mild increase in number of armored reconnaissance vehicles to detect nuclear, biological and chemical contamination
28. Mild increase in flight hours of carrier battle group to support combat operations
29. Providing a few number of frigates to support maritime surveillance
30. Deploy a few number of KC-135 to provide aerial refueling and re-supply
31. Mild increase in number of mine clearing vehicles in the AOR
32. Mild increase in number if engineering companies in the AOR
33. Drastic increase in number of missile salvos and flown sorties to support aircraft strike
34. Moderate increase in number of munitions storage facilities
35. Designating severe number of C-160 and C-130 aircraft for HA
36. Providing moderate number of F-16 in an air to ground role

Economic Actions:

37. Freeze hundreds of terrorist bank accounts
38. Severe attitude toward disrupting fund raising and recruitment
39. Moderate intend to create a unit for the surveillance of suspicious financial flows
40. Severe protection for money laundering with using electronic data processing systems
41. Mild intention to create a centre for overseeing financial transactions and payments
42. Immediate action to issue order in order to freeze enemy assets
43. Moderate changes in regulation of trade practices
44. Providing severe money aid to AOR
45. Providing severe food aid to AOR
46. Providing basic utilities

2.2. Defining Effects

In this section, the question of "to do what" is answered. Effects can be created on each one or more than one of PMSEII. Therefore; it is difficult to grouping effects according to the systems that they occur. Increasing computer power of information age provides them to be able to predict reactions caused by any action. In this study, the effects are defined according to literature review of Davis (2001). He stated in his research 9 main effects wanted to be created. We decomposed these 9 effects and for simplicity grouped them according to in which system that first they occur.

Strategic Aim: Peace and stability in the region

End States:

Populace obeys civil authority – **Political Systems**

Stop fighting; prevent future attacks and terrorists leave the country – **Military Systems**

Economic welfare in the region – **Economic Systems**

Increased life standards for the populace – **Social Systems**

Independent media operations – **Informational Systems**

Finished reconstruction of the region – **Infrastructure Systems**

Desired Effects:

Political Systems:

1. Support of neighboring countries for the central government
2. Reconciliation of central government with insurgents
3. Help of central government to bring terrorists to justice
4. Increase in number of surrendered terrorists
5. Elimination of international political support (security issues) for terrorist system
6. Increase in number of new parties

7. Local governmental control

Military Systems:

8. Dismantling the organization
9. Taking out leadership
10. Elimination of capabilities of armed forces
11. Destroying terrorist training camps
12. Confusing/diverting enemy commanders
13. Destroying safe havens and sanctuary
14. Identifying, locating, eliminating man power
15. Destroying production, storage, maintenance and distribution centers
16. Weaken internal security in adversary's military systems

Economic Systems:

17. Decrease in production which helps terrorist systems
18. Decrease in demand for exports
19. Increase in demand for imports
20. Elimination foreign aids
21. Elimination foreign investments
22. Minimization commercial interaction during war
23. Increase in consumption in domestic markets
24. After war, expanding cross-border commerce to pre-war level

Social Systems:

25. Resistance of population to insurgent influence
26. Increase in good image of our system
27. Creation of social disapproval for terrorism
28. Protection of religion, culture, history of the nation
29. Increase in social welfare after war
30. Improve in education and health care services
31. Voting of populace for against adversary's political parties
32. Start in population to form new political parties
33. Decrease in trust to enemy commander

Informational Systems:

34. Control of lines of command and communication by coalition forces
35. Media control in region
36. Decrease in support to terrorism via media
37. Limit in access to world wide web
38. Increase in enlightenment of populace

Infrastructure Systems:

39. Limited or no access to utilities during war
40. Controlling or destroying transportation channels during war
41. Controlling heavy and light manufacturing
42. Minimization physical destroy in public facilities
43. Exceed in utility output pre-war levels
44. Prevention international transportation
45. Reducing hazards for civilians
46. Increase in reconstruction after war

2.3. Integration Analysis

Effects and actions are decomposed into PMESII systems and DIME actions according to previous sections. It is impossible to decompose all possible actions and effects but the aim of this study is beyond of defining actions and effects. The relationships between actions and effects which are based on logical and historical examples are attained and this study assumes that the relationships may be assigned differently. In real EBO study, effects action matrix should be created after COAs are decided and priority effects list is prepared. Our hypothetic example and relationships of effects and actions can be seen in Figure 2. In the next step ART1 is used to complete integration analysis phase.

2.3.1. ART1. Once the interactions among the actions and systems are identified, an effect-action matrix of the interfaces is developed. An effects-action matrix $[a_{ij}]$ includes "1" , "blank" entries where an entry "1" indicates the information, material or energy link between effects i and actions j, and the direction of the link is from $j \rightarrow i$.

During the clustering of the effect-action matrix the only inputs provided to the network are the vectors representing the relationship between the effects and actions. The neural network must be self-organizing and perform in unsupervised manner. In the literature, the ART1 paradigm has not been applied to the EBO. In the following sections, the result of that ART1 is applied successfully to our hypothetic EBO example is shown.

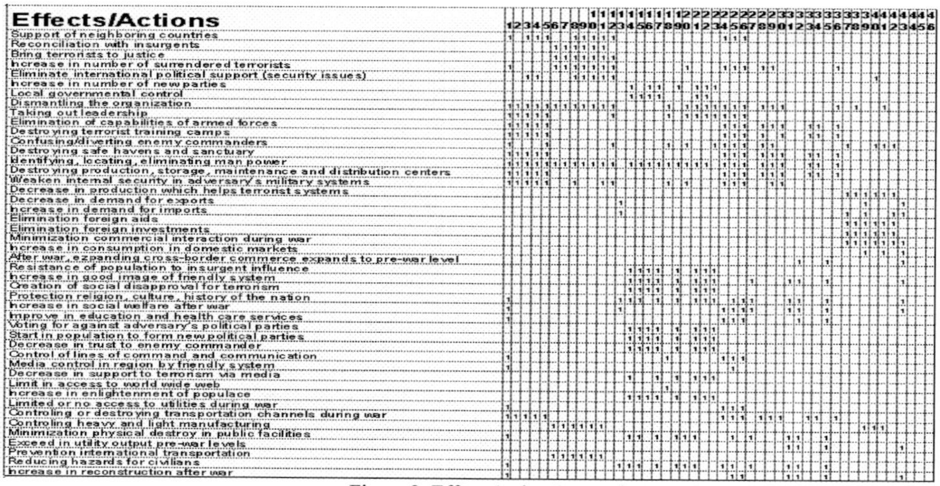

Figure 2. Effect-Action Interaction Matrix

2.3.2. Drawbacks and Improvements of ART1. Improvement steps for the ART1 algorithm should be developed to maximize effect-action integration and minimize the interaction between the action modules. The ART1 paradigm as described in the previous section may not be an effective technique for clustering the effect-action matrix, if it is applied in its basic form. It has a few drawbacks which need to be improved. In our example our matrix 46x46 is enough to show how ART1 works in the EBO. If the more input vectors are applied to the network, the stored pattern grows sparser. In order to minimize this drawback of ART1, the vigilance parameter of the network can be adjusted for the different runs of the program to obtain set of different solutions. However, the optimization still becomes difficult as the number of the input pattern increases. Another problem is being dependent on the order in which the input vectors are applied. It is also obvious that as the number of input vectors increases, the representation grow sparse and vectors with large amounts of 1's will not be classified into existing groups and will create a new category. It means that if higher vigilance parameter is used, the number of groups (number of modules) will increase (Enke et al., 2000).

2.4. Design Analysis

In this phase, the modular design is completed. ART1 architecture model is developed to allocate the actions and effects to the modules identified in the modularity matrix of the example system of systems. The application of the ART1 without improvement steps is shown in Figure 3.

Same process may be used for different case studies and for different strategic aims; hence, common modules that support various platforms can be used. The modules provide both variation and standardization. Standardization solves one of the main problems of EBO complexity which is synchronization. The modules also show that if the actions within a module applied together; it increase the possibility of reaching desired effects in that modules. As a result, modular architecting provides reduced time in the synchronization of the actions and decisions, makes strategic aim consideration easier, and also provides efficiency in the cases where there are multiple strategic aims; therefore, it facilitates the achievement of economy of national power (diplomatic, informational, military and economic). It is obvious that the modules have started to appear before improvement steps. However, the results of ART1 architecture with different parameters is not enough to complete design analysis phase. In further of this phase, modules should have their last form and architecture flow between modules and interconnect diagrams should be drawn. Those figures show the flow and transformation of national power (DIME) resulted in modules for our example.

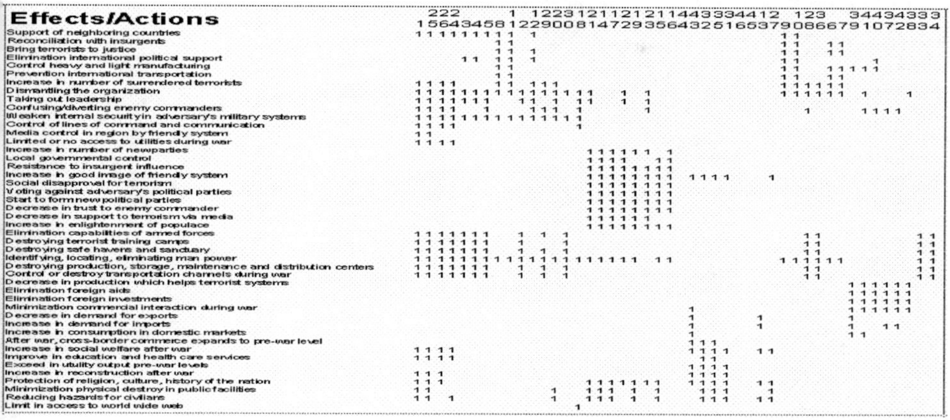

Figure 3. Modularity Matrix (Vigilance=0.2)

CONCLUSION AND FUTURE RESEARCH

An important motivation for this study was the belief that inadequate literature resource for the EBO and analysis methods to solve the EBO complexity problems need to be improved so that they can be useful in studies of computational solutions for the EBO. Therefore, in this paper, general descriptions about the EBO and methodology of it are presented that allow systems engineers to obtain in-depth understanding and to develop more exploratory analysis. A modular architecting model which composes of defining actions, defining effects, integration analysis and design analysis is developed. Then, ART1 architecture has been selected to conduct integration analysis phase. This modular architecture approach may provide economies of national power, reduced time in synchronization of actions, easier multiple strategic aim considerations.

This paper is an initial step for the aim of this research. Our next study is applying another neural network architecture, Kohonen's SOM, for clustering of the effect-action matrix. Also, in the next study design analysis will be conducted with improved insights.

REFERENCES

Cohen, S.M., Adelman, L., Tollcot, M.A., Bresnick, T.A. and Marvin, F.F., (1993), A Cognitive Framework for Battlefield Commanders' Situation Assessment, *Cognitive Technologies, Inc*., Arlington, VA 22207

Davis, K. P., (2001), "Effects Based Operations: A Grand Challenge for the Analytical Community", *RAND*, Santa Monica, CA.

Enke, D., Ratanapan, K., and Dagli, C., (2000), Larger Machine-Part Family Formation Utilizing a Parallel ART1 Neural Network, *Journal of Intelligent Manufacturing* (11), 591-604.

Rushton, J.G., and Zakarian, A.,(2000), "Modular Vehicle Architectures: A Systems Approach", *Proceedings of the 10th Annual International Conference on Systems Engineering*, July, Minneapolis, MN.

Smith, A. E., (2002)."Effects Based Operations: Applying Network Centric Warfare in Peace, Crisis, and War", *CCRP Publication Series*.

Webopedia.com, www.webopedia.com/TERM/M/modular_architecture.html, attained on May, 2007.

ANT INSPIRED ALGORITHM FOR MULTI-ROBOT COLLABORATION

SHISHIR BASHYAL
RTPIS Laboratory, Electrical and Computer Engineering Dept, University of Missouri – Rolla

G. K. VENAYAGAMOORTHY
RTPIS Laboratory, Electrical and Computer Engineering Dept, University of Missouri – Rolla

CIHAN H. DAGLI
Systems Engineering Department, University of Missouri – Rolla

ANN MILLER
Electrical and Computer Engineering Dept, University of Missouri – Rolla

ABSTRACT

There are several application scenarios where multiple robots can be used to attain goals otherwise not possible by deploying a single robot. In this study, we analyze the defensive behavior in ants and derive the inspiration for collaborative approach in multi-robot systems from them. The algorithm is then simulated in a multi-robot system to observe the effectiveness of the resulting collaboration. The finding of this study suggests that the proposed collaborative algorithm, inspired by the social defensive behavior of ants, is effective in improving multi-robot system efficiency. The implementation of the algorithm in software simulation also resulted emergence of escaping behavior in robots trapped between obstacles.

INTRODUCTION

A pack of ant-soldiers basking in the afternoon sun are alarmed by a fellow soldier hurrying towards the nest. They sense that something is wrong and are set to take action to ensure the safety of their nest. Unlike their human counterparts, they do not follow the ant to investigate the matter, nor do they wait for a command from their superior to take some action. They all take some course of action independently on their own and surprisingly, their combined effort helps them protect the nest from a possible attack. Like for many other complex problems, nature has already discovered a solution for defense and there is something for us to learn from it. The common feature of both human and ant defense mechanisms is that both are system-of-systems dealing with the safe-keeping of the society and have a means of communication to enable networking.

With robot's increasing use in different fields, it has been realized that there are several instances where multiple robots can be used to attain goals that was not possible by deploying a single robot [1]. This shift from deployment of single robot to multiple robots demands the ability to interact with functional units – strategic groupings of robots – rather than issuing commands to each individual robot [2]. Unlike in a single robot domain, multiple-robot systems do not yet have an accepted-by-all standard. The central design challenge in multi-robot system development is to develop techniques to maintain, program and execute robots without having to handle them individually [3].

Many of the concepts used in multi-robotics have been derived from a rather different field of study: socio-biology. Some of the pioneer works in this field [4, 5, 6] have derived the inspirations from ants in particular. Multi-robotics is a typical example of a network-centric system. Each individual in the team is capable of performing on its own but by forming a network among each other; they have the potential to highly improve their performance. For example in [7], the authors successfully implement a Particle Swarm Optimization (PSO) based collaborative robotic search application where a number of robots sense and share their local environmental parameters to converge to a target. The task would have been much more difficult using a single robot. With continuous development in semiconductor fabrication and increasing research in nanotechnology, it is likely that a swarm of nano-robots will be developed in near future. The need to develop effective algorithms for multi-robot collaboration is thus immediate.This study discusses the defense mechanism used by ants to defend their nest and the communication mechanism involved in the process. The network-centric defense mechanism is then simulated in a group of robots to achieve collaborative behavior.

ANT COLONIES, SOCIAL DEFENSE AND ROBOT COLLABORATION

Ants are found in virtually every terrestrial environment on earth except Antarctica [8]. In the life-rich Amazon, for example, they make up a third of the total animal biomass. Interestingly, the nest building, foraging and defense in ant colony are all executed in collaboration without any central control. The fascinating simple behaviors of the individual ants result emergence of intelligence in the colony and this fact has inspired several algorithms in computer science.

The formation of a colony in ants and the emergence of social behaviors are possible only due to their ability to communicate with others using their antennae when they are physically together or using chemical called pheromones in which case they have to be in the same territory within a time frame as the pheromone evaporates over time. The temporal nature of the communication helps the ants to come up with complex behaviors. For example, while out on foraging, ants need to find out the shortest path to the food source and this they achieve by measuring the pheromone strength while returning. If the pheromone strength is still high, they know that they took a shorter route to the food source.

This study tries to understand the behavior of ants that assists in defending their nest from enemies. Wars are common in insect colonies and ant behavior studies report that ants evaluate the need of the war in terms of resource requirement and attack other colonies only when there is scarcity of resources. It has also been established that the nature of attack differ from one species to the other. In [9], authors present a detailed analysis of defense mechanism of *Lasius niger* ant species concluding that the defense system of this ant species consists of three processes:

i. a recruiting system that allocates more workers to more valuable resources;
ii. individual ants following a simple decision rule to become more aggressive in response to increased numbers of nest-mates nearby (hence aggressive behavior reflecting the importance of each area to the colony); and

iii. variability in individual responses causing a gradual change in the proportion of fighting ants responding to a threat.

This means that the aggressiveness of an ant depends considerably in the number or ants in the neighborhood. Individual ants sense the pheromone concentration in the area and adjust their agressiveness accordingly. Different applications of multi-robot teams can benefit if the 'eagerness' of individual robots is varied with time and situation. For example, a robot with higher battery level can be set to be more eager to do tasks that consume more power just like variation in individual aggressiveness in ants. Apart from that, if the eagerness of a robot is varied based upon the number of robots in the neighborhood, robots action varies with the number of robots in the territory. Complex behaviors can emerge when this approach is adopted in a robotic team. This study investigates co-operative obstacle removal approach in which the individual robot's eagerness to remove obstacle depends on the number of robots near the obstacle. The problem formulation and proposed solution is listed below:

Problem:

A rectangular area of size 100x60 is divided into cells of size 5x5. A robot can take one step at a time to one of the position: left, right or forward. Robots originate from the leftmost column of the grid and their goal is to reach any cell in the rightmost column of the grid. Several obstacles are randomly placed in the grid. Different obstacles have different sizes varying from 1 to 4. A single robot can only displace the obstacle of unit size, two robots together can remove obstacle of size 2 units and so on. Figure 1 shows the robot field with five robots. The obstacles are assigned brightness values based on the number of robots required to displace it, the biggest obstacle requiring 4 robots is colored black.

Proposed Solution:

i. Robots always try to move forward. If there is an obstacle in the forward cell that the robot cannot remove, the robot moves to either left or right cell.

ii. The work done by a robot in moving a single step is considered to be 1 unit.

iii. The eagerness of robot **R** is set equal to the number of robots near **R**. Robots ahead of **R**, though near to **R**, are not counted while computing the eagerness as robots are not allowed to return back to help others.

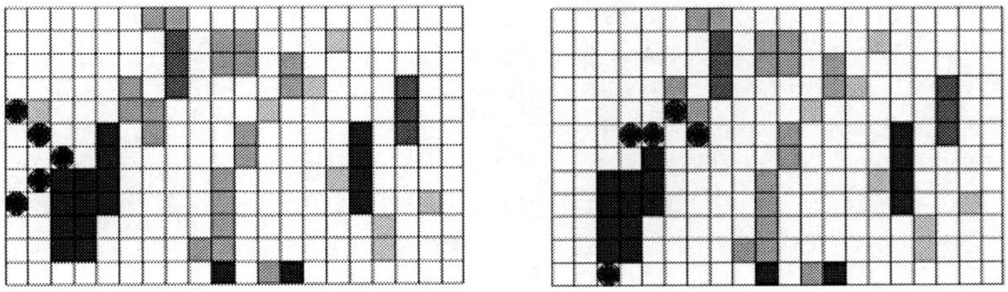

(a) (b)

Figure 1: Robots formation (a) initial and (b) as they collaborate to remove obstacles

iv. When **R** detects an obstacle in its path, it compares the obstacle size to the eagerness of the self. If the eagerness is more than or equal to the obstacle size, **R** collaborates with required number of robots near it to remove the obstacle. If the eagerness is less than the obstacle size, **R** avoids the obstacle instead of removing it.

v. The work done when removing a unit obstacle is considered to be 10 units assuming that removing an obstacle is 10 times as difficult as moving a single step.

vi. It is assumed that the removed obstacle disappears from the field rather than just displaced by a unit distance. This assumption holds true in several real implementations like robots push walls down such that they can now move over them, or robots dig a hole through the obstruction.

RESULTS AND DISCUSSION

In simulation, the number of robots in a team that start moving across the grid at the same time is varied from 1 to 5 in steps of 2. For each case, the robots are run in two modes: obstacle avoidance and obstacle removal. Once a team of robots reach the other end of the grid, second team of robots with same team size starts traversing the grid. This process is repeated 10 times for each experiment. As the field is randomly generated each time, 100 experiments are carried out for each team size and the average work done for 100 trials is reported. Figure 2(a) shows results for the robots operating in the obstacle avoidance mode where robots do not collaborate with each other. In obstacle avoidance mode, the robots try to move straight ahead but if there is an obstacle in front, the robots take a step to the left or right and then try moving forward. So even when a number of robots have moved across the grid, the obstacles remain as they are and thus the robots path length remains constant. However, since a robot is not allowed to move backward, some robots get trapped when there are obstructions in all other three directions. The trapped robots remain in the field blocking other robots from occupying the cell in the grid. As a result, paths that lead to blocked robots are eventually filled by blocked robots and new robots try a different route after detecting the blocked robots as obstacles. This way, robots coming at a later time find better routes that are less likely to be blocked and hence the average work done by individual robots decreases gradually. Figure 2(a) shows the average path length for robot teams. As more teams travel across the grid avoiding obstructions, some of the robots get trapped and unintentionally assist other robots in avoiding the trap. As a result, the average work done decreases gradually for all three team-sizes. However, robot teams with 5 robots per team discover shorter paths quicker than the teams with smaller number of robots. Figure 2(b) shows the results for collaborative obstacle removal mode. In this mode, the robots eagerness is based on the number of robots in the neighborhood. A single robot removes any obstacle that it can remove on its own but if it needs more robots in removing an obstacle, it checks if there are enough robots in the neighborhood or not. If there are enough robots in the immediate neighborhood, they collaborate to remove the obstacles. Considering the complexity in collaboration, obstacle removal is supposed to take 10 times as much effort as moving from one cell to the other. Even then, the robots greatly benefit from the collaboration as the removal of obstacles results shorter paths to other robots moving through the same path at a later time.

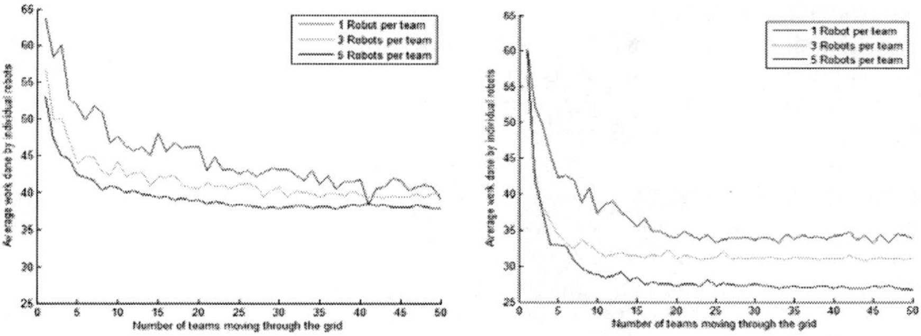

Figure 2: Average work done by robots in (a) obstacle avoidance mode and (b) collaborative obstacle removal mode

The work done by the first few teams is high as they are engaged in the obstacle removal task. However, teams arriving after about 10 teams need much less effort in getting across the grid. Teams with single robot can not remove most of the obstacles and hence there is only a slight improvement in the path length from obstacle avoidance mode to the obstacle removal mode. Teams with more number of robots can collaborate to remove most of the obstacles and hence the average work done by the team eventually reduces to 25 which is close to the minimum whereas in obstacle avoidance mode, the best average work done is two times higher than the lowest possible. Also, both figures 2(a) and 2(b) suggest that the variation in work done by teams with same number of robots in collaborative mode is less compared to that in the obstacle removal mode.

In obstacle removal mode, some robots can still get trapped when few robots are surrounded by obstacles of bigger sizes. But, when more robots arrive to the area at a later time, the eagerness of the trapped robots also increases and hence they initiate collaboration for removing the obstacle and escape successfully. This is an excellent example of an emergent behavior as the escaping behavior emerged on its own.

CONCLUSION AND FUTURE WORK

The simulation study shows that the proposed algorithm can be effectively used in multi-robot collaboration. By relating the aggressiveness in ants to eagerness in robots, the algorithm enables multi-robot teams to use simple rules achieving complex goals. This study reveals that for navigation through the grid with randomly placed obstacles, the average work done by egocentric robots, that do not remove the obstacle for others, is two times higher than the minimum possible whereas for the same problem, social robots, that collaborate to remove obstacles, end up with average work done comparable to the minimum possible after removal of all obstacles. Apart from the intended collaborative obstacle removal behavior, the ant inspired algorithm led to emergence of escaping behavior in robots where a number of robots trapped in certain situations eventually succeeded in escaping from the trap as more robots accidentally arrived in the neighborhood. With miniaturization of robots possible due to nanotechnology, it is likely that a team of robots will soon be used in several application scenarios and hence the proposed algorithm can be useful in managing the collaboration of such teams.

The implementation of the proposed algorithm in a real multi-robot team is being studied. One issue that needs to be addressed in the real implementation is the positioning system using which robots can know their relative distance from each other. Radio signal strength based indoor positioning system is being developed to enable the implementation. The effectiveness of the proposed algorithm in different collaborative tasks can also be evaluated.

REFERENCES

[1] Kartoun U., Stern H., Edan Y. 2006, Human-Robot Collaborative Learning System for Inspection, IEEE International Conference on Systems, Man, and Cybernetics, Taiwan, 2006.

[2] D. J. Bruemmer, D. D. Dudenhoeffer, J. Marble, Mixed-Initiative Remote Characterization Using a Distributed Team of Small Robots, *AAAI Mobile Robot Workshop*, Seattle, 2001.

[3] James McLurkin, Jennifer Smith, James Frankel, David Sotkowitz, David Blau, Brian Schmidt. Speaking Swarmish: Human-Robot Interface Design for Large Swarms of Autonomous Mobile Robots, AAAI Spring Symposium, March 28, 2006.

[4] David Payton, Regina Estkowski, Mike Howard. Pheromone Robotics and the Logic of Virtual Pheromones, SAB 2004 International Workshop, Santa Monica, CA, USA, July 17, 2004, Revised Selected Papers

[5] Daily, M.; Youngkwan Cho; Martin, K.; Payton, D.; World embedded interfaces for human-robot interaction; Proceedings of the 36th Annual Hawaii International Conference on System Sciences, 2003.

[6] D. J. Bruemmer, D. D. Dudenhoeffer, M. O. Anderson, M. D. McKay. A Robotic Swarm for Spill Finding and Perimeter Formation, *Spectrum 2002*, Reno, NV, August 2002.

[7] Doctor, S.; Venayagamoorthy, G.K.; Gudise, V.G., "Optimal PSO for collective robotic search applications," Congress on Evolutionary Computation (CEC2004), vol.2, pp. 1390-1395, 2004.

[8] http://www.asu.edu/news/stories/200510/20051027_hoelldobler.htm, Visited 30th May 2007.

[9] Hiroshi Sakata, Noboru Katayama; Ant defence system: A mechanism organizing individual responses into efficient collective behavior, Ecological Research 16 (3), 395–403, 2001.

[10] J. Casper and R. R. Murphy, Human-robot interaction during the robot assisted urban search and rescue response at the World Trade Center. IEEE Trans. Syst., Man, Cybern. B, vol. 33, pp. 367--385, June 2003.

[11] Burke, J., Murphy, R.R., Rogers, E., Scholtz, J., and Lumelsky, V., Final Report for the DARPA/NSF Interdisciplinary Study on Human-Robot Interaction, IEEE Systems, Man and Cybernetics Part A, vol.34, No.2, May 2004, pp. 103-112.

[12] Erol Sahin, William M. Spears. Preface in Swarm robotics: SAB 2004 international workshop, Santa Monica, CA, USA, July 17, 2004: revised selected papers.

[13] David Geer, "Small Robots Team Up to Tackle Large Tasks," IEEE Distributed Systems Online, vol. 6, no. 12, 2005.

[14] Daniel W. Palmer, Marc Kirschenbaum, Jon P. Murton, Michael A. Kovacina, Daniel H. Steinberg, Sam N. Calabrese, Kelly M. Zajac, Chad M. Hantak, and Jason E. Schatz; Using a Collection of Humans as an Execution Test bed for Swarm Algorithms; IEEE Swarm Intelligence Symposium, April 24-26, 2003.

EVENT CALCULUS AND PROBABILISTIC KNOWLEDGE BASE OF CONDITIONAL EVENTS IN COMPLEX SYSTEMS MANAGEMENT

FRANCESCO RAGO PHD
M3 Comp. LLC
Newark, DE USA

ABSTRACT

Technological environments are growing in complexity and their governance is difficult. To improve the timeliness and accuracy of non-human decision-making in a system environment we are faced with an array of crucial problems, including how to handle large amounts of incoming and uncertain information from disparate sources coded as events of different type. Artificial Intelligence has developed theories to manage events and actions (McCarthy, Kowalski) and we present an event calculus formalism featuring events and action integrated with temporal logic. We have designed and realized an algorithm that permits to a software system to be able to forecast future actions depending on events happening in a process area. The conditional relationships are calculated starting from frequency of conditional events and the resulting possible states are calculated. Each conditional event can trigger event-action modules.

INTRODUCTION

A complex system has a large number of elements capable of interacting with each other and with their environment. One of Hayek's main contributions to early systems complexity theory is his distinction between the human capacity to predict the behavior of simple systems and its capacity to predict the behavior of complex systems through modeling. He believed that complex phenomena could be modeled, but requires to capture a large set of data [1].

Distribute sensors environments are systems growing in complexity and their governance is difficult. A proactive capability has to be part of computer systems hyper-communicating in a machine-to-machine communication (M2M). Typical environments are wireless and massive sensor environments. To improve the timeliness and accuracy of non-human decision-making in a system environment we are faced with an array of crucial problems, including how to handle large amounts of incoming and uncertain information from disparate sources coded as events of different type.

Artificial Intelligence has developed theories to manage events and actions (McCarthy, Kowalski) and we presents an event calculus formalism featuring events and action integrated with temporal logic. We have designed and realized an algorithm that permits to a software system to be able to statistically forecast future actions depending on what is happening in a process area. The conditional relationship is calculated starting from frequency of conditional events and the resulting system possible states are calculated. Each conditional event can trigger event-action modules

SENSOR NETWORKS

A "sensor" is a transducer that provides functions to generate a correct representation of a sensed or controlled quantity (e.g., temperature, pressure, strain, flow, pH, etc.). A sensors network is a web of sensor nodes that collect data on sensors field, process these data and send them to other sensor nodes or to a special node called gateway, which communicates with a base station through the Internet or an Intranet. Wireless sensor network gives the characteristics of an effective, feasible, and fairly reliable monitoring system. Monitoring generates a large amount of sensor data and efficiently dealing with this large amount of data in a resource-constrained wireless sensor network is a challenge.

In the area of wireless sensor network, researchers have been working on the services that are required in most of the applications. Time synchronization between the nodes of the network is one of such service. Data from different sensor nodes can be accurately synchronized if a good time synchronization service is functional. Ganeriwal et al. [2] propose a hierarchical approach for time synchronization among nodes. It is called Timing-sync Protocol for Sensor Networks. Similarly, there are other key issues related to system deployment and operation. These are discussed in [3].

MODELING A SENSORS NET SYSTEM

Sensor measures status of systems elements and flow data between them. What it measures is a value representing a specifying system state at a specific moment. The change of state is an event in our framework. There are two main categories of events: Deterministic and Probabilistic.

A deterministic event always has the same outcome and is predictable 100% of the time.A probabilistic event is an event for which the exact outcome is not predictable 100% of the time.

We can now formalize the relationship between elements of a system using an interaction matrix. If the system S is composed in a specific representation by a set of elements $\{c_1,c_2,\ldots,c_n\}$ and each system component has internal input and outputs, We will have a matrix like:

$$V_K = \begin{pmatrix} f_{11}f_{12}\cdots f_{1n} \\ f_{21}f_{22}\cdots f_{2n} \\ \cdots\cdots\cdots \\ f_{m1}f_{m2}\cdots f_{mn} \end{pmatrix}$$

where $f(c_i,c_j,t_k)$ represents the event that describe an interaction between element a_i with element aj at the time t_k. We can assign a value v_{ijk} to the function $f(c_i,a_j,tk) = v_{ijk}$ where v_{ijk} is the specific measure of a sensor at the time

t_k of the interaction between c_i with c_j . We introduce simple examples of sensor measures:

1. $f(c_i,c_j,t_k)$: transfer of an object from site c_i to c_j., v_{ijk} is the code transmitted by a RFId tag.
2. $f(c_i,c_i,t_k)$ value of the temperature in site c_i measured at time t_k.

v_{ijk} is the formal definition of elementary event in a sensor system: something happened because a change of v_{ijk} happened.

A collection $f(c_i,c_j,t_k)$ represents complex events where c and t may change.

The matrix V_k represents time evolution of measurements of a sensor system. The series of $\{V_k\}$ is the internal representation of the measurements dynamic. The stability of the systems has a local or global meaning depending on what we want to control. We will not analyze the sensor system behavior in this report and recommend specific systems literature [4].

An event affects the elements that feed the other elements. Since we are trying to predict the effect of on event on the timing of another, we cannot consider them in isolation. What we are interested in is the time of the next event that will occur.

The events modify the operating capability of an object. When estimating the time of a next event for each element, only the current input and output values need to be considered. This is because the estimates for the times of the events are updated when they are affected by a change. The model reduces inflow into the elements affected.

This in turn affects the elements that feed the object in question.

EVENTS CALCULUS

We now introduce events in a formal way to make deduction about systems behavior. The Event Calculus was originally introduced by Kowalski and Sergot as a logic-programming framework for representing and reasoning about events [5]. The basic concepts of the model of time and change of Event Calculus are those of event and property. Properties are fluents, which hold during time periods initiated and terminated by events occurrences. An event occurrence is represented by associating the event with time point at which it occurred by means of a clause. The relation between events and properties can be defined by means of initiate and terminate clauses. The concepts of property and event can be viewed as the Event Calculus counterparts of fluent and action of Situation Calculus of McCarthy[6].

The Event Calculus can be described as a sorted predicate calculus with equality, with a sort E for events (variables e_1, e_2,..), a sort F for fluents (variables f_1, f_2,...), a sort T for timepoints (variables t, t_1, t_2,...).

A basic calculus has six predicates (other than equality):

Happens $\subset E \times T$, Happens(e_n,t_k)= $f(c_i,c_j,t_k)$. indicates that event e_k occurs at time t_k;

HoldsAt $\subset F \times T$,HoldsAt(f_i, t_k) means that fluent f_i is true at time t_k and $f_i = f(c_i,c_j,t_k)$=cost

Initiates $\subset E \times F \times T$, Initiates (e_n, f_i ,t_k) expresses that if e_n occurs at t_k it will initiate the fluent f_i;

Terminates $, \subset E \times F \times T$, Terminates(e_n,f_i ,t_k)) expresses that if e_n occurs at t_k it will respectively terminate the fluent f.

"<" is the standard order relation for time. It is a (strict) partial order, i.e. an irreflexive and transitive relation, over the set of event occurrences.

Predicates Clipped ⊂ TxFxT and Declipped ⊂ TxFxT are defined in terms of Happens, Initiates, Terminates and <. Clipped (t_{k-n}, f_i ,t_k) (respectively Declipped(t_{k-n}, f_i ,t_k)) means "the fluent f_i is terminated (respectively initiated) between times t_{k-n} and t_{k-n} ". The corresponding definitional axiomsl are:

$$Clipped(t_{k-n}, f_i, t_k) \equiv \exists e_n, t\ [Happens(e_n, t_k) \wedge$$
$$t_{k-n} \leq t \leq t_{k2} \wedge Ter\min ates(e_n, t_k, f_i)]$$

$$DeClipped(t_{k-n}, f_i, t_k) \equiv \exists e_n, t\ [Happens(e_n, t_k) \wedge$$
$$t_{k-n} \leq t \leq t_{k2} \wedge Initiates(e_n, t_k, f_i)]$$

The Event Calculus aims at modeling situations that consist of a set of events, whose occurrences over time have the effect of initiating or terminating the validity of properties, some of which may be mutually exclusive.

The Event Calculus has a meaning of temporal logic of linear discrete past time, called TL if we consider only the event logic and not the fluents. We follow the exposition in [7], tailoring the definitions somewhat towards our particular needs. The formulas are built up from the set E be a finite set of basic events, interpreted as propositional variables closed under the following formula formation rules.

TL Definition:

Every e ∈ E is a formula of temporal logic.

If $\psi \in$ TL , then their boolean combinations $e_i \vee e_j$ $\neg e_k$ are in TL. The other Boolean connectives: $\wedge, \rightarrow, \leftrightarrow$ can be defined in terms of $\neg$ and $\vee$, as usual.

If $e_i, e_j \in$ TL , then their past tense temporal combinations ● e_i and e_i**Since** e_j are in TL, where ● is spelled “previously”.

The introduction of a temporal logic for events permits to define complex logical expressions of events in a complex environment.

CONDITIONAL EVENTS PROBABILISTIC KNOWLEDGE

To represent probabilistic knowledge, we must be able to model the action effects, the relations among event variables, and the temporal nature of the decision environment. We use timed concept instances to represent actions and event variables. Probabilistic dependency statements except that we separate the logic program clauses as the preconditions in the probabilistic statements for representational clarity and inferential efficiency, represent the action effects and probabilistic relationships among the event variables. A probabilistic knowledge base is a finite set of probabilistic dependency statements and has the form of a collection of statements together with their probabilities conditioning operators in logic. Probabilistic knowledge base can be represented by rules or conditional relations symbolized typically as $(a_1 \mid e_1)$, $(a_2 \mid e_2)$,... , (read "if event e_1, then action a_1" (or equivalently, "a_1, given e_1," etc.), They may themselves represent quite complicated logical combinations of simpler events in TL, and where it may or may not be known what logical relations exist among such events. Each such rule is also assigned quantitative reliability in the form of naturally corresponding conditional probabilities. Thus, for some otherwise unspecified probability measure P, rule (a|b) is assigned value P(a|e) = P(ae)/P(e), the conditional probability of a given b, using standard Boolean and probability notation and assuming antecedent probability P(e) > 0.

Symbolically, one is considering the validity or degree of validity of the entailment schemes that is the basic element of the probabilist knowledge base:

$$G_i = [P(a_1|e_1),..., P(a_n|e_n); d_i], i = 1, 2,... ,$$

where $(P(a_1|e_1),..., P(a_n|e_n))$ can be considered the probabilistic premise set of Gi and d_i its potential conclusion called event-action module.

CONDITIONAL EVENTS PROBABILISTIC KNOWLEDGE CONSTRUCTION

The construction of the probabilistic knowledge base was realized using the following steps:

Step 1. Eliciting Control Context Information

The process begins with the analysis of the system about the control problem. It is defined the decision horizon and the list of key events.

Step 2. Defining the Action Space

In front of events, the alternative actions are listed creating an action hierarchy.

Step 3. Defining the State Space

It is identified the set of state attributes, which denote the different characteristics that constitute and distinguish the states. The state represents the values set associated to each event and the sensors measures t.

Step 4. Transition View Construction

For specific values of states a transition characteristic and a threshold are defined. A transition view is constructed for each action defined in the action space. The probabilistic dependency statements are defined for all the actions, if exit. In this case, if we can assess such transition functions directly for an action, the transition view is completed.

Step 5. The conditional probabilities are created and maintained collecting systems data. Recalculations of probabilistic dependencies are made.

DISCUSSION

We have presented the design of a probabilistic knowledge-based decision model to control sensor networks. The overall performance of the system has been evaluated with a basic result of potential disadvantage of control strategy that cannot avoid exhaustive inclusion of all the relevant variables in the knowledge base. Therefore, the influence view may become quite complex sometimes. Although it was possible to perform an abstraction on the model components to reduce the model size, a better solution is to define a set of stopping criteria which takes into account the relative significance of the available information, and the relative utility of further expanding the model.

A future research issue of this work involves incorporating a machine learning module into the system, which will learn probabilities from large data sets, so as to reduce the work on the estimation and specification of the numerical parameters involved.

CONCLUSIONS

The common characteristic of all complex systems is that they display organization without any external organizing principle. Complex systems are about adaptation and continuous improvement, but this means measures to that human or non-human can improve the system itself.

We have designed and realized an algorithm that permits to a software system to be able to decide action depending on the events in complex systems. However, a successful management will require augmenting the conceptual framework, even to the point of reshaping what one means by prediction. The challenge is the realization that many systems are not the result of a single design but an evolution and merging of designs.

REFERENCES

[1] Friedrich August von Hayek, "The Sveriges Riksbank Prize in Economic Sciences in Memory of Alfred Nobel 1974", Prize Lecture.

[2] S. Ganeriwal, R. Kumar, M. B. Srivastava, "Timing-sync Protocol for Sensor Networks", SenSys '03, 2003.

[3] S Sundresh, G. Agha, Ki. Mechitov, W. Y. Kim, and Y. Kwon, "Coordination Services for Wireless Sensor Networks", Intl. Workshop on Advanced Sensors, SHM and Smart Structures, 2003.

[4]D.H Hinrichsenand, A. J. Pritchard (2005), Mathematical Systems Theory I - Modelling, State Space Analysis, Stability and Robustness. Springer. ISBN 0-978-3-540-44125-0.

[5]R. Kowalski and M.Sergot (1986), "A Logic-Based Calculus of Events New Generation Computing", vol. 4 pp. 67–95.

[6] J. McCarthy," Situation Calculus withConcurrent Events and Narrative",11. 1995.

[7] E. A. Emerson," Temporal and modal logic", In Handbook of theoretical computer science, Vol. B, pages 995–1072. Elsevier, Amsterdam, 1990.

[8] Michiel van Lambalgen and Fritz Hamm (2005) The proper treatment of events. Blackwell Publishing, Oxford and Boston

[9] Breese, J. S.: Construction of belief and decision networks. Computational Intelligence, 8(4):624-647 (1992)

[2] Cao, C. and Leong, T. Y.: Learning Conditional Probabilities for Dynamic Influence Structures in Medical Decision Models, In Proceedings of the 1997 AMIA Annual Fall Symposium (formerly SCAMC), AMIA (1997)

[8] Pearl, J.: Probabilitic Reasoning in Intelligent Systems: Networks of Plausible Inference. Morgan Kaufmann, San Mateo, CA (1988)

Quantum Mechanical Principles of Emergence

Eric D. Smith, Missouri University of Science and Technology, erics@umr.edu
Neale R. Smith, Instituto Tecnológico y de Estudios Superiores de Monterrey

Abstract. Quantum mechanical principles are discussed as guides for emergent processes. Rules of superposition, rules of entanglement, and rules of collapse provide a process guide for similar processes in decision making, industrial distribution, and real options. The merging of genetic algorithms and quantum computing is discussed.

Introduction. In classical physics, a system is characterized by a known configuration of physical components whose positions are known exactly, and the evolution of the system over time is governed by known laws of motion and interaction. The preponderance of common experience in the world usually guides one to consider the world as purely Newtonian.

In quantum mechanics, probabilistic fields are discussed as precursors for the emergence of physical matter. Quantum mechanics can be described alternatively as matrix algebra, wave functions, transformations, or sum-of-paths [Herbert, 1985]. The various quantum descriptions can be useful for describing the emergence of physical solutions to complex problems – although distractions arise because particular quantum descriptions range from wave functions to particles with ±1 quantized characteristics.

This paper attempts to show the applicability of a few quantum principles, by generalizing the principles beyond their original context. It is an open question how much more available the quantum principles of superposition, entanglement and collapse would be if they labeled more generally as "possibilities, relations and determination," for example. The fact that Schrödinger [1935] coined the term "entanglement" (German: Verschränkung) in 1935 for his cat-in-the-box article makes it clear that many useful verbal innovations can be made.

Quantum Tic-Tac-Toe. Goff, Lehmann, and Siegel [2002] and Goff [2006] investigated a game of tic-tac-toe modified with the quantum mechanical principles of superposition, entanglement and collapse as an educational tool for teaching quantum mechanical principles. Each player's tic-tac-toe move, instead of being determined discretely (in a time-ordered alternating sequence of play with the opposing player) is entered as a superposed set of two positions – with the player's mark preliminarily not definitively appearing in either one of the positions. This is the game's rule of superposition – which could be structured in alternative ways. The next player enters her move in the same way – as a pair of possible positions. The game's rule of entanglement appears as the possible moves saturate the playing board and completely entangled cycles appear that do not allow more superposed moves; note that allowing more superposed moves would imply that there would eventually be more actualized player positions on the board than is allowed in the classical game. The game's rule of collapse states that the next player must collapse the complete cycle or cycles by choosing in which position to actualize a superposed move.

The principles of superposition, entanglement and collapse thus provide the basis for the sudden emergence of actual physical game configurations out of a set of possibilities: Superposition creates the possibilities, entanglement limits the superposition of moves, and collapse specifies when the superposed possibilities become certain.

Physical Emergence. The principles shown in the quantum tic-tac-toe game can be applied in examples as understandable as a set of boulders resting atop mountain peaks, with the boulders chained together in such a fashion – perhaps with the aid of pulleys – so that the fall of one boulder causes the fall of other boulders, at the same time precluding them from falling in certain directions – depending on the linkages.

The essential ingredients of sudden, intelligent, and organized emergence are thus: 1) the buildup of possibilities or the storing of energy, 2) the establishment of relations or linkages among the potential components, and 3) the act or decision to collapse the potentialities in a certain way. In plants, for example, photosynthesis builds up energy in the form of glucose, while complex entanglements and collapses decide the physical forms. Thus, there is a basic mechanism for harvesting energy and storing it until connected intelligence causes the energy resources to be used for the construction of a definite physical configuration.

Emerging phenomena are often unexpected and mysterious when their necessary pre-conditions are not recognized, when interacting collapse-related variables provide the possibility for widely different results to emerge, and when an unknown trigger mechanism decides the direction and time of collapse. Using simple models of emergence helps illustrate how complicated phenomena can result from the arrangement of relatively simple mechanisms.

Emergent phenomenon would not be complex without the availability of relations or connections among variables in a pre-collapse environment. A characterization of the post-collapse environment can be made by using the word "complicated" to refer to the combinatorial aspects of components interacting on a physical basis only.

Using the metaphors of quantum mechanics. The details and metaphors of quantum mechanics are a reminder to focus on the total scope of a problem. Specifically, the principles of superposition, entanglement and collapse represent a focus on the constraints of a problem – on its boundaries and edges. The following table summarizes some common methodology scopes:

Abstraction of Methodological Scope and Focus

Optimization	Fuzzy Sets	Quantum focus
Given a set of constraints, find the optimal solution within the constraints.	Given fuzzy constraints, find the best fuzzy solution.	Focus on the constraints and interrelations among the constraints, until a collapse.

Decision Making. Classical decision making, under known or uncertain probabilities, can be modeled by decision trees that are sequentially traversed from left to right as a decision path becomes decided. A focus on decision trees often creates the perception that decision making can occur only along the lines of a decision tree. However, analogously to the quantum tic-tac-toe model, mental decision making involves keeping possibilities in play and combined them into complex possible strategies that are updated as changes occur in the environment or moves are made by exterior agents.

As pertains to emergence, the structure of a classical decision trees implies that the first decision determines the value of the first parameter, after which subsequent decisions determine other parameter values. The implication for emergence is that physical structures are created only as multiple decisions determine their structure. The quantum analogue of sequential parameter determination is that first decision collapses the possibilities of the first variable only, and the second and subsequent decisions do the same for their respective variables only. Startling examples of emergence obviously are not based on such sequential determination of parameters.

Goff [2006] notes that: "Any one game in a classical game can be represented as a single path through the game tree. A quantum game has a game tree, a very much larger one in which it too is a single path, but a quantum game can also be mapped onto the game tree of its underlying classical game as a multibranched path with dead ends." Further, while the possible number of classical tic-tac-toe games is 9!=362,880, the possible number of quantum tic-tac-toe games "is about $9!^2$ or approximately 131 billion." Or, an active quantum game can be dynamically represented as a dynamically changing tree within a larger tree.

Genetic Algorithms. Genetic algorithms can be modified with rules of superposition, entanglement and collapse. These rules imply a focus not on changing the value of genetic bits, but on changing the set of constraints – by the dynamic addition, subtraction and modification of linear constraints on the solution space. As such, quantum rules represent nothing new in the field of genetic algorithms. What may be new is the quantum terminology, methodology, and concepts.

A classic tic-tac-toe game is a time-sequenced genetic algorithm. The 9-square tic-tac-toe board can be represented as a 9-bit gene: 000,000,000. A classical game can be represented with a time-sequence of 9-bit genes, or a gene growing by 9-bits in every time cycle. The genetic algorithm constraints could be formulated as follows:

A rule of superposition implies a constraint $x_l + x_j \leq 1$ (use -1's for O's).

A rule of entanglement implies a rule as to when a complete cycle of constraints exists.

A rule of collapse dictates that the entangled variables must be decided.

Ashlock [2006] commented that: "The fitness landscape in genetic programming is hard to understand. This is, in part, because the variation operators are hard to understand. There is no obvious way to compare two trees, to tell how far apart they are, or even to define what "far apart" means. A single subtree mutation can transform any tree into any other tree of the same size (if it happens at the root node). However, most subtree mutations don't change the fitness at all. The choice of operators used in evolution can have a significant impact on the kind of solutions found." Thus, a random approach is taken, and there is generally no track of tree location in genetic algorithms.

Just-In-Time. There is a similarity between quantum games with rules of superposition, entanglement and collapse, and just-in-time systems, where a changing demand environment replaces opponents making counter moves. Both involve the abstract principle that focusing on the constraints of problems allows efficiency to increase.

The basic just-in-time idea for increasing efficiency is that resources are kept available, and available for use in many places – in superposition -- while information about necessities and constraints builds up – entanglement -- until it is necessary to decide how to commit resources to a particular pattern of use -- collapse. The increase in efficiency is associated with delaying entropic increases by resource allocation for as long as possible. A just-in-time approach implies not spending any effort repeatedly solving an optimization problem, but rather updating the problem constraints until it is absolutely necessary to solve the problem and allocate resources. In logistics, a common strategy is 'postponement'. The universe also works based on delaying entropy increases, although natural scenarios often minimize the energy available. The just-in-time strategy involves a reduction of storage and holding costs of resources that are no longer energy-like and malleable, but are already configured physically.

Interestingly, the industrial theory of constraints also says that it is most important to look at the constraints of a problem. For example, work machines should have buffers that remove work constraints so that the work machines are never idle. Additionally, the activities of work machines should be “pulled” by the changing possibilities of accomplishing useful work – just as quantum tic-tac-toe players should be pulled by changing play possibilities.

Real Options. Real options may be based on quantum rules of superposition, entanglement and collapse. For example, lot buying options in a residential area may involve options that allow a buyer to: ‘Buy this lot OR that lot.’ The entrance of other buyers would entangle their options with those of the first buyer. The final location of everyone’s lots would be determined by the specific collapse of the collective entanglement. Such quantum financial options do not allow for the specificity, individuality and control that most financial agents prefer. One can imagine financial auctions, or stocks, or business deals with similar complications that most financial agents would prefer to avoid. It seems that financial agents prefer to keep fields of possibilities open in their own personal decision making, and prefer specific concrete transactions to be made at their direction. Real options analysis advocates keeping options open.

The General Problem: SAT and CSP. Constraint Satisfaction Problems (SCP) are generalizations of the Boolean satisfiability problem (SAT), in which an expression of propositional logic containing AND, OR, and NOT expressions evaluates to either TRUE or False. The problem is to decide whether there is a set of Boolean variables that make the expression TRUE. SAT was the first problem identified as NP-Complete; this was done by Stephen Cook in 1971. SAT is an important search problem in artificial intelligence. SCP is important in artificial intelligence and operations research, and is often solved with heuristics or combinatorial search methods. Sudoku is an NP-Complete CSP that has no objective function, just constraints to be satisfied. Quantum tic-tac-toe can be seen as a two SAT problems – one for each player.

Quantum Computing: A new computing method. Quantum computers have the ability to look at all the constraints and solutions of a problem simultaneously. The deterministic evolution of quantum wave functions is complemented by the possibility of a non-deterministic collapse to all particular solutions. In this way, quantum computers can look at boundary or environmental changes, while non-deterministically considering all possible solutions. D-Wave Systems [2007] recently publicly demonstrated the world's first commercially available quantum computer, and plans to deliver field-deployable systems in 2008. The design employs 16 superconducting quantum bits.

Randomness. It is known that genetic algorithms use randomness in order to search a given space. In such applications, randomness serves only to provide inputs to an evaluation equation. The necessity of quantum computing is illustrated simply by the need for random number generators. Strictly speaking, classical physics and classical computing cannot generate random numbers. Zak and Williams [1999] demonstrated that some pseudo random number generators produce correlated points in multi-dimensional spaces. The small correlations present in pseudo random numbers can become apparent in simulations such as Markov processes. Genetic algorithms rely on randomness [Brown, 2001], yet the only source of true random numbers is the quantum world.

Emergence. Interestingly, obtaining quantum circuit designs has been limited largely to finding simple solutions by hand, or by using exhaustive enumerations, but has recently been aided by the use of evolutionary genetic programming [Williams and Gray, 1999]. Lee Spector [2007] has used genetic algorithms not to alter 0/1 digits in a gene, but to change components in quantum algorithms. The components are matrixes with complex numbers that model possible manipulations made to actual quantum systems. Spector has found that genetic algorithms can effectively construct quantum algorithms where it is rather difficult for a human algorithms builder to imagine the same algorithm.

> "A "genetic programming" system is a genetic algorithm in which the chromosomes are executable computer programs. There is no sharp line between "executable computer programs" and other chromosomal encodings at a fundamental level of analysis, since the elements of any encoding could be considered "commands" in some language for which the chromosome-decoder is the "compiler" or "interpreter." But in practice so-called genetic programming systems tend to differ from other genetic algorithms in several ways. Although there are exceptions, genetic programming systems tend to use chromosome encodings that are similar in syntax and semantics to existing programming languages. They tend to allow chromosomes to vary in length (as computer programs normally do) and to incorporate hierarchical, compositional structures. And they often allow one to combine the use of special-purpose, problem-specific instructions or variables with general-purpose, commonly-used instructions (for example, + and x).
>
> As a result, a genetic programming system is not only a problem solving system but also an automatic programming system. We provide the system with information that describes what we want a program to do (primarily via the fitness function), and a successful run of the system produces (via evolution) a program that meets the desired specification. We must also provide values for a few other parameters (such as the set of instructions that may be used, population sizes, and genetic operator rates), but most of this is

straightforward and requires little expertise about the problem we are using the system to solve" [Spector, 2007, p.43].

In this way, genetic algorithms are re-arranging complex components in unexpected ways. The components embody sufficient complexity so that the resulting quantum algorithms could possibly be said to have emergent characteristics. The genetic structure is thus acting as a hierarchical overseer, working through a multi-attribute tree in order to judge the usefulness of arrangements, and preserve those that are particularly useful. Biological evolution could have proceeded similarly.

Conclusion. Quantum mechanical principles elucidate computation and problem-solving. The principles of superposition, entanglement and collapse, as well as those of satisfiability of constraints and randomness, aid model-building in the presence of emergent phenomena. Even as quantum computers will aid computation, evolutionary genetic algorithms have been found to be effective in finding useful quantum algorithms.

Acknowledgement. This research was partially supported by the ITESM Campus Monterrey research fund CAT025.

References

Ashlock, Wendy, "Mutation Vs. Crossover with Genetic Programming," in *Intelligent Engineering Systems through Artificial Neural Networks*, vol. 16, Cihan H. Dagli, Anna L. Buczak, David L. Enke, Mark Embrechts, and Okan Ersoy (Editors), ASME Press (2006), pp. 13-18.

Brown, Julian, *The Quest for the Quantum Computer*, Simon & Schuster (2001).

D-Wave, *D-Wave Systems: The Quantum Computing Company* (2007),

Goff, Allan, Dale Lehmann, and Joel Siegel, "Quantum Tic-Tac-Toe, Spooky-Coins & Magic-Envelopes, as Metaphors for Relativistic Quantum Physics", *Proceedings of the 38th AIAA/ASME/SAE/ASEE Joint Propulsion Conference & Exhibit* (2002).

Goff, Allan, "Quantum Tic-Tac-Toe: A Teaching Metaphor for Superposition in Quantum Mechanics," *American Journal of Physics* 74:11 (2006), pp. 962-973.

Herbert, Nick, *Quantum Reality: Beyond the New Physics*, Anchor Press/Doubleday (1985).

Schrodinger, E., "Die Gegenwärtige Situation in Der Quantenmechanik (the Present Situation in Quantum Mechanics)," *Naturwissenschaften*:48, 49, 50 (1935), pp. 807, 823, 844.

Spector, Lee, *Automatic Quantum Computer Programming: A Genetic Programming Approach*, vol. 7, Springer (2007).

Williams, Colin P., and Alexander G. Gray, "Automated Design of Quantum Circuits," in *Quantum Computing and Quantum Communications: Lecture Notes in Computer Science 1509*, Colin P. Williams (Editor), Springer-Verlag (1999), pp. 113-125.

Zak, Michail, and Colin P. Williams, "Quantum Recurrent Networks for Simulating Stochastic Processes," in *Quantum Computing and Quantum Communications: Lecture Notes in Computer Science 1509*, Colin P. Williams (Editor), Springer-Verlag (1999), pp. 75-88.

PART VI:

Fuzzy Systems and Machine Learning

GROWING FUZZY INFERENCE NEURAL NETWORKS: THEIR PRINCIPLE AND APPLICATION

IVETA MRÁZOVÁ
Faculty of Mathematics and Physics, Charles University
Prague, Czech Republic

JIŘÍ IŠA
Faculty of Mathematics and Physics, Charles University
Prague, Czech Republic

ABSTRACT
The article introduces a new growing fuzzy inference neural network (GFINN) based on the principle of growing neural structures. The model finds very rapidly the weights between hidden neurons and the output of the network and an original technique for the selection of significant input features supports extraction of a small set of representative, yet transparent fuzzy rules. So far, GFINN has been tested with the Boston housing data set.

INTRODUCTION

In comparison with other existing neuro-fuzzy models, the new GFINN-model should support an easier extraction of a small set of transparent fuzzy IF-THEN rules. The architecture of the model should adjust dynamically in order to be sufficiently large to yield the desired performance but at the same time also sufficiently small to offer adequate generalization. The idea of the so-called growing neural gas seems to provide these capabilities. The GFINN-model should also support an easy selection of significant input features and a quick training of the output weights.

RELATED WORKS

Most traditional models of neural networks and fuzzy inference systems (e.g. Kohonen maps (Kohonen, 2001) or the FINN-model (Nishina and Hagiwara, 1997) are rather limited by a pre-set number of rules and/or hidden neurons. Growing grids (Fritzke, 1995) enhance traditional Kohonen maps with the ability to change dynamically their structure. A similar model called growing neural gas (GNG) (Martinetz and Schulten, 1991) provides, however, a much looser structure than both the growing grid and the standard SOMs do. The topology of the output layer does not have to be two-dimensional any more. During training, new neurons can be formed, new edges can be added and old neurons and edges can be pruned.

The new GFINN-systems are inspired by classical self-organizing neural networks and by the FINN-model that almost satisfies the conditions of Jang and Sun (1993) for functional equivalence with fuzzy inference systems. FINN is an RBF-like neural network with local units arranged within a single hidden layer. Every neuron n_j in the hidden layer has assigned a reference vector $\overrightarrow{m}_j$ and a weight w_j to the output of the network. In trained FINN-networks, every hidden neuron n_j induces together with its weight w_j a fuzzy IF-THEN rule of the form:

$$IF \quad \overrightarrow{x} =_F \overrightarrow{m}_j \quad THEN \quad y = w_j.$$

To form an adequate network structure in hidden layers of RBF-like GFINN-systems, we prefer to use GNG model with an enhanced output sensitivity and prunning. Such networks will be able to grow if necessary, but will keep a stable size after the growth phase will be completed. Obviously, neurons surrounded by the majority of neurons belonging to the same class are redundant and can be pruned. We will call such a neuron n_i ε-insignificant if it holds for all of its neighbours n_j the following: $\| \overrightarrow{\eta}_i - \overrightarrow{\eta}_j \| < \varepsilon$, where $\overrightarrow{\eta}_i = (\eta_i(c_1), \ldots, \eta_i(c_k))$ denotes for n_i the vector of its class estimates to $c_1, \ldots, c_k$. Furthermore, they can be interpreted in terms of fuzzy rules easily. To simplify the fuzzy rules, a method for the selection of significant input features will be proposed later on.

THE GROWING FUZZY INFERENCE NEURAL NETWORK

The GFINN-model consists of one hidden layer and one output neuron. The hidden layer is formed by a growing neural gas structure enhanced with an improved output sensitivity and with pruning. The output y_k of the new system corresponds to the weight w_j between the hidden neuron n_j and the output neuron.

In GFINN, every neuron n_k from the hidden layer is associated with a reference vector $\overrightarrow{m}_k \in \Re^n$ and with an accumulated error variable $error_k$. With every edge e there is associated an age variable age_e. Every hidden neuron n_k has also a restriction set $\chi_k = \{\chi_k^1, \chi_k^2, \ldots\}$. This restriction set comprises all input features considered currently as significant for the respective neuron. The entire recall process can be summarized as it follows:

<u>Recall for GFINN-systems</u>

1. For the presented input pattern $\overrightarrow{x}$, compute the activation of all the hidden neurons k:
$$a_k(\overrightarrow{x}) = e^{-\frac{\|\overrightarrow{x} - \overrightarrow{m}_k\|_{\chi_k}{}^2}{2\sigma}}$$
using (restricted) Euclidean distance considered for input features from the restriction set χ_k only:
$$\|\overrightarrow{x} - \overrightarrow{m}_k\|_{\chi_k}^2 = \sum_{j \in \chi_k} (x_j - m_j)^2$$
2. Normalize the activations of all hidden neurons:
$$\nu_k(\overrightarrow{x}) = \frac{a_k(\overrightarrow{x})}{\sum_{k'} a_{k'}(\overrightarrow{x})}$$
3. Compute the output $y(\overrightarrow{x})$ of the network using the weighted average defuzzification:
$$y(\overrightarrow{x}) = \sum_k \nu_k w_k$$

The training process consists of two phases – the adjustment of the hidden layer and setting of the output weights. Both these phases are repeated iteratively until the whole system is trained. As a STOP-criterion, the number of training cycles or average approximation or classification error can be used.

COMPUTATION OF THE OUTPUT WEIGHTS

In this study we have considered two methods to find adequate output weights for GFINN-systems – the classical time-consuming least mean square

The training algorithm for GFINN-systems

1. Initialize the network with a small number of hidden neurons, e.g. two. Set the values of all the output weights to the average output value computed over the training set. Pre-set all restriction sets χ_k to include all the features: $\chi_k = \{1, \ldots, n\}$.
2. Train the hidden layer (using basically all the ideas applied before to growing neural gas). Repeat the following steps for all of the training patterns:
   (a) Take the next training pattern $(\vec{x}, \hat{y}) \in T$.
   (b) Find the nearest neuron n_{b_1} and the second-nearest neuron n_{b_2} to allow edge adjustment later.
   (c) Increment age-variable for all edges emanating from n_{b_1}:
   $$age_e(t+1) = age_e(t) + 1$$
   (d) Add the squared distance between the input $\vec{x}$ and the reference vector $\vec{m}_{b_1}$ to the local error counter of the neuron n_{b_1}:
   $$error_{n_{b_1}}(t+1) = error_{n_{b_1}}(t) + \| \vec{m}_{b_1} - \vec{x} \|^2$$
   (e) Move n_{b_1} and its direct neighbours n_k towards $\vec{x}$:
   $$\vec{m}_{b_1}(t+1) = \vec{m}_{b_1}(t) + \alpha_b \, (1 - |\hat{y} - w_{b_1}|^2) \, (\vec{x} - \vec{m}_{b_1})$$
   $$\vec{m}_k(t+1) = \vec{m}_k(t) + \alpha_k \, (1 - |\hat{y} - w_k|^2) \, (\vec{x} - \vec{m}_k)$$
   The similarity factor $(1 - |\hat{y} - w_k|^2)$ assumes $0 \leq \hat{y}, w_k \leq 1$.
   (f) If n_{b_1} and n_{b_2} are connected by an edge, reset the age of this edge to zero. Otherwise, create such an edge with an initial $age_e = 0$ value.
   (g) Remove all edges with the age larger than age_{max}. If there is a neuron without any emanating edge after this operation, remove it as well.
   (h) If the number of steps performed so far is a multiple of the integer growing parameter λ, a new neuron is inserted:
      i. Determine the neuron n_g with the largest accumulated local error $error_g$.
      ii. Find n_f, a direct neighbour of n_g, which has the largest local error $error_f$ among the neighbours of neuron n_g.
      iii. Insert a new neuron n_r halfway between n_g and n_f:
      $$\vec{m}_r = \tfrac{1}{2}(\vec{m}_g + \vec{m}_f)$$
      iv. Replace the edge between n_g and n_f by two edges: one between n_g and n_r and the other between n_r and n_f.
      v. Decrease the error variable $error_g$ and $error_f$ by a multiplicative constant α (due to 'limited forgetting'). Initialize the error variable $error_r$ with the new $error_g$ value.
   (i) Prune ε-insignificant neurons in the hidden layer.
   (j) Decrease all error variables by a multiplicative constant $d \in (0, 1)$ - the network basically 'forgets' old errors.
   (k) If the stopping criterion has not been met yet (network size, fixed number of steps, ...). Repeat from Step 2.
3. Compute the output weights, e.g. by gradient descent, LMSE or FOPT-algorithm.
4. Repeat Step 2 and 3 until the STOP-condition is met.
5. Select significant input features and alter the sets χ_k.

The algorithm for the selection of significant input features:

1. Compute the overall activation Θ_k for every hidden neuron n_k:
$$\Theta_k = \sum_{\vec{x} \in T} \nu_k(\vec{x}).$$
The most active neuron will be denoted as n_1, the least active one will be denoted as n_k. This heuristics prefers to simplify the most often used rules rather than the rarely used ones.

2. For every hidden neuron n_k, try to prune its features:
For every input feature $f \in \{1, \ldots, n\}$, check if it can be discarded from the restriction set χ_k:
   (a) Compute the output of the network $y(\vec{m}_k)$.
   (b) Temporarily remove feature f from the restriction set χ_k.
   (c) Create a set of verification patterns to each reference vector $\vec{m}_k = (m_k^1, \ldots, m_k^f, \ldots, m_k^n)$. This set contains one verification pattern $\vec{v}_{k'}$ for every other hidden neuron $n_{k'}, k' \neq k$. These patterns are identical to $\vec{m}_k$, except for the f-th feature that is replaced by the f-th element of the reference vectors for $n_{k'}$:
   $\vec{v}_{k'} = (m_k^1, \ldots, m_k^{f-1}, m_{k'}^f, m_k^{f+1}, \ldots, m_k^n)$.
   (d) For every verification pattern $\vec{v}_{k'}$, compute the output of the network $y(\vec{v}_{k'})$. If the difference $|y(\vec{v}_{k'}) - y(\vec{m}_k)|$ exceeds a given threshold δ, then the feature f cannot be pruned. If the difference of the output does not exceed the threshold, the feature f is considered to be δ-irrelevant and can be removed from the set χ_k.

error (LMSE) technique and a new approach to the computation of locally fuzzy optimum output weights (FOPT). In GFINN-systems, the normalized activations $\nu_k(\vec{x})$ of each hidden neuron n_k indicate the membership degree of the input pattern $\vec{x}$ to the pattern cluster of n_k. From this point of view, all output weights w_k can be considered independent one from another representing in principle a kind of output value attached to the respective cluster of input patterns. For the hidden neuron n_k, the locally fuzzy optimum error E_k^{FOPT} will be determined according to:

$$E_k^{FOPT} = \frac{1}{2} \sum_{(\vec{x},\hat{y}) \in T} \nu_k(\vec{x})(\hat{y} - w_k)^2 \tag{1}$$

with partial derivatives
$$\frac{\partial E_k^{FOPT}}{\partial w_k} = - \sum_{(\vec{x},\hat{y}) \in T} \nu_k(\vec{x})(\hat{y} - w_k) \tag{2}$$

where $\hat{y}$ is the desired output of the GFINN-system and w_k is the output weight attached to the neuron n_k. For local optima, we will require the partial derivatives to be equal to zero. Thus, the output weights w_k can be determined as weighted averages of the desired output values over the whole training set T:

$$w_k = \frac{1}{\sum_{(\vec{x},\hat{y}) \in T} \nu_k(\vec{x})} \sum_{(\vec{x},\hat{y}) \in T} \nu_k(\vec{x})\hat{y} \tag{3}$$

The interpretation of the rules induced by the GFINN-model can be effectively simplified when using significant characteristics in their antecedents only. Below, we will state a new method to identify such significant features. The features chosen in this way for each respective rule may, however, vary.

SUPPORTING EXPERIMENTS: THE HOUSING DATA SET

We have tested the GFINN model on the Boston housing data (Blake et al., 1998). The formed GFINN-system was also requested to provide a small set of transparent fuzzy IF-THEN rules 'explaining' its function. The data contains 506 housing patterns from the Boston Area collected during the U.S. census (1970). The goal was to predict median value of owner-occupied homes based on the following 13 attribute values (element-wise transformed into the interval $\langle 0, 1 \rangle$):

1. CRIM – per capita crime rate by town,
2. ZN – proportion of residential land zoned for lots $> 25,000$ sq ft,
3. INDUS – proportion of non-retail business acres per town,
4. CHAS – 'Charles River' variable (1 if tract bounds the river; 0 otherwise),
5. NOX – avg. ann. concentr. of nitric oxides in parts per ten million,
6. RM – average number of rooms per dwelling,
7. AGE – proportion of owner-occupied units built prior to 1940,
8. DIS – weighted distances to five Boston employment centers,
9. RAD – lower values reflect better accessibility to radial highways,
10. TAX – full-value property-tax (in \$/\$10,000),
11. PTRATIO – pupil-teacher ratio by town school district,
12. B – $1000(Bk - 0.63)^2$ where Bk is the proportion of Blacks by town. According to U.S. Census (1970), high levels of B correspond to areas with a low proportion of Blacks.
13. LSTAT – the percentage of lower status population,
14. **MEDV – median value of owner-occupied homes in \$1000's. For houses costing \$50.000 or more, the original data was censored (MEDV was set to an artificial value of \$50.000.)**

The results achieved for GFINN-systems trained by means of LMSE and FOPT are shown in Table 1 for pruning with various ε-values – we were interested in the root mean squared error (RMSE) and the mean absolute error (MAE) evaluated for 10-fold cross-validation. Both variants shared the main parameters. They grow after every third iteration ($\lambda = 3$). The learning rates start at $\alpha(0) = 0.9$ and fall according to $\alpha(t+1) = 0.997 * \alpha(t)$. α_k is adjusted according to $\alpha_k(t) = \alpha^2(t)$ and α_b according to $\alpha_b(t) = \alpha(t)$. For the neighbourhood, we have considered a constant distance threshold $d = 1$. The widths of the Gaussians were experimentally set to 0.05 and the value of δ was set to 0.001 when pruning the input neurons. LMSE usually yields slightly better results but FOPT requires much less processing time. The first five extracted rules having the highest support sound:

1. IF [$(INDUS = medium)$ AND $(CHAS = false)$ AND $(NOX = low)$ AND $(RAD = low)$] THEN $(MEDV = medium)$
2. IF [$(CRIM = low)$ AND $(ZN = medium)$ AND $(INDUS = very\ low)$ AND $(AGE = medium)$ AND $(DIS = high)$ AND $(RAD = very\ low)$ AND $(TAX = low)$ AND $(PTRATIO = high)$ AND $(B = very\ high)$ AND $(LSTAT = low)$] THEN $(MEDV = high)$
3. IF [$(CHAS = false)$ AND $(RAD = very\ high)$ AND $(B = very\ high)$] THEN $(MEDV = low)$
4. IF [$(RAD = very\ high)$ AND $(TAX = very\ high)$ AND $(B = low)$] THEN $(MEDV = low)$
5. IF $(CHAS = true)$ THEN $(MEDV = high)$

Rule 1 captures the most common housing areas with medium housing price far away from the river, with medium industrialization and low nitric oxides concentration, but still easily accessible from a highway. Rule 2 refers to areas with

ε	LMSE RMSE	LMSE MAE	LMSE Size	FOPT RMSE	FOPT MAE	FOPT Size
0.01	**0.1389**	**0.0989**	37	0.1399	0.1013	**36**
0.03	**0.1437**	**0.1018**	**25**	0.1458	0.1048	**25**
0.05	0.1482	0.1082	**21**	**0.1454**	**0.1035**	26
0.07	**0.1507**	**0.1071**	**18**	0.1580	0.1155	**18**
0.09	**0.1627**	**0.1204**	**10**	0.1629	**0.1204**	11
0.11	**0.1675**	**0.1229**	**8**	0.1711	0.1260	**8**
0.13	**0.1715**	**0.1263**	**7**	0.1740	0.1286	**7**
0.15	**0.1927**	**0.1392**	**4**	0.2368	0.1886	5
0.17	**0.2018**	**0.1462**	3	0.3583	0.3089	**2**

Table 1: Performance of GFINN-systems trained with LMSE and FOPT for different ε. 'Size' states the average size of the final network.

few black inhabitants, with a medium proportion of large mansions, away from the main employment centers, but easily accessible from a highway and with a small proportion of lower status inhabitants. Such houses are expected to be expensive. Rule 3 also describes areas with few black inhabitants, but far away from highways and the Charles river as cheap for housing. A similar relationship "far from highways ⇒ cheap" can be seen also in Rule 4 while Rule 5 sounds "If the house is by the river, it will be expensive."

CONCLUSIONS

The GFINN-model is inspired by growing neural gas enhanced with pruning and increased sensitivity to appropriate output values. The new FOPT algorithm computing the output weights yields results comparable with the optimal LMSE-method, but in much shorter time. From trained GFINN-systems, transparent fuzzy IF-THEN rules can be extracted in a simple way with irrelevant features easily pruned from the antecedents. Unfortunately, the rules extracted by GFINN-networks may vary for different initial conditions. Experiments done so far with the Boston housing data justify these observations.

ACKNOWLEDGEMENTS

This research was supported by the grant No. 358/2006/A INF/MFF of GA UK and by the program 'Information Society' under project 1ET100300517.

REFERENCES

Blake. C. L. et al., 1998, "UCI repository of machime learning databases."

U.S. Census 1970, `http://www.census.gov/prod/www/abs/decennial/1970.htm`

Fritzke, B., 1995, "Growing Grid – a Self-organizing Network with Constant Neighborhood Range and Adaptation Strength," *Neural Processing Letters*, **2(5)**, pp. 9–13.

Jang, J.-S. R. and Sun, C.-T., 1993, "Functional Equivalence Between Radial Basis Function Networks and Fuzzy Inference Systems," *IEEE Transactions on Neural Networks*, **4**, pp. 156–159.

Kohonen, T., 2001, "Self-Organizing Maps," Springer-Verlag, Berlin.

Martinetz, M. and Schulten, K. J., 1991, "A "Neural Gas" Network Learns Topologies," *T. Kohonen et al. (eds): Artificial Neural Networks*, **1**, pp. 397–402.

Nishina, T. and Hagiwara, M., 1997, "Fuzzy Inference Neural Network," *Neural Computing*, **14(3)**, pp. 223–239.

AN ADAPTIVE FUZZY CONTROL FOR A MULTI-DEGREE-OF-FREEDOM SYSTEM

S. OZCELIK & V. KALEEM
Department of Mechanical Engineering
Texas A&M University-Kingsville
Kingsville, TX, USA
sozcelik@tamuk.edu

R. CHALLOO
Department of Electrical Engineering
Texas A&M University-Kingsville
Kingsville, TX, USA
kfrc000@tamuk.edu

ABSTRACT

This work presents a comparative study of the application of fuzzy and adaptive fuzzy controller to address the problem of position control of a 3 degrees-of-freedom, linear, single-input-single-output (SISO) mass-spring-damper system. Fuzzy logic being a complementary technology most often uses verbal and linguistic information from experts. Unlike regular fuzzy controllers, adaptive fuzzy controllers adapt itself to the desired output. This paper shows how adaptation brings about better control of the system output over regular control technique. The simulation results of Adaptive Fuzzy Controller and Fuzzy Controller applied to the system and their performances are compared.

INTRODUCTION

This research work addresses the problem of position control of a linear, single-input-single-output 3-degree of freedom spring-mass-damper system. We have considered this problem because of the increasing demand for controllers, which can handle the real life uncertainties arising due to changes in the system parameters with greater accuracy. Many of today's process control techniques depend on the position control or the motion control of the actuators or the other components of the system.

The adaptive controller, which possesses the essential ability to cope with unavoidable system uncertainties as well with external environment uncertainties, is reemerging as a timely and an important class of controller design. But not all of is can be implemented to all kinds of systems; one of the main reasons may be that the system model would be very complex and highly nonlinear in nature. The most important reason that fuzzy logic has been chosen is that the adaptation algorithm can be selected such that it can completely be understood by the designer since the fuzzy logic scheme is a kind of decision making that captures the human decision making process. Therefore, the designer can put his experience and knowledge into the algorithm, much like the process of knowledge transformation between human beings. Because the designer can place his knowledge into the control algorithm, the fuzzy logic based controller can perform without the input and output training data which are necessary for neural networks based, supervised learning controllers. Zadeh

[1] laid down the fundamentals of fuzzy logic concepts. Mamdani presented the application of fuzzy logic concepts to dynamic plants [2], [3]. A more wide spread introduction of adaptive fuzzy logic products for industrial measuring and control purposes, which started in Japan have come a long way in both representation and form [4]. The adaptive fuzzy control for chaotic systems with tracking performance was presented in [5]. Use of adaptive fuzzy controllers to generate the rule base for fuzzy controller automatically was proposed in [6]. Nice treatment of adaptive fuzzy systems can be found in [7]. Recently many sources become available on fuzzy systems in general and adaptive fuzzy control in particular [8], [9], [10], [11].

PROBLEM STATEMENT

The system has three masses, four springs and one damper. In this work, both Fuzzy Logic Controller and Adaptive Fuzzy Logic Controller are used to control the position of mass1, mass2 and mass3. However, due to page limitations, only the adaptive fuzzy logic controller will be presented. For the system under consideration, the input can be applied only to mass1 but the outputs can be measured from all the three masses. This poses a challenge for the position control of mass 2 and mass3, as the control signal cannot be applied directly to these masses but is transmitted to these masses through a number of other system components. Also, what makes it more challenging is the fact that depending upon the masses whose position needs to be controlled, the relative order of the system changes from two to six. The dynamics of a Mass-Spring-Damper System with three degrees of freedom produces a sixth-order differential equation.

Neglecting the effect of gravitational force and friction the following equations of motion are obtained:

$$M_1\ddot{x}_1 = -C_1\dot{x}_1 - (K_1 + K_2)x_1 + K_2x_2 + F \qquad (1)$$

Similarly, we have

$$\begin{aligned} M_2\ddot{x}_2 &= K_2x_1 - (K_2 + K_3)x_2 + K_3x_3 && (2)\\ M_3\ddot{x}_3 &= K_3x_2 - (K_4 + K_3)x_3 && (3) \end{aligned}$$

Therefore rearranging the above equations we get;

$$\begin{aligned} \ddot{x}_1 &= \frac{1}{M_1}[F - C_1\dot{x}_1 - (K_1 + K_2)x_1 + K_2x_2] && (4)\\ \ddot{x}_2 &= \frac{1}{M_2}[K_3x_3 + K_2x_1 - (K_2 + K_3)x_2] && (5)\\ \ddot{x}_3 &= \frac{1}{M_3}[K_3x_2 - (K_4 + K_3)x_3] && (6) \end{aligned}$$

Where, M1: Mass of Mass 1, M2: Mass of Mass 2, M3: Mass of Mass 3, K1: Stiffness of the between the Support1 and Mass1, K2: Stiffness of the between the Mass1 and Mass2, K3: Stiffness of the between the Mass2 and Mass3, K4: Stiffness of the between the Mass1 and Support2, C1: Damping coefficient of the damper, x1: Displacement of Mass1, x2: Displacement of Mass2, x3: Displacement of Mass3, F: input force

DESIGN OF A DIRECT ADAPTIVE FUZZY CONTROLLER

Consider the system given by

$$x^{(n)} = f(x, \dot{x}, \ldots, x^{(n-1)}) + Bu \quad (7)$$

$$y = x \quad (8)$$

where f is an unknown function, B is an unknown positive input matrix, and u is an input and y is an output of the system. It is assume that the state vector is available for measurement. Consider the control u being composed of a basic control $u_c(\underline{x} \mid \underline{\theta})$ and a supplementary control $u_s(x)$

$$u = u_c(\underline{x}|\underline{\theta}) + u_s(x) \quad (9)$$

Here the basic control $u_c(\underline{x} \mid \underline{\theta})$ is a fuzzy logic system which is in the following form [7]

$$u_c(\underline{x} \mid \underline{\theta}) = \underline{\theta}^T \underline{\xi}(\underline{x}) \quad (10)$$

where $\underline{\xi}(\underline{x}) = [\underline{\xi_1}(\underline{x}), \cdots, \underline{\xi_M}(\underline{x})]^T$ with the fuzzy basis function being

$$\xi_l(\underline{x}) = \frac{\prod_{i=1}^{n} \mu_{F_i^l}(x_i)}{\sum_{l=1}^{m} y^{-l} \prod_{i=1}^{n} \mu_{F_i^l}(x_i)} \quad (11)$$

$\mu_{F_i^l}$ are given membership functions and $\underline{\theta} = [\theta_1, \cdots, \theta_M]^T$ is the vector of adjustable parameters. The supplementary control $u_s(x)$ is given as follows [7]

$$u_s(\underline{x}) = I^* sign(\underline{e}^T P \underline{b}_c)[\ |u_c| + b_L^{-1}(f^U + |x_m^{(n)}| + |\underline{k}^T \underline{e}|)] \quad (12)$$

with

$$I^* = \begin{cases} 1 & if \quad V_e > \bar{V} \\ 0 & if \quad V_e \leq \bar{V} \end{cases} \quad (13)$$

V_e is the Lyapunov function candidate. Adaptation of the fuzzy logic system is achieved by the following adaptation law

$$\dot{\underline{\theta}} = \begin{cases} \gamma \underline{e}^T \underline{p}_n \underline{\xi}(\underline{x}) & if \quad |\underline{\theta}| < M_\theta \\ \gamma \underline{e}^T \underline{p}_n \underline{\xi}(\underline{x}) & if \quad |\underline{\theta}| = M_\theta \ \& \ \underline{e}^T \underline{p}_n \underline{\theta}^T \underline{\xi}(\underline{x}) > 0 \\ \gamma \underline{e}^T \underline{p}_n \underline{\xi}(\underline{x}) - \alpha & if \quad |\underline{\theta}| = M_\theta \ \& \ \underline{e}^T \underline{p}_n \underline{\theta}^T \underline{\xi}(\underline{x}) < 0 \end{cases} \quad (16)$$

Details of the treatment can be found in [7]. Under this control, overall block diagram of the adaptive fuzzy control would be depicted as shown in Figure 1.

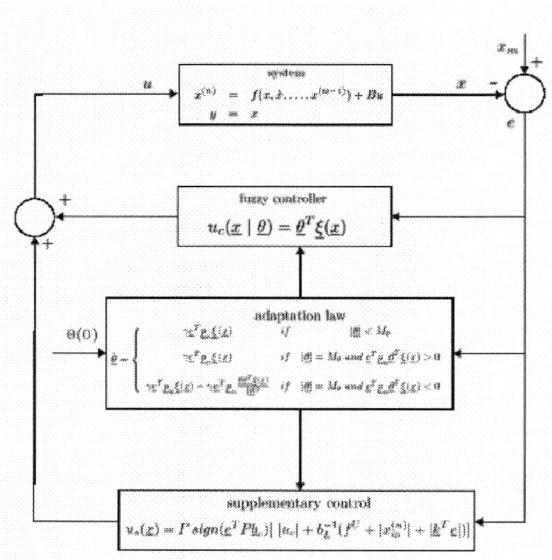

Figure 1. Adaptive Fuzzy Logic Control Scheme

Design procedure of Adaptive Fuzzy Control

1) Step 1:Off-line Preprocessing: We need to determine the desired characteristic polynomial $s^n+k_1s^{n-1}+.........+k_n = 0$, such that all the roots of are in the open left half-plane. In this research work we consider $k_1 = 2$, $k_2 = 1$so that $s^2 + 2s+1 = 0$ is stable. Then we need specify a positive definite $n \times n$ matrix Q . We define a 2×2 Q matrix for this particular problem. From the solution of Lyapunov equation

$$Q = \begin{bmatrix} 10 & 0 \\ 0 & 10 \end{bmatrix} P = \begin{bmatrix} 15 & 5 \\ 5 & 5 \end{bmatrix} K_c = \begin{bmatrix} 0 & 1 \\ -1 & -2 \end{bmatrix}$$

Then, we need to specify the design parameters M_θ, M_x, γ based on practical constrains. Here, we take $M_\theta = 30$, $M_x = 3$, $\gamma = 5$.

2) Step 2: Initial Controller Construction: We consider six fuzzy sets for each of the inputs. Then, we need to construct the fuzzy rule base, which consists of $m_1 \times m_2 ..... \times m_n$, whose IF part comprise all the possible combinations of all the fuzzy sets for all the inputs. The fuzzy basis function is of the following form:

$$\xi^{(l_1,\ldots,l_n)}(\text{x}) = \frac{\prod_{i=1}^{n} \mu_{F_i^{l_i}}(x_i)}{\sum_{l_1=1}^{m_1} \cdots\cdots \sum_{l_n=1}^{m_n} (\prod_{i=1}^{n} \mu_{F_i^{l_i}}(x_i))} \quad (20)$$

We collect ξ_(x) into a vector, then collect the maximum value of the THEN part into a vector θ, the fuzzy controller $u_c(\underline{x} \mid \underline{\theta})$ is then constructed as in (10). The supplementary controller, which will help in reducing the error, is given as

$$u_s(\underline{x}) = I^* sign(\underline{e}^T P \underline{b}_c)[\ |u_c| + b_L^{-1}(f^U + |x_m^{(n)}| + |\underline{k}^T \underline{e}|)] \tag{22}$$

$$x_m = reference\ \ input$$
$$e = error$$
$$f^U = upperbound$$

3) Step 3: On-line Adaptation: We apply the feedback controller u to the system under consideration, and use the adaptive law given by 16 to adjust the parameter vector θ. The adaptive law updates parameter vector θ, which effectively reduces the error between the desired output and the reference input. The adaptive fuzzy controller takes the input in the form of membership functions.

SIMULATION RESULTS

Before proceeding further, the following table gives the nominal plant parameter values and their respective range, which defines the upper and lower bound of their variation. The simulations were carried out for the following three cases.

Parameter	Nominal	Range
M1	2	1.4 to 2.6
M2	2	1.4 to 2.6
M3	2	1.4 to 2.6
K1	175	122.5 to 227.5
K2	450	315 to 585
K3	800	560 to 1040
K4	800	560 to 1040
C1	0.01	0.007 to 0.013

Table 1. Table of Parameters and their ranges

For comparison purposes, uncontrolled system response is obtained for the nominal case first. The simulations are carried out using both Fuzzy Logic and Adaptive Fuzzy Logic controllers designed to control the plant. All the initial conditions were set to zero.

Case	M1	M2	M3	K1	K2	K3	K4	C1
1	2	2	2	175	450	800	800	0.01
2	2.6	2.6	2.6	227.5	585	1040	1040	0.013
3	1.4	1.4	1.4	122.5	315	560	560	0.007

Table 2. Three Different Cases Considered for Simulations

CONCLUSIONS

The goal of this work was to compare the simulation results of Fuzzy Controller and Direct Adaptive Fuzzy Controller for the position control of the masses. The controllers were designed for this system and satisfactory results have been obtained. As stated earlier, due to page limitations, results from adaptive fuzzy controller are presented. The Fuzzy Inference System and Direct Adaptive Fuzzy algorithm were found to be capable of handling parametric variations and changes in the system configuration. We saw from the results that Direct Adaptive Fuzzy Controller is more effective than the non-adaptive fuzzy control.

In the following, the simulation results of the controlled system are given.

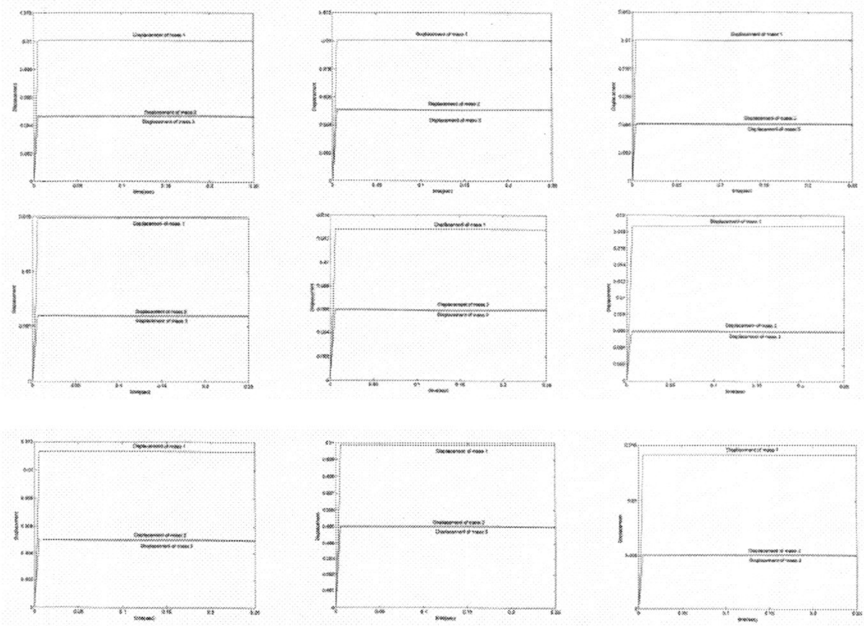

Figure 2. Positions of Mass 1(top), 2(middle), and 3 (bottom) for cases 1, 2, 3.

REFERENCES

[1] L. Zadeh. Fuzzy Sets. Informat. Control, 1965. Vol. 8, pages 338-353.

[2] E. H. Mamdani. Applications of Fuzzy Algorithms for Simple Dynamic Plant. Proc. IEE, 1974. Vol. 121, No 12, pages 1585-1588.

[3] E. H. Mamdani and S. Assilian. An Experiment in Linguistic Synthesis with a Fuzzy Logic Controller. Int. J. Man Machine Studies, 1975. Vol. 7, No 1, pages 1-13.

[4] Moon G. Joo, Jin S. Lee. Universal Approximation by Heirarichal Fuzzy system with Constraints of Fuzzy Rules. Fuzzy Sets and Systems, 2002. Vol. 130, pp.175-188.

[5] Xinping Guan and Cailian Chen. Adaptive Fuzzy Control for Chaotic Systems with H1 Tracking Performance. Fuzzy Sets and Systems, 2003. Vol. 139, Pages 81-93.

[6] L. X. Wang. Stable Adaptive Fuzzy Control of Nonlinear Systems. 31st IEEE Conf. on Decision and Control, 1992. Tucson, pages 2511-2516.

[7] L. X. Wang. Adaptive Fuzzy Systems and Control. Prentice Hall, New Jersey, 1994.

[8] Z. Kovacic and S. Bogdan. Fuzzy Controller Design: Theory and Appl. CRC Press, 2005.

[9] R. Hampel, M. Wagenknecht, N. Chaker. Fuzzy Control: Theory and Practice. Springer-Verlag, New York Inc., 2000.

[10] A. Piegat. Fuzzy Modeling and Control. Springer-Verlag, New York Inc., 2001.

[11] V. Korotkich and V. Dimitrov. Fuzzy Logic*: A Framework for the New Millenium.* Springer-Verlag, New York Inc., 2002.

INTELLIGENT CONTROL USING INTERVAL TYPE-2 FUZZY LOGIC

OSCAR CASTILLO, AND PATRICIA MELIN
Tijuana Institute of Technology, Mexico
{ocastillo, pmelin}@tectijuana.mx

ABSTRACT

Uncertainty is an inherent part of intelligent systems used in real-world applications. The use of new methods for handling incomplete information is of fundamental importance. Type-1 fuzzy sets used in conventional fuzzy systems cannot fully handle the uncertainties that are present in intelligent systems. Type-2 fuzzy sets that are used in type-2 fuzzy systems can handle such uncertainties in a better way because they provide us with more parameters. This paper deals with the design of intelligent systems using interval type-2 fuzzy logic for minimizing the effects of uncertainty produced by the instrumentation elements, environmental noise, etc.

INTRODUCTION

Uncertainty affects decision-making and appears in a number of different forms [3, 6]. The concept of information is fully connected with the concept of uncertainty. The most fundamental aspect of this connection is that the uncertainty involved in any problem-solving situation is a result of some information deficiency, which may be incomplete, imprecise, fragmentary, not fully reliable, vague, contradictory, or deficient in some other way [4]. Uncertainty is an attribute of information [21, 22]. The general framework of fuzzy reasoning allows handling much of this uncertainty, fuzzy systems employ type-1 fuzzy sets, which represent uncertainty by numbers in the range [0, 1]. When something is uncertain, like a measurement, it is difficult to determine its exact value, and of course type-1 fuzzy sets make more sense than using sets [19, 20]. However, it is not reasonable to use an accurate membership function for something uncertain, so in this case what we need is another type of fuzzy sets, those which are able to handle these uncertainties, the so called type-2 fuzzy sets [7, 23]. So, the amount of uncertainty in a system can be reduced by using type-2 fuzzy logic, because it offers better capabilities to handle linguistic uncertainties by modeling vagueness and unreliability of information [24].

Recently, we have seen the use of type-2 fuzzy sets in Fuzzy Logic Systems (FLS) in different areas of application [1, 11]. A novel approach for realizing the vision of ambient intelligence in ubiquitous computing environments (UCEs), is based on embedding intelligent agents that use type-2 fuzzy systems which are able to handle the different sources of uncertainty and imprecision in UCEs to give a good response [2]. There are also papers with emphasis on the implementation of type-2 FLS [10] and in others, it is explained how type-2 fuzzy sets let us model and minimize the effects of uncertainties in rule-base FLS [12]. In signal processing, type-2 fuzzy logic is applied in prediction of the Mackey-Glass chaotic time-series with uniform noise presence [8, 25]. In medicine, an expert system was developed for solving the problem of Umbilical Acid-Base (UAB) assessment. In industry, type-2 fuzzy logic and neural networks was used in the control of non-linear dynamic plants [2]; also we can find studies in the field of mobile robots [5, 16]. In this paper we deal with the application of interval type-2 fuzzy control to non-linear dynamic systems. It is a well known fact, that in the control of real systems, the instrumentation elements (instrumentation amplifier, sensors, digital to analog, analog to digital converters, etc.) introduce some sort of unpredictable values in the information that has been collected. So, the controllers designed under idealized conditions tend to behave in an inappropriate manner.

FUZZY LOGIC

Soft computing techniques have become an important research topic, which can be applied in the design of intelligent controllers [15], which utilize the human experience in a more natural form than the conventional mathematical approach [17, 18, 19, 20]. A FLS, described completely in terms of type-1 fuzzy sets is called a type-1 fuzzy logic system (type-1 FLS). It is composed by a knowledge base, which comprises the information given by the process operator in form of linguistic control rules. A fuzzification interface, that has the effect of transforming crisp data into fuzzy sets. An inference system, that uses the fuzzy sets in conjunction with the knowledge base to make inferences by means of a reasoning method. Finally, a defuzzification interface, which translates the fuzzy control action so obtained to a real control action using a defuzzification method [7]. In this paper, the fuzzy controller, has two input variables, which are the error $e(t)$, the difference between the reference signal and the output of the process, as well as the error variation $\Delta e(t)$,

$$e(t) = r(t) - y(t) \quad (1)$$

$$\Delta e(t) = e(t) - e(t-1) \quad (2)$$

so the control system can be represented as in Figure 1.

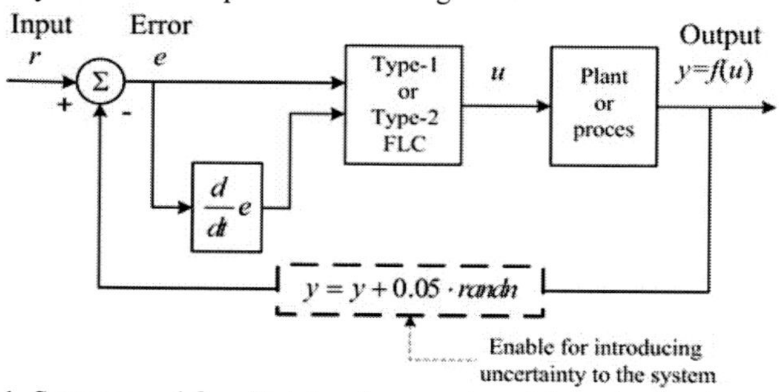

FIGURE 1. System used for obtaining the experimental results for the experiments.

A type-2 FLS is again characterized by IF-THEN rules, but its antecedent or consequent sets are now of type-2. Type-2 FLSs, can be used when the circumstances are too uncertain to determine exact membership grades such as when the training data is corrupted by noise. Similar to a type-1 FLS, a type-2 FLS includes a fuzzifier, a rule base, fuzzy inference engine, and an output processor, as we can see in Figure 2. The output processor includes type-reducer and defuzzifier; it generates a type-1 fuzzy set output (from the type-reducer) or a crisp number (from the defuzzifier) [9, 13].

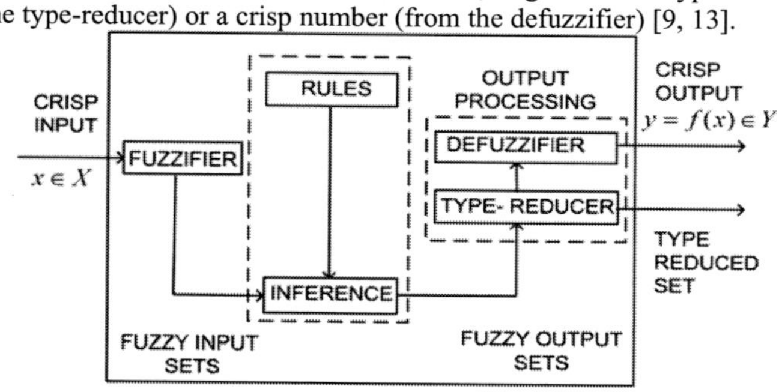

FIGURE 2. Type-2 Fuzzy Logic System

EXPERIMENTAL RESULTS

The experimental results are devoted to show comparisons in the system's response in a feedback controller when using a type-1 FLC or a type-2 FLC. A set of five experiments is described in this section. The first two experiments were performed in ideal conditions, i.e., without any kind of disturbance. In the last three experiments, Gaussian noise was added to the feedback loop with the purpose of simulating, in a global way, the effects of uncertainty from several sources. Figure 1 shows the feedback control system that was used for obtaining the simulation results. The complete system was simulated in Matlab, and the controller was designed to follow the input as closely as possible. The plant is a nonlinear system modeled with equation (3)

$$y(i)=0.2\cdot y(i-3)\cdot 0.07y(i-2)+0.9\cdot y(i-1)+0.05\cdot u(i-1)+0.5\cdot u(i-2) \qquad (3)$$

To illustrate the dynamics of this non-linear system, two different inputs are applied, first the input indicated by equation (4).

$$u(i)=\begin{cases} 0 & 1 \le i < 5 \\ .1 & 5 \le i < 10 \\ .5 & 10 \le i < 15 \\ 1 & 15 \le i < 20 \\ .5 & 20 \le i < 25 \\ 1 & 25 \le i < 30 \\ 0 & 30 \le i < 35 \\ 1.47 & 35 \le i < 40 \end{cases} \qquad (4)$$

Now, for a slightly different input given by equation (5).

$$u(i)=\begin{cases} 0 & 1 \le i < 5 \\ .1 & 5 \le i < 10 \\ .5 & 10 \le i < 15 \\ 1 & 15 \le i < 20 \\ .5 & 20 \le i < 25 \\ 1 & 25 \le i < 30 \\ 0 & 30 \le i < 35 \\ 1.4 & 35 \le i < 40 \end{cases} \qquad (5)$$

Going back to the control problem, this system given by equation (3) was used in Figure 1, under the name of plant or process, in this figure we can see that the controller's output is applied directly to the plant's input. Since we are interested in comparing the performance between type-1 and type-2 FLC systems, the controller was tested in two ways:

1. One is considering the system as ideal, that is, not introducing in the modules of the control system any source of uncertainty (experiments 1 and 2).
2. The other one is simulating the effects of uncertain modules (subsystems) response introducing some uncertainty (experiments 3, 4 and 5).

For both cases, as it is shown in Figure 1, the system's output is directly connected to the summing junction, but in the second case, the uncertainty was simulated introducing random noise with normal distribution (the dashed square in Figure 1). We added noise to the system's output $y(i)$ using the Matlab's function "randn", which generates random numbers with Gaussian distribution. The signal and the added noise in turn, were obtained with the programmer's expression (6), the result $y(i)$ was introduced to the summing junction of the controller system. Note that in expression (6) we are using the value 0.05, for experiments 3 and 4, but in the set of tests for experiment 5, we varied this value to obtain different SNR values.

$$y(i) = y(i) + 0.05 \cdot randn \tag{6}$$

The system was tested using as input, a unit step sequence free of noise, $r(i)$. For evaluating the system's response and comparing between type 1 and type 2 fuzzy controllers, the performance criteria ISE, IAE, and ITAE were used [14]. In Table 1, we summarized the values obtained in an ideal system for each criterion considering 400 units of time. For calculating ITAE a sampling time of $T_s = 0.1$ sec. was considered. For all experiments the reference input r is stable and noisy free. In experiments 3 and 4, although the reference appears clean, the feedback at the summing junction is noisy since noise for simulating the overall existing uncertainty in the system was introduced deliberately, in consequence, the controller's inputs $e(t)$ (error), and $\Delta e(t)$ contain uncertainty in the data.

In experiments 2, 4, and 5, for the type-2 FLC, as in type-1 FLC, we also selected Gaussian MFs for the inputs and for the output, but in this case we have interval type-2 Gaussian MFs with a fixed center, *c*, and an uncertain standard deviation, σ. Hence, in the type-2 FLC, for each input we defined three interval type-2 fuzzy GaussianMFs: negative, zero, positive in the interval [-10 10], as illustrated in Figures 3 and 4. For computing the output we have five interval type-2 fuzzy Gaussian MFs, which are NG, N, Z, P and PG, in the interval [-10 10]. For the type-2 FLC we used, basically, a software for type-2 fuzzy logic developed by our research group.

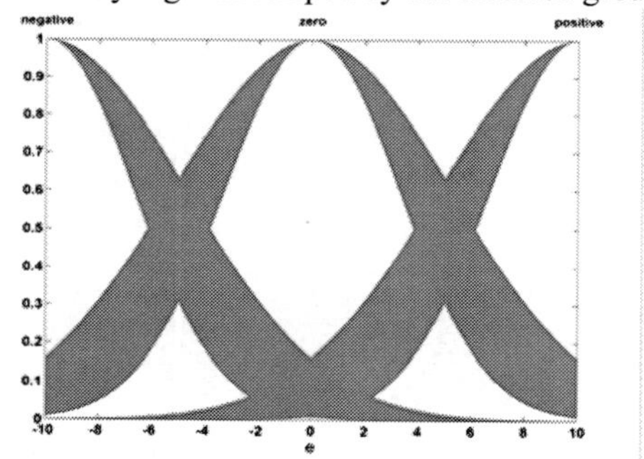

Figure 3. Input e membership functions for the type-2 FLC.

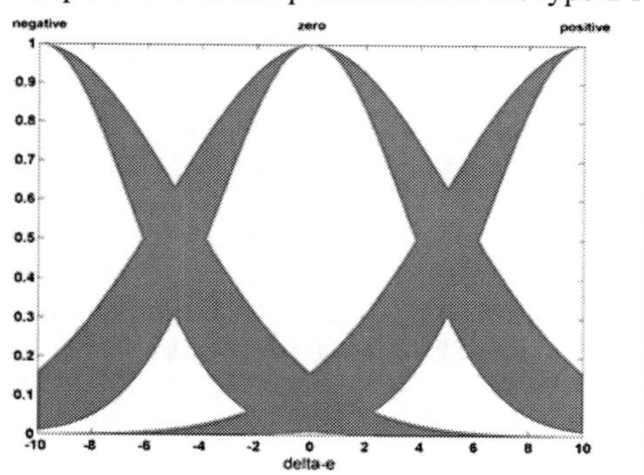

Figure 4. Input Δe membership functions for the type-2 FLC.

Experiment 1: Simulation of an ideal system with a type-1 FLC.
In this experiment, uncertainty data was not added to the system, and the system response produced a settling time of about 140 units of time; i.e., the system tends to stabilize with time and the output will follow accurately the input. In Table 3, we listed the values of ISE, IAE, and ITAE for this experiment.
Experiment 2: Simulation of an ideal system using the type-2 FLC.
Here, the same test conditions of Experiment 1 were used, but in this case, we implemented the controller's algorithm with type-2 fuzzy logic. The corresponding performance criteria are listed in Table 3, and we can observe that using a type-2 FLC we obtained the lower errors.
Experiment 3: System with uncertainty using a type-1 FLC.
In this case, equation (25) was used to simulate the effects of uncertainty introduced to the system by transducers, amplifiers, and any other element that in real world applications affects expected values. In this experiment the noise level was simulated in the range of 20 db of SNR ratio.
Experiment 4: System with uncertainty using a type-2 FLC.
In this experiment, uncertainty was introduced in the system, in the same way as in Experiment 3. In this case, a type-2 FLC was used and the results obtained with a type-1 FLC (Experiment 3) were improved.
Experiment 5. Varying the Signal to Noise Ratio (SNR) in type-1 and type-2 FLCs.
To test the robustness of the type-1 and type-2 FLCs, we repeated experiments 3 and 4 giving different noise levels, going from 30 db to 8 db of SNR ratio in each experiment. In presence of different noise levels, the behavior of type-2 FLC is in general better than type-1 FLC.

TABLE 1. COMPARISON OF PERFORMANCE CRITERIA FOR TYPE-1 AND TYPE-2 FUZZY LOGIC CONTROLLERS FOR 20 db SIGNAL NOISE RATIO. VALUES OBTAINED AFTER 200 SAMPLES.

Performance Criteria	**Type-1 FLC**		**Type-2 FLC**	
	Ideal System	Syst. with uncertainty	Ideal System	Syst. with uncertainty
ISE	7.65	19.4	6.8	18.3
IAE	17.68	49.5	16.4	44.8
ITAE	62.46	444.2	56.39	402.9

CONCLUSIONS

We have presented the study of the controllers' design for nonlinear control systems using type-1 and type-2 fuzzy logic. We presented five experiments where we simulated the systems' responses with and without uncertainty presence. In the experiments, a quantification of errors was achieved and documented in tables for different criteria such as ISE, IAE, and ITAE, it was shown that the lower overshoot errors and the best settling times were obtained using a type-2 FLC. Based on the experimental results, we can say that the best results are obtained using type-2 fuzzy systems. In our opinion, this is because type-2 fuzzy sets that are used in type-2 fuzzy systems can handle uncertainties in a better way because they provide us with more parameters and more design degrees of freedom.

REFERENCES

[1] Ching-Hung Lee, Jang-Lee Hong, Yu-Ching Lin and Wei-Yu Lai, Type-2 Fuzzy Neural Network Systems and Learning, International Journal of Computational Cognition 1 (4) (2003) 79-90.

[2] Faiyaz Doctor, Hani Hagras and Victor Callaghan, A type-2 fuzzy embedded agent to realize ambient intelligence in ubiquitous computing environments, Information Sciences 171 (2005) 309-334.

[3] G. E. P. Box and G. M. Jenkins, Time series analysis, forecasting and control, Holden Day, San Francisco, 1976.

[4] George J. Klir, Bo Yuan, Fuzzy Sets and Fuzzy Logic: Theory and Applications, Prentice Hall, New Jersey , 1995.

[5] H. A. Hagras, Hierarchical Type-2 Fuzzy Logic Control Architecture for Autonomous Mobile Robots, IEEE Transactions on Fuzzy Systems 12 (4) (2004) 524-539.

[6] Jerry M. Mendel, Computing Derivatives in Interval Type-2 Fuzzy Logic Systems, IEEE Transactions on Fuzzy Systems 12 (2004) 84-98.

[7] Jerry M. Mendel, Uncertain Rule-Based Fuzzy Logic Systems: Introduction and new directions, Prentice Hall, New Jersey, 2001.

[8] Jerry M. Mendel, Uncertainty, fuzzy logic, and signal processing, Signal Processing J. 80 (2000) 913-933.

[9] Jerry M. Mendel, On a 50% savings in the computation of the centroid of a *symmetrical* interval type-2 fuzzy set, Information. Sciences 172 (3-4) (2005) 417-430.

[10] Jerry M. Mendel, Type-2 fuzzy logic systems: Type-reduction, in: IEEE Syst., Man, Cybern. Conf., San Diego, CA, 1998, pp. 2046-2051.

[11] Jerry M. Mendel and George C. Mouzouris, Type-2 fuzzy logic systems, IEEE Transactions on Fuzzy Systems 7 (1999) 643–658.

[12] Jerry M. Mendel and Robert I. Bob John, Type-2 Fuzzy Sets Made Simple, IEEE Transactions on Fuzzy Systems 10 (2) (2002) 117-127.

[13] Jerry M. Mendel and Wu Hongwei, Properties of the Centroid of an Interval Type-2 Fuzzy Set, Including the Centroid of a Fuzzy Granule, in: Proc. FUZZ-IEEE 2005, Reno, NV, 2005, pp. 341-346.

[14] John G. Proakis, Dimitris G. Manolakis, Digital Signal Processing principles, Algorithms, and Applications, Third Edition, Prentice Hall, New Jersey, 1996.

[15] Jyh-Shing R. Jang, C.-T. Sun, E. Mizutani. Neuro-Fuzzy and Soft Computing, A Computational Approach to Learning and Machine Intelligence, Matlab Curriculum Series, Prentice Hall, New Jersey, 1997.

[16] Leslie Astudillo, Oscar Castillo, Luis T. Aguilar, Intelligent Control of an Autonomous Mobile Robot Using Type-2 Fuzzy Logic, in: Proc. Conference on Artificial Intelligence, ICAI′06, Las Vegas NV, 2006, pp. 565-570.

[17] L. A. Zadeh, Outline of a New Approach to the Analysis of Complex Systems and Decision Processes, IEEE Transactions on Systems, Man, and Cybernetics 3 (1973) 28-44.

[18] L. A. Zadeh, Similarity Relations and Fuzzy Ordering, Information Sciences 3 (1971) 177-206.

[19] L. Zadeh, The concept of a linguistic variable and its application to approximate reasoning, Part 1, Information Sciences 8 (3) (1975) 199-249.

[20] L. A. Zadeh, The Concept of a Linguistic Variable and its Application to Approximate Reasoning, Part II, Information Sciences 8 (4) 301-357.

[21] L. A. Zadeh, Toward a generalized theory of uncertainty (GTU)- an outline, Information Sciences 172 (2005) 1-40.

[22] Mamdani, E. H., Twenty years of Fuzzy Control: Experiences Gained and Lessons Learned, in: Proc. 2nd IEEE International Conference on Fuzzy Systems, San Francisco, CA, 1993, pp. 339-344.

[23] M. Mizumoto and K. Tanaka, Some properties of fuzzy sets of type-2, Information and Control 31 (1976) 312–340.

[24] M. Wagenknecht and K. Hartmann, Application of Fuzzy Sets of Type 2 to the Solution of Fuzzy Equations Systems, Fuzzy Sets and Systems 25 (1988) 183-190.

[25] Nilesh N. Karnik, Jerry M. Mendel, Applications of Type-2 Fuzzy Logic Systems to Forecasting of Time-series, Information Sciences 120 (1-4) (1999) 89-111.

FUZZY CLUSTERING REGRESSION MODEL AND A SATISFYING SOLUTION FOR AN LP PROBLEM

R. Alex
Department of Engineering and Computer Science
West Texas A&M University, Canyon, TX 79016

ABSTRACT

This work will employ methods from fuzzy regression models to build a fuzzy clustering regression model that can be used in statistical process control. Suppose that a subset D of R^n is a set of sampling points that represents n quality characteristics of K objects and a set S represents a fuzzy score space that provides ratings by experts on the characteristics of the K objects. The fuzzy ratings given by the experts can be described as a mapping $f : D \rightarrow S$. Suppose further that a clustering can be preformed on D to generate q classes based upon the experts' ratings. Let C be the set of centers of the q clustering classes. Using a special method this work will generate a core mapping f^0 from C to S based on the mapping f. Then extend the mapping f^0 to a complete mapping $f^* : F \rightarrow S$. The extension mapping is the fuzzy clustering regression model. This work will give the details of the special mapping and the extension mapping evaluations. In particular, formulation of the special mapping f^0 that involves finding a satisfying solution for a special kind of linear programming problem will be presented.

1. INTRODUCTION

Statistical Process Control is a method of studying a process from a set of small random samples from the process. The quality characteristics of each item in the sample is recorded and graphed in a chart known as Shewhart process control chart or simply a control chart. The control chart provides a graphical time series of the data from a process. There are now several distinct types and variations in use. The basic idea involved is to collect a small sample of the process to be studies. In the chart, there is a central line (CL) and two control lines referred to as the upper control line (UCL) and the lower control line (LCL). For the details of the calculation of the lines in a control chart we refer to Ott and Schilling (1990). In the chart, the central line represents an estimate of the process level, and the two control lines denote the boundaries of the normal variations. Samples are drawn from the process and the characteristic under consideration is plotted on the control chart. If any item's characteristic falls outside of the UCL and LCL, then the chart indicates the existence of an assignable cause (Ott and Schilling, 1990). Essentially, the control chart tests the hypothesis that the process is in a state of statistical control. A plotted data falling within the control lines confirms the hypothesis of the statistical control, while a point falling outside of the control lines confirms the rejection of the hypothesis. However, this elegant approach to process control has some potential flaws. Firstly, in reducing the vague observations to exact numbers, the information content of the original data also gets reduced. Secondly, the uncertainty of vague observations is essentially non-statistical. A fuzzy process control involving the construction of control chart with fuzzy numbers is presented by Cheng (2005). This work involves the task of solving a kind of linear programming problem which we call as the Lee-Tanaka's LP problem.

As an alternative, this study proposes fuzzy clustering regression model to monitor a process whose outcomes are represented by fuzzy numbers. In Section 2, we will give a satisfying solution for the Lee-Tanaka's LP problem. In Section 3, we will give the details of the extension mapping f^* and the conclusion is given in Section 4.

2. A SATISFYING SOLUTION OF LEE-TANAKA'S LP PROBLEM

Suppose that D is a subset of R^n is a set of sampling points that represents n quality characteristics of K objects. Let S represent a fuzzy score space that provides ratings by experts on the qualities of the K objects. The fuzzy ratings given by the experts can be described as a mapping $f : D \to S$. Suppose further that a clustering can be preformed on D to generate q classes based upon the experts' ratings. A general mathematical treatment of fuzzy clustering is given by Granath (1984), and fuzzy regression analysis is given by Cheng and Lee (2001) and Tanaka (1987). Let C be the set of centers of the q clustering classes are denoted as $C_1, C_2, \ldots, C_q$. Each C_j has $s(j)$ sampling points for $j = 1,2,\ldots,q$. For the center $c_j^0 \in C_j$, rate it with a triangle fuzzy number $\boldsymbol{y}_j^0 = (l_j^0, c_j^0, r_j^0) \in S$ by a mapping f^0, where c_j^0 is the mode and l_j^0, r_j^0 are the left, right boundary points of $\boldsymbol{y}_j^0$, respectively. How to get the rating $\boldsymbol{y}_j^0$? How to determine c_j^0, l_j^0, and r_j^0? The proposed mapping f^0 is like the hidden level in the FRBFN given in Cheng (2005), Cheng and Lee (2001). It also reflects some rating laws being captured by the fuzzy inference the fuzzy regression mode Alex (2006), Alex (2004). We can borrow some idea from the learning machine of neural networks to get the ratings. Tanaka and Lee transferred the learning task of fuzzy regression neural networks by means of solving a kind of linear programming, called the Lee-Tanaka's LP problem. Let c_j^0 denote the center of the class $C_j = \{x_{(1)}, x_{(2)}, \ldots, x_{(s(j))}\}$. To identify the l_j^0, c_j^0, and r_j^0, Lee and Tanaka (1999) first applied the regular least squares method to identify its mode, c_j^0, by minimizing the mean square error $\frac{1}{s(j)} \sum_k \{(c_k - c_j^0)^2 \mid x_k \in C_j\}$, where c_k is the mode of x_k and $s(j)$ is the number of sampling points in the jth class C_j. To minimize, we take the partial derivative and set it to zero and get,

$$\frac{\partial}{\partial c_j^0}\left(\sum_k \{(c_k - c_j^0)^2 \mid x_k \in C_j\}\right) = -2\left(\sum_k \{c_k \mid x_k \in C_j\} - s(j)c_j^0\right) = 0$$

Solving, we get the mode of the jth class C_j as

$$c_j^0 = \frac{\sum_k \{c_k \mid x_k \in C_j\}}{s(j)} \qquad (2.1)$$

Then a linear programming problem is formulated to identify the left and right boundary l_j^0 and r_j^0 as follows:

$$\min \quad \sum_{k=1}^{K}\sum_{j=1}^{q} f_{jk} \cdot (r_j^0 - l_j^0), \qquad (2.2)$$

$$s.t. \quad l_j^0 \le c_j^0 \le r_j^0, \quad j = 1,2,\ldots,q, \qquad (2.3)$$

$$\sum_{j=1}^{q} f_{jk} l_j^0 \le a_k, \quad k = 1,2,\ldots,K, \qquad (2.4)$$

$$\sum_{j=1}^{q} f_{jk} r_j^0 \ge b_k, \quad k = 1,2,\ldots,K, \qquad (2.5)$$

Here, we need to re-explain our understanding for the parameters f_{jk} occurring in (2.2)-(2.5).

Let $D = \{x_k = (x_{k1},\ldots,x_{kn})\} \quad (k = 1,\ldots,K)$ and $C = \{x_j^0 = (x_{j1}^0,\ldots,x_{jn}^0)\} \quad (j = 1,\ldots,q)$. For any $\mathbf{x_k} \in D$ and $\mathbf{x_j^0} \in C$, set

$$d_{jk} = \sum_i | x_{ki} - x_{ji}^0 |. \qquad (2.6)$$

$$d^\wedge = \max\{d_{jk} \mid j = 1,2,\ldots,q; k = 1,\ldots,K\}. \qquad (2.7)$$

$$h_{jk} = \frac{(d^\wedge - d_{jk})}{d^\wedge}. \qquad (2.8)$$

$$f_{jk} = \frac{h_{jk}}{\sum_{j'} h_{j'k}}. \qquad (2.9)$$

$$f_j = \frac{\sum_k f_{jk}}{K}, \qquad (2.10)$$

We can see that $f_1, f_2, \ldots, f_q$ form a group of weights. To emphasize that they are the weights, we re-write f_j as w_j. The formula (2.2) becomes

$$\min \sum_j w_j (r_j^0 - l_j^0) = \sum_j w_j r_j^0 - \sum_j w_j l_j^0 \qquad (2.2')$$

Firstly, change the order of the weights w_j such that the modes c_j are non-decreasing. That is,

$$c_1 \le c_2 \le \cdots \le c_q \tag{2.11}$$

Consider the solution $(x_1, x_2, \ldots, x_q)$ of the equation

$$\sum_j w_j x_j = 0 \quad (\sum_j w_j = 1) \tag{2.2”}$$

Proposition 2.1

Suppose we take

$P_1 = 0$;

$$P_j = \frac{w_1}{w_1 + w_2 + \cdots + w_q} + \frac{w_2}{w_2 + w_3 + \cdots + w_q} + \cdots \frac{w_{j-1}}{w_{j-1} + w_j + \cdots + w_q},$$

$2 \le j \le q$;

$$N_j = \frac{w_{j+1} + \cdots + w_q}{w_j + \cdots + w_q}, \quad 1 \le j \le q-1;$$

$$N_q = 0. \tag{2.12}$$

Then a natural solution of (2.2”) is

$$x_j = P_j - N_j,\ j = 1,2,\ldots,q \tag{2.13}$$

Set

$$l_j^0 = e_1 + \delta_1 x_j \le r_j^0 = e_2 + \delta_2 x_j \quad (j = 1,2,\ldots,q) \tag{2.14}$$

we have that

$$\sum_j w_j (r_j^0 - l_j^0) = \sum_j w_j r_j^0 - \sum_j w_j l_j^0 = e_2 - e_1 \ge 0$$

When $e_2 = e_1$, it reaches the minimum of the sum in (2.2’). Adjust the four parameters $e_1, e_2, \delta_1, \delta_2$ such that (2.14) satisfy (2.3)-(2.5) and we can get the solution of Lee-Tanaka’s LP problem. Even though it may not be optimal, it is a satisfying solution of course.

3. THE COMPLETE MAPPING

Now, let us consider the complete mapping $f^* : F \to S$. For any point $\mathbf{x}^* = (x_{*1}, \ldots, x_{*n}) \in F$, $f^*(x^*) = y^*$ is a point in the score space S, which is a triangle fuzzy number, $y^* = (l^*, c^*, r^*)$ is generated by means of the core mapping as follows:

Define

$$d_{j*} = \sum_i | x_{*i} - x_{ji}^0 |, \quad j = 1,\ldots,q, \tag{3.1}$$

where $\mathbf{x}_j^0 = (x_{j1}^0, \ldots, x_{jn}^0), \quad j = 1, \ldots, q$, are the center points of clustered classes,

$$d^\wedge = \max\{d_{j*} \mid j = 1, \ldots, q\}, \tag{3.2}$$

$$h_{j*} = \frac{d^\wedge - d_{j*}}{d^\wedge}, \tag{3.3}$$

$$f_{j*} = \frac{h_{j*}}{\sum_j h_{j*}}, \tag{3.4}$$

$$c^* = \sum_j f_{j*} c_j^0,\ l^* = \sum_j f_{j*} l_j^0,\ r^* = \sum_j f_{j*} r_j^0. \tag{3.5}$$

For an arbitrary point $\mathbf{x}^*$ in the feature space F, we can rate it by means of the procedure mentioned above.

4. CONCLUSION

This paper presents a satisfying solution to replace the ordinary solution of a kind of LP problem with a mathematical formula. The idea of satisfying solution comes from the fuzzy methodology on optimization. A satisfying solution of a LP problem is one of a set of solutions. It need not necessarily be the maximal solution, but 'very near by the maximum' with a high membership degree. That is, in fuzzy regression, we need not find the very maximum point of the LP problem; a satisfying solution is enough. Irrespective of the feasible region being empty or not, the satisfying solutions always exist.

Reference

Alex, R., 2006, "A new kind of fuzzy regression modeling and its combination with fuzzy inference," *Soft Computing Journal*, Vol. 10, No. 7, pp. 618-621.

Alex, R., 2004, "Fuzzy normal regression model and related neural networks," *Soft Computing Journal*, Vol. 8, No. 10, pp. 717 – 721.

Cheng, C.-B., 2005, "Fuzzy process control: construction of control charts with fuzzy numbers," *Fuzzy Sets and Systems*, Vol. 154, pp. 287-303.

Cheng, C.-B., Lee, E.S., 2001, "Fuzzy regression with radial basis function network," *Fuzzy Sets and Systems*, Vol. 119, pp. 291-301.

Granath, G., (1984), "Application of fuzzy clustering and fuzzy classification to evaluate the provenance of glacial till," *Journal of Mathematical Geology*, Vol. 16, pp. 283-300.

Ott, E.R., Schilling, E.G., (1990), "Process Quality Control," *McGraw-Hill Publishing Company*.

Lee, H., Tanaka, H., (1999), "Fuzzy approximations with non-symmetric fuzzy parameters in fuzzy regression analysis," *Journal of the Operations Research Society of Japan*, Vol. 42, pp. 98-112.

Tanaka, H., (1987), "Fuzzy data analysis by possibilistic linear modes," *Fuzzy Sets and Systems,* Vol. 24, pp. 363-375.

ACCUMULATED STEP DEVIATION METHOD FOR AUTOMATIC FUZZY SET GENERATION

BRUNO RENÉ SANTOS
Uninova, Monte de Caparica
Portugal

MANUEL BARATA
ISEL, Lisboa
Portugal

RITA A. RIBEIRO
Uninova, Monte de Caparica
Portugal

PEDRO SOUSA
Uninova, Monte de Caparica
Portugal

ABSTRACT
This paper presents a generic methodology for automatically generating fuzzy sets from acquired sensor data, for monitoring purposes. The novel method, denoted Accumulated Step Deviation (ASD), was devised within the scope of a European Space Agency (ESA) project, called MODI. MODI's main objective was to develop two monitoring systems: one for monitoring a drill and another for terrain detection. This project included an automatic learning module, where the ASD method was used. The ASD method is based on the traditional data preparation process, which begins with data acquisition from sensors and ends with the automatic generation of fuzzy sets to be used as input variables by the monitoring system. An illustrative example is used to discuss the behavior of the ASD method and perform a comparative study between the ASD method and a traditional Trapezoidal method.

INTRODUCTION

This work was developed within the context of the European Space Agency (ESA) project "MODI - Simulation of a Knowledge Enabled Monitoring and Diagnosis Tool for ExoMars Pasteur Payloads" (CA3 2006), which objectives are two-folded: a) to develop a monitoring system to detect faulty behaviours during the real-time drilling process; b) to develop a terrain recognition system capable of assessing which type of terrain is being drilled. Here we only address one module of the MODI, the automatic learning module, which is responsible for the automatic *fuzzification* of the sensor variables.

Hence, the focus of this paper is on proposing and discussing an automatic method, Accumulated Step Deviation (ASD), for generating fuzzy sets to be used as input variables on monitoring systems. This paper is based on previous initial work by the authors (Santos, Fonseca et al. 2006). This process is generic and can be used on any kind of phenomenon that can be perceived by sensors. We will also discuss a trapezoidal method for *fuzzifying* variables, which was part of the MODI automatic learning module, to compare both methods. To discuss the behaviour of the ASD method and compare it with the trapezoidal approach we use a small illustrative example based in one variable of the MODI project.

DATA PREPARATION

The MODI data preparation process is based on the classical data preparation paradigm (Cios, Pedrycz et al. 1998; Santos, Fonseca et al. 2006) as shown in Figure 1. In addition to the classic (crisp) paradigm a new task was added, Fuzzy Set Generation. This step is needed for the automatic construction of fuzzy membership functions that represent the nominal behaviour of each sensor variable. These fuzzy variables will later be used in the fuzzy inference system, as linguistic input variables (details about fuzzy set theory and fuzzy inference systems can be seen in (Ross 2004)).

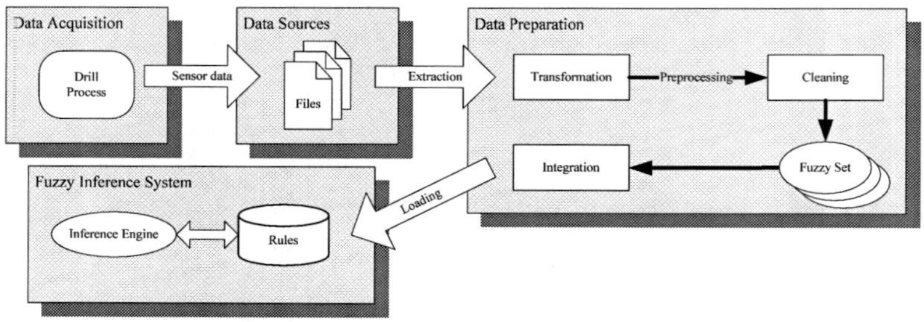

Figure 1 – Data Preparation Process Flow

Data acquisition and extraction is the phase where the drilling behaviour will be acquired from a range of sensors, used to perceive the phenomenon. In our case, the drill definitions were based on a real set-up developed by Galileo Avionica (Magnani, Ylikorpi et al. 2004) and further implemented in a simulator (CA3 2006).

In the transformation phase data is transformed from a raw state into data most suited for decision support (information). During the transformation phase the variables are transformed into a specific format using pre-defined transformations that will allow fuzzy membership functions to be generated.

The cleaning phase includes the elimination of erroneous records and checking the data fields for consistency and missing values.

With all data transformed and cleaned, it is now time to generate the fuzzy sets from the data received during the acquisition phase. This process is usually called fuzzification of input variables in inference systems (Ross 2004).

In this phase the fuzzy membership functions are generated for all variables (excluding the set points), using two methods: Trapezoidal Generation and Accumulated Step Deviation (ASD). Both methods capture the behaviour of the physical process using different approaches.

In the integration phase, all data from different scenarios will be merged, in order to define the linguistic variables corresponding to each input variable. The linguistic variables will have M elements, each corresponding to a combination of pre-defined terrains and two speeds (set points).

TRAPEZOIDAL METHOD

The trapezoidal method uses the mean and standard deviation in order to build a membership function from sensors' data. The function is calculated using as its middle value (height of the fuzzy set) the mean for the data on one scenario for a specific variable. The interval where the membership value is the highest (value of 1) is built using 10% of the standard deviation value of the data used. The minimum and the maximum values, where the membership reaches the membership value 0, are respectively calculated by the mean minus the standard deviation and the mean plus the standard deviation.

ACCUMULATED STEP DEVIATION

The Accumulated Step Deviation method (ASD) aims to incorporate on its results the distribution of samples on the data. For that we divide the scenario data in intervals with P parts to the left and P parts to the right of the middle membership value, which will be the mean value of the distribution, being P a predefined parameter. The ASD method has the following steps:

1. Identify the minimum, mean and maximum of the set. These will respectively correspond to values 0, 1 and 0 at the vertical (membership) axis.
2. Divide the range of values in P equally divided parts to the left and right of the membership function middle value, which will give us 2P parts in total, divided in length from the minimum value to the middle value for the left part and from the middle value to the maximum on the right part. This division is made separately for left bound and right bound, in order to set its last division on the middle value of the membership function.
3. Calculate the accumulated probability of one value being present from the lower (left) or upper (right) bound until the part-frontier point being calculated.
4. Obtain the total of points on the left (Total_L) and on the right (Total_R) of the mean.

Calculate each *MembershipLi* with the formula:

$$MembershipLi = \frac{\sum_{L1}^{i} count(Li)}{Total\_L} \quad (1)$$

where the value of count(Li) will be the number of case samples that fall between Li-1 and Li, being L0 = Min and LP = Mean. A similar formula is used for all Ri (right side).

Now let us assume that L is the array of value points to the left of the Mean and R the array of values to the right of the Mean. Moreover, let us assume MembershipLi,x is the x coordinate and MembershipLi,y is the y coordinate of MembershipLi, which is the left upper point of the interval where the searched value is located (the same for the right bound). With this assumptions we are now in position to define the algorithm for calculating the membership for each value.

The ASD method maintains coherence over all fuzzyset, by assuming MembershipLMin = 0, MembershipRMax = 0 and MembershipLMean = MembershipRMean = 1. On each side of the Mean the function is always monotonic and non-decreasing on the left side of Mean, and monotonic and non-increasing on the right, one of the properties of fuzzy sets (Zadeh 1965).

COMPARATIVE STUDY

In this section, both methods will be analyzed and compared. To this end, we use data obtained for the thrust sensor, during the drilling of terrain Concrete, with three different predefined set points for translational speed and rotational speed: 0.3 mm/min and 100 RPM (Rotations per minute); 0.5 mm/min and 140 RPM; and 0.7 mm/min and 180 RPM. With a sampling rate of 10 Hz during 60 seconds (603 samples) for each scenario, obtained from the drilling simulator, where it can be seen the total acquisition time of 180 seconds for the three defined scenarios.

Moreover, each scenario is divided into a training set and a testing set, where 40 seconds of data acquisition were used for training and 20 seconds for testing. Using the training data, we obtain the respective fuzzy sets, using both methods for each scenario, as shown in Figure 2 (note that each set of membership functions corresponds to a linguistic variable with three elements – one for the trapezoidal approach and the other for the ASD).

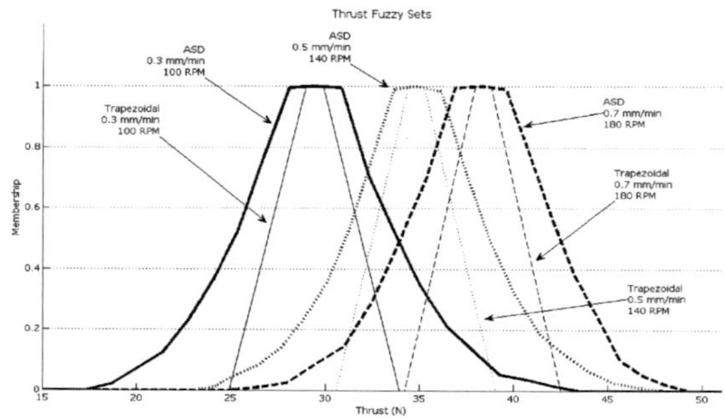

Figure 2 – Fuzzy sets for the ASD and Trapezoidal – Thrust linguistic variables

With the linguistic variables created, the next step will be to use the testing sets (20 seconds which correspond to a set of 201 samples) of each scenario, in order to validate both methods and data acquisition. For each new test sample, its membership value will be calculated and inserted on a membership value set.

In order to validate the results of our automatically generated fuzzy sets, we consider three rules with just one input and one output: *If Thrust is Nominal Then Terrain,* where Nominal corresponds to the linguistic variable Thrust for both ASD or Trapezoidal (3 elements each) and the output is a yes/no answer with a degree of confidence, given by the firing value, which is the actual membership value for the current sample. Here, our inference process starts with the calculation of the membership value of the sample value, for each of the three rules, using the following threshold IF membership ≥ 0.2 THEN belongs to Terrain-Concrete. The final aggregated result for the sample value is given by the maximum of the three membership values obtained.

A confusion matrix (Kohavi and Provost 1998) was built in order to test the performance of the classification rule system with the trapezoidal method for the 603 samples. Table 1 shows this confusion matrix, which relates the classes where each sample value belongs to (Real), with the class where the sample value was classified during the inference (Classified).

		Classified			
		Subs. 1	Subs. 2	Subs. 3	Subs. Unknown
Real	Subs. 1	119	44	5	33
	Subs. 2	54	79	51	17
	Subs. 3	22	47	103	29

Table 1 – Confusion Matrix for the Trapezoidal Fuzzy Set (Grey – correct results)

Real $Subscenario_S$ (S = [1,2,3] for the given example) is the class that represents all values that belong to subscenario S, which were used to create fuzzy sets using Train Set S; Classified $Subscenario_S$ is the class inferred by the inference system for each sample value; Subscenario Uknown is the class of all sample values where their maximum membership value was below 0.2, not being related to any predefined class.

The measure to calculate the overall performance of the classification process is Precision (Melamed, Green et al. 2003). The precision using the Trapezoidal method is 50%. It is also important to assess the average degree of confidence for the correct results, which in this case is 67%. Using the ASD method into the inference system the precision is 57% and the average degree of confidence is 89%.

In order to improve the performance of the fuzzy set generation we also applied a moving average with a moving window of 8 elements to the test data. The smoothing of the data increased the performance of both methods, with the trapezoidal method reaching 83% of well-classified sample values with an average degree of confidence of 44%, and the ASD reaches 85% of well-classified sample values with an average degree of confidence of 60%.

CONCLUSIONS

Automatic generation of fuzzy sets seems a good way to fuzzify input variables without the need of a domain expert. Further, it is a timesaving activity in terms of human intervention as well as less prone to misconceptions and errors.

According to the results obtained in the comparative study, the ASD supports a better inference performance than the Trapezoidal because it covers a wide region of the variables domain and because it has higher membership values, within the common range of the membership values. The wider coverage of the domain space is a question of tuning to be discussed with the domain expert, since it might lead to loss of precision and recognition of faulty behaviours. Regarding higher membership values, within the common range for both methods, clearly the ASD performs better because it provides more certainty in the results obtained.

In addition, the performance of both methods can be increased if we apply smoothing methods on the sensor data. Smoothing methods will remove extreme points that have lower membership values when considered in the generated membership functions. Without smoothing methods, the data domain coverage is wider, and this loss on detail can lead to loss of precision when recognizing different behaviours and loss of recognition of faulty behaviours.

In general, the performance of the ASD method with smoothed data seems the best alternative. The ASD has better precision (85%) and better degree of confidence (60%), when compared with the 82% and 44% of the Trapezoidal method. These performance results were even more obvious in the MODI inference systems, although here they are not discussed. It may be confusing to have obtained a high average degree of confidence for the ASD method with non-smoothed data (89%), but this behaviour can be explained by the very low precision value, which will remove many results by allowing only the higher valued memberships to be part of the correct results set.

In terms of algorithmic complexity, the Trapezoidal has a constant complexity, which is an advantage when compared to the ASD function that has a logarithmic complexity (caused by the binary search). If we consider a small number of points for the ASD, it will have a competing performance in comparison with the Trapezoidal method.

Concluding, the ASD showed promising results when used on monitoring fuzzy inference systems, particularly if we use smoothing methods to remove extreme points. The Trapezoidal method has a better speed performance but requires a domain expertise to specify the extreme points and this is a time consuming activity.

In the future, both methods will be used in consonance in order to achieve better speed and inference performance without direct intervention of domain experts. First, we will use the ASD to calculate the general membership function for the nominal behaviour and secondly we will approximate the result with a Trapezoidal membership function. Obviously, a final fine-tuning, supervised by the domain expert, will always be necessary to ensure accuracy of the fuzzification process.

Finally, more research must be done to ensure: recognition of instant faulty behaviours that can occur during monitoring; testing other precision measures that include the degree of confidence in monitoring systems like the MODI; merging similar fuzzy sets from a variable and from different behaviours to optimize the number of fuzzy sets used to characterize a phenomenon.

ACKNOWLEDGMENTS

The development of the generic fuzzy set generation and variable study was done under the contract 19143/05/NL/GM for the European Space Agency (ESA/ESTEC).

We first wish to thank our ESA technical officer, Mr. Raffaele Vitulli for his support during the project. Second in are in debt to both DEIMOS (sub-contractor) and Galileo Avionica (external consultant) for their contribution with testing data, respectively simulated and real.

REFERENCES

CA3. (2006). "MODI's homepage." from http://www2.uninova.pt/ca3/en/project_MODI.htm.

Cios, K. J., W. Pedrycz, et al. (1998). "Data mining methods for knowledge discovery". Boston, Kluwer Academic.

Kohavi, R. and F. Provost (1998). "On Applied Research in Machine Learning", *Machine Learning*.

Magnani, P. G., T. Ylikorpi, et al. (2004). "Deep drill (DeeDri) for Mars application", *Planetary and Space Science*, Galileo Avionica.

Melamed, I. D., R. Green, et al. (2003). "Precision and Recall of Machine Translation", *HLT-NAACL*, Edmonton, Canada, DBLP.

Ross, T. J. (2004). "Fuzzy logic with engineering applications". Hoboken, NJ, John Wiley.

Santos, B. R., P. T. Fonseca, et al. (2006). "New Data Preparation Process – A Case Study for an ExoMars Drill", *WAC - World Automation Congress*, Budapest - Hungary.

Zadeh, L. A. (1965). "Fuzzy Sets", *Information and Control*.

INTERVAL TYPE-2 FUZZY LOGIC FOR IMPROVING FEATURE EXTRACTION AND RESPONSE INTEGRATION IN MODULAR NEURAL NETWORKS FOR IMAGE RECOGNITION

OLIVIA MENDOZA[1], PATRICIA MELIN[2] AND OSCAR CASTILLO[2]
[2]UABC University, Tijuana, Mexico, omendoza@uabc.mx
[1]Tijuana Institute of Technology, Mexico {ocastillo, pmelin}@tectijuana.mx

ABSTRACT

The combination of Soft Computing techniques allows the improvement of intelligent systems with different hybrid approaches. In this work we consider two parts of a Modular Neural Network for image recognition, where a Type-2 Fuzzy Inference System (FIS 2) makes a great difference. The first FIS 2 is used for feature extraction in training data, and the second one to find the ideal parameters for the integration method of the modular neural network.

INTRODUCTION

At the moment, many methods for image recognition are available. But most of them include a phase of feature extraction or another type of preprocessing closely related to the type of image to recognize [1] [2] [3][4][5][6]. The method proposed in this paper can be applied to any type of images, because the preprocessing phase does not need specific data about the type of image [7] [8] [9]. Even if the method was not designed only for face recognition, we have made the tests with the ORL face database [10] composed of 400 images of size 112x92. There are 40 persons, with 10 images of each person. The images are taken at different times, lighting and facial expressions. The faces are in up-right position of frontal view, with slight left-right rotation. Figure 1 shows the 10 samples of one person in ORL database. To explain the proposed steps of the method, we need to separate it them in two phases: the training phase in figure 2 and the recognition phase in figure 3.

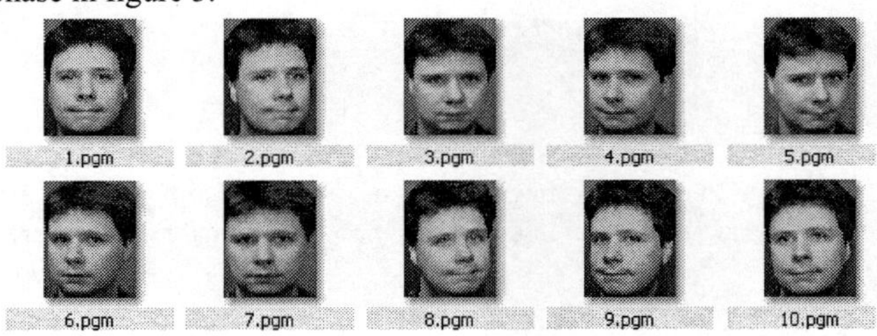

Fig. 1. Set of 10 samples of a person in ORL

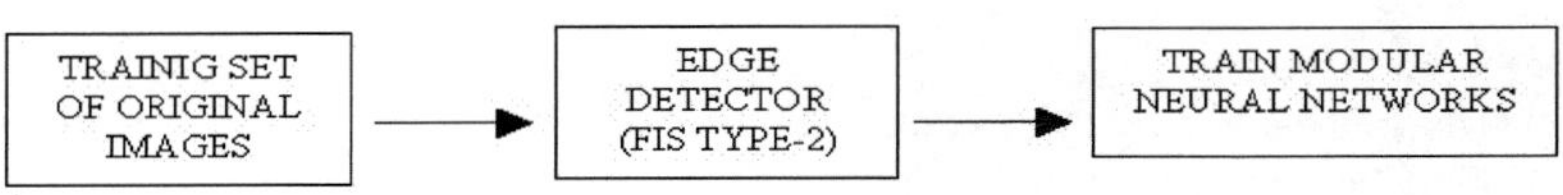

Fig. 2. Steps in Training Phase.

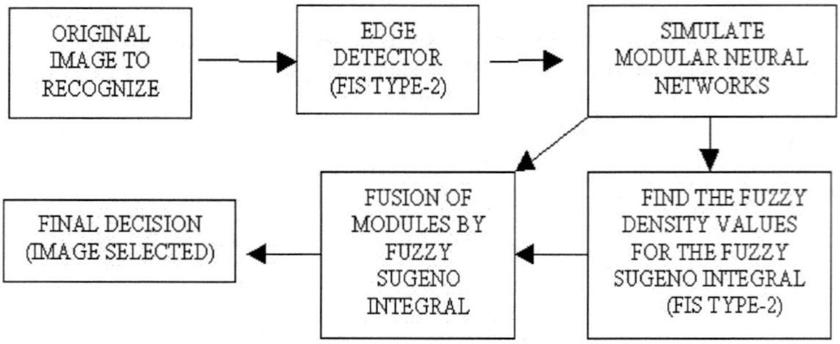

Fig. 3 Steps in Recognition Phase

TYPE-2 FUZZY LOGIC FOR EDGE DETECTION

In previous work we presented an efficient Fuzzy Inference System for edges detection, in order to use the output image like input data for modular neural networks [11]. In the proposed technique, it is necessary to apply Sobel operators to the original images, then use a Fuzzy Inference System Type-2 to generate the vector of edges that would serve like input data in a neural network. Type-2 Fuzzy Logic enables us to handle uncertainties in decision making and recognition in a more convenient way and for this reason was proposed [13].

For the Type-2 Fuzzy Inference System, 3 inputs are required, 2 of them are the gradients with respect to x-axis and y-axis, calculated with (1), to which we will call DH and DV respectively.

The Sobel edges detector uses a pair of 3x3 convolution masks, one estimating the gradient in the x-direction (columns) and the other estimating the gradient in the y-direction (rows).

$$Sobel_x = \begin{bmatrix} -1 & 0 & 1 \\ -2 & 0 & 2 \\ -1 & 0 & 1 \end{bmatrix} \qquad Sobel_y = \begin{bmatrix} 1 & 2 & 1 \\ 0 & 0 & 0 \\ -1 & -2 & -1 \end{bmatrix} \tag{1}$$

Where Sobely y Sobelx are the Sobel Operators throughout x-axis and y-axis.

If we define I as the source image, g_x and g_y are two images which at each point contain the horizontal and vertical derivative approximations, the latter are computed as (2) and (3).

$$g_x = \sum_{i=1}^{i=3} \sum_{j=1}^{j=3} Sobel_{x,i,j} * I_{r+i-2,c+j-2} \tag{2}$$

$$g_y = \sum_{i=1}^{i=3} \sum_{j=1}^{j=3} Sobel_{y,i,j} * I_{r+i-2,c+j-2} \tag{3}$$

Where gx and gy are the gradients along axis-x and axis-y, and * represents the convolution operator.

The other input is a filter that calculates when applying a mask by convolution to the original image. The low-pass filter hMF (4) allow us to detect image pixels belonging to regions of the input were the mean gray level is lower. These regions are proportionally more affected by noise, supposed it is uniformly distributed over the whole image.

The goal here is to design a system which makes it easier to include edges in low contrast regions, but which does not favor false edges by effect of noise [14].

$$hMF = \frac{1}{25} * \begin{bmatrix} 1 & 1 & 1 & 1 & 1 \\ 1 & 1 & 1 & 1 & 1 \\ 1 & 1 & 1 & 1 & 1 \\ 1 & 1 & 1 & 1 & 1 \\ 1 & 1 & 1 & 1 & 1 \end{bmatrix} \quad (4)$$

Then the inputs for FIS type 2 are: DH=g_x, DV=g_y, M= $hMF*I$, where * is the convolution operator, and de output is a column vector contains the values of the image edges, and we can represent that in graphics shown in figure 4. The Edges Image is smaller than the original because the result of convolution operation is a central matrix where the convolution has a value. Then in our example, each image with dimension 112x92 is reduced to 108x88. The inference rules and membership function parameters allow to calculate a gray value between -4.5 and 1.5 for each pixel, where the most negative values corresponds to the dark tone in the edges of the image. Then if we see the rules, only when the increment value of the inputs DH and DV are low the output is HIGH or clear (the background), in the rest of rules the output is LOW or dark (the edges). The complete set of fuzzy rules is given as follows:

1. If (DH is LOW) and (DV is LOW) then (EDGES is HIGH) (1)
2. If (DH is MEDIUM) and (DV is MEDIUM) then (EDGES is LOW) (1)
3. If (DH is HIGH) and (DV is HIGH) then (EDGES is LOW) (1)
4. If (M is LOW) and (DV is MEDIUM) then (EDGES is LOW) (1)
5. If (M is LOW) and (DH is MEDIUM) then (EDGES is LOW) (1)

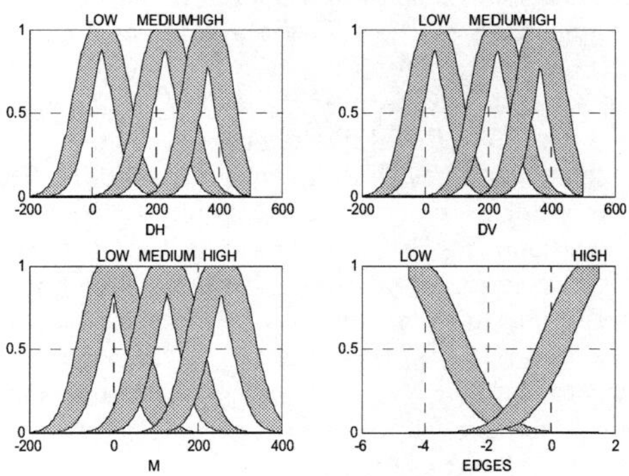

Fig. 4. Membership Function for the Type-2 FIS Edge Detector

The edge detector allows us to ignore the background color. We can see in this database of faces, different tones present for the same or another person. Then we eliminate a possible influence of a bad classification by the neural network, without losing detail in the image. Another advantage of edge detector is that the values can be normalized to a homogenous value range, independently the light, contrast or background tone in each image. At the examples in figure 5, all the edges in the images have a minimum value of -3.8 and a maximum value of 0.84. In particular for neural network training, we find these values to make the training faster: the mean of the values is near 0 and the standard deviation is near 1 for all the images.

Fig. 5 Examples of edge detection with the Type-2 FIS method.

EXPERIMENTAL RESULTS

A program was developed in Matlab that simulates each module with the 400 images of the ORL database, building a matrix with the results of the simulation of each module. These matrices are stored in the file "mod.mat" to be analyzed later for the combination of results. We can observe that in the columns corresponding to the training data, the position with a value near one is the image selected correctly. However in the columns that correspond to the test data this doesn't always happens, reason why it is very important to have a good combination method to recognize more images.

According to exhaustive tests made in the simulation matrices, we know that recognition of the images that were used for the training of the neural networks is of the 100%. Therefore the interest is focused on the recognition of the samples that do not belong to the training set, is to say samples 8,9 and 10. The parameters for the Sugeno Fuzzy Integral that will be inferred will be the Fuzzy Densities, a value between 0 and 1 for each module, which determines the rate for each module. The parameter lambda, according to the theory of fuzzy measures depends on the values of the fuzzy densities, and is calculated by searching for the roots of a polynomial. After the simulation of an image in the Neural Network, the simulation value is the only known parameter to make a decision, then to determine the fuzzy density for each module is the unique available information. For this reason we analyze the values in many simulations matrix and decide that each input to the FIS Type-2 corresponds to the maximum value of each column corresponding to the simulation of each module of each one of the 400 images. The process to recognize each one of the images is shown in figure 6.

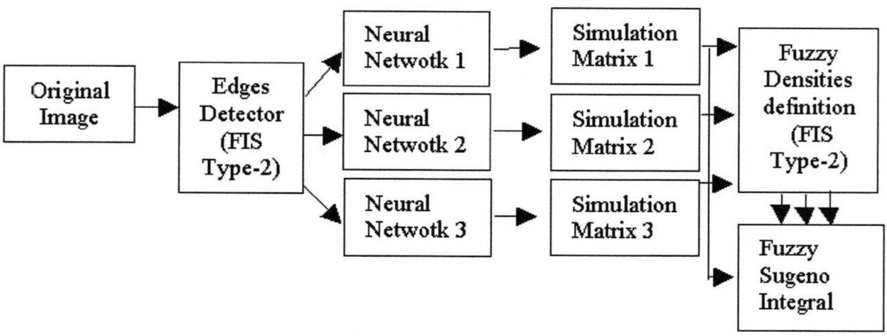

Fig. 6. Process of recognition using the type-2 fuzzy modular approach.

Then each output corresponds to one fuzzy density, to be applied for each module to perform the fusion of results later with the Fuzzy Sugeno Integral. The inference rules found fuzzy densities near 1 when de maximum value in the simulation is between 0.5 and 1, and near 0 when the maximum value in the simulation is near 0. The fuzzy rules are shown below and membership functions in Figure 7.

1. If (max1 is LOW) then (d1 is LOW) (1)
2. If (max2 is LOW) then (d2 is LOW) (1)
3. If (max3 is LOW) then (d3 is LOW) (1)
4. If (max1 is MEDIUM) then (d1 is HIGH) (1)
5. If (max2 is MEDIUM) then (d2 is HIGH) (1)
6. If (max3 is MEDIUM) then (d3 is HIGH) (1)
7. If (max1 is HIGH) then (d1 is HIGH) (1)
8. If (max2 is HIGH) then (d2 is HIGH) (1)
9. If (max3 is HIGH) then (d3 is HIGH) (1)

Although the rules are very simple, allows to model the fuzziness to rate de modules when the simulation result don't reach the maximum value 1.

However some of the images don't reach the sufficient value in the simulation of the three modules, in these cases, do not exists enough information to select an image at the modules combination, and the image is wrongly selected.

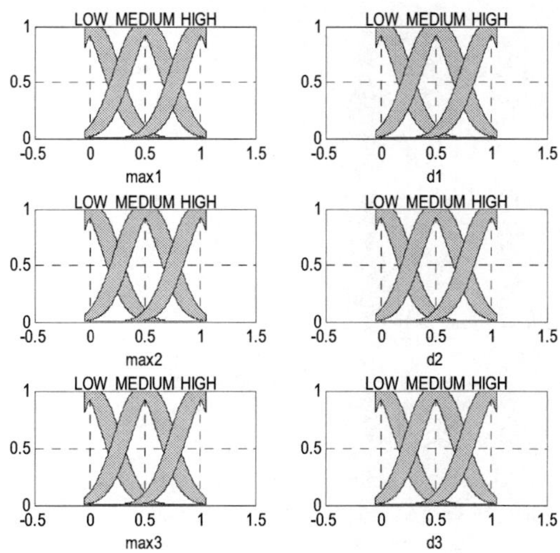

Fig. 7. Membership functions for the FIS to find fuzzy densities

In order to measure of objective form the final results, we developed a method of random permutation, which rearranges the samples of each person before the training. Once a permutation is made, the modular neural networks are trained and combined four times to obtain the sufficient information to validate the results. The average recognition rate is of 96.5%.

CONCLUSIONS

We have shown in this paper that the combination of Soft Computing techniques allows the improvement of intelligent systems with different hybrid approaches. In this paper we considered two parts of a Modular Neural Network for image recognition, where a Type-2 Fuzzy Inference System (FIS 2) help us improves the performance results in image recognition. The first FIS 2 was used for feature extraction in training data, and the second one to find the ideal parameters for the integration method of the modular neural network.

REFERENCES

[1] Renato Salinas, Luis Larraguibel, *Red Neuronal De Arquitectura Paramétrica En Reconocimiento De Rostros*, Divulgación electrónica de las Ciencias, Volumen 17, available on the web site: http://cabierta.uchile.cl/revista/17/articulos/articulos.html

[2] Patricia Melin, Oscar Castillo, *Hybrid Intelligent Systems for Pattern Recognition*, Springer-Verlag, Heidelberg, Germany, 2005.

[3] Peng Yang, Bo Du, Shiguang Shan, Wen Gao, *A Novel Pupil Localization Method Based On Gaboreye Model And Radial Symmetry Operator,* ICT-ISVISION Joint R&D Laboratory for Face Recognition, Beijing, China

[4] V.V. Starovoitov1, D.I Samal1, D.V. Briliuk1, *Three Approaches for Face Recognition*, The 6-th International Conference on Pattern Recognition and Image Analysis, October 21-26, 2002, Velikiy Novgorod, Russia, pp. 707-711

[5] J. B. Dowdalla, I. Pavlidisa, G. Bebisb, *Face Detection in the Near-IR Spectrum,* Dept. of Computer Science., University of Houston, Houston, TX 77204, bDept. of Computer Science, University of Nevada, Reno, NV

[6] Mao-Meng Chuang, Ruey-Feng Chang And Yu-Len Huang Automatic, *Facial Feature Extraction in Model*-based Coding, Journal of Information Science and Engineering 16, 447-458 (2000)

[7] Olivia Mendoza, Patricia Melín, *Sistema para el Modelado Diseño y Pruebas de Arquitecturas Neuronales Multi-Red*, International Seminar on Computational Intelligence, December 1-3, 2004 Tijuana B.C.

[8] Patricia Melin, Alejandra Mancilla, Miguel Lopez and Olivia Mendoza, *A hybrid modular neural network architecture with fuzzy Sugeno integration for time series forecasting*, accepted for publication in *Journal of Applied Soft Computing*, 2007.

[9] Olivia Mendoza, Patricia Melin, *The Fuzzy Sugeno Integral as a Decision Operator in the Recognition of Images with Modular Neural Networks*, in Edited Book: Hybrid Intelligent Systems: Design and Analysis, Springer-Verlag, pp. 299-310, 2007.

[10] AT&T Laboratories Cambridge, The ORL database of faces, information on the web page: http://www.cl.cam.ac.uk/research/dtg/attarchive/facedatabase.html

[11] Olivia Mendoza, Patricia Melín, *Sistemas de Inferencia Difusos Tipo-1 y Tipo-2 Aplicados a la Detección de Bordes en Imágenes Digitales,* International Seminar on Computational Intelligence 2006 IEEE - CIS Mexico Chapter, Tijuana, Mexico, October 9-11, 2006 Tijuana Institute of Technology (CD-rom proceedings).

[12] Jerry Mendel, *Uncertain Rule-Based Fuzzy Logic Systems: Introduction and New Directions.* NJ: Prentice-Hall, 2001.

[13] Oscar Castillo, Patricia Melin, Janusz Kacprzyk, Witold Pedrycz, *Hybrid Intelligent Systems: Design and Analysis,* Springer-Verlag, 2007.

[14] Cristiano Jacques Miosso, Adolfo Bauchspiess,*"Fuzzy Inference System Applied to Edge Detection in Digital Images*", V Brazilian Conference on Neural Networks, (Brasil, Abril de 2001).

A WEB BASED TOOL FOR ACCESSING DISTRIBUTED RELATIONAL DATABASES THROUGH MULTILINGUAL FUZZY INTERFACE

SAROJ KAUSHIK
Department of Computer Science
and Engineering, I I T Delhi,
New Delhi- 110016
INDIA
saroj@cse.iitd.rnet.in

SHIVENDRA PRASAD TIWARI
GlobalLogic India Pvt. Ltd.,
Noida (U.P.) – 201301,
INDIA
shivendra.tiwari@globallogic.com

ABSTRACT
Most of the relational database systems use SQL, which has strict syntax and semantics defined precisely to retrieve data. Sometimes, user may not be aware of SQL syntax and wants to use databases without any technical effort. In recent years, an increasing number of people have begun to realize the need for a technology to reach beyond the barriers of SQL. In the past few years, some Natural Language Interfaces to SQL have been developed but with the restriction that they are PC based and are not available for public use. Some work on fuzzy natural language interfaces have been reported by Chang, Woie (95) but none of which is available for web access. Nanda (99) proposed web based approach but was for English language. We have proposed and implemented fuzzy query system for multiple human languages. It is an interface between the user on the web and the databases spread over distributed database servers. It gives the flexibility to make fuzzy or simple query to the database in a multiple natural language. If the query is in natural language, the system parses it and converts it into equivalent SQL after resolving various tables, attributes and fuzzy terms used in the query. The system also accepts SQL query straightway. The interface is generic enough to work on all Relational Databases and is a generalized tool which can be customized for any organization and any natural language. The system has been implemented using PHP, a powerful scripting language. HTML tags have been used to design GUI and for handling connectivity with MySql Server. The system is named as MFLQ (Multilingual Fuzzy language Query) system

INTRODUCTION

The phrase "Multilingual Systems" has been used to refer to the computer programs, which permit user interaction with the computer in one or more languages, where the language can be selected dynamically, either at the time of invocation of the system or subsequently during its execution.

Another important couch used is "Fuzzy Interface" to refer to an interface which can interpret the vague instructions from the user. Fuzzy set concept was introduced by Zadeh (1965) to represent data and information processing having non-statistical uncertainty. In this paper, fuzzy logic is used to fuzzify the values of attributes of any entity in the system. For example, human can apply his intelligence in judging fuzzy term excellent in a query "list all intelligent students" but machine works on Boolean theory, therefore vague instructions can not be handled through simple if-then-else logic. Fuzzification method has to be used to solve such ambiguities. Our system also bridges the gap between natural languages and RDBMS which use complex queries to handle the data. It works as an interface between RDBMS and the user. Currently the system has been tested for three languages i.e. English, Hindi, and Telugu.

OVERALL DESIGN OF THE MFLQ TOOL

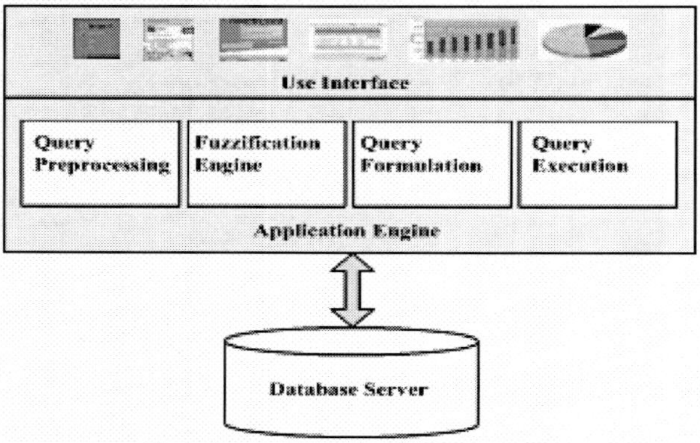

Fig.1: Layered Functional Decomposition

The system has been decomposed in three major functional layers as shown in fig.1. The User Interface is the top layer which user interacts most of the time with. It consists Language selection and query interface. The Application engine consists of the Query Processing, Fuzzification Engine, Query Formulation, and Query Execution modules. The third layer is the Database Server module, which is not necessarily in a single physical location i.e. Databases can be distributed. Each module is described briefly in the following sub sections.

A Web-Based Multilingual Fuzzy Query Interface (GUI)

Fuzzy Query interface is a user friendly GUI which provides facility to configure system for different languages. The first page of GUI contains four options viz., "General Query User", "Domain Sub-administrator", "Domain Administrator" and "Super Administrator". General Query interface can be used for triggering queries; however other all modules are useful for system administration activities. Since the major objective of the system is processing fuzzy queries intelligently, GUI explanation has been shortened.

Query Processing Module

In this module, a query is interpreted and semantics of the query is captured. The query need not be syntactically correct but should convey intended meaning. If system is unable to understand or resolve the query word(s), then it will show a spelling error. It offers suggestions to replace or ignore wrongly spelled words using LCS algorithm. It makes use of meta-dictionary stored in the system. One can modify its query if there are some words wrongly spelled.

This module tags the query words with table name, fuzzy word, numbers, condition, operators or attributes etc. The system is not only limited for English language but the concept is for language independence. It follows a very simple strategy to convert any language into English. The interesting thing is that the system has no restriction of syntax but instead it has a word to word mapping between one language to corresponding English language words stored in bilingual dictionary (language words and corresponding English words). English has been chosen as a base language to have correspondence with any language word. An administrator can create a bilingual dictionary of words from its own language and English words for the domain of database.

Query Formulation Module

Based on tagged query, the system formulates an equivqlent SQL query. The keywords are selected from the query as attributes. If some attribute comes with some value or connector or operator then this fact is taken into account while formulating a condition. Query formulation is explained later in this paper. For Example:

Input: *"I would like to see the excellent students of CSE department who are residing in hostel or age>20"*

Output: SELECT \`academic\`.\`name\` WHERE (\`academic\`.\`marks\`>=95 and \`academic\`.\`marks\` <= 100) AND \`academic\`.\`department\` like 'cse' AND \`personal\`.\`address\` like 'hostel' OR \`academic\`.\`age\`>=20

Query Execution Module

Execution module is an important part of the system. It takes the SQL query from query formulation module and directly fires SQL query to MySql and the result set is returned to user through GUI. MFLQ system is not only handles multilingual queries but it also runs SQL direct queries from the same GUI. So query could be in SQL or in natural language, the system returns same results for both semantically same queries. Joining two or more tables of data is a powerful feature found in the relational model. SQL provides facility of joining tables. The system is capable of fetching information from different tables in the database. Some sample queries in English are shown in Appendix.

DATABASE ADMINISTRATION TOOL

Database administration is a difficult task and it needs technical knowledge of database SQL queries. This system provides Look and Feel GUI using which administrator can maintain databases with fewer efforts. Super Administrator is able to authenticate some domain administrators, update, or create some new databases. There are three level of administration hierarchy in the admin module of the system as shown in fig.2.

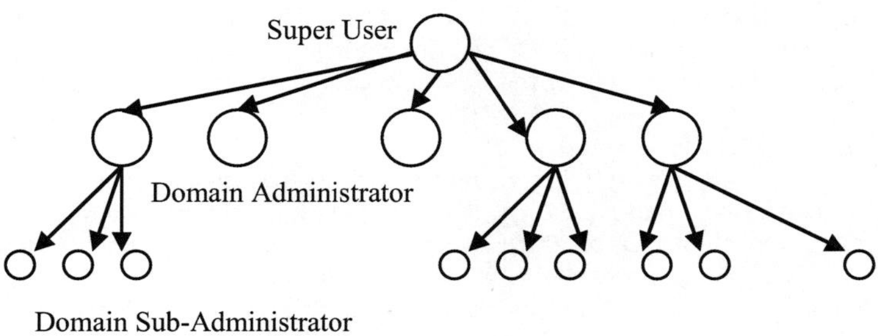

Fig.2: Administrator Hierarchy

Database Administrator Hierarchy

The system must be secured with user access control features. All users will have user names and passwords to log on to the system. The tool provides two types of users operating on the system i.e., normal user and system administrator. Normal user can choose language and use the query interface.

System Administrator is a domain expert and is able to install new languages, manages semantic graphs for registered databases. He also builds and manages fuzzy function library. He also manages users, system logs and handles system configurations. Under this category there are three types of administrators.

Super User: Super user can access all the databases in hierarchy and can authenticate or register new users. Super user is a top level authorized person who takes care of the things such as which domain administrators have to be allowed to customize this tool for their organization.

Domain Administrator (DA): DA is a chief Administrator and has full authentication for one particular domain. He can register new sub-administrators called Domain Sub-Administrators, and can manage all the databases related to that domain i.e. creation of new databases, deletion etc. as shown in fig. 2.

Domain Sub-Administrator: This is also a powerful person who is authenticated for database administration in some particular domain. However, Sub-administrator cannot authenticate other users to update databases. It means that the user information table cannot be accessed by Domain-Sub administrators.

Database operations through Look & Feel GUI

This tool has all database management capabilities such as Create New Database, Delete Databases, Create New Table, Insert data, Delete Data, Update data and Delete Tables and many other operations

Fuzzy Conditional Words and Hedge Handling

The system has a fuzzy function library, which can be updated offline time to time. Administrator can define fuzzy membership functions; fuzzy variables, hedges (very, almost, extremely etc.) and the system can calculates the corresponding conditions using fuzzy function library. Fuzzy Function definition is explained in detail in section 4.

FUZZY FUNCTION LIBRARY

Fuzzy set was proposed by Zadeh (1965). We are using the concepts of Fuzzy sets and logic. The fuzzy function library is a storehouse of all the fuzzy words in common use. The entire spectrum of fuzzy words has been split into left (rich, good, tall, old etc.), full (nearly, approximately, about etc.) and right (low, poor, short, young etc.) trapezoid functions, negatives (no, don't, not etc) and modifiers/hedges (very, pretty, much etc.) given in a book by Klir (1997). Thus depending on whether the incoming fuzzy word belongs to the set left, right or full it is mapped to a function in the Membership function library.

Membership function value

The truth-value (in fuzzy logic) or membership values (in fuzzy sets) are indicated by a value in the range [0.0,…, 1.0], with 0.0 implying absolute falseness and 1.0 indicating absolute truth. For example, in a query "Get me all GOOD students from IIT Delhi", GOOD is a Fuzzy word, which is not absolute but depends on the environment and situation that can be defined by using appropriate fuzzy membership function. Fuzzy Logic Toolbox Help MATLAB 7.0 has been used for implementation of fuzzy functions.

IMPLEMENTATION

One of the key features of the system is that it has to be deployed in a web-based environment, and it should be designed with the Client/Server model of interaction in view. The system must be able to handle concurrent queries. In order to implement the system, PHP scripts along with HTML tags have been used to design GUI and for connectivity with MySql Server. PHP is a powerful scripting language, which provides facilities to make clients side validation checks along with the connectivity with database in server side. It provides a powerful control structure to design and develop a strong logic for our application.

Converting a Natural Language Query into an SQL statement

In order to process natural language, a subset of words or phrases has been used, which hold particular meaning within the database domain. These words or phrases have been defined as keywords. Given a user input as a sequence of words in a query Q = (*w1*, *w2*... *wn*), the system approximates the meaning of Q by the meaning of the keywords present in the query. It performs natural language processing in two steps. In the first step, the system determines all the keywords contained in I. In the second step, the system uses a similarity comparison to determine the most likely meaning of each keyword contained in Q. In our system, the keywords indicate what nodes, attributes, database values, operators, and other information like fuzzy words; linguistic hedges and condition connectors are mentioned by the user. Once the set of keywords that occur in Q have been determined, the set of possible meanings for each keyword are enumerated based on how the keywords are used in the Semantic Graph and whether the keywords are used as database values.

Tokenizing and Tagging of Query: The query is tokenized based on the concept that the system has no syntactical criteria. Instead, it just searches each word for attributes, tables name, other conditional values and operators to formulate query in Meta Databases. Irrelevant words are removed. The tagged query is used by SQL formulation module as shown in fig 3. For Example:

Input: *"I would like to see the excellent students of CSE department who are residing in hostel or age>20"*
Output: *token[0][fuzzy] = "excellent"; token[1][table] = " students"; token[2][value]= "CSE"; token[3][attribute]= "department"; ... token[4][attribute] = "residing"; token[4][value] = "hostel"; token[5][connector] = "or"; token[6][attribute] = "age"; token[7][where] = ">"; token[8][value] = "20".*

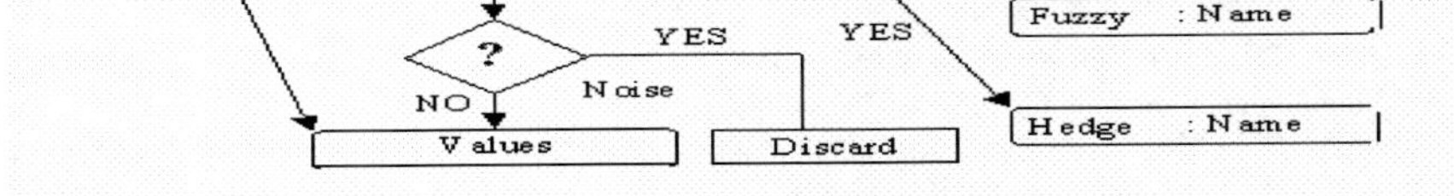

Fig. 3: Query Formulation Process

SQL Formulation

The SQL query is formulated with the help of a Semantic Graph of the database. It describes the objects and relationships between objects modeled in the database. The Semantic Graph consists of three elements: nodes, attributes, and links. The nodes are the objects in the world, the attributes are the features of the nodes, and the links describe how the nodes are related to each other. The following operations are made for SQL formulation by tagged query.

Search for attributes (select list): The 'select list' is dependent on what the user wants to return as an answer. Our system handles two types of inputs: one with explicit select lists and other with implicit select lists. Inputs with explicit select lists are those in which the user explicitly specifies which attributes are to be returned by the system. For example, ``what is the name and address of good students of CSE department?". Here name and address are explicitly mentioned. For inputs with implicit select lists, the system uses a heuristic for determining what attributes belongs in the select list. For example "Get me excellent students with their performance". Here performance is the implicit attribute. The system keeps track of all such attributes. Attributes those do not participate in the query constraints are automatically put into the select list. The assumption is that attributes used in the query constraints already have values, and those not in the query constraints need their values retrieved.

Search for tables: Search for table is a step where system searches for the entire table name in the selected database, which is needed to specify scope of the attribute names and to join tow or more tables.

Search for connectors: Connector is a symbol or a word, which is used to connect two or more conditions. If a connector comes among select list then it makes no affect but if it comes among conditions then it is placed between two conditions. There are mainly two connectors i.e. "AND" and "OR".

Search for Operators: Operator is a symbol, which comes between two operands. Here <=, >=, <, >, =, != are some of the operators which are used to make a condition.

Search for Values: If some token of symbol is not found in meta dictionary, then it is considered as a value. Every value should be associated with some attribute through an operator to formulate a condition. System searches nearest attributes of an operator to make a condition. *For Example:*

> ***Input:*** *"I would like to see the excellent students of CSE department who are residing in hostel or age>20"*
> ***Output:*** *SELECT `academic`.`name` WHERE (`academic`.`marks`>=95 and `academic`.`marks` <= 100) AND `academic`.`department` like 'cse' AND `personal`.`address` like 'hostel' OR `academic`.`age`>=20*

CONCLUSION

The MFLQ system has been designed, developed and implemented as a fuzzy natural language query tool, based on a semantic model, which takes semantic information from the database and uses a special SQL formulation methodology to formulate SQL query. It uses keywords to approximate the meaning of the user's input. The keywords are natural language labels of nodes and attributes of the semantic graph, the values contained in the database and the operators. The system uses no formal grammars or world knowledge but the most of the knowledge utilized by the system can be obtained from the databases itself. This system is most suitable for applications where the user input is short and simple, and it is necessary to support domains with a limited ontology. The system is a generalized tool which can be customized for any organization and any natural language. Currently the system has been tested for three languages i.e. English & Hindi and Telugu. It has been implemented using PHP scripts along with HTML tags to design GUI and for MySQL Server connectivity.

ACKNOWLEDGEMENT

I would like to acknowledge MHRD (Ministry of Human Resource Development) of INDIA who funded this project under thrust area scheme. I would also like to thank all those who contributed in this project during the execution phase of the project.

REFERENCES

Chang Y H et al (1995), ""Apply the Fuzzy Set Theory to the Structural Query Language of Database" Presented at International Joint Conference of CFSA/IFIS/SOFT'95 on Fuzzy Theory and Applications.

Klir G J et al (1997), "Fuzzy Sets and Fuzzy Logic Theory & Applications", Prentice Hall of India.

Nanda H & Kaushik S (1999), "Web Based access of Relational Databases using Fuzzy Language Queries", Int. Conference on Cognitive Systems, PP. 388-400.

Woei-Tzy Jong et al (1995) "Fuzzy Query Translation Techniques For Relational Database Systems" Presented at International Joint Conference of CFSA/IFIS/SOFT '95 on Fuzzy Theory and Applications.

Zadeh L A (1965), "Fuzzy sets", Information and Control, Vol 8., PP. 338-353.

Fuzzy Logic Toolbox Help MATLAB 7.0

Beginning PHP 5 by Dave W. Mercer et al. 2004, Wiley Publications.

APPENDIX

Sample English Queries and Corresponding SQL.

- Find those students along with cgpa who have cgpa greater than 9.
  SELECT `academic`.`name`, academic`.`cgpa` FROM `personal`, `academic` WHERE ( `academic`.`cgpa`>9 ) and ( `personal`.`sno` = `academic`.`sno`).
- Get me names and cgpa of civil dept students.
  SELECT academic`.`name`, `academic`.`cgpa` FROM `personal`, `academic` WHERE ( `academic`.`department` like '%civil%' ) and ( `personal`.`sno` = `academic`.`sno`).
- List the name and address of the students who reside in lucknow.
  SELECT `academic`.`name`, `personal`.`address` FROM `personal`,`academic` WHERE (`personal`.`address` like '%lucknow%') and (`personal`.`sno` = `academic`.`sno`).
- We are interested in finding the names of students and of their father or hostel.
  SELECT `academic`.`name`, `personal`.`father` FROM `personal`,`academic` WHERE (`personal`.`address` like '%hostel%') and (`personal`.`sno` = `academic`.`sno`).
- Obtain is the address of student whose date of birth is above 1980 and sno is above 8.
  SELECT `academic`.`name`, `personal`.`address` FROM `personal`,`academic` WHERE (`personal`.`dob`>1980) and (`personal`.`sno` = `academic`.`sno`).
- List the name and address of the students who are belonging to lucknow .
  SELECT `academic`.`name`, `personal`.`address` FROM `personal`,`academic` WHERE (`personal`.`address` like '%lucknow%') and (`personal`.`sno` = `academic`.`sno`).
- I want to see the good students in cse department along with their address and cgpa.
  SELECT `academic`.`name`, `academic`.`cgpa` , `personal`.`address` FROM `personal`,`academic` WHERE (`academic`.`department` like '%cse%') and (`personal`.`sno` = `academic`.`sno`) and (`academic`.`marks`>=67 and `academic`.`marks`<=78).
- Get me excellent students who reside in Delhi
  SELECT `academic`.`name` FROM `personal`,`academic` WHERE (`personal`.`address` like '%delhi%') and (`academic`.`marks'>90).

COMPUTATION OF GRADIENT AND HESSIAN IN FEED-FORWARD NEURAL NETWORKS: A VARIATIONAL APPROACH

S. LAKSHMIVARAHAN AND SUDARSHAN K. DHALL
School Of Computer Science
University Of Oklahoma
Norman, OK 73019, USA

ABSTRACT

In this paper we derive explicit expressions for the gradient and the Hessian of the objective function (which in the sum of the squared errors) with respect to the weight matrices in a general feed-forward neural network using the standard methods for computing the first-order and second-order variations of a scalar-valued function of a vector.

INTRODUCTION

Starting with the pioneering work of F. Rosenblatt in 1959, the supervised learning problem for pattern recognition using the multilayered feed-forward (artificial) neural network (NN) is usually recast as the (constrained) minimization of the sum of the squares of errors (between the output of the NN and the label assigned by the teacher) for a given set of input patterns. This minimization is performed with respect to the weights that is uniquely defined by the structure — number of layers and size of each layer — of the NN. Refer to Rosenblatt (1959) and Rosenblatt (1961) and Minsky and Papert (1988) for a historical perspective. The books by Bishop (1995), Fine (1999), and Vapnik (2000) provide a comprehensive account of the theory and applications of NN. Narendra and Parthasarathy (1990) provide a thorough view of the applications of NN in the context of identification and control of dynamical systems.

Contemporaneously and yet independent of this development, in the 1960's, the use of the second-order methods based on the combined use of the gradient and Hessian information (inverse of the Hessian or Hessian-vector product) to accelerate the rate of convergence of the unconstrained minimization algorithm was established. Since then the second-order method has become the standard workhorse for minimization and many minimization software packages based on these methods are readily available today. The books by Nash and Sofer (1996) and Dennis and Schnabel (1996) provide a comprehensive account of this part of the literature on minimization.

During the past two decades great strides have been made in the design of supervised learning algorithms for NN. First order algorithm based only on the gradient obtained using the back-propagation algorithm were first developed (Rumelhart et al. (1986) and Werbos (1990)). More recently, second-order methods have taken hold. Buntine and Weigend (1994), and Pearlmutter (1994) provide a comprehensive overview of the development and the use of second-order methods for NN.

A quick review of the literature reveals that the back-propagation method for computing the gradient and methods for computing the Hessian essentially rely on the use of the ubiquitous chain rule in differential calculus. While this enables the computation of the components of the gradient vector, Hessian matrix or the Hessian-vector product, it is often more useful to compute closed form expressions for the gradient and the Hessian matrix. For example, it is one thing to know the elements of the Hessian matrix, but it is quite another to know of its rank or its condition number.

Motivated by these considerations, in this paper we simultaneously derive closed form expressions for both the gradient vector and the Hessian matrix using the standard variational approach. While our end-results may not be new, to our knowledge the use of the variational approach to recursively uncover the structure of the Hessian has not been explored to-date.

In Section 2, a mathematical statement of the problem is given. Variational approach to computing the gradient vector and the Hessian matrix is described in Section 3. Section 4 contains concluding remarks.

STATEMENT OF THE PROBLEM

Let $\mathbf{I} = \{\mathbf{y}^k, z^k | 1 \leq k \leq L\}$ be the set of inputs (also called the **training set**) where $\mathbf{y}^k$ denotes the kth **input pattern** and $z^k \in \{-1, 1\}$ is the **label** attached to the pattern representing the membership in one of two classes.

The structure of the $(m+1)$-stage feed-forward network is such that the stages (also known as layers) are numbered from 0 through m. Stage 0 is called **input** stage and stage m is called the **output** stage. All the other stages $1 \leq r \leq m-1$ are known as **hidden** or **intermediate** stages. To simplify the notation, it is assumed that each of the first m stages, numbered 0 through $m-1$ contain the same number n of neurons. The output stage m contains only one neuron.

Let $\mathbf{u}(i)$ denote the input to stage i and $\mathbf{x}(i)$ denote the output of stage i, where $0 \leq i \leq m$. Let $\mathbf{u}(m)$ and $\mathbf{x}(m)$ denote the input and the output of stage m. The functionality of this $(m+1)$-stage feed-forward neural network is defined by the input-output relations between stages.

Input Stage: The input vector $\mathbf{u}(0)$ and the output vector $\mathbf{x}(0)$ of the input stage are initialized by the given input pattern, $\mathbf{y}^k$. Thus, $\mathbf{x}(0) = \mathbf{u}(0) = \mathbf{y}^k$, for $1 \leq k \leq L$.

Intermediate Stage r, $(1 \leq r \leq m-1)$: The input vector $\mathbf{u}(r)$ for the rth stage is given by $\mathbf{u}(r) = \mathbf{W}(r)\mathbf{x}(r-1)$, where $\mathbf{x}(r-1)$ is the output of stage $r-1$, and $\mathbf{W}(r) \in \mathcal{R}^{n \times n}$ is a weight matrix.

Let $\mathbf{u} = (u_1, u_2, \cdots, u_n) \in \mathcal{R}^n$. Define a non-linear map $\mathcal{S} : \mathcal{R}^n \rightarrow \mathcal{R}^n$, where $\mathcal{S}(\mathbf{u}) = (\mathcal{S}_1(u_1), \mathcal{S}_2(u_2), \cdots, \mathcal{S}_n(u_n))^T$ and

$$\mathcal{S}_i(u_i) = s(u_i) = \frac{e^{\alpha u_i} - e^{-\alpha u_i}}{e^{\alpha u_i} + e^{-\alpha u_i}}$$

called the sigmoid function, with $\alpha > 0$ being an arbitrary parameter.

The output $\mathbf{x}(r)$ of stage r is then given by

$$\mathbf{x}(r) = s[u(r)]$$

Output Stage m: The scalar input $\mathbf{u}(m) \in \mathcal{R}$ is given by

$$\mathbf{u}(m) = \mathbf{W}(m)\mathbf{x}(m-1) = \langle \mathbf{W}^T(m), \mathbf{x}(m-1) \rangle$$

where $\mathbf{W}(m) \in \mathcal{R}^{1\times n}$ is a row vector. The overall scalar output $\mathbf{x}(m)$ of the network is then given by

$$\mathbf{x}(m) = s[\mathbf{u}(m)].$$

Thus, if $\mathbf{x}(0) = \mathbf{y}^k$, then the overall output corresponding to this input pattern $\mathbf{y}^k$ is denoted by $\mathbf{x}^k(m)$.

Let $\overline{\mathbf{W}}(r)$ denote the vector of size n^2 obtained by concatenating the rows 1 through n of the matrix $\mathbf{W}(r)$. Let $\overline{\mathbf{W}} = (\overline{\mathbf{W}}(1), \overline{\mathbf{W}}(2), \cdots, \overline{\mathbf{W}}(m-1), \mathbf{W}(m))^T$ denote the column vector is size $(m-1)n^2+n$ obtained by concatenating the vector $\overline{\mathbf{W}}(1)$ to $\overline{\mathbf{W}}(m-1)$ and $\mathbf{W}(m)$, that is, $\overline{\mathbf{W}} \in \mathcal{R}^M$ where $M = (m-1)n^2+n$.

Let $\mathcal{J} : \mathcal{R}^M \to \mathcal{R}$ be defined by

$$\mathcal{J}(\overline{\mathbf{W}}) = \sum_{k=1}^{L} \mathbf{J}_k(\overline{\mathbf{W}}) \tag{1}$$

where

$$\mathbf{J}_k(\overline{\mathbf{W}}) = [z^k - \mathbf{x}^k(m)]^2 \tag{2}$$

Clearly, $\mathcal{J}(\overline{\mathbf{W}})$ is the sum of the squares of the difference (also called **error**) between the output $\mathbf{x}^k(m)$ induced by the input pattern $\mathbf{y}^k$ and the corresponding label z^k assigned to the pattern $\mathbf{y}^k$ by the teacher.

The goal of the learning problem may be stated as follows. Given the $(m+1)$-stage feed-forward network and a training set I, find the weight vector $\overline{\mathbf{W}}$ that minimizes $\mathcal{J}(\overline{\mathbf{W}})$.

THE VARIATIONAL APPROACH

Let $\delta\mathbf{J}$ and $\delta^2\mathbf{J}$ denote the first and second variation of $\mathbf{J}$ in (2) induced by the perturbations in the component sub-vectors $\overline{\mathbf{W}}_j$ of the weight vector $\overline{\mathbf{W}}$. Since z is a given constant, it follows that $\delta\mathbf{J}(\overline{\mathbf{W}}) = \beta(m)\delta\mathbf{x}(m)$ and $\delta^2\mathbf{J}(\overline{\mathbf{W}}) = \beta(m)(\delta^2\mathbf{x}(m) - (\delta\mathbf{x}(m))^2)$ where $\beta(m) = -2(z - \mathbf{x}(m))$ and superscript k is dropped. In view of this relation, our learning problem reduces to one of computing the first and second variations of the output $\mathbf{x}(m)$ of the NN with respect to the components $\overline{\mathbf{W}}_j$ of $\overline{\mathbf{W}}$.

In the following we develop a recursive framework for computing $\delta\mathbf{x}(m)$ and $\delta^2\mathbf{x}(m)$. Indeed, it is natural and convenient to start from the last stage.

Variation of $\mathbf{x}(m)$ with respect to $\mathbf{W}(m)$. It follows that the weight vector $\mathbf{W}(m)$ that connects the output stage $(m-1)$ with the input of stage m affects $\mathbf{x}(m)$ rather directly. Thus, perturbation in $\mathbf{W}(m)$ affects $\mathbf{x}(m)$ directly. Let $\delta = \delta_{\mathbf{W}^T(m)}$ for simplicity in notation.

Step 1A: From $\mathbf{x}(m) = s[u(m)]$, we get

$$\delta\mathbf{x}(m) = D_s[u(m)]\delta u(m)$$

and

$$\delta^2\mathbf{x}(m) = D_s^2[u(m)][\delta u(m)]^2 + D_s[u(m)]\delta^2 u(m)$$

where

$$D_s[u(m)] = \frac{\partial s[u(m)]}{\partial u} \quad and \quad D_s^2[u(m)] = \frac{\partial^2 s[u(m)]}{\partial u^2}$$

Step 1B: From $\mathbf{u}(m) = \mathbf{W}(m)\mathbf{x}(m-1) = \langle \mathbf{W}^T(m), \mathbf{x}(m-1) \rangle$, we get

$$\delta u(m) = \langle \delta \mathbf{W}^T(m), \mathbf{x}(m-1) \rangle$$

and

$$\delta^2 u(m) = \langle \delta^2 \mathbf{W}^T(m), \mathbf{x}(m-1) \rangle \; = 0$$

since $\delta \mathbf{W}(m)$ is a fixed perturbation superimposed on $\mathbf{W}(m)$ and $\delta^2 \mathbf{W}(m) = \delta[\delta \mathbf{W}(m)] = 0$.

Now, substituting Step 1B into Step 1A, and simplifying we get

$$\delta \mathbf{x}(m) = D_s[u(m)] \langle \delta \mathbf{W}^T(m), \mathbf{x}(m-1) \rangle$$

and

$$\delta^2 \mathbf{x}(m) = D_s^2[u(m)] \langle \delta \mathbf{W}^T(m), [\mathbf{x}(m-1)\mathbf{x}^T(m-1)] \delta \mathbf{W}^T(m) \rangle$$

Substituting and simplifying, from first principles, we get the gradient and the Hessian of $\mathcal{J}(\overline{\mathbf{W}})$ with respect to $\mathbf{W}(m)$, as

$$\nabla_{\mathbf{W}^T(m)} \mathcal{J}(\overline{\mathbf{W}}) = \beta(m)\mathbf{x}(m-1)$$

and

$$\nabla^2_{\mathbf{W}^T(m)} \mathcal{J}(\overline{\mathbf{W}}) = \beta(m)\Delta_1(m)[\mathbf{x}(m-1)\mathbf{x}^T(m-1)]$$

where

$$\Delta_1(m) = D_s^2[u(m)] - [D_s[u(m)]]^2.$$

Combining these with (1), we get an explicit expression for the overall gradient and the Hessian as

$$\nabla_{\mathbf{W}^T(m)} \mathcal{J}(\mathbf{W}) = \sum_{k=1}^{L} \beta^k(m)\mathbf{x}^k(m-1)$$

and

$$\nabla^2_{\mathbf{W}^T(m)} \mathcal{J}(\mathbf{W}) = \sum_{k=1}^{L} \beta^k(m)\Delta^k(m)\mathbf{x}^k(m-1)[\mathbf{x}^k(m-1)]^T \qquad (3)$$

where $\beta^k(m)$ is the $\beta(m)$ with $z = z^k$ and $\mathbf{x}(m) = \mathbf{x}^k(m)$ and $\Delta^k(m)$ is $\Delta(m)$ with $\mathbf{u}(m) = \mathbf{u}^k(m)$, where $\mathbf{u}^k(j)$ and $\mathbf{x}^k(j)$ are the input and the output of stage j induced by the input pattern $\mathbf{x}(0) = \mathbf{x}^k$.

Since $\mathbf{x}(m-1)\mathbf{x}^T(m-1)$ is a rank one outer product matrix, a necessary and sufficient condition for the overall Hessian in (3) to be of full rank is: $L \geq n$ and n of the L vectors $\mathbf{x}^k(m-1)$ for $k = 1, 2, \cdots, L$ must be linearly independent.

Variation of $\mathbf{x}(m)$ with respect to $\mathbf{W}_{i*}(k)$. From the structure of the NN, it follows that the perturbations in $\mathbf{W}_{i*}(k)$ affects only $u_i(k)$, the ith component of the input $\mathbf{u}(k)$, and hence $x_i(k)$, the ith component of the output $\mathbf{x}(k)$ of stage k. $x_i(k)$ then affects $u_j(k+1)$ and hence $x_j(k+1)$ via the weight $W_{ij}(k+1)$, for $1 \leq j \leq n$. From then on, it is a total mixing and, hence, every input and output of all stages from $(k+2)$ to m are affected. In the following, we derive the dynamics of propagation of the perturbations in $\mathbf{W}_{i*}(k)$ through the NN using a system of recurrence relations. For greater transparency, we separate the first and second variations. Refer to Lakshmivarahan and Dhall (2007) for details.

Dynamics of first variation: Let $x_{jr}(m-r)$ denote the j_r^{th} component of the vector $\mathbf{x}(m-r)$, for $1 \leq r \leq (m-k)$. Similarly, for $u(m-r)$. Also, recall that since the (first/second) variation of a vector is the vector of (first/second) variation of its components, in the following, we specify variations of a vector by those of its components. Let δ denote $\delta_{\mathbf{W}_{i*}^T(k)}$.

Stage $(m-1)$ **to** m: From $\mathbf{x}(m) = s[u(m)]$ and $u(m) = \langle \mathbf{W}^T(m), \mathbf{x}(m-1)\rangle$, we obtain the following:

$$\begin{aligned} \delta \mathbf{x}(m) &= D_s[u(m)]\delta u(m) \\ \delta u(m) &= \langle \mathbf{W}^T(m), \delta \mathbf{x}(m-1)\rangle \end{aligned}$$

Continuing likewise, we obtain

Stage k **to** $(k+1)$: From $\mathbf{x}(k+1) = D_s[u(k+1)]$ and $u(k+1) = \mathbf{W}(k+1)\mathbf{x}(k)$, it follows that

$$\begin{aligned} dtx_{j_{m-k-1}}(k+1) &= D_s[u_{j_{m-k-1}}(k+1)]\delta u_{j_{m-k-1}}(k+1) \\ \delta u_{j_{m-k-1}}(k+1) &= \langle \mathbf{W}^T_{j_{m-k-1}*}(k+1), \mathbf{x}(k)\rangle \end{aligned}$$

Stage $(k-1)$ **to** k: From $\mathbf{x}(k) = D_s[u(k)]$ and $u(k) = \mathbf{W}(k)\mathbf{x}(k-1)$, it follows that

$$\begin{aligned} \delta x_{j_{m-k}}(k) &= D_s[u_{j_{m-k}}(k)]\delta u_{j_{m-k}}(k) \\ \delta u_{j_{m-k}}(k) &= \langle \mathbf{W}_{j_{m-k}*}(k), \mathbf{x}(k-1)\rangle \\ &= \begin{cases} 0, & \text{if } j_{m-k} \neq i \\ \langle \delta \mathbf{W}_{i*}(k), \mathbf{x}(k-1)\rangle & \text{if } j_{m-k} = i \end{cases} \end{aligned}$$

By substituting back, we readily obtain the explicit expression for the gradient of $\mathbf{x}(m)$ with respect to $\mathbf{W}_{i*}(k)$, for $1 \leq i \leq n$ and $1 \leq k \leq m-1$.

Dynamics of second variation. The dynamics of the second variation can be readily obtained from above and are cataloged here for easy reference.

Stage $(m-1)$ **to** m:

$$\begin{aligned} \delta^2 x(m) &= D_s^2[u(m)][\delta u(m)]^2 + D_s[u(m)]\delta^2 u(m) \\ \delta^2 u(m) &= \langle \mathbf{W}^T(m), \delta^2 \mathbf{x}(m-1)\rangle \end{aligned}$$

Stage k **to** $k+1$:

$$\begin{aligned} \delta^2 x_{j_{m-k-1}}(k+1) &= D_s^2[u_{j_{m-k-1}}(k+1)][\delta u_{j_{m-k-1}}(k+1)]^2 \\ &\quad + D_s[u_{j_{m-k-1}}(k+1)]\delta^2 u_{j_{m-k-1}}(k+1) \\ \delta^2 u_{j_{m-k-1}}(k+1) &= \langle \mathbf{W}^T_{j_{m-k-1}*}(k+1), \mathbf{x}(k)\rangle \end{aligned}$$

Stage $(k-1)$ **to** k:

$$\begin{aligned} \delta^2 x_{j_{m-k}}(k) &= D_s^2[u_{j_{m-k}}(k)][\delta u_{j_{m-k}}(k)]^2 + D_s[u_{j_{m-k}}(k)]\delta^2 u_{j_{m-k}}(k) \\ \delta^2 u_{j_{m-k}}(k) &= \langle \delta^2 \mathbf{W}^T_{j_{m-k}*}(k), \mathbf{x}(k-1)\rangle \\ &= \langle \delta^2 \mathbf{W}_{j_{m-k}*}(k), \mathbf{x}(k-1)\rangle \\ &\equiv 0 \end{aligned}$$

Since $\delta \mathbf{W}_{j_{m-k}}(k) = 0$ for $j_{m-k} \neq i$ and $\delta^2 \mathbf{W}_{i*}(k) = 0$ as $\delta_{\mathbf{W}_{i*}^T(k)}$ is fixed. By substituting back we get explicit expressions for the Hessian of $\mathbf{x}(m)$ with respect to $\mathbf{W}_{i*}(k)$ for $1 \leq i \leq n$ and $1 \leq k \leq m-1$.

CONCLUSION

Using the standard tools from the standard variational theory, we have derived explicit expressions for the gradient and the Hessian of the sum if the squared error criterion with respect to $\mathbf{W}_{i*}(k)$ which is the ith row of the weight matrix $\mathbf{W}(k)$ that defines the interconnection between the output of the $(k-1)^{th}$ stage and the input of the kth stage. It turns out that the gradient of $\mathcal{J}(\overline{\mathbf{W}})$ with respect to $\mathbf{W}_{i*}^T(k)$ is proportional to $\mathbf{x}(k-1)$ and its Hessian with respect to $\mathbf{W}_{i*}^T(k)$ is proportional to the rank one (outer product) matrix $\mathbf{x}(k-1)\mathbf{x}^T(k-1)$. The constants of proportionality depends on the value of the first and second derivatives of the components of the mapping $\mathcal{S}$ which are the sigmoid functions.

REFERENCES

Bishop, C., 1995, *Neural Networks for Pattern Recognition*. Clarendon Press, Oxford, England.

Buntine, W., and Weigend, A., 1994, "Computing Second Derivatives in Feed-Forward Networks: A Review," *IEEE Transactions on Neural Networks*, Vol. 5, pp. 480–488.

Dennis, J. E., Jr., and Schnabel, R. B., 1996, "Numerical Methods for Unconstrained Optimization and Non-Linear Equations," *Classics in Applied Mathematics*, Vol. 16, *SIAM*, Philadelphia, PA, (378 pages).

Fine, T. L., 1999, *Feedforward Neural Network Methodology*, Springer, New York.

Lakshmivarahan, S., and Dhall, S. K., 2007, "Computation of Gradient and Hessian in Feed-Forward Neural Networks: A Variational Approach," Technical Report, School of Computer Science, University of Oklahoma, Norman, Oklahoma.

Minsky, M., and Papert, S., 1988, *Perceptrons*, (expended edition of 1969 version), MIT Press, Cambridge, MA.

Narendra, K. S., and Parthasarathy, K., 1990, "Identification and Control of Dynamical Systems using Neural Networks," *IEEE Transactions on Neural Networks*, Vol. 1, pp. 4–27.

Nash, S. G., and Sofer, A., 1996, *Linear and Non-linear Programming*, McGraw-Hill, New York, NY, (692 pages).

Pearlmutter, B., 1994, "Fast Exact Multiplication by the Hessian," *Neural Computation*, Vol. 6, pp. 147–160.

Rosenblatt, F., 1959, "Two Theorems of Statistical Separability in Perceptrons," in *Mechanization of Thought Process*: Proceedings of the Symposium held at the National Physical Laboratory, November 1958, Vol. **I**, pp. 421–456, London: HM Stationary Office.

Rosenblatt, F., 1961, *Principles of Neurodynamics: Perceptrons and the Theory of Brain Mechanisms*, Spartan Books, Washington, D. C.

Rumelhart, D. E., Hinton, G. E., and Williams, R. J., 1986, "Learning Internal Representations by Error Propagation," in *Parallel and Distributed Processing*, Vol. **I**, *Foundations* by (Ed.) D. E. Rumelhart, J. L. McClelland, and PDP Research Group, MIT Press, Cambridge, MA, pp. 318–362.

Vapnik, V. N., 2000, *The Nature of Statistical Learning Theory*, (second Ed.), Springer, New York.

Werbos, P. J., 1990, "Backpropagation Through Time: What it Does and How to Do it," *Proceedings of the IEEE*, Vol. 78, pp. 1550–1560.

BOOSTING CLASSIFICATION ACCURACY WITH SAMPLES CHOSEN FROM A VALIDATION SET

TZU-CHENG CHUANG
School of Electrical and Computer Engineering, Purdue University, West Lafayette, Indiana 47907

OKAN K. ERSOY
School of Electrical and Computer Engineering, Purdue University, West Lafayette, Indiana 47907

SAUL B. GELFAND
School of Electrical and Computer Engineering, Purdue University, West Lafayette, Indiana 47907

ABSTRACT

It is common to train a classifier with a training set, and to test it with a testing set to study the classification accuracy. In this paper, we show how to effectively use a number of validation sets obtained from the original training data to improve the performance of a classifier. The proposed validation boosting algorithm is illustrated with a support vector machine (SVM) in the application of lymphography classification. A number of runs with the algorithm is generated to show its robustness as well as to generate consensus results. At each run, a number of validation datasets are generated by randomly picking a portion of the original training dataset. At each iteration during a run, the trained classifier is used to classify the current validation dataset. The misclassified validation vectors are added to the training set for the next iteration. Every time the training set is changed, new classification borders are generated with the classifier used. Experimental results on a lymphography dataset shows that the proposed method with validation boosting can achieve much better generalization performance with a testing set than the case without validation boosting.

INTRODUCTION

Machine learning has been used in cancer prediction and prognosis for nearly 20 years (Cruz and Wishart 2006). There are several methods widely used for this purpose, such as decision tree, Naïve Bayes, k-nearest neighbor, neural network and support vector machine algorithms. New algorithms are still developed to improve the classification accuracy. One approach is to utilize feature extraction to select fewer features to train the classifier. Bagging and boosting techniques to generate different training samples are also utilized for this purpose. In this way, a number of different classifiers can be generated, and consensus techniques such as majority voting and least squares estimation-based weighting (Kim 2003) can be used to achieve better and more stable classification accuracy.

In Bagging (Breiman 1996), several classifiers are trained independently via a bootstrap method, and their results are combined together to obtain the final decision. In this procedure, a single training set $TR=\{(\mathbf{x}_i;y_i)|i=1,2,\ldots,n\}$ is used to generate K different classifiers. In order to get K different training sets and make them indepent of each other, the original training set is resampled. The new K training sets have the same size as the original dataset, but some instances may appear more than once, and some instances may not be in a new resampled training set.

The adaboost algorithm by Freund and Schapire (1994, 1996, 1997) is generally considered as a first step towards more practical boosting algorithms. A boosting algorithm defines different distributions over the training samples, and uses a weak learner to generate hypotheses with respect to the generated distributions. From the different distributions of training samples, different classifiers are generated, and they are next combined with different weights to get the final results.

Although the resampling technique of our proposed method is similar to bagging and boosting, our appoach is different since we utilize a number of validation sets obtained from the training set, and they are utilized to modify the training sets. Initially, we divide the original training data into 2 groups, one for training, and the other for validation. We use the training portion to train the classifier, and then we validate it with the validation set. The misclassified validation samples are added to the current training set to generate the next training set. At each iteration, the current validation set is regenerated as a randomly chosen part of the original training dataset with a fixed percentage. At each run, the procedure is repeated in several iterations until the validation accuracy reaches its maximum. At this point, a classifier is generated. Due to the random initialization of the training set and the validation set at each run, different independent classifiers are obtained with a number of runs. The results from different runs can be combined by a consensus rule such as majority voting to get the final results.

DATASET

Lymphography data is obtained from the UIC machine learning repository (Kononenko and Cestnik 1988). The examples in this data set use 18 attributes, with four possible final diagnostic classes. The attributes include lympho node dimension, number of nodes, types of lymphatics, etc. For convenient represantation, the attributes are transformed to integer type. There are a total of 148 samples, 2 are normal, 81 are metastases, 61 are malignant lymph and 4 are fibrosis. Because normal and fibrosis cases are scarce compared to the other two cases, we used 142 samples to classify whether the sample is metastases or malignant lymph.

SUPPORT VECTOR MACHINES

Vapnik invented SVM's with a kernel function in the 1990s (Vapnik 1992). This algorithm is initially designed for the two-class classification problem. One class output is marked as 1, and the other class output is marked as -1.

The algorithm tries to find the best separating hyperplane with the largest margin width. By getting a better hyperplane from the training samples, it is expected to get better testing accuracy. In SVM, the hyperplane of the non-separable case is determined by solving the following equation:

$$\min \frac{1}{2}\|w\|^2 + C\sum_i \xi_i$$
$$subject\ to \qquad (1)$$
$$y_i(x_i^T w + b) \geq 1 - \xi_i$$
$$\xi_i \geq 0$$

where x_i is the ith data vector, y_i is the binary (-1 or 1) class label of the ith data vector, ξ_i is the slack variable, w is the weight vector normal to the hyperplane, C is the regularization parameter, and b is the bias. It can be shown that the margin width is equal to $2/|w|$.

Usually the original data is mapped by using a kernel function to a higher dimensional representation before classification. Some common kernel functions are linear, polynomial, radial basis and sigmoid functions. In our case, we used the radial basis function given by

$$K(x, x_i) = C * \exp(-\gamma * |x - x_i|^2) \qquad (2)$$

In the experiments conducted, the SVM-Light (Joachims, 2004) software was utilized. We picked γ equal to 1 and C equal to 1 in these experiments.

TRAINING AND VALIDATION RESAMPLING TECHNIQUE

In the training phase, we initially decide the percentage of the training set as p_{train} and the percentage of the validation set as p_{vali}. We divide the original training set into 2 groups according to p_{train} and p_{vali}. These 2 initial training and validation sets do not overlap with each other. The training set is used to train the classifier, which is next validated with the validation set (Figure 1). Then, the misclassified validation samples are included in the training set to generate the next iteration training set. In the next iteration, the validation set is randomly picked from the original complete training set with percentage p_{vali}. With the new training set and the validation set, other misclassified validation samples are generated and are included in the training set to generate the next iteration training set. After several iterations, the performance of the classifier trained in this way becomes better than that of the classifier trained with all the original training set without any validation set. The iterations are stopped after reaching nearly 100% validation accuracy.

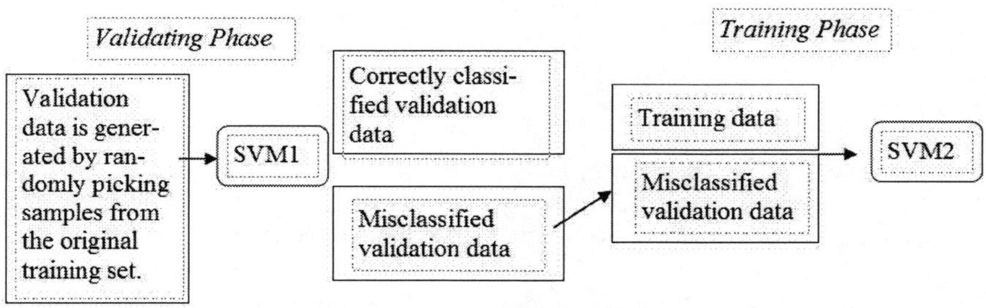

Figure 1. The misclassified validation samples are added to the training samples of the previous stage.

The proposed method emphasizes misclassified validation samples. If a misclassified sample is still misclassified the next time, it would be reemphasized, resulting in the following weighting:

$$2\times 1_{x_i,misclassified}(x_i)+1_{x_i,correclyclassified}(x_i) \quad (3)$$

where

$$1_{x_i,type}(x_i)=\begin{cases} 1, x_i -> type \\ 0, x_i -> other\ type \end{cases} \quad (4)$$

Due to the random initializations of the training and the validation set, we get different classifiers at each run. In order to get better results, we can use consensus such as majority voting between these classifiers.

EXPERIMENTS

We initially picked 50% of all the data for training and the other 50% for testing. In the training phase, we chose p_{train} equal to 0.5 and p_{vali} equal to 0.5. The results with four runs with different training and testing data are shown in Table 1. From Table 1, we can see that when the validation accuracy nearly reaches 100%, the testing accuracy also reaches its maximum. Because 100% validation accuracy means there are no misclassified validation samples, the iteration process is stopped after nearly reaching this value.

Table 1. Comparison of the testing classification accuracy between the classifier trained by all training data and the classifier trained by the proposed method. TestByAll means testing accuracy of the classifier trained by all training data. Valid means validation accuracy of that iteration. Test. means testing accuracy of that iteration.

TestByAll		0.5634		0.57746		0.59155		0.5493
iteration	Valid.	Test	Valid	Test	Valid	Test	Valid	Test
1	0.5714	0.5634	0.6286	0.5775	0.5143	0.5775	0.6000	0.5493
2	0.7143	0.4366	0.7714	0.4507	0.8286	0.4225	0.6571	0.4789
3	0.9714	0.7183	0.8571	0.4789	0.8286	0.5493	1.0000	0.5916
4	1.0000	0.6620	1.0000	0.6620	1.0000	0.6197	1.0000	0.5916
5	1.0000	0.6620	1.0000	0.6620	1.0000	0.6197	1.0000	0.5916

To test whether our proposed method is significant to improve the accuracy, we could convert the decimal value to the percentile value and then calculate Kai Square. By picking α =0.05, if the Kai square is larger than 0.3841, we can say that our proposed method is significantly different.

$$\chi^2=\sum_{i=1}^{k}\frac{(x_i-E_i)^2}{E_i} \quad (5)$$

where x_i is the percentile value of testing accuracy from our proposed method, and E_i is the percentile value of testing accuracy from the classifier which is trained by using all training data. Due to short of space, we only show 4 cases. After we take more runs, we can see that it is significantly different.

The following figure (Fig. 2) shows that the testing accuracy drops down first, and then boost to higher than initial one.

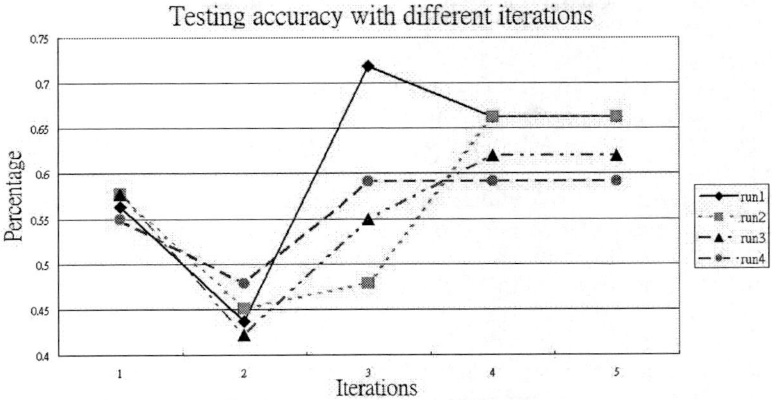

Figure 2. The testing classification accuracy varies with iterations.

In order to test for consensus results,we fixed the same training data and testing data for 3 classifiers, and then used majority voting to combine different classifier results. The results are shown in Tables 2 and 3.

Table 2. Combining 3 different classifiers by majority voting. In this training set and testing set, if we use all the training data to train the classifier, the testing accuracy is 0.5493. The three classifiers are generated from different initializations of the training set and the validation set.

Classifier	1		2		3		Consensus
Iteration	Valid.	Test.	Valid.	Test.	Valid.	Test.	Test.
1	0.54286	0.5493	0.57143	0.5493	0.6286	0.5493	0.5493
2	0.74286	0.46479	0.74286	0.46479	0.6857	0.4507	0.4507
3	1	0.60563	1	0.60563	1	0.60563	0.6056
4	0.97143	0.60563	1	0.60563	1	0.60563	0.6056

Table 3. Combining 3 different classifiers by majority voting. In this training set and testing set, if we use all the training data to train the classifier, the testing accuracy is 0.5634.

Classifier	1		2		3		Consensus
Iteration	Valid.	Test.	Valid.	Test.	Valid.	Test.	Test.
1	0.6000	0.5634	0.3143	0.4648	0.5429	0.5634	0.5634
2	0.6571	0.4507	0.7714	0.5634	0.7143	0.4507	0.4648
3	1	0.6479	0.9429	0.5775	1.0000	0.6620	0.6479
4	1	0.6479	0.9714	0.5775	1.0000	0.6620	0.6479

DISCUSSION AND CONCLUSIONS

From the results of the experiments, it is apparent that the resampling technique does generate a better training set to train the classifier, resulting in better classification accuracy.

Another approach would be to use all the training data to train the classifier, and then used the classifier to search for the misclassified vectors. However, then it is likely to get 100% training accuracy, meaning we do not know which samples to emphasize. Including validation sets works better due to this reason.

In some cases, we noticed that the testing accuracy was lower in iteration 2 or 3. However, the testing results always improved when we reached 100% validation accuracy in succeeding iterations.

In previous boosting methods, it is possible to overfit by running too many rounds. With our approach, we only add the misclassified validation samples to the training set. When 100% validation accuracy is reached, there are no more rounds to be used.

We also noticed that if the validation accuracy in the first iteration is not sufficiently high, such as better than 50%, it is a good idea to regenerate the initialization of the training and validation sets. This approach reduces the number of iterations to get the best training set.

We also considered rates of convergence. In all the experiments, the maximum validation accuracy is always reached within 5 iterations. This may take extra computation time compared to using all the training set to train one classifier, but the number of iterations to get better results is not excessive.

By using random initialization of the training set and the validation set, we can generate a number of different classifiers. The results from these classifiers can be combined, for example, by using majority voting to achieve better results. However, in our consensus experiments, the results did not improve further. This topic needs to be further investigated. We only generated 3 classifiers and then aggregated the results. It is possible that more classifiers would increase performance.

In the experiments, we chose p_{train} equal to 0.5 and p_{vali} eaul to 0.5. In order to estimate the optimal values of these parameters, it is necessary to do further research.

ACKNOWLEDGEMENT

This research was supported by NSF Grant MCB-9873139 and partly by NSF Grant #0325544.

REFERENCES

Joseph A. Cruz, and David S. Wishart, 2006, "Applications of Machine Learning in Cancer Prediction and Prognosis," *Cancer Informatics.*

Hyun-Chul Kim, Shaoning Pang, Hong-Mo Je, Daijin Kim, Sung Yang Bang, 2003, "Constructing support vector machine ensemble," *Pattern Recognition* 36 pp. 2757 – 2767.

Leo Breiman, 1996, "Bagging Predictors" *Machine Learning* 24 (2) pp.123–140.

Y. Freund and R.E. Schapire, 1994, "A decision-theoretic generalization of on-line learning and an application to boosting," In Euro COLT: European Conference on Computational Learning Theory. LNCS.

Y. Freund and R.E. Schapire, 1996, "Experiments with a new boosting algorithm," In *Proceedings 13th International Conference on Machine Learning*, pp. 146-148. Morgan Kaufmann.

Y. Freund and R.E. Schapire,1997, "A decision-theoretic generalization of on-line learning and an application to boosting." *Journal of Computer and System Sciences*,55(1):119-139.

Igor Kononenko and Bojan Cestnik, 1988, Repository of machine learning databases, http://www.ics.uci.edu/~mlearn/MLRepository.html , Irvine, CA: University of California, Department of Information and Computer Science.

B. E. Boser, I. M. Guyon, and V. N. Vapnik, 1992, "A training algorithm for optimal margin classifiers," D. Haussler, editor, 5th Annual ACM Workshop on COLT, pages 144-152, Pittsburgh, PA. ACM Press.

Thorsten Joachims, 2004, http://www.cs.cornell.edu/People/tj/svm_light/ .

AN AUTOMATED APPROACH TO GENERATE NEURAL NETWORK TOPOLOGIES

CHAITANYA VEMPATI
Department of Mechanical Engineering
The University of Texas at Austin
Austin, TX 78712
vempati.ramkumar@hotmail.com

MATTHEW I. CAMPBELL
Department of Mechanical Engineering
The University of Texas at Austin
Austin, TX 78712
mc1@mail.utexas.edu

ABSTRACT

This paper presents a novel concept of using graph grammars to generate neural network topologies. The approach involves representing the neural network as a graph and defining graph transformation rules to generate the topologies. The method starts from a small network and then uses predefined grammar rules to build the topology. Each topology generated by the graph grammar is trained using a hybrid training algorithm that uses a combination of genetic algorithms and a quasi-Newton (BFGS) method. The error of the response predicted by each topology during training, is used to further evolve the topologies. The proposed methodology appears to generate promising responses and has the potential to be a powerful process modeler.

Keywords: graph grammar, topology, Multi-Layer Feed Forward Neural Networks, Model Estimation, Learning Algorithms and Training

1 INTRODUCTION

The topology of a neural network refers to the number of nodes and their organization. These nodes can be interconnected to each other in various parallel and series combinations. These combinations give rise to a variety of architectures such as single layer neural networks, multi-layer neural networks, non-layered neural networks, recurrent neural networks etc. In current research, we focus on finding the optimum topology for non-layered feedforward neural networks.

The success of a neural network in modeling a particular problem depends upon its topology, the choice of the processing function within the nodes and the training algorithm. Simpler networks, networks with only a few hidden nodes or sparse connectivity between the nodes, do not have enough flexibility to properly capture complex input-output relationships during training. The training error such a neural network is high and doesn't decrease even after long training times. This phenomenon is commonly known as underfitting [1, 2]. On the other hand a very large and complicated neural network would not only require very large training times but would also inadvertently fit the noise in the data. A large neural network has too much flexibility in terms of the number of parameters that can be modified, so it fits even the noise without regard to underlying trends. Such networks incorrectly map data that has yet to be seen by the network even though the data follows obvious trends. The error of a network in mapping a dataset that follows the same trends as the training set but was not used for training the network is called generalization error. Such a phenomenon is called overfitting.

Therefore, it is absolutely critical for a particular problem to have the right topology since the topology of the network puts a-priori limits on complexity and performance. Although there have been some approaches to automating the generation of topologies [3-8], it is usually a trial and error process or relies on experts in the field with previous experience in the use of neural networks. But as the complexity of problem to be modeled increases, this method of trial and error becomes difficult and unmanageable.

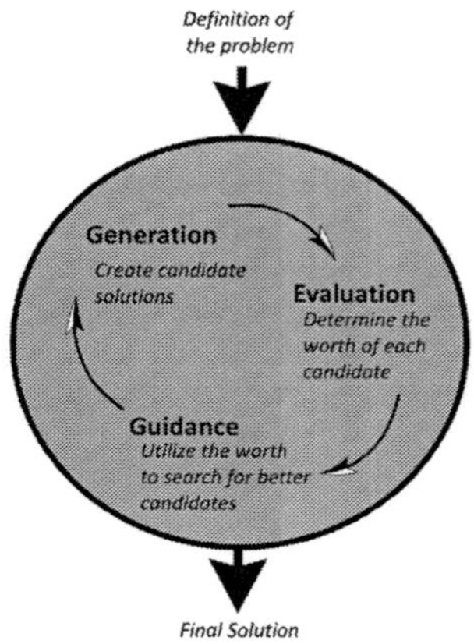

Figure 1: The method of neural network topology design

The current research presents a novel concept of using graph grammars to generate neural network topologies. The approach involves representing the neural network as a graph and defining graph transformation rules to generate the topologies. The method starts from a small network and then uses pre-defined grammar rules to build the topology. The approach is simple, efficient and has the ability to create topologies of varying complexity. The graph grammar based representation of neural networks is a flexible, robust and an intuitive approach to generating successful neural network topologies. Each topology generated by graph grammar is trained using a hybrid training algorithm that uses a combination of genetic algorithms and a quasi-Newton (BFGS) method. The error of the response predicted by the neural network, in other words the performance of each topology during training, is used to further evolve the topologies. While the methodology and the results presented in this research are still very new, they are promising and have the potential to be an automated and intelligent process modeler.

2 NEURAL NETWORK TOPOLOGY DESIGN: THE METHOD

The approach in finding the best neural network topology for a particular problem is to divide the problem into three distinct steps: generation, evaluation and guidance. ***Generation*** is the first step where a possible candidate topology or a population of topologies is created. Neural network topologies are generated by defining graph grammar or graph transformation rules. These rules have the ability to generate all possible neural networks via a recognize-choose-apply cycle (RCA) [9]. The worth of each of the candidate topologies is then ***evaluated*** by measuring their ability to properly train to a given set of data. A hybrid optimization algorithm – a combination of genetic algorithm and BFGS is utilized to train this data. After all the candidate topologies are evaluated, further changes to the topologies are defined by the ***guidance method***. A best first search approach to search through the possible topologies is presented as our guidance strategy which efficiently considers only the feasible set of neural networks. The diagram shown in Figure 1 summarizes the method and the following sections describe each step in more detail.

2.1 Generation

Generation is the first step of the method and the step where possible neural network candidate solutions are generated. The most unique aspect of the presented method lies in its use of graph grammar to generate neural network topologies. The creation of any candidate neural network begins with the seed. The seed is usually the most basic graph and contains user defined information about the problem. This forms the input to the generation step about the problem. In the case of neural networks this information is the number of input parameters and the number of responses. Hence, the seed chosen for generation of neural networks is a graph consisting of input nodes and output nodes with no connections between them. Generation now relies on a set of predefined rules to transform the seed into a solution.

Transformations from seed to a valid neural network are defined by six rules as is shown in the Appendix. Rule 1 and Rule 2 initiate the process of generating a topology. These rules ensure that there are no unconnected nodes. Rule 1 adds an arc between two unconnected nodes while Rule 2 connects two unconnected nodes by adding a node in the first hidden layer. Since the LHS of both the rules define a condition that necessitates at least one unconnected node. So this rule is applicable only when there are unconnected nodes in the graph. Rule 4 is a variation of Rule 2 and adds a node between connected nodes. Rule 4 adds a node in a new hidden layer. The LHS of Rule 4 describes a pair of connected nodes; this rule can be recognized at any point during the process of generation. The more number of times Rule 4 is called the more nodes there will be in the hidden layers and so will there be more layers. Rule 3 and Rule 5 increase the interconnectivity between the nodes. With these rules, it is easy to imagine that the entire set of possible feedforward neural networks can be captured (see Figure 2).

A complete graph or a candidate topology is generated by the application of these grammar rules iteratively via a three step process: recognize, choose and apply [9]. By applying each rule consecutively a candidate usually becomes more complex and comes closer to being a complete solution. At the start of each iteration, *recognize* is the step where all possible locations in the graph are identified where the set of rules can be applied. These locations define a list of all possible transformations that can be applied on that graph at that stage. Given the list of rule choices, the second *choose* step – is the stage where the design agent (be it human or computer) selects a particular rule that would make the best modification to the graph. The chosen rule is now *applied* on the graph and the graph is transformed to a new state.

2.2 Evaluation

To train the neural network topologies created in the generate step, we utilize a hybrid algorithm which is a combination genetic algorithms and a BFGS algorithm. A genetic algorithm is used to train the weights initially to find a good starting point for the BFGS algorithm. The BFGS algorithm uses this starting point from the genetic algorithm and quickly finds the set of weights that minimize the sum of training and generalization error. The input to the algorithm is the topology that is to be trained. The output of the algorithm is the training error and the generalization errors of the topology, in other words its worth. A genotype is a series of real numbers that correspond to weights and biases of the topology. The genetic algorithm is allowed to run for a specified period of time. After the GA is terminated, the best solution obtained by the GA is given as a starting point to the BFGS algorithm.

2.3 Guidance

The guidance step processes the error information from the evaluation step and gives direction to the generation as to which topology to pursue. The problem of finding an optimum topology is formulated as a tree search. The tree results from application of all possible rules on the seed as shown in Figure 2. Each state on the tree represents a feasible candidate topology. Even though the tree starts initially with few topologies it quickly branches out to capture a search space of all possible topologies defined by the transformations in rules. The algorithm is based on a greedy tree search. Search begins by defining a stack for storing intermediate topologies. Starting from the seed all the branches of the tree, in other words all the possible transformations, are added to the stack and trained. The training of a particular topology is completely automated using the hybrid training algorithm described earlier. The training error for a given set of data constitutes the fitness of each candidate topology. Based on this value the stack is sorted in ascending order of fitness. The candidate with the best fitness (smaller the better) or the first candidate in the stack after sorting is then removed from the stack and all its successors are expanded and then evaluated. The successors are the candidates generated by applying all the recognized rules on the first candidate in the stack. All these resulting candidates are added to the stack and the stack is sorted again.

3 IMPLEMENTATION AND RESULTS

The method described above is implemented in GraphSynth [10], which is a publicly available open-source approach to create and implement graph grammars. This allows the user to automatically define, create neural networks as graphs, implement grammar rules, generate topologies and train them. Our approach is implemented on a standard desktop computer with an AMD Athlon Dual-core 2.21 GHz processor with 2 GB RAM. All the topologies use a sigmoid activation function within each node. Table 1 shows the results of running our method on four test problems. These are standard test functions used in meta-modeling [11] that are chosen due to their complexity. The first "sine" test function is a single cycle of the sine function plus noise. The added noise makes fitting the model complicated but the resulting curve fit matches without overfitting as can be seen in Figure 3. The other three test functions have two input variables (x,y) and one output variable (z). One particular challenge in the last two test functions is the number of stationary points, this seems to correspond to the complexity of the final neural networks and the time required to train the network.

For these four test functions, the best neural networks are shown in Figure 4. Note the variety in layers, nodes and connectivity that can result from this approach. The topology generated for the Rosenbrock's function is the simplest while that generated for the Himmelblau function is the most complicated. The total time taken by the search method is essentially the sum of time taken by the method to generate, train the topologies using GA and BFGS, and store the data of results obtained. Since more rules need to be applied to generate a topology of greater complexity, more iterations are needed to generate a complex topology.

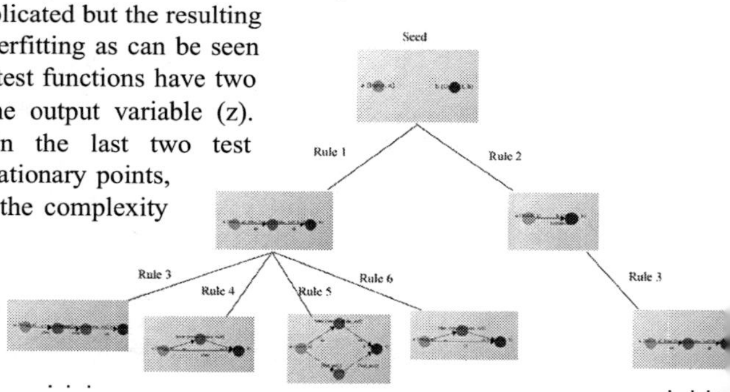

Figure 2: An illustration of the tree of solutions that is produced as a result of the grammar rules.

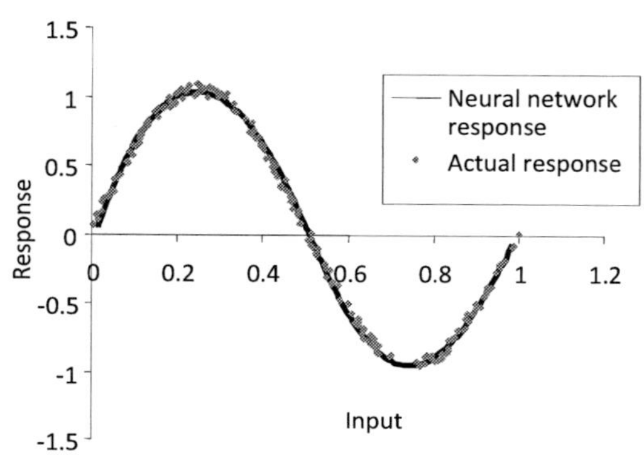

Figure 3: Neural network response vs. actual response for the sine function plus noise.

4 CONCLUSION AND FUTURE WORK

A simple and efficient method of neural network topology generation has been presented. The approach has the ability to generate both simple and complicated neural networks. Since the algorithm begins with a simple network and builds up to an optimum topology by applying rules, it generates a pruned network. Further, this methodology of generation of neural networks always ensures a feasible topology. Feasibility is guaranteed by the fact that only valid topological transformations are captured by the rules at each stage. Previous implementations of generating neural network topologies make simplifying assumptions that result in a failure to explore the complete space of possible neural networks. Another salient feature of the research is its ability to produce pruned networks. As can be observed from the results obtained almost all the topologies are already pruned. While the presented methodology has be shown to produce promising results, there are several improvements that can be made to make the entire process generating topologies faster and more efficient. Instead of training the topology for a specific time using a GA, an exit criterion can be defined in terms of the decrease in error with the number of generations. The results indicate that there is a definite mapping between topology and performance. Every problem has its own level of complexity which should be captured by the neural network topology. The current research looks very promising in capturing the right complexity and should extend well to solve more complicated problems.

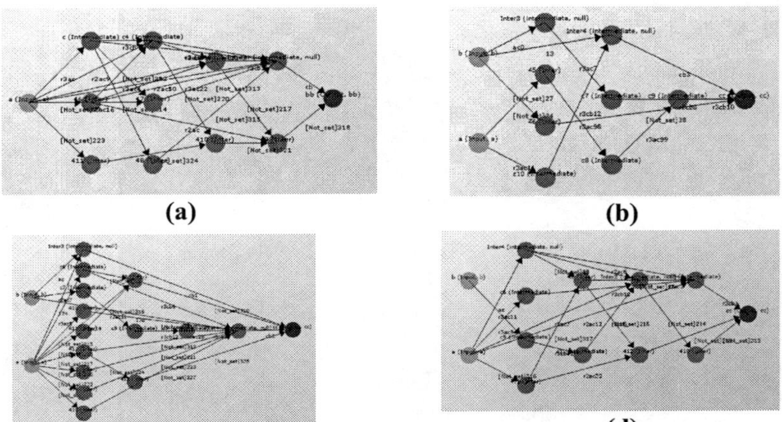

(a) (b) (c) (d)

Figure 4: The resulting optimal neural network topologies for the four test problems: (a) sine function plus noise, (b) Rosenbrock's banana, (c) Himmelblau function, (d) Rastrigin function.

Table 1: A summary of the results obtained.

Test Function	Sine	Rosenbrock	Himmelblau	Rastrigin
Number of stationary points	2	1	9	9
Training Error	0.02	0.091	0.083	0.13
Generalization Error	0.03	0.11	0.0831	0.114
Total number of candidates trained	523	872	746	569
Time for which GA was run (sec)	60	60	80	80
Number of iterations to find the optimum topology	17	15	29	22
Total time taken (approximate in hrs)	21	25	35	29

REFERENCES

[1] Andrés P.U. Structure-Adaptable Digital Neural Networks, Thesis No 2052, École Polytechnique Fédérale De Lausanne.

[2] Xin Y. 1999. Evolving Artificial Neural Networks, Proceedings of the IEEE, Vol. 87.

[3] Fahlman S. E. and Lebiere C. 1990. The cascade-correlation learning architecture, Technical Report CMU-CS-90-100, Carnegie Mellon University.

[4] Lucas, S., 1995. Growing adaptive neural networks with graph grammars. In Proc. of European Symposium on Artificial Neural Networks, pages 235-240.

[5] Kenneth O. S. and Miikkulainen R, 2002. Evolving Neural Networks through Augmenting Topologies, Evolutionary Computation 10(2): 99-127, MIT Press.

[6] Angeline P. J., Saunders G. M., and Pollack J. B. 1994. An evolutionary algorithm that constructs recurrent neural networks, Neural Computation, vol. 5, pp. 54-65.

[7] Matteucci M. 2002. ELeaRNT: Evolutionary Learning of Rich Neural Network Topologies, Technical report No. CMU-CALD-02-103, Carnegie Mellon University.

[8] Fiszelew A., Britos P., Perichisky G., and García-Martínez R. 2003. Automatic Generation of Neural Networks based on Genetic Algorithms. Revista Eletronica de Sistemas de Informacao.

[9] Campbell M.I. 2007. A Graph Grammar Methodology for Creative Systems, Artificial Intelligence, in review.

[10] Campbell M. I. 2006. The Official GraphSynth Site, http://www.graphsynth.com, University of Texas at Austin.

[11] Gilbert, E.G., 1967, "A Selected Bibliography on Parameter Optimization Methods Suitable For Hybrid Computation." Simulation, Vol. 8, No. 6, 350-352.

APPENDIX: NEURAL NETWORK GRAMMAR RULES

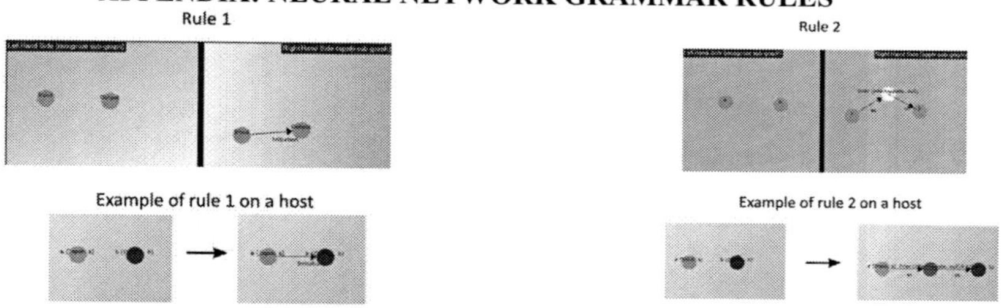

Figure A1: A Figure showing rule 1 and rule 2 with examples.

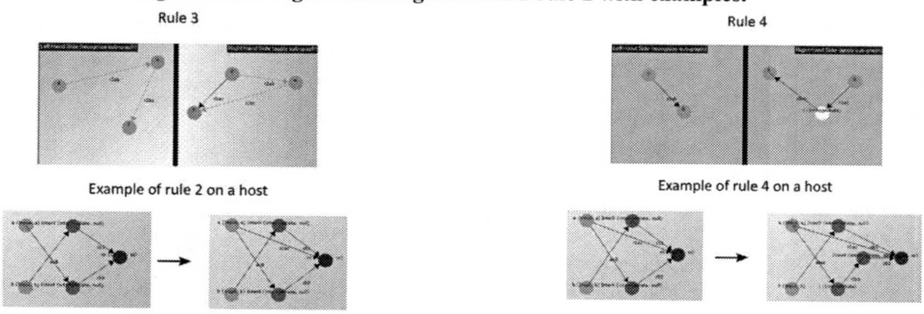

Figure A2: A Figure showing rule 3 and rule 4 with examples.

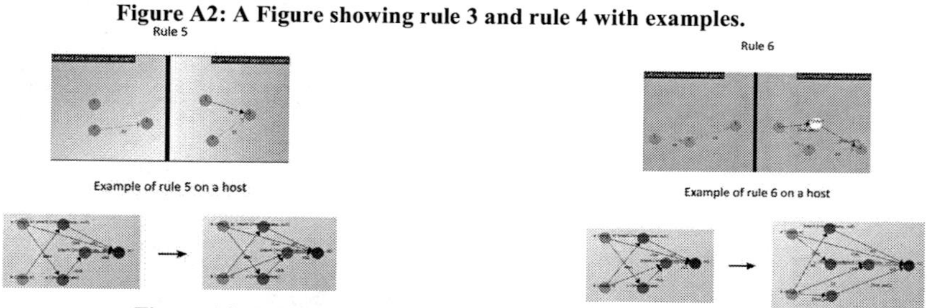

Figure A3: A Figure showing rule 5 and rule 6 with examples.

FITTING A FUNCTION AND ITS DERIVATIVE

ARJPOLSON PUKRITTAYAKAMEE, MARTIN T. HAGAN
School of Electrical and Computer Engineering, Oklahoma State University

LIONEL RAFF
Department of Chemistry, Oklahoma State University

SATISH BUKKAPATNAM
School of Industrial Engineering and Management, Oklahoma State University

RANGA KOMANDURI
School of Mechanical and Aerospace Engineering, Oklahoma State University

Abstract
This paper introduces a new procedure for gradient-based training of multilayer perceptron neural networks to simultaneously approximate both a function and its first derivatives. It is assumed that the true function values and the true derivatives are available at the training points. An algorithm is then derived to compute the gradient of a new performance function that combines both squared function error and squared derivative error. Experimental results show that the neural networks trained by the new procedure yield more accurate approximations for both the functions and their first derivatives than networks trained by standard methods. In addition, it is shown that the generalization capabilities of networks trained using this new procedure are better than those trained with early stopping or Bayesian regularization, even though no validation set is used.

INTRODUCTION

In this research, the objective is to approximate both functions and their derivatives using neural networks. Hornik *et al.* (1989) showed that feedforward neural networks are capable of approximating arbitrary functions. However, Pinkus (1999) further provided conditions under which multilayer feedforward neural networks can simultaneously and uniformly approximate a function and various partial derivatives up to order k. These conditions require the transfer functions not to be a polynomial and to be $\mathbf{C}^m$, where $m \geq k$. Similar conditions were also discussed in Ito (1993) and Li (1996).

An earlier method for approximating both functions and their derivatives was proposed in Ferrari and Stengel (2005). In a first stage, this method algebraically obtains the network parameters for approximating function values. In a second stage, the parameters are adjusted to improve the derivative approximation. In Basson and Engelbrecht (1999), the derivative approximation was performed by adding an extra output unit for each partial derivative to the regular structure of a feedforward neural networks that was used to approximate the function. The standard backpropagation procedure was then used to train the proposed network structure. Hanselmann *et al.*, (1999) created an additional network to approximate the derivatives. The derivative approximation obtained from the extra network was then combined with the network output used for function approximation using Taylor's series. In our research, a different technique was used. We will *not* modify the network structure or create an additional network for derivative approximation. Instead, we will use the same network for both function approximation and derivative approximation. In addition, unlike the algorithm in Ferrari and Stengel (2005), in our method the derivative approximation will occur simultaneously with the function approximation.

In the next section, we will introduce a new performance index, which contains both the sum square of the function errors and the sum square of the derivative errors. We will also demonstrate how to compute the gradient of this new performance index. The approximation accuracy (for both the function and its derivative) obtained using this new performance index will be investigated and compared with results obtained using standard algorithms on both simple analytical functions as well as real-world problems.

TRAINING ALGORITHM

In this section, we will introduce a new training algorithm called CFDA (Combined Function and Derivative Approximation), where the performance index includes both the sum square of the function error and the sum square of the derivative error. The new performance index can be written as

$$J = J_f + \rho J_d$$

$$= \sum_{q=1}^{Q} \sum_{k=1}^{S^M} \{g_k(\mathbf{p}_q) - a_k^M(\mathbf{p}_q)\}^2 + \rho \sum_{r=1}^{R} \sum_{q=1}^{Q} \sum_{k=1}^{S^M} \left(\frac{\partial}{\partial p_r} g_k(\mathbf{p}) - \frac{\partial}{\partial p_r} a_k^M(\mathbf{p})\right)^2 \Bigg|_{\mathbf{p} = \mathbf{p}_q} \quad (1)$$

where $g_k(\mathbf{p}_q)$ is the target function value and $a_k^M(\mathbf{p}_q)$ is the neural network output for the input $\mathbf{p}_q$. Q is the number of training data, while ρ is a scale factor that determines the importance of the derivative approximation relative to the function approximation. S^M is the number of neurons in the output layer (layer M) of the neural network, and R is the dimension of the input vector.

Note that in the performance index J the true function derivative $\partial g_k(\mathbf{p})/\partial p_r$ is compared with the derivative of the network function $\partial a_k^M(\mathbf{p})/\partial p_r$. In this way, we use the same network to approximate both a function and its derivative.

Gradient Calculation

We will use a gradient-based algorithm to minimize the performance index J. Therefore, we first need to calculate $\partial J/\partial w_{i,j}^m$ where $w_{i,j}^m$ is the weight connecting the output of the jth neuron in layer $(m-1)$ to the input of neuron i in layer m. Note further that we will only show how to compute $\partial J_d/\partial w_{i,j}^m$, since the term $\partial J_f/\partial w_{i,j}^m$ is computed using the standard backpropagation algorithm. From Eq. (1), taking the derivative of J_d with respect to $w_{i,j}^m$, we will have

$$\frac{\partial J_d}{\partial w_{i,j}^m} = -2 \sum_{r=1}^{R} \sum_{q=1}^{Q} \sum_{k=1}^{S^M} \left(\frac{\partial g_{k,q}}{\partial p_{r,q}} - \frac{\partial a_{k,q}^M}{\partial p_{r,q}}\right)\left(\frac{\partial}{\partial w_{i,j}^m}\left(\frac{\partial a_{k,q}^M}{\partial p_{r,q}}\right)\right), \quad (2)$$

Since the first factor in Eq. (2) is simply the error term, we will now focus on the computation of the second term. First note that

$$\frac{\partial}{\partial w_{i,j}^m}\left(\frac{\partial a_{k,q}^M}{\partial p_{r,q}}\right) = \frac{\partial}{\partial p_{r,q}}\left(\frac{\partial a_{k,q}^M}{\partial w_{i,j}^m}\right), \quad (3)$$

and using the fact that the term $\partial a_{k,q}^M/\partial w_{i,j}^m$ is computed as part of the standard backpropagation algorithm (Hagan, Demuth, Beale, 1994), we have:

$$\frac{\partial}{\partial p_{r,q}}\left(\frac{\partial a_{k,q}^M}{\partial w_{i,j}^m}\right) = \frac{\partial}{\partial p_{r,q}}\left(\frac{\partial a_{k,q}^M}{\partial n_{i,q}^m} \times \frac{\partial n_{i,q}^m}{\partial w_{i,j}^m}\right) = \frac{\partial}{\partial p_{r,q}}\left(\frac{\partial a_{k,q}^M}{\partial n_{i,q}^m}\right) \times a_{j,q}^{m-1} + \frac{\partial a_{k,q}^M}{\partial n_{i,q}^m} \times \frac{\partial a_{j,q}^{m-1}}{\partial p_{r,q}}, \quad (4)$$

where

$$n_{i,q}^m = \sum_{j=1}^{S^{m-1}} (w_{i,j}^m a_{j,q}^{m-1}) + b_i^m, \text{ and } a_{j,q}^m = f(n_{i,q}^m). \quad (5)$$

By combining Eq. (2) and Eq. (4) and rearranging the summations, we have

$$\frac{\partial J_d}{\partial w_{i,j}^m} = -2\left(\sum_q \left\{ a_{j,q}^{m-1} \sum_k \sum_r \left(\left(\frac{\partial g_{k,q}}{\partial p_{r,q}} - \frac{\partial a_{k,q}^M}{\partial p_{r,q}} \right) \times \frac{\partial}{\partial p_{r,q}} \left(\frac{\partial a_{k,q}^M}{\partial n_{i,q}^m} \right) \right) \right\} + \sum_r \sum_q \left\{ \frac{\partial a_{j,q}^{m-1}}{\partial p_{r,q}} \sum_k \left(\left(\frac{\partial g_{k,q}}{\partial p_{r,q}} - \frac{\partial a_{k,q}^M}{\partial p_{r,q}} \right) \times \frac{\partial a_{k,q}^M}{\partial n_{i,q}^m} \right) \right\} \right) \quad (6)$$

To compute the elements of Eq. (6), we first define

$$v_{i,q}^m \equiv \sum_k \sum_r \left(\left(\frac{\partial g_{k,q}}{\partial p_{r,q}} - \frac{\partial a_{k,q}^M}{\partial p_{r,q}} \right) \times \frac{\partial}{\partial p_{r,q}} \left(\frac{\partial a_{k,q}^M}{\partial n_{i,q}^m} \right) \right), \quad (7)$$

where $i = 1, \ldots, S^m$ and $q = 1, \ldots, Q$. Also define

$$u_{i,h}^m \equiv \sum_k \left(\left(\frac{\partial g_{k,q}}{\partial p_{r,q}} - \frac{\partial a_{k,q}^M}{\partial p_{r,q}} \right) \times \frac{\partial a_{k,q}^M}{\partial n_{i,q}^m} \right) \text{ and } z_{i,h}^m \equiv \frac{\partial a_{i,q}^m}{\partial p_{r,q}}, \quad (8)$$

where $i = 1, \ldots, S^m$, and $h = Q(r-1) + q$ in which $q = 1, \ldots, Q$ for each $r = 1, \ldots, R$. Therefore, from Eq. (7) and Eq. (8), we can rewrite Eq. (6) as

$$\frac{\partial J_d}{\partial w_{i,j}^m} = -2\left(\sum_{q=1}^{Q} \{ a_{j,q}^{m-1} v_{i,q}^m \} + \sum_{h=1}^{QR} z_{i,h}^{m-1} u_{i,h}^m \right). \quad (9)$$

Next, we need to calculate the terms defined in Eq. (7) and Eq. (8). We will begin with the term $u_{i,h}^m$ defined in Eq. (8), followed by the term $v_{i,q}^m$ in Eq. (7), and finally $z_{i,h}^m$. To compute $u_{i,h}^m$, using the terms we defined in Eq. (8), and by using the chain rule of calculus, we have:

$$u_{i,h}^m = \sum_k \left(\left(\frac{\partial g_{k,q}}{\partial p_{r,q}} - \frac{\partial a_{k,q}^M}{\partial p_{r,q}} \right) \times \sum_l \left\{ \frac{\partial a_{k,q}^M}{\partial n_{l,q}^{m+1}} \times w_{l,i}^{m+1} \times \dot{f}(n_{i,q}^m) \right\} \right). \quad (10)$$

By rearranging the summations and using Eq. (8), Eq. (10) becomes

$$u_{i,h}^m = \dot{f}(n_{i,q}^m) \sum_l \{ w_{l,i}^{m+1} u_{l,h}^{m+1} \}. \quad (11)$$

From Eq. (10) and the fact that each $n_{i,q}^M$ affects only one $a_{k,q}^M$, we have at the last layer

$$u_{i,h}^M = \sum_k \left(\left(\frac{\partial g_{k,q}}{\partial p_{r,q}} - \frac{\partial a_{k,q}^M}{\partial p_{r,q}} \right) \frac{\partial a_{k,q}^M}{\partial n_{i,q}^M} \right) = \left(\frac{\partial g_{k,q}}{\partial p_{r,q}} - \frac{\partial a_{k,q}^M}{\partial p_{r,q}} \right) \dot{f}(n_{i,q}^M). \quad (12)$$

Now, we will derive $v_{i,q}^m$. From Eq. (7) and standard backpropagation, we can express it as

$$v_{i,q}^m = \sum_k \sum_r \left(\frac{\partial g_{k,q}}{\partial p_{r,q}} - \frac{\partial a_{k,q}^M}{\partial p_{r,q}} \right) \times \left(\left(\frac{\partial}{\partial p_{r,q}} \dot{f}(n_{i,q}^m) \right) \sum_l \left\{ \frac{\partial a_{k,q}^M}{\partial n_{l,q}^{m+1}} \times w_{l,i}^{m+1} \right\} + \dot{f}(n_{i,q}^m) \sum_l \left\{ \frac{\partial}{\partial p_{r,q}} \left(\frac{\partial a_{k,q}^M}{\partial n_{l,q}^{m+1}} \right) \times w_{l,i}^{m+1} \right\} \right). \quad (13)$$

For the term $\partial \dot{f}(n_{i,q}^m) / \partial p_{r,q}$ in the above equation, by using the chain rule, we have

$$\frac{\partial}{\partial p_{r,q}}\dot{f}(n_{i,q}^{m}) = \frac{\partial}{\partial a_{i,q}^{m}}\dot{f}(n_{i,q}^{m}) \times \frac{\partial a_{i,q}^{m}}{\partial p_{r,q}}. \tag{14}$$

Thus, from Eq. (7), Eq. (8), Eq. (13), Eq. (14) and some rearrangements, $v_{i,q}^{m}$ can be computed as

$$v_{i,q}^{m} = \frac{\partial}{\partial a_{i,q}^{m}}\dot{f}(n_{i,q}^{m})\sum_{r}\left\{\frac{\partial a_{i,q}^{m}}{\partial p_{r,q}}\sum_{l}(w_{l,i}^{m+1}u_{l,h}^{m+1})\right\} + \dot{f}(n_{i,q}^{m})\sum_{l}\{w_{l,i}^{m+1}v_{l,h}^{m+1}\}. \tag{15}$$

Next, we need to calculate $v_{i,q}^{M}$ at the last layer, in order to initialize the backpropagation process defined in Eq. (15). By using Eq. (7) and Eq. (14), $v_{i,q}^{M}$ can be computed as follows

$$v_{i,q}^{M} = \frac{\partial}{\partial a_{i,q}^{M}}\dot{f}(n_{i,q}^{M})\sum_{r}\left\{\left(\frac{\partial g_{k,q}}{\partial p_{r,q}} - \frac{\partial a_{k,q}^{M}}{\partial p_{r,q}}\right) \times \frac{\partial a_{i,q}^{M}}{\partial p_{r,q}}\right\}. \tag{16}$$

To compute $z_{j,h}^{m}$, from Eq. (8) and the chain rule of calculus, we have

$$z_{j,h}^{m} = \frac{\partial a_{j,q}^{m}}{\partial p_{r,q}} = \frac{\partial a_{j,q}^{m}}{\partial n_{j,q}^{m}} \times \frac{\partial n_{j,q}^{m}}{\partial p_{r,q}} = \dot{f}(n_{j,q}^{m})\sum_{l=1}^{S^{m-1}}\left\{w_{j,l}^{m} \times \frac{\partial a_{l,q}^{m-1}}{\partial p_{r,q}}\right\}. \tag{17}$$

This equation is initialized at the input (i.e., $m = 0$). From Eq. (17) (and using the fact that $\partial a_{j,q}^{0}/\partial p_{r,q} = \partial p_{j,q}/\partial p_{r,q} = 1$ when $j = r$ and equals zero otherwise), we can initialize Eq. (17) at the input layer.

What we have shown so far is how to compute the derivatives of performance with respect to the weights. We can compute the derivatives with respect to the bias in a similar way:

$$\frac{\partial J_d}{\partial b_i^m} = -2\sum_{r=1}^{R}\sum_{q=1}^{Q}\sum_{k=1}^{S^M}\left(\left(\frac{\partial g_{k,q}}{\partial p_{r,q}} - \frac{\partial a_{k,q}^{M}}{\partial p_{r,q}}\right) \times \frac{\partial}{\partial p_{r,q}}\left(\frac{\partial a_{k,q}^{M}}{\partial n_{i,q}^{m}}\right)\right) = -2\sum_{q} v_{i,q}^{m}. \tag{18}$$

Algorithm Summary

We now summarize the steps for minimizing the performance index J_d:

1. Obtain $a_{j,q}^{m}$ and $z_{j,h}^{m}$, using Eq. (17).
2. Initialize $u_{i,h}^{M}$ and $v_{i,q}^{M}$, using Eq. (12) and Eq. (16).
3. Backpropagate $u_{i,h}^{m}$ and $v_{i,q}^{m}$, using Eq. (11) and Eq. (15). Then compute $\partial J_d/\partial w_{i,j}^{m}$ and $\partial J_d/\partial b_i^m$, using Eq. (9) and Eq. (18).
4. Using the gradient computed in Step 3, update the weights and biases using any standard gradient-based optimization technique.

Note that, from Eq. (1), there is one parameter, ρ, in the performance index that must be set. This parameter determines the relative importance of the derivative error in comparison to the function error. We propose to set $\rho = \lambda/\eta^2$ where $\eta = max(abs(derivative))/max(abs(function))$. In this way, η will account for the difference in scales between the function value and the derivative value. The same value of the parameter λ can then be used over a variety of problems. This will be tested in the next section.

SIMULATION RESULTS

In this section, we will test the approximation accuracy of the Combined Function and Derivative Approximation (CFDA) method on several univariate and multivariate function approximation problems. We will compare the results for the CFDA method with two standard algorithms that have been successfully used for function approximation in the past.

Single-Input/Single-Output Function Approximation

In this subsection, the goal is to compare the approximation accuracy obtained using the CFDA method with two other standard training algorithms on three single-input/single-output (SISO) function approximation problems. For purposes of comparison, we have chosen two standard algorithms that we have found to produce accurate training results without overfitting: 1) the Quasi-Newton BFGS (Broyden-Fletcher-Goldfarb-Shanno) optimization technique with early stopping (BFGS-ES), using 20% of the data for the validation set, and 2) the Gauss-Newton approximation to Bayesian Regularization (GNBR, Foresee and Hagan, 1997). The CDFA method can be used in combination with any gradient-based optimization algorithm. In this investigation, we have chosen BFGS quasi-Newton. The algorithms were tested on the functions in Table 1. For each function, 25 Monte Carlo trials were performed with different inital weights and data divisions. The results are shown in Table 2 and Table 3.

Table 1: List of functions

Problem	Function	Training Area	# Training Points	Testing Area	# Testing Pts	NN structure
1	$\sin(0.5\pi p)$	$-1 \le p \le 1$	40	$-0.8 \le p \le 0.8$	321	1 – 10 – 1
2	$1.2e^{-3p}\sin(8\pi p)$	$0 \le p \le 1$	200	$0.1 \le p \le 0.8$	801	1 – 60 – 1
3	$\cos(4\pi p)$	$-1 \le p \le 1$	200	$-0.8 \le p \le 0.8$	1601	1 – 40 – 1

For the CFDA-BFGS method, we first varied the penalty factor λ to determine an appropriate setting. For each problem, the median RMSE's from the 25 Monte Carlo runs obtained from CFDA-BFGS was plotted versus different values of λ. From these results, we determined that a robust value for λ would be 10^4. This value was used for all the subsequent cases. From Table 2 and Table 3 we can see that the CFDA-BFGS method provides significantly smaller errors on the test set than either BFGS-ES or GNBR. This is true for both the derivative errors and the function errors. Therefore, CFDA provides better generalization capabilities, even without using a validation set. In the next subsection, we will test the algorithm on multi-input function approximation problems.

Table 2: Approximation accuracy obtained from BFGS-ES

Problem	Median RMSE from BFGS-ES			
	Function		Derivative	
	Training	Testing	Training	Testing
1	4.3E-06	1.1E-03	7.8E-03	1.7E-02
2	5.8E-04	3.2E+00	2.5E+00	3.2E+00
3	6.4E-05	2.3E-03	4.6E-02	1.6E-01

Table 3: Approximation accuracy for GNBR and CFDA-BFGS with $\lambda = 10^4$.

Problem	Median RMSE for GNBR				Median RMSE for CFDA-BFGS			
	Function		Derivative		Function		Derivative	
	Training	Testing	Training	Testing	Training	Testing	Training	Testing
1	3.4E-07	1.9E-06	2.4E-05	3.5E-05	1.2E-08	1.2E-08	3.8E-09	1.4E-07
2	3.7E-07	4.2E-04	4.0E-02	7.8E-02	7.7E-07	3.6E-06	1.4E-06	8.0E-04
3	6.3E-07	7.6E-06	3.8E-04	5.5E-04	7.9E-08	1.1E-07	4.0E-07	2.6E-06

Multi-variate Function Approximation

In Molecular Dynamics (MD), the movements of atoms and molecules are simulated using known laws of physics to calculate the forces on individual atoms. Neural networks are applied in this field to predict the forces and the potential energy based on atomic configurations (i.e. various distances and angles between atoms, Raff, *et al.*, 2005). If the potential energy is known, the forces can be computed from the derivatives of the energy with respect to the distances and angles. Therefore, MD is an interesting application area for the use of neural networks in fitting both a function and its derivatives. We will investigate two MD problems: the Si_3 molecule, with training data obtained from the Tersoff (1989) model, and the Si_5 molecule, with data obtained with quantum calculations.

The number of data generated for each compound was 10,000 points. Approximately 1,500 points were randomly selected as the training data, while the rest of the points were used for testing. The network structure for Si_3 was $3-100-1$, while for Si_5 it was $9-45-1$. Table 4 shows the approximation accuracy comparison between the GNBR method, and the CFDA method for these two substances. As with the univariate case, the CFDA method provides improved testing errors for both the function and its derivative.

Table 4: Approximation accuracy for GNBR and the CFDA-BFGS with $\lambda = 10^4$.

Problem	RMSE from GNBR				RMSE from CFDA-BFGS			
	Function		Derivative		Function		Derivative	
	Training	Testing	Training	Testing	Training	Testing	Training	Testing
Si_3	9.9E-07	1.4E-06	1.3E-05	1.9E-05	1.0E-06	1.1E-06	4.4E-07	1.6E-06
Si_5	2.1E-04	4.2E-04	2.1E-03	2.1E-03	3.4E-04	3.5E-04	8.0E-04	8.6E-04

CONCLUSIONS

In this paper, we have presented a new gradient-based procedure for simultaneously fitting both a function and its derivative. From simulation results on both univariate and multivariate problems, we find that the approximation accuracy was significantly improved for both functions and their derivatives using the new method. In addition, the generalization ability of the new method is better than those obtained using current state-of-the-art training procedures.

ACKNOWLEDGEMENTS

This project has been funded by a grant (DMI-0457663) from the National Science Foundation (NSF).

REFERENCES

Basson, E. and Engelbrecht, A. P., 1999, "Approximation of a Function and Its Derivatives in Feedforward Neural Networks", *International Joint Conference on Neural Networks*, July 1999, Washington, Vol. 1, pp. 419-421.

Ferrari S. and Stengel, R. F., 2005, "Smooth Function Approximation Using Neural Networks", *IEEE Transactions on Neural Networks*, Vol. 16, No. 1, January 2005, pp. 24-38.

Foresee, D. and Hagan, M., 1997, "Gauss-Newton Approximation to Bayesian Learning," *Proceedings of the 1997 International Joint Conference on Neural Networks.*, vol. 3, pp. 1930 - 1935.

Gilbert, J. R., Moler, C. and R. Schreiber, 1992, "Sparse matrices in MATLAB: Design and Implementation", *SIAM Journal on Matrix Analysis and Applications*, Vol. 13, issue 1, pp. 333-356.

Hagan, M., Demuth, H. and Beale, M., 1996, *Neural Network Design*, Boston, MA: PWS Publishing.

Hanselmann, T., Zaknich, A. and Attikiouzel, Y., 1999, "Learning Functions and Their Derivatives Using Taylor Series and Neural Networks", *International Joint Conference on Neural Networks*, July 1999, Washington, Vol. 1, pp. 409-412.

Horn, R. and Johnson, C. R., 1994, *Topics in Matrix Analysis*, Cambridge University Press, New York.

Hornik, K., Stinchcombe, M. and White, H., 1989, "Multilayer feedforward networks are universal approximators," *Neural Networks*, Vol. 2, pp. 359-366.

Ito, Y., 1993, "Approximation of differentiable functions and their derivatives on compact sets by neural networks", *Math. Scient.*, 18, pp. 11-19.

Li, X., 1996, "Simultaneous approximations of multivariate functions and their derivatives by neural networks with one hidden layer," *Neurocomputing*, 12, pp. 327-343.

Pinkus, A., 1999, "Approximation theory of the MLP model in neural networks", *Acta Numerica*, 8, pp. 143-195.

Raff, L.M., Malshe, M., Hagan, M., Doughan, D.I., Rockley, M.G., and Komanduri, R., 2005, "*Ab initio* potential-energy surfaces for complex, multi-channel systems using modified novelty sampling and feedforward neural networks," The Journal of Chemical Physics 122, 084104.

Tersoff, B., 1989, "Modeling solid-state chemistry: Interatomic potentials for multicomponent systems," Phys. Rev. B 39, 5566 - 5568.

DIMENSIONALITY REDUCTION USING NEURAL NETWORKS

MOHAMMAD NAYEEM TELI
Department of Computer Science
Colorado State University
Fort Collins, CO

ABSTRACT
A multi-layer neural network with multiple hidden layers was trained as an autoencoder using steepest descent, scaled conjugate gradient and alopex algorithms. These algorithms were used in different combinations with steepest descent and alopex used as pretraining algorithms followed by training using scaled conjugate gradient. All the algorithms were also used to train the autoencoders without any pretraining. Three datasets: USPS digits, MNIST digits, and Olivetti faces were used for training. The results were compared with those of Hinton et al. (Hinton and Salakhutdinov, 2006) for MNIST and Olivetti face dataset. Results indicate that while we were able to prove that pretraining is important for obtaining good results, the pretraining approach used by Hinton et al. obtains lower RMSE than other methods. However, scaled conjugate gradient turned out to be the fastest, computationally.

1. INTRODUCTION

Dimensionality reduction is a method of obtaining the information from a high dimensional feature space using fewer intrinsic dimensions. Reducing dimensionality of high dimensional data is good for better classification, regression, presentation and visualization of data.

Recently Hinton et al. (Hinton and Salakhutdinov, 2006) used a deep autoencoder for dimensionality reduction of multiple datasets. The autoencoders are multi-layer identity mapping neural networks represented by a function $f(x) = x$, where x is a multidimensional input vector to the network. They argue that deep autoencoders could be easily trained using a gradient descent method provided the initial weights are near good solutions. They claimed that by pretraining, they were able to obtain a good set of initial weights and the fine tuning which followed the pretraining approach was able to reduce the data dimensionality very efficiently. Their results support their arguments, however, there still remain some areas which require additional studies.

Firstly, they reported deep autoencoders showed significant improvement when pretrained over the ones without pretraining (see supporting material of (Hinton and Salakhutdinov, 2006) for details). However, they studied conjugate gradient algorithm with line search for fine tuning and their results cannot be extended to other gradient based methods which do not use line search. Secondly, their pretraining approach is very complicated. It assigns a probability to every possible image via an energy function. We thought it would be pertinent to pretrain a deep autoencoder with a less complicated approach. Thirdly, one of the problems that has been identified with training multi-layer neural networks using gradient based algorithms, is the problem of local minima. Keeping this in view, we thought it would be important to train our network with a non-gradient based algorithm.

We want to further investigate some of the aspects of the dimensionality reduction using neural networks that were not explored fully by Hinton et al. In particular we want to focus on the pretraining and weight initialization of multi-layer neural networks. We hypothesize that, gradient based algorithms can train a deep autoencoder and obtain competitive results without pretraining; secondly, simpler pretraining approaches followed by fine tuning using a gradient descent method can obtain competitive results; and finally, non-gradient based algorithm can train a deep autoencoder with competitive results. In these hypotheses, competitive results would mean that the RMSE values between the input and the reconstructed datasets are lower or equal to those obtained by Hinton et al. (Hinton and Salakhutdinov, 2006).

In order to investigate our first hypothesis, we will train our deep autoencoders using scaled conjugate gradient (Moller, 1993) and steepest descent methods without pretraining. The second hypothesis will be verified by, firstly, pretraining layer by layer, using steepest descent followed by fine tuning using scaled conjugate gradient. Secondly, repeat fine tuning phase by replacing steepest descent in the pretraining phase with the correlation based non gradient algorithm, alopex (Unnikrishnan and Venugopal, 1994). To test our third hypothesis, the multi-layer neural network will be trained using alopex without pretraining.

2. BACKGROUND

Over the years many algorithms have been suggested to train autoencoders for efficient dimensionality reduction. DeMers et al. (DeMers and Cottrell, 1993), Nielsen (Nielsen, 1995), and Kambhatla et al. (kambhatla and Leen, 1997) independently observed that optimizing weights in non-linear autoencoders that have multiple layers is a very difficult task because of the choice of initial weights.

Recently Hinton et al. came up with a very efficient approach for dimensionality reduction. They argue that a gradient descent method could be used to train deep autoencoders very efficiently, if our initial set of weights were close to a good solution. They reason that autoencoders would find poor local minima with large initial weights and the gradients in the early layers would be tiny with small initial weights. This would make it infeasible to train autoencoders with many hidden layers (Hinton and Salakhutdinov, 2006). In order to overcome this, they suggested that we should pretrain our deep autoencoders first, so that we initialize the weights close to a good solution and then we could use a gradient descent method to fine tune the network. They pretrained their network using Restricted Boltzmann Machines (RBM). An RBM is a two layer Boltzmann machine without any connections between the hidden units. A Boltzmann machine is a stochastic recurrent neural network with simulated annealing dynamics. In an RBM the first layer is the visible layer, which is observable, and the second layer corresponding to the hidden layer is used for feature detection. The features learned from one RBM act as an input to train the next RBM in the group.

2.1 ALTERNATIVE ALGORITHMS

We will be using three different algorithms to address the hypotheses: steepest descent, scaled conjugate gradient and alopex. The motivation for using these algorithms is three fold: firstly, we wanted to use a very simple approach for training a deep neural network, steepest descent is one of the simplest gradient based algorithms. Secondly, they used a conjugate gradient method for fine tuning. We wanted to formulate a method which would be similar to the one used by them but at the same time we did not want to replicate the results already produced. Therefore, scaled conjugate gradient appeared to be a good alternative because it does not use the line search method unlike other conjugate gradient algorithms. Thirdly, we had to use a non-gradient based algorithm, we chose to use alopex algorithm. It uses error correlations which would avoid poor local optimum selection. The reason behind using it are two fold: firstly, it would help us know the impact of using gradient information, and secondly, this could not only be an algorithm for fine tuning but we could also see its effect as a pretraining algorithm in combination with a gradient based fine tuning method.

3. RESULTS

The main performance metric of our experiments is the error between the neural network input data and the reconstructed data at the output of the network. We also focus on the CPU time taken by the methods for MNIST dataset for which we were able to replicate Hinton et al.'s results. Each dataset was normalized in all our experiments.

3.1 USPS DATASET

USPS (Hastie et al., 2001) is a hand written zip code dataset. It has 256 features of values 0-255(the brightness value of 16x16 gray-scale image of the character). There were 7291 training samples and 2007 test samples. An eight layer autoencoder represented as, 256-1000-500-250-30-250-500-1000-256, was used for our experiment. This network was chosen based on pilot experiments.

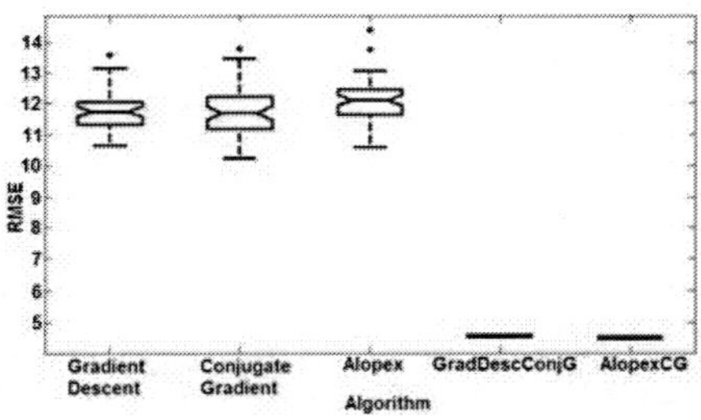

Figure 1: A comparison of the RMSE values for USPS dataset using algorithms with and without pretraining.

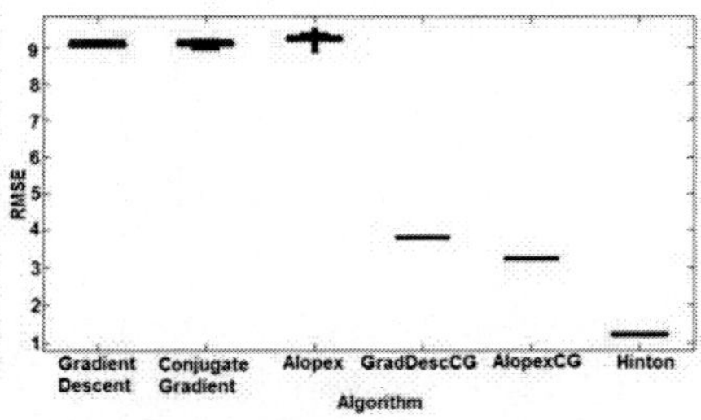

Figure 2: Comparison of test RMSE when different algorithms were used to train an 8 layer autoencoder for MNIST dataset.

A boxplot result of different combinations of the methods is presented in Fig. 1 for comparison. GradDescConjG algorithm in Fig. 1 represents pretraining using steepest descent followed by training using scaled conjugate gradient. AlopexCG is characterized by pretraining using alopex followed by scaled conjugate gradient training. This algorithm did slightly better than the one where we use steepest descent pretraining and much better than the other algorithms where no pretraining is being used. The variations in the results of the algorithms without pretraining are due to the multiple runs and the random initialization of the weights.

3.2 MNIST DATASET

MNIST (Roweis, 2007a) is a subset of the NIST digits dataset with 60000 training samples and 10000 test samples. The dimensionality of the dataset is 784. We used the same dataset and the same network, 784-1000-500-250-30-250-500-1000-784, that was used by Hinton et al. We not only ran our experiments with and without pretraining, but also reproduced results using Hinton et al.'s source code.

In order to visualize how our experiments did in comparison to each other and Hinton et al.'s approach, a boxplot is presented in Fig. 2. GradDescCG algorithm is a combination of steepest descent pretraining and conjugate gradient training. AlopexCG is alopex pretraining and conjugate gradient training. A visual comparison of Fig. 2 indicates Hinton et al.'s approach has produced the best results. Against an RMSE of 1.22 obtained using Hinton et al.'s approach we could get as close as 3.22 using conjugate gradient to train a network whose initial weights were set by alopex algorithm. However, this is not a big difference since it is over 784 intensity values. Again, the variations in the results of the algorithms without pretraining are due to the multiple runs and the random initialization of the weights.

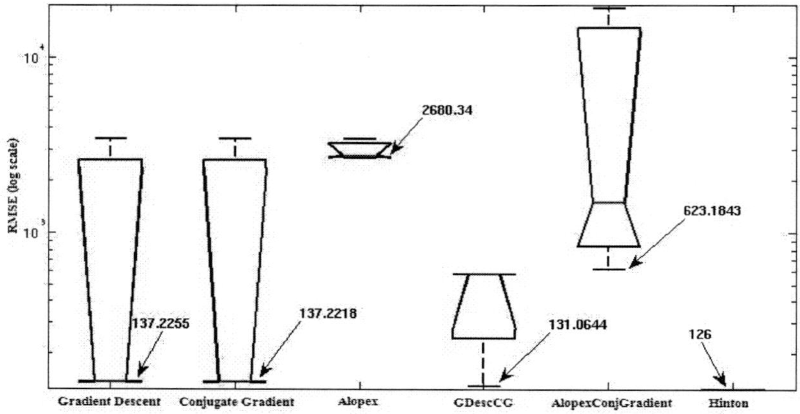

Figure 3: Comparison of test RMSE when different algorithms were used to train an autoencoder for Olivetti face dataset. GDescCG is steepest descent pretraining with scaled SCG training. AlopexConjGradient is an algorithm with alopex pretraining and SCG training.

3.3 OLIVETTI FACE DATASET

Olivetti (Roweis, 2007b) dataset was also the same as the one used by Hinton et al. It contains 10 64x64 images of each of forty different people. The authors' code was not accessible so we had to compare our results to the ones reported by Hinton et al. This dataset required some preprocessing as was done by Hinton et al. We constructed a dataset of 166000, 25x25 images from the original 400 images, by rotating (-90 to +90), scaling (1.4 to 1.8), cropping and subsampling. The rotation and scaling ranges were the same as the ones reported by Hinton et al. (Hinton and Salakhutdinov, 2006). The dataset was further divided into 41400 test images and 124600 training images. A network, 625-2000-1000-500-30-500-1000-2000-625 used by Hinton et al., was used to train this dataset. The results were obtained using algorithms without pretraining and with pretraining by steepest descent and alopex, followed by training using scaled conjugate gradient. These results are presented in Fig. 3 along with a least RMSE value reported by Hinton et al. (Hinton and Salakhutdinov, 2006). We found the least RMSE of 131.0644 against Hinton's 126. This is a very small difference over 625 intensity values. In Fig. 3 the numbers at the tails of the arrows indicate the lowest RMSE value obtained using an algorithm.

4. STATISTICAL ANALYSIS

We carried out pairwise t-tests between the different algorithms for each dataset. We used multiple comparison t-tests with pooled standard deviation and Holms method for p-value adjustment. A summary of these tests is presented in Table 1. The p-value reported is the least we could obtain using the corresponding algorithm whose difference is statistically significant than all other algorithms. For USPS dataset, steepest descent with conjugate gradient has statistically significant difference to other methods without pretraining, than alopex as a pretraining method. The two methods with pretraining for this dataset were statistically similar. For Olivetti, since we had only one value for Hinton et al. results (reported in (Hinton and Salakhutdinov, 2006)) we could not carry out any t-tests using their results. Within other methods there was no statistically significant difference between any of the approaches. For this dataset the lowest p-value is reported.

Table 1: Pairwise t-test results showing the lowest p-value of various algorithms for different datasets.

Dataset	Algorithm	p-value
USPS	Steepest Descent with Conjugate Gradient	2e-16
MNIST	Hinton	2e-16
OLIVETTI	Steepest Descent with Conjugate Gradient	0.39

We found that conjugate gradient was the fastest algorithm. We saved the CPU time using cputime command of Matlab used by each algorithm for each dataset. However, since we could compare our results with Hinton et al.'s code for only MNIST datasets, so we carried out pairwise t-test on that dataset alone. We found that conjugate gradient was significantly faster than the rest of the algorithms including Hinton et al.'s method with a p-value of 2e-16.

5. DISCUSSION

We started with three hypotheses which could be summarized as: gradient based algorithms like steepest descent and scaled conjugate gradient and non gradient based algorithm like alopex could produce competitive results in comparison to the approach used by Hinton et al. We also hypothesized that a simpler pretraining approach than RBM, using steepest descent or alopex followed by training using scaled conjugate gradient algorithm would get us competitive results when compared with Hinton et al.

Our results indicate some interesting findings. Firstly, Hinton et al.'s argument that pretraining followed by training would get better results holds true. A closer look at Fig.'s 1, 2, and 3 indicate that training followed by pretraining does indeed improve the results. Secondly, pretraining helps to find a good set of initial weights and the training that follows it tries to fine tune those weights in order to obtain better results. In Fig.'s 1 and 2, we can notice that the algorithms without pretraining have a variation around the median RMSE values while as the RMSE values obtained using pretraining and training do not show any such variation. This is because the training phase only tries to fine tune the weights such that it restricts itself to a particular search area so that it can find the global optimum.

Alopex by itself did not perform well on Olivetti dataset unlike gradient descent and scaled conjugate gradient. However, alopex performed equally well as the other methods for other two datasets without pretraining. It performed better than steepest descent when used as a pretraining algorithm for MNIST dataset, shown in Fig. 2. This leads us to an important observation. It seems that weight optimization in deep autoencoders is not entirely related to poor local optima while using gradient based algorithms.

An important finding of this research is that we were able to come up with alternative pretraining approaches, which found a low RMSE value for Olivetti face dataset, very close to Hinton et al.'s approach. We also came very close to Hinton et al.'s results for MNIST dataset. These results were achieved because steepest descent as well as alopex did well as pretraining methods. Of these two pretraining approaches, statistical results indicate steepest descent to be slightly better than alopex.

Another important result is that computationally, scaled conjugate gradient is the fastest approach. This points to a possibility of obtaining a more robust and faster training of deep autoencoders using a scaled conjugate gradient based approach with different combinations of steepest descent and alopex as pretraining algorithms. One of the possibilities would be to explore the effect of varying the number of units in the hidden layers. Another possible approach would be to train the autoencoder using Hinton et al.'s pretraining approach in combination with scaled conjugate gradient training algorithm. We could also try reversing the order of our pretraining and training algorithms.

6. CONCLUSIONS

In this research three different algorithms in different combinations for both pretraining and training were used to train deep autoencoders for three different datasets. One of the main intrinsic goals regarding pretraining was proved. Pretraining does indeed help to find a good set of initial weights and thereby help find a good solution using a gradient descent based method in a deep autoencoder. We were able to prove that other pretraining methods could perform well. However, further tuning of these other methods as pretraining and training algorithms might obtain much better results. The performance of alopex pointed towards the possibility that deep autoencoders might not be prone to local minima as we had initially thought. A more careful analysis using these algorithms could lead to more insights into training deep autoencoders and a better understanding of the weight optimization in such networks.

REFERENCES

Hinton, G, and Salakhutdinov, R., 2006, "Reducing the dimensionality of data with neural networks," *Science*, Vol. 313, pp. 504–507.

Møller, M. F., 1993, "A scaled conjugate gradient algorithm for fast supervised learning," *Neural Networks*, Vol. 6, pp. 525–533.

Unnikrishnan, K. P. , and Venugopal, K. P., 1994, "Alopex: A correlation-based learning algorithm for feedforward and recurrent neural networks," *Neural Computation*, Vol. 6, pp. 469–490.

DeMers, D., and Cottrell, G., 1993, "Non-linear dimensionality reduction," *Advances in Neural Information Processing Systems*, Morgan Kaufmann, San Mateo, CA, Vol. 5, pp. 580–587.

Hecht-Nielsen, R., 1995, "Replicator neural networks for universal optimization source coding," *Science*, Vol. 269, pp. 1860–1863.

Kambhatla, N., and Leen, T. K., 1997, "Dimension reduction by local principal component analysis," *Neural Computation*, Vol. 9, pp. 1493–1516.

Hastie, T., Tibshirani, R., and Friedman, J., 2001, "Datasets for the elements of statistical learning," *http://www-stat.stanford.edu/~tibs/ElemStatLearn/data.html*, Stanford University.

Roweis, S., 2007a, "MNIST handwritten digits,"*http://www.cs.toronto.edu/ roweis/data.html*, Computer Science Department, University of Toronto, Canada.

Roweis, S., 2007b, "Olivetti faces," *http://www.cs.toronto.edu/ roweis/data.html.* Computer Science Department, University of Toronto, Canada.

NON-DECLARATION MEASURES FOR CLASSIFIER ACCURACY

MARK A. FRIEND
Department of Operational Sciences
Air Force Institute of Technology
Wright-Patterson AFB Ohio

KENNETH W. BAUER, JR
Department of Operational Sciences
Air Force Institute of Technology
Wright-Patterson AFB Ohio

ABSTRACT
Classifier accuracy may be improved by including an option to withhold label decisions for records likely to be misidentified. In this paper we explore the efficacy of four different information theory based methods to identify such records as non-declaration records. The developed non-declaration methods are applied to an automatic target recognition problem using features drawn from high range resolution profiles generated from synthetic aperture radar data.

INTRODUCTION

Classifier performance using the Bayes optimal rule may be improved upon by withholding label decisions from those records most likely to be misidentified (Fumera, Roli, & Giorgio, 2000). Chow (1970) finds the optimal threshold for non-declarations under the assumption of perfect knowledge of class posterior distributions. Fumera et al. (2000) extends Chow's work with the use of multiple rejection thresholds and removes the assumption of perfect knowledge of class posterior distributions. Devijer & Kittler (1982) and Haspert (2000) minimize a loss function to find preferred classification and rejection rules. Laine (2005), seeks non-declaration thresholds that maximize a classification measurement of interest while meeting additional constraints imposed on the classifier by the user. Albrecht (2005) extends Laine's work and includes the identification of records a classifier has not been trained to recognize, out-of-library (OOL) records. In this paper we begin by using information theory to develop four non-declaration (NDEC) methodologies. Next we apply the methodologies to a synthetic aperture radar (SAR) automatic target recognition (ATR) problem that includes the identification of OOL records. Finally we explore the limiting effect an OOL methodology may have on improvements to classification accuracy generated by non-declaration methodologies.

APPLICATION AREA

The non-declaration methods developed are applicable to a variety of classification problems. In the SAR ATR application presented we seek to classify targets at the type and the aggregated group level. The data used for the experiment are high range resolution profiles generated from a 1ft. resolution X-band SAR radar collection on 15 stationary targets aligned in the open without concealment. For each of the targets 724 SAR images were collected over 360 degrees of aspect angle (Albrecht, 2005; Laine, 2005) with horizontal (HH) transmit and receive polarization. The resultant target chips are processed into high range resolution (HRR) profiles with an Air Force Research Laboratory process (Albrecht, 2005; Laine, 2005; Williams, Westerkamp, Gross, & Palomino, 2000). Due to the uneven collection over the aspect angles the generated HRR profiles are subsequently linearly interpolated to provide one degree increment HRR profiles. Targets are separated into three groups, hostile, friendly/neutral, and out-of-library.

TEMPLATE CLASSIFIER DESIGN

The template classifier used for our experiments is designed to use templates formed from 10 dimensional feature vectors generated from the maximum values from 10 bins of interest within HRR profiles. For each target, 24 templates are formed from the mean and covariance of 15 observations. Hughes (1968) demonstrated theoretically the loss of classification accuracy engendered by estimating parameters with insufficient training data and Landgrebe et al. have shown this to be true in practice (Fu, Landgrebe, and Philips, 1969; Shahshahani and Landgrebe, 1994). Given the size of our available data sets we chose to reduce our estimation requirements by using diagonal covariance matrices for the templates. In testing we assume prior aspect angle knowledge of ± 22.5 degrees and compute the squared Mahalanobis distance between test exemplars and 3 potential templates for each in-library target. For each test record the smallest squared Mahalanobis distance observed for each target's templates forms a 10 dimensional output vector which is subsequently processed into a posterior probability vector. These posterior probability vectors are used to make labeling and declaration decisions and the unprocessed squared Mahalanobis distance vector is used to make out-of-library labeling decisions. Test observations used in the experiment are formed by averaging sequential observations in aspect angle.

INFORMATION THEORY AND NON-DECLARATIONS

In the pattern recognition arena, information theory has been used for model selection, feature selection, classification and clutter rejection (Chang & Lu, 1995; DeVore, O'Sullivan, Anand, & Schmid, 2001; Wilson, 2007). We apply two concepts from information theory, entropy and Kullback-Liebler distance (Kullback & Liebler, 1951) to identify records more likely to be misidentified by the classifier. We first provide the definition of both measures, then show the development of 4 non-declaration methods based on these measures, and finally apply these methods to the selected application area.

ENTROPY

Given a discrete set of *n* symbols $X = \{x_1, x_2, x_3, \ldots, x_n\}$ with an associated probability mass function, $p_X\{X = x\}$, the *entropy*, $H(X)$, of this distribution is a measure of the randomness of a sequence of symbols drawn from the set (Cover & Thomas, 1991). Using this notation entropy is defined as:

$$H(X) = \sum_{i=1}^{n} p_X(x_i) \log_2\left(\frac{1}{p_X(x_i)}\right) = -\sum_{i=1}^{n} p_X(x_i) \log_2 p_X(x_i) \tag{1}$$

KULLBACK-LEIBLER DISTANCE

Kullback-Leibler distance or KL divergence captures the information not shared between distributions or the distance between two distributions. For two discrete distributions *P* and *Q* both over a real valued random variable *X* with *n* possible values KL divergence is

$$D_{KL}(P \| Q) = \sum_{i=1}^{n} p_X(x_i) \log_2\left(\frac{p_X(x_i)}{q_X(x_i)}\right) \tag{2}$$

ENTROPY BASED NDEC METHODS

The application of entropy measures applied to non-declarations is constrained in this application by a requirement to apply a non-declaration criterion developed from training data to each test observation without knowledge of other test data similarity metrics. In the entropy methods developed, target type is defined as a random variable Y and each vector of posterior probabilities is treated as a probability mass function $p_Y = \Pr\{Y = y\}$. The entropy of each output vector is a measure of the uncertainty in the posterior probability distribution over all of the target types. This uncertainty measurement forms the basis to identify records with insufficient information to make a confident classification decision.

Ent 1 Method (All Classes)

In order to utilize entropy information generated from test data we must first determine the entropy score characteristics of exemplars correctly identified by a classifier. This is accomplished by presenting training data to the classifier to create data sets of posterior probability vectors for each target type. Using the Bayes Optimal Rule the "winning" target labels are found for these posterior probability vectors. Each target's correctly labeled records are further processed by computing the entropy for correctly labeled posterior probability vectors. Finally, a threshold is created by finding a user defined quantile of the entropy scores. The resultant 1×10 vector of class entropy thresholds may then be applied to test data. In testing, the winning label is an index to select the "winning" class related entropy thresholds. The entropy of each test exemplar's posterior probability vector is compared to it's class entropy threshold determined during training. Exemplars with entropy scores exceeding their class entropy threshold are considered not to contain sufficient information to make a decision and are identified as a non-declared exemplar.

Ent 2 Method (Aggregated Classes)

This method employs the same calculations as the Ent 1 method with one slight modification. The hostile posterior probabilities and the friendly/neutral posterior probabilities are aggregated before calculating entropy scores.

KL 1 METHOD (ALL CLASSES) (PERFECT REFERENCE)

Let q_X denote the posterior probability distribution function generated across the 10 targets for each test record. Each test record has a zero entropy truth distribution, p_X. Under these assumptions, the KL distance between q_X and p_X reduces to $-\log_2 q_X(x_i)$, where i is the correct target class.

KL 2 METHOD (ALL CLASSES) (SAMPLED REFERENCE)

The KL2 method is similar in many respects to the KL1 method. Like KL1 it treats each posterior probability vector as a probability mass function q_X. However, the KL2 method does not use the simplifying assumption that p_X has zero entropy. Instead for each target class the KL2 method uses an estimated probability mass function $\hat{p}_{X,i}$ generated from the mean of posterior probability vectors from the class training records and the original KL formula shown in Eq. (2). During training, for each target class i, the correctly identified records whose posterior probability is less than $\hat{p}_{X,i}(i)$ are used to generate the KL distance threshold T_i. In testing, records with winning target index i whose posterior probability is less than $\hat{p}_{X,i}(i)$ are selected as non-declaration if $D(\hat{p}_{X,i} \| q_X) < T_i$. The effect of this additional criterion is that for a fixed quantile, the KL2 method makes fewer non-declaration decisions in comparison to the other non-declaration methods

EXPERIMENTAL RESULTS

The classification experiments performed are designed to evaluate the performance of the KL1, KL2, Ent1, and Ent2 NDEC methodologies with and without an OOL methodology present. The first experiment examines the effect NDEC methodologies have on total classification accuracy (TCA) without any OOL methodology present. Next we explore the relationship between TCA, OOL performance, and NDEC performance, and finally the last experiment is designed to evaluate the NDEC methodologies performance in conjunction with an OOL methodology.

These experiments explore potential classifier improvements in an environment where an imperfect OOL methodology results in the misidentification of some in-library records as OOL records and records with a label of OOL are not considered for non-declaration decisions. For a more parsimonious presentation, all in-library objects are considered equally important and TCA is used as the measure of effectiveness. If we define TP as the number of declared in-library objects correctly labeled, O as the number of in-library objects identified as OOL, and E as the number of declared in-library objects labeled incorrectly before non-declaration decisions are made, we may define TCA as shown in Eq. (3) and without a OOL methodology present TCA may be further simplified by the removal of the O variable.

$$TCA = \frac{TP}{TP + O + E} \tag{3}$$

NDEC METHOD PERFORMANCE WITHOUT OOL DECLARATIONS

The first and foremost performance characteristic a non-declaration method must demonstrate is the ability to select a greater percentage of mislabeled records for non-declaration than correctly labeled records. To that end each NDEC method was tested against a random selection NDEC method. TCA was found to be positively correlated with increased non-declarations for the developed NDEC methods. In contrast, the random selection NDEC method generated TCA values that fluctuated around the TCA found with all target records declared. This ability to improve TCA inversely with the number of declarations made is limited by the use of an OOL method as we shall see in the next section.

NDEC PERFORMANCE WITH OOL DECLARATIONS

Having tested the NDEC method behavior without OOL declarations present, we now let TP_{ND} equal the number of TP records identified as non-declare by a NDEC method and E_{ND} equal the E records identified as non-declare by a NDEC method. When a NDEC method and OOL method are being used together, TCA is defined as shown in Eq. (4).

$$TCA = \frac{TP - TP_{ND}}{(TP - TP_{ND}) + O + (E - E_{ND})} \tag{4}$$

In order to increase TCA, the percent decrease in TP must be less than any percent decrease in $TP + O + E$ brought about by non-declaration decisions. This relationship that must hold for TCA improvements is used in equation (5) to demonstrate the effect O has on the magnitude of TP_{ND}.

$$1 - \left(\frac{TP - TP_{ND}}{TP}\right) < 1 - \left(\frac{(TP - TP_{ND}) + O + (E - E_{ND})}{TP + O + E}\right) \rightarrow \frac{\mathrm{TP_{ND}}}{\mathrm{TP}} < \frac{E_{ND}}{O + E} \tag{5}$$

Although the magnitude of $\frac{E_{ND}}{E+O} - \frac{TP_{ND}}{TP}$ may not be directly mapped to TCA it is a strong indicator of non-declaration performance. A plot of $\frac{\mathrm{TP_{ND}}}{\mathrm{TP}}$ and $\frac{E_{ND}}{O+E}$ reveals the performance characteristics of the non-declaration methods. In Figure 1 we let the suffix E indicate the $\frac{E_{ND}}{E+O}$ measurement and T indicate the $\frac{\mathrm{TP_{ND}}}{\mathrm{TP}}$ measurement for non-declaration method. In the TCA plots we use Eq. (3) to generate an upper bound, UB, on TCA using non-declarations to reduce E to zero and a lower bound, LB, which is the classification accuracy obtained without any non-declaration decisions. These bounds demonstrate the TCA improvements made through the use of non-declarations.

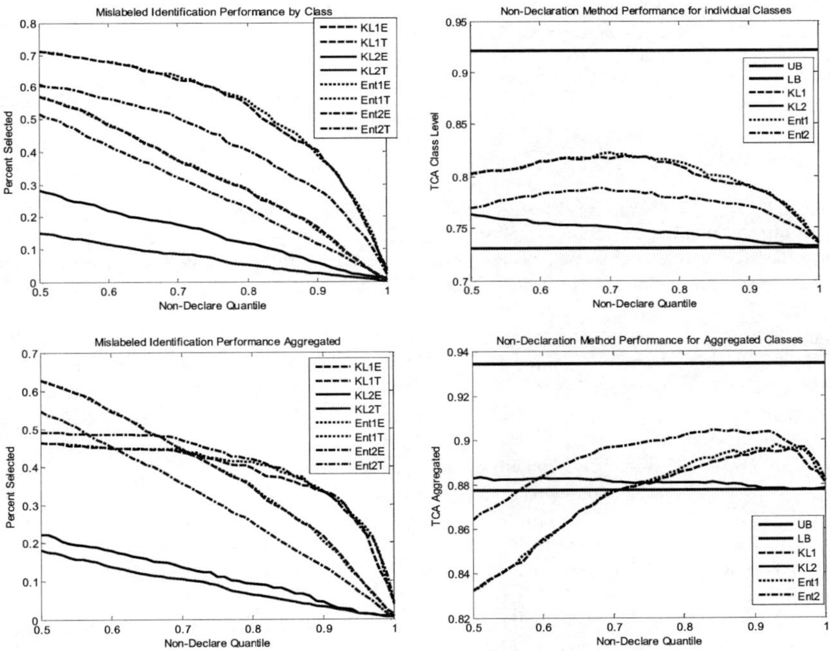

Figure 1: NDEC Method Performance, 2 Observations, Sensor HH Data

The plots in Figure 1 demonstrate the importance of matching a non-declaration method to the measure of effectiveness under consideration. In our definitions of TCA we have thus far restricted ourselves to aggregated label accuracy. It is clear from the bottom plots in Figure 1 that using records labeled correctly at the aggregate level to train Ent2 and the computation method of Ent2 garners it a performance benefit at the aggregated TCA level. However, using the same data but focusing on classification accuracy at the target class level, Ent1 and KL1 clearly dominate the Ent2 method in the top 2 plots of Figure 1.

CONCLUSIONS

Information theory based non-declaration methods may be used to improve TCA with improvement often limited only by the constraint placed on the percentage of records that must be given a label. The inclusion of an imperfect OOL methodology imposes additional limits to these improvements. Finally NDEC method choice should be a function of the measurement of effectiveness used to characterize a classifiers performance.

The views expressed in this article are those of the authors and do not reflect the official policy or position of the Air Force, Department of Defense or the U.S. Government.

NOMENCLATURE

ATR:	Automatic Target Recognition
HH:	Horizontally polarized transmitted, horizontally polarized received
HRR:	High Range Resolution
KL:	Kullback-Liebler
NDEC:	Non-Declaration
OOL:	Out-of-Library
SAR:	Synthetic Aperture Radar
TCA:	Total Classification Accuracy

REFERENCES

Albrecht, T. W., 2005, "Combat identification with sequential observations, rejection option, and out-of-library targets", PhD, Air Force Institute of Technology

Chang, K., & Lu, Y.,1995, "High resolution polarimetric SAR target classification with neural network," Proceeding of 1995 IEEE International Conference on Fuzzy Systems, Vol 3, pp. 1681-1688.

Chow, C. K., 1970, "On optimum recognition error and reject tradeoff," *IEEE Transactions on Information Theory,* Vol. IT-16, pp. 41-46.

Cover, T. M. and Thomas, J. A., 1991, *Elements of information theory,* Wiley-Interscience.

Devijver, P. A. and Kittler, J., 1982, Pattern recognition - a statistical approach. London: Prentice-Hall International.

DeVore, M. D., O'Sullivan, J. A., Anand, S. and Schmid, N., A., 2001, "Probabilistic approach to model extraction from training data," Vol. 4382, pp. 358-366.

Fu, K. S., Landgrebe, D. A. and Philips, T. L., 1969, "Information processing of remotely sensed agricultural data," *Proceedings of the IEEE,* Vol.57, Issue 4, pp. 639-653.

Fumera, G., Roli, F. and Giorgio, G., 2000, "Reject option with multiple thresholds," *Pattern Recognition,* Vol. 33, pp. 2009-2101.

Haspert, Kent J., 2000, "Optimum ID Sensor Fusion for Multiple Target Types." Institute for Defense Analysis (IDA) Document D-2451, Virginia

Hughes, G. F. 1968, "On the mean accuracy of statistical pattern recognizers," *IEEE Transactions on Information Theory,* Vol. IT-14, Issue 1

Kullback, S. and Leibler, R. A., 1951, "On information and sufficiency," *Annals of Mathematical Statistics,* Vol. 22, pp. 78-86.

Laine, T., 2005. "Optimization of automatic target recognition with a reject option using fusion and correlated sensor data," PhD, Air Force Institute of Technology

Shahshahani, B. M. and Landgrebe, D. A., 1994, "The effect of unlabeled samples in reducing the small sample size problem and mitigating the Hughes phenomenon," *IEEE Transactions on Geoscience and Remote Sensing,* Vol. 32, Issue 5, pp. 1087-1095.

Williams, R., Westerkamp, J., Gross, D. and Palomino, A., 2000, "1D high range resolution (HRR) radar," *IEEE AES Systems Magazine,* pp.37-43.

Wilson, J. N., 2007, "Landmine discrimination using the Kullback-Leibler distance," *Proceedings of the SPIE,* Vol. 6553

Predicting the Learning Performance of Artificial Intelligent Systems Using Non-Homogeneous Poisson Process Models

Wen-Li Wang, Robert Weissbach
School of Engineering
Penn State Erie, the Behrend College
(wxw18, rsw7)@psu.edu

Mei-Huei Tang
Dept. of Computer and Inf. Science
Gannon University
tang002@gannon.edu

ABSTRACT

Artificial intelligent systems can learn to adapt to environmental changes to find a better solution. Improving system performance has been a great interest of study with new objective functions and parameters being constantly applied. However, approaches for performance evaluation are often by real observation without offering a prediction capability to support decision making at run time. Prediction helps foresee the future, so that a good run can continue while a poor one can be replaced. It also assists in evaluating the efficacy of different algorithms, especially when their learning capabilities vary over time. In this paper, a statistical approach is proposed to predict run time system learning performance. We apply non-homogeneous Poisson process (NHPP) models that account for time-variant learning effectiveness to determine the future solution more precisely. Our paradigm predicts the probability for the next improvement by time. Together with the estimated extent of that improvement, the expected performance can be measured. Learning algorithms can be compared based on their performance metrics.

1. INTRODUCTION

Artificial intelligent systems, such as neural network and swarm intelligence systems [5], have the ability to learn to adapt to dynamic environments to look for reasonable solutions. Benchmarks had been focusing on the enhancement of learning algorithms to prevent local optimums and reach the global optimum. Because the random characteristic of an algorithm may yield quite different results for each run, past system performance evaluation often relies on many test runs and then computes the mean for comparison. This can be troublesome for solving optimization problems [2, 6]. It is not only time consuming to perform the tests, but also the mean may not be representative for an individual run. More importantly, dynamic learning performance in real time cannot be foreseen.

Predicting the ongoing learning performance of an algorithm can help improve the system performance. In [9], we showed that system performance can be improved many folds by terminating a poor run and then starting up a new run from scratch. However, the study did not account for time-variant learning effectiveness. By our experiments, learning becomes less effective to improve the best found solution after a period of time, because each improvement pushes closer to the global optimum. Nonetheless, an encounter of a good learning can sometimes drive several improvements consecutively. Therefore, there is a need to take into account variable improvement intensity in addition to the prediction.

The objective of this study is to cover both by using a statistical means, NHPP [3]. The inputs are the observed solutions and their improvement occurrence times, and the output is a performance metric, i.e., the expected solution for a future time of interest. NHPP has been widely used to predict reliability [4, 7, 8]. Its non-homogeneity enables the consideration of time-variant learning effectiveness. Furthermore, the independent increment of Poisson process allows us to predict the future performance using Markov models. As a consequence, the next improvement can be predicted based on the current state only. Our modeling approach performs curve fitting to the recorded observations by choosing a right class of probability distribution for NHPP. In this study, the geometric class is chosen for its best fit probability distribution for optimization problems. Based on this chosen class, the maximum likelihood estimation (MLE) is employed to compute the model parameters. The MLE [7] helps predict the probability for the next improvement by time and estimate the extent for the next improvement. They are two important factors to compute the performance metric. Learning performance for different algorithms or parameter setups can be compared by their metric values under the same time period. A case study is conducted on a swarm intelligence system to prove the concept and show the application of this model.

2. METHODOLOGY

We introduce a performance metric to compare the learning capability between different algorithms or the same algorithm with different parameter setups. From past improvement data, this metric predicts the expected future outcome to justify learning performance. Whenever a new improvement is observed, the metric can be re-computed to gain better prediction for the next. The prediction accounts for two factors: the extent of the next improvement and the chance of having an improvement at a specific time point. Both will be estimated by NHPP.

2.1 NHPP to estimate the occurrence chance of the next improvement by time

Intelligent systems improve their solutions randomly in time due to the involvement of random behavior in learning. For hard problems, predicting the number of improvements remaining can be difficult. Nevertheless, the random behavior of such an improvement process can be modeled as a Markov process. This section introduces a special case of Markov by applying the geometric class of NHPP, which considers innumerous random events, to predict the occurrence chance for the next improvement. The Poisson process of NHPP gives a good approximation of the occurrence of many real-world events, e.g., software failures. The random occurrence of improvements is similar to that of software failures to be Poisson. Because the global optimum is hard to find, the geometric class sets no limit on the number of improvements to give a closer prediction. The other strength of NHPP is its non-homogeneity, allowing the improvement intensity to vary over time. This can account for time-variant learning effectiveness.

Based on [7], the following terms are defined to develop a NHPP model for analyzing intelligent systems. Under the condition of the last known improvement time, this model predicts the occurrence chance for the next improvement at a future time point, i.e., the chance is a conditional probability.

$M(t)$: the number of recorded improvements experienced by time t.
$\mu(t)$: the expected number of improvements experienced by time t.
$L_a(t)$: the per-learning improvement probability by time t of a learning a, where $a = 1, 2, \ldots$.
$\lambda(t)$: the derivative of $\mu(t)$, the improvement intensity function, with respect to time t.
T_i': a random variable representing the ith improvement interval, where $i = 1, 2, \ldots$.
T_i: a random variable representing the ith improvement time, where $i = 1, 2, \ldots$.
ω_0: the mean of the total number of improvements remaining at time $t = 0$.
m_e: the number of recorded improvements.
t_e: the end of test time.
t_i: the time of the encountered ith improvement.

Our model is adhered to four assumptions as below. The third one assumes the improvement process has independent increments, which conforms to the history independent Markov property that the next state depends on the present state only. As a consequence, the conditional probability for the next improvement can be computed based on the condition of the last occurrence time.

1. Whenever an improvement occurs, the new learning knowledge is fed back to the intelligent system instantaneously.
2. The total number of improvements remaining in the system is a Poisson random variable with a mean of w_0 at time $t = 0$.
3. Each improvement, caused by per-learning, occurs independently and randomly in time.
4. The probability that one improvement will occur during $(t, t+\Delta t]$ is $\lambda(t)\Delta t$.

For NHPP Poisson type models, the expected number of improvements $\mu(t)$ can be stated as

$$\mu(t) = \omega_0 L_a(t) \tag{1}$$

In the geometric class, the improvement data can be used to estimate $\mu(t)$ using model parameters $\beta = (\beta_0, \beta_1)$, in which β_0, and β_1 are constants, and β_0 is a scale parameter to predict the total expected number of improvements.

$$\mu(t) = \mu(t;\beta) = \beta_0 \ln(1+\beta_1 t) \tag{2}$$

The MLE estimation is a good approach to compute β, especially when the underlying distribution of the data is specified. By [7], the conditional likelihood of β_1 given m_e improvements can be formulated as

$$MLE(\beta_1 \mid m_e) = \frac{m_e!\beta_1^{m_e} \prod_{i=1}^{m_e} \frac{1}{1+\beta_1 t_i}}{[\ln(1+\beta_1 t_e)]^{m_e}} \tag{3}$$

By computing $\frac{\partial MLE(\beta_1 \mid m_e)}{\partial \beta_1} = 0$ from (3), the point estimate $\hat{\beta}_1$ for β_1 is the solution to

$$\frac{1}{\hat{\beta}_1} \sum_{i=1}^{m_e} \frac{1}{1+\hat{\beta}_1 t_i} - \frac{m_e t_e}{(1+\hat{\beta}_1 t_e)\ln(1+\hat{\beta}_1 t_e)} = 0 \tag{4}$$

The value of $\hat{\beta}_1$ can be estimated using the Newton-Raphson method [1]. With $\hat{\beta}_1$ derived from (4), the scale estimate $\hat{\beta}_0$ for β_0 can be computed as

$$\hat{\beta}_0 = \frac{m_e}{\ln(1+\hat{\beta}_1 t_e)} \tag{5}$$

Hence, $P(t_i' \mid t_{i-1})$, the conditional improvement probability of time t_i' under the last improvement time at t_{i-1}, is derived as 1 minus the chance of not having an improvement by time $t_i'+t_{i-1}$.

$$\begin{aligned} P(t_i' \mid t_{i-1}) &= P[T_i \le t_i'+t_{i-1} \mid M(t_{i-1}) = i-1] = P[M(t_i'+t_{i-1}) \ge i \mid M(t_{i-1}) = i-1] \\ &= \sum_{j=i}^{\infty} \frac{[\mu(t_i'+t_{i-1}) - \mu(t_{i-1})]^{j-(i-1)}}{(j-(i-1))!} \exp\{-[\mu(t_i'+t_{i-1}) - \mu(t_{i-1})]\} \\ &= 1-\exp(-(\mu(t_{i-1}+t_i') - \mu(t_{i-1}))),\ i = 1, 2, \ldots \end{aligned} \tag{6}$$

For the known time t_{i-1}, if the interest is to know the additional time t_i' to have $P(t_i' \mid t_{i-1}) = p$, t_i' can be computed from Eqs. (2) and (6) as

$$t_i' = \frac{\exp\left(\frac{-\ln(1-p)}{\beta_0} + \ln(1+\beta_1 t_{i-1})\right) - 1}{\beta_1} - t_{i-1} \tag{7}$$

2.2 Predicting the next solution by count

For optimization problems, the best solution is often the maximum or the minimum. This study focuses on finding the minimal that is greater than or equal to zero. With intelligence, a search algorithm utilizes its learning knowledge to continuously look for another better solution. The learning can be effective initially to quickly find a reasonable answer. However, the effectiveness gradually degrades, resulting in the extent of improvement to become less significant. This phenomenon requires a non-linear curve fitting to predict the next solution.

In addition to predicting improvement occurrence chance, the geometric class of NHPP also gives a compelling result to fit the curve. However, the x-axis is not time but the number of improvements and the y-axis is not the cumulated number of improvements but the cumulated sum of solution values. Let s_i be the solution of the ith improvement. For n improvements, the cumulated sum of solutions $\mu(n)$ is equal to

$$\mu(n) = \sum_{i=1}^{n} s_i \tag{8}$$

Similar to Eq. (2), $\mu(n)$ can be estimated using model parameters β as below.

$$\mu(n) = \mu(n;\beta) = \beta_0 \ln(1+\beta_1 n) \tag{9}$$

By the MLE approach, the conditional likelihood of β_1 for having m_e improvements can be formulated based on Eq. (4) to be

$$\frac{1}{\hat{\beta}_1}\sum_{i=1}^{m_e}\left(\frac{s_i}{1+\hat{\beta}_1 i}\right) - \frac{\left(\sum_{i=1}^{m_e} s_i\right) m_e}{(1+\hat{\beta}_1 m_e)\ln(1+\hat{\beta}_1 m_e)} = 0 \tag{10}$$

Similarly, by Eq. (5) the scale estimate $\hat{\beta}_0$ for β_0 can be computed as

$$\hat{\beta}_0 = \frac{\sum_{i=1}^{m_e} s_i}{\ln(1+\hat{\beta}_1 m_e)} \tag{11}$$

By Eqs. (8) and (9), the solutions for the m_eth and (m_e+1)th improvements, $\hat{s}_{m_e}$ and $\hat{s}_{m_e+1}$, will be

$$\hat{s}_{m_e} = \mu(m_e) - \mu(m_e-1) = \beta_0 [\ln(1+\beta_1 m_e) - \ln(1+\beta_1(m_e-1))] \tag{12}$$

and
$$\hat{s}_{m_e+1} = \mu(m_e+1) - \mu(m_e) = \beta_0 [\ln(1+\beta_1(m_e+1)) - \ln(1+\beta_1 m_e)] \tag{13}$$

The difference between $\hat{s}_{m_e}$ and $\hat{s}_{m_e+1}$ can disclose the extent for the next improvement. However, since the statistically predicted result $\hat{s}_{m_e}$ may not be the same as the actual observation s_{m_e}, the solution for the next improvement incorporates s_{m_e} into computation. We have

$$s_{m_e+1} = s_{m_e} - (\hat{s}_{m_e} - \hat{s}_{m_e+1}) \tag{14}$$

2.3 Computing performance metric

To determine the learning capability of a running system, we propose a performance metric, P_f, considering the predicted extent of the next improvement and the occurrence probability of that improvement. The metric result can be viewed as the expected solution for a future time point, and this study prefers a small metric value that implies effective learning. Different algorithms or parameter setups can be compared by their metric values under the same time period to determine their past learning performance.

For m_e recorded improvements, the most recent event was at time t_{m_e}. Before the (m_e+1)th improvement is observed, our performance metric computes the expected result at time $t_{m_e} + t'_{m_e+1}$ by taking in s_{m_e+1} by Eq. (14) and $P(t'_{m_e+1} \mid t_{m_e})$ by Eq. (6) to be

$$P_f(t'_{m_e+1} \mid t_{m_e}) = s_{m_e}(1-P(t'_{m_e+1} \mid t_{m_e})) + s_{m_e+1} P(t'_{m_e+1} \mid t_{m_e}) \tag{15}$$

The probability $P(t'_{m_e+1} \mid t_{m_e})$ is the chance to observe the next solution s_{m_e+1} at time $t_{m_e} + t'_{m_e+1}$. Conversely, the chance of staying at s_{m_e} is equal to $1-P(t'_{m_e+1} \mid t_{m_e})$.

The number of improvements for different algorithms can be distinctive for the same duration of execution. Nevertheless, algorithms with a smaller P_f value are likely to find a better solution quicker, i.e., more effective in past learning. Regarding the same algorithm, this metric can also help judge the performance of different test runs. If a run yields a poor P_f, it may be more beneficial to terminate that run and start a new one. In [9], our threshold approach has proved this idea that significantly achieves better performance.

3. A CASE STUDY

To demonstrate our model, we studied a swarm intelligence system for the traveling salesperson problem (TSP), based on the algorithm in [2]. The system output includes the improvement occurrence time in millisecond and the solution at that time. Table 1 shows a test run, which generates 68 outputs under 280000 milliseconds, for visiting 250 cities by 25 agents. Different parameter setups can also be compared, e.g., changing the number of agents.

Table 1: An execution result of the swarm intelligence system

Format as in (t_i: improvement occurrence time in millisecond, s_i: the solution)				
(331, 24717)	(5078, 10220)	(11477, 8018)	(37605, 5617.9)	(109660, 4094.8)
(681, 21619)	(5358, 9959.9)	(12308, 7850.9)	(39248, 5519.5)	(133815, 4019)
(1012, 18961)	(5639, 9539.8)	(12579, 7555.5)	(42282, 5302.6)	(136579, 4010.6)
(1322, 17458)	(6189, 9458.2)	(13931, 7451.7)	(48862, 5165)	(136869, 3984.9)
(1603, 16696)	(6730, 9429.2)	(14492, 7363.1)	(50073, 5140.5)	(142387, 3984.4)
(1903, 15399)	(7041, 9279.7)	(14762, 6876.3)	(56302, 4971.1)	(146833, 3972.5)
(2194, 13959)	(7321, 9129.7)	(19889, 6782.2)	(57134, 4945.4)	(147104, 3959.8)
(2484, 13514)	(7852, 8916)	(20731, 6510.3)	(64745, 4813.2)	(160263, 3926.6)
(2774, 13178)	(8933, 8845.2)	(22073, 6244.2)	(69862, 4565.1)	(160833, 3881.7)
(3075, 12767)	(9504, 8758.2)	(25357, 6168.2)	(82060, 4409.5)	(179160, 3709.6)
(3355, 12565)	(10065, 8626.9)	(26169, 6056.3)	(91494, 4352)	(179470, 3620.4)
(3646, 11876)	(10626, 8555.8)	(28352, 5819.4)	(99605, 4304.9)	(267026, 3580.8)
(4217, 11473)	(10926, 8420.2)	(30805, 5693.3)	(106645, 4281.9)	
(4507, 10444)	(11197, 8418.1)	(36784, 5646.7)	(106906, 4239.6)	

The first step of our modeling process is to compute the occurrence chance of the next improvement. Suppose the interest is at the 300000th millisecond. For Eq. (4), $m_e = 68$ and $t_e = 280000$. By the MLE, β_0 and β_1 are estimated to be 14.7719 and 3.5292e-004, respectively. Applying them to Eq. (2), Fig. 1 shows the estimated $\mu(t)$ vs. the real one. By Eq. (6), the conditional probability for the next improvement to occur at the 300000th millisecond is formulated as $P(t'_{m_e+1} \mid t_{m_e}) = P(t'_{m_e+1} = 32974 \mid t_{m_e} = 267026) = 1-\exp(-(\mu(300000)-\mu(267026)) = 0.8178$.

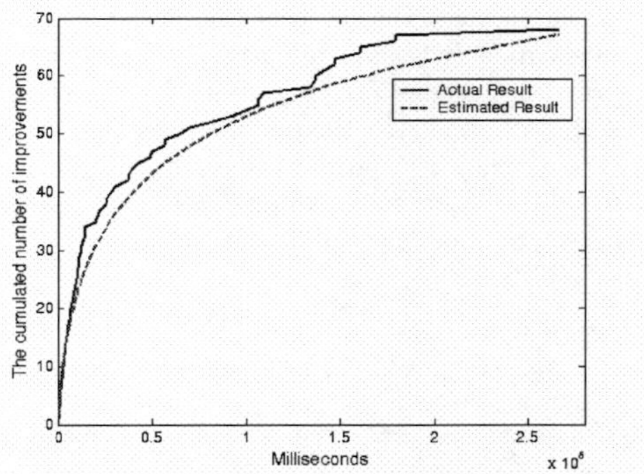

Fig. 1: Estimated $\mu(t)$ vs. the actual $\mu(t)$

The second step predicts the next solution regardless of time. By Eqs. (10) and (11), β_0 and β_1 are estimated to be 331971.4 and 0.0625, respectively. Fig. 2 shows the estimated $\mu(n)$ by Eq. (9) vs. the actual result by Eq. (8). With the estimated $\mu(n)$, $\hat{s}_{m_e}$ and $\hat{s}_{m_e+1}$ of Eqs. (12) and (13) can be computed to be 3976.1 and 3929, respectively. Accordingly, s_{m_e+1} can be computed by Eq. (14) as 3580.8–(3976.1–3929) = 3533.7.

The final step is to compute the performance metric. By Eq. (15), the result is equal to 3580.8*(1–0.8178)+3533.7*0.8178 = 3542.28. This would be the expected solution by the time at the 300000th millisecond of interest. To search for a more efficient algorithm is to explore one that can yield a lower metric value at the same time of interest. Fig. 3 shows good closeness between the performance metric values and the real solutions. The computation for the metrics follows the same time of the recorded improvements up to 300000 milliseconds.

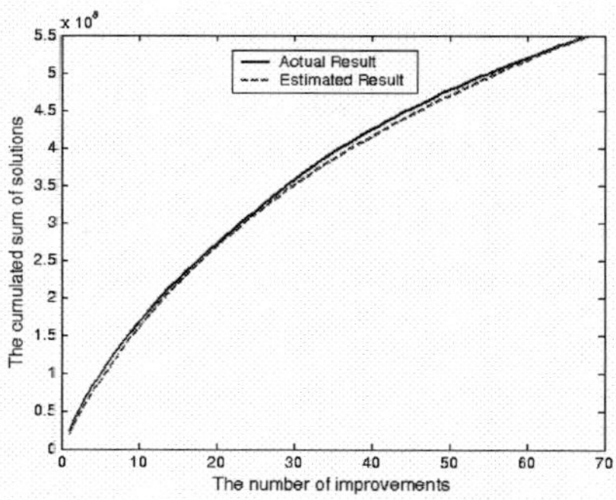

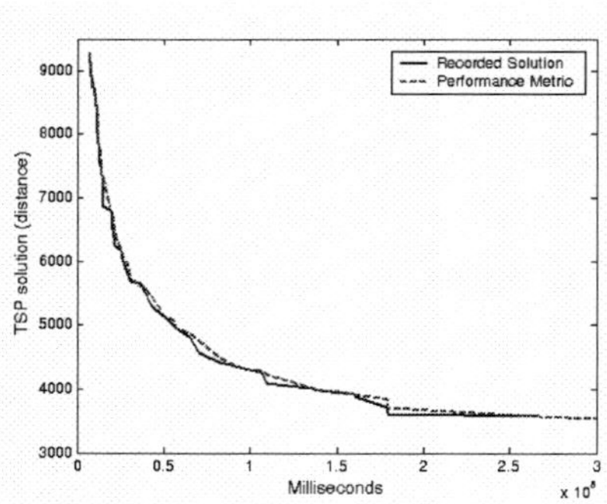

Fig. 2: Estimated $\mu(n)$ vs. the actual $\mu(n)$ Fig. 3: Recorded solutions vs. metric results over time

4. CONCLUSION:

We have introduced a performance metric to predict the future solution over time for running an intelligent system. The prediction offers an ability to justify the learning capability of different algorithms and support decision making at run time. The learning capability is determined by observing the metric value. If the value is not up to a standard, the current run is likely stocked at a local optimum and starting a new run can frequently be more effective.

The computation of the performance metric takes advantage of a statistical model, NHPP, to perform future prediction. The non-homogeneity of NHPP can account for time-variant learning effectiveness and the Poisson process allows the random occurrences of improvements to be easily modeled using Markov. By the geometric class of NHPP and the MLE approach, our model can predict the probability for the next improvement as a function of time and the possible extent of the next improvement. Finally, the metric result is to compute an expected outcome by considering both. In our case study, this paradigm demonstrates a good matching result between the predictions and the actual observations. Our future work will evaluate performance of many algorithms, guide their executions, and explore other classes of NHPP.

REFERENCES:

[1]. Deuflhard P., *Newton Methods for Nonlinear Problems (Computational Mathematics Series 35)*, Berlin, Germany: Springer, 2004.

[2]. Dorigo, M. and Gambardella, L.M., "Ant Colonies for the Traveling Salesman Problem", *BioSystems*, vol.43, pp.73-81, 1997.

[3]. Goel A. L. and K. Okumoto, "Time-Dependent Error-Detection Rate Model for Software Reliability and Other Performance Measures", *IEEE Transactions on Reliability*, R-28(3), pp. 206-211, Aug. 1979.

[4]. Huang C-Y., M. R. Lyu, and S-Y. Kuo, "A Unified Scheme of Some Non-Homogeneous Poisson Process Models for Software Reliability Estimation", *IEEE Transactions on Software Engineering*, 29(3), pp. 261-269, Mar. 2003.

[5]. Kennedy, J. and Eberhart, R. C., *Swarm Intelligence*, Morgan Kaufmann Publishers: San Francisco, 2001.

[6]. Kirkpatrick, S., Gelatt, C. D., and Vecchi, M. P., "Optimization by Simulated Annealing", *Science*, vol.220, no.4598, pp. 671-680, 1983.

[7]. Musa J. D., A. Iannino, and K. Okumoto, *Software Reliability: Measurement, Prediction, Application*, New York: McGraw-Hill, 1987.

[8]. Musa J. D. and K. Okumoto, "A Logarithmic Poisson Execution Time Model for Software Reliability Measurement", in *Proceedings of 7th International Conference of Software Engineering*, Orlando, FL, 1984, pp. 230-238.

[9]. Wang W. and M. Tang, "A Threshold Approach for Multi-Agent SI Systems to Improve Optimization Performance", in *Proceedings of Artificial Neural Networks In Engineering (ANNIE)*, 5-8 Nov., 2006, Rolla, MO, pp 19-24.

ADAPTATION IN WEIGHT SPACE THROUGH GRADIENT DESCENT FOR HOPFIELD NETAS STATIC OPTIMIZER: IS IT FEASIBLE?

GURSEL SERPEN
Electrical Engineering and Computer Science
University of Toledo
Toledo, Ohio 43606, USA

ABSTRACT

This paper reports on results of an empirical simulation study for adaptation of weights through gradient descent for a Hopfield neural network configured as a static optimizer and tested on the traveling salesman problem. Adaptation through gradient descent within the context of recurrent and non-recurrent back-propagation training was attempted in the weight space, which is high-dimensional, i.e. on the order of 1,000,000,000,000 weights for a two-dimensional node array of the Hopfield network configured for a 1000-city problem instance. Ensuing substantial empirical work, practically no noteworthy progress could be recorded for realization of the adaptation in the weight space: the adaptation algorithm failed to locate a set of weight values that established the solutions of the traveling salesman problem instances as local minima in the Lyapunov space. Accordingly, the findings in this paper tend to suggest that alternate means of adaptation schemes with a small number of freely adjustable parameters should be considered.

INTRODUCTION

The Hopfield neural network offers a true "real-time" optimization algorithm for computation of a local optimum solution of static optimization problems [Smith, 1999]. The promise is a quick and local optimum solution that also scales with the size of the problem if and when the Hopfield network algorithm is implemented in hardware so as to take full advantage of the massive degree of parallelism. Determination of weights of the Hopfield network for a given static optimization problem has been the main obstacle towards this end since techniques that pre-compute weight values are problematic at best [Abe, 1989; Abe, 1993; Abe, 1996; Serpen, 2000] and no weight learning procedure was successfully demonstrated either [Sima, 2003; Atencia et al., 2005; Bournez et al., 2006].

A theoretical framework for an adaptive Hopfield network that attains its weight vector through training for a given static optimization problem was proposed in [Serpen, 2003]. One peculiar aspect of the Hopfield network is that the weight space is incredibly high-dimensional for a typical two-dimensional topology configured for static optimization. For instance, the weight space has $O(N^4)$ dimensions, say for an N-vertex graph search problem as a generic example. Learning or training to compute a solution weight vector, i.e. through an algorithm like the gradient descent, in such a high-dimensional space poses great challenges. Accordingly, empirical simulation studies are needed to assess if in fact it is feasible to perform the intended training in such high dimensional spaces.

This paper presents, through a simulation study, an assessment of feasibility of an adaptation scheme in the weight space of Hopfield neural networks configured for static optimization. The theoretical and mathematical framework for this study is presented in the upcoming sections. Specifically, the classical Hopfield network, the adaptation scheme, and configuration of Hopfield network dynamics for static optimization and particularly for the traveling salesman problem are discussed. This is followed by the simulation study, analysis of the findings, and the conclusions.

HOPFIELD NETWORK AND ADAPTATION

A Hopfield neural network is a nonlinear dynamical system [Hopfield, 1985] with the following formal definition. Let z_i represent a node output and $z_i \in [0,1]$ with $i = 1,2,\ldots, K$, where K is the number of network nodes. Then,

$$E = -\frac{1}{2}\sum_{i=1}^{K}\sum_{j=1}^{K} w_{ij} z_i z_j + \frac{1}{\lambda}\sum_{i=1}^{K}\int_0^{z_i} f^{-1}(z)dz - \sum_{i=1}^{K} b_i z_i \tag{1}$$

is a Lyapunov function for the system of equations defined by

$$\frac{du_i(t)}{dt} = -u_i(t) + \sum_{j=1}^{K} w_{ij} z_j(t) + b_i \text{ and } z_i = f(u_i), \tag{2}$$

where w_{ij} is the weight between nodes z_i and z_j subject to $w_{ij} = w_{ji}$, b_i is the external input for node z_i, and $f(\cdot)$ is a nonlinearity, typically the sigmoid function with steepness defined by a parameter λ.

A typical static optimization problem can be mapped to a single layer and often two-dimensional node topology of the network through a partly ad hoc and problem-specific formulation of an error function [Hopfield, 1985]. This error function might be in the form of sum of linear and quadratic error terms, where each error term is associated with a particular constraint of the problem. In a generalized sense, the following format for the problem specific error function, which is defined to resemble the generic template given by the Lyapunov function in Equation 1, is pursued:

$$E(\mathbf{z}) = -\frac{1}{2}\sum_{\alpha=1}^{|S_\alpha|} g_\alpha \sum_{i=1}^{K}\sum_{j=1}^{K} d_{ij}^{\alpha}\delta_{ij}^{\alpha} z_i z_j - \sum_{\beta=1}^{|S_\beta|} g_\beta \sum_{i=1}^{K} \delta_i^{\beta} z_i \tag{3}$$

where

a. $i \neq j$;

b. the set of constraints is given by $S_\varphi = \{C_1, C_2, \ldots, C_\varphi\}$ with S_α and S_β representing sets of quadratic and linearly formulated constraints, respectively, $S_\varphi = S_\alpha \cup S_\beta$, $S_\alpha \cap S_\beta = \phi$ and $|S_\varphi| = |S_\alpha| + |S_\beta|$;

c. $g_\alpha \in R^+$ if the hypothesis nodes z_i and z_j each represent for a constraint α in S_α are mutually supporting and $g_\alpha \in R^-$ if the same hypotheses are mutually conflicting; the term δ_{ij}^{α} is equal to 1 if the two hypotheses represented by nodes z_i and z_j are related under the constraint α and is equal to 0 otherwise;

d. similarly, the term δ_i^{β} is equal to 1 or 0 to facilitate mapping of the constraint β in S_β to the network topology; and

e. the d_{ij}^{α} term is equal to 1 for all *i* and *j* under a hard constraint (which cannot be violated by definition) and is a predefined cost for a soft constraint (which can be violated in degrees), which is typically associated with a cost term in optimization problems.

The study presented in this paper will leverage the well-known NP-complete benchmark problem, the traveling salesman, from the combinatorial optimization domain. Given a list of *N* cities, the traveling salesman problem (TSP) requires a visit to each city once while minimizing the total travel distance. There are two problem constraints to satisfy: each city must be visited once, and only once, and the total travel distance should be minimum. An $N \times N$ node array may be employed as the network topology for this problem, where rows and columns represent the cities and the visiting order, respectively. In this topology, any permutation matrix, i.e. each column and row has exactly one node active, is a valid solution although most likely a locally optimal one. The error function proposed by Hopfield to map the TSP to the network topology [1985] is adopted:

$$E(\mathbf{z}) = -\frac{1}{2} g_{row} \sum_{i=1}^{N\times N} \sum_{j=1}^{N\times N} \delta_{ij}^{row} z_i z_j - \frac{1}{2} g_{col} \sum_{i=1}^{N\times N} \sum_{j=1}^{N\times N} \delta_{ij}^{col} z_i z_j - \frac{1}{2} g_{glo} \left(\sum_{i=1}^{N\times N} z_i - N \right)^2 - \frac{1}{2} g_{dis} \sum_{i=1}^{N\times N} \sum_{j=1}^{N\times N} d_{ij} \delta_{ij}^{dis} z_i z_j, \qquad (4)$$

where

$\delta_{ii}^{row} = 1$ if $row(i) = row(j)$ or $\delta_{ii}^{row} = 0$, otherwise;

$\delta_{ij}^{col} = 1$ if $col(i) = col(j)$ or $\delta_{ij}^{col} = 0$, otherwise;

$\delta_{ij}^{dis} = 1$ if $\{|col(i) - col(j)| = 1 \wedge [row(i) \neq row(j)]\}$ or $\delta_{ij}^{dis} = 0$, otherwise;

$i \neq j$; $g_{row}, g_{col}, g_{dis}, g_{glo} \in R^{-}$, d_{ij} is the distance between cities *row*(*i*) and *row*(*j*), and superscripts/subscripts *row*, *col*, *glo*, and *dis* stand for the row, column, global, and distance inhibitions, respectively. Functions *row*(*i*) and *row*(*j*) return the row and column location of nodes indexed by *i* and *j*, respectively.

Learning through experience, i.e. through prior mistakes or trials, is the only feasible option for Hopfield network as a static optimizer since no so-called training exemplars exist. Accordingly, an adaptation scheme needs to look at prior unsuccessful attempts to extract and incorporate the associated experience into the network's search algorithm. In other words, an adaptation mechanism is needed that extracts the required information from previous unsuccessful search attempts in order to incorporate it into the neural network dynamics so that it is less likely for the Hopfield network to repeat the same failed search process. Such an adaptation scheme leveraged in this study is discussed in the next subsection.

A well-detailed theoretical framework for adaptation of weights of the Hopfield network configured for static optimization was proposed in [Serpen, 2003], which will be adopted for our study herein. The Hopfield network has two sets of parameters that can be adapted: weights and constraint weighting coefficients. Gradient descent based adaptation offers a computationally and mathematically feasible option for the both [Almeida, 1987; Pineda, 1987]. Specifically, backpropagation algorithm (both non-recurrent and recurrent versions) for the weights and direct gradient descent for the constraint weighting coefficients are two first-order choices and will be employed to demonstrate the adaptation concept being assessed for feasibility.

The adaptation procedure will initially require the network to attempt computing a solution for the specific static optimization problem, in this case the traveling salesman. Once the Hopfield network converges to a fixed point which might or might not be a solution, the similarity of the fixed point, i.e. the degree to which all constraints are being satisfied, will need to be compared to a locally optimum solution. This comparison will yield an error value which can be leveraged to modify the weights (as well as constraint weighting coefficients) in accordance with an update algorithm, i.e. the gradient descent in the error space.

Starting with initial values of weights and node outputs, the process of adaptation in the weight space will proceed as follows. The network dynamics will be let converge to a fixed point. For the recurrent backpropagation, the adjoint Hopfield network computes its outputs. If the fixed point is not a locally optimum solution, outputs from the Hopfield network (and its adjoint if recurrent backpropagation) are utilized to compute the new values of weights and constraint weighting coefficients. These new values for weights are leveraged by the Hopfield network to initiate the next relaxation cycle in the overall search process.

The set of equations for adaptation of Hopfield network weights and constraint weighting coefficients were derived in [Serpen, 2003]. The adaptation algorithm employs the discretized version of Equation 2 to relax the network dynamics. The follow-up steps detailed below invoke equations and techniques presented in [Serpen, 2003]. Upon convergence to a fixed point, the error is computed. Next, if the training algorithm is recurrent backpropagation, the adjoint dynamics let settle to a fixed point. This is followed by computation of the new weight values. The next step is to update the values of constraint weighting coefficients. Consequently, upon completion of prior cycles, it is now time to launch a new relaxation and adaptation cycle pair.

SIMULATION STUDY

The simulation study was conceived with the goal of assessing feasibility of adaptation of Hopfield network dynamics in the weight space. Hopfield network dynamics were discretized and asynchronously relaxed to a fixed point. Weights of the Hopfield network were adapted using two algorithms, the standard backpropagation and the recurrent backpropagation, while the constraint weighting coefficients (CWC) were adapted through the gradient descent algorithm. Two schemes were employed to adapt the activation function slope values: a heuristic based method and a gradient descent based approach. Learning rate values were also adapted through the procedure described on page 146 of Cichocki et. al. [1990]. Accordingly, a total of four schemes as listed in Table 1 were evaluated. Pseudocode for the simulation outline is given in Figure 1.

Substantial and significant efforts were expended to facilitate the adaptation process in the weight space, which would be expected to lead the Hopfield network dynamics to one of locally minimum solutions of the traveling salesman problem. Many problem instances in the 10 to 100 cities were tested while for each problem instance exploring potentially feasible combinations of all adjustable parameters including the activation functions, and the learning rate. However, in practically all attempts, this adaptation scheme failed to guide the network towards solutions. Often the network converged to fixed points that failed to satisfy one or more constraints derived from the traveling salesman problem. In many other cases, adaptation process failed to finalize: weights were constantly being updated with no potential for stabilization in sight. The overall computational cost appeared to be substantial in some cases to the degree of being prohibitive even for small-scale problem instances.

	Weights	Constraint Weighting Coefficients	Activation Function Slopes	Learning Rates
1	BP	None	Heuristic	Cichocki
2	BP	GD	GD	Cichocki
3	RBP	None	Heuristic	Cichocki
4	RBP	GD	GD	Cichocki

Table 1. Adaptation Schemes Tested (BP: Backpropagation; RBP: Recurrent Backpropagation; and GD: Gradient Descent)

Initialization

Instantiate a Hopfield neural network topology for an N-city traveling salesman problem: determine the count and the layout of neurons, i.e. one layer two-dimensional array of ($N \times N$) neurons.

Initialize constraint weighting coefficients magnitudes randomly in the range [0.0,1.0].

Initialize weights of the network randomly in the range [-0.2,+0.2] while maintaining symmetry of the weight matrix.

Initialize the TSP inter-city distances randomly to a value in the range [0.0,1.0].

Initialize Hopfield network neuron outputs randomly in the range [0.0,1.0].

Adaptive Search

Relaxation

Relax Hopfield network dynamics until convergence to a fixed point.

Adaptation (shown for recurrent backprop – see Serpen [2003] for equations)

Setup the adjoint network and relax until convergence to a fixed point.

Update weights.

Update constraint weighting coefficients.

Adapt node activation function slopes, which implements either the heuristic that if error value for a node is large, then the sigmoid steepness should be small in the range [0.0,1.0] or a gradient descent-based approach.

Adapt learning rate values for weights: follow the procedure detailed on page 146 of Cichocki et al. [1990].

Termination Criteria

If termination criteria not satisfied (i.e. local minimum solution not found), continue with Adaptive Search.

Figure 1. Pseudocode for Simulation

The findings reported as a consequence of the simulation study might not be surprising after all. It is well established in the machine learning literature that the gradient based adaptation algorithms are likely to suffer due to ill conditioning, since the direction for most efficient descent is much more difficult to determine, as a consequence of the (extreme) high dimensionality of the search space, i.e. the space of error vs. weights. Regularization techniques, among others, are often employed to reduce the number of freely adjustable parameters to facilitate the learning process to take place. For instance, weight sharing [Nowlan et al., 1992] is leveraged to facilitate the learning and the speeding up as one option for regularization. It is also important to note the limitations of the study prior to any attempts to derive generalized conclusions, particularly those with a strong negative connotation. As a positive note, an adaptation scheme for the Hopfield network algorithm on a constraint-satisfaction, but not an optimization, problem was demonstrated to be feasible in an earlier work [Serpen et al., 1990]. The adaptation scheme has been implemented in the relatively small dimensional constraint weighting coefficient space, typically with less than ten dimensions. It is likely that this adaptation procedure in the constraint weighting coefficient space could be extended to optimization problems as well, which remains to be assessed and validated.

CONCLUSIONS

Simulation results of attempting to adapt a Hopfield network configured for a static optimization problem, the traveling salesman, indicate that the high dimensionality of the weight space makes it infeasible for gradient descent based search algorithms to succeed. The adaptation algorithms tested resulted in failure on a consistent basis. Emphasizing the limitations and constraints of the study reported, the findings coupled with the empirical and theoretical knowns in the machine learning domain in terms of training in very high dimensional spaces are expected. It is probably prudent to suggest that attempting to search a weight vector in such a high dimensional space should be resorted to as the final option. Regularization approach along with other search space dimensionality reduction methodologies might prove to be useful in this regard. Also, one can be readily tempted to experiment along the same lines reported in this study in a much lower dimensional adjustable parameter space, say less than ten dimensions , of Hopfield network dynamics.

REFERENCES

Abe, S., 1989, "Theories on the Hopfield Neural Networks," Proceedings, IEEE International Joint Conference on Neural Networks, Washington, DC, Vol. 1, pp. 557–564.

Abe, S., 1993, "Global Convergence and Suppression of Spurious States of the Hopfield Neural Networks," *IEEE Trans. on Circuits and Systems–I*, Vol. 40, pp. 246–257.

Abe, S., 1996, "Convergence Acceleration of the Hopfield Neural Network by Optimizing Integration Step Sizes," *IEEE Trans. on Systems, Man and Cybernetics B*, Vol. 26, pp. 194–201.

Almeida, L. B., 1987, "A Learning Rule for Asynchronous Perceptrons with Feedback in a Combinatorial Environment," Proceedings, IEEE International Conference on Neural Networks, San Diego, CA, pp. 609-618.

Atencia, M., Joya, G., and Sandoval, F., 2005, "Two or Three Things that We (Intend to) Know about Hopfield and Tank Networks," Proceedings, European Symposium on Artificial Neural Networks, Bruges, Belgium, pp. 25-36.

Bournez, O., and Campagnolo, M. L., 2006, "A Survey on Continuous Time Computation," New Computational Paradigms, Cooper, B., ed., Springer, Amsterdam, Netherlands.

Cichocki, A. and Unbehauen, R., 1990, "Neural Networks for Optimization and Signal Processing," Wiley, New York, NY.

Hopfield, J. J., and Tank, D. W., 1985, "Neural Computation of Decision in Optimization Problems," *Biological Cybernetics*, Vol. 52, pp. 141-152.

Pineda, F. J., 1987, "Generalization of Back-Propagation to Recurrent Neural Networks," *Physical Review Letters*, Vol. 59, pp. 2229-2232.

Serpen, G., 2003, "Adaptive Hopfield Network," Proceedings, International Conference in Artificial Neural Networks, Istanbul, Turkiye, pp. 3-10.

Serpen, G., and Livingston, D. L., 2000, "Determination of Weights for Relaxation Recurrent Neural Networks," *Neurocomputing*, Vol. 34, pp. 145-168.

Serpen, G. and Livingston, D. L., 1990, "An Adaptive Constraint Satisfaction Network," Proceedings, ASILOMAR Conf. on Signals, Systems and Circuits, Monterey, CA, pp. 163-167.

Sima, J., 2003, "Energy-based Computation with Symmetric Hopfield Nets," In Limitations and Future Trends in Neural Computation, NATO Science Series: Computer & Systems Sciences, S. Ablameyko, M. Gori, L. Goras, and V. Piuri, eds.,Vol. 186, Amsterdam: IOS Press, pp. 45-69.

Smith, K., 1999, "Neural Networks for Combinatorial Optimization: A Review of More Than A Decade of Research," *INFORMS Journal on Computing*, Vol. 11, No. 1, pp. 15-34.

Nowlan, S. J. and Hinton, G. E., 1992, "Simplifying Neural Networks by Soft Weight-Sharing," *Neural Computation*, Vol. 4, pp. 473-493.

MULTISTAGE LINEAR SVM CLASSIFICATION

QIAN XIA[1], JAMEEL SHAIKH MOHAMMED[1], OKAN ERSOY[1], HERBERT MOSKOWITZ[2]
School of Electrical and Computer Engineering [1], Purdue University,
Krannert School of Management[2], Purdue University, West Lafayette, IN 47906

ABSTRACT
Parametric models such as linear models provide useful, interpretable description of simple structure in data. However, sometimes such simple structure does not extend across an entire data set. A trade-off exists between the ease of interpretation of the model and its predictive power. We propose a Multistage Linear Support Vector Machine (MLSVM) method to address this problem. MLSVM partitions a dataset into homogenous subsets which are then individually modeled. The aggregation of linear models obtained upon termination provides the power to handle complex data while maintaining the ease of model interpretability. Empirical results show MLSVM achieves substantial improvements in training and testing accuracies as compared with a single stage linear SVM model and also circumvents the potential overfitting issue usually faced by more complex nonlinear models.
Key Words: Support Vector Machines, Multistage, Data Set Partition,

1. INTRODUCTION

In classification analysis, parametric models such as linear models are widely used in constructing the discriminant function. They provide useful, interpretable description of simple structure in data. However, such simple structure rarely extends across an entire data set. One alternative is to use nonlinear models. Unfortunately, such complex structures usually lose the power of ease of interpretation. A trade-off exists between the ease of interpretation and the predictive power of the learning algorithm.

The support vector machine (SVM) is a machine learning algorithm developed by Vapnik (1995). As many other classical methods, SVM produces a single model over the entire data set, and therefore faces the dilemma of model interpretability and the ability to handle complex data. We propose a multistage linear SVM (MLSVM) model as a new solution. The dataset is divided into a rejected and an accepted subset based on sample-margins induced by a linear SVM model. A second SVM model is then trained on the accepted subset and the rejected subset is forwarded to the next stage for further processing. The procedure proceeds until no further division is necessary or possible. The recursive partitioning of the data set naturally leads to a multistage structure and the aggregation of linear SVM models provides us the power to handle complex data while maintaining the ease of model interpretability.

MLSVM falls in the category of subspace pattern recognition, which includes the well-known tree classification methods. In some previous work on combining SVM with tree structure, Chi and Ersoy (2002) introduced a Linear Support Vector Machine Decision Tree (LSVM-DT). Bennett and Auslender (1999) introduced a support vector decision tree method for customer targeting in the database marketing. In these methods, each node is bisected based on the SVM decision boundary and the process proceeds recursively on every child node. MLSVM differs in the following ways: first, the partition criterion is margin-based rather than boundary-based. By using sample-margins we distinguish samples based on the level of confidence in making predictions for them; secondly, the multistage structure is tail-recursive – only the rejected subset is forwarded for further processing. In other words, the model structure is a chain of successive stages rather than a binary tree structure. The merits of MLSVM are demonstrated on a synthetic and a real world classification task. Experiments show that MLSVM achieves comparable predictive ability as compared to single stage linear and quadratic SVMs.

The paper is organized as follows: Section 2 explains the margin-based criterion for learning and MLSVM algorithm. In Section 3, we provide the experimental results and discussion, followed by concluding discussion in Section 4.

2. THE MLSVM ALGORITHM

The original SVM classification attempts to find the best separating hyperplane with the largest margin width and the smallest number of training errors. Sample margin is defined as the distance between the response of the sample and the discriminant hyperplane induced by the classifier (Crammer (2002)). The sample-margin produced by SVM model measures the confidence of the SVM classifier with respect to its prediction for that sample. In other words, the samples whose responses are closer to the decision boundaries are more likely to be misclassified than those further away. Based on this, the MLSVM method bisects the dataset into accepted/rejected subsets with samples "easy-to-classify" in the accepted subset and samples "difficult-to-classify" in the rejected subset. The rejected subset is then forwarded to the next stage for further processing, and a second SVM model is trained on the accepted subset. The process resembles a data filtering framework in terms of distinguishing easy and difficult samples and processes them differently. The overall model consists of a chain of successive stages, with two linear SVM models at each but the last stage.

In order to present the method further, we will discuss the criterion used for data partitioning at each stage and the rules used to stop the iterative process, followed by illustrations of the overall training and testing procedure.

2.1 Margin-based dataset partitioning

The sample-margin produced by SVM model measures the confidence of the SVM classifier with respect to its prediction for that sample. If a SVM model has high level of confidence in making prediction for a sample, it is natural to expect that the underlying model structure of that sample is well approximated by that SVM model. Furthermore, the set of samples that are well represented or approximated by the same SVM model should share a similar underlying model-structure. Therefore, by using the sample-margin information, we can pick out the subset of samples that are homogeneous in terms of the underlying model-structure and based on this "cleaned" subset, we can train a new SVM model to better capture the common underlying model-structure of the subset.

One natural way is to use the ± 1 SVM margin, and consider the samples that fall within the margin as being close to the boundary and the samples that fall outside the margin as being adequately far away from the boundary. For separable cases, it is the set of support vectors that will fall on the ± 1 margin. For nonseparable cases, this property does not hold. However, the ± 1 margin can still be a good choice for the following reasons: first, usually only a small portion of the support vectors will fall outside the margin area; secondly, the second SVM model trained on the "cleaned" dataset is expected to adjust the "bias" of the first SVM model; thirdly, widening the rejection margin tends to reject more samples and, therefore, will produce more stages during training, which can lead to potential over-fitting during testing.

2.2 Termination Rules

As the process proceeds, the set of samples that are easy-to-classify are picked out, leaving only the difficult-to-classify ones for further processing. As more stages are created, it will be more and more difficult to generate an effective partitioning because the samples left are inconsistent in nature and act like noise. On the other hand, since the partitioning is tail-recursive, there will be fewer and fewer samples left, which will increase the risk of overfitting for later stages. Here, we propose two termination rules:

Rule 1: If the number of the accepted samples is too small, then stop.

Since a second SVM model is trained on the accepted subset, it is important that the size of the accepted subset is sufficiently large. For example, the size should not be smaller than the number of explanatory variables in the model.

Rule 2: Terminate at the earliest stage which maximizes the prediction accuracy.

Cross validation can be used to evaluate the prediction accuracy, and to find the number of stages which maximizes the prediction accuracy. If there are no unique maxima, a smaller number of stages is preferred.

2.3 Training and testing procedures

The training procedure for the MLSVM method is described below:

Step 1: Set k:=1 for the first stage. Let S(0) represents the entire data set, and S(k-1) represent the rejected subset at k-1 th stage.

Step 2: Train an initial linear SVM model on the data set S(k-1) for the k th stage. Divide S(k-1) into two parts based on the ± 1 SVM margin. Then, train a second linear SVM model is trained on the accepted subset.

Step3: Check the termination conditions. If either of the two conditions is/are satisfied, then stop; otherwise, let k:=k+1and go to step 2.

During testing, trace the sample down the chain of stages until a stage that accepts the sample is founded (the first stage that accepts it), and the final classification result is given by the second SVM model of that stage.

3. EXPERIMENTS AND DISCUSSION

The MLSVM algorithm is designed to approximate the potentially complex model structure by an aggregation of linear submodels. In order to test this, we compare MLSVM with the original linear SVM model and the 2nd order polynomial SVM model. The behavior of the three methods is compared on a synthetic and a real world classification problem.

3.1 Experiments with Synthetic Data Set

We designed a simulated data set in the form of a circle shown in Figure 1. The cross points in the middle correspond to the first class (281 samples), and the circle points outside correspond to the second class (448 samples). We randomly select 600 samples as training set (228 crosses and 372 circles). The remaining 129 samples are used as testing set (53 crosses and 76 circles). Results are summarized in Table 1.

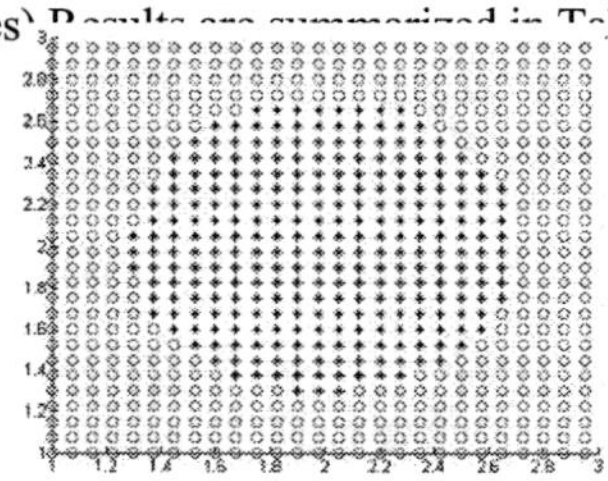

Figure 1. Synthetic data set

Table 1. Classification results with synthetic data.

	Training (# = 600)			Testing (# = 129)		
	Mis	Cor	Per	Mis	Cor	Per
Linear	228	372	62.0%	53	76	58.9%
Quadratic	3	597	99.5%	1	128	99.2%
Multistage	120	480	80.0%	30	99	76.7%

The single stage linear SVM failed to distinguish the two classes and classified all the samples as the second class. The 2nd order polynomial SVM model obtained best result in both training and testing. The MLSVM algorithm ends with 4 stages. Although, the performance of the multistage linear SVM is not as good as the 2nd order polynomial SVM, which happens to have the same form as the true underlying model, it improves substantially as compared to the single stage linear SVM. The overall training and testing accuracies are 18% and 17.8% higher than that of the single stage linear SVM, respectively. The stage wise results of the MLSVM are summarized in the table 2.

Table 2. MLSVM stage wise classification result with synthetic data.

	Training				Testing			
Stage	Total#	Mis#	Cor#	Per%	Total#	Mis#	Cor#	Per%
1	276	41	235	85.1	80	21	59	73.8
2	160	10	150	93.8	30	4	26	86.7
3	58	7	51	88.6	16	4	12	75.0
4	106	39	67	63.2	3	1	2	66.7
Sum	600	120	480	80	129	30	99	76.7

Total# - number of samples accepted; Mis# - number of samples misclassified
Cor# - number of samples correctly classified; Per% - the percentage accuracy

The stage wise training and testing accuracies are quite high, except for the last stage. This is consistent with our expectation. For previous stages, the prediction was made by the second SVM model which was trained on the "cleaned" dataset. However, for the last stage, no dataset partitioning is generated, and a single SVM model is trained on all samples available there. Since the samples forwarded to the last stage are noise like in nature, it is natural to observe a substantial decrease in training and testing accuracies for the last stage. In summary, when making predictions, the MLSVM model has higher level of confidence when the prediction is made by models at previous stages, and a lower level of confidence when the prediction is made by the model at the last stage.

3.2 Experiments with the Real World Data Set

The breast cancer dataset was obtained from the repository of a machine learning database at University of California, Irvine. This dataset has 9 attributes (3 continuous, 6 nominal) with 277 instances (196 benign cases and 81 malignant cases). We randomly select 230 samples as training set (160 benign cases and 70 malignant cases), and the rest 47 as testing set (36 benign cases and 11 malignant cases).

Results are summarized in Table 3. The MLSVM method ends with 3 stages with training and testing accuracy 8.7% and 10.6% higher than that of the single stage linear SVM model. The quadratic SVM model has the highest training accuracy, but the lowest testing accuracy, which indicates over-fitting during training. This result demonstrates that by using the MLSVM model, we obtain a substantial improvement in training and testing when compared to single stage linear model and also circumvent the risk of overfitting as compared to complex nonlinear models.

Table 3. Classification results with cancer data.

	Training (# = 230)			Testing (# = 47)		
	#Mis	#Cor	Per%	#Mis	#Cor	Per%
Linear	60	170	73.9	15	32	68.1
Quadratic	23	207	90.0	19	28	59.6
Multistage	44	186	80.9	10	37	78.7

The stage wise training and testing accuracies are summarized in Table 4. We observe similar result as with the synthetic dataset - MLSVM has higher accuracy in previous stages and lower accuracy in the last stage.

Table 4. MLSVM stage wise results with cancer data.

	Training				Testing			
Stage	Total#	Mis#	Cor#	Per%	Total#	Mis#	Cor#	Per%
1	107	13	94	87.9	23	4	19	82.6
2	56	11	45	80.4	18	4	14	77.8
3	67	20	47	70.1	6	2	4	66.7
Sum	230	44	186	80.9	47	10	37	78.7

Besides percentage accuracy, a task like the breast cancer problem should be analyzed in terms its information value in assisting diagnostic decisions. The odds of prior probability represents the diagnosing ability using only prior information, and the odds of posterior probability represents the diagnosing ability with the aid of the model response. In order to investigate informative value, we examined the relationship between odds of prior probability and odds of posterior probability. Let $\{T = t_0\}$ and $\{T = t_1\}$ represent the event of benign and malignant cases, and $\{M = m_0\}$ and $\{M = m_1\}$ be the event that the model predicts benign and malignant respectively. From Bayesian theory, we have,

$$\frac{P(T = t_0 \mid M = m_0)}{P(T = t_1 \mid M = m_0 \mid} = \lambda * \frac{P(T = t_0)}{P(T = t_1)} \text{ where } \lambda = \frac{P(M = m_0 \mid T = t_0)}{P(M = m_0 \mid T = t_1)} \qquad (9)$$

The ratio λ represents how much we gain or lose in diagnosing when taking the model response into account. When λ <1, it is better to ignore the model response; when λ >1, we gain from using the model response and the larger the value of λ is, the more benefit we obtain from it. In Table 5, we summarize the result of λ for testing for each method.

Table 5. Likelihood Ratios.

Testing	Likelihood Ratio λ	
	Y=0 (benign)	Y=1(malignant)
2nd order poly	0.88	0.63
Linear	1.02	1.09
Multistage	1.44	4.36

As shown in table 5, the 2nd order polynomial SVM model has λ's smaller than 1, which implies that the response of this model is not at all helpful. Single stage linear SVM model has λ's slightly larger than 1, which means the model is of little help. MLSVM has the largest λ's, which implies that this method is most helpful and it is especially helpful in identifying malignant cases.

4. CONCLUSIONS AND FURTHER RESEARCH

The main thrust of this report has been to propose a MLSVM approach and to discuss some of its features. MLSVM extends the original SVM method to a multistage structure. The process iteratively partitions the current dataset into an accepted and a rejected subset and builds separate linear models in each accepted subset. The aggregation of a set of linear models obtained upon termination provides the power to approximate the potentially complex nature of the dataset while maintaining the ease of interpretation.

Experiments indicate that MLSVM achieves better training and testing result than a single stage linear SVM model, and circumvents the issue of overfitting faced by a 2nd order polynomial SVM. Furthermore, early stages have higher levels of accuracy, and the last stage is expected to give worst result.

Several research directions can be further investigated. One of them is to find a better way to partition the dataset. The way we use now is the ± 1 SVM margin, which is not necessarily the best for nonlinearly separable cases. Another direction is to improve the performance of the last stage by some nonparametric methods.

References

Bachrach, R.G., Navot A., Tishby N., (2004) "Margin Based Feature Selection – Theory and Algorithms,". Proceedings of the 21th Int'l Conference on Machine Learning, Banff, Canada..

Bennet, K. P., and Auslender, L., "On Support Vector Decision Trees for Database Marketing" R.P.U. Math Report 98-100, March 1998.

Chi, H., Ersoy, O., "Support Vector Machine Decision Trees with Rare Event Detection," International Journal of Smart Engineering System Design, pp. 225-242, Vol 4, Oct 2002.

Crammer, K., Gilad-Bachrach, R., Navot, A., &Tishby, N. (2002) "Margin Analysis of the LVQ Algorithm,". Proc. 17th Conference on Neural Information Processing System.s.

Quinlan, J.R. (1990). "Introduction of Decision Trees,". In J.W. Shavlik and T.G. Dietterich (Eds.), Readings in Machine Learning. Morgan Kaufmann.

Schapire, R.E., Freund, Y., Bartlett, P., & Lee, W.S. (1998). "Boosting the Margin: A New Explanation for the Effectiveness of Voting Methods,". Annals of Statistics.

Vapnik, V.(1995). The Nature of Statistical Learning Theory, Springer-Verlag, New York.

Wu, D., Bennet, K. P., Christianini, N., and Shawe-Taylor, J., "Large Margin Trees for Induction and Transduction," ICML'99 Proceedings..

PART VII:

SMART ENGINEERING SYSTEMS

Automatically Balanced K-means for Non-photorealistic Rendering

Daniel A. Ashlock*
dashlock@uoguelph.ca

Kenneth M. Bryden†
kmbryden@iastate.edu

Balasubramaniam Karthikeyan†
balas@iastate.edu

Abstract Non-photorealistic rendering is a broad class of techniques for creating art from digital pictures. One or more digital filters is applied to create an apparent pencil sketch, watercolor, or in this study a design for stained glass. A collection of points that are the centers of a weighted Voronoi tiling are computed by using an adaptive form of the K-means clustering algorithm. A fractal model of stained glass is then run to create a stained glass texture. Tile boundaries are rendered in black. The stained glass textures are then applied within their corresponding tiles to yield a final image. The K-means algorithm is novel in mixing color and position data for the pixels and in using the variance of luminance within tiles to change the tile weights.

1 Introduction

One method of creating digital art is to start with a digital image, typically a photograph, and computationally modify it. The name for this class of techniques is *non-photorealistic rendering* [3]. These techniques can be used to render an image so that it appears to be a pen-and-ink drawing of the scene in the original image. It is also possible to generate apparent watercolors [2]. In this study a novel form of K-means clustering is used as a part of a non-photorealistic rendering pipeline to create designs for stained glass windows from digital images.

The rendering pipeline presented in this study has comparable image quality to that described in [1] but is enormously faster. The earlier pipeline optimized the segmentation of the image with an evolution strategy. The experience gained in massaging the mutation operators in that study suggested that a modification of the well-known K-means algorithm [4] might perform the optimization as well.

The remainder of this paper is organized as follows. Section 2 describes the modified K-means algorithm and specifies the rendering pipeline. Section 3 gives the details of the stained glass texture creation. Section 4 gives the results of applying the algorithm.

*Mathematics and Statistics,University of Guelph,Guelph,Ontario Canada
†Mechanical Engineering, Iowa State University, Ames, Iowa 50011

2 A Variation on K-means for Image Segmentation

Algorithm 1 gives the standard K-means algorithm which is our starting point. The typical application of K-means clustering in image processing is to cluster color values. Color values in a cluster are replaced with a single value (the cluster mean) reducing the amount of data required to store a picture. Here we will cluster data points of the form (x, y, r, g, b) which describe pixels in an image. The values x and y described the position of the pixel within the image while the r, g and b values give the pixel's color. The values for each coordinate are normalized to lie in the range $[0, 1]$. Clustering of data points of this type segments this image into groups of pixels that are close to one another both geographically and chromatically.

Algorithm 1 k-means

Input: *1) A set S of points in* $\mathbb{R}^n$
2) A desired number k of clusters.
3) A bound B on the number of cycles permitted

Output: *A category function*
$$C : S \rightarrow \{0, \ldots, k-1\}.$$

Details:
Choose k distinct points in S as initial cluster centers.
Repeat
Assign each point to the cluster whose center it is closest to, breaking ties at random.*
Recompute cluster centers as the average of all points in the cluster.
Until (no points change their cluster assignment or B cycles have occurred+)
Report the assignment of points to clusters as C.

Applying K-means clustering to all the pixels in an image can yield inconveniently large amounts of data. The rendering pipeline samples points from the image, chosen at random. The resulting cluster centers are then used as the centers of a weighted Voronoi tiling to segment the image.

A *Voronoi tiling* [5] is a division of a finite subset of the plane into convex polygonal tiles. A Voronoi tiling is generated by placing a set of *tile centers* into the plane. A *tile* associated with a given tile center p consists of all points in the plane closer to p than to any other tile center. The boundary between two tiles that meet is always a segment of the perpendicular bisector of the line segment joining the two tile centers. For many applications this is a convenient way to subdivide the plane. Images, however, often contain objects with curved boundaries. To deal with this common feature of images sets of *weighted* Voronoi tiles are used to segregate images.

A *weighted* Voronoi tiling differs from a standard Voronoi tile in that each tile center has a *weight* ω associated with it. This weight is multiplied by the distance from a tile center to a point in the plane when deciding tile membership. A large weight makes distance to a tile center more expensive, shrinking the size of the tile. A small weight make the distance to a tile center cheap, increasing the size

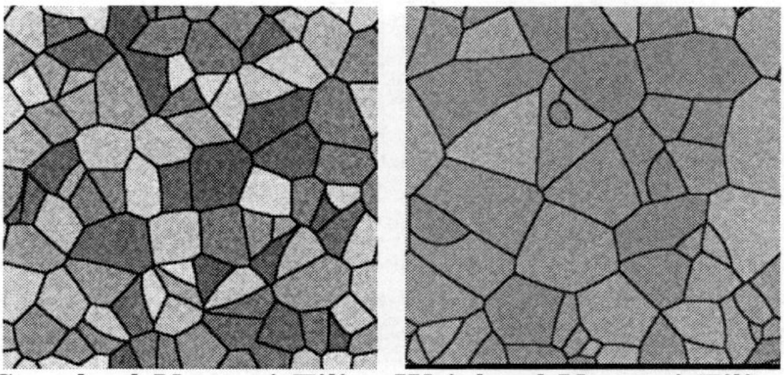

Standard Voronoi Tiling Weighted Voronoi Tiling

Figure 1: The left image is a standard Voronoi tiling of a square while the right image shows a weighted Voronoi tiling. Contrast the edge curvature in the two images. Note that when a weighted tiling is used one tile can be completely contained within another.

of the associated tile. Examples of a standard and weighted Voronoi tilings of a square are shown in Figure 1. The boundaries between weighted Voronoi tiles are no longer line segments. Examine the comparison of squared distances made for two tile centers (x_i, y_i) with weight ω_i, $i = 1, 2$ for a point (u, v) whose tile membership is to be decided:

$$\omega_1^2 \left((x_1 - u)^2 + (y_1 - v)^2 \right) < \omega_2^2 \left((x_2 - u)^2 + (y_2 - v)^2 \right)? \tag{1}$$

If $\omega_1 = \omega_2$ then all quadratic terms in the comparison cancel. If, however, $\omega_1 \neq \omega_2$ then the boundary is a quadratic curve.

We now remind the reader of a quantity that is useful in computing tile weights. The *luminance* in a red-green-blue(RGB) image is a scalar gray-scale value that is thought to capture the brightness of the image. For a pixel with color values r, g, and b, $luminance(r, g, b) = 0.299r + 0.587g + 0.114b$. Suppose that a weighted Voronoi tiling is used to segregate an image. If the sample variance of the luminance (SVL) within each tile is roughly the same then each tile contains a similar amount of information. Balancing the luminance between tiles also encourages the resulting image segregation to respect natural chromatic boundaries within the image: having tile boundaries coincide with chromatic boundaries yields a much lower SVL. Given that out K-means algorithm already approximates chromatically flat tiles, attempting to balance the SVL reinforces this desirable segmentation property.

The K-means algorithm used to create stained glass designs is modified by having it attempt to balance the luminance of the tiles. Recall that cluster means drawn from a K-means clustering of a set of sample pixels are used to segment the image in a weighted Voronoi tiling. Each tile is initially given a weight of 1.0. After each cycle of recomputing cluster centers this weight is averaged with the SVL for the pixels in cluster. This modification causes the K-means algorithm to require more iterations to converge and extends a terminal period in which a small number of points (less than ten out of 5000 in the experiments here) jump back and forth between adjacent clusters. We thus impose an upper limit of 100 iterations on the K-means algorithm. With all of the pieces in place, save

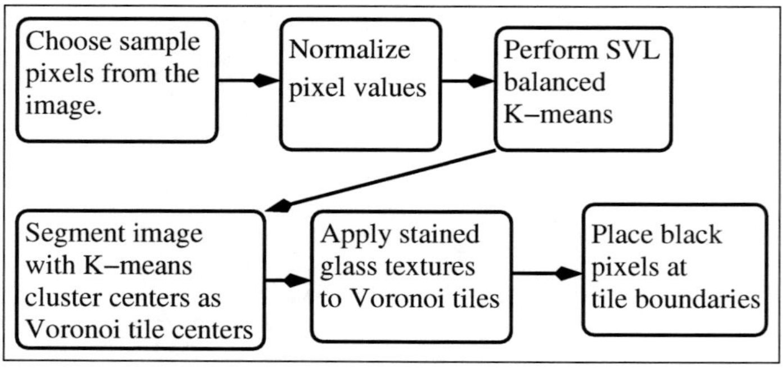

Figure 2: The stained glass rendering pipeline.

creation of stained glass textures which are described in the following section, the rendering pipeline may now be described. It is shown in Figure 2.

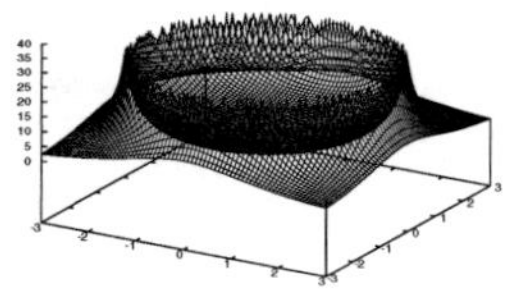

Figure 3: A plot of the crater function, used to generate the fundamental object that generates the stained glass texture.

3 Stained Glass Textures

An *iterated object placement fractal* is created by taking an object and repeatedly placing it, additively, within a drawing region. In general an *object* is a collection of vertical displacements associated with positions (x,y) within the drawing area. The object is placed at random (x,y) positions and its vertical displacement summed with any already present. The vertical direction used in this study is "more color" not upwards. Thus, while the algorithm used generates fractal landscape, the heights of the resulting relief map are exported as color intensities rather than vertical displacements. The mathematical formula used to generate the vertical displacements of the fundamental object that generates the stained glass textures is: $f(x,y) = \frac{30}{1+|R-x^2-y^2|}$. This function is called the *crater function*; a plot of the function appears in Figure 3. Details of an efficient implementation for generating stained glass textures in this fashion with a specified mean shade appears in [1].

4 Results and Discussion

The algorithm parameters used for the trials of the rendering pipeline in this study are 600 tiles evaluated with 5000 sample points. Variance amplification is

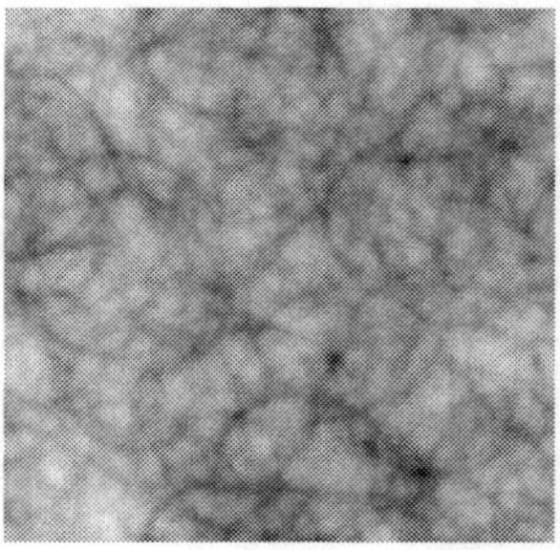

Figure 4: An example of the stained glass texture.

set to 60 with therelative importance of distance set to 0.1. These parameters were chosen by trial-and-error experimentation. Three images were used to test the rendering pipeline with the original and stained glass renderings shown in Figure 5. The second image is a photograph of three daffodils. This is a natural image with contrasting foreground and background regions as well as a chromatic grace note supplied by the green of the flower's stems. The third image is a continuous-tone varying synthetic image of a virtual planet.

The tile weights after all iteration of the K-means algorithm were completed for the Yang-yin image were equal to five decimal places. Likewise the SVL in tiles was quite small (about 10E-6) This implies that the algorithm managed to place all the tile boundaries either within one of the chromatically uniform regions or on a chromatic boundary. This image represents a control for the algorithm - if it cannot find a good segmentation of this image then it is hopeless. The resulting image does a better job on this image than the evolution strategy in [1] of placing pixels into tiles with other pixels the same color. It makes far less use of the algorithms ability to curve tile sides, resulting in a different character of stained glass rendering of the image. For some viewers the output of the ES may be more aesthetically pleasing. The results for the daffodil picture are similar in quality for the algorithm presented here and the previous ES technique. The evolution strategy required several days to finish the daffodil image required slightly less than 30 seconds on the same hardware. We note this to emphasize the enormous speed advantage of the modified K-means algorithm over the evolution strategy.

The major advantage of using a weighted Voronoi tiling is that tiles have quadratic (curved) rather than linear (strait) boundaries. When using an ES to select the tile centers and weights to minimize the variance of the SLV[1] the weights varied substantially. The weight variation resulting from the self-balancing of SLV by the K-means algorithm is far smaller. The smaller weight variation comes from the K-means placing the tile centers more effectively, in essence finding better solutions to the tile center optimization problem. This has the unexpected effect of producing better image segmentations that, in some cases, are less aesthetic as stained glass window designs (in this case the aesthetic value was judged by a small informal survey of available people). This in turn suggests that decoupling the K-means segmentation and weight assignment may restore some of the interesting irregular tiles found by the ES.

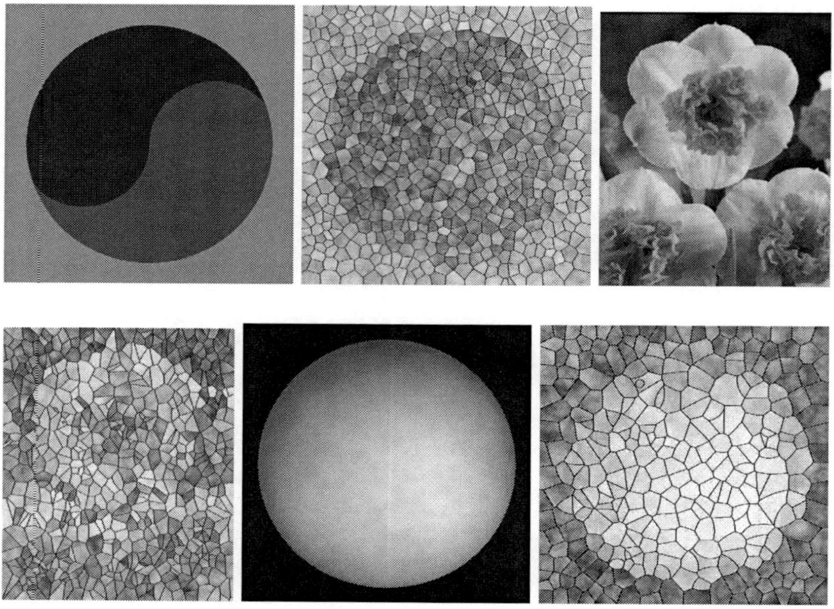

Figure 5: The original and stained glass rendering of the Yin-Yang, Daffodil, and Planet pictures.

References

[1] D. Ashlock, B. Karthikeyan, and K. M. Bryden. Non-photorealistic rendering of images as evolutionary stained glass. In *Proceedings of the 2006 Congress On Evolutionary Computation*, pages 7440–7447, 2006.

[2] Cassidy J. Curtis, Sean E. Anderson, Joshua E. Seims, Kurt W. Fleischer, and David H. Salesin. Computer-generated watercolor. *Computer Graphics*, 31(Annual Conference Series):421–430, 1997.

[3] J. Landsdown and S. Schofield. Expressive rendering: A review of nonphotorealistic techniques. *IEEE Computer Graphics and Applications*, 15(3):29–37, 1995.

[4] J. B. MacQueen. Some methods for classification and analysis of multivariate observations. In *Proceedings of 5-th Berkeley Symposium on Mathematical Statistics and Probability*, volume 1, pages 281–297. University of California Press, 1967.

[5] Okabe, Boots, Sugihara, and Chiu. *Concepts and Applications of Voronoi Diagrams*. Wiley, New York, second edition, 2000.

Lorentzian Based Model for Clustering

J. CHENG and M. R. SAYEH
Department of Electrical and
Computer Engineering
Southern Illinois University
Carbondale, IL 62901 USA

M. R. ZARGHAM
Department of Computer
Science
Southern Illinois University
Carbondale, IL 62901 USA

ABSTRACT
This paper introduces a simple adaptive dynamic system for clustering. This system is based on a set of differential equations which create valleys in its energy function for representing clusters. Each valley represents a cluster of similar input patterns. Through several examples of different pattern clusters, it is shown that the model can successfully cluster different types of input patterns.

INTRODUCTION

The goal of this paper is to develop a model which is able, in real time, to classify different patterns with similar characteristics into groups and then assigns a new pattern to one of these groups. Although the pattern clustering/recognition problem is widely investigated and many solutions have been proposed (ranging from statistical approaches to soft-computing techniques), yet more work needs to be done to make this process to respond in real time for classifying a given input pattern. The proposed method overcomes the problem of response time with being only based on ordinary differential equations and not using any IF/THEN statements.

Two of the previous work which has some resemblances to the proposed model is based on artificial neural networks a branch of soft-computing. They are Adaptive Resonance Theory (ART) (Grossberg, 1976) and Self-Organizing Maps (SOM) (Kohonen, 1982). The ART network receives an input vector (a one-dimensional array of values) and seeks to classify it with a cluster of stored patterns which is most similar to. Its best cluster is the single neuron whose set of weights (pattern) most closely matches the input vector. If a stored pattern is found that is similar to the input vector within a certain tolerance, the stored pattern is modified to become more similar to input vector. If there is no similar stored pattern, a new cluster is created by storing the input pattern. Most applications of ART can be found in (Carpenter and Grossberg, 2002).

Although ART and SOM networks and their revisions are applied to many applications, they are not flaw free and have some shortcomings. For example, these models generate some fake clusters, the storage capacity is limited and they usually use rule-based decision algorithms, which makes hardware implementation very difficult and is slow and inefficient. In order to overcome these drawbacks, we have expanded some of our previous work (Athinarayanan et al., 2002), (Sayeh et al., 1994), and developed a simpler model which is only based on ordinary differential equations (ODEs). Similar to Hopfield Network (Hopfield, 1982), these ODE's can be directly implemented in hardware through feed-back networks.

MODEL BASED ON LORENTZIAN FUNCTION

This section describes the design of a dynamic pattern clustering system. The idea behind this system is that the input patterns are stored on the surface of a dynamic system. Comparing the proposed system with ART network, the important difference is that the dynamics of our system is solely described by the ODEs, as oppose to the ART dynamics where in addition to differential equations IF/THEN logical statements are also included. This makes the true analysis of ART impossible. The SOM has only the weight dynamics in a very linear simple version and more importantly, similar to ART, includes IF/THEN logical statements.

To identify clusters, a local energy function is formulated to create a single valley centered at each input pattern or similar set of input patterns, which is called *V*-energy function. For example, in one-dimensional space, given an input pattern, say u, *V*-energy function is to create a valley, say w_1, under the pattern u. Originally, the weight w_1 (referred to as a ghost which is randomly placed) may be far away from u. Later w_1 is referred to as the center of a cluster which includes u. Similarly, in cases where there are more than one input pattern, the valleys are created for patterns that are close to each other.

Generally, the *V*-energy function is constructed to represent M valleys centered at locations $W_j = (w_{1j,}, w_{2j,} \ldots w_{Nj,})^T$, for $j = 1$ to M, representing M clusters for P input patterns. Considering an N dimensional space for input patterns, V is represented as:

$$V = \sum_{p=1}^{P} \left[\gamma - \sum_{j=1}^{M} x_{jp} \right]^2$$

Here:

$$x_{jp} = \frac{1}{1 + \lambda \, \| U_p - W_j \|^2}$$

is a Lorentzian function, where P is the number of input patterns, and ||.|| is the Euclidean norm.

The constant γ encourages a number of valleys (clusters) to be formed and the factor $1/\lambda$ approximately reflects the radius of the generated valleys. The parameter $x_{jp}\,(for\ j = 1,\ldots, M)$, which has value between 0 and 1, represents the depth of each valley. As distance between U_p and W_j decreases, the value of x_{jp} move toward 1 indicating W_j as a cluster for U_p.

To represent the performance of this model, first we derive the dynamics.

$$\frac{\partial V}{\partial w_{ij}} = \sum_{P=1}^{P}\left[\gamma - \sum_{j=1}^{M} x_{jp}\right] x_{jp}^{2}(u_{ip} - w_{ij})$$

where:
$$x_{jp} = \frac{1}{1 + \lambda\left(\sum_{i=1}^{N}(u_{ip} - w_{ij})^{2}\right)}$$

Thus, dynamics of *Ws* can be expressed as:

$$\frac{dW_{ij}}{dt} = -\tau_w \sum_{p=1}^{P}\left[\gamma - \sum_{j=1}^{M} x_{jp}\right] x_{jp}^{2}(u_{ip} - w_{ij})$$

where τ_w, is related to the learning rate.

PERFORMANCE OF THE MODEL ON TWO EXAMPLES

Example1: This example demonstrates simulation results of our model on two set of two dimensional input patterns as is shown in Fig. 1. Figure 2 (a) demonstrate the set of obtained clusters after 1,000 iterations for λ equal to 20, γ equal to 2, and τ_w equal to 5. As can be observed from Fig. 2, two clusters were formed. Figure 2 (b) is the final contour image of the data set represented in Fig. 1 after 1000 iterations. The contour surface of this image represents the *V*-energy function in terms of different input values. There are two valleys representing two different clusters. Each generated valley is represented by a set of nested rings with different shades representing different levels of valley's depth. The small circles represent the final position of *Ws*; the ones with grey color represent the center of the clusters.

Example 2: This example represents the performance of our model on patterns with higher dimensionality. In preparing these patterns, we utilized 33 of the input patterns that were used by Carpenter and Grossberg in the ART-2 model (Carpenter and Grossberg, 1988). These patterns represent waveforms with different feature characteristics. Each pattern was scanned and processed by Photoshop to form a twenty dimensional vector. To construct such a vector, each pattern was sampled at twenty equally spaced points. The sampled values were bounded to 1 and 10.

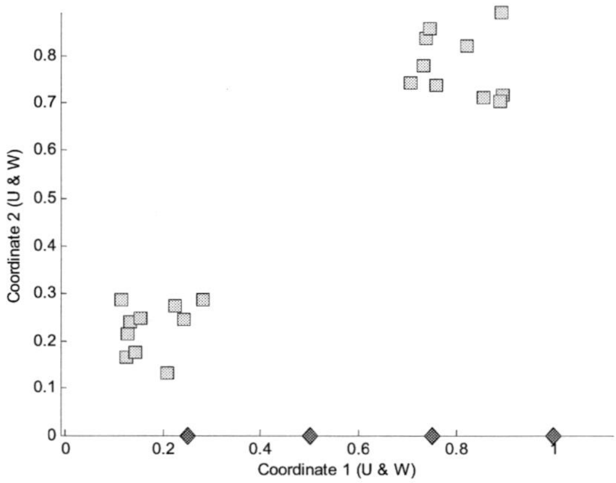

Figure 1.Two classes of simple clusters; there are 20 input patterns (represented by squares) and 4 *W*s (represented by the diamonds) positioned at the bottom.

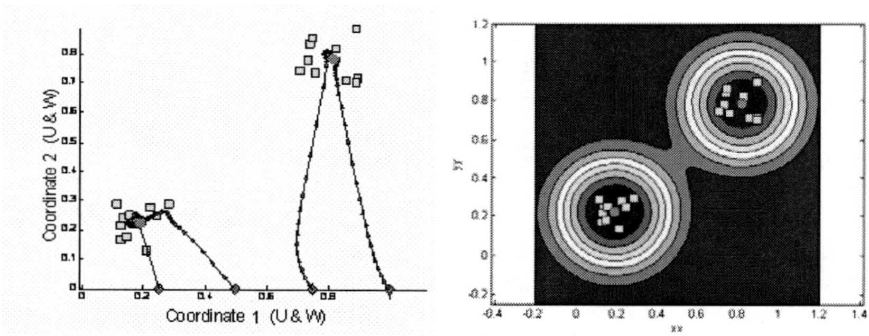

(a) Moving path and final position of *W*s (b) Contour representation

Figure 2. Final positions of *W*s and contour representation of the *V*-energy function.

To obtain the performance of the proposed model on these patterns, we generated 30 twenty dimensional *W*s. Each component of these weights was randomly generated by having a value between 1 and 10. Figure 3 demonstrates the set of obtained clusters after 100,000 iterations for λ equal to 20, γ equal to 7, and τ_w equal to 10. As observed from the figure, six clusters were formed. These clusters are represented with different border patterns. Within every cluster the distance of each pattern from the corresponding *W* is less than 5. The similar patterns were successfully categorized into same clusters. The patterns with less similarity were not formed clusters as can be seen in Fig. 3 with white borders.

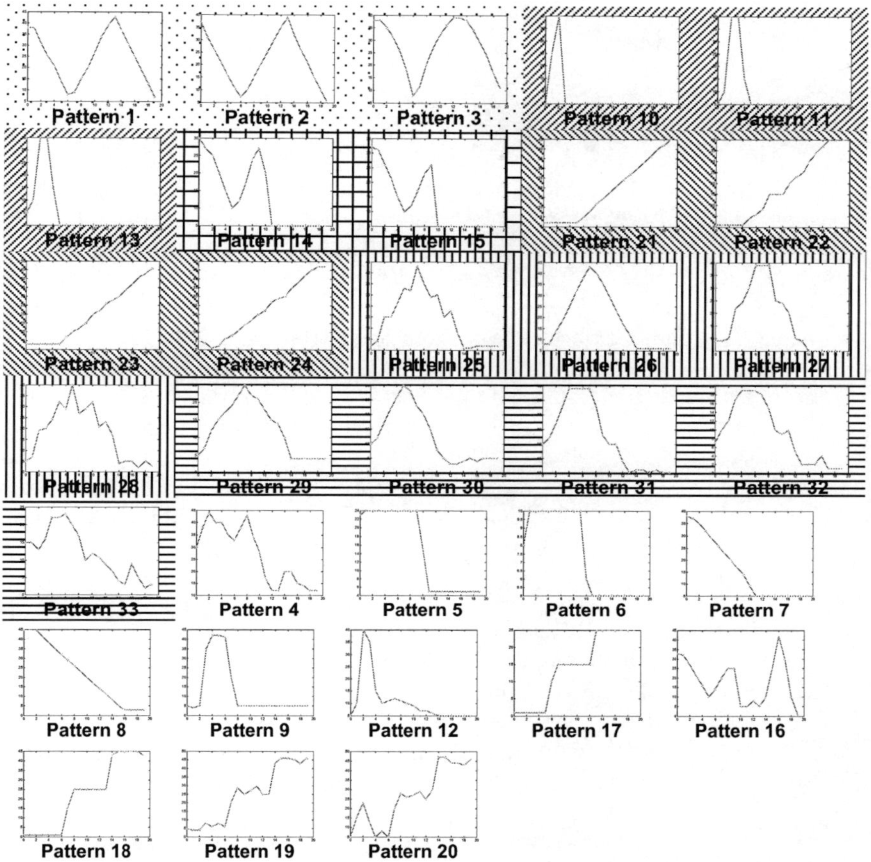

Figure 3. Six different clusters are formed as indicated with same pattern borders. Clusters with single element are marked with white border.

CONCLUSIONS

The main purpose of this paper was to develop a model which is able, in real time, to classify different patterns with similar characteristics into groups and then assigns a new pattern to one of these groups. This model is realized with a set of ODEs that can be easily implemented on the analog type devices to take advantage of high-speed electronics or photonics technologies.

REFERENCES

R. Athinarayanan, M. R. Sayeh and Dale A. Wood, " Adaptive Competitive Self-Organizing Associative Memory," *IEEE Transactions on Systems, Man, and Cybernetics*, Vol. 32, No. 4, pp. 161-471, July 2002.

G. A. Carpenter and S. Grossberg, "Adaptive Resonance Theory," in *The Handbook of Brain Theory and Neural Networks, Second Edition* Michael A. Arbib, Editor Cambridge, Massachusetts: MIT Press, 2002.

G. A. Carpenter and S. Grossberg, "The ART of Adaptive Pattern Recognition by a Self-Organizing Neural Network," IEEE Computer, Vol. 21, no. 3, pp. 77-88, March 1988.

S. Grossberg, "Adaptive Pattern Classification and Universal Recoding, II: Feedback, Expectation, Olfaction, and Illusions," *Biological Cybernetics*, Vol. 23, pp. 187-202, 1976.

J. J. Hopfield, "Neural Networks and Physical Systems with Emergent Collective Computational Abilities," Proceedings of the National Academy of Sciences of the U.S.A., Vol. 79, pp. 2554-2558, 1982.

T. Kohonen , "Self-organized Formation of Topologically Correct Feature Maps," *Biological Cybernetics*, Vol. 43, pp. 59-69, 1982.

M. R. Sayeh , R. Athinarayanan, and M.R. Zargham, "Operation of an Associative Memory"," *Pattern Recognition*, Vol. 27, No. 12, pp. 1815-1821, 1994.

ASSOCIATION-BASED IMAGE RETRIEVAL

ARUN KULKARNI
Computer Science Department
The University of Texas at Tyler, Tyler, TX 75799
Arun_kulkarni@UTTyler.edu

ABSTARCT
Typical content-based image retrieval (CBIR) system captures image features that represent image properties such as color, texture, or shape of objects in the query image and try to retrieve images from the database with similar features. In this paper, we propose a new approach for image storage and retrieval called association-based image retrieval (ABIR). We try to mimic human memory. We use a generalized bi-directional associative memory (GBAM) to store associations between feature vectors that represent images stored in the database. As an illustration, we have considered a database with three sets of images. The results of our simulation are presented in the paper.

INTRODUCTION

The rapid growth in the number of large-scale image repositories in many domains such as medical image management, multimedia libraries, document archives, art collection, geographical information the systems, law enforcement management, environmental monitoring, biometrics, and journalism has brought the need for efficient CBIR systems. Many ideas from fields such as the computer vision, database, image processing, and information retrieval are used in CBIR. Effective retrieval of image data is an important building block for general multimedia information management. For an image to be searchable, it needs to be indexed by its content. Earlier CBIR systems used keyword annotation for indexing images in the database. The keyword annotation method involves a large amount of manual effort. Furthermore, the keyword annotation depends upon human interpretation of image content, and it may not be consistent. To overcome these difficulties automatically extracted low-level features such as color, texture, and shape features are used for image indexing. Many CBIR systems with low-level features have been developed. Swain and Ballard (1991) used color histograms for indexing images. Chang and Kuo (1998) proposed the use of texture features via the wavelet transform. Shape features based on Fourier descriptors, moment invariants also have been used in conjunction with color, and texture features in CBIR systems (Shekhar and Chaudhuri, 2006). Krishnapuram et al. (2004) have proposed a fuzzy approach for CBIR. It may be noted that CBIR does not rely on describing the content of the image in its entirety. It may be sufficient that a retrieval system presents similar images in some user-defined sense. Smelders et al. (2000) consider the description of image content in two steps. The first step deals with image processing operations that transpose the image data array into another spatial data array that can include methods over local color, local texture, or local geometry. The second step is to extract invariant features. Rui et al. (1998)

proposed a relevance feedback based interactive retrieval approach. In their approach during the retrieval process, the user's high-level query and subjectivity are captured by dynamically updated weights based on the user's feedback. In this paper, we propose a new approach for image retrieval called association-based image retrieval (ABIR). We try to mimic the human brain. Association is one of the fundamental characteristics of the human brain. The human memory operates in associative manner; that is, a portion of recollection can produce an associated stream of data from the memory. The human memory can retrieve a full image from a partial or noisy version of the image as the query image. Furthermore, given a query image as the input, the human brain can recall associated images that have been stored in the past. The human memory can respond to abstract queries. For example, if we see an image of a person, we can recall images of his house, spouse, and car. The associative storage and retrieval mechanism is not explored in the present CBIR systems. We propose architecture for the ABIR system that is based on a generalized bi-directional associative memory (GBAM). The rest of the paper is organized as follows: In Section 2, we provide the overview of CBIR systems. In Section 3, we describe the architecture of the proposed ABIR system, and introduce generalized bi-directional associative memory models. In Section 4, we present results of our simulation, and Section 5 deals with conclusions.

BACKGROUND

Stages in a typical CBIR system include annotation, preprocessing, and feature extraction. In order to store an image in the database, it is first annotated. In the feature extraction stage, features based on attributes such as the color, texture, or shapes are extracted. Most CBIR systems use color histograms to compare the query image with images in the database. Often color histograms alone are not sufficient to retrieve desired images from the database, because a single color histogram may represent multiple images in the database. Many CBIR systems use texture or shape features in conjunction with color features. In order to retrieve similar images from the database the feature vector obtained from the query image is compared with feature vectors of images in the database. The most commonly used similarity function is the Minkowski distance that is given by Equation (1).

$$d\left(\mathbf{f}_q,\mathbf{f}_d\right)=\left[\sum_{j=1}^{n}\left|f_{qj}-f_{dj}\right|^2\right]^{\frac{1}{2}} \tag{1}$$

where n is the number of features. The query image is compared with images in the database and the images in the database are ranked based on the similarity measure. Images that are similar to the query image are retrieved and displayed. While it is feasible to retrieve a desired image from a small collection by exhaustive search, more effective techniques are needed with a large database. The most commonly used technique for efficient retrieval is the indexing technique. Other features of a typical CBIR system include defining query feature space and displaying query results. The first component of the query space is the selection of a subset of images from the large image archive. In order to reduce the query space images in the database can be clustered into a small number of categories.

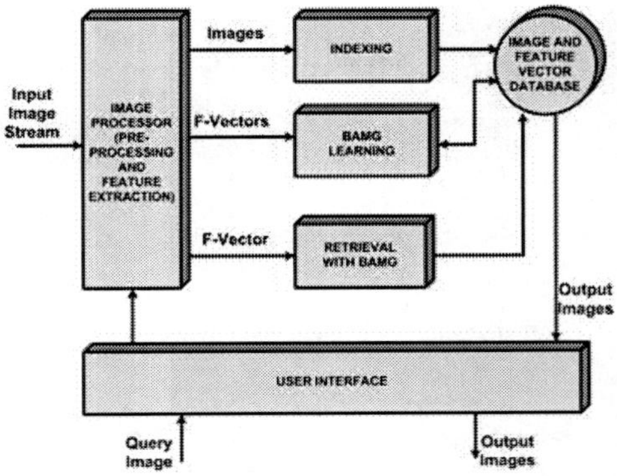

Figure 1. Architecture for an association-based image retrieval system

METHODOLOGY

Architecture for the proposed ABIR system is shown in Figure 1. There are two data paths in the system. One deals with storing a set of images in the database, while other path deals with a query image. All images stored in the database are passed through the image-processing block that deals with preprocessing and feature extraction. The preprocessing stage deals with geometric and radiometric corrections, mapping the image from red, blue, green (RBG) color space to hue, saturation, and intensity (HIS) color space. Extracted features represent color, texture, and shape characteristics of objects in the image. The indexing stage deals with creating the index table that contains feature vectors and pointers to the corresponding image locations. For faster retrieval, images in the database can be grouped in predefined categories. The user can select a query image via the user interface. The query image undergoes the same processing as the stored images. The feature vector obtained from the query image is used as the input to the GBAM, which produces the feature vectors of associated images as the output. The associated images are then retrieved from the databases and presented to the user via the user interface. The most important feature of the proposed ABIR system is that we use associative memory for retrieval of images. The GBAM is used to store association between feature vectors of associated images. The basic functions of the GBAM are to store associative pairs through a self-organizing process and to produce an appropriate response pattern on receipt of the associated stimulus input pattern. The stimulus and response patterns represent feature vectors corresponding to the query image and output images. Many associative memory models have been proposed to simulate human memory. However, the potential of these models has not been explored for CBIR. Linear associative memories have been studied extensively by Kohonen (1988). Bidirectional associative memories (BAMs) have studied by Kosko (1988). A two-layer network with feedback that simulates a BAM is shown Figure 2. The network is designed to store pairs of stimulus vectors $\{\mathbf{x}_1, \mathbf{x}_2, \ldots \mathbf{x}_n\}$ and response vectors $\{\mathbf{y}_1, \mathbf{y}_2, \ldots \mathbf{y}_n\}$. In an auto-associative network, the response vector $\mathbf{y}_i$ and the corresponding stimulus vector $\mathbf{x}_i$ are the same. In a hetero-associative memory, the stimulus and the response vector are not identical. Associative memories are often able to produce correct response patterns even though stimulus patterns are distorted or incomplete. Conventional BAMs are used to store are retrieve pairs of stimulus and response patterns. However, if the number of associated inputs and/or outputs is more than two; that is, instead of pairs of vectors, if we want to store triplets or quadruplets of vectors, then we need to use generalized BAMs. Pairs of vectors $(\mathbf{x}_i, \mathbf{y}_i)$ can be stored with the BAM by summing bipolar correlation matrices.

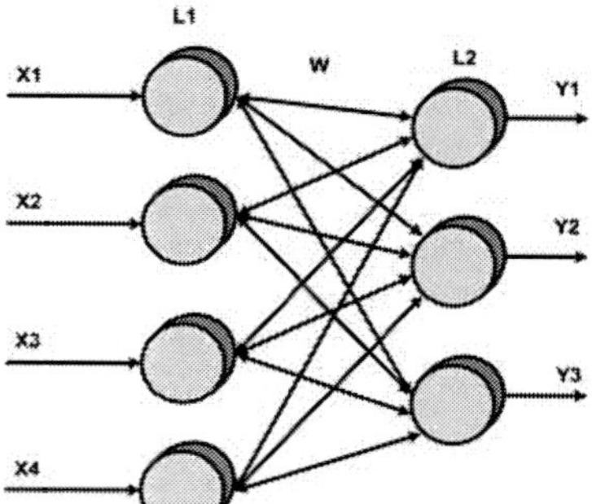

Figure 2. Bi-directional associative memory

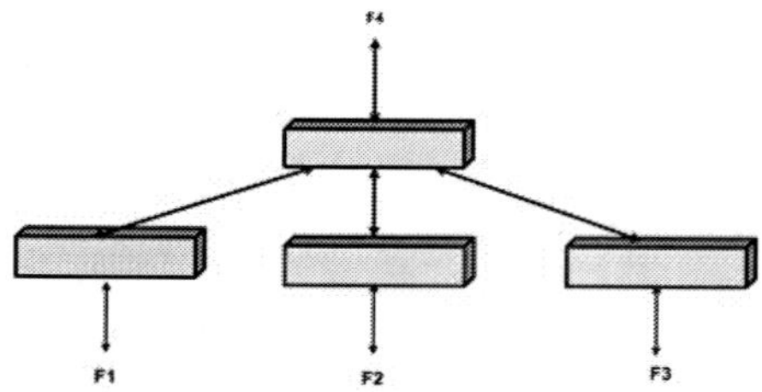

Figure 3. GBAM- Tree structure

If the input vectors are orthonormal then the recall is perfect.

$$\mathbf{x}_i \mathbf{x}_j = \begin{cases} 1 \text{ for } i = j \\ 0 \text{ for } i \neq j \end{cases} \tag{2}$$

If the input vectors are not orthonormal, then the output vector may contain cross talk. In a dual BAM, feedback is achieved with $\mathbf{W}^{\mathrm{T}}$ and is given by

$$\mathbf{W}^{\mathrm{T}} = \sum_{i=1}^{N} \left(\mathbf{y}_i \mathbf{x}_i^{\mathrm{T}}\right)^{\mathrm{T}} = \sum_{i=1}^{N} \mathbf{x}_i \mathbf{y}_i^{\mathrm{T}} \tag{3}$$

If we assume a nonlinear transfer function for neurons in the BAM, then the recalled output is a nonlinear function of a transformed input vector and is given by

$$\mathbf{y}_i = F(\mathbf{W}\mathbf{x}_i) \tag{4}$$

With the feedback the input vector $\mathbf{x}_i$ can be estimated as

$$\mathbf{x}_i = F(\mathbf{W}^T \mathbf{y}_i) \tag{5}$$

The simplest transfer function for the BAM is a step function. The stable reverberation corresponds to the system energy local minimum. When the BAM neurons are activated, the network quickly evolves to a stable state of two-pattern reverberation or a non-adaptive resonance. Humpert (1990) has suggested generalization of the BAM that can store associated multiple input/output patterns. We suggest three models of the GBAM with topologies such as the bus, ring, and tree. The GBAM with tree topology is shown in Figure 3.

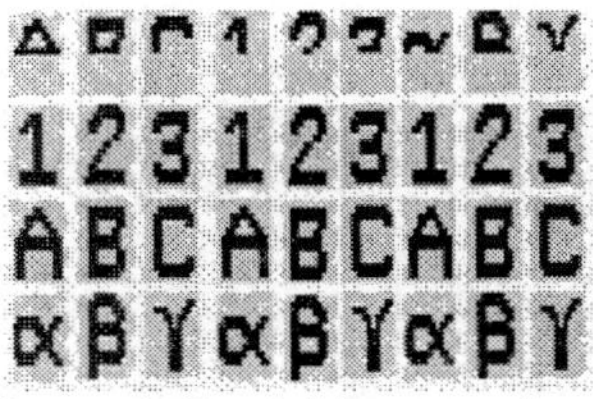

Figure 4. Partial input and recalled images

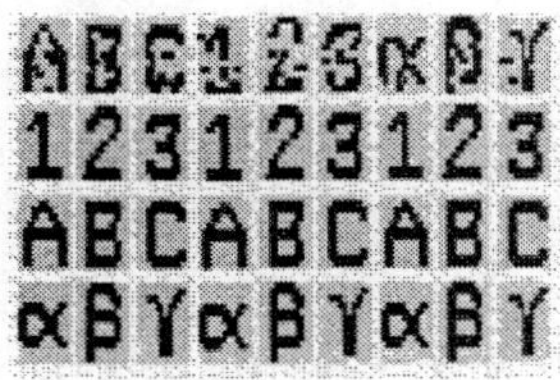

Figure 5. Noisy input and recalled images

Figure 6. Query and retrieved images

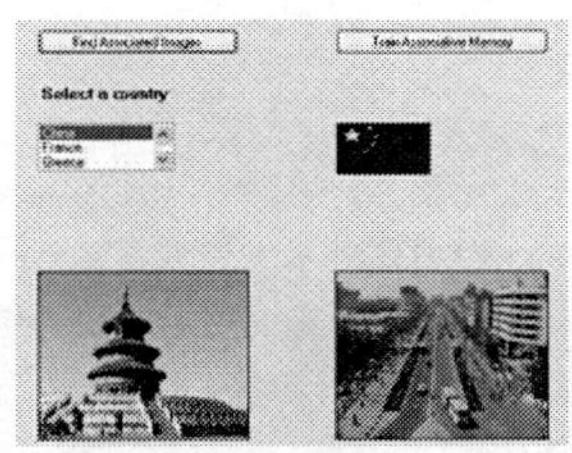

Figure 7. Query and retrieved images

RESULTS

We have developed software to simulate the ABIR system. In the first example we have consider three sets of images (Kulkarni, 2001). The first set contained images of numbers {1, 2, 3}, the second set contained images of characters {A, B, C}, and the third set contained images of Greek characters {α, β, γ}. Each character was represented by a 12x12 matrix. Each image was represented by a feature vector of 144 elements. These feature vectors are stored in the GBAM. During retrieval, any one image (partial or noisy) from any set was used as the query image, and the corresponding images from the other sets were retrieved. The retrieved images are shown in Figures 4 and 5. The first row shows the query images and subsequent rows show corresponding retrieved images. The GBAM stores and recalls feature vectors. In the present simulation, we used a tree topology for storing three sets of images.

In the second example, the database contains three sets of images. The first set contains images of flags of various countries, the second set contains images of monuments, and the third set contains images of outdoor scenes. We used the

GBAM to retrieve images from the database. We assigned simulated feature vectors to index images in the database. The retried images are shown in Figures 6 and 7.

CONCLUSIONS

In this paper we have suggested a new approach for image retrieval. We have suggested an association-based image retrieval (ABIR) system. CBIR systems retrieve images based on low-level features, whereas in ABIR systems expert's knowledge in the form of associations between images is encoded and used to retrieve images. The suggested approach is useful in many practical applications such as biometrics and automatic target recognition.

References

Chang, T. and Kuo, C. C. J., 1998. Texture analysis and classification with tree-structured wavelet transform, *IEEE Transactions on Image Processing*, vol. 2, no. 4, pp 429-441.

Humpert, B., 1990. Bi-directional associative memory with several patterns, *Proceedings of International Joint Conference on Neural Networks,* vol. 1, San Diego, pp. 741-750.

Kohonen, T., 1988. *Self-Organization and Associative Memory.* Springer-Verlag, Berlin.

Kosko, B. 1988. Bi-directional associative memories, *IEEE Transactions on Systems, Man and Cybernetics* vol. 18, pp 49-60.

Krishnapuram, R. Medasani, S. Jung, S., Choi, Y., and Balasubramanian, R., 2004. Content-based image retrieval based on a fuzzy approach, *IEEE Transactions on Knowledge Engineering,* vol. 16, no. 10, pp 1185-1199.

Kulkarni, A. D., 2001. *Computer Vision and Fuzzy-Neural Systems*. Prentice Hall, Upper Saddle River, NJ.

Rui, Y, Huang, T. S., Ortega, M., Mehrotra, S., 1998. Relevance feedback: A power tool for interactive content-based image retrieval. *IEEE Transactions on Circuits and Systems for Video Technology*, vol 8, no. 3, pp 644-655.

Smeulders, A. W., Worring, M., Santini S., Gupta, A. and R. Jain, R., 2000. Content-based image retrieval at the end of the early years. *IEEE Transactions on Pattern Recognition and Machine Learning,* vol. 22, no. 12, pp 1349-1380.

Shekhar R. and S. Chaudhuri, S., 2006. Content-based image retrieval in presence of foreground disturbances, *Proceedings of EE, Vision Signal Processing*, vol. 153, no. 5, pp 625-638.

Swain M. J.and Ballard, D. H., 1991. Color indexing, *International Journal of Computer Vision*, vol. 7, no. 1, 1991, pp 11-32.

ADAPTIVE GRAYSCALE MORPHOLOGICAL OPERATORS FOR IMAGE ANALYSIS

Z. CHENG
Dept. of Soil & Atmospheric Sci.
University of Missouri
Columbia, MO 65211

S.H. ANDERSON
Dept. of Soil & Atmospheric Sci.
University of Missouri
Columbia, MO 65211

ABSTRACT

Grayscale morphological operators are often used for image structure analysis. Normally these operators use the maximum (dilation) or minimum (erosion) of nearest neighbor values. This study introduces the use of an adaptive structural element for use as a grayscale morphological operator in image analysis. The structural element introduces changes with grayscale pixel value. Simple and more complex synthetic sets are presented along with their dilation and erosion analyses. This adaptive operator permits calculation of the image surface area with changing resolution. The surface area is estimated from the adaptive operator through dilation or erosion with differencing of constructed images. Results illustrate the ability to use a signature or multi-fractal function to show effects of particle distributions. The rate of change of the surface area is used to estimate the multi-fractal dimension (signature) as a function of changing resolution. Results appear promising for use in evaluating size and distribution of particles. Future applications of this method include quantifying and analyzing image features in geologic materials.

INTRODUCTION

Image analysis often extracts features of interest, evaluates them and classifies them to provide information on differences among images. These techniques can be used to identify specific image objects. One major challenge in image analysis is to find the best method to express the features of interest. Analysis technique selection depends on the investigator's interests and needs. Image analysis methods include texture analysis and segmentation {[1] and [2]}. These researchers have illustrated methods for granulometry, size distribution and spatial size distribution of objects. Their studies used probability descriptors of shapes and textures to describe image objects.

Further techniques are needed in evaluating images which can handle variable grayscales. The current study focuses on introducing a new method for grayscale image analysis, in which features of object structures are calculated based on the change in their properties with changing resolution (not resolution of imaging equipment but differences in image filtering). The area of the gray level surface is measured for several resolutions. This gray-scale surface area decreases at coarser resolutions since fine details that contribute to the area disappear. Signatures of the image are computed from the rate of decrease in gray level surface area compared to change in resolution. A description of the method used to calculate the rate of decrease was introduced a few years ago [3].

The objective of this paper is to provide an overview of the theory of the use of an adaptive grayscale morphological operator. In the next section (I), the mathematical definitions for signatures will be presented. In the subsequent section (II), procedures for computing the signature or feature measurements for an image by the adaptive morphological operators will be shown.

I. MATHEMATICAL DEFINITIONS OF SIGNATURES

A. Generated Upper and Lower Surfaces. In this section, computational definitions of the signatures will be developed while the next section (II) will illustrate more details of the method. We begin with an image containing grayscales referred to as function *g*. This grayscale image is represented by $g(i, j)$ where *i* and *j* are image pixel locations. To generate two new surfaces both a distance ε from the original surface, we consider all points in *3-D* space a distance $+/-\varepsilon$ from the original by covering the original surface both above and below. The new surfaces are referred to as the upper $[u_k(i, j)]$ and lower $[l_k(i, j)]$ surfaces where *k* increases by integer values and the distance from the original is $k*\varepsilon$. The initial upper and lower grayscale surfaces for $k=0$ are equal to the original surface

$$u_0(i, j)=l_0(i, j)=g(i, j). \tag{1}$$

The new upper and lower surfaces for $k>0$ are obtained using Eqs. (2) to (4). Upper surfaces are defined by Eq. (2) and constrained by Eq. (3)

$$u_k(i,j)=max\{u_{k-1}(i,j)+\varepsilon, u_{k-1}(m,n)\} \tag{2}$$

$$|(m,n)-(i,j)|\leq 1 \tag{3}$$

where (m, n) are pixel locations a distance 1 pixel from location (i, j) and taken to be the four nearest neighbors for pixel location (i, j). Thus, the maximum function operates for five pixels. Lower surfaces are defined by Eq. (4) and constrained by Eq. (3)

$$l_k(i,j)=min\{l_{k-1}(i,j)-\varepsilon, \; l_{k-1}(m,n)\} \tag{4}$$

The minimum function also operates on five pixels. The maximum and minimum operators smooth the new surfaces to some extent relative to the original. In essence, the above relations illustrate that we are taking a maximum/minimum in a radius around a point.

B. Surface Area Estimation. The area of the image can be obtained using three different approaches. One of these is using the space occupied between two upper surfaces (u_k and u_{k-1}). The volume between these upper surfaces is calculated from the following:

$$v_U(k) = \Sigma\Sigma[u_k \; (i,j) - u_{k-1} \; (i,j)]. \tag{5}$$

The summation is over (i, j) for all pixels. The surface area $[A_U(k)]$ of the image as a function of *k* is then obtained:

$$A_U(k)= v_U(k)/\varepsilon. \tag{6}$$

Another method for estimating surface area uses the space occupied between two lower surfaces (l_k and l_{k-1}). The volume between these lower surfaces is calculated from the following:

$$v_L(k) = \Sigma\Sigma[l_{k-1} \; (i,j) - l_k \; (i,j)]. \tag{7}$$

The surface area $[A_L(k)]$ of the image as a function of *k* is then obtained:

$$A_L(k)= v_L(k)/\varepsilon. \tag{8}$$

The third method is an average of the upper and lower surface estimates. The surface area $[A(k)]$ as a function of *k* is:

$$A(k)=[A_U(k) + A_L(k)]/2. \tag{9}$$

C. Mathematical Definition of Signatures. The theoretical area $A(k)$ of the gray scale surface behaves according to the expression

$$A(k) = Fk^{2-D} \tag{10}$$

where $A(k)$ is the area of the gray-scale surface representation of the image, and *F* is a constant signifying the surface area when the dimension is two. The multi-fractal function $SIG(k)$ which we will refer to as a signature in this paper is the measurement of an image surface area, where *k* is the integer, changing from 1 to the upper limit. To determine the signature $SIG(k)$, $A(k)$ is plotted vs. *k* on a log-log scale for all *k*, where the signature $SIG(k)$ is computed for each *k* by finding the slope of the best fit straight line through three points: *{[log(k -1), log (A(k -1))], [log(k), log (A(k))], [log(k +1), log (A(k +1))]}*. The signature will be determined using

$$SIG(k) = 2 - Slope(k) \tag{11}$$

where the slope is a function of *k*. The features of the multi-fractal function or *SIG(k)* will have a constant value for images which are fractal and variable values for non-fractal images. The definitions for three different estimates of signatures are similar but use different surface area estimates: $A_U(k)$ for SIG_U, $A_L(k)$ for SIG_L and $A(k)$ for SIG_C where SIG_U, SIG_L, and SIG_C are the upper, lower and combined signatures, respectively.

II. PROCEDURES TO ESTIMATE SIGNATURES USING IMAGE PROCESSING

A. Adaptive Grayscale Morphological Operators for Dilation and Erosion. To understand the operations used to obtain upper and lower grayscale surfaces, the derived surfaces can be considered as obtained by dilation for an upper surface and erosion for a lower surface. These techniques come from mathematical morphological analysis [4]. The basis comes from set theory where the main *Set A* can be the overall grayscale image and *Set B* is a structuring element used to filter the image. The image is filtered using the structuring element; and each time *Set A* is filtered with *Set B*, it becomes a new surface *(Set A')* that is used for subsequent filtering.

The structuring element shown in Fig. 1a is a common filter used in dilation. The dilation procedure is to weight the center and four nearest neighbors by a factor of 1 and to select the maximum gray-scale value and replace the center pixel with this new value. This is done for each pixel in the image.

We start with a grayscale surface that is to be dilated. The dilation is done by use of a structuring element illustrated in Figure 1b. However, the structuring element for the operation is not a constant element in this case. Work in this study suggests a deviation from Eq. (2) in that the center element is no longer weighted equally to its nearest neighbors. This results in an adaptive morphological operator rather than a regular grayscale operator.

An example of the use of the adaptive morphological operator with a simple binary image (Fig. 2a) is illustrated in Fig. 2b. It is noted that the original black/white binary image becomes a grayscale image as the adaptive dilation operator is used. The center of the binary image is unchanged since it is already at the maximum value (white is relative maximum and black is relative minimum). Only the outer edges of the center section are altered during dilation. This operation tends to increase the values for pixels on the edge of objects in the image that are at a higher value relative to their surroundings. This means that they will have no effect when dealing with a maximum operator inside the object since it is at a maximum.

The erosion of image surfaces is similar to dilation processes. The structuring element used for erosion is illustrated in Figure 1c. An example of the use of the adaptive erosion operator is illustrated in Fig. 2c. The original binary image (Fig. 2a) changes to a grayscale image with erosion. The outside sections remain unchanged since these pixels are already at the minimum value; only the edges of the inside image section are altered during erosion. This operation tends to reduce the values of pixels within the center edges of objects in the image that are at a higher value relative to their surroundings. However, outside the object the operator has no effects since these pixel values are already at the minimum value.

The dilation and erosion operations provide nonlinear image transformations that locally modify geometric objects in images. The minimum and maximum filters are performed and result in the grayscale erosion and dilation. These operations use a structuring element to interact with the set and extract information from images or image objects. Applications of the adaptive structuring element can establish both sizes and perimeters of particles and voids.

B. Surface Area Determined from Image Differencing. Figure 3 shows the images obtained by differencing with the upper surfaces (Fig. 3a), with lower surfaces (Fig. 3b), and combined upper and lower surfaces (Fig. 3c). Image differences with dilation (Fig. 3a) characterize the progression of each step of the outside boundary of the original object (Fig. 2a). Note that saturation of a 15 x 15 image occurs at k=8 which is the point at which the migration of the object moves to the edge of the image. Saturation is the integer value at which differences in images are equal to ε. Differences with erosion (Fig. 3b) characterize the progression of the inside boundary of the original object. Saturation occurs with erosion at k=4 which is the point at which the shrinkage of the object moves down to one pixel. Image differences with a combined erosion and dilation (Fig. 3c) characterize the progression of both the outside and inside boundaries of the object at the same time.

0	1	0
1	1	1
0	1	0

a

0	1	0
1	1+ ε/g	1
0	1	0

b

0	1	0
1	1- ε/g	1
0	1	0

c

Fig. 1. Structuring elements for dilation and erosion used in derivation of surfaces. (a) Element used for regular dilation and erosion, (b) Element used for dilation in this study, and (c) Element used for erosion in this study. ε is defined as the distance of separation for the new surface with the previous surface and g is the grayscale value of the previous surface.

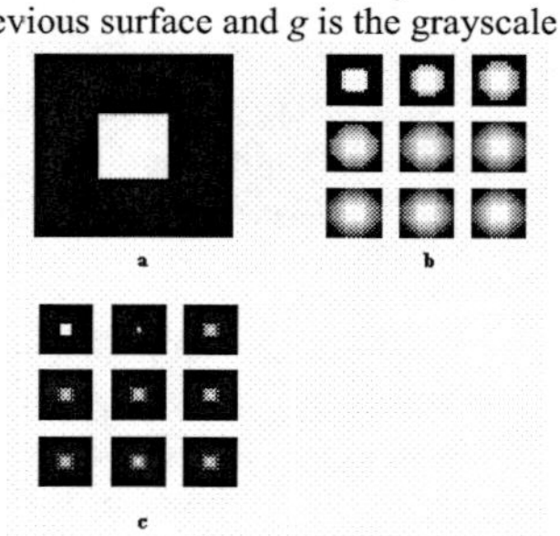

Fig. 2. Illustration of new images created by dilation and erosion using adaptive operators. (a) Original image for creating new surfaces, image size is 15 x 15 pixels. (b) Illustration of new images created by dilation, order of sub-figures is from left to right and top to bottom with k=1 through k=9 with image sizes of 15 x 15 pixels for sub-figures. (c) Illustration of new images created by erosion, order of sub-figures is from left to right and top to bottom with k=1 through k=9 with image sizes of 15 x 15 pixels for sub-figures.

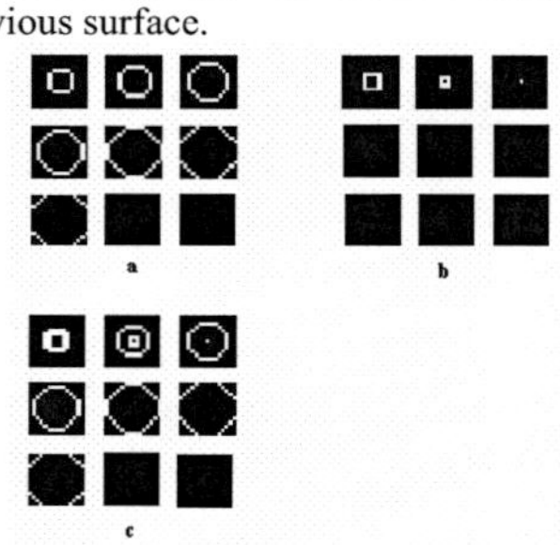

Fig. 3. Illustrations of images created by differencing. Prior image iteration is subtracted from new dilated or eroded image created by the adaptive operator with an original binary image (Fig. 2a). (a) Image differences created with dilated images. (b) Image differences created with eroded images. (c) Image differences created from combined (addition) dilated and eroded images. Order of sub-figures is from left to right and top to bottom. For dilation, the first subfigure is dilated image at k=1 minus original image, the second sub-figure is image at k=2 minus image at k=1, the third sub-figure is dilated image at k=3 minus image at k=2, etc. The same pattern was used for erosion and combined images.

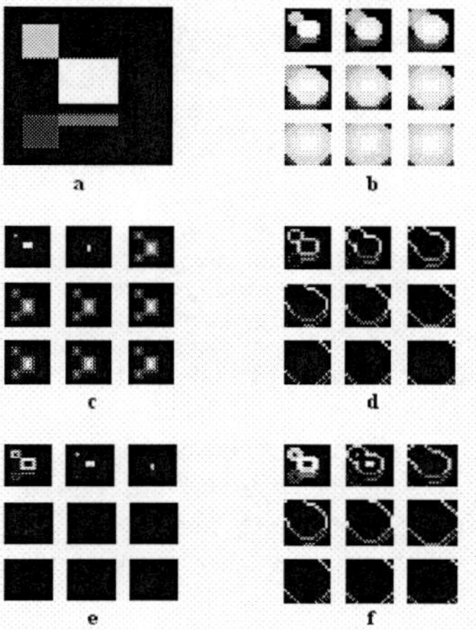

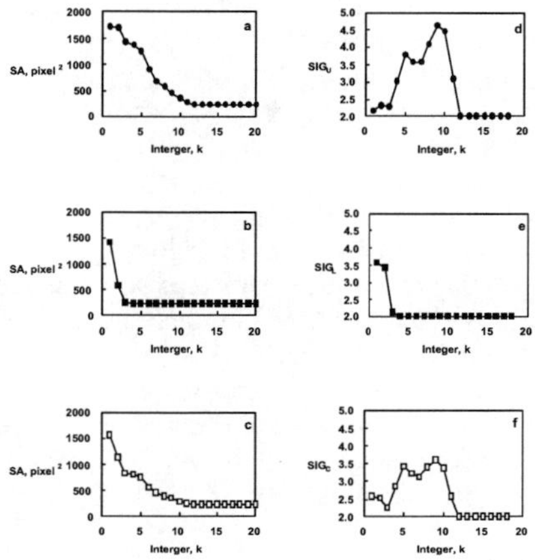

Fig. 4. Illustration of changes in images due to use of the adaptive operators with a gray-scale image. (a) Original gray-scale image with 6 levels and image size of 15 x 15 pixels, (b) images created by dilation, (c) images created by erosion, (d) images created by difference of prior image from new dilated image, (e) images created by difference of prior image from new eroded image, (f) images created by difference of prior image from combined new eroded and dilated image. Order of sub-figures is from left to right and top to bottom with k=1 through k=9 and image size of sub-figures is 15 x 15 pixels.

Fig. 5. Estimated surface area (SA) as a function of k for the (a) upper, (b) lower, and (c) combined surfaces corresponding to Figs. 4d-f, respectively. Signatures (SIG) determined for the (d) upper (SIG_U), (e) lower (SIG_L), and (f) combined (SIG_C) surfaces from the surface area curves in Fig. 5a-c.

The surface area of the images created by differencing is the summation of all of the pixel values of that image divided by ε {Eqs.(5) through (8)} for the upper and lower surfaces. For the kth value, the surface area characterizes features of the embedded object within an image. At small values of *k* or high frequency, the surface area is dependent on fine details of the object. At higher values of *k* or low frequency, the surface area is dependent on coarse details of the object.

In Figure 4, the same processes as illustrated in Figures 2 and 3 were performed for an image that contained five objects and six levels of grayscale (Fig. 4a). Dilation and erosion processes of the image are shown in Figs. 4b and 4c. Images created by differencing operations for the dilated and eroded images are illustrated in Figs. 4d and 4e. The differencing operation with dilated images tends to merge objects together and remove the differences between objects. This differencing operation with eroded images tends to make objects in the image smaller, separates objects that are touching, and removes isolated pixels (Fig. 4e). The differences for combined dilated and eroded images are shown in Fig. 4f.

The estimated surface areas as a function of *k* for the upper, lower, and combined images created by differencing in Figures 4e-f are illustrated in Figures 5a-c. All estimates of surface area decrease in value as a function of *k*, due to the loss of details with lower frequency. For the original image in Fig. 4a, the estimate of surface area for the eroded or lower image decreased more rapidly than the dilated or upper image (Fig. 5b). This was due to the rapid saturation of the eroded images at k=4 (Fig. 4e). Figures

5a-c show that the measured surface area characterizes the following parameters: (i) the perimeters of individual objects, (ii) the sizes of objects in the image, and (iii) the measured surface areas.

C. Signatures Computed from the Change in Surface Area. Signatures for the upper, lower and combined images are shown in Figures 5d-f. These signatures were computed from the surface area curves. Data indicate that the upper signature characterizes the spatial distribution of objects, the lower signature characterizes the size of objects, and the combined signature is a combination of both. Thus, these curves are helpful in describing different features of image objects.

For non-fractal surfaces such as a binary image and when the scale range and magnification factors are appropriate, the SIG_C, SIG_U, and SIG_L are positive and increase to a peak (the highest value of D), then decrease to 2.0, corresponding to the global value when the scales become comparable with the surface heights. The shapes of these signature curves depend on the black vs. white area and their distributions, where the black area represents the pixels with low value (pore) and the white area represents the pixels with high value (solid).

CONCLUSIONS

This study illustrates the details of using an adaptive grayscale morphological operator for evaluation of two synthetic data sets. Simple and more complex synthetic sets are presented along with their dilation and erosion analyses. The adaptive operator calculates the image surface area with changing resolution. The surface area is estimated from the adaptive operator through dilation or erosion with differencing of constructed images. Results illustrate the ability to use a signature or multi-fractal function to show effects of particle distributions [5]. The rate of change of the surface area is used to estimate the multi-fractal dimension (signature) as a function of changing resolution. Potential applications of using these signature curves exist for evaluating materials containing pores such as earth materials [6]. The upper signature describes pore features through dilation processes while the lower signature describes solid features through erosion processes. The combined signature describes a combination of both pores and solids. The adaptive morphological operators used in this study have promise as a new technique for image analysis of earth and soil materials [7].

REFERENCES

[1] G. Ayala and J. Domingo, "Spatial size distributions: Applications to shape and texture analysis", *IEEE Trans. Pattern Analysis and Machine Intelligence*, Vol. 23, no.12, pp. 1430-1442, Dec. 2001.

[2] R.C. Gonzalez and R.E. Woods, *Digital image processing*, chapter 8, pp. 558-560, Addison-Wesley, 1992.

[3] S. Peleg, J. Naor, R. Hartley, and D. Avnir, "Multiple resolution texture analysis and classification", *IEEE Trans. Pattern Analysis Machine Intelligence*, vol. 6, no. 4, pp. 518-523, 1984.

[4] J. Serra, *Image analysis and mathematical morphology*. Academic Press, 1982.

[5] F. Albregtsen, B. Nielsen, and K. Yogesan, "Fractal dimension, only a fraction of the truth?", *Proc. 11th Int'l Conf. Pattern Recognition,* IEEE Comput. Soc. Press, vol. 3, pp. 733-736, 1992.

[6] S. Talibuddin and J.P. Runt, "Reliability test of popular fractal techniques applied to small two-dimensional data sets", *J. Applied Physics*, vol. 76, pp. 5070-5078, 1994.

[7] Z. Cheng, S.H. Anderson, C.J. Gantzer, and Y. Chu, "Fuzzy logic for predicting soil hydraulic conductivity using CT images", *Intelligent Engineering Systems Through Artificial Neural Networks* vol. 11, pp. 307-312, 2001.

A DESIGN METHOD FOR STATE OBSERVERS FOR TIME-DELAY PLANTS

KOU YAMADA
Graduate School of Engineering, Gunma University
1-5-1 Tenjincho, Kiryu, Japan

MASAHIKO KOBAYASHI
Department of Mechanical System Engineering, Gunma University
1-5-1 Tenjincho, Kiryu, Japan

ABSTRACT
In this paper, we examine a design method for state observers for time-delay plants. State observers are initially proposed by Luenberger and are used to estimate the state variable of the plants. Several papers on design methods of state observer have been published. However, no paper examines the parametrization of all state observers for time-delay plants. In this paper, we propose the parametrization of all state observers and linear observers for time-delay plants.

Introduction

In this paper, we examine a design method for state observers for time-delay plants. State observers are initially proposed by Luenberger(Luenberger, 1964; Luenberger, 1966; Luenberger, 1971). State observers are used to estimate the state variable of the plants when the state variable of the plant is unavailable. Several papers on design methods of state observer have been published. Recently the parametrization of all state observers (Goodwin and Middleton, 1989) and of all linear observers (Ding, Guo, Frank, 1994) are obtained. However, no paper examines the parametrization of all state observers for time-delay plants.

In this paper, we propose the parametrization of all state observers for time-delay plants. First, the problem considered in this paper is described. Next, the parametrization of all state observers for time-delay plants are proposed. Next, using the obtained parametrization of all state observers for time-delay plants, we present a design method of state observer with unknown input for time-delay plants. Finally, we present the parametrization of all state observers for time-delay plants.

Problem formulation

Consider the time-delay plant written by

$$\left\{ \begin{array}{rcl} \dot{x}(t) & = & Ax(t) + Bu(t-L) \\ y & = & Cx(t) + Du(t-L) \end{array} \right. , \tag{1}$$

where $x \in R^n$ is the state variable, $u \in R^p$ is the input, $y \in R^m$ is the output, $A \in R^{n \times n}$, $B \in R^{n \times p}$, $C \in R^{m \times n}$, $D \in R^{m \times p}$, $L > 0$ is the time-delay. It is assumed that (A, B) is stabilizable and (C, A) is detectable. The transfer function from u to y in Eq.(1) is denoted by

$$y(s) = G(s)e^{-sL}u(s), \tag{2}$$

 where

$$G(s) = C(sI - A)^{-1}B + D \in R^{m\times p}(s). \tag{3}$$

When the state variable $x(t)$ is not available, in many cases, the state estimator named the state observer is used. The state observer estimates the state variable $x(t)$ in Eq.(1) using available measurement $y(t)$ and $u(t-L)$. Since the available measurements of the plant in Eq.(1) are $u(t-L)$ and $y(t)$, the general form of the state observer for Eq.(1) is written by

$$\xi(s) = F_1(s)y(s) + F_2(s)e^{-sL}u(s), \tag{4}$$

where $F_1(s) \in R^{n\times m}(s)$, $F_2(s) \in R^{n\times p}(s)$, $\xi(s) = \mathcal{L}\xi(t)$ and $\xi(t) \in R^n(t)$. In the following, we call the system in Eq.(4) state observer, if

$$\lim_{t\to\infty}(x(t) - \xi(t)) = 0 \tag{5}$$

is satisfied. The problem considered in this paper is to obtain the parametrization of all state observers in Eq.(4) for the time-delay plant in Eq.(1).

Parametrization of all state observers

In this section, we clarify the parametrization of all state observers in Eq.(4) for the time-delay plant in Eq.(1).

The parametrization of all state observers in Eq.(4) for the time-delay plant in Eq.(1) is summarized in the following theorem.

Theorem 1 *The system in Eq.(4) is state observer if and only if $F_1(s)$ and $F_2(s)$ are written by*

$$F_1(s) = (sI - A + BK)^{-1}BX(s) + Q(s)\tilde{D}(s) \tag{6}$$

and

$$F_2(s) = (sI - A + BK)^{-1}BY(s) - Q(s)\tilde{N}(s), \tag{7}$$

respectively, where $K \in R^{m\times n}$ makes $A - BK$ have no eigenvalue in the closed right half plane, $\tilde{N}(s)$ and $\tilde{D}(s)$ are coprime factors of $G(s)$ on RH_∞ satisfying

$$\begin{aligned} G(s) &= \tilde{D}^{-1}(s)\tilde{N}(s) \\ &= N(s)D^{-1}(s), \end{aligned} \tag{8}$$

$X(s) \in RH_\infty$ and $Y(s) \in RH_\infty$ are functions satisfying

$$\begin{aligned} &\begin{bmatrix} Y(s) & X(s) \\ -\tilde{N}(s) & \tilde{D}(s) \end{bmatrix}\begin{bmatrix} D(s) & -\tilde{X}(s) \\ N(s) & \tilde{Y}(s) \end{bmatrix} \\ &= \begin{bmatrix} I & 0 \\ 0 & I \end{bmatrix} = \begin{bmatrix} D(s) & -\tilde{X}(s) \\ N(s) & \tilde{Y}(s) \end{bmatrix}\begin{bmatrix} Y(s) & X(s) \\ -\tilde{N}(s) & \tilde{D}(s) \end{bmatrix} \end{aligned} \tag{9}$$

and $Q(s) \in RH_\infty^{n\times m}$ is any function.

Proof of this theorem requires following lemma.

Lemma 1 *The eight functions satisfying Eq.(9) are written by*

$$\begin{bmatrix} D(s) & -\tilde{X}(s) \\ N(s) & \tilde{Y}(s) \end{bmatrix} = \begin{bmatrix} I & 0 \\ D & I \end{bmatrix} + \begin{bmatrix} -K \\ C-DK \end{bmatrix} (sI-A+BK)^{-1} \begin{bmatrix} B & F \end{bmatrix} \quad (10)$$

and

$$\begin{bmatrix} Y(s) & X(s) \\ -\tilde{N}(s) & \tilde{D}(s) \end{bmatrix} = \begin{bmatrix} I & 0 \\ -D & I \end{bmatrix} - \begin{bmatrix} -K \\ C \end{bmatrix} (sI-A+FC)^{-1} \begin{bmatrix} B-FD & F \end{bmatrix}. \quad (11)$$

(Nett, Jacobson, Balas, 1984)

Lemma 2 *Suppose that* $A(s) \in RH_\infty^{n\times m}$, $B(s) \in RH_\infty^{q\times m}$, $C(s) \in RH_\infty^{p\times m}$, rank $\begin{bmatrix} A^T(s) & B^T(s) \end{bmatrix}^T = r$. *The equation written as*

$$X(s)A(s) + Y(s)B(s) = C(s) \quad (12)$$

has a solution $X(s) \in RH_\infty^{p\times n}$ *and* $Y(s) \in RH_\infty^{p\times q}$ *if and only if there exists* $U(s) \in \mathcal{U}$ *to satisfy*

$$\begin{bmatrix} A(s) \\ B(s) \\ C(s) \end{bmatrix} = U(s) \begin{bmatrix} A(s) \\ B(s) \\ 0 \end{bmatrix}. \quad (13)$$

When a pair of $X_0(s) \in RH_\infty^{p\times n}$ *and* $Y_0(s) \in RH_\infty^{p\times q}$ *is a solution to Eq.(12), all the solutions are given by*

$$\begin{bmatrix} X(s) & Y(s) \end{bmatrix} = \begin{bmatrix} X_0(s) & Y_0(s) \end{bmatrix} + L(s) \begin{bmatrix} W_1(s) & W_2(s) \end{bmatrix}, \quad (14)$$

where $W_1(s) \in RH_\infty$ *and* $W_2(s) \in RH_\infty$ *are rational functions satisfying*

$$W_1(s)A(s) + W_2(s)B(s) = 0 \quad (15)$$

and

$$\text{rank} \begin{bmatrix} W_1(s) & W_2(s) \end{bmatrix} = n + q - r, \quad (16)$$

$L(s) \in RH_\infty^{p\times(n+q-r)}$ *is an arbitrary stable rational function (Vidyasagar, 1985).*

Using above-mentioned lemmas, Theorem 1 is proved.
(Proof) First, necessity is shown. That is, we show that if the system in Eq.(4) satisfies Eq.(5), then $F_1(s)$ and $F_2(s)$ in Eq.(4) are written by Eq.(6) and Eq.(7), respectively. The system in Eq.(1) is rewritten by

$$\begin{cases} D(s)\hat{x}(s) &= e^{-sL}u(s) \\ y(s) &= N(s)\hat{x}(s), \end{cases} \quad (17)$$

 where $\hat{x}(s)$ satisfies

$$x(s) = (sI - A + BK)^{-1} B\hat{x}(s). \tag{18}$$

Substituting Eq.(17) into Eq.(4), we have

$$\xi(s) = F_1(s)N(s)\hat{x}(s) + F_2(s)D(s)\hat{x}(s). \tag{19}$$

When the state space description of $F_1(s)$ and $F_2(s)$ are written by

$$\begin{cases} \dot{x}_{f1}(t) = A_{f1}x_{f1}(t) + B_{f1}u_{f1}(t) \\ y_{f1}(t) = C_{f1}x_{f1}(t) + D_{f1}u_{f1}(t) \end{cases} \tag{20}$$

and

$$\begin{cases} \dot{x}_{f2}(t) = A_{f2}x_{f2}(t) + B_{f2}u_{f2}(t) \\ y_{f2}(t) = C_{f2}x_{f2}(t) + D_{f2}u_{f2}(t), \end{cases} \tag{21}$$

respectively,

$$\begin{aligned} & x(t) - \xi(t) \\ = \; & \mathcal{L}^{-1}\left[\left\{(sI - A + BK)^{-1}B - F_1(s)N(s) - F_2(s)D(s)\right\}\hat{x}(s)\right. \\ & + (sI - A + BK)^{-1}x(0) - C_{f1}(sI - A_{f1})^{-1}x_{f1}(0) \\ & \left. - C_{f2}(sI - A_{f2})^{-1}x_{f2}(0)\right] \end{aligned} \tag{22}$$

is satisfied. From the assumption that Eq.(5) is satisfied, A_{f1} and A_{f2} have no eigenvalue in the closed right half plane, that is, $F_1(s) \in RH_\infty$ and $F_2(s) \in RH_\infty$. In addition,

$$(sI - A + BK)^{-1}B - F_1(s)N(s) - F_2(s)D(s) = 0 \tag{23}$$

is satisfied. From Lemma 2, all of $F_1(s)$ and $F_2(s)$ hold Eq.(23) are written by Eq.(6) and Eq.(7), respectively, where $Q(s) \in RH_\infty^{n\times m}$. In this way, necessity has been proved.

Next, sufficiency is proved. That is, we show that if $F_1(s)$ and $F_2(s)$ are written by Eq.(6) and Eq.(7), then Eq.(4) satisfies Eq.(5). $x(t) - \xi(t)$ is written by

$$\begin{aligned} & x(t) - \xi(t) \\ = \; & \mathcal{L}^{-1}\left[\left\{(sI - A + BK)^{-1}B - F_1(s)N(s) - F_2(s)D(s)\right\}\hat{x}(s)\right. \\ & + (sI - A + BK)^{-1}x(0) - C_{f1}(sI - A_{f1})^{-1}x_{f1}(0) \\ & \left. - C_{f2}(sI - A_{f2})^{-1}x_{f2}(0)\right]. \end{aligned} \tag{24}$$

Since $F_1(s)$ and $F_2(s)$ are written by Eq.(6) and Eq.(7), after simple manipulations,

$$(sI - A + BK)^{-1}B - F_1(s)N(s) - F_2(s)D(s) = 0 \tag{25}$$

is satisfied. In addition, from Eq.(24),Eq.(25), $F_1(s) \in RH_\infty$ and $F_2(s) \in RH_\infty$, we have

$$\lim_{t\to\infty} (x(t) - \xi(t)) = 0. \tag{26}$$

We have thus proved Theorem 1.

In this section, using the result in the preceding section, we present a design method of state observer with unknown input for time-delay plants.

In some cases, $u(t)$ and $x(t)$ are unavailable. In this case, in order to estimate state $x(s)$, state observers use only $y(t)$. Such state observer is called state observer with unknown input. Using Theorem 1, when $m \geq p$ and $G(s)$ is of minimum-phase, the state observer with unknown input is designed as

$$\xi(s) = F_1(s)y(s), \tag{27}$$

where

$$F_1(s) = (sI - A + BK)^{-1} BX(s) + Q(s)\tilde{D}(s), \tag{28}$$

$$Q(s) = (sI - A + BK)^{-1}BY(s)\hat{N}(s) \tag{29}$$

and $\hat{N}(s)$ is stable rational function satisfying

$$\hat{N}(s)\tilde{N}(s) = I. \tag{30}$$

Parametrization of all linear observers

In this section, we present the parametrization of all linear observer to estimate

$$v(t) = Px(t) \tag{31}$$

for time-delay plants in Eq.(1), where $P \in R^{m_P \times n}$. Since the available measurements of the plant in Eq.(1) are $u(t-L)$ and $y(t)$, the general form of the linear observer for Eq.(1) is written by

$$\xi_P(s) = F_{1P}(s)y(s) + F_{2P}(s)e^{-sL}u(s), \tag{32}$$

where $F_{1P}(s) \in R \in R^{m_P \times m}(s)$, $F_{2P}(s) \in R^{m_P \times p}(s)$, $\xi_P(s) = \mathcal{L}\xi_P(t)$ and $\xi_P(t) \in R^{m_P}(t)$. In the following, we call the system in Eq.(32) linear observer, if

$$\lim_{t \to \infty} (v(t) - \xi_P(t)) = 0 \tag{33}$$

is satisfied. The problem considered in this section is to obtain the parametrization of all linear observers in Eq.(32) for the time-delay plant in Eq.(1).

The parametrization of all linear observers in Eq.(32) for the time-delay plant in Eq.(1) is summarized in the following theorem.

Theorem 2 *The system in Eq.(32) is linear observer if and only if $F_{1P}(s)$ and $F_{2P}(s)$ are written by*

$$F_{1P}(s) = P(sI - A + BK)^{-1} BX(s) + Q(s)\tilde{D}(s) \tag{34}$$

 and

$$F_{2P}(s) = P\left(sI - A + BK\right)^{-1} BY(s) - Q(s)\tilde{N}(s), \tag{35}$$

respectively, where $K \in R^{m\times n}$ *make* $A - BK$ *have no eigenvalue in the closed right half plane,* $\tilde{N}(s)$ *and* $\tilde{D}(s)$ *are coprime factors on* RH_∞ *of* $G(s)$ *satisfying Eq.(8),* $X(s) \in RH_\infty$ *and* $Y(s) \in RH_\infty$ *are functions satisfying Eq.(9) and* $Q(s) \in RH_\infty^{n_P\times m}$ *is any function.*

(Proof) Theorem 2 is proved using similar procedure of proof of Theorem 1. Therefore the detail of proof of 2 is omitted on account of limitted space.

Conclusion

In this paper, we proposed the parametrization of all state observers and all linear observers for time-delay plants.

References

D. G. Luenberger: Observing the state of a linear system, IEEE Trans. Mil. Electron, MIL-8, pp.74-80(1964)

D. G. Luenberger: Observers for multivariable systems, IEEE Trans. on Automatic Control, AC-11, pp.190-197(1966)

D. G. Luenberger: An introduction to observers, IEEE Trans. on Automatic Control, AC-16, pp.596-602(1971)

G. C. Goodwin and R. H. Middleton: The class of all stable unbiased state estimators, Systems & Control Letters, 13, pp.161-163(1989)

X. Ding, L. Guo and P. M. Frank: Parametrization of linear observers and its application to observer design, IEEE Trans. on Automatic Control, AC-39, pp.1648-1652(1994)

C. N. Nett, C. A. Jacobson and M. J. Balas: A connection between state-space and doubly coprime fractional representation, IEEE Trans. on Automatic Control, AC-29, pp.831-832(1984)

M. Vidyasagar: Control system synthesis–A factorization approach–, MIT Press, London, (1985)

MODIFIED PID CONTEROLLER FOR NON-MINIMUM PHASE TIME-DELAY PLANTS

KOU YAMADA
Department of Mechanical System Engineering, Gunma University
1-5-1 Tenjincho, Kiryu, Japan

TAKAAKI HAGIWARA
Department of Mechanical System Engineering, Gunma University
1-5-1 Tenjincho, Kiryu, Japan

YOSUKE SHIMIZU
Department of Mechanical System Engineering, Gunma University
1-5-1 Tenjincho, Kiryu, Japan

ABSTRACT

PID controller structure is the most widely used one in industrial applications. However no method has been published to guarantee the stability of PID control system for any time-delay plants and the admissible sets of P-parameter, I-parameter and D-parameter to guarantee the stability of PID control system are independent from each other. In this paper, we propose a design method for modified PID controllers such that the modified PID controller make the closed-loop system for any time-delay plants stable and the admissible sets of P-parameter, I-parameter and D-parameter are independent from each other.

Introduction

PID (Proportional-Integral-Derivative) controller is most widely used controller structure in industrial applications (Datta, Ho, and Bhattacharyya, 2000). Its structural simplicity and sufficient ability of solving many practical control problems have contributed to this wide acceptance. If there exists a stabilizing PID controller, the parametrization of all stabilizing PID controller is considered in Yang(1994), Ho, Datta and Bhattacharyya(1997) and Datta, Ho and Bhattacharyya(2000). However the method in Yang(1994), Ho, Datta and Bhattacharyya(1997) and Datta, Ho and Bhattacharyya(2000) remains a difficulty. The admissible sets of P-parameter, I-parameter and D-parameter in Yang(1994), Ho, Datta and Bhattacharyya(1997) and Datta, Ho and Bhattacharyya(2000) are related each other. That is, if P-parameter is changed, then the admissible sets of I-parameter and D-parameter change. From practical point of view, it is desirable that the admissible sets of P-parameter, I-parameter and D-parameter are independent from each other. Yamada and Moki initially tackle this problem and propose a design method of PI controllers for any minimum-phase systems such that the admissible sets of P-parameter and I-parameter are independent from each other (Yamada and Moki, 2003). Yamada expands the result in Yamada and Moki(2003) and propose a design method of PID controllers for minimum-phase plant such that the admissible sets of P-parameter, I-parameter and D-parameter are independent from each other (Yamada, 2005). Yamada and Hagiwara(2006) gave a design method for PID controllers for unstable plants. However, no paper examines a design method for modified PID controllers for any time-delay plants.

In this paper, we propose a design method for modified PID controllers such that the modified PID controller makes the closed-loop system stable for any time-delay plants and the admissible sets of P-parameter, I-parameter and D-parameter to guarantee the stability of closed-loop system are independent from each other.

Problem formulation

Consider the closed-loop system written by

$$\left\{ \begin{array}{rcl} y & = & G(s)e^{-sL}u \\ u & = & C(s)(r-y) \end{array} \right. , \tag{1}$$

where $G(s)e^{-sL}$ is the single-input/single-output plant, $G(s) \in R(s)$ is strictly proper function and L is the time delay. $G(s)$ is assumed to be coprime and to have no zero on the origin. $C(s)$ is the controller, $u \in R$ is the control input, $y \in R$ is the output and $r \in R$ is the reference input.

When the controller $C(s)$ has the form written by

$$C(s) = a_P + \frac{a_I}{s} + a_D s, \tag{2}$$

then the controller $C(s)$ is called the PID controller (Yang, 1994; Ho, Datta and Bhattacharyya, 1997; Datta, Ho and Bhattacharyya, 2000), where $a_P \in R$ is the P-parameter, $a_I \in R$ is the I-parameter and $a_D \in R$ is the D-parameter. a_P, a_I and a_D are settled so that the closed-loop system in Eq.(1) has desirable control characteristics such as steady state characteristic and transient characteristic. For easy explanation, we call $C(s)$ in Eq.(2) the conventional PID controller.

The problem is to propose a design method for modified PID controllers $C(s)$ to make the closed-loop system in Eq.(1) stable for any time-delay plant $G(s)e^{-sL}$ such that the admissible sets of P-parameter a_P, I-parameter a_I and D-parameter a_D to guarantee the stability of closed-loop system are independent from each other.

The basic idea

In order to design modified PID controllers $C(s)$ that can be applied to any time-delay plants $G(s)e^{-sL}$, we adopt the parametrization of all stabilizing Smith predictors for time-delay plants $G(s)e^{-sL}$. According to Yamada and Takenaga(2007), the parametrization of all stabilizing Smith predictors $C(s)$ for time-delay plants $G(s)e^{-sL}$ is written by

$$C(s) = \frac{C_f(s)}{1 - C_f(s)G(s)e^{-sT}}, \tag{3}$$

where $C_f(s)$ is given by

$$C_f(s) = \frac{1}{G_u(s)}\left(\bar{G}_u(s) + \frac{Q(s)}{G_u(s)}\right), \tag{4}$$

$G_u(s)$ is biproper minimum-phase function of $G(s)$ that is, when $G(s)$ is factorized as

$$G(s) = G_u(s)G_s(s), \tag{5}$$

$G_u(s)$ is the unstable biproper minimum-phase function and $G_s(s)$ is the stable function, $\bar{G}_u(s) \in RH_\infty$ satisfies

$$\bar{G}_u(s_i) = \frac{1}{G_s(s_i)\, e^{-s_i T}}, \tag{6}$$

$s_i (i = 1, \cdots, n)$ denote unstable poles of $G(s)$, and $Q(s) \in RH_\infty$ is any function. On the parametrization of all stabilizing Smith predictors $C(s)$ in Eq.(3) for $G(s)e^{-sL}$, the Smith predictor $C(s)$ in Eq.(3) includes free-parameter $Q(s)$. Using free-parameter $Q(s)$ in Eq.(3), we propose a design method of modified PID controllers $C(s)$ to make the closed-loop system in Eq.(1) stable and to be able to apply to any time-delay plant $G(s)e^{-sL}$. In order to design the modified PID controllers $C(s)$, the free parameter $Q(s)$ in Eq.(3) is settled for $C(s)$ in Eq.(3) to have the same characteristics to conventional PID controller $C(s)$ in Eq.(2). Therefore, next, we describe the role of conventional PID controller $C(s)$ in Eq.(2) in order to clarify the condition that the modified PID controller $C(s)$ must be satisfied. From Eq.(2), using $C(s)$, the P-parameter a_P, the I-parameter a_I and the D-parameter a_D are decided by

$$a_P = \lim_{s\to\infty} \left\{ -s^2 \frac{d}{ds}\left(\frac{1}{s}C(s)\right)\right\}, \tag{7}$$

$$a_I = \lim_{s\to 0} \{sC(s)\}, \tag{8}$$

$$a_D = \lim_{s\to\infty} \frac{d}{ds}\{C(s)\}. \tag{9}$$

Modified PID controller

In this section, we describe a design method of the free parameter $Q(s)$ in Eq.(3) to makes the Smith predictor $C(s)$ in Eq.(3) works as a modified PID controller.

1. Modified P controller

   The modified P controller $C(s)$ satisfying Eq.(7) is written by Eq.(3), where

   $$Q(s) = \lim_{s\to\infty} \left\{a_P G_u^2(s) - G_u(s)\bar{G}_u(s)\right\}. \tag{10}$$

   Since $Q(s)$ in Eq.(10) is included in RH_∞ independence of a_P, the controller $C(s)$ in Eq.(3) with Eq.(10) makes the closed-loop system in Eq.(1) stable for any time-delay plant $G(s)e^{-sL}$ independence of a_P.

2. Modified I controller

The modified I controller $C(s)$ satisfying Eq.(8) is written by Eq.(3), where

$$Q(s) = \frac{q_0 + q_1 s}{\tau_0 + \tau_1 s}, \tag{11}$$

$$\begin{aligned} q_0 &= \tau_0 G_u(0)\left(\frac{1}{G_s(0)} - \bar{G}_u(0)\right), \\ q_1 &= \tau_1 G_u(0)\left(\frac{1}{G_s(0)} - \bar{G}_u(0)\right) \\ &\quad + \tau_0 \left\{ \frac{2 \left.\frac{d}{ds}\left(G_u(s)\right)\right|_{s=0}}{G_s(0)} - \left.\frac{d}{ds}\left(G_u(s)\bar{G}_u(s)\right)\right|_{s=0} \right\} \\ &\quad + \frac{\tau_0}{G_s^2(0)}\left\{ TG(0) - \frac{1}{a_I} - \frac{d}{ds}\left.\left(G(s)\right)\right|_{s=0} \right\}, \end{aligned}$$

$\tau_i \in R > 0 (i = 0, 1)$. From $\tau_i > 0 (i = 0, 1)$, $Q(s)$ in Eq.(10) is included in RH_∞ independence of a_I. This implies that the controller $C(s)$ in Eq.(3) with Eq.(11) makes the closed-loop system in Eq.(1) stable for any time-delay plant $G(s)e^{-sL}$ independence of a_I.

3. Modified D controller

The modified D controller $C(s)$ satisfying Eq.(9) is written by Eq.(3), where

$$Q(s) = a_D \lim_{s\to\infty} \left(G_u^2(s)\right) s. \tag{12}$$

Since $Q(s)$ in Eq.(12) is improper, $Q(s)$ in Eq.(12) is not included in RH_∞. In order for $Q(s)$ to be included in RH_∞, Eq.(12) is modified as

$$Q(s) = a_D \lim_{s\to\infty} \left(G_u^2(s)\right) \frac{s}{1 + \tau_D s}, \tag{13}$$

where $\tau_D \in R > 0$. From $\tau_D \in R > 0$ in Eq.(13), $Q(s)$ in Eq.(13) is included in RH_∞ independence of a_D. This implies that the controller $C(s)$ in Eq.(3) with Eq.(13) makes the closed-loop system in Eq.(1) stable for any time-delay plant $G(s)e^{-sL}$ independence of a_D.

4. Modified PI controller

The modified PI controller $C(s)$ satisfying Eq.(7) and Eq.(8) is written by Eq.(3), where

$$Q(s) = \frac{q_0 + q_1 s + q_2 s^2}{\tau_0 + \tau_1 s + \tau_2 s^2}, \tag{14}$$

$$\begin{aligned} q_0 &= \tau_0 G_u(0)\left(\frac{1}{G_s(0)} - \bar{G}_u(0)\right), \\ q_1 &= \tau_1 G_u(0)\left(\frac{1}{G_s(0)} - \bar{G}_u(0)\right) \end{aligned}$$

$$+\tau_0\left\{\frac{2\left.\dfrac{d}{ds}\left(G_u(s)\right)\right|_{s=0}}{G_s(0)}-\left.\frac{d}{ds}\left(G_u(s)\bar{G}_u(s)\right)\right|_{s=0}\right\}$$

$$+\frac{\tau_0}{G_s^2(0)}\left\{TG(0)-\frac{1}{a_I}-\left.\frac{d}{ds}\left(G(s)\right)\right|_{s=0}\right\},$$

$$q_2 = \tau_2\lim_{s\to\infty}\left\{a_P G_u^2(s)-G_u(s)\bar{G}_u(s)\right\},$$

$\tau_i \in R > 0 (i = 0, 1, 2)$. From $\tau_i > 0 (i = 0, 1, 2)$, $Q(s)$ in Eq.(14) is included in RH_∞ independence of a_P and a_I. This implies that the controller $C(s)$ in Eq.(3) with Eq.(14) makes the closed-loop system in Eq.(1) stable for any time-delay plant $G(s)e^{-sL}$ independence of a_P and a_I.

5. Modified PD controller

The modified PD controller $C(s)$ satisfying Eq.(7) and Eq.(9) is written by Eq.(3), where

$$Q(s) = q_0 + q_1 s, \tag{15}$$

$$q_1 = a_D\lim_{s\to\infty} G_u^2(s),$$

$$q_0 = \lim_{s\to\infty}\left\{a_P G_u^2(s)-G_u(s)\left(\bar{G}_u(s)+2a_D s^2\frac{d}{ds}(G_u(s))\right)\right\}.$$

Since $Q(s)$ in Eq.(15) is improper, $Q(s)$ in Eq.(15) is not included in RH_∞. In order for $Q(s)$ to be included in RH_∞, Eq.(15) is modified as

$$Q(s) = q_0 + \frac{q_1 s}{1+\tau_D s}, \tag{16}$$

where $\tau_D \in R > 0$. From $\tau_D > 0$ in Eq.(16), $Q(s)$ in Eq.(16) is included in RH_∞ independence of a_P and a_D. This implies that the controller $C(s)$ in Eq.(3) with Eq.(16) makes the closed-loop system in Eq.(1) stable for any time-delay plant $G(s)e^{-sL}$ independence of a_P and a_D.

6. Modified PID controller

The modified PID controller $C(s)$ satisfying Eq.(7), Eq.(8) and Eq.(9) is written by Eq.(3), where

$$Q(s) = \frac{q_0+q_1 s+q_2 s^2}{\tau_0+\tau_1 s+\tau_2 s^2}+q_3 s, \tag{17}$$

$$q_0 = \tau_0 G_u(0)\left(\frac{1}{G_s(0)}-\bar{G}_u(0)\right),$$

$$q_1 = \tau_1 G_u(0)\left(\frac{1}{G_s(0)}-\bar{G}_u(0)\right)$$

$$+\tau_0\left\{\frac{2\left.\dfrac{d}{ds}\left(G_u(s)\right)\right|_{s=0}}{G_s(0)}-\left.\frac{d}{ds}\left(G_u(s)\bar{G}_u(s)\right)\right|_{s=0}\right\}$$

$$
\begin{aligned}
& +\frac{\tau_0}{G_s^2(0)}\left\{TG(0)-\frac{1}{a_I}-\frac{d}{ds}\left(G(s)\right)\bigg|_{s=0}\right\} \\
& -\tau_0 a_D \lim_{s\to\infty}\left\{G_u^2(s)\right\}, \\
q_2 &= \tau_2 \lim_{s\to\infty}\left\{a_P G_u^2(s)-G_u(s)\left(\bar{G}_u(s)+2a_D s^2\frac{d}{ds}(G_u(s))\right)\right\}, \\
q_3 &= a_D \lim_{s\to\infty} G_u^2(s)
\end{aligned}
$$

and $\tau_i \in R > 0 (i = 0, 1, 2)$. Since $Q(s)$ in Eq.(17) is improper, $Q(s)$ in Eq.(17) is not included in RH_∞. In order for $Q(s)$ to be included in RH_∞, Eq.(17) is modified as

$$
Q(s) = \frac{q_0 + q_1 s + q_2 s^2}{\tau_0 + \tau_1 s + \tau_2 s^2} + \frac{q_3 s}{1+\tau_D s}, \tag{18}
$$

where $\tau_D \in R > 0$. From $\tau_D > 0$ and $\tau_i > 0 (i = 0, 1, 2)$, $Q(s)$ in Eq.(18) is included in RH_∞ independence of a_P, a_I and a_D. This implies that the controller $C(s)$ in Eq.(3) with Eq.(18) makes the closed-loop system in Eq.(1) stable for any time-delay plant $G(s)e^{-sL}$ independence of a_P, a_I and a_D.

Conclusion

In this paper, we proposed a design method of modified PID controllers such that the modified PID controller makes the closed-loop system for any time-delay plants stable and the admissible sets of P-parameter, I-parameter and D-parameter are independent from each other.

References

A. Datta, M.T. Ho, and S.P. Bhattacharyya: Structure and Synthesis of PID Controllers, Springer-Velag, London, (2000)

J. Yang: Parameter Plane Control Design for a Two-tank Chemical Reactor Systems, Journal of the Franklin Institute, 331B(1), pp.61–76, (1994)

M.T. Ho, A. Datta, and S.P. Bhattacharyya: A linear programming characterization of all stabilizing PID controllers, Proceedings of the American Control Conference 1997, (1997)

K. Yamada and T. Moki: A design method for PI control for minimum phase systems, Intelligent Engineering Systems Through Artificial Neural Networks, 13, (2003)

K. Yamada: Modified PID controllers for minimum phase systems and their practical application, Proceedings of The 2005 Electrical Engineering/Electronics, Computer, Telecommunication, and Information Technology (ECTI) International Conference, II, pp.457–460, (2005)

K. Yamada and T. Hagiwara: A design method of modified PID controller for unstable plants, Intelligent Engineering Systems Through Artificial Neural Networks, Vol.16, pp.753-758, (2006)

K. Yamada and H. Takenaga: The parametrization of all stabilizing Smith predictors for time-delay plants, submitted for publication

SMART SEMI-ACTIVE CONTROL OF FLOOR-ISOLATED STRUCTURES

TZU-KANG LIN
National Center for Research on Earthquake Engineering
PEI-YANG LIN
National Center for Research on Earthquake Engineering

KUO-CHEN CHANG
National Taiwan University

ABSTRACT
A semi-active control system for structures with sensitive equipments is proposed in this paper. While technology of structure control in regular buildings gradually matured and is implemented practically over the last two decades, hospitals and facilities with electronic devices are still vulnerable to seismic loadings. To solve this problem, a hybrid floor isolation system comprising a bidirectional roller pendulum system (RPS) and a controllable magnetorheological (MR) damper is proposed to ameliorate the response of structures under external excitation. For the best performance of the control system, two smart control algorithms for the MR damper are developed. The controllers are designed to automatically switch between displacement-oriented or acceleration-oriented objectives, and the capability of the structure under earthquake can be largely improved. Theoretical analysis is carried out, and the results have demonstrated the feasibility and advantage of the proposed control system on floor-isolated structures.

INTRODUCTION

Equipment that is vital to the operation of critical facilities such as hospitals, information storage databases, and military command and control centers is often at risk of sustaining damage in the event of major ground motions. Furthermore, manufacturers of sensitive electronic equipment are facing with the possibility of their products being damaged in modest or severe seismic events. With such hazards in mind, many facilities have installed floor isolation systems to alleviate potential damage that could occur in a seismic temblor. Commonly, rolling pendulum systems (RPS) are used to isolate the substructure from the superstructure. These devices are very cost effective when contrasted with their rubber bearing counterparts.

A new concept of smart structures has recently been proposed and widely discussed. Structures with self-modification capability are called adaptive or smart structures. The smart structure technology can avoid loss of life and damage to the structure and its contents, particularly for large structures containing hundreds of people. To implement the concept of smart structure, neural networks, which are used for the learning ability, were then applied to active control. A series of active control theory based on neural networks has been developed over the last ten years (He Y.A. and Wu J., 1998) (Liut D.A. et al, 1999) (Kim J.T. et al, 2000) (Hung S.L. et al, 2000), and (Brown A. S., and Yang T. Y., 2001). For example, Ghaboussi and his coworkers developed a neural network-based control rule for nonlinear structures and successfully verified it on a small shaking table (Ghaboussi J. and Joghataie A., 1995) (Bani-Hani K. and Ghaboussi J., 1998a) (Bani-Hani K. and Ghaboussi J., 1998b) (Bani-Hani K. et al, 1999a), and (Bani-Hani K. et al, 1999b). Fuzzy logic was also integrated into another system for better robustness (Hung S.L. and Lai C.M., 2001). To get the best result with limited resources, the placement of the sensor/actuators on civil structures was also considered by genetic algorithms (Abdullah M.M. et al, 2001). Moreover, numerical and experimental studies have been conducted by Lin *et al* (2006a) and (2006b) to evaluate the effectiveness of neural network controllers implemented in active structural control systems. These researchers created an effective and stable active control method that exhibits superior performance in comparison with classical linear quadratic gain (LQG) methods.

In the current study a RPS floor isolation system is combined with magnetorheological (MR) dampers to form a new smart system for superior mitigation of bidirectional seismic excitations. The performance of two different control methods are applied to a floor-isolated structure and evaluated. The first control method is a neural network-based controller termed NEURO-MR. The second control method, termed COC-VG, consists of a linear quadratic regulator (LQR) with clipped optimal control and variable gains.

OVERVIEW OF THE BENCHMARK STRUCTURE AND CONTROL DEVICES

A single-degree-of-freedom system shown in figure 1 is used as the numerical model to develop the smart control algorithms for the MR dampers. The structure is assumed to be seated on a roller pendulum system, and the period of the system is 2.8 sec. Considering about the influence of the RPS system in different directions, the mass of the system in the longitudinal direction is 17.3 tons, and the mass in the latitudinal direction is set to 16 tons. Two MR dampers are used to control the response of the structure and the capacity are 20KN and 17KB, respectively. Two control criteria of the superstructure displacement are set to evaluate the developed controllers.

1. Displacement of the structure should be controlled beneath 12 cm under the El Centro Earthquake of PGA 700 gal, which represents far-field earthquake.
2. Displacement of the structure should be controlled beneath 10 cm under the TCU 068 Earthquake of PGA 700 gal, which represents near-fault earthquake.

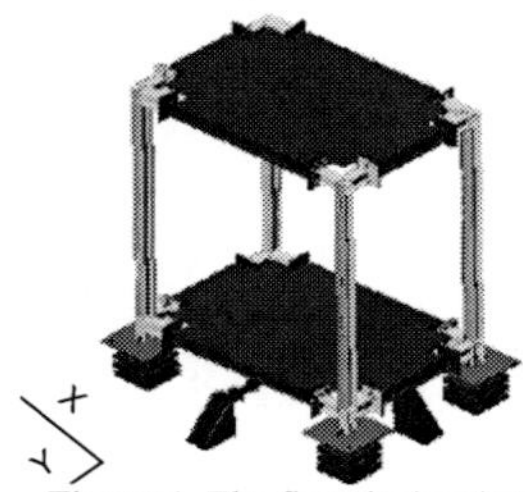

Figure 1. The floor-isolated structure

OUTLINE OF THE PROPOSED CONTROLLERS

An overview and discussion of the two control methods (NEURO-MR and COC-VG) employed for operating of the dampers is discussed. Each control method in this study attempts to reduce displacement and acceleration responses, in order to protect the superstructure and the stored items as the objectives shown in figure 2. In many cases, the displacement of the platform relative to the base motion can be alleviated most when command voltages sent to the MR dampers are maximum; this is referred to as 'passive-on' in the following text. However, due to an effective increase in structural stiffness, the corresponding acceleration response is often aggravated during 'passive on' operation. Conversely, structural acceleration responses can be kept to a minimum when the voltage signals sent to the MR dampers are turned off, which is referred as the 'passive-off' case. Thus, intelligent management of voltage levels is required to reduce both displacement and acceleration responses effectively. These methods emphasize reduction of acceleration responses and ensure that the maximum base drift remains below the displacement limit of the MR dampers and RPS system. The basic theories of these controllers are described in the following two sections.

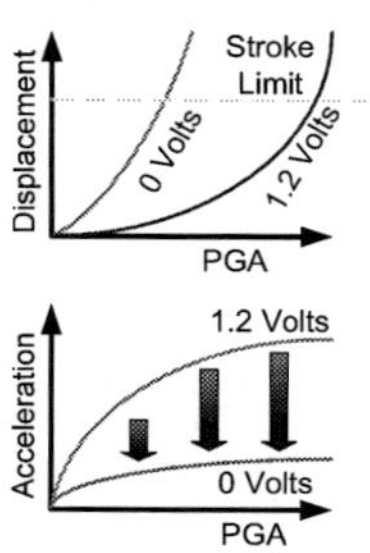

Figure 2. Objectives of the NEURO-MR and COC-VG

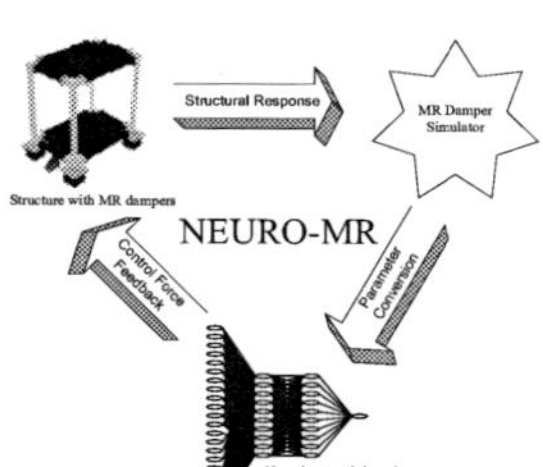

Figure 3. The NEURO-MR control system

3.1 NEURAL-NETWORK-BASED CONTROLLER

The first controller to be described is the NEURO-MR controller. This concept was initially proposed by Lin et al (2006a) (2006b) and is described in figure 3. In order to identify optimal control signals, the NEURO-MR controller requires an accurate model of the MR dampers. By utilizing a simplified MR damper simulator, the NEURO-MR controller can rapidly calculate the optimal command voltage for the MR dampers.

Structural responses at the current time step and the previous two steps are considered by the control criterion J described below. Specific weights are assigned to each time step to represent different key factors of displacement, velocity, and acceleration. Therefore, a more effective and smooth reduction of structural responses can be achieved.

The control criterion J is defined as follows:

$$J = W_1 Z_{k+1}^T Q Z_{k+1} + W_2 Z_{k+2}^T Q Z_{k+2} + W_3 Z_{k+3}^T Q Z_{k+3} + W_4 A_{k+1}^2 + W_5 A_{k+2}^2 + W_6 A_{k+3}^2 \qquad (1)$$

where

$$Z_k = [D_k V_k]^T \qquad (2)$$

$$Z_{k+1} = [D_{k+1} V_{k+1}]^T \qquad (3)$$

$$Z_{k+2} = [D_{k+2} V_{k+2}]^T \qquad (4)$$

D_k, V_k and A_K are the displacement, velocity and acceleration of the isolated structure at state space Z_k with time step k, and D_{k+1} V_{k+1} and A_{K+1} are the displacement, velocity and acceleration of the structure at time step $k+1$. Criterion Q, which comprises the mass and stiffness of the system, is a weighting matrix used to evaluate the total energy of the structure. W_1, W_2, W_3, W_4, W_5 and W_6 are coefficients used to designate the relevant importance of the different time steps. Weights W_4, W_5 and W_6 correspond to acceleration inputs and are adjusted to minimize either displacement or acceleration responses as needed.

The objective of the smart control system is to minimize the acceleration response of the isolated floor with due consideration being given to displacement limitations of the MR dampers and the RPS system. To impart robustness to the NEURO-MR controller, three different training cases with various PGA levels are used. The first training case consists of the 1940 El Centro NS component record and a white noise signal, which are normalized to 150 and 50 gal levels, respectively. These two excitations are used to represent small intensity ground excitations, and the control criterion focuses on reducing acceleration response in this training case. Moreover, the 1940 El Centro NS component record and a white noise signal are normalized to 400 and 100 gal, respectively, to represent moderate intensity ground motions. In this case, the control criterion is altered to concentrate on reducing both the displacement and acceleration of the superstructure. Finally, the 1940 El Centro NS component record and a white noise signal are normalized to 700 and 200 gal, respectively, to represent strong intensity ground motions. The control criteria are altered again to focus on reducing displacement of the superstructure. By this arrangement of training patterns, the NEURO-MR controller adaptively switches between acceleration-oriented and displacement-oriented control. Namely, the acceleration of the system can be minimized, while taking into account limitations of the MR damper stroke and RPS devices.

Displacement, velocity, acceleration responses of the structure from the previous three time steps as well as the command voltages to the MR dampers in the previous three time steps are chosen as the input nodes of the NEURO-MR network, and the output of the NEURO-MR network is the optimal MR damper voltage signal. A total of 48,000 patterns are collected and used to train the NEURO-MR controller. After a trial and error process, the network structure of the controller is chosen to consist of four layers with nodal configuration of 12-7-7-1. Inputs of the network are composed of current, previous structural responses, and previous control voltages while the output is the current command voltage to one MR damper.

3.2 LQR/CLIPPED OPTIMAL CONTROLLER WITH VARIABLE GAINS

In the past, active control of structures has been used to resist earthquake excitations in numerical and experimental studies. Various control algorithms have been proposed including LQR (Bryson *et al,* 1975; Abel-Rohm *et al,* 1980), predictive control, modal control, sliding mode control, H∞ control, as well as others. Direct feedback control of measured quantities can be used with many of these controllers. In this section, a LQR/clipped optimal controller algorithm, COC-VG, is used with output weighting, and variable gains is described.

The first controller, *G1*, is designed to control the base isolation system when the base drifts are small, while the second controller, *G2*, is designed to control the base isolation system when the base drifts become large. Three stroke levels ($0 < x1 < x2 < xlimit$) are defined, and interpolations are made between each level as follows:

$$\begin{cases} |x| \le x_1 & G = G_1 \\ x_1 < |x| < x_2 & G = G_2 * \dfrac{x - x_1}{x_2 - x_1} + G_1 * \dfrac{x_2 - x}{x_2 - x_1} \\ |x| \ge x_2 & G = G_2 \end{cases} \qquad (5)$$

In this study, the MR dampers are used to provide adaptive control of the floor isolation system. Since the COC-VG algorithm specifies a force to be applied to the structure and the MR damper is a semi-active device, an inverse transformation of the ideal force to the correct voltage level is required. Thus, a clipped–optimal controller is used to calculate a suitable command voltage in each time step.

Figure 4 shows a flowchart of the clipped optimal control algorithm. The desired control force, calculated by COC-VG, is used to compare desired and estimated MR damper forces.The two voltage levels that show the best performance are used to interpolate the command voltage, according to the desired control force as follows:

$$V = V_j * \frac{F_{desired} - F_{MR\_V_i}}{F_{MR\_V_j} - F_{MR\_V_i}} + V_i * \frac{F_{MR\_V_i} - F_{desired}}{F_{MR\_V_j} - F_{MR\_V_i}} \qquad (6)$$

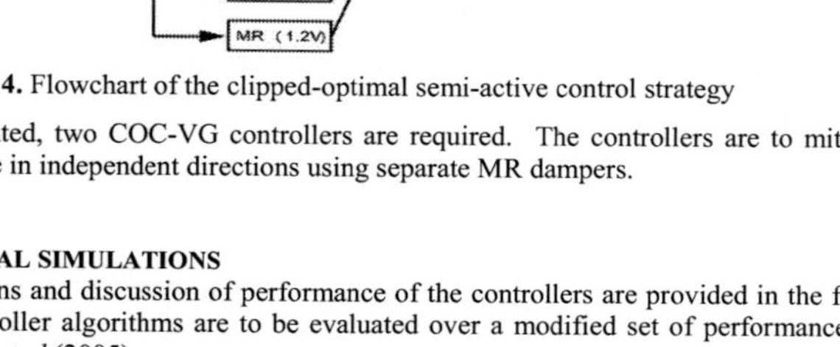

Figure 4. Flowchart of the clipped-optimal semi-active control strategy

When implemented, two COC-VG controllers are required. The controllers are to mitigate the response of the superstore in independent directions using separate MR dampers.

RESULTS OF NUMERICAL SIMULATIONS

Numerical simulations and discussion of performance of the controllers are provided in the following section. Generated controller algorithms are to be evaluated over a modified set of performance metrics proposed by Narasimhan *et al* (2005).

Results of numerical simulations reflect the considered optimization objectives as described in figure 2 where J_2 represents peak base drift of the structure, and J_3 represents peak absolute acceleration of the isolated structure. The NEURO-MR and the COC-VG algorithms attempt to provide significant consideration to superstructure acceleration levels, while ensuring that the base displacements do not exceed MR damper stroke lengths or RPS limitations. These algorithms typically perform well with respect to acceleration responses, especially at lower PGA levels.

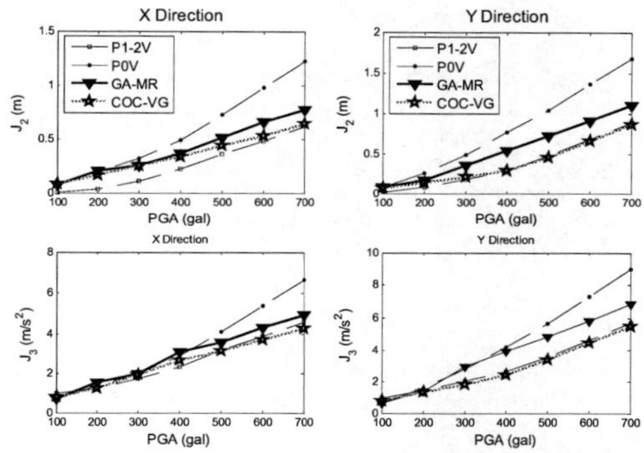

Figure 5. Comparative results: Chi-Chi (TCU0-68)

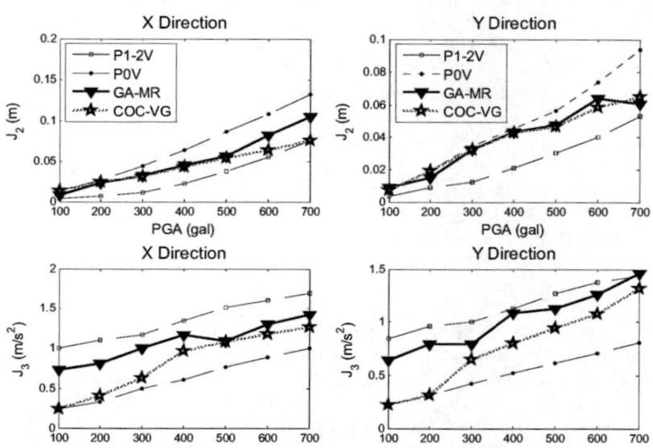

Figure 6. Comparative results: Chi-Chi (Station: TCU-129)

To graphically interpret results for all excitation levels, figures 5 and 6 show controller performance over all excitation levels for the two specific earthquake records with respect to selected metrics. As can be observed, performance of the semi-active controllers is generally bounded by the passive-off and passive-on cases; however, in several instances, semi-active controllers show superior performance with respect to these passive operational levels. In figure 6, results from simulations utilizing the record of station TCU-129 under the Chi-Chi earthquake are summarized. With respect to displacement response, the performances of the two semi-active cases are similar. This shows the intelligent operation of the controllers since it is able to effectively minimize accelerations in far-field excitations, and yet give more consideration to displacement control for near-field excitations.

SUMMARY AND CONCLUSION

Theoretical analysis of semi-active control systems for floor-isolated structures is proposed in this paper. To meet the vibration requirement of high-tech instruments, the acceleration response of the floor-isolated control system should be alleviated as much as possible both during micro vibration and large PGA values when the stroke of the MR dampers are kept in their designed displacement.

Two different methods including NEURO-MR and COC-VG are used to develop the control algorithms for a floor-isolated structure. To demonstrate the performance of these controllers, control criteria proposed by Narasimhan *et al.* (2005) for smart base isolated benchmark building are used for comparison. Theoretical analysis have shown that these methods can effectively alleviate the accelerations of structures during micro vibration and suppress the displacement of the isolated floor under large PGA values to extend the service range of the isolated structures. The smart characteristics of these systems can give more protections to the vulnerable high-tech equipments. Experimental verifications of the smart system will be carried out in the next paper.

ACKNOWLEDGMENTS

Mr. David Shook in Zachry Department of Civil Engineering, Texas A&M University, College Station, TX, U.S.A. is due much appreciation for providing assistance in finishing this research.

REFERENCES

He Y.A. and Wu J., 1998. Control of Structural Seismic Response by Self-recurrent Neural Network (SRNN). Earthquake Engineering & Structural Dynamics; Vol.27 No.7, 641-648.

Liut D.A., Matheu E.E., Singh M.P. and Mook D.T., 1999. Neural-Network Control of Building Structures by a Force-matching Training Scheme. Earthquake Engineering & Structural Dynamics; Vol.28 No.12, 1601-1620

Kim J.T., Jung H.J. and Lee I.W., 2000. Optimal Structural Control Using Neural Networks. Journal of Engineering Mechanics, ASCE; Vol.126 No.2, 201-205.

Hung S.L., Kao C.Y. and Lee J.C., 2000. Active Pulse Structural Control Using Artificial Neural Networks. Journal of Engineering Mechanics, ASCE; Vol.126 No.8, 839-849.

Brown A. S. and Yang T. Y., 2001. Neural Networks for Multiobjective Adaptive Structural Control. Journal of Structural Engineering; Vol. 127, No. 2, 203-210.

Ghaboussi J. and Joghataie A., 1995. Active Control of Structures Using Neural Network. Journal of Engineering Mechanics, ASCE; Vol.121 No.4, 555-567.

Bani-Hani K. and Ghaboussi J., 1998. Nonlinear Structural Control Using Neural Networks. Journal of Engineering Mechanics, ASCE; Vol.124 No.3, 319-327.

Bani-Hani K. and Ghaboussi J., 1998. Neural Networks for Structural Control of a Benchmark Problem, Active Tendon System. Earthquake Engineering & Structural Dynamics; Vol.27 No.11, 1225-1246.

Bani-Hani K., Ghaboussi J. and Schneider S.P., 1999. Experimental Study of Identification and Control of Structures Using Neural Network. Part 1 Identification. Earthquake Engineering & Structural Dynamics; Vol.28 No.9, 995-1018.

Bani-Hani K., Ghaboussi J. and Schneider S.P., 1999. Experimental Study of Identification and Control of Structures Using Neural Network. Part 2 Control. Earthquake Engineering & Structural Dynamics; Vol.28 No.9, 1019-1039.

Hung S.L. and Lai C.M., 2001.Unsupervised Fuzzy Neural Network Structural Active Pulse Controller. Earthquake Engineering & Structural Dynamics; Vol.30 No.4, 465-484.

Abdullah M.M., Richardson A. and Hanif J., 2001. Placement of Sensors/actuators on Civil Structures Using Genetic Algorithms. Earthquake Engineering & Structural Dynamics; Vol.30 No.8, 1167-1184.

Lin, T.K., Chang K.C., Chung L.L. and Lin Y.B., 2006. Active Control with Optical Fiber Sensors and Neural Networks, Part I: Theoretical Analysis. Journal of Structural Engineering ASCE; Vol. 132, No. 8: 1293-1303.

Lin, T.K., Chang K.C., and Lin Y.B., 2006.Active Control with Optical Fiber Sensors and Neural Networks, Part II: Experimental Verification. Journal of Structural Engineering ASCE; Vol. 132, No. 8: 1304-1313.

Bryson A E and Ho Y C, 1975. Applied optimal control, John Wiley and Sons, New York, N. Y.

Abel-Rohm M., Quintana V. H., and Leipholz H. H., 1980. Optimal control of civil engineering structures J. of Engr. Mech. 106(6) 57–73

Narasimhan S, Nagarajaiah S., Johnson E. A., and Gavin H. P., 2005. Smart base-isolated benchmark building. Part I: Problem definition Structural Control and Health Monitoring 13 573-588

FUZZY LOGIC CONTROL OF A SMART BASE-ISOLATION SYSTEM WITH MR DAMPER

PEI-YANG LIN
National Center for Research on
Earthquake Engineering, Taiwan

TZU-KANG LIN
National Center for Research on
Earthquake Engineering, Taiwan

ABSTRACT

In this study, fuzzy logic control of a smart base-isolation system with semi-active controlled MR damper is presented. The Fuzzy-Logic control is used to calculate the optimal command to the semi-active controlled MR damper. A large-scaled, one-story building structure equipped with this smart base-isolation system is tested on the shaking table in NCREE. Different earthquake excitation rerecords and various PGA levels are tested on the shake table. The test result shows that the application of fuzzy-logic control cans effective control the base-isolation system. Both the acceleration and displacement are well controlled; moreover, it is adaptive to various kinds and intensity of excitations.

INTRODUCTION

Low power consumption, high reliability and fail-safe operation make the semi-active control technique one of the more promising approaches for mitigation of seismic responses in civil engineering structures. Currently, magnetorheological (MR) dampers are being widely studied for their potential use as semi-active control devices [Dyke][Spencer][Chang]. An MR damper resembles an ordinary linear viscous damper but the cylinder of the damper is filled with special fluid that contains tiny polarizable particles. The fluid state can be changed drastically from liquid to solid and vise versa by adjusting the magnitude of an applied magnetic field produced by a coil that is wrapped around the piston head of the damper. When no current is supplied to the coil, the damper behaves as an ordinary viscous damper. On the other hand, when current is sent through the coil, the fluid inside an MR damper becomes a semi-solid, the yield strength of which depends on the level of current applied. Since control commands simply adjust the parameters of a MR damper that is placed in a real structure, control instability never occurs and only a small amount of energy is required. Therefore, MR dampers are reliable and fail-safe. This study combined the rolling pendulum system and the semi-active controlled MR damper to the smart base-isolation system. The rolling pendulum system provides the suitable restoring forces, while the semi-active controlled MR damper provides the optimal damping force. The command voltage to the MR damper is calculated by Fuzzy-Logic control. Various earthquake excitations with several PGA levels are tested on the shake table to validate the stability, efficiency, and adaptively of this smart base-isolation system. All the test result shows that the smart base-isolation system is stable, reliable, fail-safe and effective. And most important thing, it is adaptive to different intensity of excitation as compared to the traditional passive base-isolation system.

DESIGN OF FUZZY-LOGIC CONTROL

The semi-active controlled MR damper is used in the hybrid isolation system; its resistance can be rapidly adjusted by the command voltage. In this study, the Fuzzy-Logic control is used to calculate the online command voltage to the MR damper. The Fuzzy-Logic control is designed according to the feedback signals and control object. The Fuzzy-logic control maps an input space to an output space by means of if-then rules, see Fuzzy Logic Toolbox Users Guide. Control components of the input signal are transformed into linguistic values through a fuzzification interface at each time step. Use of a fuzzy controller is advantageous in that performance is not overly sensitive to changes in the input signal. For output the mapped linguistic values are transformed into numerical values through a defuzzification interface

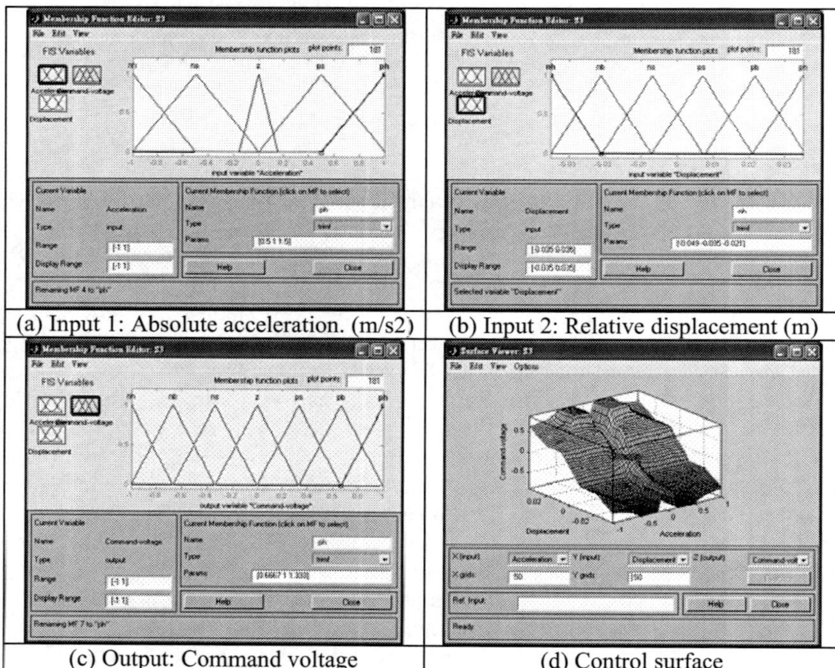

(a) Input 1: Absolute acceleration. (m/s2)
(b) Input 2: Relative displacement (m)
(c) Output: Command voltage
(d) Control surface

Figure 1: Input and output membership functions of semi-active control surface "S3"

Design of a Fuzzy-Logic controller is separated into three parts: (1) use a fuzzy inference system (FIS) editor to define the number of input and output variables and choose the type of inference to be used; (2) define membership functions for the input and output variables; and (3) define if-then rules. In this study, a trial and error process results in the use of two inputs (displacement and acceleration) and one output (voltage) variable. Next, triangular membership functions and the range of their variables are defined for each input and output variable. Finally, the if-then rules are edited so as to connect each input and output

For the semi-active controller "S3", the absolute acceleration and relative displacement are selected as inputs, and the output is the command voltage. The number of membership functions used for the inputs are five and six, while seven membership function are used for the output. Figure 1 (a), (b) and (c) show the result of applying the inputs and output membership functions over the whole range of the input variables for semi-active control case "S3". Figure 1 (d) shows the control surface of semi-active control case "S3". The design approach for semi-active control case "S3" is to control both the absolute acceleration and the relative displacement.

SHAKE TABLE TEST

In this study, a base-isolated structure with a rolling pendulum system and a 20 kN / 300 mm MR damper is tested on a shake table. Figure 2 shows a schematic drawing and test photo of the experimental setup. The shaking table controlled by the MTS system delivers the desired earthquake excitation. Responses of the system are measured by numerous accelerometers, load cells, thermal couples and LVDTs, and are sent to the data acquisition system. Some transducer readings are also feed back to the semi-active control system (PC/Simulink). Through a semi-active control surface (pre-loaded in PC/Simulink) and the feedback signals, the command voltage can be calculated online. The command voltage is converted to command current by a voltage control current source (VCCS) and fed to the magnetorheological damper in order to optimally control the hybrid base-isolation system.

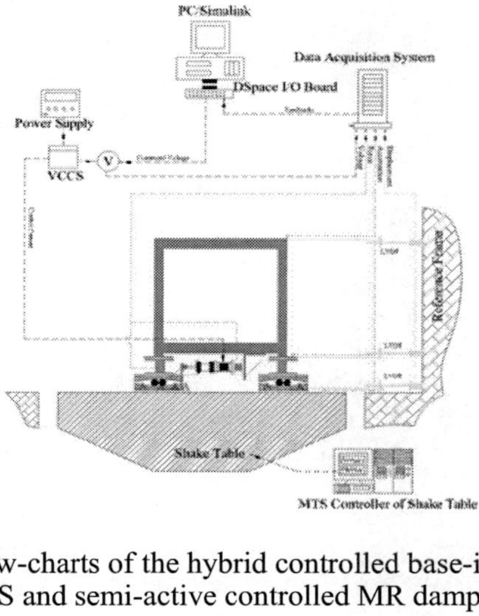

Figure 2: Control flow-charts of the hybrid controlled base-isolation system with RPS and semi-active controlled MR damper.

The floor plane of the upper structure is 2m by 3m, the floor height is 3m. The mass of each floor level are 6 tons (including the structural weight and the mass blocks). As results, the total mass of the upper structure is about 12 tons. The natural period of the rolling pendulum system is designed as 2.77 sec. Ends of the Magnetorheological damper are securely attached to the top surface of the shake table and the bottom of the isolated structure. The command voltage is sent from the "dSPACE" control system with one BNC line to the voltage controlled current system (VCCS). The VCCS converts the command voltage to the command current and sent to the MR damper. The power supply provides the 24volts power to the VCCS. All the displacement, velocity, acceleration, command voltage, damper force, shear and axial force are measured and recorded by suitable sensors and "Pacific" data acquisition system in NCREE. Eight excitation cases, including El Centro NF (1970 array#6), El Centro FF (1940), three Chi-Chi earthquake records (1999) from three different station TCU075, TCU076 and TCU082, another earthquake record from station TCU075 (Far-field case), Kobe and random excitations, are tested in this shaking table test. Four passive control cases (with four different command levels), three semi-active control cases and fixed-base case are tested in this study.

Figure 3: 3D drawing and photo of the smart base-isolation system with rolling pendulum system (RPS) and MR damper in the shake table test.

Figure 4 shows the maximum responses of four passive control cases (P-off, P-04V, P-08V and P-max) and the semi-active control case "S3" under various PGA levels El Centro FF (1940) excitations. The higher command voltage will leads the smaller stroke, but higher acceleration. Two passive control cases ("P-08V" and "P-max") and the semi-active control case "S3" can protect the base-isolation system up to 700 gals El Centro FF (1940) excitation. Among the three cases, only the semi-active control case "S3" can reduce the acceleration more in small intensity of excitation. That means the smart base-isolation system is more adaptive to different excitation intensity as compared to the passive cases.

The traditional passive isolation system designs the system according the design earthquake of spectrum. As results, the control effect of the isolation system is good under the design PGA level. When the excitation is minor, the control effect (isolated acceleration responses) decreased. It shows that the passive control is not adaptive to different intensity of excitations, while the semi-active control is adaptive to a large range of excitation.

Figure 5 shows the damper command, force time histories and hysteretic loops of the semi-active control “S3” under 700 gals El Centro FF (1940) excitation. The hysteresis loops in semi-active control case “S3” varied with the command voltage. It represents the designed control object “reduce the acceleration with the constraint of the stroke ”. Comparing the hysteresis loops, the semi-active control system vary the MR damper’s characteristic according the excitations, as results, the smart base-isolation system is more adaptive to different kinds and intensity of excitations.

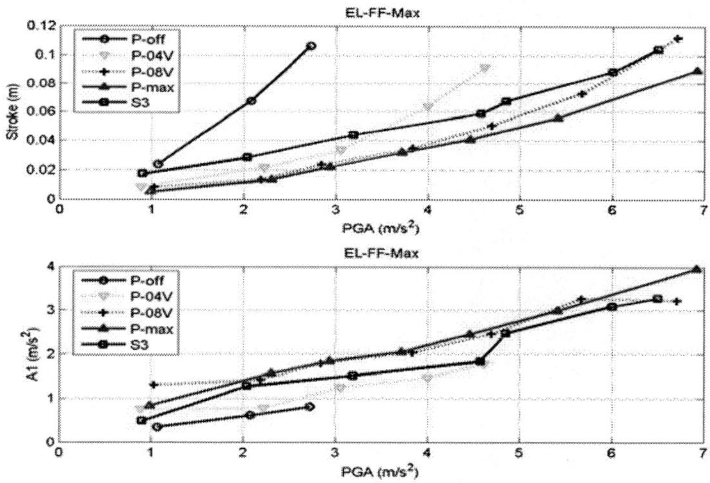

Figure 4: Comparison of the maximum responses of four passive control cases (P-off , P-04V, P-08V and P-max) and the semi-active control case “S3” under various PGA levels El Centro FF (1940) excitations.

CONCLUSIONS

The shake table test of smart base-isolation system shows the great benefit of the smart damper. The most different between the base-isolation system with passive damper and semi-active control Magnetorheological damper is that the semi-active control system is adaptable to various kinds and intensity of excitations. While the passive control system can only focus on some cases (P-max: intensive earthquakes; P-off: small earthquakes). Also, this study provides evidence of full-scale, real-time control of augmenting a common base-isolation system with a smart Magnetorheological damper that is modulated with a fuzzy controller.

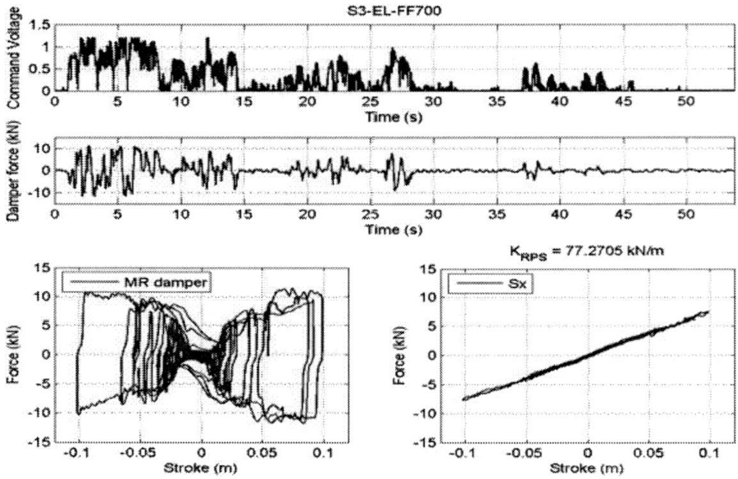

Figure 5: Comparison of the damper command and force time history responses and hysteretic loops of the semi-active control case "S3" under 700 gals El Centro FF (1940) excitation.

REFERENCES

Dyke, S. J., Spencer, Jr. B. F., Sain, M. K., and Carson, J. D.(1996) "Modeling and control of magnetorheological dampers for seismic response reduction," Smart Mat. and Struct., 5, pp. 565-575.

Spencer, Jr. B. F., Dyke, S. J., Sain, M. K., and Carson, J. D.(1997), "Phenomenological model of Magnetorheological dampers," J. Engrg. Mech and Mech., ASCE, 123(3), pp. 230-238.

Spencer, Jr. B. F., Dyke, S. J., Sain, M. K., and Carson, J. D. (1998), "An experimental study of MR dampers for seismic protection," J. Engrg. Mech and Mech., Smart Materials and Structures: Special Issue on Large Civil Structures, 7(5), pp. 693-703.

Chang, C. C. and Roschke, P. (1998), "Neural network modeling of a magnetorheological damper," Journal of Intelligent Material Systems and Structures, 9, pp. 755-764.

"Fuzzy Logic Toolbox Users Guide", The Math Works, Inc.

Lin, Pei-Yang, Roschke, P. N., Loh, C. H. "Hybrid base-isolation with magnetorheological damper and fuzzy control", Structural Control and Health Monitoring Volume 14, Issue 3, April 2007, Pages: 384-405.

Kim, Hyun-Su, Roschke, P. N., Lin, Pei-Yang, Chin-Hsiung Loh, "Neuro-fuzzy model of hybrid semi-active base isolation system with FPS bearing and MR damper", Engineering structures 28, 2006, PP947-958

Lin, Pei-Yang, Roschke, P. N. and Loh, C. H. , "System Identification and Real Application of the smart Magneto-Rheological Damper", 20th IEEE International Symposium on Intelligent Control (ISIC'05), 27-29 June 2005, Cyprus

A MICROCONTROLLER-BASED NEURAL NETWORK FOR DTMF DECODING

THOMAS L. HEMMINGER
DAVID R. LOKER
SHOULING HE
School of Engineering
Penn State Erie, The Behrend College, Erie, PA 16563 USA
e-mail: hemm@psu.edu, drl3@psu.edu, and sxh@psu.edu

ABSTRACT
In an earlier paper we demonstrated that reliable dual-tone multifrequency (DTMF) decoding can be accomplished using feedforward neural networks. Many possible solutions for DTMF decoding have been successfully employed, including filter banks and FFT algorithms. These designs often increase the parts count and complexity of the systems. Software designs can require manipulation of complex numbers, resulting in a heavy burden for integer-based microprocessors. This paper extends our work by addressing additional system noise for the PC implementation, and by introducing a microcontroller-based decoder using an inexpensive PIC 18F452.

INTRODUCTION

A common signaling method in telecommunications is known as dual tone multi-frequency (DTMF) dialing. In telephone applications, both low-band and high-band frequencies are simultaneously generated for each pushbutton of the standard keypad. Multi-channel DTMF decoding is performed by multiplexing channels on a digital signal processor (DSP) such that a large number of channels can be decoded simultaneously (Bagchi, S. and Mitra S. K., 1995). Hardware-based approaches (Becker, R. and Mulder J., 1991) employ filter banks for the detection of frequencies, and are implemented as custom circuits. These outputs are then applied to detectors for reconstruction of the appropriate key. Fourier Transform algorithms have been used in DSP applications (Bagchi, S. and Mitra S. K., 1995), (Mock, P., 1985), and filter banks have also been applied (Shatnawi, A. M. et al., 1997), (Brandao, A. L. and Romano J. M. T., 1990), (Valenzuela, R. A., 1990), but can be inefficient due to the high computational load. Artificial Neural Networks offer several advantages for this application. ANNs are intrinsically parallel and fast, and they are robust for detecting signals in the presence of noise (Masters, T., 1994), (*Neural Network Theory, Technology and Applications*, 1996), (Daponte, P. and Grimaldi D., 1998). An inefficient example of an ANN on a DSP for DTMF decoding can be found in (Daponte, P et al., 2000). This paper discusses an approach to develop an improved neural network-based implementation that can be used on a PC or a microcontroller.

DTMF STANDARDS AND PREVIOUS DECODING METHODS

Some standard parameters for DTMF signals are summarized in Table 1. Most systems employ an 8 kHz sampler, common to telephone systems. The minimum frequency separation in a DTMF encoder is 73 Hz, meaning that any algorithm should have a resolution greater than this. An FFT routine with 128 frequency bins would require about 900 complex computations yielding a resolution of 62.5 Hz. This meets the criterion stated above but it is clearly a heavy burden for most microcontrollers. There are alternative methods of spectral estimation that focus on specific frequency bands, requiring a lower number of floating point calculations. DTMF decoding lends itself to this idea since there are only seven distinct frequencies of interest.

Table 1. CCITT standards for DTMF signals

Parameter	Specification
SNR	> 24 dB
Twist	< 4 dB
Frequency distortion	< 1.5%
Dynamic Range	> 25 dBm
Silence time interval	> 45 ms
Required tone duration (T)	45 ms < T < 55 ms
Processing time	< 100 ms

PROPOSED ARCHITECTURE

Our approach is similar to that presented in (Daponte, P et al., 2000) except that we have been able to substantially reduce the computational overhead. We decode the signal by using pattern recognition in the time domain via the feedforward neural network. An important issue in processing these patterns is to force the signals to appear phase independent. This can be accomplished by making certain that the patterns are input to the network at the proper point in the waveform. We call this the characteristic point T_C and designate it as the maximum input value. To determine T_C, we require that the algorithm observe the input signal during the silence time interval to provide an unambiguous starting point. A small replica of the input vector is copied, the mean square value is computed, and the result compared against the noise level. Once the beginning of a signal has been detected it is necessary to determine where the actual peak of the signal is located. This is found by determining the minimum number of samples that must contain at least one signal peak. This worst-case duration is designated T_D and to establish its value we refer to eqn 1. A DTMF signal can be represented as:

$$\begin{aligned} s(t) &= A\cos(\omega_1 t) + A\cos(\omega_2 t + \phi) \\ &= 2A\cos[(\omega_1 + \omega_2)t/2 + \phi/2]\ \cos[(\omega_1 - \omega_2)t/2 - \phi/2] \end{aligned} \tag{1}$$

where A and ϕ are the magnitude and phase respectively. The lower frequency can be thought of as a modulating signal determining the worst-case period of the signal. This period T_D is equal to $4\pi/(\omega_1 - \omega_2)$. From table 2 we see that the key corresponding to '*' (941 Hz and 1209 Hz) yields the lowest difference frequency (268 Hz) and, hence T_D = 7.5 ms or 60 samples. The peak value should occur within this range. A simple test for the maximum value was conducted over this range to determine the characteristic point of the signal T_C. We employ aliasing to reduce the number of processing points. There are only seven frequencies of interest, so undersampling can be used if care is taken to retain the identities of each. Details can be obtained in (Loker, D. and Hemminger T., 2006). For this work, a nominal sample rate of 8 kHz was employed. These signals were undersampled at selected integer values of two through eight with a value of six being determined as optimal. We call this the *skip interval* and it corresponds to a 1.333 kHz sample rate. From table 3 we see that the key corresponding to '#' (392 Hz and 144 Hz) yields the lowest frequency separation; hence T_D = 8.1 ms or 11 samples in this case. We employed a PC to address the effects of additive noise, twist, and frequency distortion. This was necessary because the large values of distortion required the use of a larger network to produce suitable results. The microcontroller was used to decode signals with additive noise only, because we were able to employ a smaller network which kept the processing time within reason.

Table 2. Original DTMF signals and their differences. Lowest frequency separation indicated by shaded block.

		Column Frequencies		
		1209	1336	1477
Row Frequencies	697	512	639	780
	770	439	566	707
	852	357	484	625
	941	268	395	536

Table 3. This is the result of undersampling the values in table 2 by a factor of six. The rows and columns have the same keypad meanings as in the earlier tables. Lowest frequency separation indicated by shaded block.

		Column Frequencies		
		124	3	144
Row Frequencies	636	512	633	492
	563	439	560	419
	481	357	478	337
	392	268	389	248

We found that $T_D = 14$ samples was sufficient, yielding zero error for all practical conditions. To include frequency distortion and twist (Table 1) in the PC implementation it was necessary to increase T_D to 24 samples. Another very important consideration is the number of neural network inputs. One criterion was that the network contain only one layer of hidden nodes to reduce the computational burden on a microcontroller. In addition, all nodes were specified to be sigmoidal to restrict the output values. Initially, a network with 14 inputs, one hidden layer with 18 nodes, and 12 output nodes (14:18:12) was developed. This worked reasonably well for high SNRs, but degraded with increasing noise. It was decided to decrease the SNR to a level of 10 dB to correspond with other work in this area. It was found that a minimum of 20 input nodes provided near 100% accuracy in detecting DTMF tones at this noise level. During this time we reduced the number of nodes in the hidden layer, which reduced training time, with no degradation of performance. Although this configuration worked well, it was insufficient with the additional frequency distortion components mentioned earlier. As a result, the final network format was 30:12:12 for the PC configuration and 20:12:12 for the microcontroller with additive noise only. The phases of the DTMF signals are considered random with a uniform distribution. Therefore, for the training set we held the phase of one signal component constant while varying the phase of the other. We found 64 phase shifts on the second signal to be appropriate, corresponding to 5.6^o for each shift. Uniformly distributed random values of frequency distortion for both the low-band and the high-band tones, constrained to be within +/-0.6%, were used for training. To run the simulations, a set of signal vectors was created, corresponding to all 12 keypad entries. The vectors were separated by a silence time interval equivalent to 45 ms, additive noise was added to correspond to each SNR employed in the testing, and all signals were normalized. The parameters are acquired only during the interval where a signal candidate exists as to enable the use of local signal characteristics, in this case thirty samples. To test the results we created a test vector **V** containing all 12 DTMF tones with uniformly random phases. Each signal was separated by 45 ms intervals of silence time as specified by the CCITT standards summarized in Table 1. A summary of the procedure follows:

- **V** is searched sequentially via mean square value to locate the starting point of the signal, i.e., until the input reaches a fixed level above the background noise.
- Once the starting point has been found, flag F is set to allow for only one signal evaluation during this interval to avoid repeats or echoes.
- The next $T_D = 14$ (24 with PC) values are searched for the characteristic point T_C.
- After T_C is found, a copy of the next 20 (30) values is normalized using Eqn. 3.
- These values are sent to the neural network for evaluation.
- No action is taken on the input until it drops down to that of the background noise indicating the current signal has ended, at which time F is reset and the procedure continues to search for the next signal contained in $\mathbf{V}$.

To test the procedure, the algorithm was repeated 1000 times with a new set of DTMF tones for each iteration, resulting in 12000 evaluations per experiment. To verify the procedure, three independent tests were performed.

THE MICROCONTROLLER

Much of this research was originally conducted on a PC (Loker, D. and Hemminger T., 2006). The current project extends this work and includes implementation on a PIC18F452 microcontroller. We used a 10Mhz oscillator with an embedded C program separated into five functions,

1) Main function: All facilities used in this implementation are initialized here. They are Timer1, the CCP Compare module and interrupt service routine, as well as the LCD display. After the initialization, a while-loop runs indefinitely to detect the start point of each key press, searches for the maximum point of 14 additional samples, then selects the input pattern.

2) CCP_1 interrupt service routine: This timing routine is invoked whenever the parameter, CCP_1, equals Timer1. CCP_1 is initialized to a value of 1875. A 10Mhz oscillator is used here, meaning that Timer1 counts at a 10Mhz/4 = 2.5Mhz rate. When Timer1 is equal to CCP_1 (1875), it means 1875x0.4 = 750us, has elapsed, corresponding to the 1.33 kHz sample rate. Meanwhile, a window that contains the five-point sample sequence will move one step ahead. This is the level detector sequence. If the start point is detected, the AD converter will record the input data sequence to be processed with the neural network.

3) Input pattern normalization.

4) Neural network: The neural network is implemented with this subroutine.

5) Initialization of LCD and output to LCD: These functions are used to initialize the LCD module and send control instructions and data to be displayed to the module.

PC RESULTS

This uses a 30:12:12 configuration. For the first test, Gaussian noise was added to **V** yielding specific SNR values. The procedure was repeated for each SNR shown in figure 1(a) ranging from 24 dB to 10 dB. For reliable DTMF signal decoding, the system needs to yield zero errors. This is accomplished by restricting the SNR to a range of 24 dB down to 14.8 dB, providing a significant level of reliability. The second test measured the performance of the DTMF decoder when various twist (phase distortion) factors were used, ranging from 0 to 5dB. The results are shown below in Figure 1(b). As can be seen in the figure, values of twist up to 5dB resulted in very low error rates. When the twist was constrained to within 2dB, this resulted in zero error.

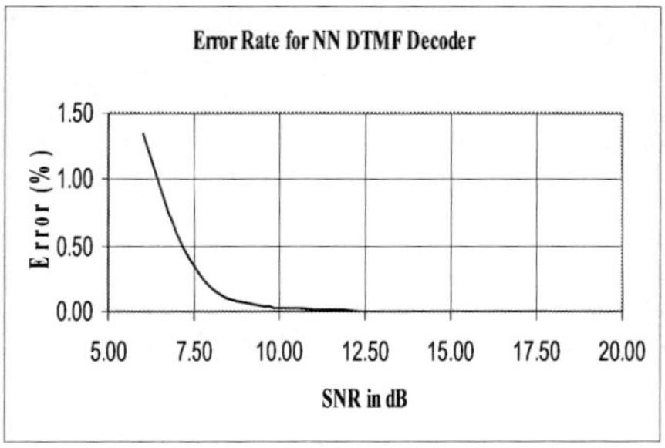

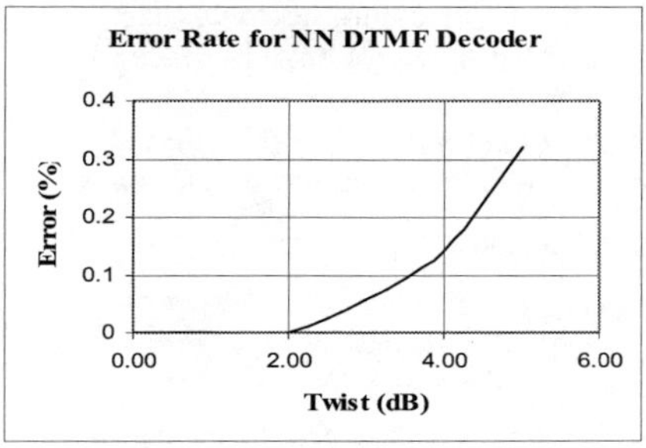

(a) (b)

Figure 1. A plot of error versus SNR for the PC DTMF decoder (a). A plot of error versus twist for the PC DTMF decoder (b).

The third test measured the performance of the DTMF decoder when various frequency distortion factors were used, ranging in value from 0 to 0.5%. Based upon each frequency distortion factor, uniformly distributed random values were used to determine the frequencies of both the low-band and the high-band tones. The results are shown below in Figure 2(a). As can be seen in the figure, values of distortion less than 0.5% resulted in very low errors. When the values were constrained to less than 0.3% the result was zero error.

MICROCONTROLLER RESULTS

The microcontroller was employed to decode DTMF signals with additive noise only. To cover all of the parameters addressed in the PC tests would have required about 320 ms per code symbol, which is well beyond the 100 ms upper bound. This reduced-parameter test was accomplished with the 20:12:12 format. Here, the processing time was about 200 ms, but we are in the process of acquiring a 40 MHz oscillator which will reduce the processing time to well below 100 ms. The same tests as done before with additive noise were conducted on the microcontroller system yielding the results presented in figure 2(b).

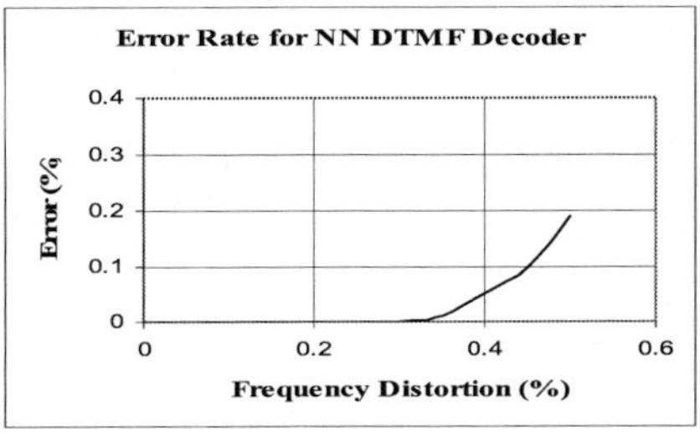

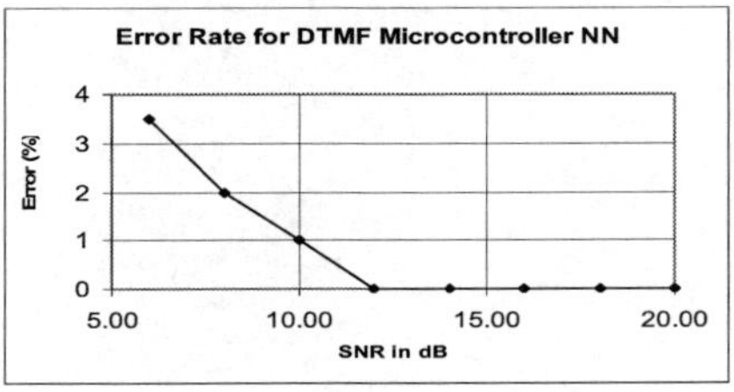

(a) (b)

Figure 2. A plot of error versus frequency distortion for the PC DTMF decoder (a). A plot of error versus SNR for the microcontroller DTMF decoder (b).

CONCLUSION

We will continue to work with the microcontroller in order to get the processing time within the CCITT constraints. The (30:12:12) network requires approximately 528 floating point multiplications and additions. This, along with additional overhead translates into approximately 320 ms of processing time. The smaller network was faster but not within the necessary parameters. The processing time can be significantly reduced if an alternate PIC with hardware multiply is employed. Alternatively, lookup tables and integer approximations can be explored along with faster clock speeds.

REFERENCES

Bagchi, S. and Mitra S. K., "An Efficient Algorithm for DTMF Decoding Using the Subband NDFT," *Proceedings of the IEEE Symposium on Circuits and Systems*, Seattle, WA, pp. 1936-1939, April-May 1995.

Becker, R. and Mulder J., "SIGFRED: A Low-Power DTMF and Signaling Frequency Detector," IEEE J. Solid-State Circuits, vol. 26, no. 7, pp. 1027-1037, 1991.

Shatnawi, A. M., A. I. Abu-El-Haiji, and A. M. Elabdalla, "A Digital Receiver for Dual Tone Multifrequency (DTMF) Signals," *Proceedings of the IEEE Conference on Instrumentation and Measurement Techniques*, Ottawa, Canada, pp. 997-1002, May 1997.

Brandao, A. L. and Romano J. M. T., "An Adaptive Notch Filter for Frequency and Power Estimation with Applications in Telephone Signaling Codes," *Proceedings of the IEEE International Symposium on Telecommunications* ITS'90, New York, NY, pp. 312-316, 1990.

Mock, P., "Add DTMF Generation and Decoding to DSP-uP Design," *EDN*, pp. 543-557, 1985.

Valenzuela, R. A., "Efficient DSP Based Detection of DTMF Tones," *Proceedings of the IEEE Global Telecommunications Conference*, GLOBECOM '90, San Diego, CA, pp. 1717-1721, Dec. 1990.

Masters, T., *Signal and Image Processing with Neural Networks*, Wiley, 1994.

Neural Network Theory, Technology and Applications, IEEE Technology Update Series, 1996.

Daponte, P. and Grimaldi D., "Artificial Neural Networks in Measurement," *Measurement J.*, vol. 23, no. 2, pp. 93-115, 1998.

Daponte, P., Grimaldi D., and Michaeli L., "Neural Network and DSP Based Decoder for DTMF Signals," *IEE Proc.-Sci. Meas. Technol.*, vol.147, no. 1, pp. 34-40, 2000.

Loker, D. and Hemminger, T., "A Neural Network Based DTMF Decoder," Artificial Neural Networks in Engineering, 2006.

SVM BASED PATTERN CLASSIFIER THROUGH CLUSTERING

NATACHA GUEORGUIEVA
Computer Science, City University of New York
Staten Island, NY 10314

IREN VALOVA
Computer and Information Science, University of Massachusetts
N. Dartmouth, MA 02747

KNUT SIEM
Computer Science, City University of New York
Staten Island, NY 10314

ABSTRACT
The concept of Support Vector Machines (SVMs) for classification and regression has been introduced by Vapnik in 1995. The classification algorithm proposed by Vapnik was intended as binary classification for linearly separable input data and the solution was to find an Optimal Separating Hyperplane (OSH). Very often the input data is not linearly separable. SVMs employ a technique commonly known as the kernel trick to address this problem. In this paper, we propose a pattern classifier topology which includes: a) Pattern generator able to produce a set of two- and three-dimensional patterns with desired characteristics; b) An incorporated Graphical User Interface (GUI) in order to perform learning by using different types of kernels and compare the results; c) We introduce an approach for preliminary clustering of the training set in order to avoid some limitations of traditional SVM algorithms; d) Visualization for two and three-dimensional case to facilitate the interpretation of the results. As demonstrated in the experiment section, the fitting ability of the kernel depends greatly on the characteristics of the input as well as on the parameter selection. The proposed classifier was tested on generated and well-known benchmark datasets.

INTRODUCTION

Support Vector Machines (SVMs) are a set of related supervised learning methods, for which solving classification and regression tasks is formulated as quadratic programming (QP) problems [1, 2]. SVMs may be defined as a classification method that determines that maximum-margin hyperplane. In the case of basic linear classification, a SVM creates a maximum margin hyperplane that lies in a transformed input space [3]. Given binary choice training examples (labeled either `yes' or `no'), a maximum-margin hyperplane divides the `yes' and `no' examples, such that the distance from the closest examples, i.e. the margin, to the hyperplane is maximized. Because of their area of implementation, SVMs compete with Neural Networks as tools for solving pattern recognition problems.

We propose an approach for constructing SVM based pattern classifiers through initial clustering. The calculation of training set centroids allows us to reduce the size of the learning set and to connect the SVM with Pattern Selection (PS).

Separable Input Data

In the case of two linearly separable classes, assuming that the training vectors $x_i \in \Re^n, i = 1,2,...,l$ and the output vector $y \in \mathfrak{R}^l$ such that $y_i \in \{-1,+1\}$, the decision function is

$$y(x) = sign(\mathbf{w}^T \mathbf{x} + b) \tag{1}$$

When the classes are separable the following holds:

$$y_k \left[\mathbf{w}^T \cdot \mathbf{x}_k + b\right] \geq 1, \quad k = 1,2,..., \; l \tag{2}$$

As is noted in [4], SVM formulations are done within a context of convex optimization theory which leads to the suggestion to formulate a constrained optimization problem in the primal weight space and then formulate the Lagrangian. With the conditions of optimality, the problem can be solved in the dual space of Lagrange multipliers (henceforth known as support values. The solution of problem P is the following: maximize the margin by

$$\begin{aligned} &\min_{w,b} J_P(\mathbf{w}) = \frac{1}{2}\mathbf{w}^T \mathbf{w} \\ &\text{such that} \quad y_k \left[\mathbf{w}^T \cdot \mathbf{x}_k + b\right] \geq 1, \quad k = 1,...,N \end{aligned} \tag{3}$$

This gives the Lagrangian

$$L_P(\mathbf{w}, b; \alpha) = \frac{1}{2}\mathbf{w}^T \mathbf{w} - \sum_{k=1}^{N} \alpha_k \left(y_k \left[\mathbf{w}^T \mathbf{x}_k + b\right] - 1\right) \tag{4}$$

with Lagrange multipliers $\alpha_k \geq 0$ for $k = 1,...,N$.

Minimizing L_P with respect to $\mathbf{w}, b$ one obtains the Karush-Kuhn-Tucker (KKT) conditions:

$$\begin{cases} \dfrac{\partial L_P}{\partial \mathbf{w}} = 0 \rightarrow \mathbf{w} = \sum_{k=1}^{N} \alpha_k y_k \mathbf{x}_k \\ \dfrac{\partial L_P}{\partial b} = 0 \rightarrow \sum_{k=1}^{N} \alpha_k y_k = 0 \end{cases} \tag{5}$$

resulting in the following decision function

$$y(x) = \mathbf{sign}\left(\sum_{k=1}^{N} \alpha_k y_k \mathbf{x}_k^T \mathbf{x} + b\right) \tag{6}$$

Substituting $\mathbf{w}$ in L_P with the expression obtained after minimization determines the following QP problem as a dual problem in the Lagrange multipliers α_k

$$\begin{aligned} &\max_{\alpha} J_D(\alpha) = -\frac{1}{2}\sum_{k,l=1}^{N} y_k \, y_l \, x_k^T \, x_l \, \alpha_k \alpha_l + \sum_{k=1}^{N} \alpha_k \\ &\text{such that} \quad y_k \left[\mathbf{w}^T \cdot \mathbf{x}_k + b\right] \geq 1, \quad k = 1,...,N. \end{aligned} \tag{7}$$

Solving this problem is the essence of training a linear classifier with separable input data.

Nonseparable Input Data

In order to separate hyperplanes for the case of non-separable input data, a slack variable, ξ, is introduced in the constraints.

$$\begin{Bmatrix} \mathbf{w}^T \cdot \mathbf{x}_i + b \geq +1 - \xi_i & \text{if } y_k = +1 \\ \mathbf{w}^T \cdot \mathbf{x}_i + b \leq -1 + \xi_i & \text{if } y_k = -1 \end{Bmatrix} \tag{8}$$

From this the primal weight space optimization problem can be constructed

$$\min_{w,b,\xi} J_P(\mathbf{w},\xi) = \frac{1}{2}\mathbf{w}^T\mathbf{w} + c\sum_{k=1}^{N}\xi_k$$
$$\text{such that} \quad y_k\left[\mathbf{w}^T \cdot \mathbf{x}_k + b\right] \geq 1 - \xi_k, \quad \xi_k > 0, \quad k = 1,...,N \tag{9}$$

where c is a positive real constant. The Lagrangian then becomes

$$L_P(\mathbf{w},b,\xi;\alpha,\nu) = J_P(\mathbf{w},\xi) - \sum_{k=1}^{N}\alpha_k\left(y_k\left[\mathbf{w}^T\mathbf{x}_k + b\right] - 1 + \xi_k\right) - \sum_{k=1}^{N}\nu_k\xi_k \tag{10}$$

with Lagrange multipliers $\alpha_k \geq 0$, $\nu_k \geq 0$ for $k = 1,...,N$, $\alpha_k \geq 0$ for $k = 1,...,N$.
Minimizing L_P with respect to $\mathbf{w}, b, \xi$ one obtains

$$\begin{Bmatrix} \dfrac{\partial L_P}{\partial \mathbf{w}} = 0 \rightarrow \mathbf{w} = \sum_{k=1}^{N}\alpha_k y_k \mathbf{x}_k \\ \dfrac{\partial L_P}{\partial b} = 0 \rightarrow \sum_{k=1}^{N}\alpha_k y_k = 0 \\ \dfrac{\partial L_P}{\partial \xi} = 0 \rightarrow 0 \leq \alpha_k \leq c, \quad k = 1,...,N \end{Bmatrix} \tag{11}$$

resulting in the classifier

$$y(x) = \mathbf{sign}\left(\sum_{k=1}^{N}\alpha_k y_k \mathbf{x}_k^T\mathbf{x} + b\right) \tag{12}$$

One of the main advantages of the SVM over other networks is that its training is performed through the solution of a linearly constrained convex QP problem. In this way only a global (not necessarily unique) minimum exists which can be found by efficient algorithms performing approximation in a finite number of steps. At the same time SVM based pattern classifiers have some limitations as: a) The number of Support Vectors (SV) is relatively high and in the cases of high degree of class overlapping some non-informative pairs (one pattern located at the wrong side of the border plus one pattern of the opposite class very near to it) are selected. There is not an immediate extension of standard SVM formulation to multi-class problems as well [5, 6].

On the other hand, pattern classifiers based on traditional neural networks (NNs) topology experience difficulties in adapting to the problem complexity and their complexity becomes unmanageable. As a result, NNs very often have problems with generalization through under- and/or overfitting [5].

KERNELS

This section is an overview of some of the kernels than can be used in SVM environments. Three constant parameters are used, a degree d , a multiplier γ and a scalar term c_0 . Kernel functions used in SVM are typically an equivalent of an inner product in a higher-dimensional feature space if they satisfy Mercer's conditions [5]:

$$K(\mathbf{x},\mathbf{x}_k) = \sum_m \alpha_m \varphi_m(\mathbf{x})\varphi_m(\mathbf{x}_k), \quad \alpha_m \geq 0 \tag{13}$$

$$\int\int K(\mathbf{x},\mathbf{x}_k)g(\mathbf{x})g(\mathbf{x}_k)d\mathbf{x} \ \ d\mathbf{x}_k > 0, \quad g \in \boldsymbol{L}_2 \tag{14}$$

Here we present some of the SVM kernels used in the shown experiments which satisfy the Mercer's conditions

- Polynomial kernels provide a simple nonlinear mapping to the feature space:

$$K(\mathbf{x}_k,\mathbf{x}_l)=(\gamma \ \ \mathbf{x}_k\mathbf{x}_l+c_0)^d \tag{15}$$

- The sigmoid function/MLP is the kernel equivalent of a multilayer perceptron with a single hidden layer.

$$K(\mathbf{x}_k,\mathbf{x}_l)=tanh(\gamma \ \ \mathbf{x}_k\mathbf{x}_l+c_0) \tag{16}$$

The γ parameter is called the scale and c_0 the offset. The support vectors correspond to the first layer and the Lagrange multipliers to the weights.

- Radial Basis Functions (RBFs) are popular kernel choices, and the Gaussian variant [4] in particular has received much attention.

$$K(\mathbf{x}_k,\mathbf{x}_l)=e^{-\gamma\|\mathbf{x}_k-\mathbf{x}_l\|^2} \tag{17}$$

The γ parameter controls the standard deviation of the Gaussian model.

- Symmetrical function kernel of type

$$K(\mathbf{x}_k,\mathbf{x}_l)=\frac{1}{\gamma\|\mathbf{x}_k-\mathbf{x}_l\|^2+1} \tag{18}$$

SVM PATTERN CLASSIFIER WITH PRE-CLUSTERING

In order to avoid the high computational effort for designing standard SVM and existing difficulties in solving a multi-class problem we propose a preliminary selection of critical patterns by means of clustering which includes the following steps:

- Clustering the patterns of each class.
- Making decision about critical clusters.
- Select patterns in each cluster according to their proximity to the border.

In our experiments we use a modified frequency sensitive competitive learning (FSCL) with reinforcement learning and Kohonen's Learning Vector Quantizer (LVQ) for clustering [7]. We applied the Kohonen's LVQ following the rules given below:

- If the two closest centroids to pattern x_k are z_i and z_j, where z_i corresponds to the x_k's class and z_j corresponds to a different class

$$z_i(k+1)=z_i(k)+\lambda[x_k-z_i(k)] \tag{19}$$

$$z_i(k+1)=z_i(k)-\lambda[x_k-z_i(k)] \tag{20}$$

where x_k is in a 20%-width window.

- If the two closest centroids are of the same class as x_k

$$z_{i,j}(k+1)=z_{i,j}(k)-\eta[x_k-z_{i,j}(k)] \tag{21}$$

- The other cases exclude changes.

The second step - making decision about critical centroids is the most important one. The principal that we used is the following one: when two centroids of different classes are the nearest in both senses, both are included in the first group of critical centroids.

The third step - selecting patterns in each cluster according to their proximity to the border was performed in the following way: a) For each centroid of each class determine the nearest centroid of the other class. b) Select patterns in accordance to their proximity to the middle point between these centroids. c) Repeat the steps in the same order for each centroid of the second class not selected in the above process.

EXPERIMENTS

The proposed pattern classifier is an extension of LIBSVM [8] and is developed in Java. The training pattern sets are shown in Fig. 1.

Random Data	Partial Iris Data	Spiral Data
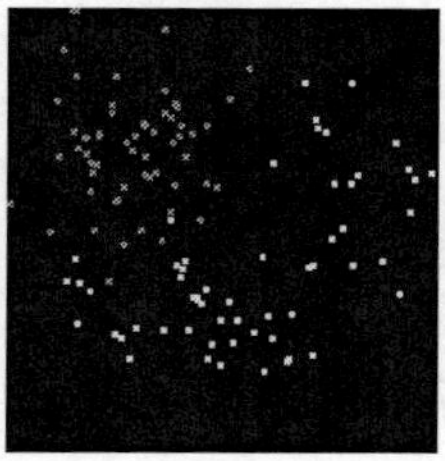	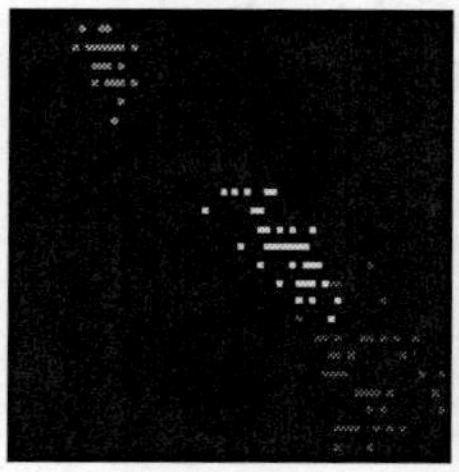	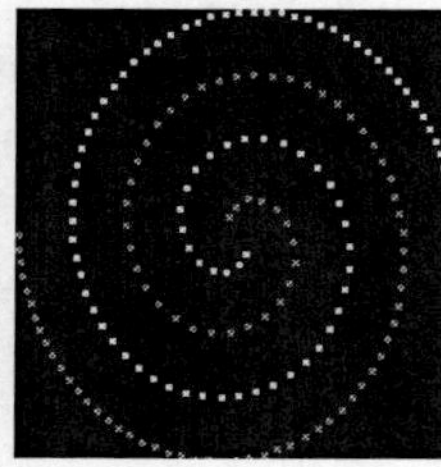

Figure 1. Training pattern sets

We propose a pattern generator able to produce patterns with different dimensionality and given number of classes. Fig. 2 demonstrates some experimental results with random data and different kernels.

RBF	**Sigmoid/MLP**	**Symmetrical Function 1**
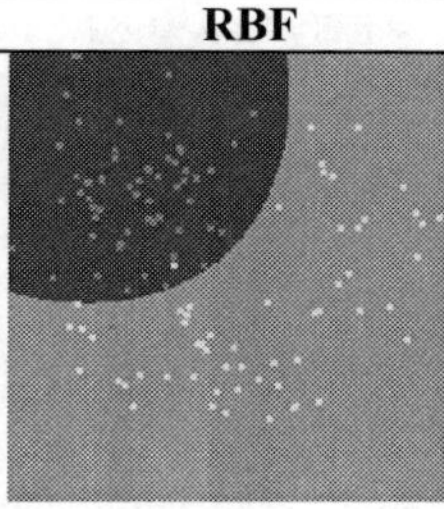	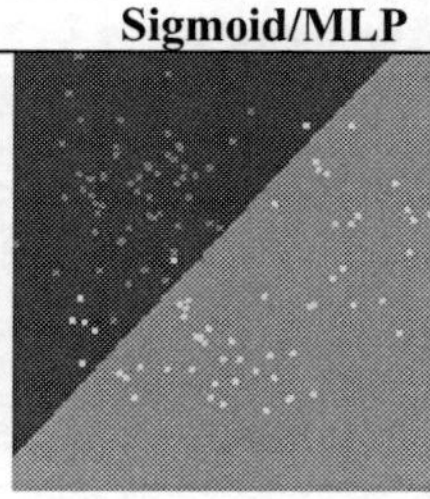	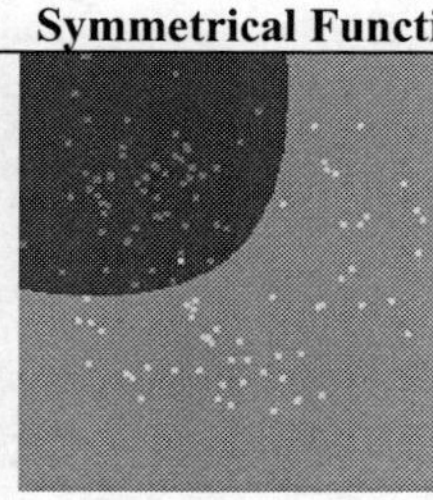
$\gamma = 2, c = 5$ 2 errors (1.66%)	$\gamma = 0.01$ 8 errors (6.66%)	$\gamma = 2, c = 5$ 2 errors (1.12%)

Figure 2. Experimental results with a generated pattern set

The iris training data set has three classes each of them containing 50 patterns. Fig. 3 shows partial experimental results with different kernels.

RBF	**Polynomial Kernel:** degree $d = 3$, $c_0 = 1.0$	**Symmetrical Function 1**
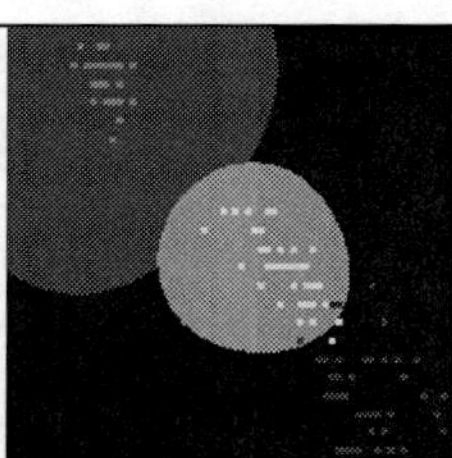 $\gamma = 10, c = 10$ 6 errors (4.00%)	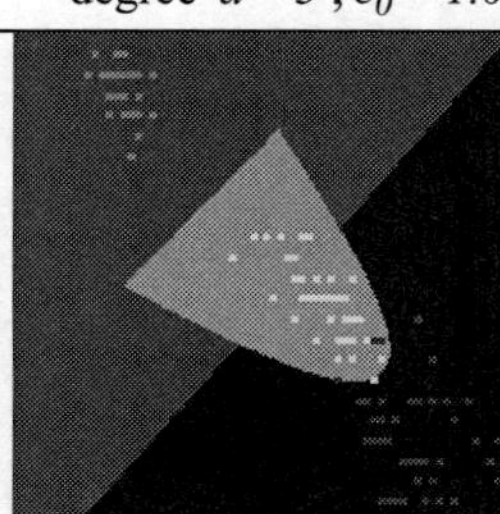 $\gamma = 10, c = 10$ 5 errors (3.33%)	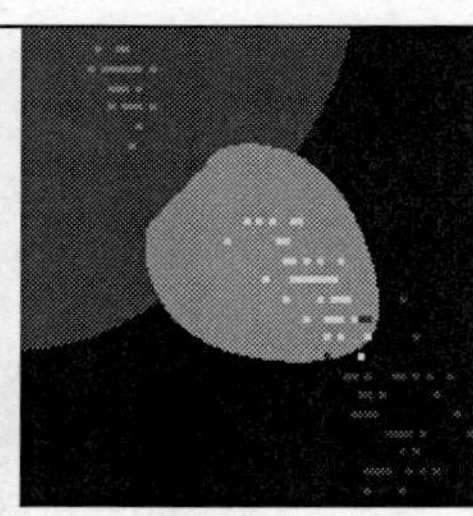 $\gamma = 10, c = 10$ 5 errors (3.33%)

Figure 3. Experimental results with Iris data

The spiral problem is a well-known, hard classification problem. All three of the kernels used in this experiment have a cost $c = 3$ and the Sigmoid/MLP kernel has $c_0 = 3$. The best results are shown in Fig. 4.

RBF	Sigmoid/MLP	Symmetrical Function 1
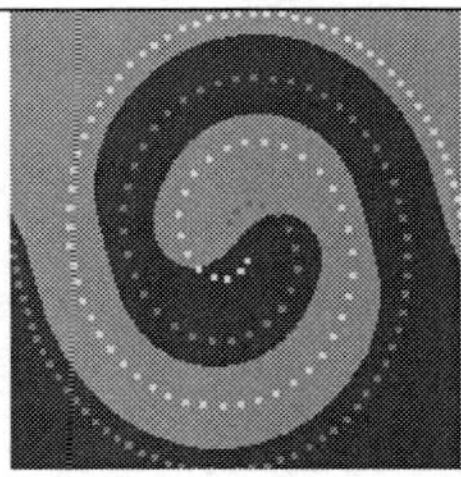	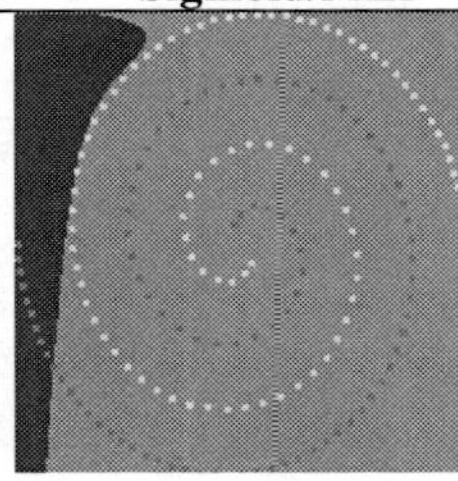	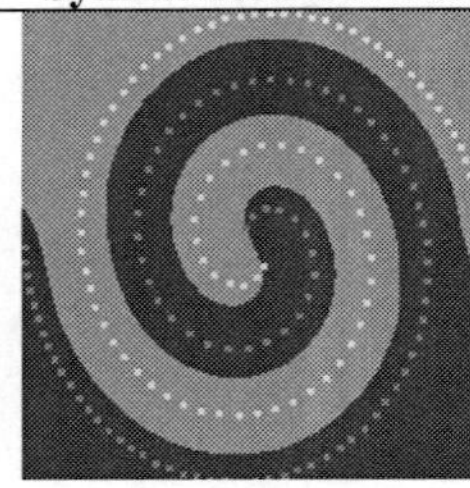
$\gamma = 20$ 10 errors (5.61%)	$\gamma = 20$ 79 errors (44.30%)	$\gamma = 20$ 2 errors (1.12%)

Figure 4. Experimental results with spiral data

CONCLUSIONS

The random generated pattern was more or less successfully classified by all the three kernels in that part of the experiment. The Polynomial kernel provides simple decision boundaries and work effectively, but fails dramatically, just like the Sigmoid/MLP kernel, when applied to the spiral pattern. As expected, the results produced by the Gaussian RBF kernel and the Symmetrical Function are similar. This is because both kernels are inversely proportional to the square of the norm of the two input vectors.

As demonstrated in the experiments, the fitting ability of the kernel depends greatly on the characteristics of the input. This means that choosing one kernel and expecting it to be the best for all applications is rather naïve. Parameter selection is also an issue; leaving the process of preparing a machine for classification yet one of much trial and error. Future challenges lie in automating this task.

REFERENCES

[1] Boser B., Guyon I, Vapnik V., A Training Algorithm for Optimal Margin Classifiers. In Fifth Annual Workshop on Computational Learning Theory. Pittsburgh, ACM, pp. 144--152, 1992.

[2] Vapnik V., Golowich S., and Smola A., "Support vector method for function approximation, regression estimation and signal processing," in Advances in Neural Information Processing Systems. Cambridge, MA: MIT Press, 1997, vol. 9.

[3] Vapnik V., "The support vector method of function estimation," in Nonlinear Modeling: Advanced Black-Box Techniques, J. A. K. Suykens and J. Vandewalle, Eds. Boston, MA: Kluwer, 1998, pp. 55–85.

[4] Suykens, J.A.K., T. Van Gestel, J. De Brabanter, B. De Moor, J. Vandewalle, Least Squares Support Vector Machines. World Scientific, Singapore, 2002, ISBN 981-238-151-1.

[5] Hsu, C.-W. and C.-J.L. Lin, A Comparison of Methods for Multiclass Support Vector Machines. IEEE Transactions on Neural Networks. 13(2): p. 415-425, 2002.

[6] Rossi, F. and N. Villa. Classifcation in Hilbert Spaces with Support Vector Machines. Proceeding of the International Symposium on Applied Stochastic Models and Data Analysis (ASMDA05). Brest, France: ASMDA, 2005.

[7] Kohonen T., The self-organizing map, Proc. IEEE, vol. 78, pp. 1464–1480, 1990.

[8] Chang, C.-C. and C.-J.L. Lin, LIBSVM: a library for support vector machines, 2001.

CONFIDENCE FRAMEWORK IN CLASSIFICATION

NATHAN J. LEAP
Dept. of Operational Sciences
Air Force Institute of Technology
Wright-Patterson AFB, Ohio

KENNETH W. BAUER, JR
Dept. of Operational Sciences
Air Force Institute of Technology
Wright-Patterson AFB, Ohio

ABSTRACT

There is no universally accepted methodology to determine how much confidence one should have in a classifier output. This research proposes a framework to determine the level of confidence in an indication from a classifier system where the output is a measurement value. The level of indication confidence is comprised of "classifier confidence" and "exemplar confidence." Classifier confidence is estimated using the exemplars in a test set and is reflected in quantities such as classification accuracy, average entropy, and confidence regions about these quantities. A classifier exhibits exemplar confidence through the magnitude of its posterior probability estimate for that exemplar. Classifier confidence and exemplar confidence are combined to form the level of confidence for a given classifier indication from a given classifier. In this paradigm, posterior probabilities are essentially adjusted based upon the confidence in the underlying classifier. This paradigm is applied to synthetic data as well as two real-world data sets.

INTRODUCTION

There is an extensive literature regarding the subject of classification confidence, but there seems to be no current methodology that is universally accepted. One approach to measuring confidence is to devise a measure whose range is the set of all real numbers ([1], [2], and [3]). These measures have no absolute meaning since there is no lower or upper bound on the measure; thus, the analyst must be intimately familiar with the measure to understand what the magnitude of the measure means. Another approach to developing a confidence measure is to limit the range of the confidence measure to the closed interval [0,1] ([4], [5], [6], [7], [8], [9], [10], [11], [12], and [13]). This limitation allows for an absolute interpretation of the measure. As stated in [8], this means that an analyst need not know the inner workings of the measure since it has absolute meaning. By definition, any measure that treats confidence as a probability falls into this category. Yet another approach to developing a confidence measure is to further limit it to be a binary value ([14], [15], and [16]).

Richards *et al.* [8] describe the basic need for a confidence measure. Often, classifier systems are judged as a whole by using measures such as true positive rate, false positive rate, and Receiver Operating Characteristic (ROC) curves. It is also desirable to be able to assess the confidence one might place in a given indication by a given classifier system. Much of the current literature uses either a posterior probability or something similar to a posterior probability to measure confidence in a given declaration ([5] and [7]). However, there are problems with simply using posterior probabilities. Roberts [1] states that the problem in using posterior probabilities as a measure of confidence is that they sum to 1; this is a problem because if a "rogue" data point is evaluated, it is treated as if it is a "genuine" data point and thus is classified with "apparent confidence" into one of the output classes. In the automatic target recognition (ATR) paradigm, this is the problem with out-of-library targets. Richards *et al.* [8] states that using a posterior probability is troublesome because this requires specification of the prior probabilities which are often unknown. Ross and Minardi [9] state that posterior probabilities are particulary difficult for classification systems to estimate. Because of these problems, researchers continue to search for a better confidence measure for a single declaration. This paper outlines the beginning of a new confidence paradigm and discusses applying the paradigm to two real-world data sets. Finally, some conclusions about the paradigm are provided.

CONFIDENCE PARADIGM

There is no universally accepted methodology to determine how much confidence one should have in a classifier output. This research proposes a framework to determine the level of confidence in an indication from a classifier system where the output is a measurement value. The level of indication confidence is comprised of "classifier confidence" and "exemplar confidence." Classifier confidence is estimated using the exemplars in a test set and is reflected in quantities such as classification accuracy, average entropy, and confidence regions about these quantities. A classifier exhibits exemplar confidence through the magnitude of its posterior probability estimate for that exemplar. In addition, perhaps some contextual information is available to supplement the classifier. Classifier confidence and exemplar confidence are combined to form the level of confidence for a given classifier indication from a given classifier. In this paradigm, posterior probabilities are essentially adjusted based upon the confidence in the underlying classifier.

CLASSIFIER CONFIDENCE

Ultimately, we seek a confidence measure for each declaration; that is, we want to know how confident we are in any given declaration. We attribute our confidence in the declaration to two factors: the strength of the indication (i.e., the magnitude of the posterior probability estimate as afforded by the classifier) and our confidence in the classifier itself. As a starting point, we will reflect confidence in a classifier through point estimates of its accuracy and entropy. Further, we also calculate a joint-confidence region about these estimates and use its area as an indication of our uncertainty in these performance measures due to finite test samples. The following describes a methodology to form a confidence measure for a classifier.

In information theory, the amount of uncertainty in a sequence of observations can be measured using entropy [17]. Here, we are interested in the amount of uncertainty in the posterior probability estimates since these are often used to make declarations. Entropy for a given exemplar is symmetric about the value 0.5. This means that the posterior probability vectors [0.2, 0.8] and [0.8, 0.2] will produce the same entropy. This poses a small problem. Consider the problem where two classifiers were attempting to produce a declaration for the same exemplar. If one classifier produced the posterior probability vector [0.2, 0.8] and the other [0.8, 0.2], by simply examining the entropy, we would be equally confident in both. However, assuming that the classification threshold for both classifiers was between 0.2 and 0.8, one would correctly classify the exemplar while the other would incorrectly classify the exemplar. Here, both classifiers would have the same measure of entropy for this exemplar, but one would be correct and the other incorrect. This forces us to combine entropy with some other measure that accounts for this type of error. Since this is simply classification error, it is logical to combine entropy with classification accuracy (CA).

THE BONFERRONI BOX

Average entropy and classification accuracy are now supplemented with the construction of a joint confidence region about their point estimates. The area of this region is reflective of our uncertainty in these performance measures. In this fashion, we link classifier confidence to the specific data used to train and test the classifier. A straight-forward and conservative confidence region is given by use of the Bonferroni inequality, and in two dimensions, the confidence region is rectangular. In three dimensions, the confidence region is a box. Thus, we need to estimate the standard error of both the classification accuracy and the average entropy.

Under the assumption of normality, the half-width of a 100(1-α)% confidence interval around classification accuracy is given by $HW[CA] = t_{\alpha/2,n-1}\hat{\sigma}_{CA}$ where $\hat{\sigma}_{CA}$ is the standard error of classification accuracy. A similar confidence interval can be formed around the average entropy, $\overline{E}$. The half-width of this interval is given by $HW[\overline{E}] = t_{\alpha/2,n\text{-}1}\ \hat{\sigma}_E\ /\sqrt{n}$ where $\hat{\sigma}_E$ is the sample standard deviation of E_i, $i=1,...,n$.

A joint confidence region that yields an overall confidence of at least (1-α) is the Bonferroni interval that is obtained by using $t_{\alpha/4,n\text{-}1}$ for both intervals [18]. The two intervals combined form a "Bonferroni box" around the point estimate $(CA,\ \overline{E})$.

One complication in the Bonferroni box is that it is possible (and quite likely) that it extends to values greater than one or less than zero in either direction. This is problematic because we can never observe values for classification accuracy or average entropy less than zero or greater than one. Thus, we set the maximum of any confidence interval value to one and the minimum confidence interval value to zero. This restricts the Bonferroni box to only feasible values.

CONFIDENCE FUNCTION

Since we are interested in a single measure of confidence for the classifier, one way to combine the three is to examine a plot where CA is on the x-axis and average entropy is on the y-axis and the Bonferroni box is displayed around the point. In such plots, the ideal point would be the point (1,0) and a corresponding area of zero for the Bonferroni box. This corresponds to perfect classification accuracy, zero average entropy, and no error in either estimate. That is, the closer a classifier is to the point (1,0) on the plot with a small Bonferroni box, the more confidence we would have in that classifier. If we make the area of the Bonferroni box into a third dimension, then the desired point would be (1,0,0).

Hence, the three-dimensional distance (i.e. the Euclidean distance between (1,0,0) and the point (classification accuracy, average entropy, area of the Bonferroni box)) divided by $\sqrt{3}$ is in the interval [0,1]. We propose to map this value into a "confidence" through the use of a fifth order polynomial

In order to fit a specific fifth order polynomial, we want to define points for six categories: supremely confident, extremely confident, very confident, moderately confident, extremely not confident, and supremely not confident. These categories correspond to the following six points: (0, 1), (0.1, 0.97), (0.25, 0.95), (0.65, 0.25), (0.95, 0.01), and (1, 0).

Now, for any classifier, we can calculate the 3-D distance and map this value into a confidence score. First, this gives us a new way to compare classifiers. That is, this is a new measure that can be calculated for a set of classifiers in order to choose which classifier is preferred. Second, this gives us classifier confidence that we can use to combine with exemplar confidence in order to derive an overall confidence score for every declaration. It should be noted that this paradigm can only be used for systems that output a posterior probability or some other measure of class membership. Some fusion schemes such as some of those used in [19] have output labels with no associated posterior probability or other measurement of class membership.

COMBINING CLASSIFIER AND EXEMPLAR CONFIDENCE

The methodology for combining classifier confidence and exemplar confidence is very similar to the methodology for assigning classifier confidence. With classifier confidence, we calculated a three-dimensional distance from an ideal point where the three dimensions were classification accuracy, average entropy, and area of the Bonferroni box around the point in the space defined by classification accuracy and average entropy. In the methodology to combine classifier and exemplar confidence, we will measure the distance from an ideal point in two dimensions. The first dimension is the posterior probability for a given exemplar. The second dimension is the classifier confidence. For a given indication, the desired point is (1, 1). Thus, one can measure how far any point is from the ideal point. This distance is bounded by $\sqrt{2}$.

We need a methodology to map the normalized two-dimensional distance into overall confidence. The shape of this curve should look similar to that of the classifier confidence. Again, a fifth order polynomial seems to provide us with an adequate curve so we will need six points and will use the same six category names. These categories correspond to the following six points: (0, 1), (0.05, 0.98), (0.25, 0.93), (0.5, 0.5), (0.9, 0.01), and (1, 0).

Finally, we now have a paradigm that we can use to examine how much confidence a user should have in any given indication. This confidence is a function of exemplar confidence as well as classifier confidence. The next step is to define a way to test this paradigm, and this work is currently underway.

APPLICATION TO REAL-WORLD DATA

This section details the process of testing the confidence framework developed thus far. Since real data sets are being used and replications cannot be arbitrarily generated, bootstrapping is employed. In this process, a random selection of the data is taken to train the classifier, and the remaining data is left for validation.

For a given classifier family and a given data set, the following process was followed. First, the data set was randomly split into training data and validation data, and a classifier was trained from a family of classifiers. For this classifier, the classification accuracy, average entropy, entropy standard deviation, and set of posterior probabilities were calculated for both the training set and the validation set. Then, confidence scores were generated. ROC curves were built by applying thresholds to the posterior probability estimates, and ROC-like curves were built by applying thresholds to the confidence scores. This process was then repeated 750 times for different random samples of the data set, and an average ROC curve was calculated for both the training set and the validation set and for both methodologies (i.e. posterior probabilities and confidence scores).

From the average training ROC curves, a threshold space was chosen where the confidence scoring outperformed the posterior probability and these thresholds were applied to the validation set. Thus, a region of interest resulted where thresholding the confidence scoring outperformed thresholding the posterior probabilities. The idea was to examine whether the trends hold when independent validation data was employed.

This process was implemented on two real-world data sets: ripley crab data and University of Wisconsin breast cancer data [20]. In both cases, a feed-forward neural network was used as the classifier and was trained using a back-propagation training algorithm. The results show that on average, using the simple post-processing scheme that involves basically adjusting the posterior probability estimates improves classification performance at a significant region on the ROC curve.

CONCLUSIONS

There is no universally accepted methodology to calculate the confidence one should have in a given classifier indication. There is a wide variety of literature on the subject, and most methods depend on employing posterior probability estimates. In this paper, a paradigm is developed to measure how much confidence one should have in a given indication from a given classifier.

In this paradigm, the level of indication confidence is a combination of classifier confidence and exemplar confidence. Classifier confidence is a function of a specific test set and is reflected in quantities such as classification accuracy, average entropy, and the confidence regions about these quantities. A classifier exhibits exemplar confidence through the magnitude of its posterior probability estimate for that exemplar. Classifier confidence and exemplar confidence are combined to form the level of confidence for a given classifier indication from a given classifier.

This paradigm was implemented on Monte Carlo simulation data throughout the development. In addition, it was implemented on two real-world data sets. It was shown that using this simple post-processing which involves adjusting the posterior probabilities, there is an improvement in classification performance.

REFERENCES

[1] Roberts, S. J., 1996, "Assessing the Confidence of Classification in Artificial Neural Networks," IEE.

[2] Jaeger, S., 2004, "Using Informational Confidence Values for Classifier Combination: An Experiment with Combined On-Line/Off-Line Japanese Character Recognition," Proceedings of the 9th International Workshop on Frontiers in Handwriting Recognition, IEEE.

[3] Ho, T. K., Hull, J. J., and Srihari, S. N., 1992, "On Multiple Classifier Systems for Pattern Recognition," IEEE.

[4] Melnik, O., Vardi, Y., and Zhang, C., 2004, "Mixed Group Ranks: Preference and Confidence in Classifier Combination," IEEE Transactions on Pattern Analysis and Machine Intelligence, 26(8).

[5] Wan, E. A., 1990, "Neural Network Classification: A Bayesian Interpretation," IEEE Transactions on Neural Networks, 1(4).

[6] Parker, D. R., Gustafson, S. C., and Ross, T., 2005, "Receiver Operating Characteristic and Confidence Error Metrics for Assessing the Performance of Automatic Target Recognition Systems," Optical Engineering, 44(9).

[7] Weintrab, M., Beaufays, F., Rivlin, Z., 1997, "Neural-Network Based Measures of Confidence for Word Recognition," IEEE.

[8] Richards, J. A., Jr., W. J. B., and Bray, B. K., 2003, "An informative confidence metric for ATR," Proceedings of SPIE, 5095.

[9] Ross, T., and Minardi, M. E., 2004, "Discrimination and Confidence Error in Detector Reported Scores," Algorithms for Synthetic Aperture Radar Imagery XI, Proceeding of SPIE, 5427.

[10] Huang, Y. S., and Suen, C. Y., 1993, "Combination of Multiple Classifiers with Measurement Values," IEEE.

[11] Tubbs, J. D., and Alltop, W. O., 1991, "Measures of Confidence Associated with Combining Classification Results," IEEE Transactions on Systems, Man, and Cybernetics, 21(3).

[12] Zhang, X., Gupta, N., and Gupta, R., 2006, "Pruning Dynamic Slices with Confidence." PLDI.

[13] Thomas, I. L., and Allcock, G. M., 1984, "Determining the Confidence Level for a Classification," Photogrammetric Engineering and Remote Sensing, 50pp. 1491-1496.

[14] Black, M., and Franklin, M., 2004, "Perceptron-Based Confidence Estimation for Value Prediction," IEEE.

[15] Bailey, T. C., Everson, R. M., Fieldsend, J. E., 2007, "Representing Classifier Confidence in the Safety Critical Domain: An Illustration from Mortality Prediction in Trauma Case," Neural Computing and Applications, 16(1).

[16] Delany, S. J., Cunningham, P., and Doyle, D., 2005, "Generating Estimates of Classification Confidence for a Case-Based Spam Filter," Lecture Notes in Artificial Intelligence, 3620 pp. 177-190.

[17] Duda, R., Hart, P., and Stork, D., 2001, "Pattern Classification," John Wiley and Sons, New York, NY.

[18] Wackerly, D.D., III, W.M., and Scheaffer, R.L., 2002, "Mathematical Statistics with Applications," Duxbury Press, Pacific Grove, CA.

[19] Nathan J. Leap, Paul P. Clemans, Kenneth W. Bauer, J., Paper accepted for publication on 05 Jan 07, "An Investigation of the Effects of Correlation and Autocorrelation on Classifier Fusion and Optimal Classifier Ensembles," International Journal of General Systems-Intelligent Systems Design.

[20] D.J. Newman, S. Hettich,C.L.Blake, and C.J. Merz, 1998, "UCI Repository of Machine Learning Databases."

A NEW HYBRID ALGORITHM FOR POLYPHASE CODES DESIGN

S.P SINGH
ECE Dept, MGIT
Hyderabad -75, India
singh_spgahlot@rediffmail.com

K SUBBA RAO
ECE Dept, College Of Engg, (O.U)
Hyderabad -7, India
kakarlasubbaraoa@yahoo.com

ABSTRACT

A new Hybrid Algorithm (HA) is proposed as a statistical technique for obtaining approximate solutions to combinatorial optimization problems. The proposed algorithm is a combination of Threshold Accepting (TA) and Hamming Scan algorithms. It combines the good methodologies of these algorithms like global minimum converging property of TA algorithm and fast convergence rate of Hamming scan algorithm. In this paper HA is used to design 32-phase sequences, which have good autocorrelation properties. We have found Barker codes up to length 23 with maximum alphabet size of only 32. The sequences of lengths from 24 to 289 have correlation properties better than well known Frank codes. The synthesized 32-phase sequence sets are promising for practical application to radar and spread spectrum communication. The convergence rate of HA is shown to be good.

I INTRODUCTION

Sequences with low aperiodic autocorrelation sidelobe levels are useful for channel estimation, radar, and spread spectrum communication applications. Sequences achieving the minimum peak aperiodic autocorrelation sidelobe level one (i.e. the Barker condition) are called Barker Sequences.

The aperiodic autocorrelation function (ACF) of sequence S of length N is given

$$A(k) = \begin{cases} \sum_{n=0}^{N-k-1} s_n s^*_{n+k} \; ; & 0 \le k \le N-1 \\ \sum_{n=0}^{N+k-1} s_n \, s^*_{n-k} \; ; & -N+1 \le k \le 0 \end{cases} \qquad \dots \quad (1)$$

If all the sidelobes of the ACF of any polyphase sequence are bounded by

$$|A(k)| \le 1, \quad 1 \le |k| \le N-1 \qquad \dots \quad (2)$$

then the sequence is called a generalized Barker sequence

In 1953 [1] Barker introduced binary sequences for lengths N = 2,3,4,5,7,11, and 13, fulfilling the condition in (2). The binary Barker can be regarded as a special case of polyphase Barker sequences. If the sequence elements are taken from an alphabet of size M, consisting of the Mth roots of unity.

$$S_m = \exp\left\{2\pi i.\frac{m}{M}\right\} =: \exp(\, i\phi_m) \qquad 0 \le m \le M-1 \qquad .. \quad (3)$$

the sequence is alternatively named as M-phase Barker sequence.

In 1965, Golomb and Scholtz [2] first investigated generalized Barker sequences and presented six phase Barker sequence of lengths N ≤ 13. In [3], 60-phase Barker sequences upto length 18 were reported. Polyphase Barker sequences of lengths up to 31 were given in [4-6], wherein, at length 30, an alphabet size of 7200 had to be used. In [7] , polyphase Barker sequences of lengths 32-36 were presented, wherein, at length 35, an alphabet size of 11520 was necessary. In [8], polyphase Barker sequences of lengths 37-45 were presented, wherein, an alphabet size of 120. Recently, in [9] polyphase Barker sequences of lengths 46-63 were presented, wherein, an alphabet size of 2000 had to be used. However, polyphase Barker sequences for larger lengths require larger alphabets and the possibility for exhaustive search diminishes. The synthesis of polyphase codes with good correlation properties can be regarded as a nonlinear multivariable optimization problem, which is usually difficult to tackle. The Threshold Accepting (TA) technique, introduced by Dueck. G & Scheuer [10] proved to be an efficient and powerful tool to find optimal or near optimal solutions for complex multivariable nonlinear functions. The concept of Hamming scan algorithm has been employed for obtaining the pulse compression sequences at larger lengths with good correlation properties [11,12]. This algorithm has fast convergence rate but has demerit viz., the tendency to be stuck with local minima. The HA has global minimum estimation capability of TA algorithm and fast convergence rate of Hamming scan algorithm.

Binary code is one of the most commonly used radar pulse compression signals due to the easy signal generation and processing [13]. However, polyphase signal has larger main lobe-to- peak sidelobe ratio over binary signal of the same code length. In addition, polyphase waveforms have a more complicated signal structure and thus, are more difficult to detect and analyze by an enemy's electronic support measures (ESMs). With the maturity of digital signal processing, the generation and processing of polyphase signals has become easy and less costly. In this paper, HA have been used for the design of thirty- two phase-coded sequences having good correlation properties.

II. THIRTY- TWO PHASE SEQUENCES DESIGN

A thirty-two phase coded sequence of length N bits is represented by a complex number sequence

$$\left\{ s(n) = e^{j\phi_m(n)} \ , \quad n = 1, 2, ..., N \right\} \qquad .... \quad (4)$$

where $\phi_m(n)$ is the phase of nth bit in the sequence and lies between 0 and 2π. If the number of the distinct phases available to be chosen for each bit in a code sequence is M, the phase for the bit can only be selected from the following admissible values:

$$\phi_m(n) \in \left\{ 0, \ \frac{2\pi}{M}, \ 2\frac{2\pi}{M}, ..., \ (M-1)\frac{2\pi}{M} \right\} \qquad ....(5)$$

Considering a polyphase coded sequence S with code length N, and distinct phase number M, one can concisely represent the phase values of S with the following 1 by N phase matrix:

$$S = [\phi_m(1) \ , \phi_m(2), ..., \phi_m(N)] \qquad ... \quad (6)$$

where all the elements in the matrix can only be chosen from the phase set in (5).

A more practical approach to design polyphase sequences with properties in (2) is to numerically search the best polyphase sequences by minimizing a cost function that measures the degree to which a specific result meets the design requirements. For the design of thirty-two phase coded sequences used in radar and communication, the cost function is based on the sum of squares of autocorrelation sidelobe peaks. Hence, from (1) the cost function can be written as,

$$E = \max \left\{ \left| A(k)_{k \neq 0} \right| \right\} \qquad ... \quad (7)$$

the minimization of cost function in (7) generates thirty-two phase sequences that are automatically constrained by (2).

III. DISCRIMINATING FACTOR (DF)

The discriminating factor (DF) is defined by Golay as follows.

$$DF = \frac{A(0)}{\max_{k \neq 0} |A(k)|} \quad \ldots (8)$$

The denominator is the peak sidelobe value and is related to the L∞ norm of the sidelobes.

IV THRESHOLD ACCEPTING ALGORITHM

The SA technique, introduced by Kirkpatrick et al [14] proved efficient and powerful tool to find optimal or near optimal solutions for complex multivariable nonlinear functions. The major advantage of the SA algorithm over the traditional "greedy" optimization algorithms is the ability to avoid becoming trapped in local optima during the search process.

A modification of Simulated annealing is called Threshold Accepting (TA) has been proposed [10] and claimed to be 'overwhelmingly superior' to it. Simulated annealing has a probabilistic acceptance rule that a perturbation is accepted, even if ΔE is positive, with a probability exp $(-\Delta E/T_i)$ [10] where ΔE is the change in cost due to mutation. T_i is the control parameter, which by analogy is known as the system "temperature." Normally, the temperature T_i is slowly decreased from a large value to a very small one during the annealing process. In Threshold accepting method, the acceptance rule is simpler. The perturbation is accepted as long as $\Delta E > T_i$. The remaining features of the simulated annealing method are retained.

V. HAMMING SCAN ALGORITHM (HSA)

The HSA is a traditional greedy optimization algorithm, which searches in the neighborhood of the point in all directions to reduce the cost function and has fast convergence rate. The basic difference between Genetic Algorithm (GA) and HSA is that GA uses random but possibly multiple mutations. The Mutation is a term metaphorically used for a change in an element in the sequence. For example if a phase value of a polyphase sequence is ϕ_m $(1 \leq m \leq M)$, it is replaced with phase ϕ_i, i =1, 2, M, $i \neq m$, and the cost for each ϕ_i, is evaluated. If the cost is reduced due to a change in phase value, the new phase value is accepted; otherwise, the original phase value is retained. This process is recursively applied to the entire sequence. The HSA mutates all the elements in the given sequence one by one and looks at all the first order-Hamming neighbors of the sequence. Thus, Hamming scan performs recursively local search among all the Hamming-1 neighbors of the sequence and selects the one whose objective function value is minimum.

VI. THE HA FOR POLYPHASE SIGNAL DESIGN

The HA is a combination of both TA and HSA. It combines the good methodologies of the two algorithms like global minimum converging property of TA algorithm and fast convergence rate of HSA. The demerit of HSA is that it gets stuck in the local minimum point because it has no way to distinguish between local minimum point and a global minimum point. Hence, it is sub-optimal [11-12]. The drawback in TA is that it has a slow convergence rate. The HA overcomes these drawbacks. The flowchart of HA for optimizing the orthogonal polyphase code set is shown in Fig.1.

The computational cost for searching the best polyphase, code of length N, and distinct phase number M through an exhaustive search, i.e., minimizing (7), is of the order of M^N grows exponentially with the code length and set size. Therefore, the numerical optimization of polyphase sequence is an NP-complete problem.

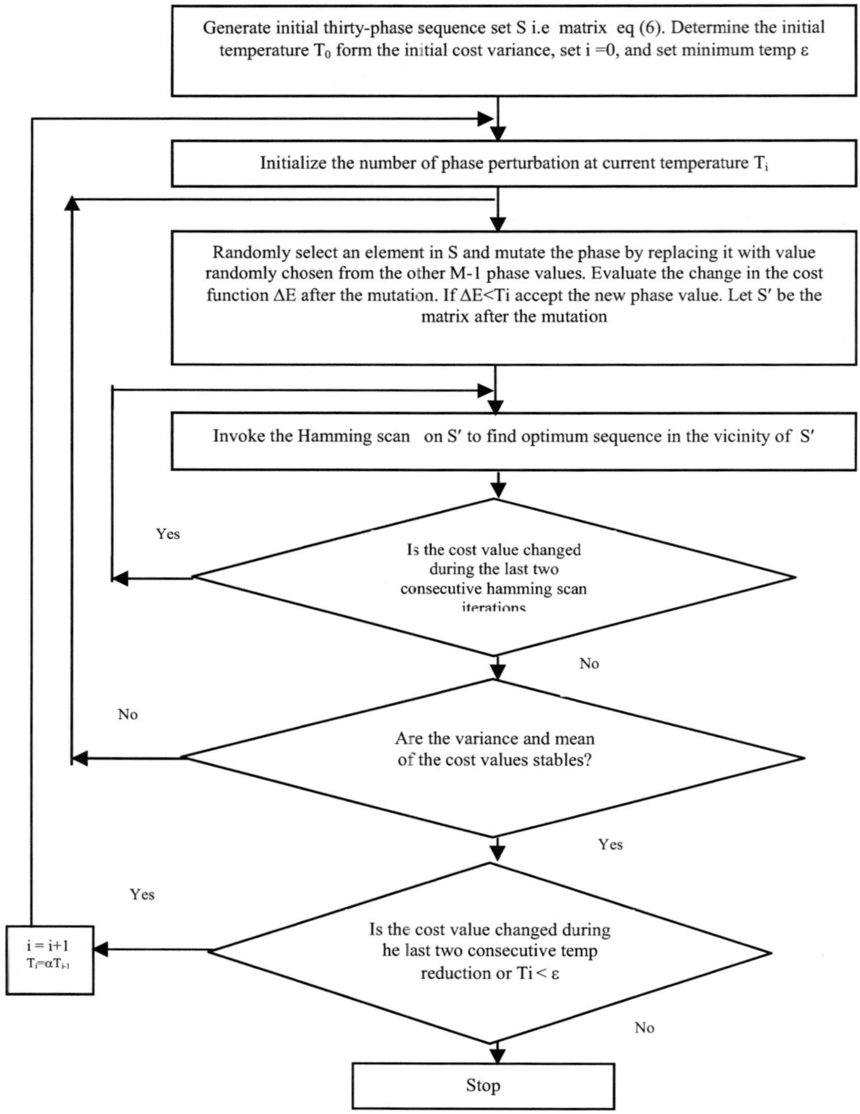

Fig. 1 Flow chart of Hybrid Algorithm

To successfully implement the HA, one needs to determine few key parameters or criteria for the annealing process. These are the starting temperature, the rate of decrement of temperature, i.e., cooling schedule, the determination of equilibrium condition at each temperature, and the annealing stopping criterion. In this work, the starting temperature T_0 is decided based on the standard deviation σ of the initial cost distribution by setting.

$$T_0 = 15\,\sigma \quad \text{.... (9)}$$

From the initial temperature T_0, the system temperature is systematically reduced according to

$$T_{i+1} = \alpha T_i \;\; (0 < \alpha < 1) \quad \text{.... (10)}$$

where α is a constant and chosen to be 0.95 in this design. At a temperature T_i ($i > 0$), the alphabets of the sequence are constantly "mutated" and accepted if $\Delta E > T_i$ until the cost function reaches equilibrium state. Then, the temperature is reduced to T_{i+1} according to (10), and the "mutations" are repeated until a new equilibrium state is reached at the updated temperature.

The annealing process is stopped if no "mutated" phase is accepted during three consecutive temperature reductions or $T < \varepsilon$, where ε is the minimum stopping temperature. The value of ε is chosen as 0.01.

VII. DESIGN RESULTS

Thirty–two phase sequences are designed using the HA. The length of the sequence, N, is varied from 2 to 289. The cost function for the optimization is based on (7). In this work all, the autocorrelation sidelobe peaks (ASP) are single realizations obtained using Pentium - IV, processor. Table I shows the synthesized results. In table I, columns 1, 4 & 7 show sequence length, N, columns 2, 5 & 8 show autocorrelation sidelobe peaks (ASP), columns 3, 6 & 9 show the Discrimination factor (DF) of the sequences. We have found nearly Barker codes up to length 23 (except length 22) with maximum alphabet size of only 32. The sequences of lengths from 24 to 289, have good correlation properties. Table II shows comparison of ASP of the synthesized sequences with Frank codes for different values of N. As shown in the table-synthesized sequences, have better correlation properties than Frank codes [15].

Table: I Correlation Properties (M = 32, N varied from 2 to 289)

Sequence Length (1)	ASP (2)	DF (3)		Sequence Length (4)	ASP (5)	DF (6)		Sequence Length (7)	ASP (8)	DF (9)
2	1.00	2.0		14	1.00	14.00		40	1.76	21.78
3	1.00	3.0		15	1.00	15.00		49	2.01	24.32
4	1.00	4.0		16	1.01	15.83		64	2.14	29.88
5	1.00	5.00		17	1.00	17.00		81	2.44	33.09
6	1.00	6.00		18	1.03	17.36		100	2.62	38.11
7	1.00	7.00		19	1.00	19.00		121	3.35	36.27
8	1.00	8.00		20	1.02	19.51		144	3.51	40.94
9	1.00	9.00		21	1.02	20.56		169	3.82	44.15
10	1.00	10.00		23	1.04	22.12		196	4.42	44.32
11	1.00	11.00		24	1.06	22.65		225	4.69	47.92
12	1.00	12.00		25	1.10	22.72		256	5.03	50.89
13	1.00	13.00		36	1.55	23.15		289	5.06	57.06

Table II : Comparison between synthesized 32-phase sequences and Frank codes

Sequence Length	25	36	49	64	81	100	169	196	225	289
Frank codes ASP	1.6	2.0	2.2	2.6	2.8	3.2	4.14	4.49	4.78	5.4
Synthesized ASP	1.2	1.4	1.9	2.1	2.4	2.6	3.82	4.42	4.69	5.0

It may be observed that as the length, N, increases, the DF also increases, which is the conformity with other findings. We have optimized/ maximized the Discrimination factor (DF) by minimizing the peak sidelobe.

VIII CONCLUSIONS

An effective HA has developed for the design of thirty-two phase-coded sequences used in radar systems and Spread spectrum communications for significantly improving performance of the system. This new approach combines the TA algorithm and the Hamming Scan algorithm and provides a powerful tool for the design of polyphase sequences. With the proposed optimization algorithm, thirty-two phase sequences are designed with different code lengths, N. We have found Barker codes up to length 23 with maximum alphabet size of only 32. The sequences of length from 24 to 289 have good correlation properties and better than Frank codes. The convergence rate of the algorithm is shown to be good.

REFERENCES

[1] Barker, R.H: "Group synchronizing of binary digital system" in Jackson, W, (Ed): *Communication theory* (Butterworths, London, 1953), pp. 273-287.

[2] Golomb, S. W, and Scholtz, R. A, "Generalized Barker sequences", *IEEE Trans, Inf. Theory,* 1965, IT-11,(4), pp. 533-537.

[3] Zhang., N, and Golomb, S. W, " Sixty phase Generalized Barker sequences", *IEEE Trans, Inf. Theory*, 1989, IT-35,(4), pp. 911-912.

[4] Bomer, L and Antweiler,M., " Polyphase Barker sequences". *Electronics letters* Vol. 34, No 16, 1989 25, pp 1577-1579.

[5] Friese, M, and Zottmann, H: " Polyphase Barker sequences upto length 31 ", *Electronics letters*, 1994,30,(23), pp 1930-1931.

[6] Zhang., N, and Golomb, S. W, " 7200 - phase Generalized Barker sequences", *IEEE Trans, Inf. Theory*, 1996, IT-35,(4), pp. 1236-1238

[7] Friese, M, " Polyphase Barker sequences upto length 36 ", *IEEE Trans, Inf. Theory*, 1996, IT-42,(4), pp. 1248-1250.

[8] A.R Brenner., " Polyphase Barker Sequences upto length 45 with small alphabets" *Electronics letters* Vol. 34, No 16, Aug 1998, pp 1576-1577.

[9] Peter Browein and Ron Fergusion " Polyphase sequence with low autocorrelation", *IEEE Trans, Inf. Theory,* 2005, IT -51, No 4, pp1564-1567.

[10] Dueck. G & Scheuer. T., "Threshold accepting: A general purpose optimization algorithm appearing superior to simulated annealing", *J. Computational physics,* 90, pp. 161-175, 1990.

[11] Moharir.P.S, Singh.R.and Maru. V.M., "S-K-H algorithm for signal design", *Electronics letters,* Vol 32, no 18, pp.1642-1649, Aug 1996.

[12] Moharir.P.S and Maru. V.M and Singh.R., "Bi-parental Product algorithm for coded waveform design in radar", *Sadhana*, Vol.22, no.5,pp 589-599, Oct. 1997.

[13] E. C. Farnett and G. H. Stevens, "Pulse compression radar," *Radar Handbook, Second ed.* New York: McGraw-Hill, 1990,

[14] S. Kirkpatrick, C. D. Gelatt, and M. P. Vecchi, "Optimization by simulated annealing," *Science,* Vol. 220, pp. 671–680, May 1983.

[15] R. L. Frank, "Polyphase codes with good non-periodic correlation properties," *IEEE Trans. Inform. Theory*, Vol. IT-9, pp. 43–45, Jan. 1963.

SYNTHESIS OF NON PERIODIC TERNARY SEQUENCES BY COMBINATORIAL OPTIMIZATION

N. RAJAIAH
ECE Dept, College of Engineering
(O U) Hyderabad-7, INDIA
rajaiah53@yahoo.com

K.SUBBA RAO
ECE Dept, College of Engineering
(O U) Hyderabad-7, INDIA
kakarlasubbarao@yahoo.com

ABSTRACT- Spread spectrum multiple access communications, and pulse compression radars, requires a set of sequences having peaky auto- correlation properties as well as minimum possible cross-correlation properties. Design of sequences with such properties is a non-linear multivariable optimization problem, for which modern optimization techniques should be applied. in this paper we have used combinatorial optimization using the techniques like hamming-scan, back-tracking to design sets of aperiodic ternary sequences with good auto-correlation, and cross-correlation properties. Some of these designed sequences are presented, and their properties are better than m- sequences.

1. INTRODUCTION

In code division multiple access (cdma) spread spectrum communication, number of users simultaneously transmits information over a common channel using different code sequences referred as spreading sequences (simon, 1985 *et al)*. in most cdma systems signals from different users are received asynchronously at base station receiver. in order to achieve low level of interference among competing users of the channel, reliable message synchronization and better spectrum spreading, the code sequences need to have (i) negligible cross-correlation at all lags for all pairs of sequences and (ii) individually peaky auto-correlations. such sequences also find applications in pulse-compression radars.

The measure of good sequences for CDMA system in recent times is the mean square values of aperiodic Cross-Correlations (CC) (Karkkainen, !992 & Pursely,1977). Hence CDMA system requires sequence sets with lowest possible average aperiodic CC values. Though good analytical design procedures based on powerful number theoretical results are available for sequences with optimal periodic auto and cross-correlations, but not if the sequences are taken in the aperiodic sense. Thus the problem of signal design can be assumed as multi objective non-linear optimization problem (Deng, 1996).One of the objectives being the maximizing auto-correlation of individual sequences and the other being minimizing the cross-correlation for all the pairs of sequences.

M-sequences, Gold sequences and Kasami sequences are some examples of spreading signals. These are periodic binary sequences cannot be used in applications where secrecy is of most importance. The **D**iscriminating factors (**D**) of these sequences are also not good. Hence it is proposed to design non-periodic non-binary sequences (ternary) by combinatorial optimization using techniques such as Hamming-scan, back-tracking.

2. ALGORITHMS FOR SIGNAL DESIGN

Many Global optimization techniques are proposed in the literature, such as Genetic algorithm (Goldberg, 1989), Simulated annealing (Boulanagar, 1996 *et al),* Evolutionary approach (Natarajan,2005 *et al*), and Tabu search (Boulanagar,1996 *et al*). Many modifications to these algorithms are also suggested. An algorithm called Eugenic algorithm is developed recently (Sing, 1996 *et al),* and is used to list ternary sequences with high merit factors. It supplements chance by the notion of locally complete search. One component of this algorithm is Hamming scan.

2. 1 HAMMING SCAN ALGORITHM

Hamming scan algorithm is more efficient though it is sub-optimal. It looks at all Hamming neighbors by mutating the elements of the sequence set and picks up the one with the largest cost function. If it is better than the original sequence set the algorithm recursively continues from there as long as an improvement exists. Mutation means changing the element of the sequence.

2.2 BACK TRACKING

Back tracking algorithm is invoked to improve the sequence set further where Hamming scan finds no better Hamming neighbors. Let this location be L. Back track then looks at the best H, Hamming neighbors though all of them are inferior (the algorithm takes unfavorable search with respect to cost function) and starts Hamming scan from each of them separately by C times where C is called climb. During C runs of Hamming scans if a better sequence set is found the algorithm shifts to that point, and improves the sequences by recursive backtracking. If the improvement ceases before C runs of the Hamming scan for a particular starting sequence set the climb is terminated.

3. OPTIMALITY CRITERION

Let S_i *and* S_j be two binary sequences of length N belongs to sequence set size M.

$$S_i = \left(S_i(0)\, , S_i(1), \ldots\ldots, S_i(N-1)\right) \tag{1}$$

$$S_j = \left(S_j(0), S_j(1), \ldots\ldots, S_j(N-1)\right) \tag{2}$$

The aperiodic auto- correlation of the sequence S_i is

$$r(k) = \sum_{n=0}^{N-1} S_i(n) S_i(n+k) \qquad 0 \le k \le N-1 \tag{3}$$

The aperiodic cross-correlation (Pursely, 1977) for the pair of sequences S_i *and* S_j is

$$c_{ij}(k)=\begin{cases}\sum_{n=0}^{N-1+k} S_i(n-k)S_j(n) & 1-N\le k\le -1\\ \sum_{n=0}^{N-1+k} S_i(n)S_j(n+k) & 0\le k \quad N-1\\ 0 & elsewhere\end{cases} \tag{4}$$

.

In pulse compression techniques the peakyness of the auto-correlation of a sequence is characterized by Discriminating factor (D) (Golay, 1977), defined as

$$D=\frac{r(0)}{\max_{k\neq 0}|r(k)|} \tag{5}$$

Since there are M users in the set, it is a good measure to study minimum D among the set as given by

$$D=\{\min(D_i) \quad i=1,2,...,M\} \tag{6}$$

We try to maximize this minimum value of D, which is one of the requirements.

Pursely (1997) showed that the average SNR at the out put of a Binary Phase Shift Keying (BPSK), asynchronous, Direct Sequence (DS), Spread Spectrum Multiple Access (SSMA) system for the jth user as a function of the Average Interference Parameter (AIP) for the M other users and the power of the additive white Gaussian noise present in the channel, in the form

$$SNR_j=\left(\frac{N_0}{2E_b}+\frac{1}{6N^3}\sum_{i=1,i\neq j}^{m} r_{ij}\right)^{-1} \tag{7}$$

where E_b is the bit error energy, N_0 is the noise power spectral density and N is the sequence length. The AIP, r_{ij}, is defined as

$$r_{ij}=2\mu_{ij}(0)+\mu_{ij}(1) \tag{8}$$

which is approximated by Karkkainen (1992) as

$$r_{ij}=2\sum_{k=1-N}^{N-1}|c_{ij}(k)|^2 \quad i=1,2,.,M,\ j=1,2,.,M,\ i\neq j \tag{9}$$

In order to minimize the interference, the parameter r_{ij} should be as small as possible for all $i \neq j$. Hence the Energy (ERG) in the cross-correlation of every pair of sequences is computed and maximum among all pairs is ascertained as follows:

$$ERG=\underset{i,j}{Max}\left[\frac{1}{2N-1}\sum_{k=1-N}^{N-1}|c_{ij}(k)|^2\right] \quad i=1,2...M;\ j=1,2,....M;\ i\neq j \tag{10}$$

Minimization of ERG is taken to be one of the other requirements. Since one component of the objective function (D) is to be maximized and the other (ERG) to be minimized, this situation is well matched to the application of the Simonization algorithms (Simon, 1956). Simon (1956) suggested that it is not wise to exclusively and rigorously optimize any quantitative measure one should only “satis fice”. This has also been recommended by others (Moharir, 1996 *et al*), thus gives relief from the difficulty that no optimization procedure can guarantee to yield the global optimum. The combined cost function is defined as

$$CF = \left(\frac{D_{am} - D_{bm}}{D_{bm}} \right) - \alpha \left(\frac{ERG_{am} - ERG_{bm}}{ERG_{bm}} \right) \tag{11}$$

where D_{bm}, D_{am} are the minimum Discriminating factors before and after mutation

ERG_{bm}, ERG_{am} are the maximum energies before and after mutation

α is the appropriate weightage factor, which characterizes the importance of the cross-correlation, and is taken as unity.

4. TERNARY SPREADING SIGNAL DESIGN

Initially the ternary sequence sets are chosen by random choice. The set size M and code length N in each sequence set is chosen similar to m-sequences. For every sequence in the set, the individual Ds, and pair wise CCs and also ERG in CCs are calculated. Maximum and minimum Ds, maximum and minimum ERGs and maximum cross-correlation are ascertained. These values are shown in table 1. Hamming scan is made to operate on these sequence sets. For ternary sequences (-1, 0, 1) the alphabets can be mutated in two possible ways: -1 $\rightarrow$ 0 or 1, 0 $\rightarrow$ -1 or 0 and 1 $\rightarrow$ -1 or 0, resulting two strands of Hamming neighbors. Better neighbor of these two strands is selected. When Hamming scan ceases to improve the cost function, the back tracking algorithm is applied. The results of the optimized sequence sets after successive Hamming scan and backtracking are shown in table 2. From these tables it is clear that, Hamming scan and back tracking considerably improve the cost function. For example the initial values of minimum D ($D_{\min}$), maximum ERG ($ERG_{\max}$) and maximum CC ($C_{\max}$) for a sequence set of size M=48 and length N=511 are 6.46, 142.87 and 72 respectively. With the Hamming scan and backtracking, $D_{\min}$ is increased to 12.10. $ERG_{\max}$ and $C_{\max}$ are reduced to138.82 and 65.

Table.1 Initial minimum Discriminating factors and maximum energy in cross- correlation of randomly selected sequence sets

M	N	$D_{\min}$	$ERG_{\max}$	$C_{\max}$	F= $C_{\max}$ / N
2	7	1.66	1.07	2	0.28
2	15	2.25	4.31	4	0.26
6	31	3.00	8.34	9	0.29
6	63	2.68	16.72	15	0.23
18	127	3.71	39.47	30	0.23
16	255	4.65	71.16	43	0.16
48	511	6.46	142.87	72	0.14
60	1023	8.78	272.69	104	0.10

Table. 2 Results of the optimized sequences with successive Hamming scan and backtracking

M	N	D_{min}	ERG_{max}	C_{max}	F= C_{max} / N
2	7	4	1.07	2	0.28
2	15	4.5	2.06	3	0.20
6	31	5.75	7.27	9	0.29
6	63	5.42	12.81	13	0.20
18	127	7.16	33.99	29	0.22
16	255	8.63	61.47	37	0.14
48	511	12.10	138.82	65	0.12
60	1023	14.79	261.01	93	0.09

Table. 3 Comparison of optimized sequences with the m –sequences

M	N	m-sequences		optimized sequences	
		C_{max}	F= C_{max} / N	C_{max}	F= C_{max} / N
2	7	5	0.71	2	0.28
2	15	9	0.60	3	0.20
6	31	11	0.35	9	0.29
6	63	23	0.36	13	0.20
18	127	41	0.32	29	0.22
16	255	95	0.37	37	0.14
48	511	113	0.22	65	0.12
60	1023	383	0.37	93	0.09

584 CONCLUSION

Signal design for spread spectrum applications with optimal correlation properties is most difficult and challenging optimization problem. In this paper we have attempted to solve such problem by combinatorial optimization, and presented families of aperiodic ternary sequences. The results show that Hamming scan and backtracking algorithms are more efficient. Comparison of optimized sequences with m-sequences also shows that the optimized sequences are better in terms of C_{max} and F. Hence they are more suitable for spread spectrum military communications.

ACKNOWLEDGE MENTS

The authors wish to acknowledge the help rendered by late Dr P.S.Moharir

REFERENCES:

Boulanager,C., Lequepeys ,J.R.,and Herault, L.,1996, "New spreading binary sequences obtained by Simulated Annealing and Tabu Search," Proceedings of ICPMSC'. PP 277-280. HongKong.

Deng,H.,1996, "Synthesis of Binary Sequences with Good Auto and Cross-Correlation Properties by Simulated Annealing" IEEE Trans on Aerospace and electronics systems Vol.32,No1,PP.98-107

Goldberg, D.E.,1989," Genetic Algorithm in Search Optimization, and Machine Learning," Reading, Addison-Wesley.

Golay,M.J.E.,1997, "Sieves for low Auto Correlation Binary sequences", IEEE Trans. on IT, Vol.23, No. 1, PP. 43-51.

Karkkainen K.H.,1992, "Mean-square Cross-correlation as a performance Measure for Spreading Code Families. "IEEE 2nd International Symposium on Spread Spectrum techniques and Applications(ISSTA '92) pp. 147-150.

Karkkainen, K.H., and Leppanen, P.A.,1991, "Comparison of the performance of some Linear spreading code Families For Asynchronous DS/SSMA systems", Proc.IEE Military Communications Conference,(MILCOM),Nov 4-7, pp.784-790.

Mohrir P.S., Maru V.M., & Singh, R.,1996, "Simonization for signal design", Sadana23: 351-358.

Natarajan, B.,.Das, S., and.Stevens, D.,2005," An Evolutionary Approach to Designing Complex Spreading Codes for DS-CDMA," IEEE Trans.Wireless.Commun,Vol.4 No. 5, PP. 2051-2055.

Pursely, M.B.,1997 " Performance Evaluation of Phase Coded Spread-Spectrum Multiple-Access Communication-Part1: System Analysis" IEEE Transactions on Communications, Vol. COM-25, No.8, August. pp. 795-799.

Proakis, J.G., and Salehi,M.,. "Communication systems engineering ", 2nd Ed,.Pearson Education Asia.

Simon. H.A.,1956, "Rational Choice and the structure of the environment", Psychol. Rev. 63, PP. 129.

Simon, M.K., Omura,K.,.Scholtz, R.A., and.Levitt, B.K.., 1985 " Spread Spectrum Communications" Vol.1,Rock ville,MD:Sci.Press.

.Sing,,R., Moharir,P.S., and Maru,V.M., 1996,"Eugenic algorithm based search for ternary pulse compression sequences," JIETE, Vol 42, PP. 11-19.

INDEPENDENT COMPONENT ANALYSIS FOR SEPARATING WATER VAPOR SPECTRUM FROM TERAHERTZ SPECTRA

GUANGYIN ZENG
Department of Decision Sciences & Engineering Systems,
Rensselaer Polytechnic Institute,
Troy, NY 12180

ALBERT REDO-SANCHEZ
Center for Terahertz Research,
Rensselaer Polytechnic Institute,
Troy, NY 12180

XI-CHENG ZHANG
Center for Terahertz Research,
Rensselaer Polytechnic Institute,
Troy, NY 12180

MARK J. EMBRECHTS
Department of Decision Sciences & Engineering Systems,
Rensselaer Polytechnic Institute,
Troy, NY 12180

ABSTRACT

Water vapor absorption significantly distorts the spectra of compounds that are measured in atmospheric air and makes it difficult to identify the compounds' terahertz fingerprints. To get clean spectra, it's necessary to measure the compounds in nitrogen environment or to measure another reference spectrum in air. But both methods face problems in real-life application scenarios, in which compounds are measured in atmospheric air with unknown and varying humidity. In this paper, Independent Component Analysis (ICA) is used to separate the water vapor absorption features from the Terahertz spectra of compounds that are measured in atmospheric air. The cleaned Terahertz spectra using ICA are compared with the reference spectra measured in nitrogen, which do not contain water vapor absorption features. Results show that ICA can effectively remove water vapor features from the compounds with only a single measurement.

INTRODUCTION

In the electromagnetic spectrum one can distinguish microwaves, infrared, visible light and X-rays. Terahertz (THZ) radiation lies between the microwave and infrared frequencies, at a frequency of about a trillion cycles per second. Its waves (T-rays) can be used to create pictures and transmit information.

Terahertz time-domain spectroscopy (THz-TDS) has been recognized as a powerful, non-destructive and non-contact sensing technique [1]. Terahertz (THz) radiation can penetrate through most of dry-dielectric materials such as paper, plastic, leather, and wood [2-4]. Many chemical materials, such as Explosives and Related Compounds (ERC), show resonant peaks due to intermolecular or phonon vibrations within the THz region [5-7]. Several sensing and imaging modalities using pulsed THz wave have been developed in the past ten years [8]. All these properties indicate that Terahertz sensing technology is an excellent tool for ERC detection. However, Terahertz spectra measured in atmospheric air suffer from the distortion caused by water vapor absorption. The performance reduces as the standoff distance increases.

Traditionally, the way to remove the effects of water vapor absorption is to measure the compound in a nitrogen environment so that there is no water vapor absorption. But this approach is not feasible for standoff distance sensing because there is little control over the environment. Another method is to measure the spectra of the compounds and the spectrum of a reference substrate, such as a mirror, and then perform a relative measurement by dividing the compound spectra by the reference spectrum. However, this method is also not realistic in a real-world scenario because it will be very difficult to place a reference substrate at the same location (distance and orientation) as the compound to be measured.

Several research efforts have been proposed to deal with the water vapor absorption distortion problem in standoff sensing measurements. A recent one is to use Terahertz Air-Breakdown-Coherent-Detection (THz ABCD) [9, 10]. However, this approach is technically complex because it requires precise control over the laser pulse profile, timing and phase.

The use of advanced signal processing techniques has gained little attention in the Terahertz community, which has been mostly focused on developing better emitters, better detectors, reducing acquisition time, and building a library of spectra of compounds. This paper shows the use of Independent Component Analysis (ICA) for eliminating the effect of water vapor absorption. ICA tries to find independent components underlying the observed data [11,12]. It can be seen as a nonlinear extension of Principal Component Analysis (PCA) [13]. Unlike PCA, which finds uncorrelated components by solving the eigenvalue problem, ICA finds components that are mutually independent.

The spectra of compounds measured in atmospheric air contain water vapor absorption features and the absorption features of the compounds, which are independent from each other. Hence, ICA can be used to separate them and remove the water vapor features out from the compounds' spectra. ICA offers the benefit that it does not require a prior measurement of the atmospheric air as a reference, which enables the practical use of Terahertz spectroscopy in practical environment. Two explosives, hexahydro-1,3,5-trinitro-1,3,5-triazine (RDX) and tetramethylene-tetranitramine (HMX), have been tested using ICA.

TERAHERTZ SPECTRA DATA

The Terahertz spectra are acquired with a common Terahertz time-domain spectroscopy setup [3]. Basically, a femto-second laser beam is split into a pump beam and a probe beam. The pump beam generates a Terahertz pulse through the excitation of a biased photoconductive antenna. The probe beam scans the Terahertz pulse by changing the delay between them with a mechanical delay stage. The modulation of the Terahertz pulse over the probe beam is detected by a ZnTe crystal through optical rectification and two balanced photo-diodes. The data (waveform) is acquired in time-domain and the frequency information is obtained via a Fourier Transform.

Figure 1 shows the time-domain waveforms and power spectra of atmospheric air and nitrogen. The sampling interval is 0.067 picoseconds and each waveform has 1001 data points, which results in a frequency resolution of 14.6 GHz. Rectangular windowing method is used in the Fast Fourier Transform. The critical frequency is 7.5 THz but the bandwidth of the system reduces the trustable range of data to 0.4 ~ 2.5 THz. This range has been used for ICA analysis. Data outside this range can be considered as noise.

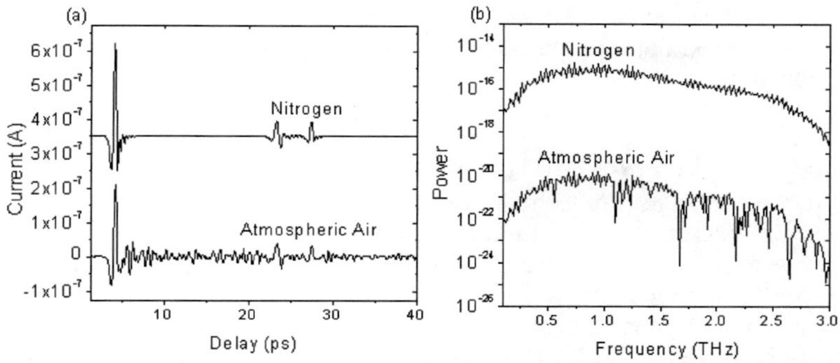

Figure 1. Time-domain waveforms (a) and power spectra (b) of atmospheric air and nitrogen gas. The baselines of waveform and power spectrum of nitrogen have been offset for clarity.

Six waveforms have been measured: atmospheric air, nitrogen gas, RDX in atmospheric air, RDX in nitrogen gas, HMX in atmospheric air, and HMX in nitrogen gas. The humidity ranges from 20% to 40% in atmospheric air measurements. The spectra of RDX and HMX are shown in the results section.

ICA ALGORITHMS

Let's start assuming that there are m observed variables $(x_1, x_2, \ldots x_m)$, which are linear mixtures of h independent sources $(s_1, s_2, \ldots s_h)$. Using matrix notation, the model can be expressed as:

$$X = AS \tag{1}$$

where $S \in R^{h \times n}$, $X \in R^{m \times n}$, $A \in R^{m \times h}$, $h \leq m$. The statistical model in Eq. (1) is commonly referred to as the ICA model. The goal is to find both the transform matrix A and the source S based only on the observation X. This problem is also called Blind Source Separation (BSS) problem [14, 15]. The BSS problem can be solved by finding a transform matrix W that satisfies:

$$S = WX \tag{2}$$

where $S \in R^{h \times n}, W \in R^{h \times m}, X \in R^{m \times n}$. The mixing matrix *W* and the source *S* can be determined by applying the Fast Fixed Point ICA algorithm developed by Hyvärinen et al [11]. Let's Denote $W = [w_1, w_2, \ldots w_n]^T$, and $S = [s_1, s_2, \ldots s_n]^T$, where $w_i, s_i \in R^{1 \times n}$. Note that it is possible to recover the observed signal *X* using the estimated *S* and *W* by taking a pseudo inverse as shown in Eq. (3). For simplicity, the number of signals is assumed to be equivalent to the number of independent sources, i.e. *m=h*. If this is not true, the dimensionality of signals can always be reduced by performing a PCA analysis.

$$\begin{aligned} X &= (W^T W)^{-1} W^T S \\ &= (W^T W)^{-1} (w_1^T s_1 + w_2^T s_2 + \ldots w_i^T s_i + \ldots w_n^T s_n) \end{aligned} \tag{3}$$

We can filter noise or irrelevant components out from the observed signals. In Eq. (3), if weight vector w_i is set to 0, then the effect of source s_i is eliminated from the recovered mixture X. If there are several independent components to be filtered out, we only need to set their corresponding weight vectors to zero simultaneously.

ICA has been applied to a variety of fields such as signal processing [16], financial market data mining [17,18], telecommunication [19], and image denoising [20].

ICA FOR EXTRACTING THE WATER VAPOR SPECTRUM

The spectrum in atmospheric air is considered as a linear mixture of two independent sources, the spectrum in nitrogen and the spectrum of water vapor. ICA analysis is applied to separate the water vapor spectrum from this spectrum and a spectrum in pure nitrogen gas.

In order to compare the results obtained by ICA, a reference water vapor spectrum is obtained using THz-ABCD method [19]-[20], which can provide a spectrum with a higher resolution (190 MHz). The water vapor spectrum obtained by this method is shown in Figure 2a.

The extracted water vapor spectrum by applying ICA on the spectra data in previous section is shown in Figure 2d. The resolution is lower (14.6 GHz) than that of the reference spectrum (190 MHz). To get a better comparison, the reference spectrum is interpolated to match the resolution of the spectrum obtained by ICA. The interpolated water vapor spectrum is shown in Figure 2c, in which some peaks of the original reference water spectrum disappear after the interpolation due to lower frequency resolution.

A longer scan is acquired in order to improve the frequency resolution of the spectrum. This scan has 3001 points and the sampling interval is also 0.067 picoseconds, which results in a frequency resolution of 3.65 GHz. The extracted water vapor spectrum using ICA is shown in Figure 2b.

Comparing the extracted water vapor spectra with the reference spectra, it can be seen that the positions of peaks of the spectra match almost perfectly. It can also be seen that the spectrum in Figure 2b is more similar to the original reference spectrum in Figure 2a than the spectrum in Figure 2d. This leads to the conclusion that improving the frequency resolution results in an extracted water vapor spectrum that is closer to the reference spectrum.

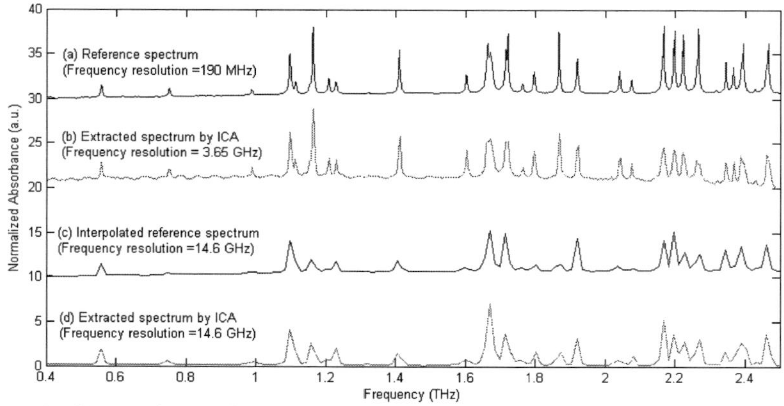

Figure 2. Comparison of the extracted water vapor spectrum with a reference water vapor absorption spectrum. The spectra have been normalized and offset for clarity.

ICA FOR REMOVING WATER VAPOR FEATURES FROM RDX AND HMX

In this section, ICA is used to remove the water vapor absorption features from the spectra of RDX and HMX measured in atmospheric air. The spectra of these compounds have three components: the spectra of the compounds, the spectrum in nitrogen, and the water vapor absorption spectrum, which are considered independent from each other. In order to eliminate the setup artifacts and the spectrum in nitrogen, we first perform a preprocessing by dividing the spectra of the compounds by a reference spectrum in nitrogen. This reference spectrum can be measured before and independently.

There are two water vapor spectra that can be used for ICA filtering, the interpolated reference water vapor spectrum and the extracted water vapor spectrum by ICA. Using the ICA filtering algorithm, we can separate the water vapor component from the spectra of compounds. The results are shown in Figure 3.

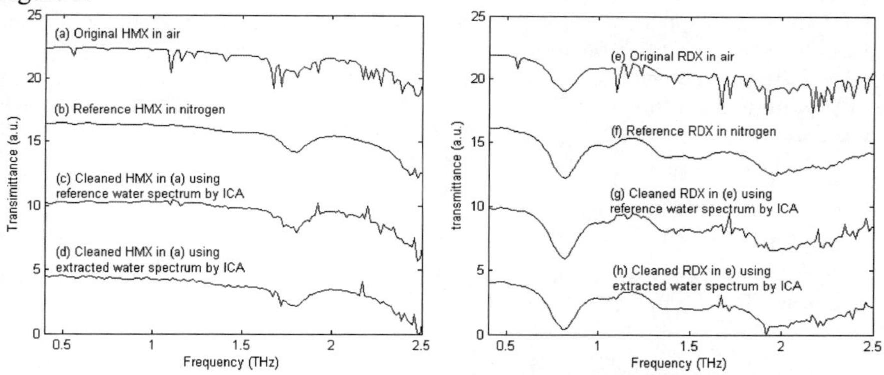

Figure 3 ICA separation of water vapor from HMX and RDX. The spectra have been offset for clarity.

From the above figure, it can be seen that most of water vapor peaks have been removed by ICA below 1.5 THz. There are some peaks remaining above 1.5 THz. But their amplituds are much smaller. Now it is possible to identify the peak of HMX at 1.8 THz based on Figure 3c or 3d, which is difficult to see in the original spectrum in Figure 3a. For RDX, the resulting spectra in Figure 3g and 3h using ICA cleansing are also better than the original spectrum in Figure 3e.

CONCLUSIONS

This paper has demonstrated that ICA is able to separate the water vapor features from the spectra of compounds measured in atmospheric air. The resulting spectra are very similar to the spectra obtained by measuring the compounds in nitrogen environment (no water vapor absorption).

The Terahertz spectra of compounds usually have to be measured in nitrogen environment in order to reduce the effects of water vapor absorption. But this is unrealistic in practice if the compounds are to be measured from a standoff distance, where the humidity of air changes quickly and it is not feasible to purge the water vapor content. This limitation can be reduced by using the ICA filtering methods introduced in this paper.

The spectrum in atmospheric air is considered as a linear mixture of two independent spectra, the spectrum in nitrogen and the spectrum of water vapor. ICA is used to separate these two independent spectra. Comparison of the extracted water spectrum and a reference water spectrum obtained from another experimental setup shows that ICA can effectively extract water vapor spectrum without measuring it directly. This result provides a new method to measure the spectrum of water vapor spectrum and separate it from other compounds.

In this paper, RDX and HMX have been tested by the ICA separation technique. Results show that ICA can successfully separate water vapor from the compounds. The practical benefit of this technique for standoff distance sensing is that it allows us to get a clean spectrum of the compound with only one measurement. There is no need to measure a reference spectrum by replacing the compound with a reference surface, such as a mirror, at the same location of the compound, which is not feasible from the practical point of view. ICA can obtain a clean spectrum of the compound using the waveform of the compound and the spectrum of water vapor that could be measured separately.

Future issues include improving the quality of the cleaned spectra and seek for more applications of ICA in Terahertz field. For example, ICA can be used to determine the content of different materials. It can also be applied to quality control process in sensing applications.

ACKNOWLEDGEMENT

The authors would like to thank Jian Chen, Nicholas Karpowicz, and Jianming Dai for providing the data.

REFERENCES

1. H. Zhong, J. Xu, X. Xie, Y. Tao, R. Reightler, E. Madaras, and X.-C. Zhang, "Nondestructive defect identification with terahertz time-of-flight tomography," *IEEE Sens. J.*, **5**, pp. 203-208 (2005).
2. B. Ferguson, and X.-C. Zhang, "Materials for terahertz science and technology," *Nature Mater.*, **1**, pp. 26-33 (2002).
3. D.M. Mittleman, R.H. Jacobsen, R. Neelamani, R.G. Baraniuk, and M.C. Nuss, "Gas sensing using terahertz time-domain spectroscopy," *Appl. Phys. B*, **67**, pp. 379-390 (1998).
4. D.M. Mittleman, M. Gupta, R. Neelamani, R.G. Baraniuk, J.V. Rudd, and M. Koch, "Recent advances in terahertz imaging," *Appl. Phys. B*, **68**, pp. 1085-1094 (1999).
5. Y. Chen, H. Liu, Y. Deng, D. Veksler, M. Shur, X. -C. Zhang, D. Schauki, M. J. Fitch, and R. Osiander, "Spectroscopic characterization of explosives in the far infrared region," *Proc. SPIE 5411*, **1**, pp. 1-8 (2004).
6. M.C. Kemp, P.F. Taday, B.E. Cole, J.A. Cluff, A.J. Fitzgerald, and W.R. Tribe, "Security applications of terahertz technology," *Proc. SPIE*, **5070**, pp. 44-52 (2003).
7. F. Huang, B. Schulkin, H. Altan; J.F. Federici, D. Gary, R. Barat, D. Zimdars, M. Chen, and D.B. Tanner, "Terahertz study of 1,3,5-trinitro-s-triazine by time-domain and Fourier transform infrared spectroscopy," *Appl. Phys. Lett.*, **85**, pp. 5535-5537 (2004).
8. D. Arnone, "Terahertz pulsed imaging and spectroscopy for chemical detection and security," *The Joint 30th International Conference on Infrared and Millimeter Waves*, **1**, pp. 198 (2005).
9. J. Dai, X. Xie, and X.-C. Zhang, "Detection of broadband terahertz waves with laser-induced plasma in gases," *Phys. Rev. Lett.*, **97**, 103903 (2006).
10. Xu Xie, Jianming Dai, and X.-C. Zhang, "Coherent control of THz wave generation in ambient air," *Phys. Rev. Lett.*, **96**, 075005 (2006).

11. A. Hyvärinen, and E. Oja, "A fast fixed point algorithm for independent component analysis," *Neural Computation*, **9**, 7, pp. 1483-1492 (1997).
12. A. Hyvärinen, and E. Oja, "Independent Component Analysis: Algorithms and Applications," *Neural Networks*, **13**, 4-5, pp. 411-430 (2000).
13. K. Pearson, "Principal Components Analysis," *The London, Edinburgh, and Dublin Philosophical Magazine and Journal of Science*, **6**, 2, pp. 559 (1901).
14. A. Bell, and T. Sejnowski. "An Information-maximization Approach to Blind Separation and Blind Deconvolution," *Neural Computation*, **7**, pp. 1129–1159 (1995).
15. G. Burel, "Blind Separation of Sources: A Non-linear Neural Algorithm," *Neural Networks*, **5**, pp. 937-947 (1992).
16. R. Vigário, "Extraction of Ocular Artifacts from EEG Using Independent Component Analysis," *Electroenceph.Clin. Neurophysiol.*, **103**, 3, pp. 395–404 (1997).
17. K. Kiviluoto, and E. Oja, "Independent Component Analysis for Parallel Financial Time Series," *In Proc. Int. Conf. on Neural Information Processing (ICONIP'98)*, **2**, pp. 895–898, Tokyo, Japan (1998).
18. A.D. Back, and A.S. Weigend, "A First Application of Independent Component Analysis to Extracting Structure from Stock Returns," *Int. J. on Neural Systems*, **8**, 4, pp. 473–484 (1997).
19. T. Ristaniemi, and J. Joutsensalo, "On the Performance of Blind Source Separation in CDMA Downlink," *In Proc. Int. Workshop on Independent Component Analysis and Signal Separation (ICA'99)*, 437–441, Aussois, France (1999).
20. A. Hyvärinen, "Sparse Code Shrinkage: Denoising of Nongaussian Data by Maximum Likelihood Estimation," *Neural Computation*, **11**, 7, pp. 1739–1768 (1999).

MODELING TRANSITION METAL NANOCLUSTERS FOR HYDROGEN STORAGE CAPACITY USING ARTIFICIAL NEURAL NETWORKS

WILLIAM O. GRIFFIN

JERRY DARSEY*

Department of Chemistry, University of Arkansas at Little Rock, Little Rock, Arkansas

ABSTRACT

Chemistry has often sought to explain bulk properties by looking at the molecular features that drive an overall system. This leads to chemical relationships, which predict whether such a property can be predicted based on the features of the smaller molecules. By correctly modeling these small molecules, the orbital energies, bond angles, lengths, and correct crystal structure are discovered using highly accurate *ab initio* molecular modeling techniques. The molecular orbital energies give an accurate representation of the bond the hydrogen and metal form in a metal hydride complex. Artificial neural networks using feed-forward back-propagation architecture reveal the relationship in this data to the percent hydrogen storage capacity of the metals in the bulk system. They do this by a self-consistent procedure adjusting the weight of the input according to the error of the preceding calculation. We are studying numerous transition element clusters to represent the bulk property of hydrogen adsorption. From these studies, the most promising potential materials will be further studied experimentally.

INTRODUCTION

Hydrogen is seen as a future fuel for powering the vehicles in the near future. In order for the hydrogen to be utilized, it must be "fill-able" in a car like the cars of today, and usable as gasoline's convenience is utilized. One of the solutions for a future fuel in replacing the diminishing oil supply is hydrogen. This simple fuel when burned produces water vapor, a product much more favorable to some than carbon dioxide.

Corresponding Author
Email: jadarsey@ualr.edu
Chemistry Department SCLB 451
2801 South University Avenue
Little Rock, AR 72204
(501) 569-8828

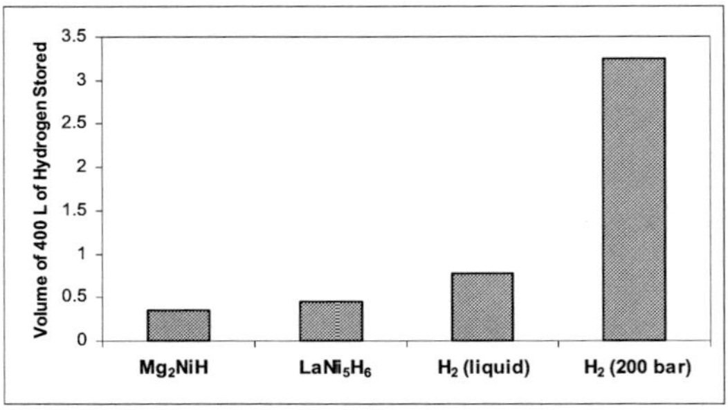

Figure 1 – The volume comparison of (from left to right) two metal hydrides, cryogenically stored liquid hydrogen, and hydrogen under pressure.

Metal hydrides are a complex between a metal unexposed to air and water that would bind hydrogen atoms, storing them for later use (MacKay 1966). Not only is this efficient as seen in figure 1, but it also is safer than storing hydrogen in pressurized tanks. Hydrogen is reactive and flammable as a diatomic gas (H_2). The risk of explosions in hydrogen vehicles would be too great considering our propensity for car accidents. Ideally, the hydrogen should be released or produced on a need to use basis, instead of available and reactive all at the same time.

Under most circumstances, the metal hydrides are an ideal solution to the energy needs of our vehicles. The problems that the metal hydrides face are the recharging time of a metal hydride cell, the release kinetics of the hydrogen, and the weight of the metal. There are many types of metal hydrides with varying properties (MacKay 1966). The spectra of what's available will expectantly lead to an optimum combination that would bring the best metal hydride mixture of low weight, good release kinetics, and fast recharge time.

METHODS

Due to the complicated problems associated with the metal hydride conundrum, we applied a solution that would take this large amount of data available for metal hydrides and metals and see if a prediction could be made based on the information given to a neural network as has been done previously for metal hydrides (Zeng and Chen 1997) as well as other materials. (Darsey et al 2000) A neural network takes the information given, where correlations are not readily apparent and can transfer the input through a sigmoidal function to an output. Artificial neural networks are especially good at tackling large amounts of data (Sumpter et al. 1994).

The information that we wanted correlated was the percent weight hydrogen of the metal hydrides, which we used as an output. The inputs are HOMOs and LUMOs. HOMOs and LUMOs are the Highest Occupied Molecular Orbital and Lowest Unoccupied Molecular Orbital of the structure, which is derived from the atomic orbitals of the atoms that comprise the nanocluster. Each of the HOMOs and LUMOs energy values are found from correct modeling of the molecule, which gives values for the inputs of the neural network. (Darsey 2002)

The program we used to get the inputs for our neural network approach is a molecular modeling technique called Gaussian 03. This is used to model the lowest energy structure of a molecule or nanocluster of metals. It is based on an *ab initio* approach to calculating the molecular orbital energies. Small nanoclusters of metals have interesting properties, and we have found that at the level of two to three atoms the bulk properties of the metals can often be elucidated from the HOMOs and LUMOs.

Ab initio modeling is a form of molecular modeling based on physical constants (Planck's constant, mass of an electron, the speed of light) and the elemental characteristics of the atoms (number of protons, neutrons, and electrons) (Grant and Richards 2003). Then, through a solution to the Schrödinger equation based on some assumptions, the orbital energies are solved for the molecule.

Seventeen nanoclusters were chosen based on their overall adsorption properties as metal hydrides. Pure metal hydrides of only one kind of metal atom were modeled in a two-atom cluster as well as a 3-atom cluster. In other words, all the cluster sets were of two or three atoms. This has also been seen from previous studies as important because it has been shown that more atoms seem to give more information in the nanoclusters considered (Mitchell et al. 2005).

The network was set to learn from the matrix of 16 molecules in a cross-validated leave one out training, so that the learning could be tested on the one left out. When the network finds an area that is in a low error range, we take and evaluate the training at that many training cycles with respect to the other 15 tested inputs from their respective cross-validated nets. The second network was made with 11 hidden nodes as well, but with 14 inputs instead of the 16. It was cross-validated as explained above.

Table 1: The Percent hydrogen storage capacity of metal hydrides used in this study.

Metal	Percent Weight of Hydrogen
Zr	2.16%
V	3.81%
Ti	3.98%
Pd	0.72%
Mg	7.66%
TiFe	1.86%
Mg_2Ni	3.60%
ZrNi	1.85%
$ZrMn_2$	1.77%
$ZrCr_2$	1.82%
TiCo	1.45%

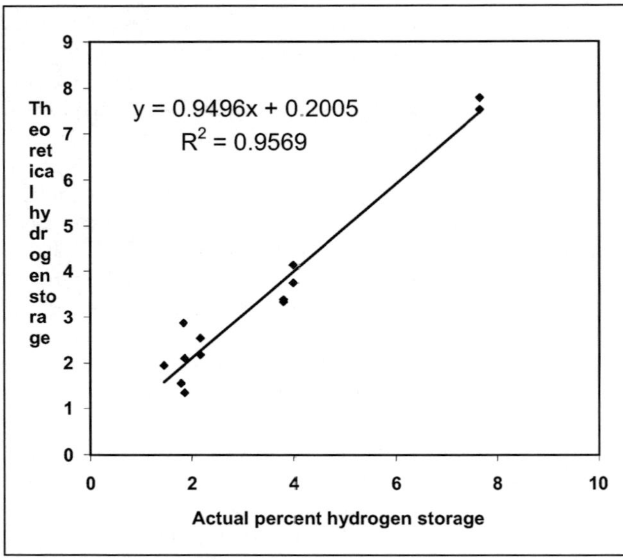

Figure 2 – The correlation of the cross-validated data. The R^2 value shows a greater than 95% fit to the prediction.

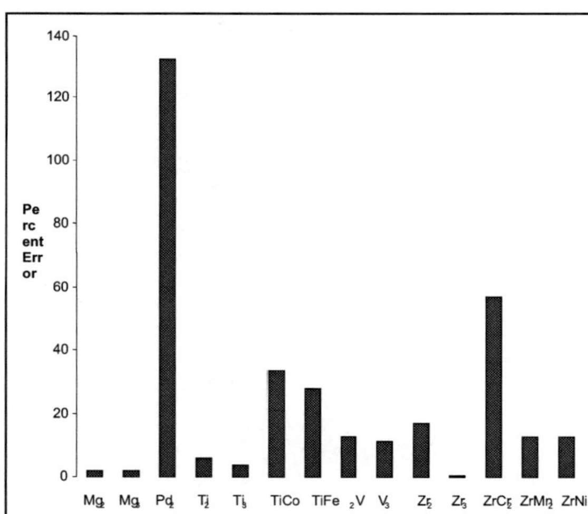

Figure 3 – The Neural Network produced a low percent error on all metals trained except Pd_2, possibly because there were not enough similar compounds in the training set. After confirming this with using the q test, we removed it from the prediction set.

RESULTS

The neural network had 34 inputs and 11 nodes in the hidden layer. It trained and learned well using about 250,000 training cycles. There were two outlying points with this training. One was high above and one beneath the linear fit of all the other points. Both removals were confirmed with a q test to validate their removal. The network was trained without the samples and there was one outlier. A q test was again to justify the removal, which was confirmed. Its point was removed and the fit was better than the previous correlation.

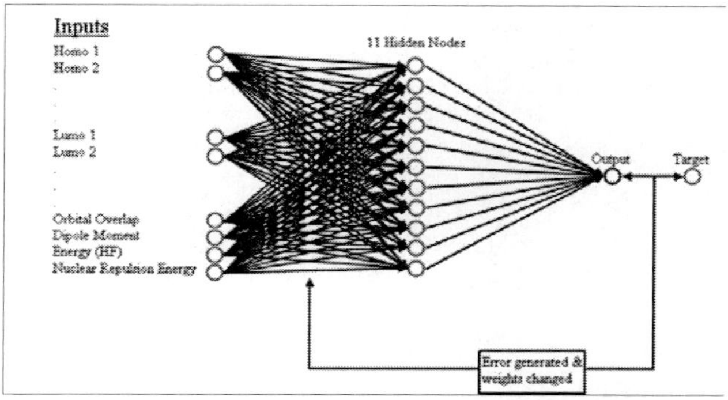

Figure 4 – The inputs consisted of HOMOs, LUMOs, the total energy, the nuclear repulsion energy, and the dipole moment calculated from Gaussian 03.

CONCLUSIONS

It will be possible to model the percent weight hydrogen and hopefully other factors that can be used to find better materials for hydrogen storage, looking at hydrogen binding and release factors that can be used for future materials investigations. Once the successful neural network predictions are validated, we hope to be able to model new materials that could be used to test new compounds *in silico* before evaluating them in a laboratory.

REFERENCES:

Darsey, J.A.; Sumpter, B.G.; Noid, D.W. "Correlating Physical Properties of Polymeric Materials and Organic Precursors of Polymers Using Artificial Neural Networks," *Smart Engineering Systems Design*, Vol. 2, (2000), pp. 283-298.

Darsey, J.A. "Neural Networks," In *Encyclopedia of Smart Materials*, Schwartz, Mel. Ed; John Wiley & Sons Inc: New York, NY, March, (2002), pp. 682-690.

Grant, G.H.; Richards, W.G. "Quantum Mechanics," In *Computational Chemistry*; Compton, R.G.; Oxford Chemistry Primers No. 29: Oxford University Press; New York, (2003); pp 16-17.

MacKay, K.M. "Introduction," In *Hydrogen Compounds of the Metallic Elements*; Bunbury, H.M., Ed.; General and Industrial Chemistry Series; E. &F.N. Spon Ltd: London, (1966).

Mitchell, N.C.; Darsey, J.A.; Buzatu, D.A., "Artificial Intelligence Modeling of Materials' Bulk Chemical and Physical Properties," *Proceedings of the IPSI-2005 CAMBRIDGE Conference* [CD-ROM]; Cambridge, Massachusetts, USA, July 2005.

Sumpter, B. G.; Getino, C.; Noid, D.W. "Theory and applications of neural computing in chemical science," Annual *Reviews in Physical Chemistry*, 1994, Vol. 45, pp. 439-81.

Zeng, W.; Chen, N. "Artificial neural network method applied to enthalpy of fusion of transition metals," Journal *of Alloys and Compounds*, 1997, Vol. 257,Issue 1-2, pp. 266-267.

Gaussian 03, Revision C.02, Frisch, M. J.; Trucks, G. W.; Schlegel, H. B.; Scuseria, G. E.; Robb, M. A.; Cheeseman, J. R.; Montgomery, Jr., J. A.; Vreven, T.; Kudin, K. N.; Burant, J. C.; Millam, J. M.; Iyengar, S. S.; Tomasi, J.; Barone, V.; Mennucci, B.; Cossi, M.; Scalmani, G.; Rega, N.; Petersson, G. A.; Nakatsuji, H.; Hada, M.; Ehara, M.; Toyota, K.; Fukuda, R.; Hasegawa, J.; Ishida, M.; Nakajima, T.; Honda, Y.; Kitao, O.; Nakai, H.; Klene, M.; Li, X.; Knox, J. E.; Hratchian, H. P.; Cross, J. B.; Bakken, V.; Adamo, C.; Jaramillo, J.; Gomperts, R.; Stratmann, R. E.; Yazyev, O.; Austin, A. J.; Cammi, R.; Pomelli, C.; Ochterski, J. W.; Ayala, P. Y.; Morokuma, K.; Voth, G. A.; Salvador, P.; Dannenberg, J. J.; Zakrzewski, V. G.; Dapprich, S.; Daniels, A. D.; Strain, M. C.; Farkas, O.; Malick, D. K.; Rabuck, A. D.; Raghavachari, K.; Foresman, J. B.; Ortiz, J. V.; Cui, Q.; Baboul, A. G.; Clifford, S.; Cioslowski, J.; Stefanov, B. B.; Liu, G.; Liashenko, A.; Piskorz, P.; Komaromi, I.; Martin, R. L.; Fox, D. J.; Keith, T.; Al-Laham, M. A.; Peng, C. Y.; Nanayakkara, A.; Challacombe, M.; Gill, P. M. W.; Johnson, B.; Chen, W.; Wong, M. W.; Gonzalez, C.; and Pople, J. A.; Gaussian, Inc., Wallingford CT, 2004.

NEURAL NETWORK BASED FAILURE PREDICTION MODEL FOR COMPOSITE HYDROGEN STORAGE CYLINDERS

J. Chen, J.Hu, V.G.K. Menta and K. Chandrashekhara
Department of Mechanical and Aerospace Engineering, University of Missouri-Rolla, MO 65409
and
William Chernicoff
US Department of Transportation, Washington, DC 20590

ABSTRACT
Composite high-pressure cylinders have potential application for hydrogen storage in automotive and transportation systems. Safe installation and operation of these cylinders is of primary concern. A neural network model has been developed for predicting the failure of composite storage cylinders subjected to thermo-mechanical loading. A backpropagation Neural Network model is developed to predict composite cylinder failure. The inputs of the neural network model are the laminate thickness, winding angle, and temperatures. The output of the model is the failure pressure. The finite element model of the cylinder is based on laminated shell theory accounting for transverse shear deformation and geometric nonlinearity. A composite failure model is used to evaluate the failure under various thermo-mechanical loadings. The neural network is trained using failure results of simulation under different thermal loadings and lay-up. The developed neural network model is found to be quite successful in determining the failure of hydrogen storage cylinders.

INTRODUCTION

The composite high-pressure cylinder is made with a high molecular weight polymer or aluminum liner that serves as a hydrogen gas permeation barrier. A filament-wound, carbon/epoxy composite laminate over-wrapped outside of the liner provides the desired pressure load bearing capacity (Liang, 2002). The cylinder is capable of sustaining pressures of 5000 psi or higher by taking advantage of high modulus, high strength and low specific weight of modern high performance composite (Takeichia, 2003). To design composite high-pressure cylinders with the most possible safety, reliability and minimum weight considerations, the failure of composite structures under various mechanical and thermal loadings need to be well understood.

To account for complex composite wall structure and environmental temperature influence, a comprehensive finite element model is developed and implemented in commercial finite element code ABAQUS (Mitlitsky, 2000) to analyze the failure of the composite cylinder. Due to a large number of parameters such as varying thermal loads, winding angles, cylinder dimensions and lay up configurations, it is a tremendous task to optimize the cylinder design and predict cylinder failure pressure through case-by-case finite element analysis simulation. A backpropagation Neural Network (NNk) model is employed to predict the failure pressure using the results obtained from a few finite element simulation cases. Three sets of simulation results with various winding angles and thermal loadings are applied for Neural Network training. The trained NNk model can be used as a tool to predict the failure pressure of the hydrogen storage cylinder under a given set of loads.

FINITE ELEMENT SIMULATION OF COMPOSITE HYDROGEN CYLINDER

A typical structure scheme of hydrogen storage cylinder is shown in Fig. 1. The inner aluminum liner is subject to mechanical pressure and temperature. In many current designs, a glass/epoxy layer is placed over the carbon/epoxy laminate to provide impact and damage resistance. The doubly curved shell theory accounting for out of plane shear deformations and geometric nonlinearity is used for the analysis of composite hydrogen storage cylinders.

The composite hydrogen storage cylinder is modeled and meshed using ABAQUS (2001) commercial Finite Element Analysis code (Fig. 2). A laminated shell element (S8R), based on doubly curved shell theory, including membrane bending and transverse shear effects, is used for modeling the cylinder. In order to estimate the failure pressure of the cylinder, it is necessary to include failure criterion. Tsai-Wu failure theory is utilized to check and report the ply-by-ply laminate failure by using user subroutine UVARM. In addition, temperature dependent material properties are incorporated in the model by using user subroutine USDFLD so that, at each integration point, the material properties are determined by the given temperature. The orientation of each element at every ply is handled by subroutine ORIENT according to winding pattern. The comprehensive model is then solved by using ABAQUS/standard solver with geometric nonlinearity considered. The dimensions considered in the present cylinder analysis are based on a typical design from literature (Mitlitsky, 2000). The outer radius of the cylinder is taken as R_{out} = 0.235 m and inner radius R_{in} = 0.22 m (Fig. 1). The pressure bearing carbon/epoxy laminate consists of 24 plies with a total thickness of 28 mm. The protective glass/epoxy layer and liner are 2 mm and 2.5 mm thick respectively. The cylinder is subjected to an internal pressure that gradually increases until the first ply failure occurs. To manufacture closed cylinders, two types of winding patterns are usually used: hoop winding and helical/polar winding. The thickness ratio R (total thickness of helical laminate/total thickness of hoop laminate) affects the failure pressure of the cylinder and a range of 0.1 to 2.0 is considered. Winding angle of laminae also affects the failure pressure of the cylinder and a range of 10° to 30° for helical winding and 89° for hoop winding has been considered based on the manufacturing feasibility. The plies in the protection laminate are oriented as 45° angle ply.

FAILURE MODEL FOR COMPOSTIE HYDROGEN STORAGE CYLINDERS BY FEEDFORWARD BACKPROPAGATION NEURAL NETWORK

Feedforward backpropagation Neural Network (Hornik, 1989) is used to predict the failure pressure of hydrogen cylinder. The schematic of the NNk is shown in Fig. 3. The relationship of failure pressure and the inputs (thickness ratio R, temperature inside of the cylinder T_{in}, temperature outside of the cylinder T_{out}, and filament winding angle θ) is modeled by a two-layer (hidden layer and output layer) network. Each layer consists of a number of processing units, known as neurons (Fig. 3). To obtain an easy training and a robust NNk model, inputs are scaled to a desirable range by a designed transfer function f_{in}^{T} before entering the input layer. Inputs are then passed through weighted connections to the hidden layer and then to the output layer. The output P' is, finally, scaled back to failure pressure P by a designed transfer function f_{out}^{T}. The number of neurons in the hidden layer and the characterizing weights and biases in the model are determined by training the NNk.

The input Transfer function

$$f_{in}^{T}(I_i) = I_i \cdot a_i + b_i \text{ and } I_i' = f_{in}^{T}(I_i) \qquad (1)$$

where,

I_i are input variables (corresponding to $\{R, T_{in}, T_{out}, \theta\}$ respectively)

I_i' are normalized input variables

$$a_i = \frac{n_{\max} - n_{\min}}{\vec{I}_i^{\max} - \vec{I}_i^{\min}},\ b_i = n_{\min} - Min\{a_i \cdot \vec{I}_i\},\ i = 1,\ 2, 3, \text{and } 4$$

[$n_{\min}$ $n_{\max}$] is the desired scale range (Here, taken as [-1 1])

and $\vec{I}_i$ are input training pattern vectors.

The activation function in the hidden layer is the Log-sigmoid function

$$f^{h}(x) = \frac{1}{1+e^{-x}} \qquad (2)$$

The activation function in the output layer is the Pureline function

$$f^{o}(x) = x \qquad (3)$$

The relationship of normalized input and output of NNk

$$P' = \sum_{j=1}^{N} W_{1j}^{o} f^{h}\left(\sum_{i=1}^{4} W_{ji}^{h} I_i' + {}^{h}b_j\right) + {}^{o}b \qquad (4)$$

where,

N is number of neurons in the hidden layer

W_{1j}^{o} are weights in the second (output) layer

W_{ji}^{h} are weights in the hidden layer

b^{o} is bias in the second (output) layer

b_j^{h} are biases in the hidden layer

and P' is the normalized output burst pressure.

The output Transfer function

$$f_{out}^{T}(I_i) = (P' - b)/a \text{ and } P = f_{out}^{T}(P') \qquad (5)$$

where,
P is the final output burst pressure

$$a = \frac{n_{\max} - n_{\min}}{\vec{P}^{\max} - \vec{P}^{\min}}, \; i = 1,\ 2, 3, \text{and}\ 4$$

$$b = n_{\min} - Min\{a_u \cdot \vec{P}\}, \; i = 1,\ 2, 3, \text{and}\ 4$$

and $\vec{P}$ is the output training pattern vector.

Training consists of providing a set of known input-output pairs (or patterns) to the network. During the backward propagation, weights in each node are iteratively adjusted based on the errors using the gradient descent equations. The procedure is repeated until convergence is achieved. In this study, the input-output sets are obtained from finite element simulation results. The model is trained in MATLAB NNk tool box. The fast training Levenberg-Marquardt algorithm is adopted for the training process.

CURVE FITTING

The network is trained by using the simulation results from three cases shown in Table 1. There are 12 neurons used in the neural network (NNk) model and the learning parameter is set to 0.01. After convergence is achieved, the network model is capable of predicting the failure pressure for a given cylinder thickness ratio (R), temperature distribution (T_{in} and T_{out}) and winding pattern (θ). To evaluate the performance of NNk prediction, six test cases (shown in Table 2) are studied. The test data from ABAQUS simulation results, for winding angles not included in the training (15° and 25°), are used to compare the predicted results from the trained neural network. The performance of the NNk is then illustrated in Fig. 4 and Fig. 5, and the maximum errors are reported for each case in Table 2. The test case 1 (T40-40) implies T_{in}=40°C and T_{out} = 40°C. It can be seen that the simulation results are in good agreement with the results predicted by NNk model.

CONCLUSIONS

A comprehensive finite element model is developed for composite hydrogen cylinder. Taking parameters determining the failure behavior of cylinder as inputs and failure pressure as output, a 2-layed backpropagation neural network model is developed and used to predict the failure pressure. The performance of the trained neural network is then evaluated by comparing the predicted values with test cases. The results are in good agreement.

ACKNOWLEDGEMENT

The project is sponsored by the US. Department of Transportation and University Transportation Center.

REFERENCES

1. Liang, C., Chen, H. and Wang, C., 2002, "Optimum Design of Dome Contour for Filament-wound Composite Pressure Vessels Based on a Shape Factor," Composite Structures, Vol. 58, pp. 469–482.
2. Takeichia, N., Senoha, H., Yokotab, T., Tsurutab, H., Hamadab, K., Takeshitac, H., Tanakaa, H., Kiyobayashia, T., Takanod, T. and Kuriyamaa, N., 2003, "Hybrid Hydrogen Storage Vessel, A Novel High-pressure Hydrogen Storage Vessel Combined with Hydrogen Storage Material," International Journal of Hydrogen Energy, Vol. 28, pp. 1121 – 1129.
3. Mitlitsky, F., Weisberg, A., H. and Myers, B., 2000, "Vehicular Hydrogen Storage Using Lightweight Tanks," Proceedings of the U.S. DOE Hydrogen Program Review NREL/CP-570-28890.
4. ABAQUS, 2001, ABAQUS Theory Manual.
5. Hornik, K., Stinchcombe, M., and White, H., 1989, "Feedforward Networks are Universal Approximators," Neural Networks Vol. 2, pp. 359-366.

Table 1: Lay-up configurations for various winding patterns

Case No.	Lay-up pattern	Description
1	$[\pm 20^\circ/\pm 89^\circ]_6$	24 layers with total thickness of 28 mm
2	$[\pm 10^\circ/\pm 89^\circ]_6$	T_{in}=[25 50 75 100 120 140] °C
3	$[\pm 30^\circ/\pm 89^\circ]_6$	T_{out}=[25 50 75 100 120 140] °C

Table 2: Testing cases and maximum error of prediction

Testing case	Inputs			Max. Error
	T_{in} (°C)	T_{out} (°C)	θ (Degree)	(%)
1	40	40	15	1.32
2	30	80	15	-1.54
3	130	40	15	1.23
4	40	40	25	1.11
5	30	80	25	-2.40
6	130	40	25	-1.00

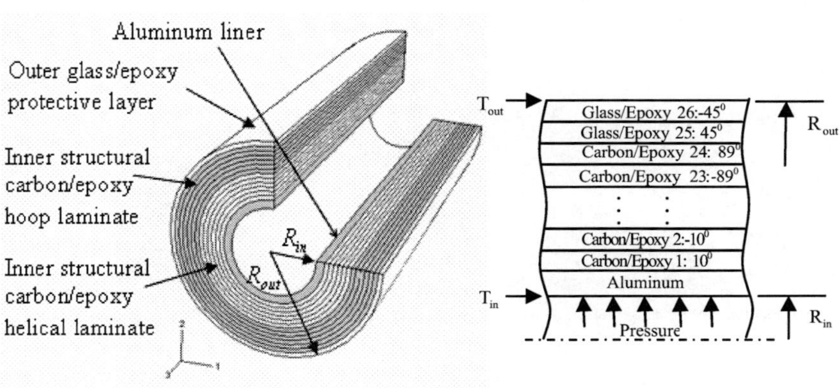

Figure 1 Structure scheme of hydrogen storage cylinder

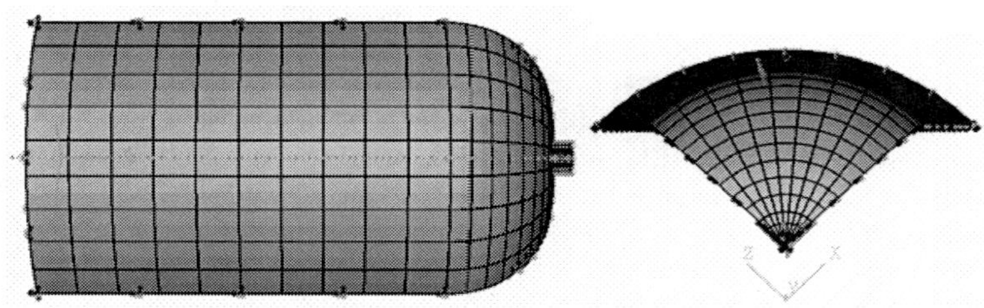

Figure 2 Finite element simulation of hydrogen cylinder

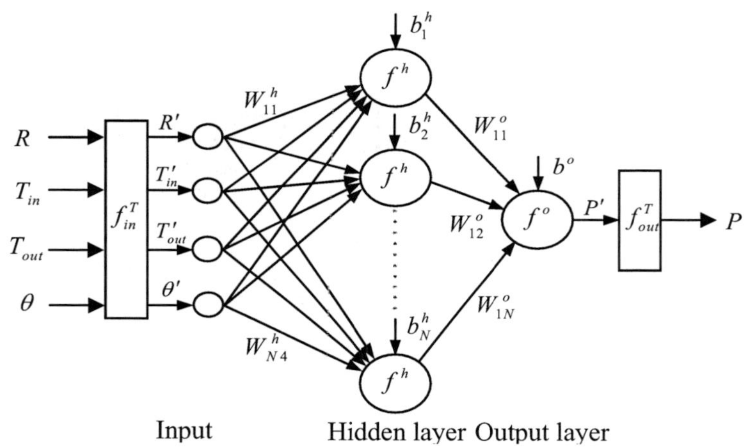

Figure 3 Feedforward Back-propagation Neural Network Architecture

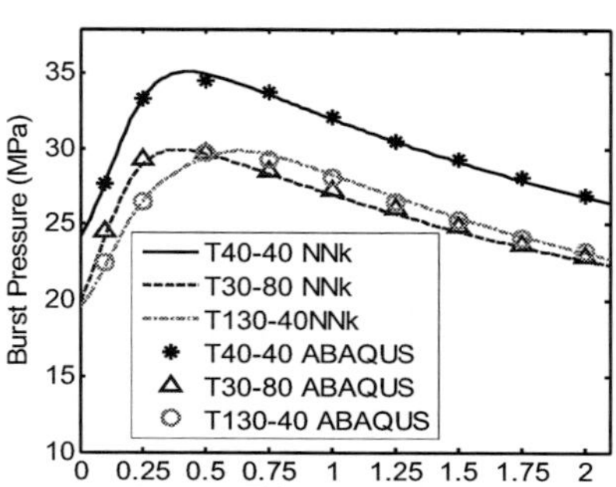

Figure 4 Comparison of burst pressures from NNk and ABAQUS at θ = 15°

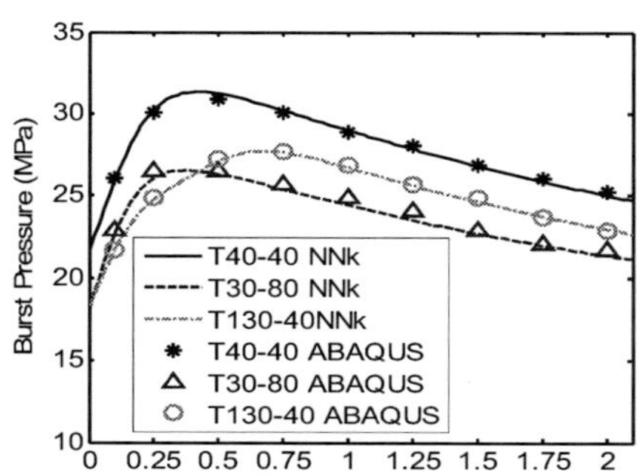

Figure 5 Comparison of burst pressures from NNk and ABAQUS (θ = 25°)

PATTERN INDEPENDENT TOTAL STATIC LEAKAGE ESTIMATION OF COMPLEX GATES IN ADVANCED CMOS PROCESSES

Hussam Al-Hertani, Dhamin Al-Khalili and Come Rozon
Electrical and Computer Engineering Dept.
Royal Military College of Canada

ABSTRACT
Estimation of static leakage current is an increasingly important issue for VLSI designers, as it can contribute significantly to the total power dissipation in circuits based on advanced CMOS technologies. In this paper, two new artificial feedforward neural network (AFNN) based approaches for high level, pattern independent static leakage current estimation in CMOS ultra deep submicron (UDSM) complex gates, are presented: one is regression-based and the other classification-based. In MatLab, the 'Levenberg-Marquardt' algorithm was used for training both AFNNs. To evaluate the AFNNs, the 'Leave-one-out cross validation' scheme was utilized. The average percentage error for the regression-based and classification-based approaches was 8.9% and 10.6% respectively.

INTRODUCTION

As process technologies move into the nanometer domain, increasing transistor counts, frequencies and decreasing channel lengths are contributing to an ever-increasing trend in total circuit power dissipation. Specifically, at such small channel lengths, second order effects cause significantly large static leakage currents that are becoming the major source of power dissipation. The importance of leakage current to the power dissipation paradigm means that designers must be able to accurately estimate the magnitude of leakage current in their design as well as try to minimize it. Currently the most comprehensive way of estimating total leakage current is through running BSIM4 based netlists through SPICE simulations. The disadvantage of this is that it's very computationally intensive, especially when estimating leakage of large circuits containing millions of transistors.

The main objective of the work presented in this paper is to develop a high-level estimation methodology capable of estimating static leakage dissipation in complex gates and other larger components. For basic logic gates with few inputs, pattern dependent leakage estimation techniques maybe feasible, however for complex gates and other larger circuits, pattern dependent leakage estimation becomes increasingly difficult with exponentially increasing run times. The proposed pattern independent leakage estimation methodology is capable of fast leakage estimation in CMOS complex gates exhibiting small run times without losing too much accuracy. Estimation of leakage in larger circuits composed of many such complex gates can then be accomplished simply by adding the average total leakage due to each complex gate component in the circuit.

Similar work has been reported in (Cao, 2002) and (Cao et al, 2005), however in both approaches a regression / classification model was used to estimate the leakage caused by a single input vector for a particular circuit i.e. one artificial feedforward neural network (AFNN) is used for input pattern dependent estimation based on a single circuit topology. This work however, creates a methodology whereby a single instance of an AFNN or other pattern recognition / regression algorithms maybe used to estimate (pattern independently) the average total leakage of any CMOS complex gate in a particular technology. The metric used to characterize the static leakage associated with a single complex gate is the average total static leakage (ATSL). This metric is the mean of the sum of static leakage for every input for a particular complex gate.

Two new approaches to this pattern independent high-level static leakage estimation methodology are introduced; (i) a regression-based approach and, (ii) a classification-based approach. Both approaches are based on the development of a dataset (feature extraction) and both are implemented using AFNNs.

The next section will provide background information on sources of static leakage current. The following section will address feature selection and extraction. Sections 4 and 5 will introduce the regression and classification based approaches respectively. Section 6 will describe the simulation setup and present the results and finally a conclusion is presented.

2. STATIC LEAKAGE CURRENTS MECHANISMS

In this section a brief overview of the major static leakage currents is provided. A more comprehensive review of all static leakage current mechanisms can be found in (Mahmoodi et al, 2003).

Subthreshold Current

As transistor devices are scaled down, the voltage supply V_{dd} must be also scaled appropriately. This has a negative impact on performance since the delay parameter T can be expressed as:

$$T \propto \frac{V_{dd}}{\left(V_{dd} - V_{th0}\right)^{\partial}} \quad (1)$$

where V_{th0} is the threshold voltage and ∂ is a technology dependent parameter between 1 and 1.5. If the voltage supply is scaled down without scaling the threshold voltage appropriately, the gate drive term $(V_{dd} - V_{th0})^{\partial}$ will cause a reduction in performance. In order to alleviate this problem the threshold voltage has to be scaled along with the supply voltage. However, this systematic reduction of the threshold voltage causes subthreshold current to increase exponentially as demonstrated by equation (2). Subthreshold current is the main static leakage current component in the 130 - 90nm technologies. The expression for subthreshold current can be given by:

$$I_{SUB} = \mu_0 C_{ox}^{'}\left(\frac{W}{L_{eff}}\right)v_t^2 \exp(1.8)\exp\left(\frac{\left(V_{gs} - V_{th0} - \gamma^{'} V_s + \eta V_{ds}\right)}{n v_t}\right)\left(1 - \exp\left(V_{ds} / v_t\right)\right) \quad (2)$$

where V_{gs} is the gate to source voltage, V_{ds} is the drain to source voltage, $v_t = kT/q$ is the thermal equivalent voltage, V_{th0} is the zero bias threshold voltage, $\gamma^{'}$ is the linearized body effect coefficient, η is the DIBL coefficient, $C_{ox}^{'}$ is the gate oxide capacitance and μ_0 is the zero bias mobility.

For a stack of OFF transistors, the subthreshold current across the stack is significantly less than the subthreshold current across a single OFF transistor due to the stacking effect.

Gate tunneling current

As transistor devices shrink, it becomes necessary to reduce the gate oxide thickness of the devices. As the gate oxide thickness (t_{ox}) is reduced, the electric field across the gate oxide increases for a given gate voltage. This causes electrons to tunnel through the gate oxide layer. This phenomenon is called direct tunneling or gate tunneling. Gate tunneling leakage current can be very large for thin oxide layers (t_{ox} < 20Å). For NMOS transistors, gate tunneling current density becomes a significant leakage component at the 90nm node.

Gate tunneling is most significant when the device is ON and is in strong inversion i.e. when the gate terminal is connected to a high potential such as V_{dd}, while both the drain and source terminals are connected to a zero potential. Gate tunneling current in PMOS is approximately one order of magnitude lower than the gate tunneling current in NMOS and can be neglected in most cases.

Gate induced drain leakage

Gate induced drain leakage (GIDL) is caused due to the existence of a high electric field at the drain junction of MOS transistors. In NMOS transistors, GIDL takes place when the drain is at a higher potential than the gate. This causes significant band bending in the drain (band-to-band tunneling), allowing electron-hole pair generation through avalanche multiplication and band-to-band tunneling. GIDL has a significant impact in UDSM processes. The simplified expression used for calculating GIDL (Choi et al, 2003) is shown below:

$$I_{GIDL} = A \cdot E_s \cdot \exp\left(\frac{B}{E_s}\right) \qquad (3)$$

where $E_s = \left(V_{gd} - V_{fb} - 1.2\right) / t_{ox}$, V_{gd} is the gate-to-drain voltage, V_{fb} is the flat-band voltage and A and B are process dependent constants. Equation (3) illustrates the strong dependence of GIDL on both V_{gd} and t_{ox}.

3. FEATURE SELECTION

As with most patten recognition based approaches, the first step is to create a dataset. Ideally this dataset is composed of many row entries. Each entry (row) in the dataset represents a vector of features associated with a single complex gate. Therefore, each column in this dataset represents a single feature (variable). The feature selection process involves chosing input and output features(variables) that best describe the formulation of the problem. Input features that describe CMOS complex gate components could include the number of transistors, a transistor width measure, the number of inputs (fan in) as well as the transistor netlist. The output variable is a static leakage metric. The key here is to choose features that could be used by the AFNN to adequately represent a particular component.

For the purposes of this paper it was found that the following input features represent the complex gates adequately; (i) effective width of the NMOS network (NMOS_EW), (ii) effective width of the PMOS network (PMOS_EW) and (iii) a simple circuit level netlist of the NMOS pull-down network.

The effective width of the NMOS/PMOS networks is calculated by collapsing each network into a single transistor that posesses an equivalent worst-case load (resistance) to that of the original NMOS/PMOS network.

Each transistor in the netlist is assigned two numbers; one for the drain and one for the source. Transistors are listed in order from the top left most transistor, moving down the first stack and then onto the top transistor on the next stack and so on. For example, the netlists of the complex gates shown in Figs. 1(a) and 1(b) are "122334403020" and "1223302440" respectively. Note that the PMOS network is complementary to that of the NMOS and its topology does not contain any additional 'uniqueness' not available in the NMOS netlist. ATSL was used as the output feature (for the regression based approach). This metric is derived primarily from SPICE simulations (using Predictive Technology Model's (PTM) BSIM4 45nm (Zhao et al, 2006) model files) for each complex gate considered in the dataset.

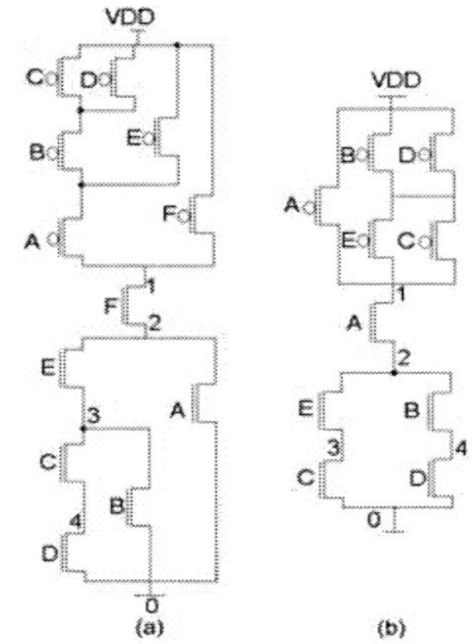

Figure 1. Schematic for the (a) 7_generic and (b) 9_generic complex gates found in Table. 1

4. REGRESSION BASED APPROACH

The regression-based approach attempts to find a function capable of relating input variables/features listed in section 3 to the output variable ATSL. Recall that ATSL was obtained by simulating the complex gates in SPICE *'a priori'*, and typically includes all sources of static leakage modeled by the 45nm BSIM4 PTM process. These primarily include subthreshold current, gate tunneling leakage and gate induced drain leakage. Once the dataset is complete, the AFNN can then be trained to estimate the leakage of an unknown complex gate (not provided in the dataset) based on its features (same as those chosen in the dataset). The dataset should generally contain a large number of complex gate entries to improve the leakage estimation process.

To demonstrate this concept, a dataset composed of 23 complex gates was developed. Note that each gate considered has its own feature vector in the table with its corresponding output variable. Also note that the netlist in each feature vector describes the topology of the NMOS section of each gate as demonstrated in Fig.1. The objective is to train an ANN to estimate the ATSL of a complex gate directly from its equivalent PMOS/NMOS effective width and netlist. The complete dataset is shown in Table 1.

5. CLASSIFICATION BASED APPROACH

The classification-based approach attempts to create classifier rules capable of estimating the leakage of any complex gate by categorizing it into a class of other complex gates exhibiting similar leakage levels. Notice that the input features for this approach are identical to those used in the regression approach. The output variable however is not ATSL, but a class number derived from clustering the feature vectors based on their ATSL value. To demonstrate this classification-based approach, first the dataset used for regression (See Table 1.) is clustered into 5 classes based only on the ATSL of each observation, using the K-means clustering algorithm. After experimenting with the number of classes, it was found that a group of 5 classes was adequate since it exhibited a reasonable estimation error. Once the observations are clustered into each class, the ATSL for all observations within a class is averaged to develop the 'mean class leakage average' (MCLA). In the regression-based approach, the ANN was trained to directly estimate leakage power of a complex gate. In the classification-based approach however, the neural network is trained to 'classify' an unknown complex gate into a class whose leakage characteristic is closest to it. The newly classified complex gate is then assigned the MCLA of the class it just joined as its ATSL estimate (See Table 1).

Table 1. Dataset for the 23 complex gates. NMOS_EW, PMOS_EW and the NMOS netlist are the input features. ATSL is the output feature for the regression approach. Class number is the output feature for the classification approach..

Name	NMOS_EW	PMOS_EW	NMOS Netlist												ATSL (nA)	Class	MCLA (nA)
1_Aoif2201	0.225	0.450	1	2	2	0	1	3	3	0	0	0	0	0	5.273	5	5.381
2_Oaiff2201	0.225	0.450	1	2	2	0	1	2	2	0	0	0	0	0	5.230	5	5.381
3_Blf00001	0.338	0.300	1	2	2	0	1	0	0	0	0	0	0	0	4.864	2	4.931
4_Blf00101	0.150	0.675	1	2	2	0	1	2	0	0	0	0	0	0	4.603	4	4.558
5_generic	0.135	0.750	1	2	2	3	3	0	2	0	0	0	0	0	4.848	2	4.931
6_generic	0.375	0.270	1	0	1	2	2	0	1	2	0	0	0	0	4.819	2	4.931
7_generic	0.139	0.731	1	2	2	3	3	4	4	0	3	0	2	0	5.301	5	5.381
8_generic	0.366	0.277	1	2	2	3	3	0	1	2	1	3	1	0	5.112	2	4.931
9_generic	0.113	0.900	1	2	2	3	3	0	2	4	4	0	0	0	4.667	4	4.558
10_generic	0.450	0.225	1	0	1	3	3	0	1	3	3	0	0	0	4.563	4	4.558
11_generic	0.150	0.675	1	2	2	0	1	2	0	0	0	0	0	0	4.604	4	4.558
12_generic	0.338	0.300	1	0	1	2	2	0	0	0	0	0	0	0	4.864	2	4.931
13_generic	0.263	0.386	1	2	2	0	1	2	2	0	2	0	0	0	5.722	5	5.381
15_generic	0.169	0.600	1	2	2	0	1	2	1	2	0	0	0	0	4.544	4	4.558
16_generic	0.300	0.338	1	2	2	3	3	0	1	0	0	0	0	0	5.069	2	4.931
17_Norf201	0.450	0.225	1	0	1	0	0	0	0	0	0	0	0	0	3.971	3	3.985
18_Nor301	0.675	0.150	1	0	1	0	1	0	0	0	0	0	0	0	3.004	1	2.591
19_Nor401	0.900	0.113	1	0	1	0	1	0	1	0	0	0	0	0	2.172	1	2.591
20_Nanf201	0.113	0.900	1	2	2	0	0	0	0	0	0	0	0	0	3.998	3	3.985
21_Nanf301	0.075	1.450	1	2	2	3	3	0	0	0	0	0	0	0	3.030	1	2.591
22_Nanf401	0.056	1.80	1	2	2	3	3	4	4	0	0	0	0	0	2.156	1	2.591
23_Invf101	0.225	0.450	1	0	0	0	0	0	0	0	0	0	0	0	4.369	4	4.558

6. SIMULATIONS AND RESULTS

For both approaches, a neural network with three layers was setup in MatLab. The input layer has the same number of neurons as inputs (15). The hidden layer has 15 neurons and the output layer has a single neuron for the regression based approach (ATSL) and 5 neurons for the classification based approach (5 classes). The 'Levenberg-Marquardt' algorithm was used for training the ANN.

To truly evaluate the performance of both approaches one must test for two things; (i) how well did the neural net memorize the observations in the dataset (memorization) and (ii) how well can the trained neural net estimate leakage due to a complex gate (observation) it hasn't seen before (generalization). The error metric used for both of these (observation and generalization) was the mean square error (MSE). The neural network in both approaches did well as far as memorization goes with a worst-case MSE of 1e-10nA. To test for generalization a 'Leave-one-out cross validation' scheme was utilized. In this scheme only 22 out of the 23 observations in the dataset are used for training. Once training is completed, the single observation not used in the training is evaluated. This is repeated 23 times so that every-time a different observation is left out of the training and the other 22 are used in training. During every testing phase, the observation evaluated was not seen during the training phase, and hence evaluates the neural nets generalization ability rather than its memorization ability. The MSE for the 'Leave-one-out cross validation' scheme was 0.2066nA for the classification-based scheme and 0.1760nA for the regression-based scheme. The average percent error was also calculated. For the regression-based and classification-based approaches it was 8.9% and 10.6% respectively.

CONCLUSIONS

This paper developed a new high level pattern independent methodology for total static leakage estimation. This methodology removes the complexity associated with pattern dependent static leakage estimation and decreases run times especially in larger circuits. Regression based and classification based approaches were introduced.

For the small dataset used, both the regression and classification approaches for high-level leakage estimation fared well with an average percentage error of 8.9% and 10.6% respectively. A larger dataset with more complex gates may yield even less error.

REFERENCES

Cao L, 2002 *Circuit power estimation using pattern recognition techniques*, IEEE/ACM International Conference on Computer Aided Design. ICCAD 2002. 10-14 Nov. 2002 Page(s):412 – 417

Cao W.Q.Y, Yuanyuan Yan, Xun Gao, 2005, *Neural Network Macro-model for High-Level Power Estimation of CMOS Circuits*, ICNN&B '05. International Conference on Neural Networks and Brain Volume 2, 13-15 Oct. 2005 Page(s):1009 – 1014

Mahmoodi-Meimand H., Roy K., Mukhopadhyay S., 2003, "Leakage current mechanisms and leakage reduction techniques in deep-submicrometer CMOS circuits," in Proceedings of the IEEE, Feb 2003, vol. 91, pp. 305–327.

Choi Y K et al, 2003, "Investigation of gate-induced drain leakage (GIDL) current in thin body devices: Single-gate ultra-thin body,symmetrical double-gate, and asymmetrical double-gate mosfets," Jpn. J. Appl. Phys, vol. 42, pp. 20732076, 2003.

W. Zhao, Y. Cao, 2006, "New generation of Predictive Technology Model for sub-45nm design exploration," pp. 585-590, ISQED, 2006. http://www.eas.asu.edu/~ptm/

AN APPROACH TO THE TOOL WEAR MODEL CONSTRUCTION USING ACOUSTIC SIGNALS OF CUTTING PROCESS

VALERIY I. DUBROVIN
ASSOCIATE PROFESSOR OF SOFTWARE DEPARTMENT, ZAPORIZHZHYA NATIONAL TECHNICAL UNIVERSITY, UKRAINE

YURIY N. VNUKOV
SCIENTIFIC WORK PRO-RECTOR, HEAD OF ENGINEERING TECHNIQUES DEPARTMENT, ZAPORIZHZHYA NATIONAL TECHNICAL UNIVERSITY, UKRAINE

SERGEY I. DYADYA
ASSOCIATE PROFESSOR OF METAL-CUTTING MACHINS AND TOOLS DEPARTMENT, ZAPORIZHZHYA NATIONAL TECHNICAL UNIVERSITY, UKRAINE

YURIY S. AFONIN
ASSISTANT OF SOFTWARE DEPARTMENT, ZAPORIZHZHYA NATIONAL TECHNICAL UNIVERSITY, UKRAINE

TATYANA V. MANILO
STUDENT OF SOFTWARE DEPARTMENT, ZAPORIZHZHYA NATIONAL TECHNICAL UNIVERSITY, UKRAINE

ABSTRACT

A way of tool wear model definition is presented, which is the combination of wavelet packet transform for acoustic signal processing and artificial neural network for tool wear estimation.

An acoustic signal is chosen as an essential informational source of the process or the object under consideration. Measurement acoustic signals are processed using wavelet transform, which decomposes the acoustic signal into approximation and detail components at different levels. Wavelet transform decomposition is used to gather time-frequency-based information from acoustic signals. The artificial neural network is proposed as a tool of relation establishment between the results of wavelet decomposition and parameters of interest.

The results indicate that wavelet transform is an effective tool in the analysis of acoustic signals by providing information relative to the process under question.

The performance of the method is demonstrated by means of numerical and experimental investigations.

The presented method implementation is prospective in on-line tool wear diagnostic system.

INTRODUCTION

One of the central aspects in effective cutting process providing is before-the-fact revelation of tool wear factor. As the tool wear affects the acoustic signal features it is advisable to apply the acoustic methods of the tool wear diagnostics.

The purpose is the approach consideration and analysis of the tool wear modeling.

The task of tool wear diagnostics using acoustic signal is to determine the tool wear factor in the presence of acoustic signal of cutting process. So the target (output) attribute is tool wear factor and the input data are the characteristics of acoustic signal. There is need for proper model construction for relation establishment between input and output data.

The tool wear model construction using acoustic signal includes the next steps:

(1) Acoustic signal processing and further detection of informative signal characteristics.

(2) Correlation establishment between detected characteristics and tool wear.

The packet wavelet transform as a main mathematical tool for the task of signal processing and the artificial neural network for match making between wavelet transform results and tool wear factor are proposed to use.

FAST AND PACKET WAVELET TRANSFORM

Wavelet transform is a powerful mathematical tool for the task of analysis of inherently nonstationary acoustic signals.

The wavelet transform basis construction implies the usage of scale transformation and shifts (Chui, 1992). In particular because of scaling the wavelet transform is good for identification of signal features variations at different frequency bands, and by means of shifts the wavelet transform is able to analyze the signal characteristics over the whole signal interval of interest (Daubechies, 1992).

The result of WT are approximation and detail coefficients, are defined as Eq.1, 2:

$$c_{j,k} = \frac{1}{\sqrt{2^j}} \sum_{t=0}^{L} s(t)\phi\left(\frac{t-2^j k}{2^j}\right), \qquad (1)$$

$$d_{j,k} = \frac{1}{\sqrt{2^j}} \sum_{t=0}^{L} s(t)\psi\left(\frac{t-2^j k}{2^j}\right), \qquad (2)$$

where $c_{j,k}$, $d_{j,k}$ – approximation and detail coefficients, respectively; j - scale parameter; k - shift parameter; s(t) - initial signal; L - signal length; φ - scaling function; ψ - wavelet function.

The modifications of wavelet transform are fast wavelet transform (FWT) and packet wavelet transform (PWT) (Hernandez, 1995).

Using FWT the fast iterative calculations are realized (*Jaideva, 1999*), as a result of it the approximation and detail coefficients can be found using approximation coefficients of the previous level (Eq.3,4).

$$c_{j+1,k} = \sum_n h_n c_{j,2k+n} , \qquad (3)$$

$$d_{j+1,k} = \sum_n g_n c_{j,2k+n} , \qquad (4)$$

where h_n, g_n – wavelet-sequence and scaling sequence, n – sequences length.

As the coefficients $c_{j,k}$ at the first level are the time-discrete values of signal (Eq.5)

$$c_{0,k} = s(k) , \qquad (5)$$

The scheme of such calculations is shown in Fig.1a.

Using PWT the analysis tree is decomposed for both approximation and detail coefficients of any level (Fig.1b). Such decomposition allows to pick out both low- and high-frequency components of signal.

ACOUSTIC SIGNAL PROCESSING USING PACKET WAVELET TRANSFORM

Signal processing is a process of signal transformation and its informative characteristics evaluation.

The packet wavelet transform for acoustic signal analysis is chosen, because the decomposition operation can be applied to the both low- and high-frequency components (Hernandez, 1996), which can be the source of tool wear condition information under consideration.

Figure 2 shows the PWT results of two signals. Signals measured under different tool wear conditions. The first signal corresponds to the relatively low tool wear value concerning the second signal.

At the 4th level there are 16 windows, which show the decomposed signal results corresponding different frequency bands.

The root-mean-square (RMS) values of wavelet coefficients in each window (RMS-1, RMS-2,…,RMS-16) determine the signal energy level in appropriate frequency band (Yao, 1999).

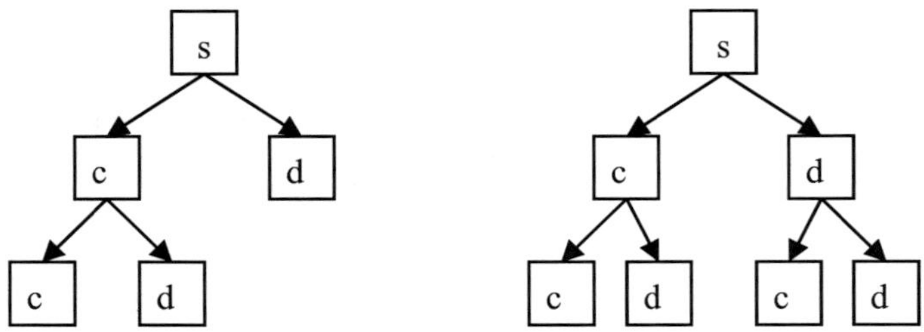

(a) (b)

Figure1 – Wavelet transform scheme, (a) - FWT, (b) - PWT

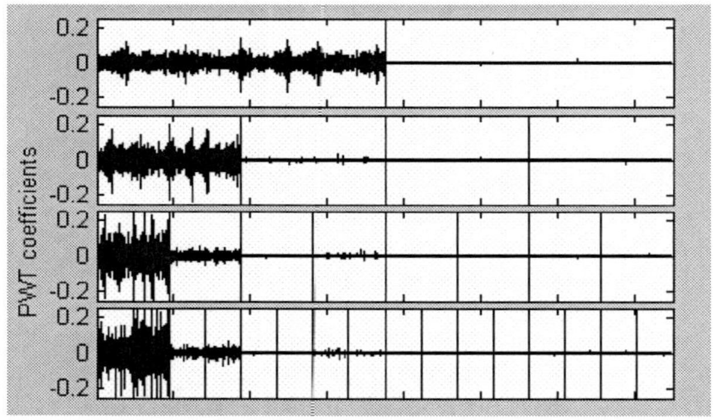

(a)

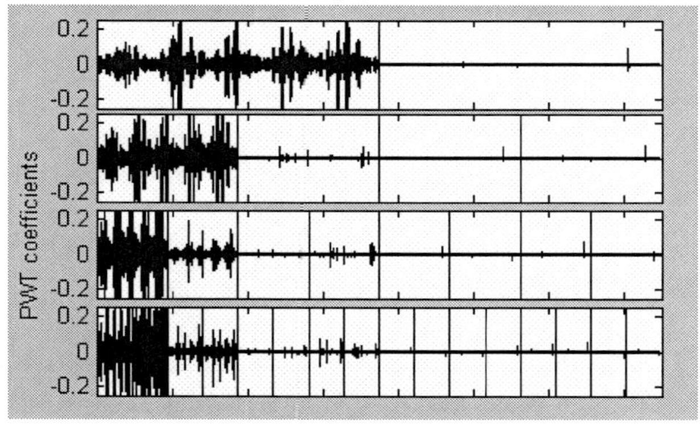

(b)

Figure 2 – PWT coefficients as a result of acoustic signal processing (a) – tool wear<0.2mm, (b) – tool wear>0,2mm

RMS values of wavelet coefficients in windows that are the most sensible to the tool wear can be used as informative characteristics. In this case among them the RMS-1, RMS-2, RMS-3, RMS-4, RMS-7, RMS-8 is selected.

TOOL WEAR ESTIMATION USING ARTIFICIAL NEURAL NETWORK

The artificial neural network (ANN) for relation establishment between informative characteristics of acoustic signal and tool wear factor is used, because ANN is good for complicated dependencies approximating.

The ANN with following structure was used for model construction: two layers of neurons, the first consists of 6 neurons (it corresponds the number of informative characteristics), and the second includes 1 neuron with sigmoid activation functions. The ANN was trained by Levenberg-Marquardt method (Haykin, 1999).

As a result of neural network training the model presented in Table1 was obtained. The table shows neural-network model parameters, such as weight ratios and threshold valuations for each neuron.

The figure shows the high accuracy of the tool wear evaluation, which is about 97%. The results of test sample testing are graphically presented in Fig.3.

CONCLUSION

The examined approach to the model construction embodies the combination of packet wavelet transform for acoustic signal processing and artificial neural network for diagnostic parameters estimation.

Tool wear model, constructed on basis of this approach shows the high accuracy of tool wear evaluation. So the conclusion is drawn about effectiveness of proposed method.

The integration of presented method into online tool wear diagnostic system is quite prospective for further tool wear detection in real time.

Table1 – Neuronet model of tool wear estimation

Layer	1						2
Neuron number in layer	1	2	3	4	5	6	1
Weight ratios	13.3759	-3.018	-5.2558	3.4364	-9.009	2.804	-7.4461
	1.0645	0.92168	72.9234	-32.0328	133.1599	0.11889	-1.9372
	0.054435	-0.00956	-24.4895	-4.1369	-15.2134	-0.03587	58.6086
	0.066867	0.012423	-42.8313	1.7003	14.5045	0.019599	12.1296
	-0.03743	0.29665	23.0538	-68.7071	-63.4904	-0.01181	-48.686
	-0.01810	-0.04274	-1.8792	-0.50713	25.3585	-0.00397	-2.0132
Threshold valuations	433757	-4.9022	20.7449	-20.9038	21.848	6.0621	-2.6703

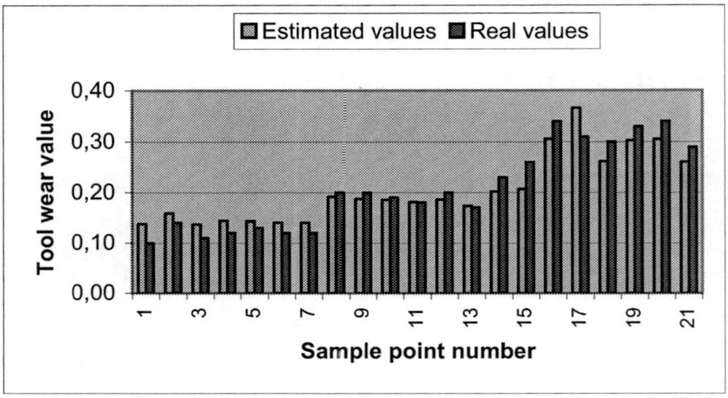

Figure 3 – Comparative results of real and estimated values

REFERENCES

1. C.K. Chui, 1992, "*An Introduction toWavelets*", Academic Press.
2. I.Daubechies, 1992, "*Ten Lectures on Wavelets*", CBS-NFS Regional Conferences in Applic. Mathematics, 61, SIAM.
3. E. Hernandez, X. Wang, G.Weiss, 1995, "*Characterization of Wavelets, scaling functions and wavelets associated with multiresolution analy*ses", Washington University in St. Louis, Preprint.
4. Jaideva C. Goswami, Andrew K. Chan, 1999, "*Fundamentals of wavelets. Theory, algorithms and applications*", Wiley Interscience.
5. E. Hernandez, X. Wang, G.Weiss, 1996, "*A First Course on Wavelets*", Washington University in St. Louis, CRC Press.
6. Y.X. Yao, X. Li, Z.J. Yuan, 1999, "Tool wear detection with fuzzy classification and wavelet fuzzy neural network", *International Journal of Machine Tools & Manufacture*, 39, pp.1525–538.
7. S. Haykin, 1999, "*Neural Networks*", A Comprehensive Foundation, 2nd edition, Prentice Hall.

OUTPUT- AND INPUT-RESPONSE SURFACES GENERATED FROM AN ARTIFICIAL NEURAL NETWORK FOR AN EMPERICAL TO SEMI-MECHANISTIC MODEL: A HEURISTIC APPROACH

WILLIAM A. YOUNG II
Department of Industrial and
Systems Engineering
Ohio University
Athens Ohio 45701, USA

GARY R. WECKMAN
Department of Industrial and
Systems Engineering
Ohio University
Athens Ohio 45701, USA

ABSTRACT
Artificial Neural Networks (ANNs) are primarily used in academia, due to their ability to model complex non-linear systems. Though ANNs have been used to solve practical problems in industry, they are not typically used in non-academic environments. It is believed that ANNs are not used because they are not very well understood, hard to implement, or the fact that they have a reputation of being a "black-box" model. Surfaces can be analyzed to provide evidence that the relationships "learned" by ANNs are in fact valid and can be used to gain additional knowledge of the system. In addition, equations can be derived that represents the constructed surfaces. This paper extends this form of ANN knowledge extraction technique to derive a semi-mechanistic equation.

INTRODUCTION

Since the estimates obtained from an Artificial Neural Network (ANN) are not transparent, decision makers often rely on predicative models based on linear or non-linear regression rather than ANNs. This is rather surprising since ANNs have historically outperformed their regressive counterparts. However, the major benefit of a regressive model is that the derivation of how the estimate was formed is very transparent. Thus, decision makers can verify and validate the relationships that the regressive model is using to model a system. However, if regression techniques are blindly used, the mathematical factors and constants determined may not use the inputs to the system correctly. These models are often called empirical because the model is derived from sample data, and typically do not use known relationships of the system. Problems can arise in determining factors and constants if an insignificant input is highly correlated with a desired response. Mechanistic equations are preferred over empirical ones, because they are based on scientific understanding of the system.

This paper describes a heuristic methodology to formulate a semi-mechanistic model based on ANN knowledge extraction. The foundation of the semi-mechanistic model comes from an ANN, because it is believed to model system interactions more effectively than methods used in classical statistics.

2. BACKGROUND

The first type of model is data driven and is called an empirical model. The second, and most difficult to derive, is a mechanistic model. A mechanistic model is preferred over an empirical model, because it can provide more realistic predictions, and the results are often more meaningful since the model is based on laws of physics or chemistry. They also do not suffer from the idiosyncrasies within a data driven model, but require that the fundamentals of the physical or chemical properties of the system are known.

ANNs are used in many engineering environments because of their ability to “learn” and “adapt” to non-linear systems by mathematically imitating very basic human-like intelligence. Though using of ANNs to solve practical problems has increased dramatically in the last decade (Meireles, Almeida, & Simões, 2003), they are still viewed by many as a “black-box” model. Some researchers suggest that ANN models are incapable of learning physical relationships within a system through empirical data (Lange, 1999). However, it is becoming apparent that extracting knowledge from a trained ANN can lead to new discoveries of relationships that lie within a system (Browne, Hudson, Whitley, Ford, Picton, & Kazemian, 2003). Thus, there is a need to develop knowledge extraction techniques further for a trained ANN.

The most popular form of knowledge extraction technique are decision trees such as C4.5, CART (Breiman, Friedman, Olshen, & Stone, 1984), and TREPAN (Craven, 1996). Rule based methodologies also exist and are of the form of “If-Then” rules, where is a subset of M-of-N algorithms (Cristea, Cristea, & Okamoto, 1997). Non-symbolic methods have been developed that determine the importance of input variables and network nodes (Garson, 1991). Researchers have also developed a stochastic method to determine the importance of input variables within an ANN (Nath, Rajagopalan, & Ryker, 1997). Another popular method is called a ‘sensitivity about the mean’ analysis, which is a procedure that records the changes to the output as a single input value is varied while holding the rest of the input values to their sample means.

3. METHODOLOGY

The object of this methodology is to produce an equation that has the form shown in Equation 1, where functions f_1 through f_n are determined iteratively. The premise of this methodology determines an equation for f_1, which has more influence to the output than the preceding equation such as f_2. Thus, each function that appears from left to right has a smaller amount of influence towards the output respectively.

$$z = \omega_1 \cdot f_1(x_1, y_1) + \omega_2 \cdot f_2(x_2, y_2) + \omega_3 \cdot f_3(x_3, y_3) + \cdots + \omega_n \cdot f(x_n, y_n) + \varepsilon_{n+1} \tag{1}$$

The heuristic used to determine the functions in Equation 1 starts by training an ANN to model an output response with input parameters. Sensitivity analysis about the mean is preformed to screen for input variables to ensure that f_1 has more significance to z than the preceding functions in the equation. The two most sensitive variables (x_1 and y_1) based on the sensitivity analysis are used to determine f_1 in the first iteration. An output-response surface is generated using each input’s sample range and optimization techniques are used to fit a surface equation to the output-response surface based on minimizing the sum of squared error. Thus, the first iteration of this heuristic method results in an equation that has the form $f(x_1,y_1)$. Each iteration of this heuristic determines the preceding equations f_2 through f_n.

The equation $f(x_1,y_1)$ is then compared to the desired response (z). This equation $f(x_1,y_1)$ is scaled, based on the values obtained from the sensitivity analysis. A scaling factor (ω_1) was used in this methodology so each input variable would be used to explain a portion of the desired response. The scaling factor limits the amount of variation that can be explained by the individual functions. Completing the first iteration of determining functions and parameters in Equation 1 is shown in Equation 2.

$$z = f_1 + \varepsilon = \omega_1 \cdot f(x_1, y_1) + \varepsilon_1 \tag{2}$$

Once $f(x_1,y_1)$ and ω_1 are determined in the first iteration, $f(x_2,y_2)$is to be determined in the second iteration. The second iteration is preformed similarly to the first iteration after the inputs used to find $f(x_1,y_1)$ are removed. An additional ANN is trained with the resulting inputs to model ε, a sensitivity analysis is performed to generate an additional output-response surface. This resulting surface is then fit with an equation. This equation is compared to $\varepsilon 1$ and then scaled similarly to $f(x_1,y_1)$. This process repeats until there are no more input parameters left to explain the output response. The formulation of the semi-mechanistic equation is shown in Equation 3.

$$MPG = MPG_{1st\ Iteration} + MPG_{2nd\ Iteration} + MPG_{3rd\ Iteration} + \varepsilon_4 \tag{3}$$

This methodology utilizes a dataset obtained online from University of California Irvine's Machine Learning Repository website (Newman, Hettich, Blake, & Merz, 1998). This dataset contains 398 instances (or samples) of nine attributes which include five continuous values (mpg, displacement, horsepower, weight, and acceleration), three multi-valued discrete values (cylinders, model year, and origin), and one string (car name) value. The original dataset was modified for this particular research example by limiting origin to a discrete value of '1', which represented 245 American made automobiles manufactured between 1970 to 1982. Both car name and model year were both omitted.

4. RESULTS

To baseline the proposed heuristic methodology of creating a semi-empirical model by analyzing input-response surfaces from an ANN, linear and non-linear regression was performed on the seventy-five percent of the data used for training and cross-validation from the ANN, leaving twenty-five percent left for testing of the two methods. A comparison of the twenty five percent of the data remaining will be used to benchmark the method proposed in this paper, with the following models.

NeuroSolutions®, which is a software product developed by NeuroDimensions®, was used in training the ANN for this research (NeuroDimension, 2006). The final network design of 5-4-4-1 was determined after experimenting with many of the network parameters and using 60% of the sample data for training, 15% for testing, and 15% for cross-validation. This experiment included varying the number of processing elements (1 through 15), the number of hidden layers (1 through 3), the training algorithm (momentum, conjugant gradient, and Levenberg-Marquardt), and the threshold function types (linear, sigmoid, and hyperbolic tangent). The combination of parameters that produced the best generalization (or the lowest normalized mean squared error on the cross-validation dataset) was a network that used the Levenberg-Marquardt learning algorithm, two hidden layers with hyperbolic tangent functions, and an output layer that was a pure linear function. The combination of these parameters produced a coefficient of determination (or R2) of 82.1%.

Building output-response surfaces (Young, Weckman, & Marvel, 2007) starts by modeling the ANN structure, and then determining which input is varied. The first iteration starts by identifying the two variables that cause the largest range effect to the output based on sensitivity about the mean analysis. The results from this analysis are found in Figure 1a.

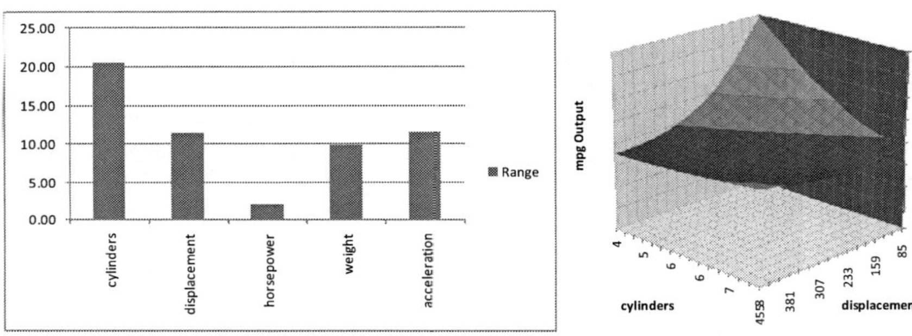

Figure 1: a) Sensitivity about the Mean b) 1st Iteration Response

From this figure, cylinders and displacement are the two most influential. After identifying these variables, the output-response surface is generated and is shown in Figure 1b. After the output-response surface is determined, optimizing techniques are used to determine a three-dimensional equation that best fits the surface based on sum of squared error. The equation that best predicts the surfaces shown in Figure 1b is the "Power E" equation (see . Equation 4). The value of Scale (0.580) is determined by the results of the sensitivity found in Figure 1a. The effective range of cylinders and acceleration is 20.58 and 11.58 respectively. The sum of the effective range of all of the input variables in this figure is 55.44. The scale is then calculated as a ratio of the sum of the range of the variables used in the surface equation (cylinders and displacement) and the total sum. Thus, the ratio is 32.15/55.44, or 0.58. The value of It. Rsq. Is 0.748.

$$MPG_{1st\ Iteration} = 0.58 * \left(160.706 * (cylinders^{-0.382} * displacement^{-0.253})\right) \quad (4)$$

The second iteration starts by eliminating the first two variables (cylinders and acceleration) from the MPG database. A second ANN is then trained with the remaining inputs to "learn" the difference of the actual and the result from the first iteration, or ε1, shown in Equation 4. The two most sensitive input variables to model ε1 are horsepower and weight and an output-response surface is generated. The surface equation that best fits the output-response was a "Power

D" equation with a value of R2 of 0.91. The Scaling factor ω is 0.80 and an It. Rsq. value of 0.76.

$$MPG_{2nd\ Iteration} = 0.799*\left(164.203*\text{horsepower}^{-0.448}\text{-}1.132*\text{weight}^{0.301}\right) \quad (5)$$

The third iteration continues with the iterative process resulting in only one input remaining which is acceleration. A third ANN is then trained and output-response curve is shown in Figure 1a2. From this curve, the relationship of ε2 and weight s is linear (see Equation 6).

$$MPG_{3rd\ Iteration} = 1 * (7.81 - \frac{acceleration}{2.95}) \quad (6)$$

The resulting equation after all iterations were completed is shown in Equation 7. Table 1 presents the benchmark results for the proposed methodology of the semi-mechanistic model against other methods including ANN, linear regression, and non-linear regression. This table shows the metrics commonly used to validate a model, such as R2, mean squared error (MSE), standard deviation (std.) and average percent error (Avg. %Error).

$$MPG = \begin{matrix} 0.580 * \left(160.706 * (cylinders^{-0.382} * displacement^{-0.253})\right) \\ +0.799 * (164.203 * horsepower^{-0.448} - 1.132 * weight^{0.301}) \\ +1.0 * \left(7.81 - \frac{acceleration}{2.95}\right) \end{matrix} \quad (7)$$

Table 1: Benchmark of Methods

Metric	ANN	Semi-Mechanistic	Non-Linear Regression	Linear Regression
Rsq.	82.1%	79.0%	78.8%	74.0%
Avg. %Error	4.4%	5.2%	2.0%	29.6%
MSE	34.84	26.14	31.40	63.92
std.	6.17	5.81	6.04	6.78

As anticipated, the ANN model was the most accurate in terms its R2 value. The ANN model was able to outperform the semi-mechanistic and non-linear regression model by about 3% and was nearly 8% better than the linear regression model. In terms of Avg. %Error, the non-linear regression model was able to outperform any other model benchmarked in this study. The semi-mechanistic equation was able to outperform the other models in two related categories, which included MSE and std.

CONCLUSIONS

The approach this methodology takes into creating a semi-mechanistic is based on a logic similar to bottoms-up and parametric based estimation theory. Several traits that this method provides are desirable in a system model. One trait is to allow sensitive parameters more influence on the estimate than less sensitive variables. It also provides a method that processes a summation of functions that can be verified and explained. The scaling procedure is an attempt to allow part of the desired output signal to be modeled by less sensitive variables and attempts to limit over-fitting input parameters. Another benefit of this model is the use of modeling interactions from input parameters.

The semi-mechanistic equation was not the most accurate in terms of R2, but it did perform rather well in other metrics such as MSE, Avg. %Error, and std. The equation presented from non-linear regression had comparable results, but lacked reasoning as to how the equation was fit, other than the fact that parameters were adjusted to reduce the sum of squared error.

A limitation of this model results from the fact that ANN models have large degrees of freedom when arriving at final weights and biases of a system; it is likely that many sets of weights can arrive at identical outputs (Schmitz, Aldrich, & Gouws, 1999). Thus, many models may have to be examined before a model that represents true physical meaning is used in practice. This limitation could be further hindered if expert knowledge about the system is not known. The results of the model seem to be reliable, which can be explained, and relatively easy to implement. One of the most important aspects of this methodology seemed to be the ability of the semi-mechanistic equation to better generalize the system than other models that were benchmarked. The other model appeared to have suffered from over-fitting the parameters, which could have explained such a difference in MSE.

REFERENCES

Breiman, L., Friedman, H., Olshen, R., & Stone, C. (1984). Classification and regression trees. Beltnont, CA: Wadsworth.

Browne, A., Hudson, B., Whitley, D., Ford, M., Picton, P., & Kazemian, H. (2003). Knowledge extraction from neural networks. Proceedings of the 29th Annual Conference of the IEEE Industrial Electronics Society, (pp. 1909-1913). Roanoke, Virginia.

Craven, M. (1996). Extracting eomprehensiblc models from trained neural networks. Ph.D. Dissertation, University of Wisconsin, Madison, Wisconsin.

Cristea, A., Cristea, P., & Okamoto, T. (1997). Neural networks knowledge extraction. Electrotech in Engineering , 42 (2), 477-491.

Garson, D. (1991). Interpreting neural-network connection weights. AI Expert , 6 (2), 47-51.

Lange, N. (1999). New mathematical approaches in hydrological modeling -- an application of artificial neural networks. Physics and Chemistry of the Earth, Part B: Hydrology, Oceans and Atmosphere , 24 (1), 31–35.

Meireles, M., Almeida, P., & Simões, M. (2003). A comprehensive review for industrial applicability of artificial neural networks. IEEE Transactions on Industrial Electronics , 50 (1), 585-601.

Nath, R., Rajagopalan, B., & Ryker, R. (1997). Determining the Saliency of input variables in neural network classifiers. Computers Operations Rsearch , 24 (4), 767-773.

NeuroDimension. (2006). Intelligent Software Solutions. Retrieved 2007, from http://www.nd.com/

Newman, D., Hettich, S., Blake, C., & Merz, C. (1998). UCI repository of machine learning databases. (I. University of California, Producer, & Dept. of Information and Computer Sciences) Retrieved April 5, 2007, from ftp://ftp.ics.uci.edu/pub/machine-learning-databases/auto-mpg/

Schmitz, G., Aldrich, C., & Gouws, F. (1999). ANN-DT: An algorithm for extraction of decision trees from artificial neural networks. IEEE Transactions on Neural Networks , 10 (6), 1392–1401.

Young, W. & Weckman, G. (2007). Generating Variable Relationships by Optimization Techniques through Sensitivity Analysis from a Neural Network, Industrial Engineering Research Conference, Nashville, Tennessee, May, 2007

N ADAPTIVE METHODOLOGY FOR PREDICTING THE UPPER BOUND IN WWW TRAFFIC

CLAUDE F. TURNER, PH.D.
Bowie State University
Computer Science Department
14000 Jericho Park Rd
Bowie, Maryland 20715

ANTHONY JOSEPH, PH.D.
Pace University
Computer Science Department
163 William Street
New York, New York 10038

ABSTRACT

This article presents a methodology for predicting the upper bound of website traffic volume. The method employs a pre-processing scheme that transforms the data into a form that renders it more tractable for analysis and estimation. The output of the processing scheme is presented to two separate prediction methods—a dynamic artificial neural network scheme and a linear prediction scheme—to predict the future upper bound of the traffic. The method is applied to the Calgary University data set and a comparative analysis of the two prediction models is given. The results illustrate the promise of such methods for effective short-term traffic upper bound predictions in a bursty traffic environment, which can be useful for Web server overload prevention.

INTRODUCTION

The ability to predict the number of website visits (or hits) is important for many reasons. Zhao and Schulzrinne (2003) stated that "to provide expected quality of services, a Web site needs to predict its future traffic volume: long term predictions (e.g., in months or years) are useful for capacity planning, while short term predictions (e.g., in seconds or minutes) are useful for overload prevention." To prevent a website from becoming overwhelmed by highly bursty requests leading to server crashes or degradation of service quality, it is sometimes necessary, for a server to take "needed actions in advance when it anticipates an approaching peak load which is likely to exceed its capacity or a preset threshold," such as, switching some of the load to other servers (Zhao and Schulzrinne, 2003). For such actions to be taken, dynamic load shedding and migration are needed, and short term traffic predictions become necessary. However, the self-similarity of World Wide Web (WWW) traffic (Crovella and Bestavros, 1997) makes prediction in such an environment challenging.

As discussed in Zhao and Schulzrinne (2003), for overload prevention only an upper bound of the future traffic volume need to be predicted to determine whether the future traffic volume will exceed the server's capacity or a preset threshold. As long as the future traffic volume is below the predicted upper bound, the exact volume is unimportant in this context. Further, over-estimation is less of a penalty than under-estimation for overload prevention; a false alarm only incurs unnecessary overheads, but a miss prediction of excess traffic can cause the server to overload.

The goal of this article is to present a methodology for predicting the upper bound of website traffic volume. The method first employs a pre-processing scheme that transforms the data into a form that renders it more tractable for analysis and estimation.

The output of the preprocessing algorithm is then presented to two separate prediction methods -- a dynamic neural network scheme and a linear prediction scheme -- to predict its future upper bound. The method is applied to the Calgary University data set and a comparative analysis of the two prediction methods is given.

BACKGROUND

Two prediction models that will be used to predict the upper bound for WWW traffic: linear prediction (LP) and an artificial neural network (NN).

Linear prediction is defined as $\hat{x}_n = \sum_{k=1}^{q} a_k x_{n-k}$, where $(a_k)_{k\in\{1,2,\ldots,q\}}$ are the predictor coefficients and q denotes the order of the predictor. The main objective of linear prediction is to determine the coefficients $a = (a_k)_{k\in\{1.2,\ldots,q\}}$. This is usually accomplished by imposing some kind of minimization criterion on the difference, $|x_n - \hat{x}_n|$, between the predicted value of the process, $\hat{x}_n$, and its actual value, x_n. One of the popular criteria is the *Minimum Mean Square Error* (MMSE) (Haykin, 2001; Box, 1994), which seeks to minimize the Root Mean Square Error, $E|x_n - \hat{x}_n|^2$. This minimization leads to a set of equations that may be solved efficiently using the Levinson-Durbin recursive algorithm (Levinson, 1947; Durbin, 1960). Another criterion, the *Weighted Least Square Error* (WLSE) (Covella, 1997), seeks to minimize a weighted sum of the difference $\Phi(a) = (1/2)\sum_{i=1}^{k} \alpha_i [a^T u_i - x_i]^2$ where $(\alpha_i)_{i\in\{1,2,\ldots,k\}}$ are the weights, T denotes the transpose, and $u_i = (x_{i-1}, \cdots, x_{i-q})$. The prediction is then given by $\hat{x}_k = a^T u_k$.

Neural networks are crude but insightful exemplars of the brain. They can be broadly classified into static and dynamic networks. Dynamic networks may be either time-lagged feedforward networks (TLFNs) or recurrent networks (Haykin, 1999). Recurrent networks have feedback loops that lead to more intricate designs, improved learning capabilities, and dynamic operations; TLFNs are extensions multilayer feedforward networks and as such may be viewed as universal approximators -- they have short-term memory in time delay elements and are commonly trained using the backpropagation algorithm (Haykin, 1999, Principe et al, 2000). The definition of the backpropagation algorithm adopted in this study is the one posited by Bishop (1995), where backpropagation relates to the efficient calculation of the derivatives of the error function with respect to the network weights and the weight adjustments using the derivatives are computed using any of the many optimization techniques including as gradient descent, Newton's method, and Levenberg-Marquardt algorithms.

METHODOLOGY

The main objectives of this study is to (1) determine the performance of two well known prediction methods' ability to predict the upper bound on World Wide Web traffic and (2) provide a comparative analysis of the results of the two methods. The data are first preprocessed to make them more tractable for analysis. The output of the preprocessing is a sequence consisting of the maximums over the partitioned difference between the training data and a shifted version of itself. It is presented to the prediction methods to determine future values of the upper bound of the traffic. Finally, performance metrics are computed, and a comparative analysis of the prediction models performance is given.

DATA PRE-PROCESSING ALGORITHM

As stated in the Matlab online documentation for the Simulink Toolbox (http://www.mathworks.com), "When dealing with empirical data, it is often useful to remove outliers, smooth, detrend, or otherwise treat the data to make it more tractable for analysis and estimation purposes." In light of this principle, we analyzed the differences, d, in the data rather than the empirical data, x. We accomplished this by first partitioning the data into intervals, $I_k = x_{(k-1)p+1}, x_{(k-1)p+2}, \ldots, x_{kp}$, of length p. This length may also be referred to as the *window size* or *prediction length*. For ease of computation, it is assumed that the length n of x is a multiple of p. If this is not the case, the data is appropriately truncated. This results in an updated value of the length of x which equals $\lfloor n/p \rfloor * p$. After the data is partitioned, the difference, $d = (d_k)_{k=1,\cdots,(\lfloor n/p \rfloor-1)}$, is obtained by subtracting adjacent intervals, $d_k = I_k - I_{k-1}$, of the partitioned x. Since our interest lies in the maximum number of visits over a specified time interval rather than on individual visits, we take the absolute value of d followed by its maximum over each d_k to obtain $M = (M_j)_{j=1,\cdots,(\lfloor n/p \rfloor-1)}$. The sequence, $\widetilde{M}$, obtained by repeating each value in M p times in a manner that preserves the order in M is given formally as $\widetilde{M}_j = M_{\lfloor j/(p+1) \rfloor}, j = 1, \cdots, mp$. It is an extension of M with length equal to the length of d, and is actually the sequence of desired maximums to be predicted over their prescribed windows, each of length p.

METRICS

This section introduces some of the metrics that will be used in our comparative analysis. Let $\widehat{M}$ denote the prediction of M. By introducing the sets $O = \{k | \widehat{M}_k > M_k, k = 1, \ldots, m\}$ and $U = \{k | \widetilde{M} > d_k, k = 1, \ldots, mp\}$, the percentage of predicted maximums that are overestimations may be represented mathematically as $e_0 = card\{O\}/card\{M\} \times 100\%$, where *card{X}* denotes the cardinality (the number of elements) in set X. Similarly, the percentage of predicted maximums that are underestimations may be represented mathematically as $e_u = card\{\bar{O}\}/card\{M\} \times 100\%$, where $\bar{O}$ represents the complement of the set *O*. The ratio of the sum of the differences between the overestimations and their actual maximums to the sum of their actual maximums may be represented mathematically as $O_r = \sum_{k\in O}[\widehat{M} - M_k]/\sum_{k\in O} M_k$. Finally, the ratio of the sum of the differences between the underestimated values and their predicted maximums to the sum of the underestimated values is given by $U_r = \sum_{k\in U}\left[d_k - \widehat{\widetilde{M}}_k\right]/\sum_{k\in O} M_k$.

EXPERIMENTAL RESULTS AND DISCUSSION

This section provides performance results of the linear prediction and neural network models for the Calgary dataset obtained from a University of Calgary WWW server that is publicly available at the Internet Traffic Archive (http://ita.ee.lbl.gov/html/traces.html). Each line gives the date and time of occurrence of the visit as well as other information.

Linear Prediction

The linear prediction coefficients were obtained via the WLSE (LP-WLSE) with an optimal prediction order of q=5 for the dataset. Examination of the percentage of the differences, d, that lie below their predicted maximum over window sizes from 1 through 15 showed that the percentages of the number of samples that lie below their predicted maximum increases over window sizes from 1 through 8, peaks at 9 with a value of 92%, and then generally (but not strictly) decline for the remainder of window sizes (10 through 15). Over this same range of window sizes, the number of maximum values in

(M_i) that are available for the prediction model decreases as the window size increases. Indeed at p=9, this number is $38 = \lfloor 353/9 \rfloor - 1$, since this length is given by $\lfloor n/p \rfloor - 1$. This latter results illustrates that although the window size of 9 might yield desirable results for the LP model, it does not yield a data size (38 samples) that is practical for the neural network training and testing nor for comparisons between the LP and NN methods. It would be ideal to have the values of the predicted maximum equal or exceed the actual maximum values. The predicted maximum often lies above (overestimation) or below the real maximum (underestimation). For the objective this study of capacity prediction overestimation is favored over underestimation. A predicted maximum of 100% means that each data value of the original differences is either less than or equal to its predicted maximum.

Neural Network

The 175 maximum values of website hits were normalized to within ± 1 by removing the mean value of the dataset, 1159.2, and dividing by the maximum value of the dataset, 5271, to produce a new dataset. The neural network model used was of the two-input one output single-layer focused TLFN with eight taps per input. The network is described by $o(n) = \phi(\sum_{k=1}^{8}(\sum_{m=1}^{2} w_m(k)u(n-k)) + b_o)$, where o is the output, u is the input, w is the weights, b is the bias weight, and ϕ is the tanh activation function. The network contained a total of 17 weights: 8 weights per input plus the bias. This suggests that the dataset is small since for a mean square error of 10% on the testing data, the training set size need be 170 samples. Nonetheless, with a dataset of this size, it is still possible to find an oversized network model that generalizes well because the backpropagation learning algorithm may converge in a sufficiently good suboptimal solution and because of the universal approximation property of the backpropagation algorithm trained feedforward networks (Haykin, 1999; Lawrence et al, 1997). Besides, Jagric (2003) used approximately nine years of monthly data (105 samples with 56 used for training) to show that neural network models yielded better forecasts of business cycles for the transitional Slovenia economy than the "National Bureau of Economic Research (NBER) leading indicator [linear] model." Levenberg-Marguardt learning algorithm was used in the experiments because of its desirable properties: relatively faster speed of convergence on small- to medium-sized regression problems (MathWorks, 2003) and its relatively more accurate training on these problems. The many factors the influence the choice of a desirable learning algorithm include differentiability, problem size, data noisiness, robustness, local minima, problem type, step size, and error convergence goal (Reed and Mark II, 1999; MathWorks, 2003). The network model was trained under supervised learning with about 70% of the data allocated to the training subset and the remainder (51 samples for the multi-step ahead prediction and 52 samples for the 1-step ahead prediction) allocated to testing. The termination criterion on network training was a 1000 epochs maximum or the change in the mean square error was less than a threshold of 0.0001. Weight update was done in batch mode and the best weights obtained during training were used in the testing phase. The network weights were purposefully minimized to make them more comparable to the coefficients used in the linear prediction model. The out-of-sample forecast of the test subset of maximum values of website hits were done for 1-step ahead, 2-step ahead, 3-step ahead, and 4-step ahead predictions. The results of these forecasts in terms of mean square error (mse) and correlation (r) were mse = 0.0234 and r = 0.643, mse = 0.2227 and r = 0.661, mse =

0.0211 and 0.719, and mse = 0.0242 and r = 0.658 respectively. They were done in 27, 16, 26, and 41 epochs respectively.

Comparisons

The comparative analysis of the linear prediction and neural network models are specifically provided for correlation coefficient, root mean square error, percentage of predicted maximums that are overestimations, percentage of predicted maximums that are underestimations, the ratio of the sum of the differences between the overestimations and the sum of their actual maximums, and the ratio of the sum of the differences between the underestimated values and their predicted maximums to the sum of the underestimated values for the two methods. Since the neural network model uses 70% percent of the data for training, we only examine the two models performance for the latter 30% of the data, and since the linear predictor was only done for 1-step ahead prediction, this is the only case comparatively analyzed.

The results of the neural NN approach overall performance were better than the LP approach. In terms of root mean square error and correlation, the results were rms=0.153 and r=0.643 to linear the prediction's rms=0.290 and r=0.511. As previously discussed, overestimation is preferable to underestimation, thus with a higher overestimate percentage (e_o=48%) and lower underestimate percentage, e_u=34%, the NN approach is again favored over e_o=33% and e_u=42%, respectively, for the LP approach. The same argument holds for the over- and underestimation ratios—o_r= .604 and u_r=.221 compared to o_r=.380 and u_r=0.254 for the LP model. Figure 1 gives a comparison of the plots for the original difference, d, and its predicted maximums for the neural network and linear prediction approach.

Although the neural network approach generates better overall performance, one limitation of this approach is the relatively large amount of data that is needed for training. For a real-time adaptive system, this could pose a challenge since a significant amount of training will need to be done offline at various times. Because of the recursive manner in which the WLSE generates its weights, this might be less of a challenge for the linear prediction approach.

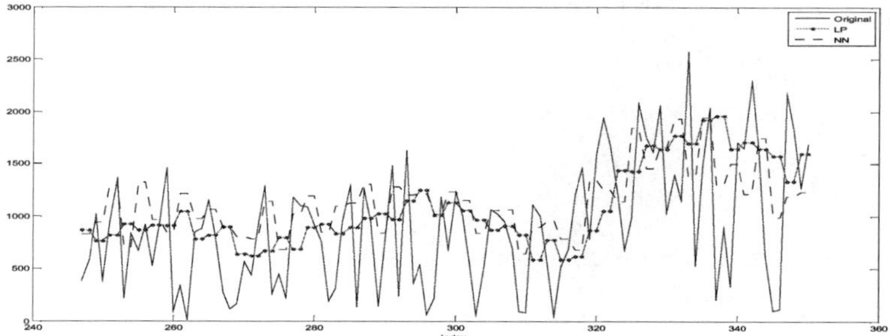

Figure 1: Comparison—Difference, NN, LP-WLSE, p = 2

CONCLUSION

This work presented a methodology for predicting the upper bound of website traffic volume. The method employs a pre-processing scheme that transforms the data into a form that renders it more tractable for analysis and estimation. The output of the processing scheme is presented to two separate prediction methods -- a focused time-delay neural network scheme and a linear prediction scheme -- to predict the future upper bound of the traffic. The method has been applied to the Calgary University data set and a comparative analysis of the two prediction models has been established. The comparison revealed the superiority of the dynamic neural network model over the linear prediction model. The results illustrate the promise of such methods for effective short-term traffic bound predictions in bursty traffic environments, which can be useful for Web server overload prevention.

REFERENCES

Bishop, C., 1995, *Neural Networks for Pattern Recognition*, Clarendon Press, Oxford, pp. 140-163, 253-294.

Box, G., Jenkins, G.M., 1994, *Time Series Analysis: Forecasting and Control*, 3rd Ed., Prentice Hall, Upper Saddle Hill, pp 46-180.

Coffman, E., Jelenkovic, P., Nieh, J., and Rubenstein, D., 2001, "The Columbia Hotspot Rescue Service," Technical Report, Columbia University, February.

Crovella, M.E., and Bestavros A, 1997, "Self-Similarity in World Wide Web Traffic: Evidence and Possible Causes," *IEEE/ACM Transactions on Networking*, 5(6):835–846.

Durbin, J., (1960). "The fitting of time series models," *Rev. Inst. Int. Stat.*, Vol. 28, pp. 233-243.

Haykin S., 1999, *Neural Networks, A comprehensive Foundation*, 2nd Ed., Prentice Hall, Upper Saddle Hill, pp. 1-38, 50-84, 117-143, 156-245, 635-660, 790.

Haykin, S., 2001, *Adaptive Filter Theory*, 4th Ed., Prentice Hall, Upper Saddle Hill, pp 4-7

Jagric, T., 2003, "Forecasting with Leading Economic Indicators – A Neural Network Approach," *Business Economics*, October, pp.42-54.

Lawrence, S., Giles, C., and Tsoi, A., 1997, "Lessons in Neural Network Training: Overfitting may be Harder than Expected," *Proceedings of the 14th National Conference on Artificial Intelligence, AAAI-97*, AAAI Press, Menlo Park, California, pp. 540-545.

Levinson, N., (1947). "The Wiener RMS Error Criterion in Filter Design and Prediction," *J. Math. Phys.*, v. 25, pp. 261-278.

MathWorks, 2003, MATLAB and Toolbox Documentation: Neural NetworkToolbox, ver. MATLAB 6.5.1 (R13) Service Pack 1, Natick, MA.

Mozinga, R.A., and Miller, T.W., 1980, *Introduction to Adaptive Arrays*, John Wiley & Sons, New York, pp 319-347.

Principe, C., Euliano, N., and Lefevre, W., 2000, *Neural and Adaptive Systems: Fundamentals through Simulations*, John Wiley & Sons, New York, pp. 429-524.

Reed, R. and Mark II, R., 1999, *Neural Smithing: Supervised Learning in Feedforward Artificial Neural Networks*, MIT Press, Cambridge, pp. 155-183.

Strobach, P., 1990, *Linear Prediction Theory, A Mathematical Basis for Adaptive Systems*, Springer Verlag, New York, pp 33-35

Zhao, W., Schulzrinne, H., 2003, "Predicting the Upper Bound of Web Traffic Volume Using a Multiple Time Scale Approach," (Poster) The Twelfth International World Wide Web Conference, May 20-24, Budapest, HUNGARY, http://www2003.org/cdrom/papers/poster/p243/index.html

A NEW APPROACH TO PLANT LOCATION PROBLEM: THE ELECTRE-TRI TECHNIQUE APPLICATION

ORHAN TURKBEY
Dept. of Industrial Eng.,
Fac. of Eng. and Arch.,
Gazi University,
Ankara, Türkiye
turkbey@gazi.edu.tr

ZEHRA ISIK
The Ministry of Transportation.,
Ankara, Türkiye
i_zehra@hotmail.com

YESIM KALENDER
Dept. of Industrial Eng.,
Fac. of Eng. and Arch.,
Gazi University,
Ankara, Türkiye
fyesim@gazi.edu.tr

ABSTRACT
The methodology of electre method requires the construction of an outranking relation, concordance and discordance indices and analysis of the results obtained from derived outranking relations. The electre method can especially be used in the solution of various problems as a method capable of turning data into a quantitative form in problems that are dominated by qualitative data. However, a detailed description of the problem, determination of the basic constraints, emphasis of important points and determination of a basic objective are required. In this study, Electre-Tri technique, one of the Electre techniques which is a multiple criteria assignment method that assigns project options to predefined categories, is examined. General information on the technique, its application method, the application of a selected example first manually and then via the prepared software and in conclusion interpretation of results obtained constitute the main content of this study.
Key Words: Multiple criteria decision analysis, Electre Tri technique, Plant location problem.

1. INTRODUCTION

Electre, which is a decision making technique is abbreviation of words of "Elimination Et Choix la Realite" meaning "elimination and selection towards the reality" (Lourenco and Costa, 2002). The Electre Tri method was developed by the LAMSADE (Laboratoire d'Analyse et Modélisation de Systèmes pour l'Aide à la Décision, Université Paris-Dauphine France).

The methodology of this technique requires forming of an outranking relation, forming concordance and discordance indices (including relative importance of each criteria within the work) and analyzing the obtained results from created outranking relations.

There are 6 main versions of Electre technique defined as I, II, III, IV, Tri and 1S. In each situation project options are evaluated in terms of different criteria and every criterion becomes evaluation measurement according to certain point of view of the user's choice. Electre techniques can be applied in several areas that require decisions such as selection of government projects, construction of equipment plans, selection of a machine-device-car-plane. Electre technique can be used in quantitatively dominant areas as qualitative converter solution technique (Lourenco and Costa, 2002).

Nevertheless, in order to use this technique in a healthy fashion for these purposes; decision maker should be able to refer to certain expert group when needed and "should make detailed definition of the problem, determine basic limitations, focus on important points and determine a basic aim most importantly". Later on, decision maker members should establish an *"interdisciplinary council"* which includes experts on several branches regarding attribute, extent and importance of the subject. In the solution of the problem determined by decision maker, interdisciplinary council will follow through those steps of Electre technique;

i. Determination of alternatives related to problem,
ii. Determination of important criteria for the problem,
iii. Evaluation of each alternative according to standards,
iv. Application of Electre technique for solution,

like in the most of the multi criterion approaches (Mousseau et al 1999).

Four different criterion are used in Electre techniques; real criterion, semi-real criterion, gap criterion and pseudo (nick/imitation) criterion. However, only first and last of them are generally used in Electre technique (Roy, 1996).

In a multi-criterion problem, usually without regarding knowledge level of the decision maker and knowledge quality related with criteria used for each option, it is said that "*option a outrank option b*", if option *a* is as good as option *b* and there is no opposing idea.

2. HISTORICAL REVIEW

Electre technique is developed as a multi-criterion decision making system by Bernard Roy (1968). Electre-Tri technique is a multi-criterion assignment technique that assigns project options to predefined categories (Roy, 1996; Mousseau et al. 1999).

According to Lourenco and Costa (2002) Electre-Tri technique belongs to Electre method group (Yu, 1992a; Roy and Bouyssou, 1993) and takes several criterion as base and is designed specially for classification of activity set of problem according to arranged category (Dias et al., 2002). These categories are characterized in an arranged fashion for whole set according to predefined rules named as reference activities (Lourenco and Costa, 2002).

At late 70s, for ranking problems trichotomy procedure which accepts outranking as a base is proposed (Moscarola, 1977; Moscarola and Roy, 1977; Roy, 1981). After a few years, Electre A is developed as a new technique in order to assist big bank companies that face with acceptance or rejection of credit requirements and this was applied in 10 sectors that show balance sheet activity of them (Figueira et al 2005). In 1992, a technique named as Electre-Tri, accepting previous studies as base, was developed (Yu, 1992a,b). This technique is the most successful method of application for ranking problems that require multiple criteria decision aiding (see: e.g. Damart et al., 2002; Dimitras et al., 2001; Gabriel, 1994; Georgopoulou et al., 2003; Mavrotas et al., 2003; Merad et al., 2004; Raju et al., 2000).

Aubert and Reiffers (2003) used multiple criteria Electre-Tri technique that makes solution depending on outranking relations in order to compare economical performances of Middle East and North Africa information economies and to determine which country has which performance level. Lyra et al. (2001) present how to classify knowledge-system based project portfolio for companies by using Electre-Tri technique, which is one of multiple criteria decision methods.

Mousseau V., Slowinski R. and Zielniewicz P. (2000) used an application of banking sector in order to demonstrate feasibility of Electre-Tri software 2.0.

Additionally, Electre-Tri technique was used in realization of TIRS (Transport Infrastructure Regional Study in the Balkans), which is a very large scale motorway Project and of Paris underground Project (Roy and Hugonnard, 1982). Again Electre-Tri technique; is used in several areas such as global planning problems, water source planning, water refinement systems, agriculture, watering and geographic planning and transportation (Lourenco and Costa, 2001; Joerin at al, 2001; Dias et al. 2002).

3. A COMPUTER AIDED APPLICATION TO PLANT LOCATION PROBLEM

In this section, computer program application of Electre-Tri technique, whose processing and manually application was defined before, will be performed. Algorithmic portion of the program was coded in C++ programming language and logic of codes depends on flow chart of the algorithm (Figure 3.1). Codes which are prepared by application of series and loop rules sequentially apply algorithm steps and make comparisons according to obtained numerical results, then give assignment results according to selected sample.

After inputing *a*, *g*, *b* numbers and λ value, by using "*create*" button, data input matrices such as evaluation matrix, profile definition and preference parameters are formed. Codes forming these matrixes are given in Appendix 1.

Main objective of the assignment procedure is to determine which categories that the alternatives will be assigned to. The procedure has two assignment procedures; optimistic and pessimistic.

According to the algorithm, it is necessary to obtain the same results in optimistic and pessimistic assignment procedures to assign an alternative (factory) to a category (candidate plant location). Otherwise, the assignment is not done.

we obtain the results regarding which alternative will be assigned to which category. Code for finding assignment results is given in Appendix 6 and obtained results are presented in Table 3.9.

Table 3.9. Assignment results

Alternatives	Pessimistic assignment results	Optimistic assignment results	General result
a_1	C_3	C_3	C_3
a_2	C_1	C_3	No assignment
a_3	C_2	C_2	C_2

4. CONCLUSIONS

In this study, in order to explain the application sense of the Electre-Tri technique best, the technique is applied to the problem of assigning new factories to suitable locations. The main objective handled in the problem is "Solution of the problem that fits best with the enterprise's benefits".

As stated before in a detailed way, 3 factories (a_i: alternatives) producing same products with different capitals and capacities are evaluated with 5 criteria defined as energy supply (g_1), transportation (g_2), raw material supply (g_3), manpower (g_4) and market (g_5). This evaluation is based on logic of forming outranking relations in order to enable comparison of the alternatives and categories (C_j: candidate factory locations) with the profiles (b_h) that define their limits. By applying the assignment procedures, alternatives are assigned to suitable categories in accordance with the outranking relations.

First, manual application and then the computer program application of the example case is performed. Since in large scaled problems, manual application becomes a very tedious and a long process, the program prepared in JAVA programming language for eliminating this hindrance can be easily applied in large scaled problems as well.

When we observe the results regarding the example examined, we see that factory a_1 is assigned to C_3 and factory a_3 is assigned to C_2 establishment location. Factory a_2 is assigned to C_3 plant location with the optimistic method and to C_1 plant location with the pessimistic method. The difference between these two results is because a_2 is both incomparable with b_1 and b_2 profiles so factory a_2 could not be assigned to any candidate plant locations.

VARIABLES OF THE PROBLEM TO BE SOLVED

a_i	Alternative
b_h	Profile
c	General concordance index
c_j	Partial concordance index
C	Category
d_j	Discordance index
g_j	Criterion
λ	Cutting value

REFERENCES

Aubert, J. E., Reiffers, J. L., 2003, "Knowledge economies in the Middle East and North Africa: toward new development strategies"/edited by Jean-Eric

Aubert, Jean-Louis Reiffers, Washington, D.C.: World Bank, c.2003. HD 30.2 W67 2002 (Rutherford).

Damart, S., Mousseau, V., Sommerlatt, I., 2002, "Du mode d'implication d'acteurs multiples dans cadre de l'utilisation d'un modéle d'affectation multicritére: Analyse au regard d'une application á la tarification des transports publics", INFOR 40 (3), 199–222.

Dias, L., Mousseau, V., Figueira, J. and Cl´ımaco, J., 2002, "An aggregation/disaggregation approach to obtain robust conclusions with ELECTRE TRI", EJOR, 138(2), 332–348.

Dimitras, A. I., Zopounidis, C., Hurson, C., 2001, "Assessing financial risks using a multi-criteria sorting procedure: The case of country risk assesment", Foundations of Computing and Decision Sciences, 29, 97-109.

Figueira, J., Greco, S., Ehrgott, M., 2005, "Multiple Criteria Decision Analysis: State of the Art Surveys", Springer Science+Business Media, Inc., New York.

Gabriel, V., 1994, Méthodologie pour la planification de production de sistéma d'observation de la terre par satellites: Aspects algorithmiques et multicritéres. Ph.D. thesis, Universite Paris-Dauphine.

Georgopoulou, E., Sarafidis, S., Mirasgedis, S., Zaimi, S., Lalas, D., 2003, "A multiple criteria decision-aid approach in defining national priorities for greenhouse gases emissions reduction in the energy sector", EJOR, 146, 199-215.

Joerin, F., Theriault, M., Musy, A., 2001, "Using GIS and outranking multicriteria analysis for land-use suitability assessment", I. J. of Geographical Information Science 15(2), 153-174.

Lourenco, R. P., Costa, J. P., 2001, "Multiple criteria decision making in the new millennium", Lecture Notes in Economics and Mathematical Systems, 507, 132-141.

Lourenco, R. P., Costa, J. P., 2002, "The LinearTri software: sorting solutions in Multiple Objective Linear Integer Programming Problems", 56th meeting of the European Working Group "Multiple Criteria Decision Aid", Coimbra, Portugal.

Lyra, G., Almeida, A.T., Costa, A.P.C.S., 2001, "ELECTRE multicriteria aid on priorities assignment of information systems", In Information Resources Management Association International Conference, Toronto-Canada, pp. 681.

Mavrotas, G., Diakoulaki, D., Capros, P., 2003, "Combined MCDA-IP approach for project selection in the electricity market", Annals of Oper. Res., 120(1), 159–170.

Merad, M., Verdel, T., Roy, B., Kouniali, S., 2004, "Use of multi-criteria decision-aids for risk zoning and management of large area subjected to mining-indiced hazards", Tunnelling and Underground Space Technology, 19, 125–138.

Moscarola, J., 1977, "Aide á la décision en présence de critéres multiples fondée sur une procédure trichotomique: Méthodologie et application", Thése de doctorat de 3e cycle, Université Paris V René Decartes.

Moscarola, J., Roy, B., 1977, "Procédure automatique d'éxamen de dossiers fondée sur une ségmentation trichotomique en présence de critéres multiples", RAIRO Recherche Opérationnelle, 11 (2), 145–173.

Mousseau, V., Slowinski, R., Zielniewicz, P., 1999, "ELECTRE TRI 2.0a Methodological Guide and User's Manual", Document no. 111, LAMSADE, Université Paris-Dauphine, France, Institute of Computing Science, Poznan University of Technology, Poznan, Poland.

Mousseau, V., Slowinski, R., Zielniewicz, P., 2000, "A user-oriented implementation of the ELECTRE TRI method integrating preference elicitation support", Computers & Operations Research, (27), 757–777.

Raju, K., Duckstein, L., Arondel, C., 2000, "Multicriterion analysis for sustainable water resources planning: A case study in Spain", Water Resources Management, 14(6), 435–456.

Roy, B., 1968, "Classement et choix en présence de points de vue multiples (la méthode ELECTRE)", *RIRO, 2e année,* no. 8, 57-75.

Roy, B., 1996, Multicriteria Methodology for Decision Aiding: Nonconvex Optimization and its Applications, Kluwer Academic Publishers, Link Dordrecht.

Roy B., Bouyssou D., 1993, "Aide Multicritére á la Décision: Méthodes et cas", Economica, Paris.

Roy, B., Hugonnard, Jean-Christophe, 1982, "Ranking of suburban line extension projects on the Paris metro system by a multicriteria method", Transportation Research, 16A (4), 301-312.

Yu W., 1992a, "Aide multicritére à la décision dans le cadre de la problématique du tri: concepts, méthodes et applications", Ph. D. Thesis, Université Paris-Dauphine.

Yu, W., 1992b, "ELECTRE TRI: Aspects méthodologiques et manuel d'utilisation", Document du LAMSADE no.74, Université Paris-Dauphine.

APPENDICES

Appendix 1: Code written for screen and forming of data input matrices.
Appendix 2: Code written for computing concordance indices.
Appendix 3: Code written for computing discordance indices.
Appendix 4: Code written for computing credibility indices.
Appendix 5: Code written for determining preference relations.
Appendix 6: Code written for obtaining assignment results.

NEURAL NETWORKS AS A USEFUL TOOL FOR REAL-TIME FACIAL EXPRESSION RECOGNITION

ONDREJ LINDA
Graduate student
Computing and Information Sciences,
Kansas State University,
Manhattan, Kansas, USA
olinda@ksu.edu

SRIVARDHAN CHANDRAPATI
Graduate student
Mechanical and Nuclear Engineering,
Kansas State University,
Manhattan, Kansas, USA
csri@ksu.edu

AKIRA TOKUHIRO
Associate Professor,
Mechanical Engineering,
University of Idaho,
Idaho Falls, Idaho, USA
tokuhiro@uidaho.edu

ABSTRACT
Human beings communicate by rather skillful projection and reception of facial and voiced-expressions, as well as gestures (hand, posture, etc.). In order to develop "tools" by which humans can interact with a computer, we need to develop a software tool that is capable of processing, for example a wide range of facial expressions (FEs) in time. Facial expressions, in their variety and in addition, digital quality when captured presents itself as nontrivial challenge in applied biometrics. In this paper we present results to date on development of a real time and computationally "light" facial expression recognition software tool. Contrary to past studies reporting elaborate processing and FE classification methods, we undertook an approach extracting a small number of facial feature points, and one that realistically contained noise. We additionally proposed a vectoral descriptor and amplitude for a given FE, relative to a reference, neutral FE image. Moreover we trained and tested across FE databases that were ethnically/culturally and gender-wise different. Our results to date opened for consideration the following in brief, that: 1) there are similarities/differences across FE database, 2) vectoral descriptors and amplitude per FE appear effective, 3) "happy" and "surprise" FEs are well classified across different database, while 4) "angry-distress" and "fear-surprise" pairs are linked by misclassification across databases; that is, there is some evidence that these are ethnicity/culture specific and therefore misinterpreted.

INTRODUCTION

During one-on-one communication it is widely accepted that significant amount of information is conveyed through non-verbal communication, such as gestures or facial expressions (Mehrabian, 1968). Developing a software tool for automatic facial expression recognition would significantly improve the communication between human and computer not only during normal use, but also when communicating with an handicapped person.

Amongst a wide range of functional skills, we often take our ability to recognize and project facial expressions for granted, this in fact is a challenge when tried to achive using computers. The range of human expressions and the cognitative state to which the expression is attached is vast, even a simple expression such as a smile is amongst individuals. Additionally, there is some evidence of differences amongst different ethnic groups and cultures. However, as humans are able to distinguish between basic expressions/emotions irrespective of ethnicity/culture, it is reasonable to assert that there exists a common functional modality for each expression. Discovering this modality and developing a tool to accurately extract an inherent "pattern" from an expression may be key to building robust facial expression recognition software.

Automated facial expression recognition has been proposed by a number of investigators; a notable, detailed survey is that given by Fasel and Luettin (2003) and also by Pantic and Rothkrantz (2003). However, most of the approaches are far from being applicable in a real time environment except perhaps for that developed by Littlewort et al. (2003). This is mainly due to the limitation on computing power; that is, considerable computing is needed for detailed image preprocessing and feature point extraction. In the present work, our objective is to develop a useful real time application that will be capable of a relatively accurate classification of the facial expressions. Because each captured frame needs to be processed as quickly as possible, we plan to limit the processing time per image. This poses a number of challenges. First of all we cannot extract a large number of (facial) feature points. Instead we search for the most distinguishable points in the face which can consistently be tracked spatiotemporally (frame-to-frame if needed) with high accuracy. Secondly, we exercise the option to limit the search for feature points if an initial "pass" does not easily reveal them. Thus, at the onset we expect a lower overall accuracy in facial expression recognition and from a learning point of view (from a dataset), we acknowledge the existence of noise in our data.

It is here at the pre-processing stage that we significantly differ from Littlewort et al. (2006) where they began with the whole face region to get 92,160 possible features which they later reduced to a minimum of 500 features using machine learning techniques. Under such circumstances they found an optimum set of facial features, but without specifics on what features are being learned and are important. Based on Bassili's (1978) experiment, we instead choose to select points on the face.

In the present work we will introduce results to date of work toward developing a real-time facial expression recognition tool. We used a feed-forward Artificial Neural Network (ANN) trained with back propagation training using gradient descent with momentum rule, to adapt to the characteristics of our facial expression image database. The ANN was built using the neural network toolkit in MATLAB™. In this paper we present our results on classifying captured facial expressions from snapshot and video frames, extracted from ethnically different databases.

LEARNING

One of the most common techniques for facial expression classification is via Artificial Neural Network (ANN). Based on our previous work on facial expression biometrics (Chennamsetty and Tokuhiro, 2004; Chennamsetty, 2005), we sought to use ANN to perform (quasi) real-time FE-classification.

There are two basic approaches to classifying expressions, one being identifying the presence of each (facial) action unit (AU) as defined by FACS (Ekman and Friesen, 1978) and then classifying based upon combinations of AUs, and the other, classifying the FE into one of the six basic emotions as defined by Ekman (1992). Here, we chose to classify the given face into one of the six basic expressions: Anger, Distress, Fear, Happy, Sad and Surprise. The tracking of AUs requires high resolution and standardization of images, which is ultimately far from reality. We also recognize that *precisely* defining a FE is difficult (perhaps irrelevant) and it is often unrealistic (questionable) to seek a "pure" FE. In fact, as FEs are highly spatiotemporal, each given expression can be a mixture of several FEs. Thus, ideally the ANN should output its "degree of belief/confidence" in how much a given FE is similar to one of our six expressions. A feed-forward ANN trained with back propagation learning rule is capable of giving such a response (Kobayashi and Hara, 1992).

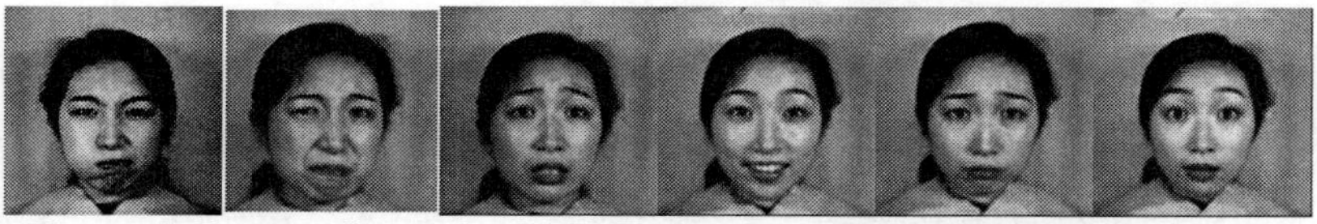

Figure 1. Images from the JAFFE database. You can see the six basic expressions that we are trying to classify. From left to right: Angry (An), Distress (Di), Fear (Fe), Happy (Ha), Sad (Sa), Surprise (Su).

Figure 2. Frames from the video sequences from the MMI database. From left to right: Angry, Distress, Fear, Happy, Sad, Surprise.

INPUT DATA

For our experiments we used two facial image databases. The first, known as the JAFFE database, consists of 213 images of 10 Japanese females (Lyons, 1998). The second database, known as MMI, consists of video sequences of 15 different subjects (Pantic et al., 2005). The MMI subjects are mostly young white males and females. Each video sequence starts with a neutral face and then slowly transitions to an expression and back to a neutral face.

Feature Points Extraction

Although some may contest Johansson's (1973) point light experiment and Bassili's (1978) experiments based on Johansson's work, there is evidence that extraction of only the key facial features or points are enough for classification of FEs. Further, as we are cognizant of in-the-field application issues, we decided to locate 17 of the most distinguishable feature points on the eyebrows, eyes, nose and mouth, as shown in the Fig. 3. To locate these points we used an algorithm consisting of three main stages; 1) *Face region estimation,* 2) *Feature region detection,* and 3) *Feature point detection.* First, face region location is estimated using a seeded region growing method. The face region information is then passed to the feature region detection algorithm that identifies key points of the eyes, mouth, nose and eyebrow using a Particle Swarm Optimization-based search. The feature points associated with these regions are defined based on the assumption that the feature is continuous at least along its edge and separated from other features, even if large variations appear within the feature itself. This is accomplished by using the bounding box function built in Matlab™.

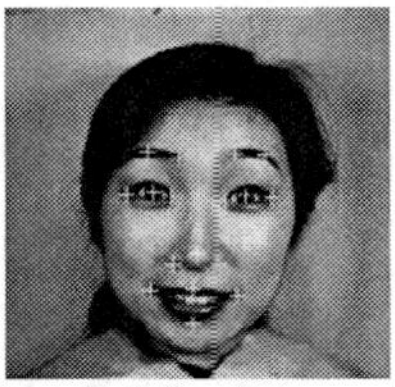

Figure 3. Face with located feature points that we select on the face.

Descriptors

In Fig. 3, the absolute coordinates of the feature points only tell us where they are located in the image. Rather one needs to construct 'a descriptor' that describes the position of the feature points with respect to the other points, for a given FE. This descriptor is also important in order to compensate for head movement and differences in placement of the face in the field-of-view (frame). In other words, we want to extract information about the feature points' movement on the face. We constructed a descriptor 'vector' consisting of 42 different distances, angles and vectors among the points. This descriptor vector served as the input vector for the ANN.

Tracking changes in expression

Rather than processing each FE image as single entity, in the present work we focused our interest on changes in FE between two images; that is, what Cohen et al. (2003) called static processing. As FEs constantly change, we sought to extract information like whether the corners of the mouth move relative to one another or whether the eyebrows are rising relative to the mouth. Our goal throughout was to classify the expression into one of the six accepted FE groups. Therefore we generated a descriptor vector for a given FE image and compared it with the descriptor vector of the neutral FE for the subject. In this manner we derive the change in descriptor vector relative to the neutral vector FE for the subject. For the JAFFE database we have three neutral FE images for each subject. So we took each of these as a reference neutral descriptor and calculated difference vectors between these reference vectors and the FE vector of interest. Where-as in the MMI video-frame sequence we first split the video into individual frames and took the first FE frame (always neutral) in the sequence as the reference and then calculated the relative changes between subsequent frames relative to this reference. We thus generated a sequence of vectoral descriptors tracking the spatio-temporality of FEs.

Target function

In order to train the ANN we constructed a training set containing labeled instances. In the case of the JAFFE database we assigned the same apriori expression label from the database to the derived descriptor relative to the reference neutral FE. However, for the MMI database, lacking apriori labels, we devised an approach as follows. Here we sought to attach one of the six FE labels to each frame from the video sequence. The question then is what label is assigned to each FE frame, as the expression changes in time from an initial, neutral expression. Further, in some frame-to-frame sequence, we had to assign incrementally different FE labels; that is, at some point, one frame for instance, was 'neutral' and the next 'happy', even if it represented the first of several sequential images expressing 'happy'. We recognize that this incremental labeling introduces noise into the data and may influence the training exercise.

EXPERIMENTAL RESULTS

In considering ANN methods, we initially used a feed forward NN with back propagation training rule. We began with the JAFFE database as our training set and the pre-processed video sequences from the MMI database as our testing set, since our eventual goal is to run this on real-time video. As a comparison we also constructed 10 pairs of training and testing sets from the JAFFE database. In each pair there are 9 subjects in the training set and the tenth was used for cross validation. Splitting the data in this way for each subject gave us 10 cross validation sets. By tabulating the result for each cross validation set, we were able to assess and compare the performance of the ANN of the JAFFE and the MMI database.

Training and testing on the JAFFE database

We first present our results of training the ANN on the JAFFE database and testing it on the JAFFE database cross validation sets as was described above. We present these results first, because we anticipate that testing the ANN on a non-JAFFE images would potentially yield either misclassification of FEs or a strong indication of the goodness of a minimal feature point set that we used per FE. In fact, this baseline should yield expectedly *good* results. The following results were achieved by using ANN with one hidden layer containing 30 neurons. We trained 10 different networks and then averaged the results.

Table 1. True positive rate for ANN trained and tested on the JAFFE database.

TP	Angry	Distress	Fear	Happy	Sad	Surprise
58.43%	57.33%	42.00%	50.89%	74.17%	44.22%	90.00%

As shown in Table 1, the average, overall accuracy of the trained ANN was 58.43%, but ranges from a low of 42% for "distress" to as high of 90% for "surprise". This average figure of merit (FOM) is disappointing when compared to higher figures reported by other studies. However, we emphasize the limited number of feature points extracted per FE and additionally, noise contained in the data. In fact, other studies when tested with FE images beyond its training database show a comparable true positive rate. Thus, in the task of classifying six different FEs, we conditionally accepted this figure of merit. In fact, this figure is consistent with the drop in accuracy of FE recognition by human subjects (Bassili 1978). Also the table shows accuracies for each particular FE. From above we can clearly see that our methodology is suited for classifying FEs with clear movement of feature points, i.e. "happy" and "surprise" relative to the neutral FE. We assert that these FEs are well-suited for the ANN to learn and classify. On the other hand, FEs such as "angry, distress, fear and sad" are "weakly" characterized by our descriptor and therefore not well-suited for the ANN to learn. Another possible source of the lower figure of merit is the inherent noise , that could mask any pattern in our training or testing data.

Training on the JAFFE database and testing on MMI database

It is impractical to assume that we will always have associated training dataset available for a FE which we want to analyze. Therefore to evaluate the ability of our ANN-based application to recognize FEs on unseen (external) faces, we first trained the ANN with the JAFFE database and then tested images taken from the MMI database. In fact, not only are the training/test sets different, they span ethnic and cultural characteristics. Thus, again using the same ANN, we averaged the results of 10 networks. A validation-based early termination was used while training the ANN; this method uses a sample validation set that is used to test the network after every epoch of training and stops furthur training upon increase in the error for the validation set, even if the maximum number of training epochs has not been reached. In this case we used a sample of the MMI database as the validation set so as to improve performance. This also prevents the ANN from overfitting, which is when the ANN begins to memorize the training set rather than identifying a trend.

Table 2. True positive rate for ANN trained on the JAFFE database and tested on the MMI database.

TP	Angry	Distress	Fear	Happy	Sad	Surprise
55.96%	57.38%	17.22%	13.05%	90.80%	50.24%	92.00%

Table 2 shows that the average accuracy achieved on a different database was 55.96%, which is comparable to the accuracy we achieved using the JAFFE database. This is also comparable to the nonlinear AdaSVM developed by Littlewort et al. (2006) who reported impressive results for FE recogniton on one database; however, this figure of merit dropped to ~60% when tested on another database.

One can again see that "surprise(Su)"and "happy(Ha)" FEs consistently produce higher figures of merit and support the view that these FEs are easiest to train on and classify. Interestingly, the ANN is some 15% better at recognizing the happy expression in the MMI than in JAFFE on which it was trained. Though limited, we learned through our own FE database development that subjects from some ethnicities and cultures (here Asians) do not express a sense of "happy" with the same "intensity" as from other backgrounds. However, the vectoral descriptor appears to be consistent for both JAFFE and MMI. We can also see that "angry" and "sad" FEs are still relatively difficult for the ANN to classify; that is, the descriptor for these MMI FEs is not particularly distinct. Interestingly the true positive rates for FEs "distress" and "fear" exhibited a significant decrease relative to Table 1. In fact, in the case of "fear", the rate is worse than that achieved by random guess. So overall the performance of the ANN trained on the JAFFE and tested on the MMI database appears to be consistent for Angry, Happy, Sad and Surprise expressions and (provisionally) suspect for Distress and Fear, relative to Table 1.

INTERPRETATION OF OUR RESULTS

We briefly present observation where there was a notable decrease in figure of merit with respect to Distress and Fear expressions on the MMI database. The consistency in the decrease suggests that indeed there may be some potential ethnic and cultural constraints (within a group) and equally, bias (beyond a group) when training and testing on different database.

Lower FOM on Distress and Fear FEs

We observed from Table 1 and 2 that the figure of merit resulting from the ANN were consistent except for the FEs, "distress" and "fear". To investigate the potential source of these results, we generated a simple "confusion matrix", Table 4, for ten networks of our ANN. Here the top *row* FE labels represent the classification FE and the leftmost column FE labels represent the true (known) classification. Thus the diagonal elements signify a (desired) one-to-one "correct" correspondence; the off-diagonal elements per column signify incidence of misclassification. The asterisk (*) is discussed below.

Table 4. The confusion matrix of ANN trained on the JAFFE database and tested on the MMI database.

true\classif.	Angry	Distress	Fear	Happy	Sad	Surprise
Angry	30	5	5	0	2	0
Distress	26*	6	1	0	2	1
Fear	0	4	2	1	8	31*
Happy	0	1	2	44	2	1
Sad	3	6	3	0	23	6
Surprise	0	0	0	0	3	47

For for all FEs "Angry, Distress, Fear, Happy, Sad and Surprise", we see that one element (box) contains the largest number of incidents. In fact, "Angry, Happy, Sad and Surprise" are correctly classified with conditions to be discussed. Further, by the low number of incidents on the diagonal, both "Distress and Fear" are inconclusively classified. That is, these two FEs learnt from JAFFE training set is weakly correlated to the same FEs from the MMI test set, and thus misclassified as other FEs, except surprise. At the least, "distress and fear" is not taken as "surprise" by the paired JAFFE-MMI datasets.

Next, we noted that a large number of incidents of "distress and fear (column)" test examples were misclassified as "angry and surprise(row)" respectively. Here, the network is simply not able to distinguish between "fear and surprise" and "angry and distress" FEs; that is, even though human experience may indicate a measure of confidence to the contrary (that one can indeed distinguish Fear, Surprise, Angry and Distress), from a image processing and ANN perspective, these FEs "look" similar and are paired. Moreover, there were not recorded incidents of "surprise" misclassified as "fear" and only a few cases of "angry" misclassified as "distress". Based on this, the vectoral descriptor corresponding to "Distress and Fear" learned via ANN/JAFFE appears to be similar to the descriptor corresponding to the MMI FE for "Angry and Surprise" respectively.

Finally, from an application perspective we suggest the following ethnic and cultural inferences. As noted, the JAFFE database used here for ANN training consists of young to middle-aged (20-35 years old) Japanese female subjects. On the other hand, the MMI database for training, consists of diverse (gender, age) subjects but contains in majority, white (Caucasian, European) men and women. There are individual subjects (see Fig. 2) with different ethnic/cultural appearance. This leads us to the partial interesting conclusion that ethnic, cultural and gender differences between these two databases can lead to misclassification and expectedly, a lower figure of merit. These results may also suggest that, although six basic FEs are widely accepted as common across ethnicities and cultures, the subtleties associated with FEs that are either mixed FEs and/or lower in amplitude/"intensity" (not a distinct vectoral descriptor) can be misclassified when the training occurs on a ethnicity/culture-specific database and testing on another specific or non-specific database. Further, results from our work suggests that the FEs, "distress-fear" and "fear-surprise", are prone to a paired misclassification when training occurs on JAFFE (Japanese) and testing on MMI (mostly Caucasian/European) FE datasets.

CONCLUSIONS TO DATE

Human beings communicate by many means but in spoken conversation with one or more subjects, the voiced and facial expressions (FEs) are dominant modes in conveying information. In the present work we sought to process facial expressions images and classify them into one of six standard FEs; that is, "happy, sad, surprised, fear, distress and angry". A seventh neutral expression was taken as a reference for both of the two FE databases we used in the study. To maintain minimal computational effort, we used a limited number of facial feature points (per FE image) and accepted a measure of noise as well.

In spite of these conditions and constraints, we showed that the ANN conditionally performs in a reasonable manner. That is, although the overall figure of merit (FOM) of our method does not attain a high value reported by previous investigators, we note that these higher FOMs are achieved at the expense of off-line and computationally-intensive FE recognition, image processing and classification. From the present objective applications in the field, quasi-real time acquisition, processing and classification of lower quality facial images are anticipated. Our results to date indicate the following for consideration, mainly that:

- Training with an ethnicity/culture specific (JAFFE) facial expression database and testing with different database reveals some consistencies and differences among FEs.
- Generating vectoral descriptors for a given FE and tracking change in its amplitude, relative to a reference, neutral FE is a simple way of analyzing spatiotemporal changes in FEs.
- The "happy" and "surprise" FEs are consistently recognized and classified across these FE database via use of vectoral descriptors.
- The "angry-distress" and "fear-surprise" FE pairs are evidently linked by misclassification across different FE databases; that is, there is some evidence that these are ethnicity/culture specific.

ONGOING WORK

In our research we are working toward improving the performance of our algorithm. And also improve the performance of the ANN using a larger database of ethnically/culturally diverse facial images or video sequences. We are also investigating initial indications that differences in FEs (and thus emotions) between different ethnicities and cultures can indeed be revealed by the systematic misclassification as indicated by our "confusion matrix". Finally, in order to reduce computational time while simultaneously improving figure of merit, we are developing a means to quasi-simultaneously process the voiced expression along with the FE since the voice also conveys a measure of the state of emotion (Ramamohan and Dandapat, 2006).

REFERENCES

[1] Bassili, N. J., 1978, "Facial Motion in perception of faces and of Emotional Expressions", *Journal of Experimental Psychology : Human Perception and Performance*, Vol. 4, No. 3, pp. 373-379.

[2] Chennamsetty, N. and Tokuhiro, A., 2004, "Facial Expression Classification Using Learning Vector Quantisation Networks", ANNIE 2004, Smart Engineering System Design, St. Louis, MO, USA, Nov. 7-10.

[3] Chennamsetty, N., 2005, "Development of Automatic Facial Expression Recognition System Using Gabor Wavelets and Learning Vector Quantisation Networks", M.S. Thesis, University of Missouri-Rolla.

[4] Cohen, I., Sebe, N., Garg, A., Chen, L. and Huang, T. S., 2003, "Facial expression recognition from video sequences: Temporal and static modeling.", *Computer Vision and Image Understanding*, Vol. 91, No. 1-2, pp. 160-187.

[5] Ekman, P. and Friesen, W. V., 1978, "Facial Action Coding System Manual", Palo Alto: Consulting Psychologists Press.

[6] Ekman, P., 1992, "Are there Basic Emotions?", *Psychological Review*, Vol. 99, No. 3, pp.550-553.

[7] Fasel, B. and Luettin, J., 2003, "Automatic Facial Expression Analysis: A Survey", *Pattern Recognition*, Vol. 36, pp. 259--275.

[8] Gunnar Johansson, 1973, "Visual Perception of Biological Motion and an order for its Analysis", *Perception and Psychophysics*, Vol. 14, No 2, pp. 201-211
[9] Kobayashi, H. and Hara, F., 1992, "Recognition of Six Basic Facial Expressions and their Strength by Neural Network", Proceedings, International Workshop Robot and Human Communication, pp. 381-389.
[10] Littlewort, G., Stewart, M., Fasel, I., Chenu, J. and Movellan, J., 2003, "Analysis of Machine Learning Methods for Real-Time Recognition of Facial Expression from Video", Technical Report, Machine Perception Laboratory, Institute for Neural Computation, University of California, San Diego.
[11] Littlewort, G., Bartlett, M., Fasel, I., Susskind, J. and Movellan, J., 2006, "Dynamics of facial expression extracted automatically from video", Image and Vision Computing, Vol. 24, No. 6, pp. 615-625.
[12] Lyons, M. J., Akamatsu, S., Kamachi, M. and Gyoba, J., 1998, "Coding Facial Expressions with Gabor Wavelets", Proceedings, Third IEEE International Conference on Automatic Face and Gesture Recognition, Nara Japan, IEEE Computer Society, pp. 200-205.
[13] Mehrabian, A.,1968, "Communication without Words", *Psychology Today*, Vol. 2, No. 4, pp. 53-56.
[14] Pantic, M. and Rothkrantz, L. J. M., 2003, "Automatic Analysis of Facial Expressions: The State of the Art", *IEEE Transactions on Pattern Analysis and Machine Intelligence*, Vol. 22, No. 12, pp. 1424-1455.
[15] Pantic, M., Valstar, M. F., Rademaker, R. and Maat, L., 2005, "Web-based Database for Facial Expression Analysis", Proceedings IEEE International Conference on Multimedia and Expo, Amsterdam, The Netherlands.
[16] Ramamohan, S. and Dandapat, S. 2006, "Sinusoidal Model-Based Analysis and Classification of Stressed Speech", *IEEE Transactions on Speech and Audio Processing*, Vol. 14, No. 3, pp. 737-746

Subject Index

A

B

C

J

K

L

M

N

O

P

Author Index

A

B

C

D

E